Glass Capillary Chromatography in Clinical Medicine and Pharmacology

Glass Capillary Chromatography in Clinical Medicine and Pharmacology

edited by

Halvor Jaeger

L.A.B. Gesellschaft für pharmakologische
Untersuchungen m.b.H. & Company
Neu-Ulm
Federal Republic of Germany

MARCEL DEKKER, INC. *New York and Basel*

Library of Congress Cataloging in Publication Data
Main entry under title:

Glass capillary chromatography in clinical medicine and pharmacology.

Includes index.
1. Chromatographic analysis. 2. Biomolecules—Analysis. 3. Drugs—Analysis. I. Jaeger, Halvor, [date]. [DNLM: 1. Chromatography, Gas—methods.
QD 272.C44 G549
QP519.9.C47G53 616.07'56 84-22972
ISBN 0-8247-7103-6

MARCEL DEKKER, INC.
270 Madison Avenue, New York, New York 10016

Current printing (last digit):
10 9 8 7 6 5 4 3 2 1

PRINTED IN THE UNITED STATES OF AMERICA

Preface

This book represents an effort to provide practicing medical, clinical, pharmaceutical, and biological scientists with practical and relevant state-of-the-art information in the field of capillary gas chromatography. Although a number of excellent books that describe and discuss the latest developments in gas chromatography are available, many are written at a highly technical level for use by chromatographic theorists. Others are written on an instructional level with only minor emphasis on specific types of application. In this volume, it is the intent of the editor to bring together contributions from world leaders in the fields of clinical medicine, pharmacology, and analytical and biological chemistry, all of whom are also renowned chromatographers, and in so doing to bridge the gap between theory and practice. Part I provides background theory and a description of the state of the art without emphasizing unnecessary and esoteric technical terminology and theoretical concepts that are not useful or interesting to those whose areas of expertise lie in other scientific disciplines. In Part II each chapter deals with a particular scientific problem and, because in each case the scientific problem is of central importance and capillary gas chromatography is the tool that facilitates the solution, readers are provided with the most practical and realistic insight into the kinds of chromatographic problems wherein their interests lie.

The editor is grateful to a number of people for their direct input, support, and cooperation. Needless to say, the majority of the time and effort necessary for the creation of this edition came from the authors of the individual chapters. For their work and their invaluable contributions, the editor is deeply appreciative. The editor is also grateful to Dr. Werner Gielsdorf and Dr. James A. Settlage for the support, advice, and many hours that they contributed in the coordination of this effort. Much help was also provided by Frau Gisela Brugger and all the staff at L.A.B. GmbH.

Halvor Jaeger

Contributors

Douwe D. Breimer, Ph.D. Department of Pharmacology, Subfaculty of Pharmacy, University of Leiden, Leiden, The Netherlands

Albertus G. de Boer, Ph.D.* Department of Pharmacology, Subfaculty of Pharmacy, University of Leiden, Leiden, The Netherlands

J. R. Carlin, B.Sc. Department of Animal Drug Metabolism and Radiochemistry, Merck Sharp & Dohme Research Laboratories, Rahway, New Jersey

Jeffrey A. Corkill, Ph.D.† College of Pharmacy and Allied Health Professions, and Barnett Institute of Chemical Analysis, Northeastern University, Boston, Massachusetts

Andre P. De Leenheer, Ph.D. Professor, Laboratoria voor Medische Biochemie en voor Klinische Analyse, Rijksuniversiteit Gent, Gent, Belgium

Hans Ditschuneit, Prof. Dr. Med. Department of Internal Medicine, University of Ulm, Ulm, Federal Republic of Germany

Wolfgang Dünges, Ph.D. Fachbereich Geowissenschaften der Universität, Mainz, Federal Republic of Germany

Hartmut Frank, Ph.D. Institut für Toxikologie, Eberhard-Karls-Universität, Tübingen, Federal Republic of Germany

Simon J. Gaskell, Ph.D. Tenovus Institute for Cancer Research, Welsh National School of Medicine, Cardiff, Wales

---

Current affiliations:

*Department of Pharmaceutical Technology and Biopharmacy, Subfaculty of Pharmacy, University of Leiden, Leiden, The Netherlands
†Assistant Professor, Department of Chemistry, Eastern Washington University, Cheney, Washington

C. F. Gelijkens, Ph.D. Laboratoria voor Medische Biochemie en voor Klinische Analyse, Faculteit van de Farmaceutische Wetenschappen, Rijksuniversiteit Gent, Gent, Belgium

Emilio Gelpí, Ph.D. Analytical Neurochemistry Unit, Instituto de Química Bio-Orgánica (CSIC), Barcelona, Spain

Werner Gielsdorf, Ph.D. L.A.B. Gesellschaft für pharmakologische Untersuchungen m.b.H. & Co. (Laboratory for Pharmacological Research), Neu-Ulm, Federal Republic of Germany

Roger W. Giese, Ph.D. College of Pharmacy and Allied Health Professions, and Barnett Institute of Chemical Analysis, Northeastern University, Boston, Massachusetts

K. Halpaap-Wood, Ph.D. Institute for Lipid Research, Baylor College of Medicine, Houston, Texas

Marjorie G. Horning, Ph.D. Institute for Lipid Research, Baylor College of Medicine, Houston, Texas

Yasuyuki Hoshika, Dr. Eng.* Aichi Environmental Research Center, Aichi, Japan

Elsa Ilias Chemist, L.A.B. Gesellschaft für pharmakologische Untersuchungen m.b.H. & Co. (Laboratory for Pharmacological Research), Neu-Ulm, Federal Republic of Germany

Halvor Jaeger, M.D. L.A.B. Gesellschaft für pharmakologische Untersuchungen m.b.H. & Co. (Laboratory for Pharmacological Research), Neu-Ulm, Federal Republic of Germany

Walter G. Jennings, M.S., Ph.D. Department of Food Science and Technology, University of California, Davis, California

Johannis P. Kamerling, Ph.D. Department of Bio-Organic Chemistry, State University of Utrecht, Utrecht, The Netherlands

Hans U. Klör, Ph.D. University of Ulm, Ulm, Federal Republic of Germany

Wilfried A. Koenig, Dr. Institut für Organische Chemie, Universität Hamburg, Hamburg, Federal Republic of Germany

Simon H. Kuttab, Ph.D. College of Pharmacy and Allied Health Professions, and Barnett Institute of Chemical Analysis, Northeastern University, Boston, Massachusetts

K. Lertratanangkoon, Ph.D. Institute for Lipid Research, Baylor College of Medicine, Houston, Texas

Hartmut M. Liebich, Ph.D. Medizinische Universitätsklinik, Tübingen, Federal Republic of Germany

Dieter Lutz, Dr. Chemist, L.A.B. Gesellschaft für pharmakologische Untersuchungen m.b.H. & Co. (Laboratory for Pharmacological Research), Neu-Ulm, Federal Republic of Germany

---

Current affiliation:

*Department of Hygiene, Shinshu University School of Medicine, Nagano, Japan

Neil K. McCallum, D.Phil. Chemistry Division, Department of Scientific and Industrial Research, Petone, New Zealand

Wilhelm Nebelung, M.D. University of Ulm, Ulm, Federal Republic of Germany

Carl D. Pfaffenberger, Ph.D. Professor and Director, Division of Chemical Epidemiology, Department of Epidemiology and Public Health, University of Miami School of Medicine, Miami, Florida

Johannes Rasper, Ph.D. L.A.B. Gesellschaft für pharmakologische Untersuchungen m.b.H. & Co. (Laboratory for Pharmacological Research), Neu-Ulm, Federal Republic of Germany

Pat. J. F. Sandra, Dr. Sc. Laboratorium voor Organische Scheikunde, Rijksuniversiteit Gent, Gent, Belgium

Eugen Schlegel, Ph.D. L.A.B. Gesellschaft für pharmakologische Untersuchungen m.b.H. & Co. (Laboratory for Pharmacological Research), Neu-Ulm, Federal Republic of Germany

James A. Settlage, Ph.D. L.A.B. Gesselschaft für pharmakologische Untersuchungen m.b.H. & Co. (Laboratory for Pharmacological Research), Neu-Ulm, Federal Republic of Germany

Cedric H. L. Shackleton, Ph.D. Children's Hospital Medical Center of Northern California, Oakland, California, and The University of California, San Francisco, California

Wanda G. Stillwell, Ph.D.* Research Assistant Professor, Institute for Lipid Research, Baylor College of Medicine, Houston, Texas

Cristina Suñol Analytical Neurochemistry Unit, Instituto de Química Bio-Orgánica (CSIC), Barcelona, Spain

W. J. A. VandenHeuvel, Ph.D. Department of Animal Drug Metabolism and Radiochemistry, Merck Sharp & Dohme Research Laboratories, Rahway, New Jersey

Nicholas P. E. Vermeulen, Ph.D. Departments of Pharmacology and Pharmaceutical Analysis, Subfaculty of Pharmacy, University of Leiden, Leiden, The Netherlands

Johannes F. G. Vliegenthart, Ph.D. Department of Bio-Organic Chemistry, State University of Utrecht, Utrecht, The Netherlands

R. W. Walker, B.Sc. Department of Animal Drug Metabolism and Radiochemistry, Merck Sharp & Dohme Research Laboratories, Rahway, New Jersey

---

Current affiliation:

*Consultant Chemist, Applications Laboratory, Delsi/Nermag, Houston, Texas

Contents

Glass Capillary Chromatography in Clinical Medicine and Pharmacology

I

CLINICAL MEDICINE

1

Introduction

Halvor Jaeger and James A. Settlage / *L.A.B. Gesellschaft für pharmakologische Untersuchungen m.b.H. & Co. (Laboratory for Pharmacological Research), Neu-Ulm, Federal Republic of Germany*

During the two decades following the introduction of capillary columns to the field of gas chromatography, the technique enjoyed a slow but significant growth as witnessed by the literature. Until recently though, capillary columns have been largely confined to relatively few laboratories where specialists solved particularly challenging separation problems. In the meantime the vast majority of gas chromatographic work was done with packed columns. But now there are indications that the scales may tip the other way. Jennings has written that capillary chromatography is experiencing a renaissance (Chap. 3). We might, on the other hand, say that the child has finally reached adolescence and is presently undergoing a characteristic growth spurt.

In any case the last few years have been a period of remarkable expansion for this technique. While this development has paralleled a growth in all areas of analytical chemistry and has been further fueled by the increasing number of separation problems which can be solved only by the higher resolving power and greater inertness of capillary columns, it is also clear that the growth has partially stemmed from an increasing tendency for laboratories to phase in capillary columns for routine work and to phase out packed columns. This transition is not likely to occur rapidly, nor is it ever likely to be complete, but the trend itself speaks clearly and firmly on the side of the newer technique.

For the practicing scientist in any field confronted with the need to develop a laboratory system for carrying out gas chromatographic separations, it will be instructive to consider some of the factors which have contributed to the recent increase in popularity of capillary chromatography and, on the other hand, those factors which have prevented the present growth spurt from occurring much earlier and are still effective in slowing that growth.

It is not difficult to document the inherent advantages of capillary columns in terms of resolution, inertness, and detection limits. There is simply no way to challenge these. Less definitive but nevertheless sound arguments have been made for the potential economy of capillary columns versus packed columns. Furthermore, high-quality capillary columns and a large resevoir of technical know-how have been available to use for a number of years. Whey then, if arguments for capillary columns are valid, are most GC separations still conducted with packed columns or, to put it the other way, what has happened in recent years to elicit the keen interest that was formerly lacking? Some factors are clear. Packed columns have always been less expensive to purchase or, alternatively, less difficult to produce in the laboratory. They require much less skill to use and are far more forgiving of such abuses as inadequate sample cleanup, contaminated carrier gas, improper heating, careless injection, and a long list of others. In a profit-oriented lab or in a lab where GC serves only a secondary purpose and is used by workers whose expertise is confined to other areas, for example in a synthesis lab, these are very real advantages. As a result, while capillary columns were still in their infancy, the ease and economy with which packed columns could be implemented were already making that technique extremely popular. It followed naturally that chromatographic instruments, including ovens, gas supplies, inlets, detectors, and even integrators were designed specifically for packed columns. At the same time methodologies ranging from the monitoring of synthetic reactions to U.S. EPA protocols were being written for the implementation of packed columns. Such a body of hardware and information generated over more than 20 years will certainly possess a formidable inertia.

This by itself provides a tremendous barrier against a rapid transition to capillary usage. But there is in addition a purely human aspect to the problem. Practicing chromatographers who have become experts in the use of packed columns find themselves forced into the position of novices when they undertake their first study with a capillary column. The natural reluctance that this creates coupled with the much greater demands of the new technique and often with the difficulties of adapting available equipment has surely led to many false starts. Capillary columns that are not properly installed and properly used may very well produce chromatograms that are not significantly better than those generated by packed columns, and practitioners who obtain inferior results from their first efforts might well be inclined to conclude that capillary columns are not justified except in cases where they are absolutely required.

But the fact is that in most analytical situations capillary columns are at least justified simply because they are inherently capable of generating more information with a higher level of reliability. Comparison of chromatograms of identical samples generated by each method clearly demonstrates this. Often peaks which appear to represent a single substance in a packed column chromatogram are shown to include two or more substances when eluted from a capillary column. Furthermore, compounds which do not elute at all from a packed column may chromatograph quite well on a fused silica or even a glass capillary column. Even the apparent advantage of larger injection volumes on packed columns is usually more than offset since the broader solute bands generated on packed columns deliver a lower concentration of sample per unit of time to the detector in spite of the larger injection volume.

The point here is not to reiterate the basic superiority of capillary columns but to make a case for their use even in studies where packed columns demonstrate "reliable" results.

The problem is that there is a degree of uncertainity in any chromatographic study, and while test runs made with standards may generate reproducible results, the reliability of subsequent runs is always jeopardized by changes in the column which may cause variations in the way the solutes of interest adsorb or degrade on the column, by fluctuations in the baseline, especially when measuring near the detection limit, and by extraneous substances coeluting with coumpounds of interest. The justification for capillary columns is that they increase the credibility of any study by minimizing these sources of error. And when measurements are made, the increased sensitivity of the capillary column often provides greater precision because measurements are made further above the detection limit, where coefficients of variation become prohibitively large.

The growing realization of these facts has surely been a factor in the recent shift in favor of capillary columns. But perhaps no single factor has had more influence than the recent introduction of fused silica. These new columns possess all of the advantages of the glass capillary, are more inert than all but the most scrupulously prepared glass, and, perhaps most important, their felxibility greatly facilitates handling and installation and obviates the need to straighten ends.

Fused silica columns and other state-of-the-art developments are discussed in Chapter 3, but it is worth emphasizing some of those developments which especially enhance and further justify capillary chromatography. A major opportunity offered by fused silica columns which is often overlooked is the possibility of connecting them directly into most detector cells, hence optimizing sensitivity, preventing possible sample loss on active surfaces and avoiding any peak distortion.

This advantage applies especially to mass spectrometers where the use of glass columns or, of course, packed columns necessitates passage of the sample through a long and often oversized transfer line which contains dead volumes and which butts with but does not enter the source block.

On-column injection is reviewed by several authors in this volume and is recommended for its quantitative reproducibility and for the fact that in connection with fused silica columns it provides an optimally inert chromatographic system from sample container to detector. Such a system makes it possible to chromatograph compounds which might not survive conditions required for split or splitless injection. An additional advantage not frequently mentioned is in method development. Troubleshooting and then optimizizing a quantitative analytical method involves adjustment of a large number of parameters especially in the case of GC-MS. On-column injection minimizes potential chromatographic problems and provides reliable and constant chromatographic conditions while other aspects of the method such as detector variables and solvent systems are optimized. Later, the system can be converted to split or splitless as desired or needed.

Finally, the advent of bonded liquid phases is worthy of mention. Capillary columns can easily be destroyed by a single on-column injection, if the material injected contains a large amount of nonvolatile residue. Even split or splitless injection often results in a small amount of residue finding its way onto and eventually throughout the column. The result is a loss of efficiency until the column becomes useless. The washability of bonded-phase columns often makes it possible to rejuvenate a column, thus helping to remove another barrier against their use.

The recent developments discussed have both facilitated and encouraged the implementation of capillary chromatography, and a trend is clearly developing, but there is still a great deal more that can and will be done which is likely to further accelerate the transition. The current high interest in capillaries will motivate a parallel investment in research and development by the makers of chromatographic equipment. Today's instruments that are advertized as being "designed for capillary columns" are usually shipped with 1/4-inch fittings on detectors and inlets and a conversion kit complete with the necessary adapters and a hanger for the capillary column.

(This often does little to prevent the column from contacting oven walls and thus generating local hot or cold spots in the column.) Detectors are invariably designed for packed-column effluents. Even column inlets are often not adequately modified to provide for proper split or splitless injection. In spite of these deficiencies most systems can be made to yield excellent results but, on the other hand, several exciting possibilities for development exist.

First, detectors with cell volumes commensurate with capillary columns are likely to be developed which will provide detection limits orders of magnitude below the present levels. Second, a whole new generation of bonded liquid phases tailored for a variety of separation problems is already beginning to appear and can be expanded almost without limit.

Third, it is likely that recent developments in effluent splitting and switching and even recycling will mature and find use in an increasing number of applications. Finally, there is reason to hope that inlet systems will be developed which will provide injection of a wide range of sample types and require no more skill than present-day packed-column inlet systems.

As developments in the application of capillary columns continue to occur, the range of possible uses grows at an even faster pace. It is hoped that the reader will find in this volume a fair representation of their present application in the medical field. Efforts have been made to update this material throughout the editorial and publication process so as to provide the reader with the latest possible developments.

2

Gas Chromatographic Theory

Walter G. Jennings / *Department of Food Science and Technology, University of California, Davis, California*

I. INTRODUCTION

The dictionary defines "chromatography" as the separation of mixtures into their constituents by preferential adsorption by a solid. Few scientists would agree with this definition. More precisely, chromatography is the separation of a mixture into its constituents by repetitively partitioning those components between two phases; usually one of those phases is mobile and the other stationary. The word itself was suggested by Tswett [1], who developed a liquid-solid adsorption chromatography system for his studies on plant pigments. Paper and thin-layer chromatography are other examples of liquid-solid chromatography. The simplest single-transfer example of liquid-liquid chromatography is a separatory funnel extraction, extended to repetitive separations by the Craig countercurrent distribution apparatus. Where the mobile phase is gaseous, two types of chromatography are possible, gas-solid (adsorption) chromatography, and gas-liquid chromatography. The latter, also known as gas-liquid partition chromatography and to which we will confine our attention, is the more powerful technique and has become known simply as "gas chromatography" or "GC."

Gas chromatography has become the world's most universally used analytical technique. Its wide acceptance and rapid growth can be largely attributed to the facts that it affords good powers of separation, that analysis times are relatively short, and that sensitivities are reasonably high. Because of these advantages, many methods have been developed for the production of volatile derivatives, permitting gas chromatographic analysis of many higher-boiling or relatively nonvolatile substances such as amino acids, steroids, and even metals.

The judicious exploitation of this powerful tool is contingent upon the selection of the best conditions for a given analysis, and the utilization of newer methods of sample introduction, separation, and detection; realization of this goal requires some exposure to gas chromatographic theory. And as is the case with practically every technical field, this exposure involves a new and specialized vocabulary. The glossary at the end of this chapter summarizes the definitions and symbols that will be used.

II. BASIC APPARATUS

The gas chromatograph consists of three basic components: (1) the column, which contains the stationary liquid phase and which is continuously swept by the mobile gas phase (carrier gas); (2) an inlet by which the sample to be separated is introduced onto the front of the column; and (3) a detector, which generates a signal for each eluant emerging from the column. These various components will be considered in Chap. 3, but our introduction to gas chromatographic theory requires some distinction of the different column types.

The two extremes are packed columns and wall-coated open-tubular (WCOT) columns. The former have larger sample capacities and are more forgiving of deficiencies in instrument design and operator technique, but the latter are capable of achieving much more complete separations, shorter analysis times, and higher sensitivities; in addition, they can provide a more inert system. Porous-layer open-tubular (PLOT) columns offer a compromise between these two extremes. Support-coated open-tubular (SCOT) columns are merely PLOT columns that use a solid support material (rather than fused glass, etc.) to produce the porous layer. Whiskered columns, produced by growing elongated crystals (usually sodium fluoride) inside a glass capillary before the coating is applied, are really a special form of PLOT columns. These various column types are shown schematically in Figure 2.1.

III. THEORY

A. Basic Considerations

The purpose of the column, of course, is to permit partitioning of sample constituents between the stationary liquid phase and the moving gas phase; this is encouraged by having the liquid phase in the form of a thin film, with a large surface area exposed to the moving gas phase. As a solute (sample component) is injected onto the column, the molecules partition between the liquid and the gas phase in accordance with the *distribution constant*, K_D:

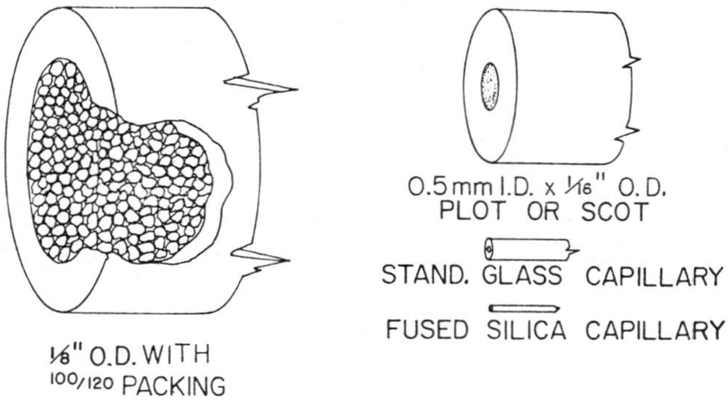

Figure 2.1 Schematic representation of various column types, approximately to scale. Standard glass capillaries are about 0.8 mm OD and 0.25–0.32 mm ID; flexible fused silica capillaries are about 0.20–0.32 mm ID, with a wall thickness of about 0.025 mm.

$$K_D = \frac{\text{concentration in liquid phase}}{\text{concentration in gas phase}} = \frac{c_L}{c_G} \tag{1}$$

This is a true equilibrium constant, specific for that solute and that liquid phase at that temperature. The continuously moving carrier gas conducts those solute molecules in the gas phase to virgin liquid phase, where they again partition in accordance with K_D; those solute molecules that were in the liquid phase overcome the dispersive forces and emerge from solution to the gas phase, where they advance and continue the partitioning process.

Whenever a solute is in the gas phase, it is carried through the column at the same rate as the carrier gas (i.e., the moving gas phase). The molecules of a solute that is not strongly attracted to the liquid phase emerge to the carrier gas stream more frequently than do the molecules of a solute that has a higher affinity for the liquid phase. Hence the first solute has a shorter *retention time* than does the second, and separation is achieved. The degree of separation is, of course, also affected by broadening of the solute bands as they progress through the column; excessive band broadening leads to interdiffusion of these separating solute bands and at least partially negates the separation process.

One of the factors accounting for the much higher degree of separation that can be achieved in an open-tubular column as compared to a packed column is that the latter experiences a higher degree of band broadening due to its multiplicity of flow paths (Figure 2.2). In an open-tubular column, there is essentially a single flow path, and this contribution to band broadening is absent. Another factor that affects band broadening is the uniformity of the liquid film phase thickness. If the film is uniformly thin, then all molecules of a given solute experience approximately the same degree of retardation with each solution and revaporization step. In the case of a nonuniform film thickness, a molecule dissolving in a globule or thick patch of liquid phase will be retarded more than one dissolving in a thin portion of the film, and band broadening results. In packed columns, the liquid phase is distributed over a granular support, and where these particles contact each other, there is a tendency to form thicker areas of liquid phase.

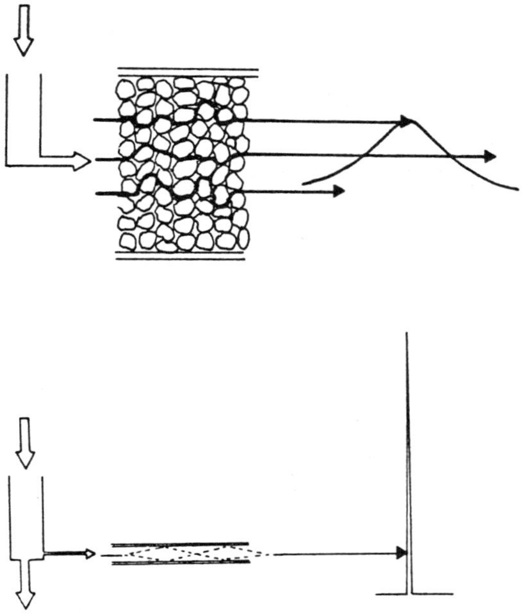

Figure 2.2 Comparison of the flow paths in packed and open-tubular columns. The total injection is normally delivered to the packed column, which, partly because of a multiplicity of flow paths, generates a relatively broad, low-intensity peak. Although only a fraction of the injection may be delivered to the open-tubular system, the smaller quantity delivered to the detector arrives as a tighter band, resulting in more sample per unit time, which generates a sharp, high-intensity peak.

Thus, the efficiency of a column can be related to the degree of band broadening it engenders in a solute, relative to the length of time that solute has been in the colum. If two columns exhibit the same retention time for a given solute, that column in which a lower degree of band broadening occurred has achieved a higher efficiency. This is illustrated in Figure 2.3.

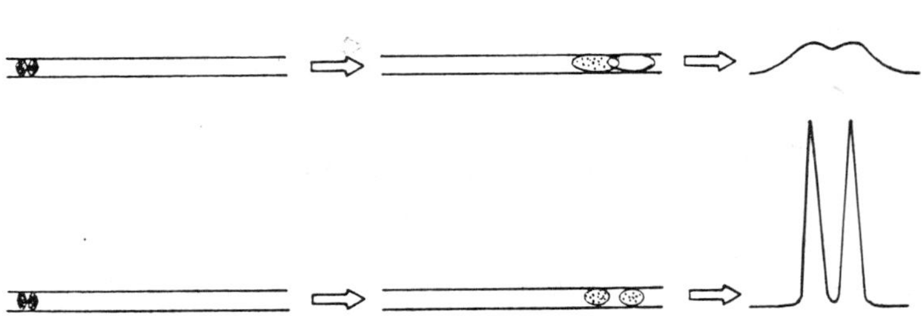

Figure 2.3 Schematic representation of columns with low (top) and high (bottom) efficiencies. Assuming identical solutes, liquid phase, and temperature, the relative retentions are the same in both cases; the solutes in the low-efficiency column have suffered a greater degree of longitudinal diffusion, resulting in an incomplete separation as they are delivered to the detector.

As a solute emerges from the column, the shape of the peak generated by the detector is affected by (1) the width of the emerging solute band, and (2) the concentration of solute in the gas phase, that is, K_D. If K_D is large, most of that solute is in the liquid phase and the concentration in the gas phase is low; solute molecules emerge to the gas phase to reestablish K_D only slowly, as their predecessors are swept to the detector. A larger volume of carrier gas (i.e., a longer time) is required to conduct all of that solute to the detector, which, in consequence, generates a broad, low-intensity peak. If K_D is small, the concentration in the gas phase is high, the emerging solute is conducted to the detector at a higher concentration in a shorter time, and a tall, sharp peak is produced. Under isothermal conditions, the K_D for a given solute is constant, and with the exception of overloaded conditions (see below), column efficiencies can be related to peak widths and retention times. Where the column temperature is increased (usually in a linear fashion) during the course of a run (temperature programming), this situation becomes more complex.

B. Theoretical Plate Numbers

Provided the sample was properly injected and the capacity of the column was not exceeded (i.e., the solute chromatographed in accordance with its equilibrium K_D), a Gaussian peak should result. For isothermal determinations, column efficiencies are normally defined in terms of *theoretical plates*, n:

$$n = \frac{(t_R)^2}{\sigma} \tag{2}$$

where t_R is the retention of the solute measured from the point of injection to the peak maximum, and σ is the standard deviation of the peak (Figure 2.4). Both values must be measured in the same units. The *peak width at half height* ($w_{.5}$) can be measured with some precision (an optical micrometer is normally used), and in a Gaussian peak, this amounts to 2.35σ (Figure 2.4). Hence

$$n = \frac{(t_R)^2}{(w_{.5}/2.35)^2} = 5.54\left(\frac{t_R}{w_{.5}}\right)^2 \quad \text{(Figure 2.5)} \tag{3}$$

Column efficiencies are also described in terms of *theoretical plates per meter*, but more frequently the inverse of this is used, and we speak of the *height equivalent to a theoretical plate*, h:

$$h = \frac{L_{(mm)}}{n} \tag{4}$$

A certain volume of gas is required to carry even a nonsorbed component (one that spends no time in the liquid phase, e.g., air) through the column. This is known as the gas holdup volume of the column. Under conditions of constant flow, volumes are proportional to times, and particularly with compressible fluids such as gases, it is usually simpler to accurately measure times than volumes. It is for this reason we normally use *retention times* rather than *retention volumes* to describe the retardation of a solute, and the term *gas holdup time* is employed in a similar manner. Because the more

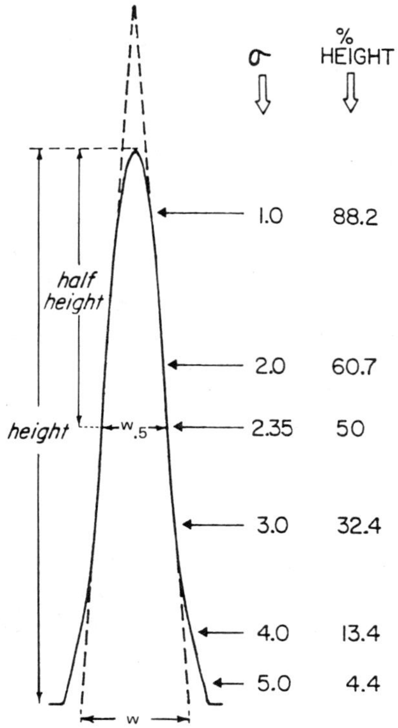

Figure 2.4 Characteristics of a Gaussian peak.

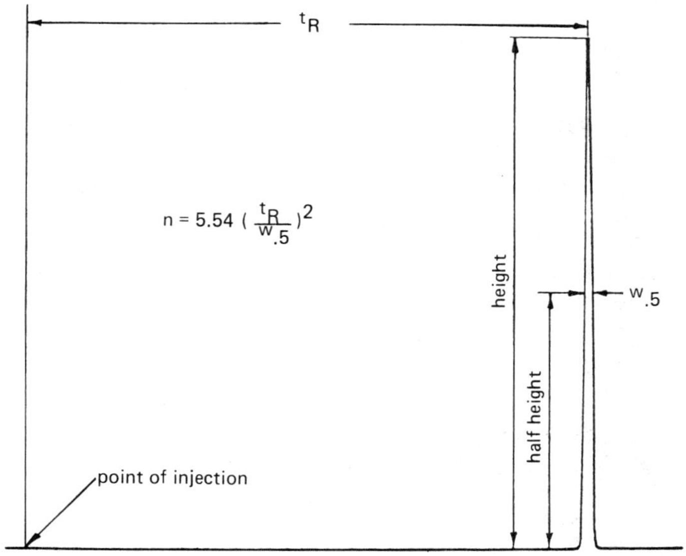

Figure 2.5 Method of calculating the number of theoretical plates. The height equivalent to one theoretical plate is obtained by dividing the column length in millimeters by n.

sensitive detectors required for high resolution gas chromatography are generally insensitive to air, the leading edge of a methane peak is commonly used to measure the gas holdup time; the symbol t_M is used to denote this value.

As mentioned earlier, whenever a solute is in the gas phase, it is swept toward the detector at the same rate as the carrier gas. Hence every solute spends the same time (equivalent to t_M) in the gas phase. Inasmuch as the total retention time of a solute is t_R, the time that solute spends in the liquid phase, that is, the *adjusted retention time* (t'_R), is:

$$t'_R = t_R - t_M \quad \text{(Figure 2.6)} \tag{5}$$

Solutes separate on the basis of the difference in the degree of attraction which the liquid phase has for them; separation is maintained (hopefully), but it is neither generated nor improved upon while those solutes are in the gas phase. Because theoretical plate measurements are based on the total retention (t_R), they embrace both t_M (which does not affect separation) and t'_R (which does affect separation). In capillary columns, t_M is proportionately much larger. In other words, not all of the theoretical plates embraced by n are effective; the number of *effective theoretical plates*, N, is defined as:

$$N = 5.54\left(\frac{t_R - t_M}{w_{.5}}\right)^2 = 5.54\left(\frac{t'_R}{w_{.5}}\right)^2 \quad \text{(Figure 2.7)} \tag{6}$$

Figure 8 shows that, regardless of whether we deal with theoretical or effective theoretical plates, there is a certain elasticity to these numbers that is apparently related to the retention characteristics of the solutes used as test compounds. This solute characteristic is the partition ratio, the ratio of the amount (i.e., the weight; *not* the concentration) of that solute in the liquid phase to the amount in the gas phase. These values are, of course,

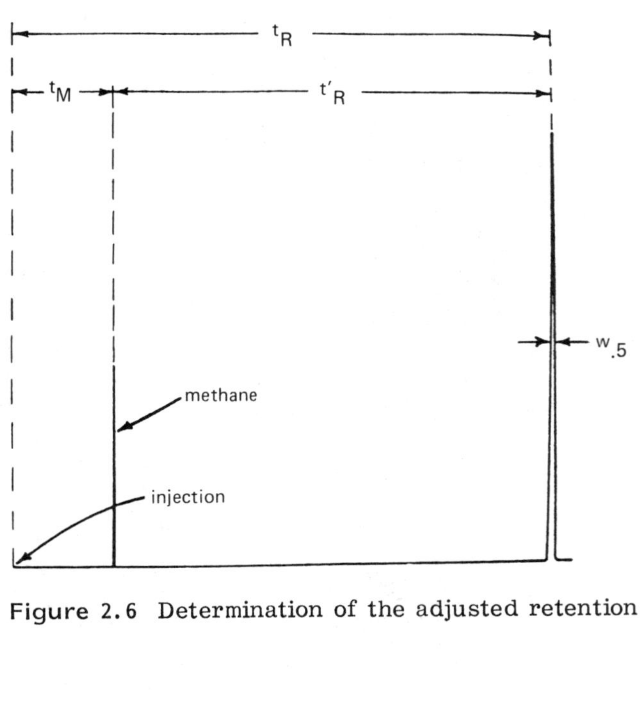

Figure 2.6 Determination of the adjusted retention time, t'_R.

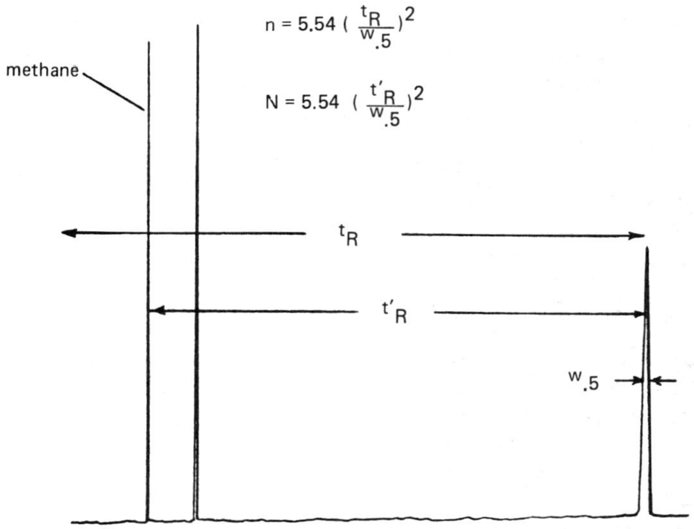

Figure 2.7 Method of determining the effective theoretical plate number, N.

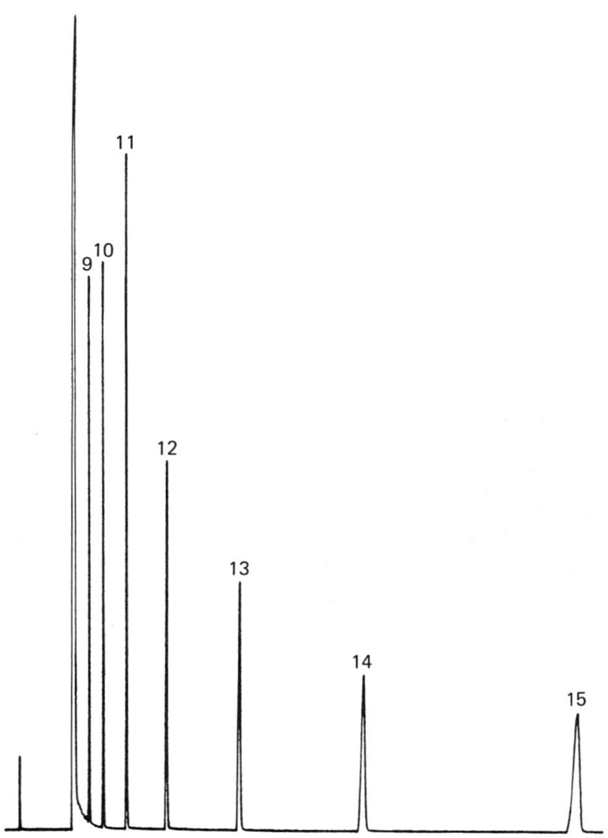

Figure 2.8 Effect of the partition ratio (t) on theoretical and effective theoretical plate numbers.

proportional to the time that solute spends in the liquid phase and the time it spends in the gas phase. Hence, the *partition ratio*, k, is defined as

$$k = \frac{t_R - t_M}{t_M} = \frac{t'_R}{t_M} \qquad \text{(Figure 2.6 and 2.8)} \qquad (7)$$

Figure 2.9 shows the theoretical maximum possible plate numbers as a function of the partition ratio.

The maximum possible efficiency that could in theory be obtained from a column of a given radius with a perfect coating is reflected by the *theoretical minimum height equivalent to a theoretical plate*, h_{min} [2]:

$$h_{min} = r_o\left(\frac{11k^2 + 6k + 1}{3(1 + k)^2}\right)^{1/2} \qquad (8)$$

where r_o is the radius of the capillary, and k is the partition ratio of the test compound. Comparison of this value with that obtained from (4), above, permits calculation of what has been termed the *coating efficiency* (CE) of the column:

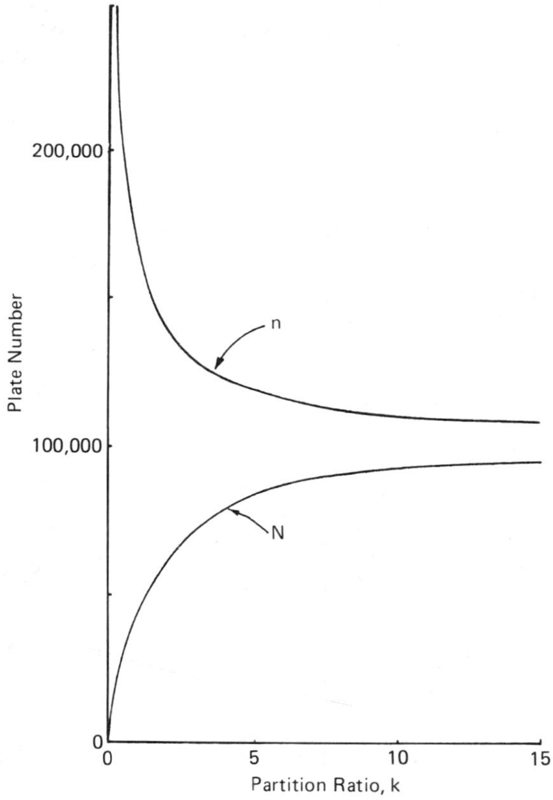

Figure 2.9 Plots of the theoretical maximum plate numbers as functions of the partition ratio.

$$\% \text{ CE} = \frac{h_{min}}{h} \times 100 \tag{9}$$

It should be emphasized that Equation 8 is applicable only to smooth-bore WCOT columns; it has no validity with PLOT, SCOT, or whisker columns, and its application to deeply etched WCOT columns is questionable.

In an attempt to eliminate the dependence of plate numbers on partition ratios, Kaiser [3,4] took advantage of the linear relationship that exists between peak widths and half height and partition ratios in an isothermal separation. The validity of this relationship has been challenged (e.g., Franken [5], Guichon [6], and Smuts et al. [7]), but others argue that discrepancies are minor (e.g., Smuts et al. [8] and Nilsson [9,10]). Using linear regression to calculate the best-fit line for the experimental data, the widths at half height that correspond to solutes with partition ratios of zero (air) and 10.0 are then calculated; these are termed b_o and b_{10}, respectively. From Equation 7, it can be seen that the adjusted retention time of a solute $t'_R = kt_M$. The adjusted retention time of that hypothetical solute with k = 10 is therefore $10t_M$. The *real plate number* (n_{real}) can therefore be defined as:

$$n_{real} = 5.45\left(\frac{10t_M}{b_{10} - b_o}\right)^2 \tag{10}$$

This value is, of course, independent of k, and falls between the values for n and N. All of the efficiency measurements discussed to this point are designed for isothermal determination and have no validity when used under conditions of temperature programming.

C. Phase Ratio

A major distinction between packed and open-tubular columns lies in the relative availability of gas and liquid phase, that is, the *phase ratio*, β. This characteristic of the column is defined as:

$$\beta = \frac{\text{volume of gas phase}}{\text{volume of liquid phase}} \tag{11}$$

and in a WCOT column (for all practical purposes)

$$\beta = r_0/2\, d_f \tag{12}$$

where r_o is the *radius of the capillary* and d_f is the average thickness of the liquid-phase film. In packed columns, typical β values range from about 5 to 25, and in WCOT columns, they normally range from about 100 to 300, and can be as high as 1500 in extreme cases.

Note, by referring back to Equation 1, that the distribution constant K_D is based on solute concentrations. Concentrations can be expressed as the amount (i.e., weight) of solute per unit volume of solvent:

$$K_D = \frac{\text{amount of solute in liquid phase/volume of liquid phase}}{\text{amount of solute in gas phase/volume of gas phase}}$$

$$= \frac{\text{amount of solute in liquid phase}}{\text{amount of solute in gas phase}} \times \frac{\text{volume of gas phase}}{\text{volume of liquid phase}}$$

The first fraction was defined as the partition ratio, k (Eq. 7), and the second is the phase ratio, β (Eq. 11). Hence,

$$K_D = \beta k \tag{13}$$

This fundamental relationship will be used again and again, especially in Chap. 3, as we explore special injection procedures, and parameters such as film thickness, column diameter, column length, temperatures, and program rates and carrier gas velocity. Through Equation 13, we can explore the relationships between these variables and separation efficiencies, analysis times, and the temperatures to which chromatographing solutes will be subjected, which can be critically important, for example, for thermally labile compounds.

D. Measurements of Component Separation

1. Relative Retention

The relative retention of two solutes is determined by dividing the adjusted retention of the second by the adjusted retention of the first; thus, if any degree of separation is achieved, the relative retention is greater—never less—than one:

$$\alpha_{A,B} = \frac{t'_{R\,(B)}}{t'_{R\,(A)}} \tag{14}$$

Relative retentions are affected by differences in the degrees of attraction that the liquid phase has for the solutes in question. Nonpolar solutes such as the paraffin hydrocarbons will exhibit larger relative retentions in a nonpolar liquid phase such as a methyl silicone SE 30 than they will in a more polar liquid such as the polyethyleneglycol Carbowax 20 M. Polar solutes such as alcohols will exhibit smaller relative retentions in the apolar liquid phase and larger relative retentions in the polar liquid phase. For a given change in temperature, the vapor pressure of a larger homologue is affected to a larger degree than is the vapor pressure of its smaller relative. Hence, relative retentions vary inversely with temperature; thus, at lower column temperatures relative retentions are larger, but peaks, as discussed earlier, are broader.

Relative retentions do not in themselves constitute a measure of component separation because they give no indication of peak sharpness. Figure 2.3 illustrates two very different degrees of separation, although relative retentions are the same in both cases. As the efficiency of the system (i.e., peak sharpness) increases, separations can be accomplished at lower relative retention values. The low plate numbers generated by packed columns required a vast array of liquid phases, which the chromatographer selected on the basis of the relative retentions of the compounds to be separated. Modern capillary systems can generate such high efficiencies that less selective (i.e., those producing lower relative retentions) liquid phases that fail to resolve those solutes on a packed column may be eminently suitable to their separation in a capillary system.

2. Resolution

Resolution takes into account both the separation of the peak maxima and the sharpness of those peaks. It is defined as twice the distance between those

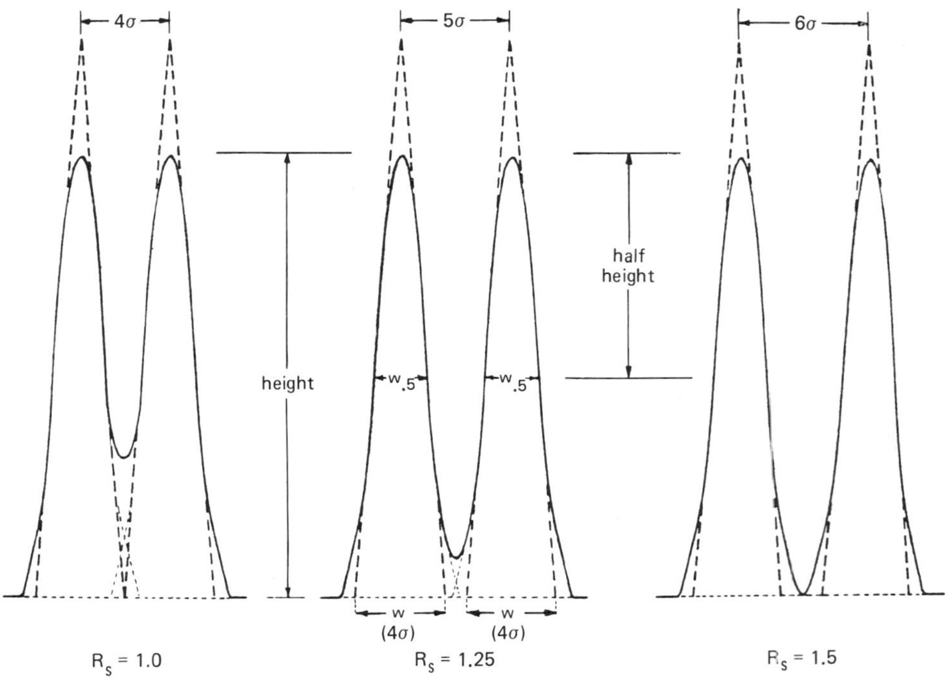

Figure 2.10 Resolution of two Gaussian peaks.

peaks divided by the sum of their widths (idealized) at base line:

$$R_s = \frac{2(t_{R(B)} - t_{r(A)})}{w_{(A)} + w_{(B)}} \qquad \text{(Figure 2.10)} \qquad (15)$$

It is usually simpler to measure peak widths at half height with good precision. In a Gaussian peak, base width is 4.0σ, and the width at half height is 2.355σ. Hence $w_{.5} = 0.589w$; twice this value is 1.18, and we can say:

$$R_s = 1.18\frac{t_{R(B)} - t_{R(A)}}{w_{.5(A)} + w_{.5(B)}} \qquad (16)$$

Qualitatively, a resolution of 1 will achieve about 95% separation, and a resolution of 1.5 will yield essentially baseline separation.

The number of theoretical plates required for any desired degree of separation can be determined from the relationship

$$n_{req} = R_s^{\,2}\left(\frac{\alpha}{\alpha - 1}\right)^2\left(\frac{k + 1}{k}\right)^2 \qquad (17)$$

where k represents the partition ratio of the second component. The use of this relationship in the selection of columns may be facilitated by remembering that, assuming uniform column efficiency, the number of theoretical plates increases as a square root function of the column length; that is, a fourfold increase in L is necessary to double n (valid only at the optimum carrier gas velocity; see below). Because some suppliers list theoretical, and others specify effective theoretical plates, it is also useful to realize that

$$n = N\left(\frac{1 + k}{k}\right)^2$$

hence, (18)

$$N = \left(\frac{k}{k + 1}\right)^2$$

3. *Separation Number*

The separation number is based on the separation of two consecutive members of a homologous series; the normal (straight-chain) paraffin (saturated) hydrocarbons are most widely used. The concept was originally suggested by Kaiser [11], who called it the Trennzahl, and the German abbreviation is still used. The *separation number*, TZ, is defined as:

$$TZ = \frac{t_{R(B)} - t_{R(A)}}{w_{.5(A)} + w_{.5(B)}} - 1 \qquad \text{(Figure 2.11)} \qquad (19)$$

where A and B in this case represent two members of a homologous series possessing n and n + 1 carbon atoms, respectively (i.e., differing by a single methylene unit). The separation number is, therefore, closely related to resolution. This system has been used both for measuring component separation and for describing column efficiency. It has been applied to both isothermal and programmed temperature conditions. Values become larger as compounds of longer retention are used as test substances, and as the temperature (or program rate) of the determination is decreased. This latter effect is over and above the increase in retention associated with the lower temperature [12].

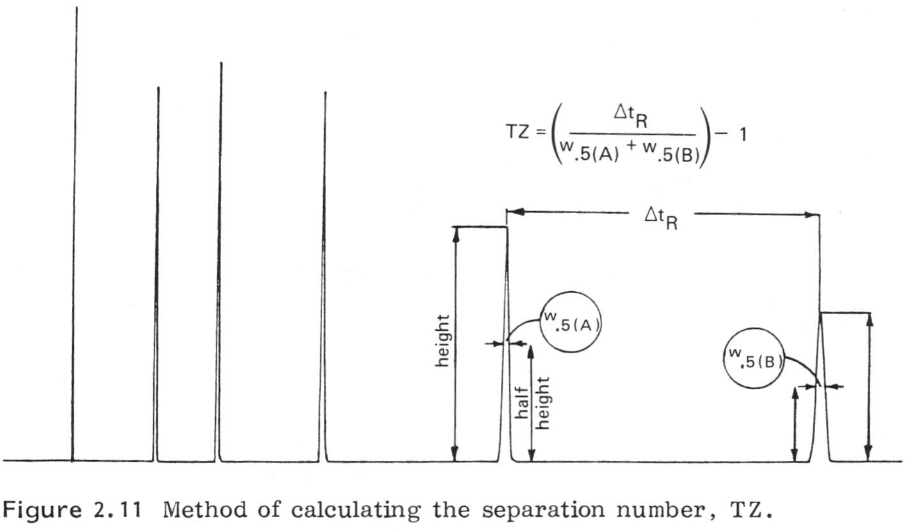

Figure 2.11 Method of calculating the separation number, TZ.

E. Carrier Gas Considerations

The efficiency of a given column, as reflected by the number of theoretical plates (or the magnitude of h) that it generates on a given solute at a given temperature, is strongly influenced by the velocity of the carrier gas flowing through the column. This velocity, of course, is the same velocity at which the solute molecules are forced to move whenever they are in the gas phase.

At any given temperature, the degree of dispersion of solute molecules in the gas phase is time dependent; thus, a greater degree of diffusion occurs during a long analysis (low carrier gas velocity) than during a short analysis (high carrier gas velocity). This diffusion influences separation efficiency in two ways which have opposing effects on the value of n (or h).

As a solute traverses the column, the establishment of its equilibrium distribution constant at each point of partitioning is encouraged by rapid transverse diffusion, that is, the free and unhindered movement of those solute molecules from the liquid phase into the gas phase and (more important in WCOT columns) back into the liquid phase. Diffusion in the longitudinal direction, however, would lead to excessive band broadening. The interrelationship of these and other factors that influence colum efficiency is expressed by the van Deemter equation [13]. As applied to open-tubular columns, this is known as the Golay equation [2]. In its simplest form, this states that

$$h = B/\bar{u} + (C_G\bar{u} + C_L\bar{u}) \tag{20}$$

where h represents the height equivalent to a theoretical plate; B, C_G, and C_L are constants for that column under those conditions and represent, respectively, longitudinal diffusion, the resistance to solute molecule transfer from the carrier gas to the liquid phase (closely related to a resistance to transverse diffusion), and the resistance to solute molecule transfer from the liquid phase to the carrier gas; $\bar{u}$ is the average linear carrier gas velocity. In WCOT columns the liquid phase film is very thin, and the gas volume is disproportionately large; this makes C_L vanishingly small compared to C_G, and

$$h \simeq B/\bar{u} + C_G\bar{u} \tag{21}$$

The smaller h, the more efficient the column. At low velocities (i.e., long analysis times) longitudinal diffusion becomes limiting; the $B/\bar{u}$ term dominates and h is large. At fast velocities, the opportunity for longitudinal diffusion is decreased, but now there is insufficient time for the solute molecules to diffuse transversely through the large volume of gas phase to the liquid phase before they are swept past the next available partitioning site; although $B/\bar{u}$ is now small, $C_G\bar{u}$ becomes limiting and h is large. Obviously, there must exist some optimum velocity at which the sum of these two terms is minimal; this is reflected by the van Deemter plot (Figure 2.12).

Solute diffusion is influenced both by the diffusivity of the carrier gas (see below) and by the diffusivity of the solute molecules. Small solute molecules diffuse more rapidly in a given carrier gas than do larger solute molecules. Hence the value of $\bar{u}_{opt}$ reflected by the van Deemter plot will vary with the diffusion characteristics of the test compound; as the size of the molecule (which usually parallels the partition ratio) increases, the optimum gas velocity decreases. Because the larger solute molecule diffuses more slowly, a longer time (i.e., a lower gas velocity) is required for transverse diffusion, and a longer time is permissable before the degree of solute dispersion resulting from longitudinal diffusion (band broadening) becomes limiting.

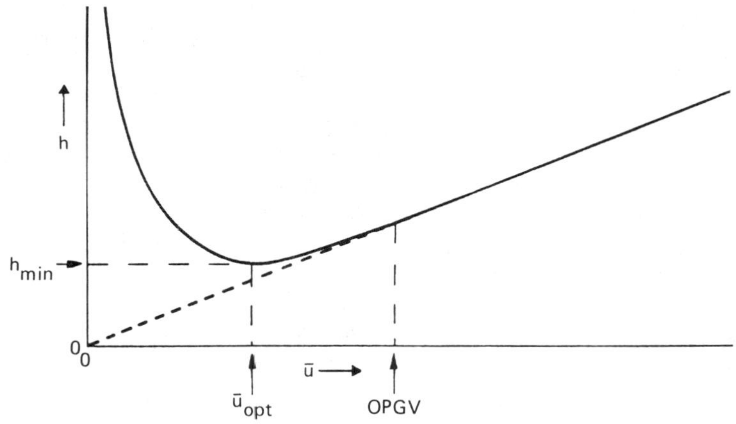

Figure 2.12 Theoretical van Deemter (Golay) curve for an open-tubular column.

One practical observation that emerges from these facts related to temperature programming (see below). Capillary systems are usually run at constant pressure drop, and as the column temperature increases, the velocity of the carrier gas decreases. The use of devices to counteract this velocity decrease, by continually increasing the carrier gas pressure during temperature programming, is periodically suggested. However, the size of molecule most affected by the chromatographic process increases as the temperature rises; the concurrent decrease in carrier gas velocity tends to be self-optimizing, and the wisdom of increasing the gas pressure to negate this effect is questionable.

Figure 2.13 illustrates van Deemter curves obtained with three different carrier gases, all other conditions being constant [14]; differences are

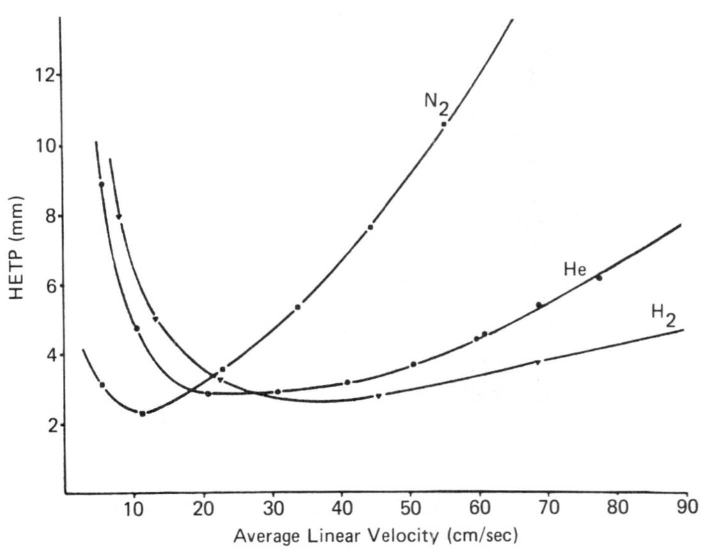

Figure 2.13 Experimental van Deemter curves obtained on a 50 m × 0.25 mm glass capillary with different carrier gases. (From Ref. 14.)

consistent with the diffusivities of those gases and bear out our earlier rationalizations; shorter analysis times (i.e., higher gas velocities) are preferred under conditions of higher diffusivity. As reflected by Figure 2.14 [15], this has great practical significance.

It should be noted that while $\bar{u}_{opt}$ is used to verify column efficiency as specified by the manufacturer, most experienced practitioners prefer to operate at carrier gas velocities one and one-half to two times $\bar{u}_{opt}$ because, as the velocity is increased above $\bar{u}_{opt}$, the increase in h is more than offset by the decrease in analysis time. This fact is stressed by Figure 2.15, which plots the efficiency per unit time as a function of the carrier gas velocity, and which stresses the vast superiority of hydrogen as carrier gas. The advantages and disadvantages of the various carrier gases and precautions to be taken with hydrogen have been discussed elsewhere [16].

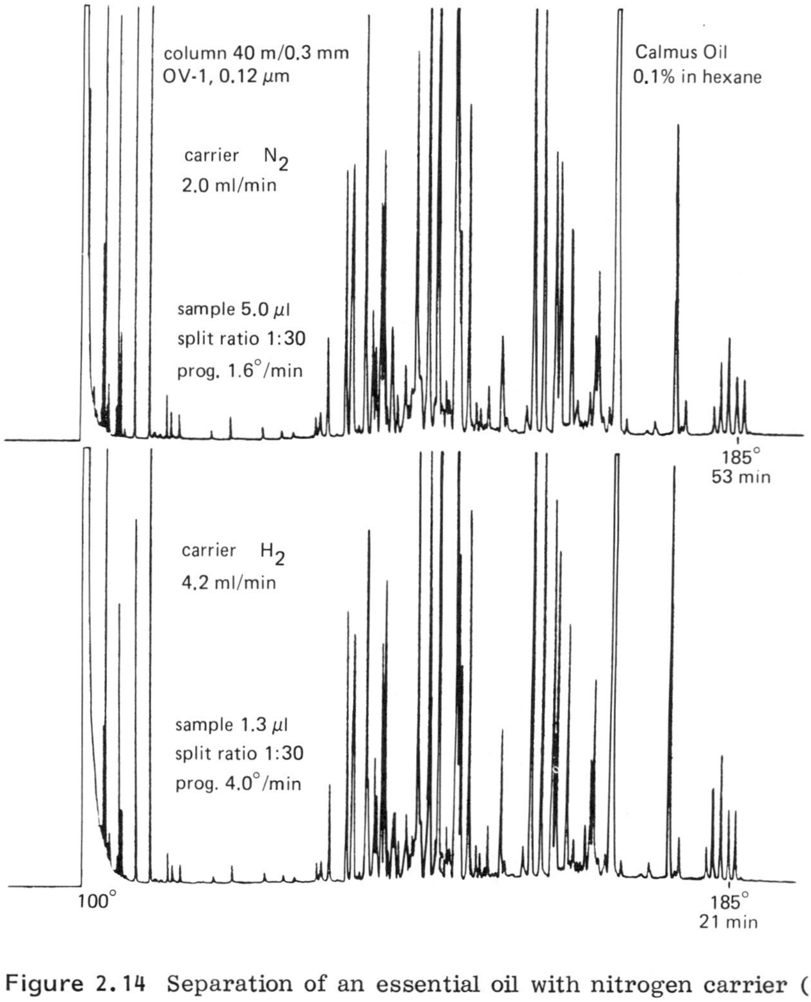

Figure 2.14 Separation of an essential oil with nitrogen carrier (analysis time, 53 min) and hydrogen carrier (analysis time, 21 min). Because of its higher optimum velocity, the velocity of the hydrogen carrier was 2.1 times that of the nitrogen carrier; a faster program rate was also used in the second case. (From Ref. 15.)

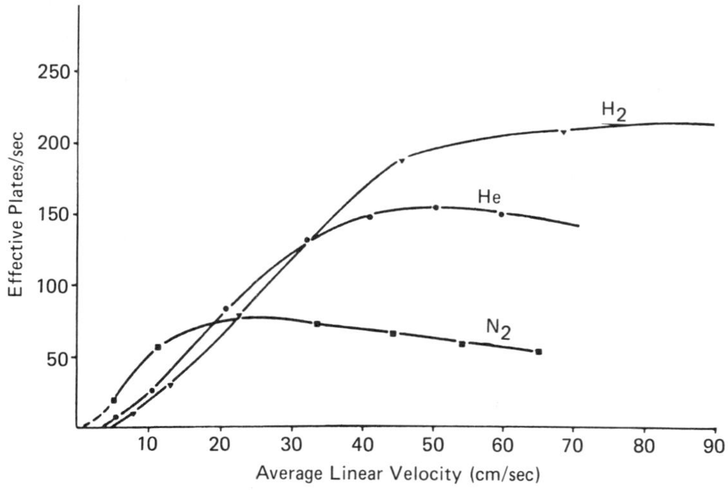

Figure 2.15 Plots of efficiency per unit time as a function of the average linear carrier gas velocity for different carrier gases. (From Ref. 14).

F. Temperature Programming

It was pointed out earlier that the temperature at which a given solute is eluted from the column to the detector has a profound effect on peak shape. As the temperature increases, the higher vapor pressure of the solute changes the c_L/c_G ratio, the steady flow of carrier gas delivers the solute to the detector at a higher concentration for a shorter time, and the detector generates a narrow, high-intensity peak. It was also pointed out that higher temperatures have an adverse effect on relative retentions. Because the improvement in peak shape does not compensate for the decrease in relative retention, component separation (as measured by resolution or the separation number) is in most cases affected adversely by temperature increases.

Temperature programming—beginning the chromatographic process at a low temperature and increasing the temperature at some regular rate until the compounds are eluted (or to some higher value which is then held until elution occurs)—allows the analyst to realize both benefits. Relative retentions are larger during at least part of the analysis, but by the time each solute reaches the end of the column, the temperature is higher. At these higher temperatures, compounds elute to produce narrow, high peaks that result in improved sensitivities. The selection of program parameters and their effects on separation efficiency, analysis time, thermal exposure, and elution orders are beyond the scope of our discussion here, and the interested reader is referred to other sources [16].

G. Retention Systems

1. *General Considerations*

The prospects of using chromatographic retentions as a criterion of compound identification have attracted much interest since the earliest days of gas chromatography. It must be realized, however, that while a discrepance between two retentions (under the same chromatographic conditions) does establish that the two substances are different, a coincidence of retentions does not

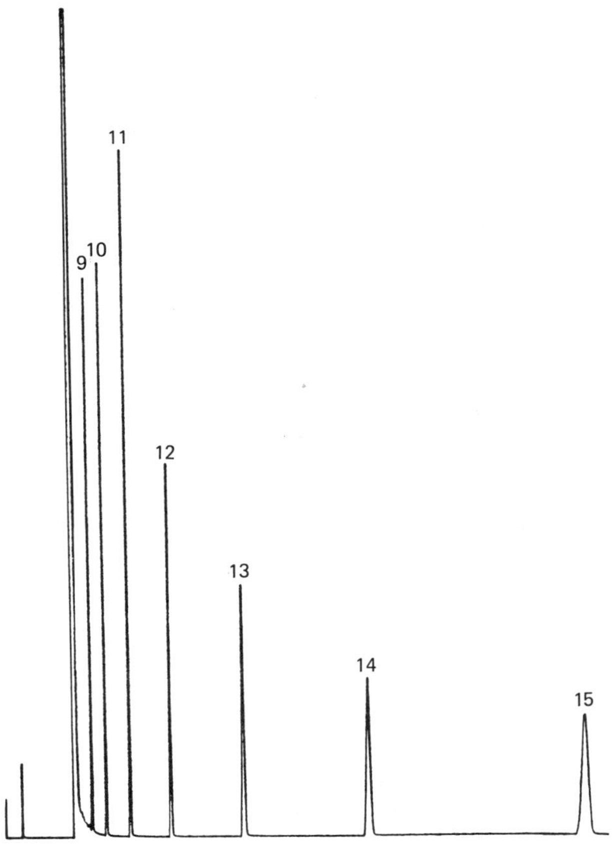

Figure 2.16 Isothermal chromatogram of a homologous series of compounds: C_9 through C_{15} *n*-paraffin hydrocarbons at 90°C.

establish that the two materials are one and the same. It is apparent that the degree of reliability that can be assigned to retentions is related to the separation efficiency of the system; a high-resolution system can distinguish between compounds of similar retention that would cochromatograph to yield a single peak on a low system. By resorting to the use of two high-performance columns possessing different chromatographic properties, the probability of incorrect assignments based on retention data can be greatly reduced.

Figures 2.16 and 2.17 show the separation of homologous compounds under various temperature conditions, and Figures 2.18 through 2.22 show graphs that search those data. Note that for isothermal determinations, a straight line is obtained when the logarithms of the adjusted retention times of the members of a homologous series are plotted as functions of the carbon numbers (or molecular weights). Results of the programmed temperature runs exhibit curvature when either the adjusted retention times themselves or their logarithms are plotted as functions of the carbon numbers. Nevertheless, point-to-point departure from a straight line is not severe. These relationships have been utilized in establishing several retention systems.

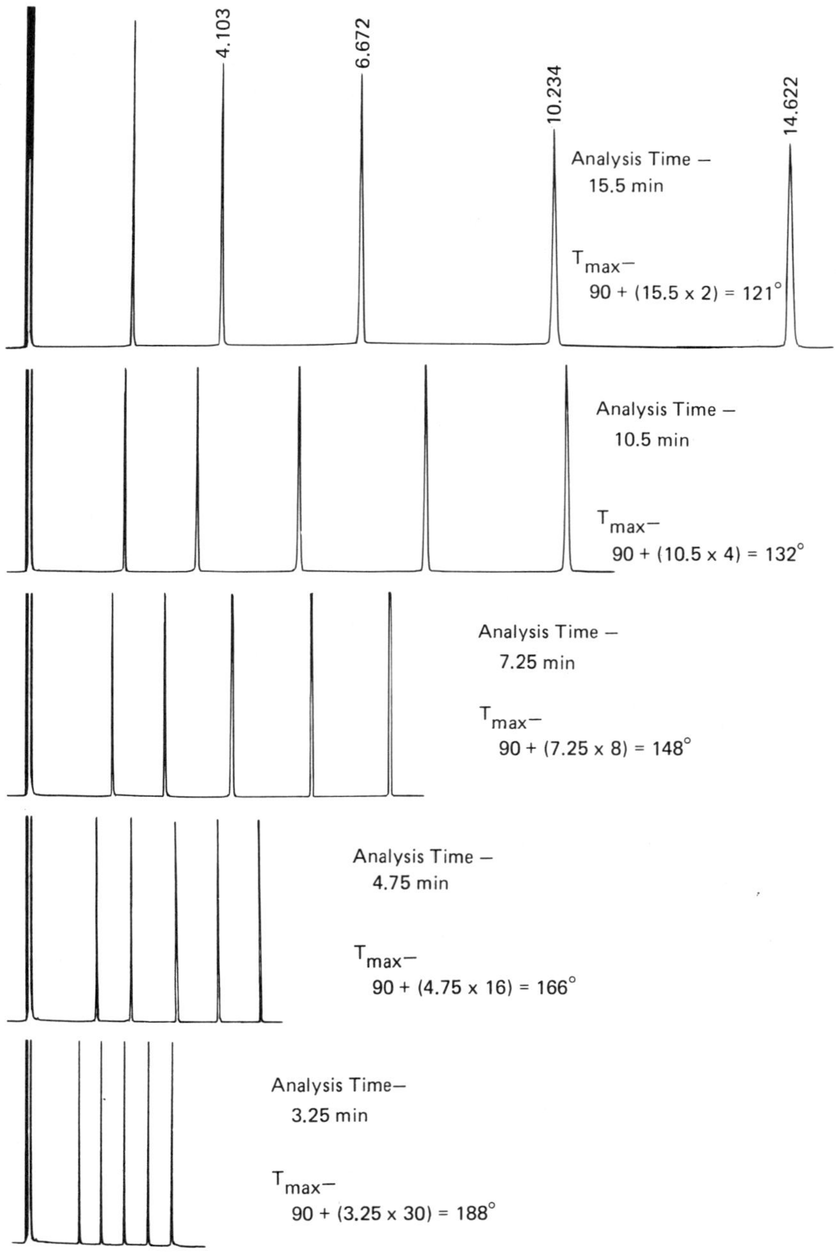

Figure 2.17 Programmed temperature runs of the C_9 through C_{15} *n*-paraffin hydrocarbons. Each run programmed from 90° at the indicated rate: (1), 2°/min; (2) 4°/min; (3) 8°/min; (4) 16°/min; (5) 30°/min. Note the effect on analysis time, peak shape, component separation, and final elution temperature.

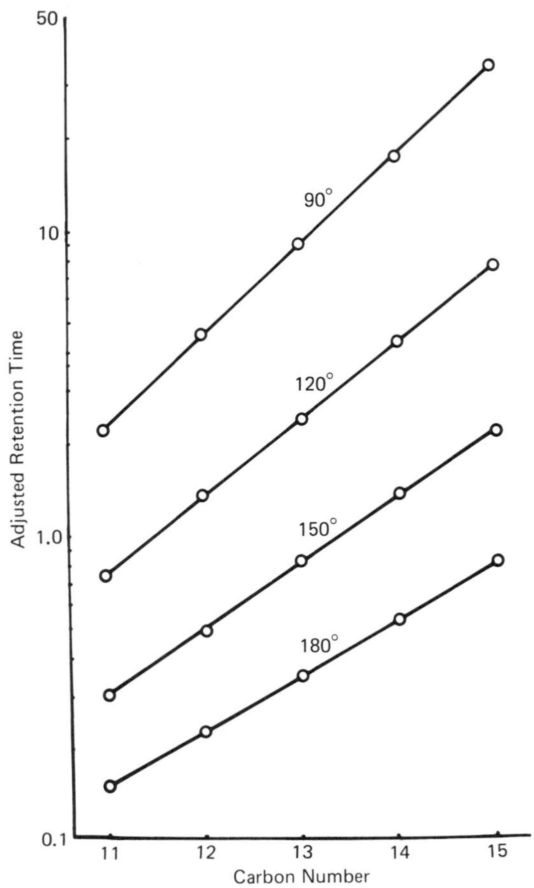

Figure 2.18 Semilogarithmic plots of the adjusted retention times as functions of the carbon number for the *n*-paraffin hydrocarbons at various isothermal temperatures.

2. *Calculation of the Gas Holdup Time, t_M*

While the gas holdup time can be conveniently measured by methane injection, it can also be calculated by reiterative techniques. The various proposals for this have been reviewed [17]. The popular Peterson and Hirsch approach employs three members of a homologous series of chain lengths C_n, C_{n+x}, and C_{n+2x} (n and x are integers greater than zero), whose isothermal retention times we will represent by X, Y, and Z, respectively. The gas holdup time, t_M, can be calculated from the relationship:

$$t_M = \frac{Y^2 - XZ}{2Y - (X + Z)} \qquad (22)$$

3. *Kovats Retention Index System*

This widely used system [18] takes advantage of the relationship that is illustrated in Figure 2.18. It assigns to each *n*-paraffin hydrocarbon a retention index 100 times the carbon number; thus, hexane has a retention index of 600 and octane a retention index of 800 on any liquid phase at any temperature. The retentions of all other compounds are described as being equivalent to those of hypothetical hydrocarbons (usually) possessing a mixed number of carbon atoms. In other words, the adjusted retention time of a compound having a retention index of 1450 would fall halfway between the adjusted retention times of tetradecane and pentadecane chromatographed under the same conditions, when plotted on a semilogarithmic scale (Figure 2.22). More specifically, the *retention index*, I, of a compound, A, is defined as:

$$I_b^a = 100\,N + 100\,n\,\frac{\log t'_{R(A)} - \log t'_{R(N)}}{\log t'_{(R_{N+n})} - \log t'_{R(N)}} \tag{23}$$

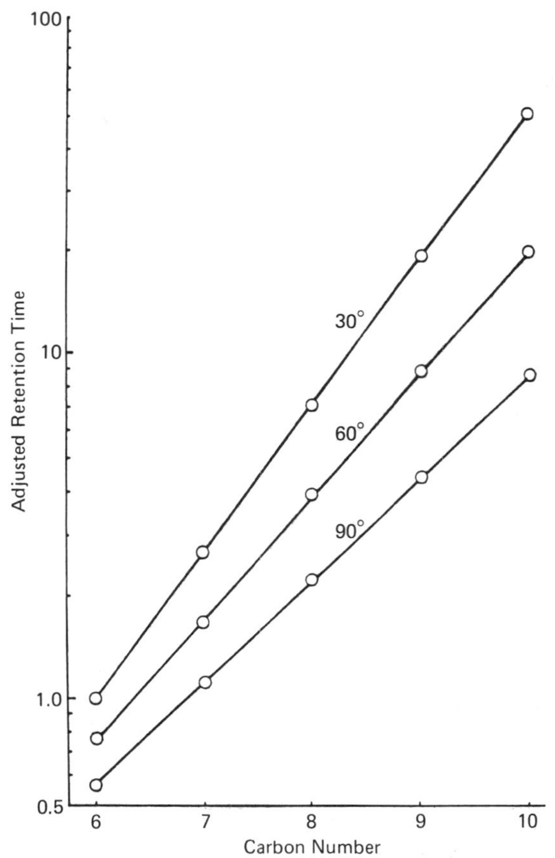

Figure 2.19 Semilogarithmic plots of the adjusted retention times versus the carbon number for a series of primary alcohols.

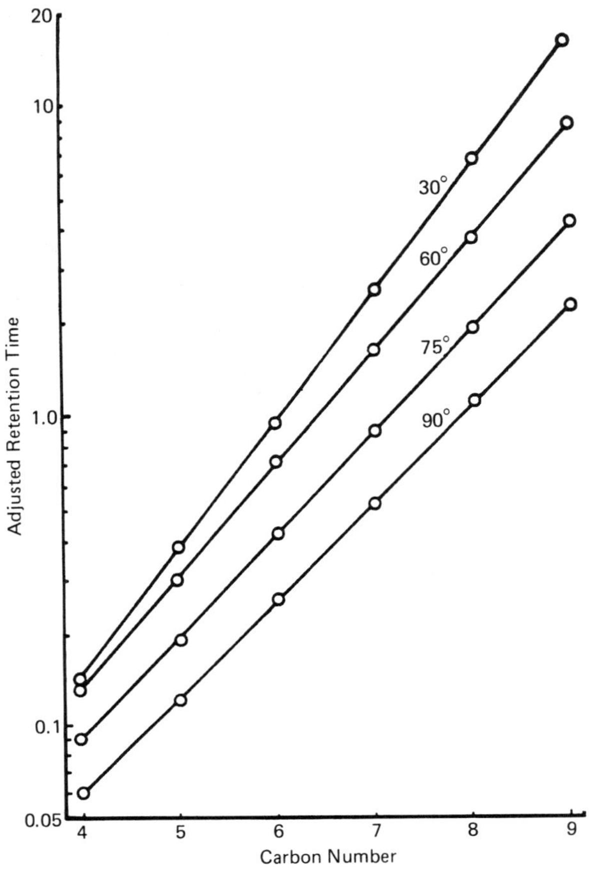

Figure 2.20 Semilogarithmic plots of the adjusted retention times as functions of the carbon number for a series of methylketones (butan-2-one through nonan-2-one).

where I is the retention index of compound A on liquid phase a at temperature b, N represents the number of carbons in the hydrocarbon standard whose retention is less than that of A, and n represents the additional number of carbons possessed by the hydrocarbon standard whose retention exceeds that of A. For greater precision, the retention of the unknown compound should be bracketed by those of the standards.

Retention indices are affected by the temperature of determination to the degree that the temperature change affects the vapor pressure of the solute (and the interactions between the solute and the liquid phase) differently for the compound in question than for the hydrocarbon standards. As the polarity of the liquid phase, and the polarity, size, and functionality of the solute increase, the magnitude of the shift becomes larger. Several other retention systems, using different reference compounds, have also been proposed [e.g., 19–21].

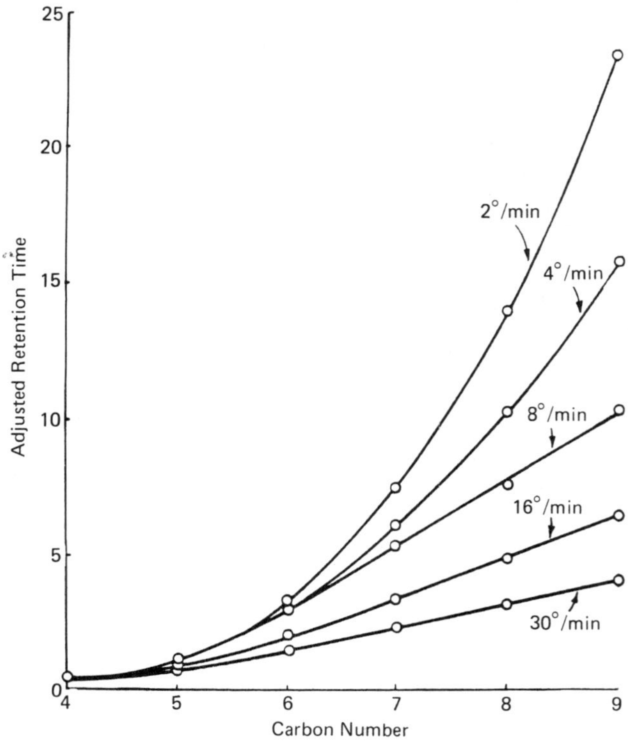

Figure 2.21 Plots of the adjusted retention times as functions of the carbon number for temperature-programmed separations of the normal alcohols.

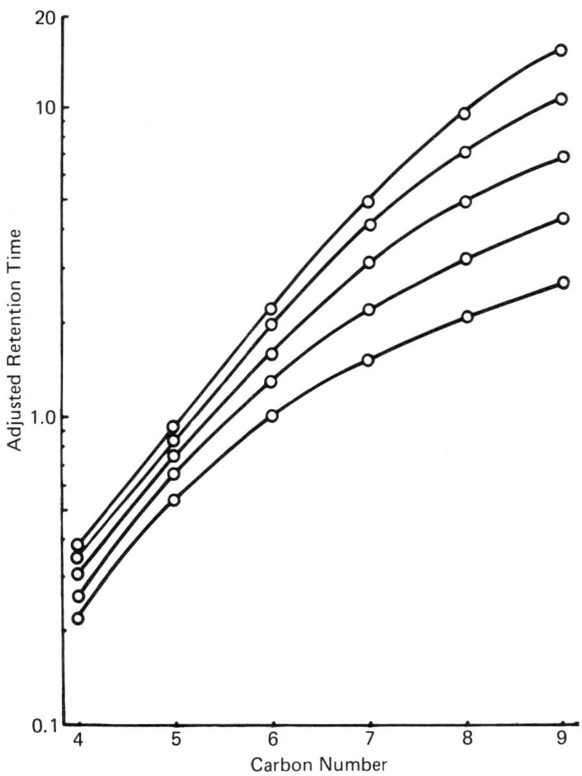

Figure 2.22 Plots of the logarithms of the adjusted retention times as functions of the carbon number for temperature-programmed separations of the methyl ketones.

GLOSSARY

A confusing multiplicity of symbols and terms are used in the field of gas chromatography. The terminology used in Chaps. 2 and 3 complies with or is analogous to the recommendations of the Commission on Analytical Nomenclature of the International Union of Pure and Applied Chemistry [22].

$\alpha_{A,B}$ Relative retention. The ratio of the adjusted retention times of two components; by convention, α is greater than unity. Hence,

$$\alpha_{A,B} = \frac{t'_{R\ (B)}}{t'_{R\ (A)}}$$

β Phase ratio. The volume of the column occupied by gas relative to the volume occupied by liquid phase:

$$\beta = V_G/V_L$$

d_f Film thickness. Average thickness of the liquid phase film, usually measured in micrometers (μm)

h Height equivalent to a theoretical plate (HETP). Column length (in millimeters) divided by the theoretical plate number, n.

H Height equivalent to an effective theoretical plate (HEETP). Column length (in millimeters) divided by the effective theoretical plate number, N.

I Retention index. A number describing the retention characteristics of a solute with reference to those of the *n*-paraffin hydrocarbons.

k Partition ratio. The amount of a solute in the liquid phase relative to the amount in the gas phase, which is equivalent to the time spent in the liquid phase relative to the time spent in the gas phase: $k = t'_R/t_M$. This term is preferred to the term "capacity ratio," and this symbol is preferred to the symbol k' used by some authors.

K_D Distribution constant. The ratio of concentrations of a solute in the liquid and gas phases, $K_D = c_L/c_G$. This term is preferred to the term "partition coefficient."

L Length of the column. Limited to that portion of the column containing liquid phase (i.e., excluding connections and manifold lines).

n Theoretical plate number. A unitless number used to express column performance.

N Effective theoretical plate number. Similar to the theoretical plate number, except that this takes into account the gas holdup volume.

n_{real} Real plate number. Used to express system performance, this number is based on calculations extending over a range of partition ratios.

R_s Peak resolution. The separation of two solutes, calculated on the basis of the distance between the two peak maxima and the sharpness (i.e., widths) of those peaks.

r_o Radius of the capillary.

t_M Holdup time. The time (or recorder chart distance) required for the passage of a nonsorbed substance (e.g., air) through the system.

t_R Retention time. The time (or recorder chart distance) from the point of injection to the peak maximum. This measurement has no validity with a peak that exhibits asymmetry.

t'_R Adjusted retention time. The time (or recorder chart distance) from the point of injection to the peak maximum, minus the holdup time (or distance). Equivalent to the time spent in the liquid phase.

TZ Separation number. A measure of column efficiency based on the separation of two consecutive members of a homologous series.

$\bar{u}$ Average linear carrier gas velocity. Equivalent to $L_{(cm)}/t_{M(sec)}$.

$\bar{u}_{opt}$ Optimum gas velocity. That velocity indicated by the minimum of the van Deemter plot.

$w_{.5}$ Width of the peak at half height.

REFERENCES

1. M. Tswett, *Proc. Warsaw Soc. Nat. Sci., Biol. Sect. 14* No. 6 (1903).
2. M. J. E. Golay, in *Gas Chromatography 1958* (D. H. Desty, ed.), Butterworths, London, 1958, p. 36.
3. R. E. Kaiser, *Chromatographia 8*:491 (1975).
4. R. E. Kaiser, *Chromatographia 9*:337 (1976).
5. J. J. Franken, *Chromatographia 9*:643 (1976).
6. G. Guichon, *Chromatographia 11*:249 (1978).
7. T. W. Smuts, T. S. Buys, K. de Clerk, and T. G. du Toit, *HRC&CC 1*:41 (1978).
8. T. W. Smuts, T. S. Buys, K. de Clerk, and T. G. du Tuit, *HRC&CC 2*:456 (1979).
9. O. Nilsson, *HRC&CC 2*:191 (1979).
10. O. Nilsson, *HRC&CC 2*:720 (1979).
11. R. Kaiser, *Z. Anal. Chemie. 189*:1 (1961).
12. W. Jennings and K. Yabumoto, *HRC&CC 3*:177 (1980).
13. J. J. van Deemter, F. J. Zuiderweg, and A. Klinkenberg, *Chem. Sci. 5*:271 (1956).
14. T. A. Rooney, L. H. Aetmayer, R. R. Freeman, and E. H. Zerenner, *Am. Lab. 11* No. 2:81 (1979).
15. K. Grob and G. Grob, *HRC&CC 2*:109 (1979).
16. W. Jennings, *Gas Chromatography with Glass Capillary Columns*, 2nd edition, Academic Press, New York, London, 1980.
17. R. Kaiser, *HRC&CC 1*:115 (1978).
18. E. Kovats, *Adv. Chromatogr. 1*:229 (1965).
19. E. P. Woodfard and C. M. van Gent, *J. Lipid Res. 1*:88 (1960).
20. H. van den Dool and P. D. Kratz, *J. Chromatogr. 11*:463 (1963).
21. J. Novak and J. Ruzickova, *J. Chromatogr. 91*:79 (1974).
22. *Pure Appl. Chem. 37*:445 (1974).

3

The State of the Art

Walter G. Jennings / *Department of Food Science and Technology, University of California, Davis, California*

I. INTRODUCTION

Gas chromatography is a prime example of a subject for which the state-of-the-art consideration is especially necessary today. Developments in column technology, inlet devices, and injection techniques and advances in detector design and data processing all testify to the fact that gas chromatography is enjoying an Age of Renaissance. In this chapter we will examine developments in these areas only to the degree that they have a bearing on glass capillary gas chromatography in clinical analysis; in that sense, this coverage should not be considered complete. The interested reader can glean additional information from the references cited.

II. INLETS

A. General Considerations

Chromatographic efficiency is related to the width of the solute bands as they are eluted to the detector (see Chap. 2); this width is a function of (1) the width of the band (i.e., the length of column occupied by that band) as it begins the chromatographic process, and (2) the degree of longitudinal diffusion (i.e., band broadening) that the solute experiences as it traverses the column and proceeds to the detector. Even an efficient column delivers broad peaks if the initial band width is large; hence the injection is critically important to both column efficiency and component separation. A variety of injection processes that achieve narrow starting bands merit brief consideration here; more detailed discussion is available elsewhere [1].

B. Inlet Splitters

The simplest and most widely used method of sample introduction utilizes an inlet splitter. Components to be detected should exist at levels of 0.01 to 10% of the sample injected. The inlet splitter is a heated chamber, swept with a relatively large volume of gas (typically 50–300 cm^3/min), most of which is discharged to atmosphere via a restrictor. A small portion of the gas flow (1–2 cm^3/min) is directed through the column and constitutes the carrier gas. The injected sample is rapidly vaporized, mixed and split. Because of the high gas flow through the vaporization chamber, the vaporized sample endures in the splitter only momentarily. Consistent with this short time, a narrow sample plug is directed to the column, while the large gas flow through the splitter flushes the major portion of the injection from the inlet; this ensures that the sample on the column is followed by pure carrier gas. Many samples are injected in a low-boiling solvent, which comprises the major portion of the injection. As the solvent vaporizes, it occupies a much larger volume, and unless that volume can be accommodated in the injection chamber, areas such as the inner face of the septum and even the carrier gas line may become contaminated. This can lead to large solvent tails that obliterate a major portion of the chromatogram. For this reason, most inlet splitters are constructed with an inner volume of 2–5 cm^3.

Considerable attention has been directed to design characteristics and their effect on sample linearity, that is, the degree of fidelity with which the sample split to the column reflects the composition of the sample injected [e.g., 2–4]. Except in special circumstances [5], the temperature of the splitter must be sufficiently high to ensure rapid vaporization of the injection; with thermally labile compounds, this may be a disadvantage. In addition, it is usually advisable that all sample-contact surfaces be of glass or quartz. A typical inlet splitter is shown in Figure 3.1.

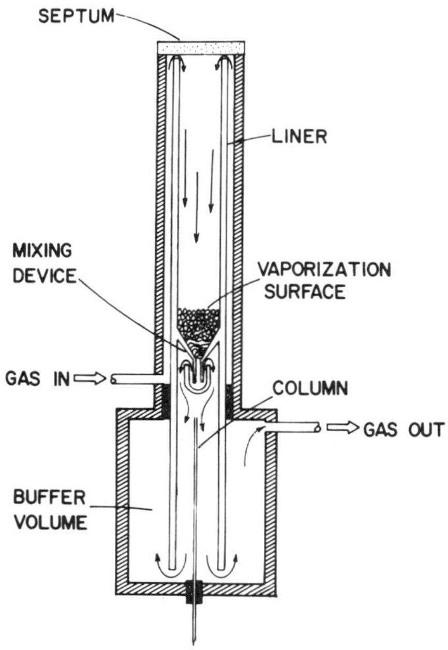

Figure 3.1 Schematic representation of a typical inlet splitter. The injected sample is rapidly vaporized, thoroughly mixed (in this case by flow inversions) with the relatively large stream of gas, expanded, and a small portion split to the column; the buffer volume minimizes pressure surges that accompany sample vaporization.

C. Splitless Injection

This injection mode requires that the sample be injected in a suitable volume of solvent, under conditions where that solvent vaporizes in the inlet and recondenses to a liquid on the first portion of the column [5,6]. The shape assumed by that plug of condensed solvent results in a short segment where the column phase ratio continuously decreases in the direction of migration [7]. In accordance with Equation 13 in Chap. 2, the front edge of each band moving into that section must decelerate at a faster rate than the rear of the band, and band narrowing results [8]; this phenomenon is known as the solvent effect.

Splitness injection requires an inlet similar to that represented by Figure 3.2. The sample-containing volume is usually about 1 cm^3. All of the gas flow into the inlet is directed to the column during the period of sample introduction; this normally requires 1–2 min. This longer "injection time" permits the sample to be vaporized more slowly, and the inlet temperature can be lower with a splitless than with a split mode. Because the flow capacity of the column is limited, an excessively long time would be required to transport all of the sample from the inlet to the column, and the solvent tail would be excessively large and interfere with the analysis. Consequently, an alternate flow path that permits purging of the vaporization chamber is normally provided.

Solutes that have boiling points lower than that of the solvent undergo a reverse solvent effect [9] in splitless injection, and their peak shapes can be

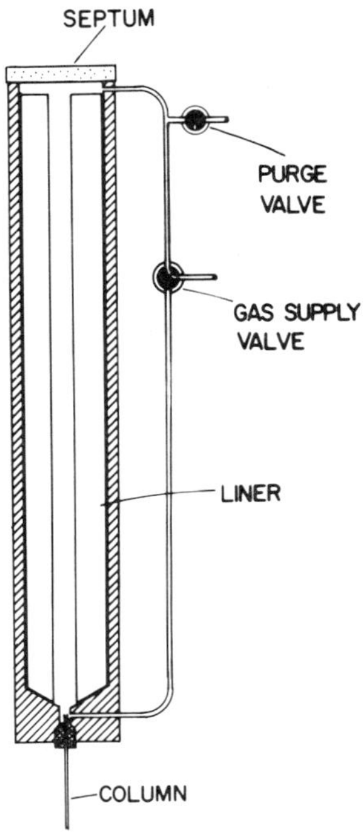

Figure 3.2 Schematic representation of a typical splitless inlet. The sample is injected with the purge valve closed and carrier gas introduced at the top; after a predetermined time, the carrier gas is introduced through the bottom line and the purge valve is opened to flush the last residues from the system.

seriously degraded. Splitless injection is usually employed for trace component analysis (concentrations less than 0.1%), and because the solutes must be high-boiling with respect to the solvent, and the solvent must be high-boiling with respect to the initial column temperature, it is not suited to the analysis of low-boiling components. On-column solvent condensation can lead to localized degradation of the liquid-phase coating, and have a deleterious effect on column life. Recent developments in bonded-phase fused silica columns (see below) promise to alleviate this problem.

D. On-Column Injection

To ensure the lowest vaporization temperatures possible (i.e., vaporization at column temperature), on-column injections have long been used for thermally labile compounds with packed and large-bore open-tubular columns. Schomburg et al. [10] devised a rather exotic method for accomplishing on-column injection on small-bore capillaries, but the equipment and technique requirements were relatively stringent. Grob and Grob [11] suggested a simpler method that involved the use of a standard microliter syringe fitted with a 32-gauge needle that can be inserted into a 0.32-mm capillary column. These fine needles are so flexible that they do not have sufficient rigidity to pierce the septum, and other methods have been devised. One [12] involves a septum-free valved inlet, and another [1] uses a detached pilot needle to pierce the septum and to guide the fine needle through the septum and into the column.

At the present time, on-column injections require the use of wider-bore (0.32 mm) columns; while they offer considerable promise in the analysis of biologicals and other thermally labile compounds, they also lead to more rapid column deterioration. Again, recent developments in bonded-phase columns may circumvent this latter difficulty.

E. Other Injection Modes

Several other injection mechanisms have been suggested. These include headspace techniques, with [13,14] and without [15] prior enrichment, on-column cold trapping, a falling needle injector [16], and capsule insertion [17]. Another system, recently described by Rijks et al., utilizes thermal focusing to achieve a narrow solute band (Figure 3.3).

F. Injection Technique

The operator's injection technique influences to a very large degree the results obtained. With split injections, the sample should be drawn up into the glass syringe barrel (empty needle) and the needle wiped. The needle is

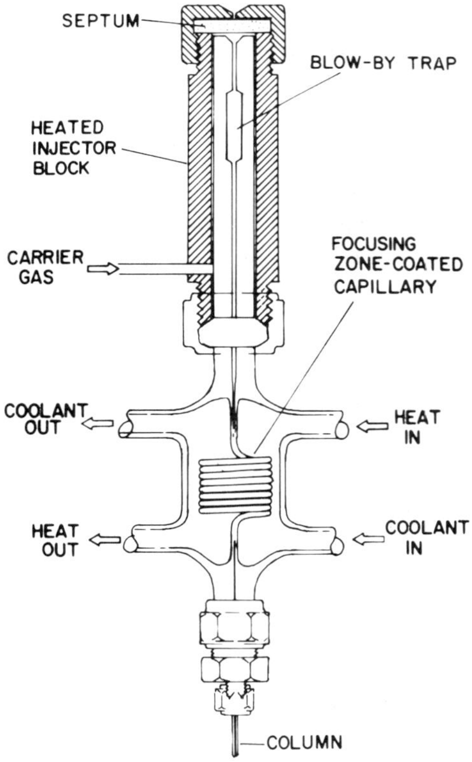

Figure 3.3 Schematic representation of a thermally focused inlet. The temperature gradient which exists along the trapping capillary in both the concentration and volatilization steps causes the rear of the sample plug to decelerate more slowly that the front of the plug; band narrowing results. (From Ref. 18.)

inserted through the septum as deep as possible, the plunger depressed, and the syringe withdrawn, all in one fast, smooth, and continuous motion. Broken segments of column, lodged in the splitter flow path, can lead to doublets for each peak [1]. With splitless injection, the empty needle technique is also recommended, but better results are obtained if, after insertion into the heated inlet, the needle is allowed to warm up for 3–4 sec before the sample is injected [19].

Most operators report that on-column injections give better results when the empty needle technique is used; early suggestions recommended a slow, prolonged injection [11,12], but as additional experience has been gained with this new technique, most workers feel that a normal swift depression of the syringe plunger gives improved performance [21,22].

III. COLUMNS

A. General Considerations

While each component of the system can be regarded as critically important in that a maladjustment or defect in any one can lead to inferior results, the column, which is really the heart of the chromatograph, is of paramount importance. The use of an inferior column, or the improper installation of a good column, forces even the best chromatographic system to produce shoddy results.

The cross-sectional dimensions of the column should be uniform throughout its length; while some experimental high-aspect ratio columns possess an oval or rectangular cross section [e.g., 23–25], in most cases the cross-sectional dimension can be expressed as the diameter or radius of the column. It is also important that the liquid phase endures as an absolutely uniform thin film; a single globule or thickened patch will exercise a drastic and adverse effect on column efficiency. With smooth-bore WCOT columns, the coating efficiency (Chap. 2) indicates how well these two conditions have been met.

B. Materials of Construction

Some liquid phases are repelled by glass, and a variety of treatments have been explored to encourage wetting of the surface and spreading of the liquid phase (see below). Contact angle relationships have been changed by etching or roughening the surface [26–28], or by the inclusion of microparticles [29–33]. While this has a tendency to stabilize the film, it can also result in point-to-point variations that have an adverse effect on the uniformity of the film thickness. Consequently, much of the early work in column technology emphasized separation efficiencies and theoretical plate numbers. It soon became evident, however, that another parameter was also important: A reputation for inertness notwithstanding, some compounds failed to negotiate some glass capillary columns, and others produced malformed or tailing peaks that attested to substrate interaction.

Glass, of course, is essentially fused silicone dioxide, but a variety of impurities may be present, and a number of addends may be used during manufacture to impart special properties to the finished product. Because glass is a high-viscosity, supercooled fluid rather than a solid, ions from these addends will continue to migrate through the matrix. Removal of ions at the surface has, therefore, only a temporary effect. Common soda-lime (soft) glass employs a fluxing agent that produces a finished product high in sodium and calcium ions; smaller amounts of aluminum, iron, and magnesium are also usually present. Borosilicate glass (Pyrex) contains boron addends, whose primary purpose is a finished product with a lower coefficient of thermal expansion. High-density metals such as lead are added to other glasses to increase the refractive index and produce crystal. Naturally occurring fused silicon dioxide is known as quartz, and large deposits that occur in Brazil are mined commercially to produce bars, ingots, and tubing of fused quartz. This material is characterized by 50–100 ppm of metallic impurities, usually dominated by magnesium and iron. A purified quartz which rarely contains less than 10 ppm of metallic impurities is also commercially available. The term "fused silica" is usually reserved for high-purity synthetic glass, whose metal content can be less than 1 ppm, produced from high-purity silicon tetrachloride.

Glass possesses a high-energy surface, and while it normally has fewer adsorptive sites than the equivalent metal surface, those that it does have are extremely tenacious. In consequence, the glass capillary column can offer an active surface whose characteristics may be modified by the nature and availability of impurities (or addends) in the glass. Surface pretreatments such as silanization [34,35], the use of stabilizing addends (such as, particulate matter [29–33]), surface-active agents [36–37], or the initial application of a more polar liquid phase [38], have all been explored. One productive route entails blocking surface access by the deposition of a more inert layer such as barium carbonate [39–41].

Most early efforts in column manufacture utilize soda-lime glass, which is more easily etched by hydrochloric or hydrofluoric acid. The surface is slightly alkaline and especially at higher temperatures, degradation of silicone-type liquid phases can be demonstrated on such columns [42]; some solutes may also undergo reaction. Several column test mixtures have been suggested whose behavior might yield some information as to the suitability of a given column for qualitative or quantitative analysis of compounds possessing a variety of functional groups [43–45].

Borosilicate glass has also been widely used, and while it produces more stable silicone columns [44,46,47], the surface cannot be regarded as inert [48]. As with the soda-lime glass, a variety of treatments to produce more inert borosilicate columns have also been explored, but none has been entirely successful.

Dandeneau and Zerenner [38] investigated the chromatographic properties of several glasses. They reported that columns produced from an uranium glass gave generally inferior results, but columns made from a lead-potash glass yielded excellent chromatographic results except for sulfur-containing compounds.

Quartz (presumably the naturally occurring material) has been reported to be unsatisfactory as a column material, in that the surface is sufficiently active to catalyze a number of reactions, including liquid-phase degradation [40,49]. Columns drawn from high-purity fused silica, however, are much more inert and give superior results on test compounds ranging from cyclic amines, alcohols, and free fatty acids to chloro- and nitrophenols [38]. Even so, some surface deactivation treatment is usually required; regardless of the glass utilized, the efficacy of that treatment can be affected by the types of compounds analyzed and by the temperature to which the column is subjected. In many cases, upper temperature limits are dictated more by the thermal lability of the deactivation treatment rather than by the characteristics of the liquid phase. Columns that have developed activity can sometimes be pacified in situ [48].

At the time of this writing, remarkable developments are occurring with fused silica columns. Columns whose deactivation will endure prolonged exposure to 300°C and shorter exposure to temperatures up to 350°C are appearing on the market; bleed rates, even at these elevated temperatures, are remarkably low (see below). Some success has been achieved in chemically bonding the liquid phase to the fused silica surface, producing columns capable of long-term, high-temperature operation. In some cases such columns have been filled with solvents (e.g., dichloromethane), sealed, heated overnight, cooled, and drained without changing their characteristics. This of course introduces exciting possibilities, including micro open-tubular HPLC.

C. Column Installation

The general areas of converting packed column chromatographs to capillary capability and the installation of glass capillary columns have been covered in detail elsewhere [1]. In brief, the outlet end of the capillary column must terminate in a high-velocity gas zone. This condition is normally achieved by positioning the end of the column as close as possible to the zone of detection, and introducing hydrogen (for flame ionization detection, FID) and make-up or moderating gas (as required) at a position such that it sweeps the end of the column with a high-velocity gas stream. With split-mode, the inlet end should project up into the high-velocity gas zone; with splitless injection, the inlet end should be at the lowest point in the inlet. Columns produced from conventional glasses are made in a coiled configuration, and short sections at each end may have to be flame-straightened to permit proper column installation. This can lead to the generation of active sites that may cause degradation and tailing of sensitive compounds unless those sites are pacified [1]. A major advantage of the thin-wall, flexible, fused silica columns is that they are inherently straight; the required length of column is merely freed from the coil and inserted into the detector and inlet as required.

In most cases, graphite ferrules are preferred method of achieving column connections, but care should be exercised to ensure that this material (or any other elastomer) does not come in contact with the flow stream. In that event, adsorptive tailing is frequently observed.

D. Column Selection

Because they offer higher maximum operating temperatures, greater stability, more inert conditions, easier installation, and less fragility, fused silica columns, which represent state-of-the-art development, are generally superior and should be employed. At a given coating efficiency, the number of theoretical plates offered by the column increases with column length, although the relationship is linear only at the optimum linear carrier gas velocity (see below). The separation efficiency per unit time, the optimum practical gas velocity, and the range of solute partition ratios for which conditions can be optimized all decrease with increasing column length. Hence, columns should be as short as possible, consistent with the number of theoretical plates required for a given separation. In other words, columns of high coating efficiency are preferred because the necessary number of theoretical plates can be delivered by a shorter column, which results in shorter analysis times, the prospect of optimal conditions over a wider range of retention times, and (in programmed mode) lower elution temperatures, which can be important in the analysis of thermally labile compounds.

Liquid phases are selected with a view to their stability under the conditions of analysis and the degree of separation that they achieve, that is, the relative retentions of the substances to be separated (see Chap. 2). Analytical conditions that may affect column stability include the nature of the materials injected (i.e., silanizing reagents or peculiar solvent requirements), the temperature of the analysis, and the mode of sample injection. These considerations have been thoroughly reviewed [1]. Other ways in which column length, phase ratio, and liquid phase influence the conditions to which the sample components are subjected are considered under selection of parameters (Section IV.D).

IV. OPERATIONAL PARAMETERS

A. General Considerations

With any given chromatographic system, several parameters that can have a profound effect on the analysis are under the control of the operator. The skilled chromatographer varies these to maximize separation efficiency in one case, or to sacrifice superfluous efficiency in the interest of shorter analysis times in another case. If detectability is a problem, conditions favoring maximum peak height may be preferred at the expense of separation efficiency, and steps can be taken to improve the survival of thermally labile compounds. Understanding the interrelationships between these variables can help in the attainment of superior analytical results.

B. Carrier Gas Choice and Velocity

The shapes of van Deemter curves produced by various carrier gases were explored in Chap. 2, where it was pointed out that both the flatness of the curve and the rate of the optimum linear velocity increase in the order nitrogen < helium < hydrogen. Faster carrier gas velocities produce shorter analysis times (see below), and flatter van Deemter curves mean that the carrier gas velocity can be optimized over a greater range of partition ratios (i.e., a longer section of the chromatogram). This is especially important under conditions of temperature programming, because the average linear carrier gas velocity decreases as column temperature increases. Another factor influencing the flatness of the van Deemter curve is column length; shorter columns produce flatter curves [50–52].

Although the highest column efficiency (i.e., the largest number of theoretical plates) is obtained at the optimum linear carrier gas velocity, most analyses should be performed at velocities that are 50–100% higher than this value because the loss in efficiency is more than compensated for the decreased analysis time.

C. Program Parameters

If the components of a sample have boiling points that do not exceed a range of about 50°C, it is usually possible to effect their separation under isothermal conditions, where separation efficiencies are highest. With samples that represent a wide range of boiling points, isothermal separations are rarely practical. At a temperature where the lower boiling components are suitably separated, the gas phase concentrations of higher boiling solutes are so low that they are delivered to the detector at low concentrations over extended times to produce

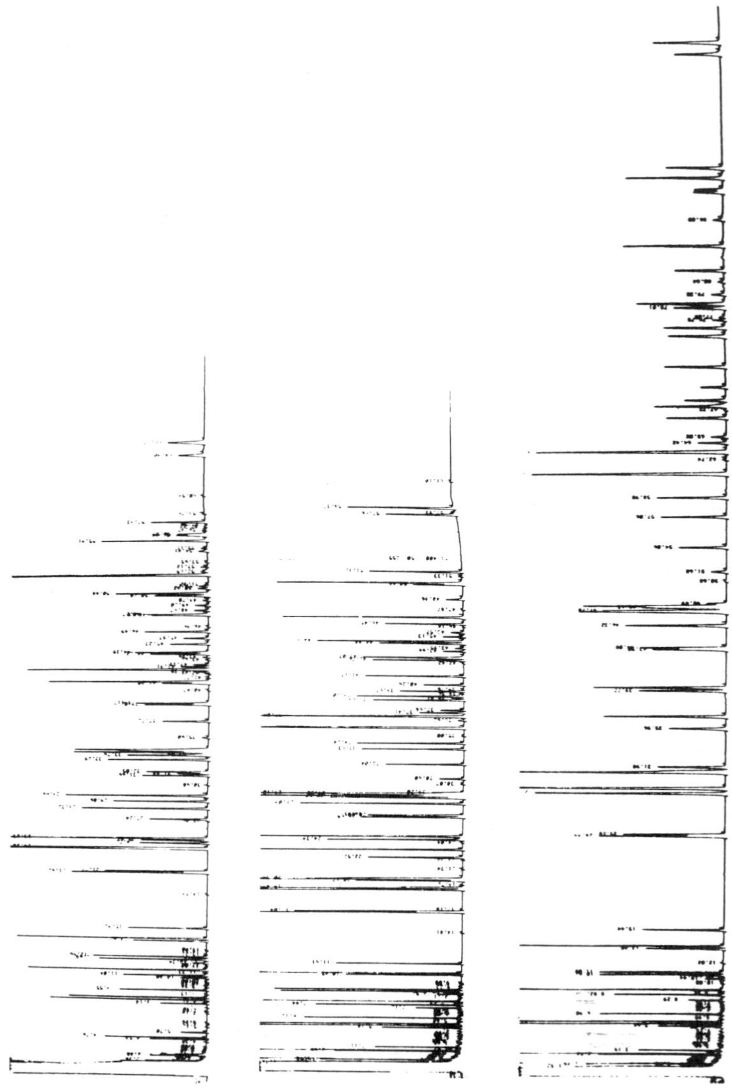

Figure 3.4 Effect of changing carrier gas velocity at constant program rate (top vs center), and effect of changing program rate at constant carrier gas velocity (center vs bottom). Initial temperature, 30°C; final temperature, 280°C; other conditions as shown:

Figure		Program rate (°/min)	$\bar{u}_{initial}$ (cm/sec)	$\bar{u}_{final}$ (cm/sec)
Top	Helium	4	29	19
Center	Hydrogen	4	62.5	40
Botton	Hydrogen	2	62.5	40

broad, low-intensity peaks; this not only results in prolonged analysis times, but also affects sensitivity. The use of higher temperatures will correct this problem, but the lower boiling solutes are then inadequately separated. As explored in Chap. 2, both difficulties can be corrected by temperature programming. For a simple single-ramp program, program parameters include initial temperature, initial hold time, program rate, final temperature, and final hold time. Initial temperatures should be low enough to achieve adequate separation of the low-boiling components; an abundance of low boilers may make advisable an initial hold time. The upper temperature limit should be sufficiently high to ensure suitable gas phase concentration as high boilers are delivered to the detector. The initial and final temperatures must be selected with reference to the temperature limits of the liquid phase. If the upper limit is reached before all components have eluted, a final hold time can be observed until this takes place. Program rates are chosen with attention to the analysis time and separation efficiency; higher rates lead to shorter analysis times and lower separation efficiencies. Considerable experimentation may be necessary in the selection of program parameters.

D. Selection of Parameters

It is important to realize that changing one parameter invariably causes changes in others. Increasing the rate of temperature programming, for example, causes a simultaneous decrease in carrier gas velocity. Under isothermal conditions, the analysis time is directly proportional to the carrier gas velocity, but this is not always the case under conditions of temperature programming. This is because those components with larger partition ratios spend most of their time in the liquid phase until higher temperatures have been reached; hence they are influenced by the higher carrier gas velocity only during a relatively short portion of the program when they are delivered to the detector. With higher-boiling compounds, the program rate has a more profound effect on analysis time in a temperature-programmed mode. Both parameters will require attention in the optimization of an analysis. These points are illustrated by Figure 3.4.

Equation 13 from Chapter 2 can be of immense benefit in the selection of chromatographic parameters. Starting from the relationship $K_D = \beta k$ and expanding each term, we can see that

$$\frac{c_L}{c_G} = \frac{r_o}{2d_f} \times \frac{t'_R}{t_M} = \frac{r_o \bar{u} t'_R}{2d_f L}$$

Hence,

$$t'_R = \frac{2c_L L d_f}{c_G r_o \bar{u}}$$

Bearing in mind that both separation efficiencies and analysis times vary directly with t'_R, the influence of a number of parameters can now be estimated. An increase in c_L or decrease in c_G (i.e., a more retentive liquid phase, lower column temperature, or lower program rate), an increase in column length or in liquid film phase thickness, or a decrease in the column radius or in the carrier gas velocity (assuming this remains above $\bar{u}_{opt}$) would result in better component separations and longer analysis times. At a given program rate, longer analysis times may mean that compounds are being subjected to higher

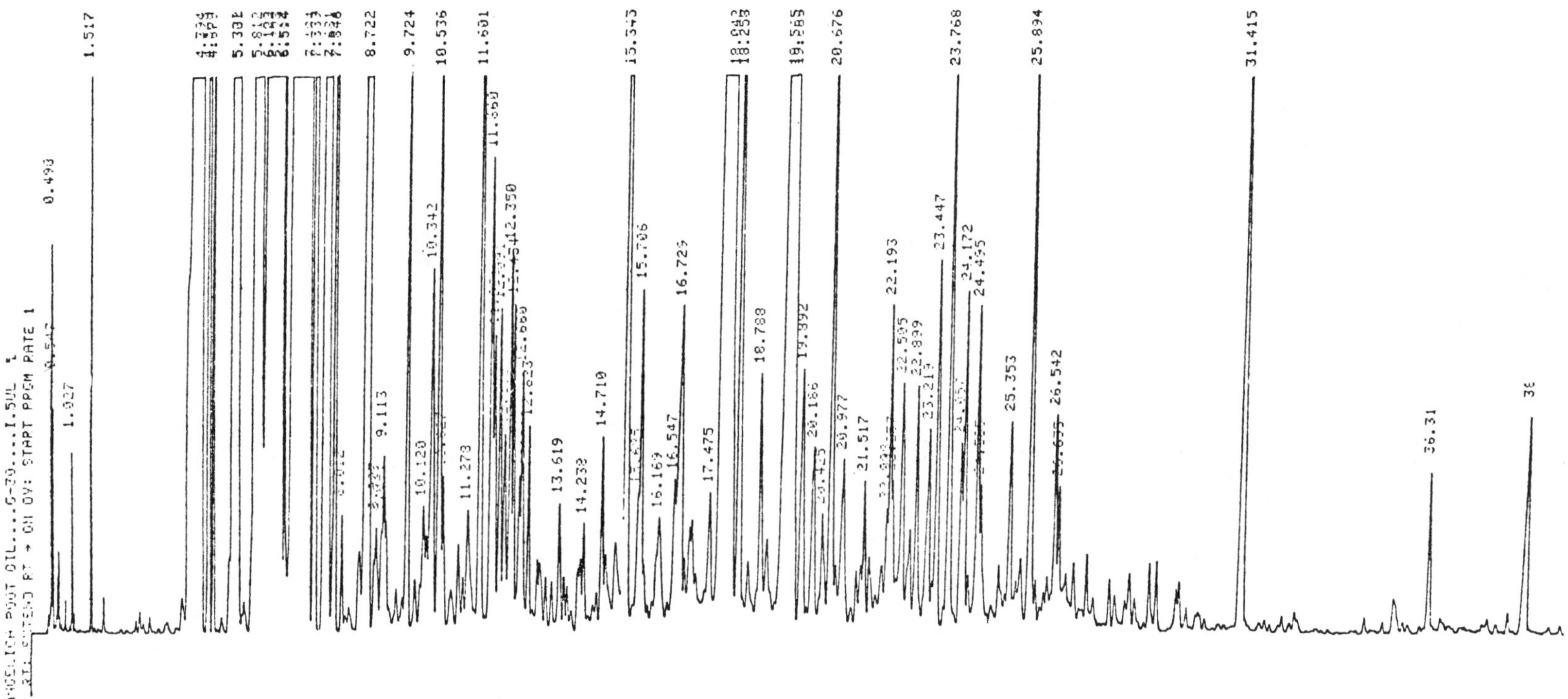

Figure 3.5 Separation of angelica root oil on a 20-m fused silica capillary coated with SE 54; 65–275°C at 4°/min.

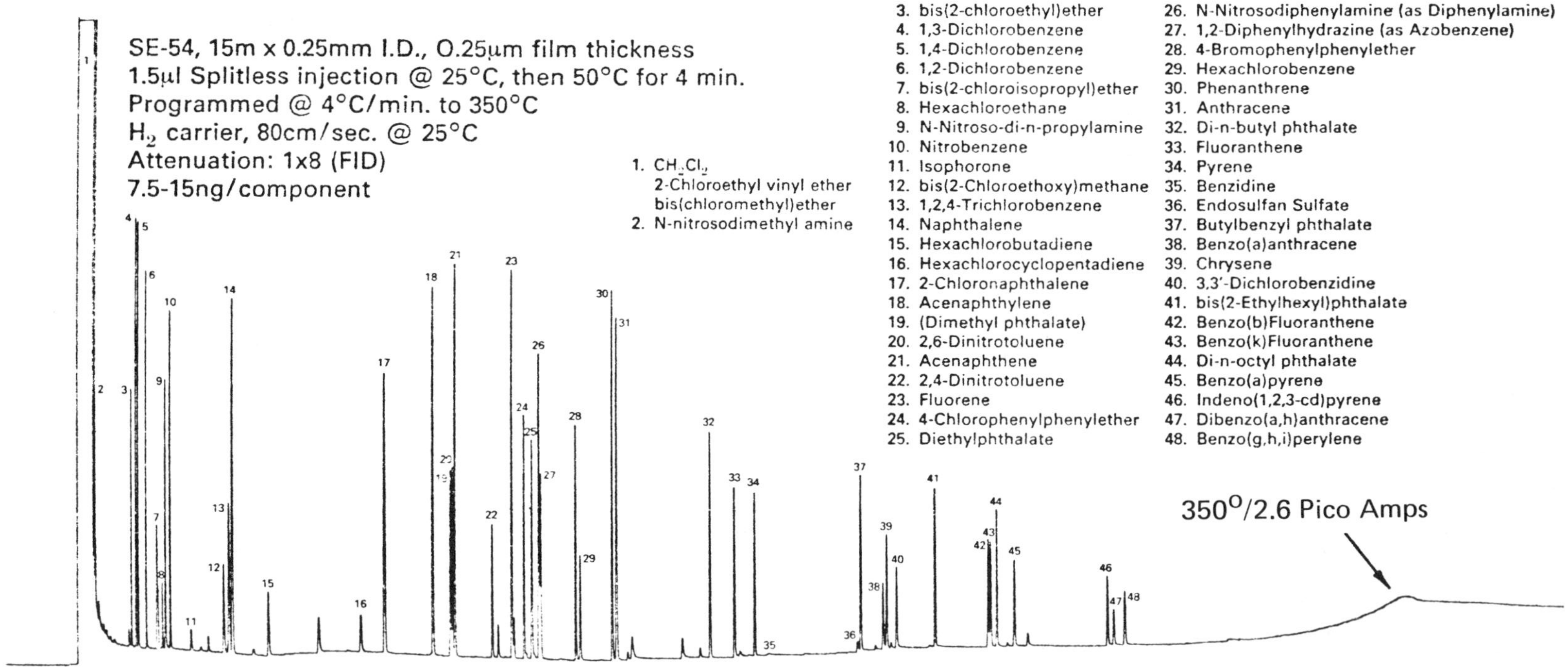

Figure 3.6 Separation of EPA base-neutral extractables on a 15-m fused silica column coated with SE 54. Note the very low bleed (attenuation 1 × 8; signal 2.6 pico amp) at 350°C. (Courtesy of J&W Scientific, Inc., Ranch Cordova, CA.)

temperatures, but there are complicating factors here (see below). Conversely, a less retentive liquid phase, higher temperature, shorter column, decreased liquid film phase thickness, larger diameter column, or higher carrier gas velocity would lead to shorter analysis times and decreased separation efficiencies. For thermally labile compounds, the most productive routes usually entail the use of less selective liquid phases, shorter- and/or larger-diameter columns, thinner liquid phase films, and higher carrier gas velocities; increasing the rate of temperature programming results in compounds being exposed to higher temperatures, but for shorter periods of time.

V. CONCLUSIONS

Because they frequently deal with complex mixtures of larger-molecular-weight compounds that may be prone to decompose or to interact with components of the analytical system, those involved in clinical analysis should be especially interested in recent developments in column technology and improved methods of sample introduction and detection.

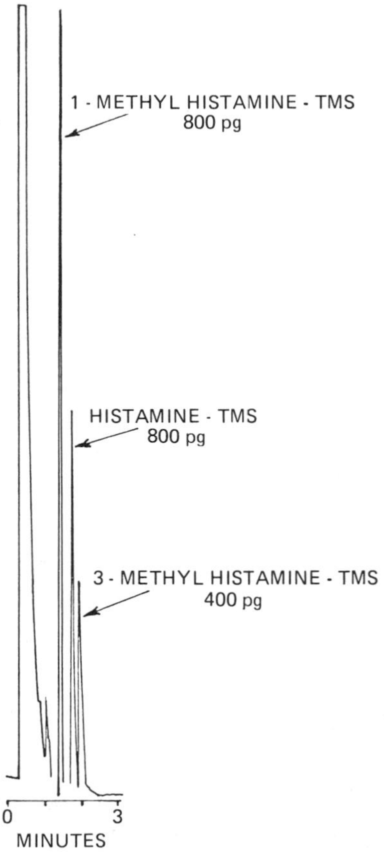

Figure 3.7 Rapid analysis of histamine and metabolites, using a short capillary and a nitrogen detector. (From Ref. 53.)

```
BILE ACIDS; JAEGER SAMPLE, 10 M SE 54 FUSED SILICA COLUMN; 8 UL COLD TRAP
LIST

OVEN TEMP=100°C   SETPT=100°C   LIMIT=405°C
   EQUIB TIME = 0.20 MIN

OVEN TEMP PROFILE:       (ANNOTATION OFF)
   INITIAL VALUE = 100°C
   INITIAL TIME = 5.00 MIN
   LEVEL 1
      PRGM RATE = 30.00°C/MIN
      FINAL VALUE = 220°C
      FINAL TIME = 0.00 MIN
   LEVEL 2
      PRGM RATE = 2.00°C/MIN
      FINAL VALUE = 300°C
      FINAL TIME = 5.00 MIN
EXIT
```

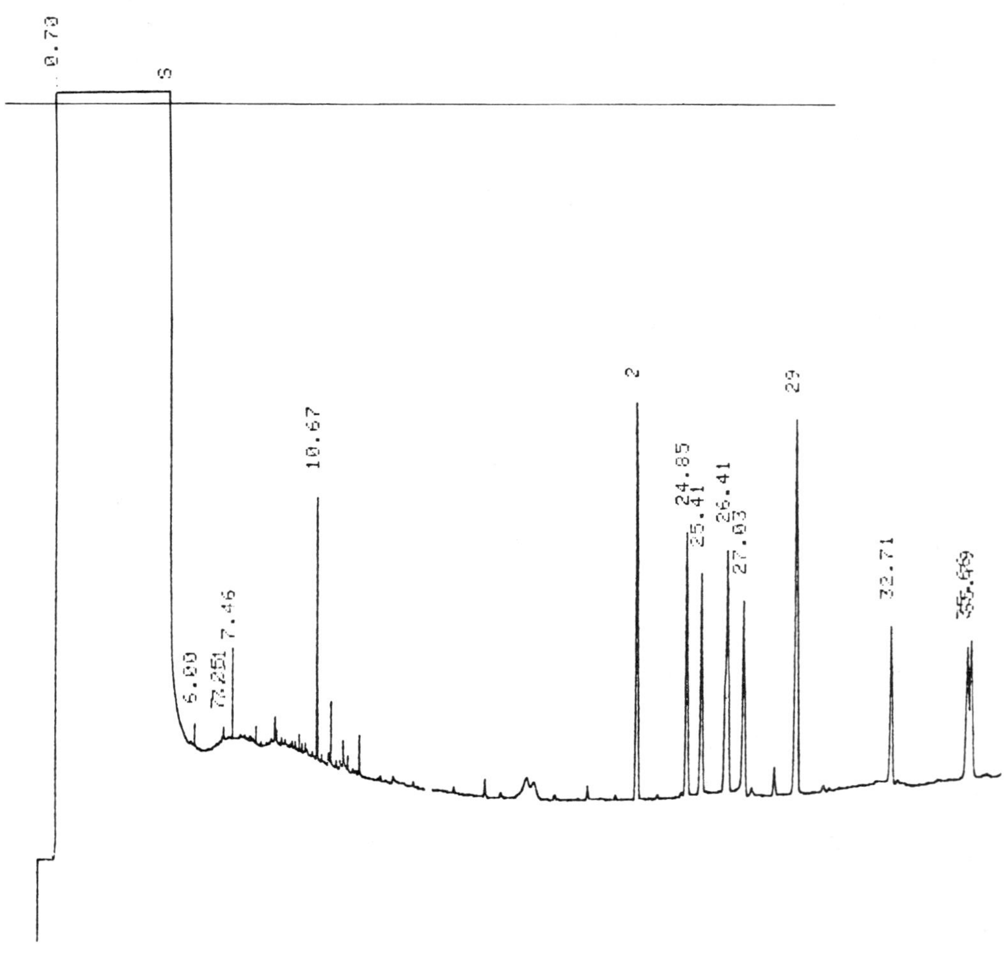

Figure 3.8 Separation of a standard mixture of bile acids on a 20-m fused silica column coated with SE 54. Cold-trapped (splitless sans solvent) injection. Column held at 100°C during period of splitter deactivation; when splitter activated, column programmed rapidly to 220°, and from 220° to final temperature at 4°/min.

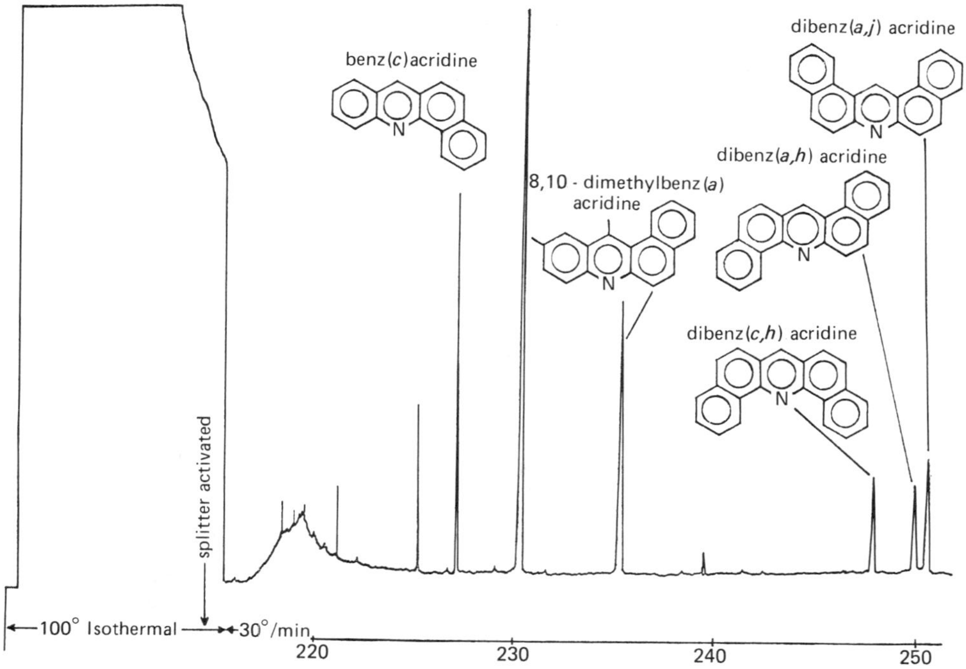

Figure 3.9 Separation of azaarenes on a 20-m fused silica column coated with SE 54. Cold-trapped (splitless sans solvent) injection. Column conditions same as in Figure 3.8.

Figure 3.5 illustrates the superior resolution that can be achieved with an opentubular glass capillary column produced from high-purity fused silica; Figure 3.6 shows the low bleed rates exhibited by such columns; even at elevated temperature. In Figure 3.7, a glass capillary column coupled with a selective detector demonstrates rapid and high sensitivity analysis of several biogenic amines, and Figures 3.8 and 3.9 show other analyses related to clinical testing.

Properly installed in a well-designed system, the more inert glass capillary (and, in particular, the fused silica capillary) is capable of producing superior separations of higher sensitivity in shorter times. On a meter-for-meter basis, the state-of-the-art columns not only outperform the outmoded packed column in every respect but are also less expensive; it would appear that the days of the packed column are numbered.

GLOSSARY

A confusing multiplicity of symobls and terms are used in the field of gas chromatography. The terminology used in Chapters 2 and 3 compiles with or is analagous to the recommendations of the Commission on Analytical Nomenclature of the International Union of Pure and Applied Chemistry [54].

$\alpha_{A,B}$ Relative retention. The ratio of the adjusted retention times the two components; by convention, α is greater than unity. Hence

$$\alpha_{A,B} = \frac{t'_{R\ (B)}}{t'_{R\ (A)}}$$

β Phase ratio. The volume of the column occupied by gas relative to the volume occupied by liquid phase:

$$\beta = V_G/V_L$$

d_f Film thickness. Average thickness of the liquid phase film, usually measured in micrometers (μm).

h Height equivalent to a theoretical plate (HETP). Column length (in millimeters) divided by the theoretical plate number, n.

H Height equivalent to an effective theoretical plate (HEETP). Column length (in millimeters) divided by the effective theoretical plate number, N.

I Retention index. A number describing the retention characteristics of a solute with reference to those of the *n*-paraffin hydrocarbons.

k Partition ratio. The amount of a solute in the liquid phase relative to the amount in the gas phase, which is equivalent to the time spent in the liquid phase relative to the time spent in the gas phase: $k = t_R/t_M$. This term is preferred to the term "capacity ratio," and this symbol is preferred to the symbol k' used by some authors.

K_D Distribution constant. The ratio of concentrations of a solute in the liquid and gas phases, $K_D = c_L/c_G$. This term is preferred to the term "partition coefficient."

L Length of the column. Limited to that portion of the column containing liquid phase (i.e., excluding connections and manifold lines).

n Theoretical plate number. A unitless number used to express column performance.

N Effective theoretical plate number. Similar to the theoretical plate number, except that this takes into account the gas holdup volume.

n_{real} Real plate number. Used to express system performance, this number is based on calculations extending over a range of partition ratios.

R_s Peak resolution. The separation of two solutes, calculated on the basis of the distance between the two peak maxima and the sharpness (i.e., width) of those peaks.

r_o Radius of the capillary.

t_M Holdup time. The time (or recorder chart distance) required for the passage of a nonsorbed substance (e.g., air) through the system.

t_R Retention time. The time (or recorder chart distance) from the point of injection to the peak maximum. This measurement has no validity with a peak that exhibits asymmetry.

t'_R Adjusted retention time. The time (or recorder chart distance) from the point of injection to the peak maximum, ninus the holdup time (or distance). Equivalent to the time spent in the liquid phase.

TZ Separation number. A measure of column efficiency based on the separation of two consecutive members of a homologous series.

$\bar{u}$ Average linear carrier gas velocity. Equivalent to $L_{(cm)}/t_{M(sec)}$.

$\bar{u}_{opt}$ Optimum gas velocity that velocity indicated by the minimum of the van Deemter plot.

$w_{.5}$ Width of the peak at half height.

REFERENCES

1. W. Jennings, *Gas Chromatography with Glass Capillary Columns*, 2nd Edition, Academic Press, New York, London, 1980.
2. W. G. Jennings, *J. Chromatogr. Sci. 13*:185 (1975).
3. G. Schomburg, H. Behlau, R. Dielmann, F. Weeke, and H. Husmann, *J. Chromatogr. 142*:87 (1977).
4. C. Watanabe, H. Tomita, and N. Sato, *HRC&CC 2*:481 (1979).
5. K. Grob and G. Grob, *Chromatographia 5*:3 (1972).
6. K. Grob and K. Grob, Jr., *J. Chromatogr. 94*:53 (1974).
7. K. Grob and K. Grob, Jr., *HRC&CC 1*:57 (1978).
8. W. G. Jennings, R. R. Freeman and T. A. Rooney, *HRC&CC 1*:275 (1978).
9. R. J. Miller and W. G. Jennings, *HRC&CC 2*:72 (1979).
10. G. Schomburg, R. Dielmann, H. Husmann, and F. Weeke, *Chromatographia 10*:383 (1977).
11. K. Grob and K. Grob, Jr., *J. Chromatogr. 151*:311 (1978).
12. K. Grob, *HRC&CC 1*:263 (1978).
13. W. G. Jennings and M. Filsoof, *Agric. and Food Chem. 25*:440 (1977).
14. K. Yabumoto, M. Yamaguchi, and W. G. Jennings, *J. Food Chem. 3*:7 (1978).
15. B. Kolb, P. Pospisil, T. Borath, and M. Auer, *HRC&CC 2*:283 (1979).
16. M. L. J. van der Berg, Doctoral thesis, Tech. Univ. Delft, The Netherlands, 1975.
17. H. H. Buser and H. M. Widmer, *HRC&CC 2*:177 (1979).
18. J. A. Rijks, J. Drozd, and J. Vovak, in *Advances in Chromatography, 1979* (A. Zlatkis, ed.), Chromatography Symposium, Univ. of Houston, Texas, 1979, p. 195.
19. K. Grob and G. Grob, *HRC&CC 2*:109 (1979).
20. K. Grob, *HRC&CC 1*:263 (1978).
21. T. A. Rooney, Personal communication (1979).
22. S. Trestianu, Personal communication (1979).
23. H. D. Pappendick and J. Baudisch, *J. Chromatogr. 122*:443 (1976).
24. G. DuPlessis and P. Torline, *Chromatographia 10*:624 (1977).
25. R. Dandeneau, Personal communication (1979).
26. K. Grob, *Helv. Chim. Acta 48*:1362 (1965).
27. G. Alexander and G. A. F. M. Rutten, *Chromatographia 6*:231 (1973).
28. G. Alexander, G. Garzo, and G. Palyi, *J. Chromatogr. 99*:81 (1974).
29. G. Nota, G. C. Goretti, M. Armenante, and G. Marino, *J. Chromatogr. 95*:229 (1974).

30. E. Schulte, *Chromatographia 9*:315 (1976).
31. C. Watanabe, and H. Tomita, *J. Chromatogr. Sci. 13*:123 (1975).
32. R. C. M. de Nijs, G. A. F. M. Rutten, J. J. Franken, R. P. M. Dooper and J. A. Fijks, *HRC&CC* 2447 (1979).
33. H. T. Badings, J. J. G. van der Pol, and J. G. Wassink, *HRC&CC 2*: 297 (1979).
34. K. D. Bartle and M. Novotny, *J. Chromatogr. 94*:35 (1974).
35. K. Grob, G. Grob, and K. Grob, Jr., *HRC&CC 2*:31 (1979).
36. L. D. Metcalf and R. J. Martin, *Anal. Chem. 39*:1204 (1976).
37. E. J. Malec, *J. Chromatogr. Sci. 9*:318 (1971).
38. R. Dandeneau and E. Zerenner, *HRC&CC 2*:351 (1979).
39. K. Grob and G. Grob, *J. Chromatogr. 125*:471 (1976).
40. K. Grob, G. Grob, and K. Grob, Jr., *Chromatographia 10*:181 (1977).
41. K. Grob, Jr., G. Grob, and K. Grob, *HRC&CC 1*:149 (1978).
42. A. Venema, L. D. G. v.d. Ven, and H. v.d. Steege, *HRC&CC 2*:405 (1979).
43. G. Schomburg, H. Husman, and F. Weeke, *Chromatographia 10*:580 (1977).
44. G. Schomburg, *HRC&CC 2*:461 (1979).
45. K. Grob and G. Grob, *HRC&CC 1*:302 (1978).
46. H. W. Dürbeck, I. Büker, and W. Leyman, *Chromatographia 11*:372 (1978).
47. G. Schomburg, R. Dielmann, H. Borwitzky, and H. Husmann, *J. Chromatogr. 167*:337 (1978).
48. R. C. M. de Nijs, J. J. Franken, R. P. M. Dooper, J. A. Rijks, H. J. J. M. de Ruwe, and F. L. Schulting, *J. Chromatogr. 167*:231 (1978).
49. K. Grob, *Chromatographia* 7:94 (1974).
50. K. Yabumoto and W. VandenHeuvel, *J. Chromatogr. 140*:197 (1977).
51. W. Jennings, J. A. Settlage, and R. J. Miller, *HRC&CC 2*, 441 (1979).
52. W. Jennings, J. A. Settlage, R. J. Miller, and O. G. Raabe, *J. Chromatogr. 186*:189 (1979).
53. Fong-Yi Lieu, and W. Jennings, *HRC&CC* 3:89 (1980).
54. *Pure Appl. Chem. 37*:445 (1974).

4

Considerations in Quantitative Analysis

Walter G. Jennings / *Department of Food Science and Technology, University of California, Davis, California*

James A. Settlage / *L.A.B. Gesellschaft für pharmakologische Untersuchungen m.b.H. & Co. (Laboratory for Pharmacological Research), Neu-Ulm, Federal Republic of Germany*

I. INTRODUCTION

Quantitation is influenced by, among other things, the method of sample preparation; while critically important, a detailed discussion of those procedures is beyond the scope of our considerations here. We will begin with the assumption that appropriate preparation procedures have been completed and the analyte material resides in a vial ready for analysis, and we will devote our attention to considerations of those factors that influence our measurement of its quantitative composition,

The analyst would like to feel (1) that everything in the analyte survived storage in the vial (or other container) to the time of analysis; (2) that a representative quantity of each solute in the analyte began the chromatographic process (i.e., actually reached the beginning of the column); (3) that everything beginning the chromatographic process finished the chromatographic process; and (4) that the signal delivered by the detector to the data processing equipment was treated so that it gives a precise quantitative representation of the amount of that solute reaching the detector. While we have made remarkable progress, we have not yet attained, or in all probability will we ever attain, these goals for all solutes. In fact, for any solute there exists a concentration limit below which the chromatographic system ceases to be inert and/or the counting mechanisms ceases to be precise.

II. SOURCES OF ERRORS

With respect to the analyte, materials in the vial may react with each other, with air, or with trace contaminants in the vial; compositional changes may result from absorption of particular solutes into the elastomeric closure on the vial, or adsorption to the column wall. The transfer step itself, in which a "representative" sample is taken from the vial and conducted to the inlet of the gas chromatograph, also presents opportunities for compositional changes that are usually ignored by all but the most critical investigators.

A major source of error with respect to the latter is inherent to the sampling syringe itself. The syringe barrel and needle offer new opportunities for adsorption and reaction of sensitive components, particularly where a metal syringe needle is employed. Indeed, the syringe can be viewed as a microchromatographic system in its own right: what comes out of the syringe is not always what went in.

The type of inlet employed to deliver the sample to the column can also exercise a significant effect on the composition of the sample that actually reaches the column and begins the chromatographic process. While split and "splitless" injectors can be made to work well for heat-stable solutes of the proper volatility range, on-column injection is preferred for heat-labile and/or higher boiling solutes. At the present time, the best results are usually achieved by using a deactivated, fused silica syringe needle with a suitably designed on-column injector so that the sample contained in the fused silica needle can be deposited directly in the column. Because the needle is transparent and of very small bore, precise calibration is possible, and the analyst can ascertain with considerable accuracy the size of the injected sample.

Not everything that begins the chromatographic process actually finishes the chromatographic process. While it is now possible to produce open-tubular columns that are much more inert than was possible just a few years, our demands have outpaced these advances. There was a time when the alkyl phenols were regarded as demanding compounds; as we discovered and developed new methods of column deactivation, the alkyl phenol ceased to be a problem, and we turned our attention to the more acidic halogenated and/or nitro phenols. Today we are concerned with the analysis of reactive organonitrates and highly reactive drugs and metabolites, whose analysis would have been considered impossible even a few years ago.

In spite of the remarkable progress that has been made in the area of column deactivation (see Chap. 3), no component of the system, including the vial, the sampling syringe, the inlet, or the column, can be considered as totally inert with respect to all solutes. Inertness toward a given solute or group of solutes is not adequately demonstrated by massive injections of the solute in question; if the injection is sufficiently large, it may result in a chromatographic signal, even in an active system. Consequently, it is wise to test the system for activity. The activity can reside in the syringe, in the inlet, in the column, or in the transfer tubing (if any) connecting the end of the column with the zone of detection.

One of the better ways for determining the level of activity existing in a system was explored by Schomburg et al. [1] and by Dandeneau and Zerenner [2]. A solution of one of the most demanding of the test solutes whose quantitation is desired and a compound that can safely be assumed to be immune to these solute-system interactions (paraffin hydrocarbon?) is prepared in a suitable solvent, thoroughly mixed, and diluted 1:100 (or other suitable amount). The dilute solution is thoroughly mixed, and another dilution, perhaps 1:100, is prepared.

This process is repeated until the final solution contains a lower concentration of the solute in question that is anticipated in the solutions that will ultimately be subjected to quantitative measurements. The ratio of the two solutes should be the same in each of the above solutions. Beginning with the most dilute solutions, an aliquot of each of the solutions is then injected at a suitable sensitivity setting, which may be changed between samples. If the system responds similarly to both the active and the inactive component, the ratios of the two peak areas should remain constant over the entire concentration range. If the system (syringe, inlet, column, etc.) has a demand capacity for the active solute, this is readily apparent, since the ratio of the two solutes reaching the detector varies as a function of the amount in the sample. The level at which this ratio begins to changes represents the lower limit for reliable quantitative measurement.

Figure 4.1, from Dandeneau and Zerenner [2], illustrates the application of this technique. In the sample shown, one series of test solutions explores the suitability of the system for the quantitative analysis of 3,5-dimethylphenol (and presumably, similar compounds); it is assumed that *p*-dimethoxybenzene is immune to the solute-system interactions being evaluated. The ratio of the peak areas (active solute/inactive solute) is seriously affected at low concentrations, which establishes that something in this system makes it unsuitable for

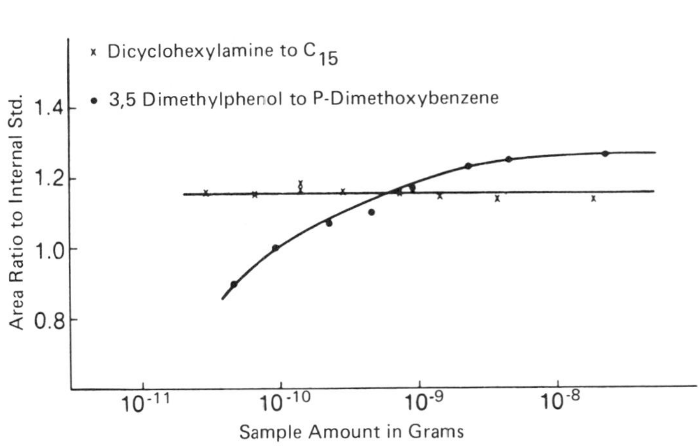

Figure 4.1 Method of checking the activity of a system toward a given solute by measuring the ratio of the active solute/inactive solute in a series of dilutions of the same solution. The concentration of the active solute in the most dilute solution should be below the minimum level desired for quantitative measurements (From Ref. 2.)

the quantitative analysis of phenols of this type below injected amounts of 10^{-7} g. This does not condemn the column (or the syringe, or the inlet, or?) to the wastebasket; it only establishes that this system should not be used for low-level phenol analysis, unless some corrective action is first taken. Analysis of the other set of solutions, containing dicyclohexylamine (active) and pentadecane (inactive), indicates that the system is suitable for the quantitative analysis of this amine down to levels at least as low as 5×10^{-11} g.

It should be noted that a change in the solute peak ratio does not necessarily testify to an activity fault in the system. Peak ratio changes can also result from decomposition of one of the solutes, from a nonlinear split injection, or from a nonlinear detector response. Where the solute ratio does remain constant, the reliability of the results, at least within the limits of these dilutions, has been established. Where the ratio does change, one should be aware of the other possible causes, as some other corrective action may be required.

III. METHODS OF QUANTITATION

A. General Considerations

While not all of the methods considered below can be regarded as satisfactory for use in quantitative analysis, they form a logical progression, and in many cases a comprehension of one assists in the understanding of the use, as well as advantages and disadvantages, of a method considered subsequently. At the same time, it is not our purpose here merely to fill space; antiquated approaches such as excising peaks from a chromatogram so they may be weighed, or even the use of a disk integrator, will not be considered. We will assume the availability of modern electronic integration equipment. At the same time, the user must be aware of the fact that while such devices print results that pretend great precision and appear beyond challenge, they can be no better than the quality of the signal supplied by the detector, the analog-to-digital conversion, and the algorithms with which the integrator manipulates those data. All such measurements should be made with the detector optimized and operating in a plateau region. With an FID, for example, it is possible to operate the detector under conditions where miniscule changes in gas flows affect the sensitivity of the detector response; quantitation would be hopeless under such circumstances.

B. Area Normalization

In this method, the area of each component is reported as a percent of the total area of all peaks:

$$\text{Area \% component X} = \text{Area peak X/Total area all peaks} \times 100$$

These results reflect the relative quantitative composition of a multicomponent sample, but only if all components are eluted, all components are detected, all components are unambiguously measurable (i.e. separation is adequate), and components generate the same detector response per unit of weight. In most cases detector responses are not equivalent and response factors must be determined.

C. Response Factor Determination

A standard solution is prepared containing known weights of each of the substances to be measured; a standard reference compound may be included. The solution is injected, preferably under conditions where the components are completely resolved, and subjected to integration. The response factor (RF) for each component is equal to its weight percent divided by its area percent. As an example, assume were prepare a solution containing 0.187 g X, 0.162 g Y, and 0.194 g Z in a suitable solvent whose weight is neglected. The total solute weight is the sum of the above, or 0.543 g, and the weight percents of each are 34.44, 29.83, and 35.73, respectively. Assume that the peaks are well separated in the resulting chromatogram; and that they are well removed from the solvent tail so that the integrated peak areas represent only the areas of those solutes and do not include an unknown amount of solvent. We would then determine the area percent of each component, and divide that into the weight percent for that component to obtain the response factor:

Component	WT %	Area units	Area %	Response factor
X	34.44	3250	30.63	1.12
Y	29.83	4270	40.25	0.74
Z	35.73	3090	29.12	1.23
	100.00	10610	100.00	

In subsequent efforts at quantitation, the area percent of component X multiplied by the response factor 1.12 should equal the weight percent of component X reaching the detector. Quantitation via response factors has an appealing simplicity and seems quite direct, but without further calibration it provides little absolute quantitative information and assumes that response factors are constant over the range of concentrations being measured. When this method is chosen, response factors must be rechecked against the possibility of variations in the detector.

D. Quantitation by Means of Calibration Curves (i.e., External Standards)

In this approach, which is most commonly employed with gas samples or headspace injections, a separate solution is prepared for each solute that is to be quantified later; the solutions should be at or slightly above the upper concentration limit that will be comsidered. From above the upper concentration limit that will be considered. From these, a series of dilutions is prepared that will permit the injection of known weights of each solute. Plots are then constructed of the detector response (i.e., integrated peak area) versus the weight injected (Figure 4.2). In use, the graph is consulted to determine the weight of a particular solute that would produce an observed detector response under those conditions. The data represented in Figure 4.2 meet the first criterion: the individual curves must pass through the origin, that is, there should be no response at zero concentration.

The second criteria, linearity, is also met; it can be seen from the figure, however, that ther is some departure from straight lines. This type of calibration provides, on the one hand, a graph for measuring over a certain concentration range and, on the other hand, an indication of the precision which can be expected. Precision can be determined by calculating the standard deviation of the results from a series of injections at the same concentration and/or

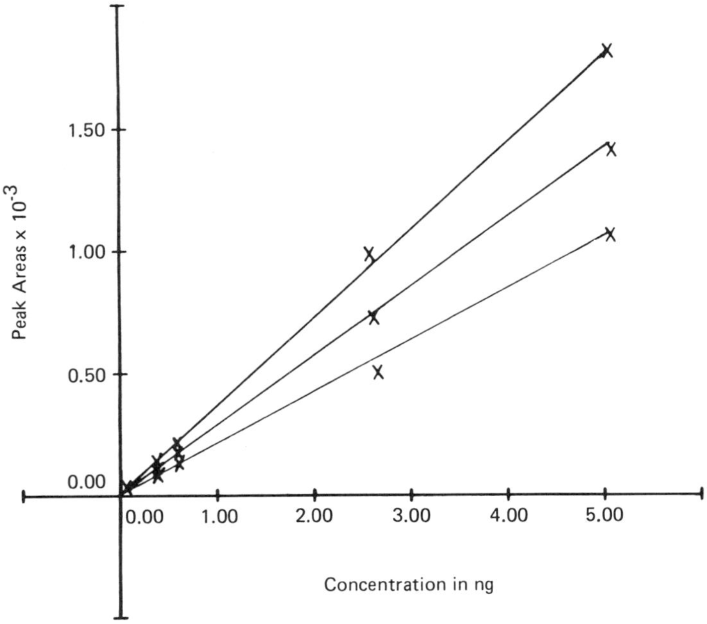

Figure 4.2 Calibration curves for measurement of three substances. Each substance is injected at five concentrations and the curves are determined by linear regression.

by calculating the coefficient of correlation for the entire curve. Formulas for these calculations can be found in most textbooks on statistics and are preprogrammed into many hand calculators.

Use of this calibration method requires careful control of the injection volume and a constant, though perhaps unknown, split ratio. Furthermore, it assumes that sources of variability, such as adsorption, evaporation, decomposition, differential splitting, etc., affect the analyte solutions to the same degree as they do the calibration solutions. For this reason, all samples should be dissolved in the same solvent and when analytes are contained in a complex matrix (e.g., extraction residues), all solutions should be as dilute as possible. Finally, since detector response may vary with time, the calibration must be checked periodically during a series of analyses, or before and after a single analysis.

It should be added that when the analytes to be quantified are obtained, for example, by extraction from a measured volume of material, such as water, plasma, fruit tissue, etc., validation of the quantitative method requires that the calibration standards be added to the same type of material and subjected to the same extraction procedure as the analytes.

This will, of course, provide myriad possibilities for variability and may preclude meaningful quantitation by this method.

E. Internal Standards

This is probably the most satisfactory method from a general purpose viewpoint A stock solution is prepared containing a carefully measured weight of the analyte substance. From this, by successive dilutions, a series of solutions is prepared. The concentrations of these solutions should be calculated so that small aliquots (perhaps 100 μl) can be added to a given volume of substrate (perhaps 2 ml of urine) to create a set of samples spiked over the full range of interest. A second stock solution is prepared containing the internal standard.

It is not essential to known the exact concentration of this solution, but it may be useful for later diagnostic purposes; it should at least be approximately established to provide concentrations in the mid-range of the expected analyte concentrations or, if the internal standard is to provide a carrier effect, perhaps 3–5 times higher. Each calibration sample is then spiked with an equal aliquot of this solution, and the samples are subjected to the predetermined preparation procedure. As the samples are chromatographed, analyte and internal standard peaks are measured, and area or height ratios (analyte/internal standard) are recorded. These ratios are then plotted against the respective weight concentration of analyte. Figure 4.3 shows a calibration curve for the measurement of codeine in human plasma. One milliliter portions of blank plasma were spiked with codeine at concentrations ranging from 1 ng/ml to 200 ng/ml and with a methyl analogue of codeine at about 20 ng/ml. It can be determined from the curve that a peak ratio of 1.00 represents a plasma concentration of 130 ng/ml. Therefore, when unknown plasma samples spiked with the same amount of internal standard are measured, their codeine concentrations can be obtained by multiplying the peak ratio by 130.

Quantitation via internal standards has two major advantages from a chromatographic point of view. The size of the injection need not be known, and long-term detector stability is not usually critical. It is still possible to experience inaccuracies due to discrimination in sampling (e.g., component

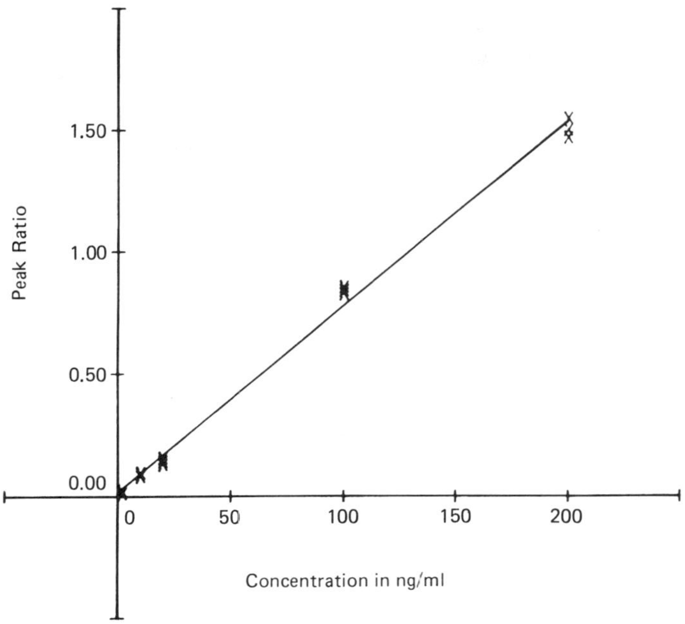

Figure 4.3 A calibration curve for the quantitative determination of codeine using an internal standard. Five concentrations are used, and each concentration is extracted and injected three times. The curve is then determined by linear regression.

adsorption by the syringe needle) or injection (e.g., a nonlinear split), or anything that affects the accuracy of the peak area measurements, such as an inaccurate integrator (not unknown) or inadequate solute-solute or solute-solvent separations.

The major advantage of quantitation via an internal standard, however, is that errors occurring at each step, whether positive or negative, tend to be self-compensating because the gain or loss affects the analyte and the internal standard proportionately. This advantage derives from and is highly dependent on a close chemical similarity between analyte and internal standard. In general, the greater the similarity between these two substances, the better the precision will be for the overall measurement process. On the other hand, increased similarity places greater demands on the separation system. Here high-resolution capillary columns provide still another advantage over packed columns, which is often overlooked.

IV. CONCLUSIONS

This last statement serves to illustrate a point. Creatively employed, capillary columns offer more than increased plate numbers and better deactivation than can be provided by packed columns. Capillary GC often provides opportunities for fundamentally different approaches to chromatographic problems, and detection limits may be orders of magnitude lower than those achievable with packed columns. With these advantages, however, come new challenges.

Quantitation of a capillary separation may seem more difficult sometimes because of the presence of a peak or peaks incompletely resolved from the analyte. In the corresponding packed column chromatograms, these peaks may not appear, but the substances they represent are still there and contribute to the peak measurement. Good quantitation will require resolution of the analyte from the interfering substances, which can sometimes be accomplished by optimizing the chromatographic conditions and at other times may require the use of another liquid phase.

A final note pertains to the challenge presented by increased sensitivity. Quantitation at the picogram or femtogram level is often limited by the inability to obtain a zero value for a blank sample. Every effort must be made to ensure that the current injection does not contain a detectable quantity of the previous analyte.

REFERENCES

1. G. Schomburg, *HRC&CC 2*:46I (1979).
2. R. Dandeneau and E. Zerrener, *HRC&CC 2*:351 (1979).

5

Prechromatographic Microliter Techniques in Capillary GC Trace Analysis*

Wolfgang Dünges / *Fachbereich Geowissenschaften der Universität, Mainz, Federal Republic of Germany*

I. INTRODUCTION

Microliter techniques are used as prechromatographic procedures for capillary GC in two ways (see Table 5.1). It is obvious that their role differs in microliter/ppm analyses (respectively in microliter/ppb analyses) and in milliliter/ppb analyses; however, these techniques are equally valuable in both cases. At first, it may seem surprising that until quite recently (see [1]) no effort was undertaken to investigate systematically this important aspect of organic trace analysis. In my opinion the difficulties of the micro range are evident to everyone who works with microliter amounts. Nevertheless, the possibilities were obscured and became apparent only after long experimental work, through which the reasons for the failure of conventional operations were elucidated.

---

*This paper is dedicated to Professor Dr. L. Horner on the occasion of his seventieth birthday.

Table 5.1 Range of Application for Prechromatographic Microliter Techniques

	μl/ppm analyses	ml/ppb analyses
Primary sample (example)	5–100 μl (50 μl of an enzyme preparation)	0.2–10.0 ml (5 ml of drinking water)
Solutes	Phenols (1 ppm each)	Phenols (1 ppb each)
Dosage volume for capillary GC	2 μl	2 μl
Detection limit	1 ng	1 ng
Volume of the extract[a]	150 μl	15 ml
Enrichment for quantitative determinations	3-fold	3000-fold
Conventional milliter operations (end-volume)	Not feasible	Extracting, concentrating, drying (200 μl)[b]
Microliter operations	Extracting, concentration, drying, derivatization	Concentrating further and derivatization
End volume (i.e., dosage sample)	5 μl	5 μl

[a]e.g., with ethyl acetate, recovery of solutes is 90%.
[b]If this is the definite end-volume, the degree of enrichment is still insufficient. A 50-fold greater amount of the primary sample must be used in this instance.

It should be kept in mind that microliter experiments can be carried out only if all conditions are optimal and all working instructions (see [1]) are strictly adhered to, just as, for example, in the preparation and coating of capillary columns, where similar conditions must be met. In this chapter, recent experiments from the author's lab are described. Exact working details are provided in Ref. 1b. Because of the paucity of systematic investigations on all prechromatographic procedures, the literature on this subject is equally scarce. Exceptions are an excellent review by E. Reid [2] and several monographs on derivatizations [3–5]. If information on a particular method of sample pretreatment is needed, the reader should consult one of the surveys [6-10].

II. EXPERIMENTAL TECHNIQUES

A. Extractions

Extractions play a pivotal role in prechromatographic experiments, expecially if biological material is to be examined. Incorrect extraction procedures lead to low yields of recovery and to erroneously positive or negative results; they even may lead to severe damage to capillary systems as, for example, if biological components such as lipids are transferred into the column. In some instances, purification of extracts has been recommended, using reextraction procedures. It is, however, preferred that impurities be avoided in the first step. In earlier experiments we developed extraction techniques for small amounts of biological fluids. The technique of *extraction with slow stirring* is carried out as follows:

Ten to fifty μl of whole blood, plasma, or a microsomal suspension are covered with the same volume of an organic solvent. Both phases are stirred with a glass or a PTFE rod at less than 300 rpm, so that the clear-cut separation of the layers is maintained, and no mixing of the phases is observed in the form of droplets. Mixtures of 1 part of acetone with 1 part ether or of ethyl acetate have been found to be best suited for biological samples, and give higher recoveries from buffer solutions as well.

High yields of phenols and barbiturates were obtained with only minimum coextraction of lipids or proteins. These extraction procedures, which are meticulously described in Ref. 1, are the basis of glass capillary GC determinations of these compounds, either with derivatization (barbiturates [15]) or without (phenols [1]). The method of stirring out, which dates back to P. L. Kirk's work [11], gave highly satisfactory results in our investigations, and is strongly recommended as a basis for microliter extractions of biological samples with other solvents as well.

In more recent experiments we have introduced a different approach, namely, *extraction with complete water transfer* [13]. In this method an aqueous sample is mixed with an organic solvent. It is assumed that during the reduction of the aqueous phase, equilibrium conditions prevail and Nernst's law of distribution ($c_{org.\ ph.}/c_{water} = k$) is valid. By shifting the volume ratio to the side of the organic phase until $v_{org.\ ph.}/v_{water}$ approaches infinity, the solutes are carried into the organic liquid, even if the coefficient of distribution is very small. The following conditions must be met: (1) The solute should be extractable to a finite amount; and (2) the equilibrium of dissociation of ionizable compounds (i.e., acids and bases) should be established rapidly.

The validity of the method was investigated using the acidic, neutral, and basic drugs which are listed in Table 5.2. These drugs were extracted from a phosphate buffer (pH 7.4) with ethyl acetate/acetone (1:1) until complete transfer of water was accomplished. The yield was in all instances higher than 80% [13]. The principle of the method is illustrated in Figure 15.1. The practicability of the technique depends on several prerequisites:

1. Only water and organic solutes should be transferred into the organic solvent. Inorganic salts, lipids, and proteins should stay behind.
2. Reliable procedures for drying the extracts should be available by which losses of the solutes may be minimized.
3. Since the volumes of the extracts are larger than usual, reliable methods for enrichment must be used whereby the organic traces to be determined are retained insofar as is possible.
4. During concentration procedures, impurities from solvents are enriched as well as solutes; therefore, organic liquids of the highest purity should be available.

These last two problems have been solved previously: Special procedures for the purification of acetone and ethyl acetate have been published [1], and concentration under partial reflux (see below, and [1]) is the required

Table 5.2 Results of Enrichment (A) and Extraction (B) Procedures

	A[a] (%)	B[b] (%)
Debrisoquine	95	95
4-Hydroxydebrisoquine	95	95
Diazepam		80
Carbamazepine	52	90
Phenprocoumon	30	90
Phenylbutazone	13	90

[a](A) Solutions (3 ml ethyl acetate with 10 μg of each of the drugs) were taken to dryness in an air stream; temperature, 80°C.
[b](B) Extraction with complete water transfer into ethyl acetate/acetone (see [13]). Concentrated under partial reflux. Ethyl acetate removed by Methanol as described in the Text.

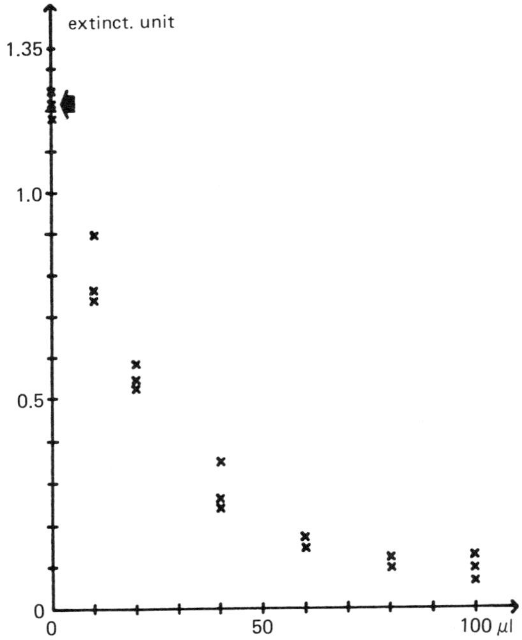

Figure 5.1 Variation of extraction yield with amount of aqueous phase. The amounts of the solute (fenclozic acid, 10 μg) and of the organic phase (200 μl ethyl acetate/acetone, 1:1) were kept constant. Each of the two liquids had been presaturated with the other one; thus no change in the amounts of the phases was observed after extraction (3 min at 300 rpm). Arrow: Extraction with complete water transfer. (From Ref. 13.)

enrichment procedure. Drying of the extracts has been done either azeotropically (e.g., with ethyl acetate), or by freezing out (see [13]), or by combining both techniques. Of many investigated solvent systems, ethyl acetate/acetone (1:1) has been found to be particularly well suited for the extraction of the buffer, and acetone alone for the extraction of plasma [13]. The new method appears to be a valuable tool, especially for capillary work [13].

B. Volume Reduction of Solutions Under Partial Reflux

Volume reduction of solutions under partial reflux down to 5 μl, has recently been developed for the preconcentration of dilute organic solutions [1]. The technique, which is based on a method for IR spectroscopy [14], is described in Figure 5.2. Its principles may be summarized as follows:

1. The solvent is never chased completely.
2. Vaporized solutes are washed out of the gaseous phase by the condensing solvent.
3. Solutes which have been left behind on the glass walls from the receding solutions are redissolved by the refluxing liquid.

The method has been evaluated for 60 chemically different compounds in several solvents [1]. Solutes with a boiling point of 175°C or higher are retained nearly quantitatively in acetone. A stream of solvent vapor constantly leaves the vessels, and therefore, humidity cannot enter from ambient air. Dry end-concentrates from solvents result, which cannot be desiccated by azeotropic

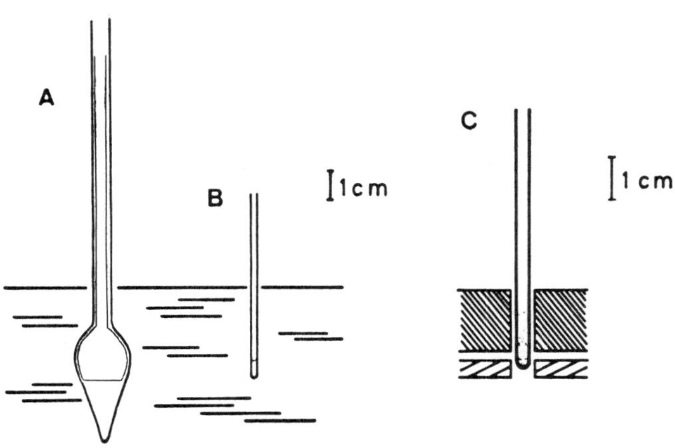

Figure 5.2 Concentration under partial reflux (A,B) and reaction under total flux (C). Volume of the solutions reduced from 1 ml to 30 μl (A) and subsequently to 5–7 μl (B). The bulbs are immersed in a water bath (at 90°C, for ethyl acetate). The tubes act as an air condenser. After the last trace of the liquid layer vanishes, the vessels are transferred immediately into ice. The velocity of evaporation and the shape and dimensions of the vessels are critical (see [1]). (C) 7-μl reflux reactions are carried out under vibrations and by cooling the tubes with a metal cooler. Heat is applied from an air bath. After application of a boiling chip and of the needed reagents, reaction is carried out with the end-concentrates in the same vessel as (B).

distillation, as can acetone. Volume reduction under partial reflux—a cornerstone in our prechromatographic methodology—is easily performed when optimal experimental conditions are reached and maintained reliably.

C. Displacement of Solvents in the Vapor Phase

Because of the special conditions required for working in the micro range, small amounts of compounds having relatively high "macro boiling points" could be lost if their solutions were taken to dryness (see Table 5.2). Up to now this step was essential if a solute was to be transferred from one solvent into another having a lower boiling point. This task can now accomplished by the displacement of solvents in the vapor phase in the following way:

1. Evaporation of the primary solvent under partial reflux, for example, 1 ml ethyl acetate (b.p. 77°C), in a water bath at 90°C, until the liquid is no longer visible.
2. Rapid addition of the lower boiling solvent (such as 100 μl of methanol) into the hot vessel, which is kept in the water bath. Concentration under partial reflux follows. The procedure is then repeated.
3. After the disappearance of the last methanol layer, quick transfer of the vessel into ice.

In the example described above, ethyl acetate could not be detected by UV absorption; thus less than 5% was present. All compounds shown in Table 5.2 were nevertheless retained to more than 95%.

The procedure, which is based on the controlled conditions shown in Figure 5.2, was also proven to be of great value in sample preparation, since it allows one to choose freely the optimum medium for each part of the analysis, including (1) extraction; (2) drying (i.e., freezing out or azeotropic distillation); (3) derivatization; and (4) measurement (optical, chromatographic, or other). To cite an example (see also [13]); the extraction of plasma was performed with acetone; ethyl acetate was then added, and the extract was concentrated and dried. Subsequently, either optical measurements were performed in methanol after the ethyl acetate was removed, or derivatizations were carried out in acetone (see [15]), which had displaced ethyl acetate.

D. Microliter Reflux

Reactions with complete reflux are the method of choice for derivatizations in the microliter range, as has been demonstrated previously [1]. In amounts less than 50 μl and under atmospheric pressure, these reactions can be performed only with a *microrefluxer* [16] (see also [1]), whose features include the following:

1. Constant precise temperature of the reaction mixture is guaranteed by the boiling solvent.
2. Smooth reflux is reached reliably as equilibrium between evaporation and condensation is achieved and maintained more easily in smaller masses.
3. Because local overheating is avoided with microliter reflux, the concentrations of reagents and solutes are equal in all parts of the reaction mixture, and side reactions are thus minimized.

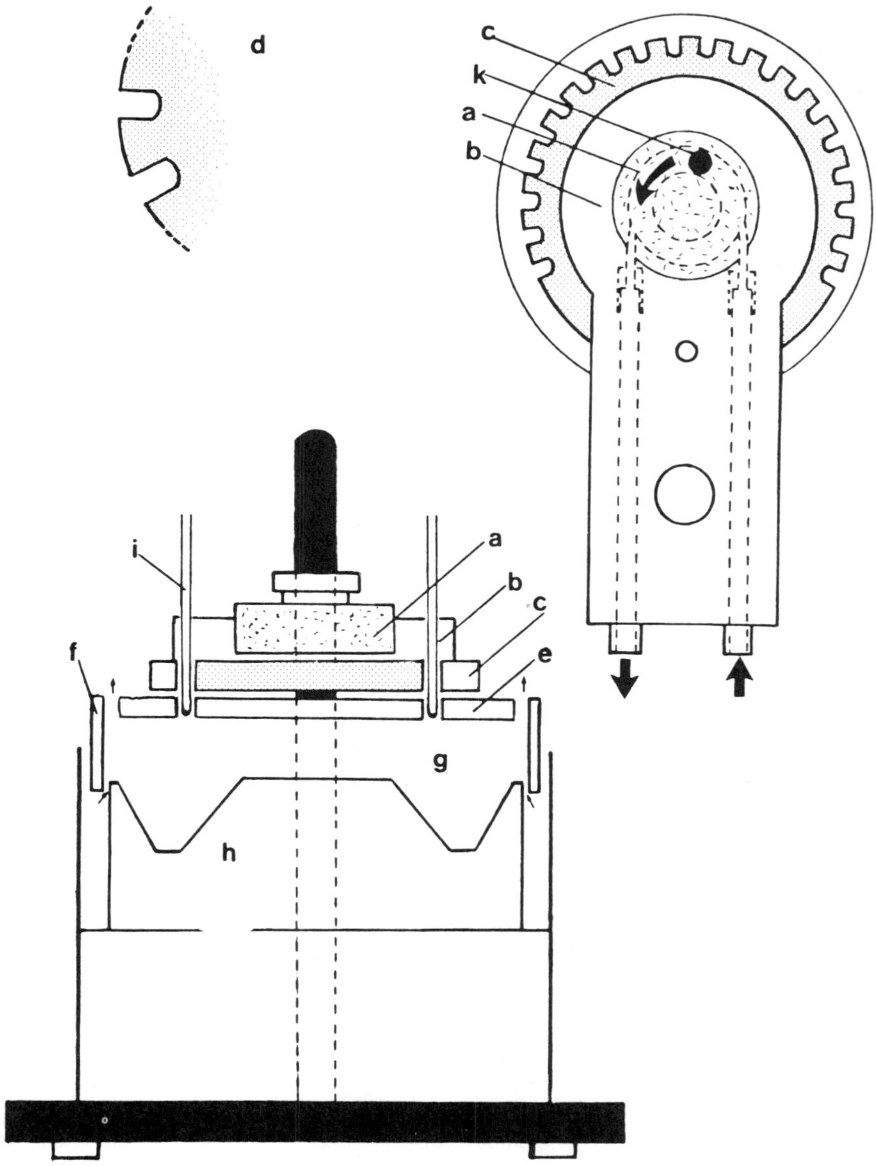

Figure 5.3 A new model of the microrefluxer. The principle was first shown in [16]; for earlier versions, see [1]. a, Water-driven vibrator with metal ball, k; b, metal block with holes for the flow of water; c, metal holder plus cooler for 7-μl vessels [d, for 1-ml vessels]; e, cover of air bath, g; f, glass side-walls of g; h, heater. The streaming water causes strong vibrations and therefore, the formation of droplets in the cooled zone is not possible, and a constant refluxation results for μl amounts. Liquids of low and high b.p. are fluxed reliably (see Table 5.3). This device, including micro glassware, can be obtained from Forschungsinstitut Berghof, Tübingen, West Germany.

It should be noted that the development of our derivatization methods [1,15] was made possible only with the aid of the microrefluxer, which permitted a large series of reactions to be performed in a relatively short time span. The microrefluxer is thus extremely important in the scheme of our microliter techniques. All derivatizations, be they in microliter/ppm or in milliter/ppb analyses (see Table 5.1), can now be performed under optimal conditions, that is, in 5-μl end-concentrates.

In collaboration with G. Maassen of Tübingen, an improved version of this device has been developed (Figure 5.3) which has the following properties:

1. The actions of the vibrator are stronger and can be transmitted equally to all vessels.
2. The distance between heated zone and cooled zone (see Figure 5.2C) is minimized; therefore, the amount of vaporized solvent is kept small.
3. Because of the stronger agitations, reflux vessels with a slightly smaller inner diameter can be used. Evaporation of low-boiling liquids is reduced.

Results from Table 5.3 indicate the applicability of the technique to many kinds of solutions. It is noteworthy that 7-μl samples were heated under reflux for 30 min with no problems, and even though relatively viscous liquids such as water or *n*-butyl alcohol were employed.

E. Auxiliary Techniques

Microliter methods could be performed only if elaborate auxiliary techniques were available. These include: (1) purification and drying of solvents; (2) storage of microliter amounts of dilute organic solutions; (3) pressure reactions of aqueous or organic solutions; (4) cleaning of microglassware; and (5) recognition of water in 7-μl dosage samples. It was thus necessary

Table 5.3 Reflux Experiments with the New Microrefluxer (Figure 5.3): Weight Loss with Different Solvents[a]

	Temperature of the air bath, °C	Average weight loss, % (standard deviation)
Methylene chloride (b.p. 40°C)	65	13.8 (0.836)
Acetone (b.p. 56°C)	70	10.2 (0.816)
Methanol (b.p. 65°C)	77	5.2 (0.408)
Ethyl acetate (b.p. 77°C)	94	4.0 (1.636)
Water (b.p. 100°C)	138	2.4 (1.032)
n-Butyl alcohol (b.p. 118°C)	173	0.2 (0.707)

[a]Six vessels containing 6–7 μl were heated for 30 min.

to develop these special techniques along with the required tools right from the very beginning. In Ref. 1, these techniques have been described together with detailed working instructions. In the following, the handling of milligram amounts of hygroscopic salts is discussed (1) as an example of how in the microliter range, parameters have to be considered which can be neglected in working with larger amounts; and (2) since this technique has been used recently to solve a problem in a derivatization reaction for capillary GC trace analysis. Small amounts of salts adsorb water more rapidly than large ones because of the increased surface-to-mass ratio. On the other hand, micro amounts can be handled more easily if proper precaution is taken. A simple micro hood (see [1]) has been found to be very helpful in this regard.

This device can be assembled easily from a glass funnel with a maximum diameter of 40 cm and appropriate sheets of plastic. A stream of air is passed through its neck with a velocity of up to 40 ml/min. Equilibrium values of humidity are reached in the micro hood within about 20 min since it is practically void of dead volume. These values correspond closely to those of the predried streaming air in the following investigations (i.e., 4%), and this despite an ambient air humidity of 72%. These values could then be maintained with a reduced air flow of 10 ml/min.

We investigated the methylation of diphenylhydantoin using Claisen's carbonate method (see Chap. 29, this volume) [15]. It was assumed that the water content of alkali carbonates, which are basic reaction partners, could play a role here. We therefore determined the change of weight when dry and moist Cs_2CO_3 and K_2CO_3 are exposed to ambient and to dry air. Moist salts did not change in weight under streaming, dry air to the extent that the dry carbonates did (see Figure 5.5), especially if one considers the amount of time normally required for their addition into reaction vessels (i.e., less than 1 min).

Since it was now clear that the water content of alkali salts was not altered during the preparation of the reaction mixtures, reliable investigations of the reaction conditions could be carried out. To 7-μl volumes containing the other reactants [15], 0.5–1.0 mg of different carbonates were added. Results using glass capillary GC after 30 min of reaction with the microrefluxer are shown in Figure 5.4. It can now be stated:

1. Alkali carbonates with a water content of more than 10% (A',B') lead to complete *N,N*-methylation of phenobarbital (a) and to the bis-(methyl) product (b), and to the monomethyl derivatives of diphenylhydantoin (c).
2. If salts are used which have been desiccated thoroughly (A,B), only the peaks of the *N,N*-permethylated compounds (a,b) are detected, indicating the quantitative conversion of both phenobarbital and diphenylhydantoin.
3. Peaks of additional side-products (like +E, Figure 5.1 of [15]) could not be discerned. This was confirmed by comparison with blank chromatograms, where the small side-peaks (x) showed up in the same intensity.

Results from the literature [17] discussed in [15] are similar to those seen in Figure 5.4B'. It seems highly possible that they, too, were achieved with moist carbonate.

Because of the importance of a reliable derivatization method for capillary trace determinations, the influence of the water content on the outcome of the carbonate method should be investigated on a broad scale. Earlier

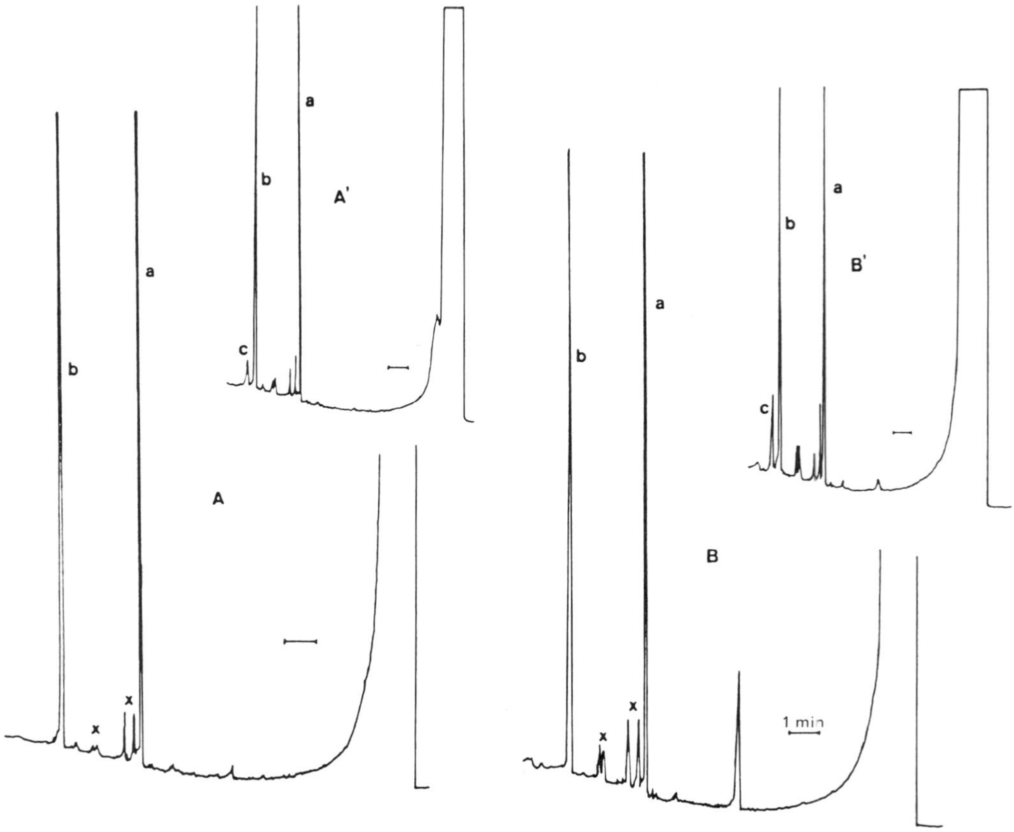

Figure 5.4 Gas chromatograms of 1-μl aliquots from 7-μl reactions of phenobarbital+diphenylhydantoin in acetone with different carbonates and methyl iodide. A,A', Cs_2CO_3; B,B', K_2CO_3. A,B, each salt desiccated at 250°C, 15 torr, over P_2O_5. A',B', each salt contained about 15% of water as determined by weight loss after heating. The salts were added under controlled conditions (see text). a, *N,N*-dimethylphenobarbital, b, *N,N*-diphenylhydantoin, c, *N*-methyldiphenylhydantoin. For GC conditions, see Figure 5.2. (From Ref. 15.)

methylation experiments with dihydroxybenzenes, dicarboxylic acids or drugs of greater complexity did not give unequivocal results, and therefore, they should be repeated under the described conditions as well.

CONCLUSIONS

Recently, organic and biomedical trace analysis has made enormous progress due to improved capillary GC. But even if further advances in the development of instruments cannot be foreseen, it seems unlikely that some day a black box will be invented which, upon being fed a raw primary sample, will deliver exact qualitative and quantitative information on the interesting organic traces. Thus, for the time being, all results from instrumental

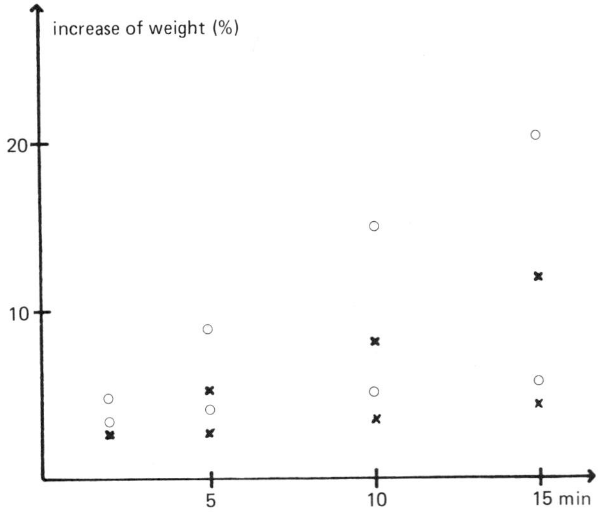

Figure 5.5 Relative increase of weight under controlled conditions (lower graphs) and in ambient air (upper graphs). 1–2 mg amounts of dry (see Figure 5.4) Cs_2CO_3 (crosses) and of K_2CO_3 (circles) were kept for different times under a micro hood (4% relative humidity) or one the laboratory bench (72% humidity).

analyses are based on sophisticated prechromatographic techniques, such as microliter procedures.

ACKNOWLEDGMENTS

The experiments have been carried out in the laboratories of W. Dosch, Fachbereich Geowissenschaften der Universität Mainz, to whom I am greatly indebted.

I thank F. P. Ott, Norheim, and W. Christmann, Mainz, for helpful discussions, and B. Dünges, P. Dünges, and B. Soltau for their excellent technical assistance.

This work has been generously supported from the Doktor-Robert-Pfleger-Stiftung, Bamberg, Federal Republic of Germany.

REFERENCES

1a. W. Dünges, *Prä-chromatographische Mikromethoden*. A. Hüthig, Heidelberg, (1979).
1b. W. Dünges, *Pre-chromatographic Micromethods*, 2nd ed. A. Hüthig, Heidelberg, in preparation, 1984.
2. E. Reid, *Analyst 101:*1–18 (1976).
3. J. D. Micholson, Analyst *103*:1–28, 193–227 (1978).
4. K. Blau and G. King, eds., *Handbook of Derivatives for Chromatography*, Heyden and Son, London, 1977.

5. D. R. Knapp, *Handbook of Analytical Derivatisation Reactions*, J. Wiley & Sons, New York, 1979.
6. E. Reid, ed. *Assay of Drugs and Other Trace Compounds in Biological Fluids*, North Holland Publishers, Amsterdam, 1976.
7. E. Reid, ed., *Blood, Drugs, and Other Analytical Challenges*, Ellis Horwood Publishers Chichester and J. Wiley & Son, New York, 1978.
8. E. Reid, ed., *Trace-Organic Sample Handling*, Ellis Horwood Publishers, Chichester and J. Wiley & Sons, New York, 1981.
9. K. Tsuji ed., *GLC and HPLC Determinations of Therapeutic Agents*, 3 Parts, M. Dekker, Inc., New York, 1978–79.
10. H. S. Hertz and S. N. Chesler, eds., *Trace Organic Analysis: A New Frontier in Analytical Chemistry*, National Bureau of Standards Spec. Publ. #519, Washington, D.C., 1979.
11. P. L. Kirk, *Adv. Chromatogr.* 5:79–125 (1968).
12. W. Dünges, R. Langlais, and R. Schlenkermann, *J. High Res. Chromatog. Chrom. Comm.* *2*:361–365 (1979).
13. W. Dünges, J. Bode, H. Hoffmann, M. Müller, and B. Soltau, in Proceedings of the 4th International Symposium on Capillary GC, "Hindelang 1981," (R. E. Kaiser et al., eds.), A. Hüthig, Heidelberg, Basel, New York, 1981, pp. 765-778.
14. K. Beyermann, A. Kessler, and P.-W. Ungerer, *Fres. Z. Analyt Chemie 251*:289-293 (1970).
15. W. Dünges, Chapter 29, this volume.
16. W. Dünges, *Analyt. Chem. 95*:963–964 (1973).
17. R. De Sagher, J. Pocius, and J.-P Thenot, *J. Chromatogr. 156*:43-53 (1978).

6

Capillary GLC of Amino Acids

Hartmut Frank / *Institut für Toxikologie, Eberhard-Karls-Universität, Tübingen Federal Republic of Germany*

Halvor Jaeger / *L.A.B. Gesellschaft für pharmakologische Untersuchungen m.b.H. & Co. (Laboratory for Pharmacological Research), Neu-Ulm, Federal Republic of Germany*

I. INTRODUCTION

Amino acids as inherent constituents of living matter represent more than 50% of the dry weight of a cell, mainly as building blocks of proteins. These not only constitute structural elements of the cell architecture, but also function as biocatalysts, as messenger substances, or as vehicles for the selective transport of biological fuels. Amino acids also participate in the cell chemistry in their free forms. Some occur as intermediates in metabolism, serve as nutrients, neurotransmitters, or precursors for biosynthesis of cell constituents [1–3]. Their presence in sedimentary rocks has been considered as indication of primordial life [4].

Structural features common to all of them are an amino group, a carboxyl group, and a hydrogen atom attached to the same carbon atom, and a fourth substituent of variable structure which confers to each amino acid its particular chemical property. From the twenty amino acids commonly found in proteins (Table 6.1), the fifteen neutral amino acids with aliphatic or aromatic side chains are compounds soluble in water as zwitter ions with isoelectric points

Table 6.1 Structures of Protein Amino Acids: $H_2N-\underset{R}{CH}-COOH$

Amino acid	Side chain (R)	Abbreviation	One-letter symbol [5]
Glycine	$-H$	Gly	G
Alanine	$-CH_3$	Ala	A
Valine	$-CH(CH_3)_2$	Val	V
Isoleucine	$-\overset{*}{C}H(CH_3)(C_2H_5)$	Ile	I
Leucine	$-CH_2-CH(CH_3)_2$	Leu	L
Serine	$-CH_2OH$	Ser	S
Threonine	$-\overset{*}{C}HOH-CH_3$	Thr	T
Cysteine	$-CH_2-SH$	Cys	C
Methionine	$-(CH_2)_2-S-CH_3$	Met	M
Proline	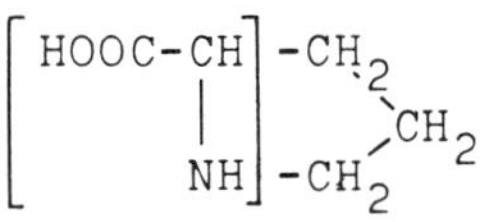	Pro	P
Phenylalanine	$-CH_2-$ 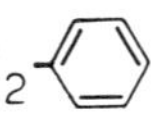	Phe	F
Tyrosine	$-CH_2-$ 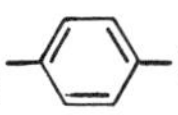 $-OH$	Tyr	Y
Tryptophan	$-CH_2-$ 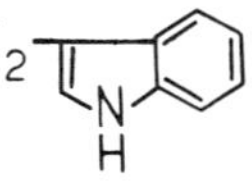	Trp	W
Asparagine		Asn	N
Glutamine	$-(CH_2)_2-CONH_2$	Gln	Q
Aspartic acid	$-CH_2-COOH$	Asp	D
Glutamic acid	$-(CH_2)_2-COOH$	Glu	E
Lysine	$-(CH_2)_4-NH_2$	Lys	K
Arginine	$-(CH_2)_3-NHC(=NH)NH_2$	Arg	R
Histidine	$-CH_2-$ 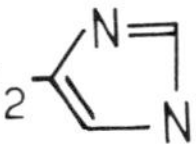	His	H

between pH 6 and 7. The others carry an additional carboxyl group or nitrogen containing function in the side chain, as the acidic amino acids aspartic and glutamic acid, or the basic arginine, histidine, and lysine.

Amino acids exhibit considerable differences in their chemical properties and reactivities; even the fifteen neutral amino acids are distinctly dissimilar due to their different side chains. While six of them have relatively inert aliphatic and aromatic hydrocarbon moieties, others contain reactive functional groups: hydroxyl groups in serine, threonine, and tyrosine, sulfur in cysteine, cystine, and methionine, an indole ring in tryptophan, or carboxamide groups in asparagine and glutamine. Other amino acids may occur in biological samples, the most common being β-alanine, α-aminobutyric acid ($R = -CH_2-CH_3$), γ-aminobutyric acid (GABA), citrulline ($R = -(CH_2)_3-$ $-NH-CO-NH_2$), homocysteine ($R = -CH_2-CH_2SH$), homoserine ($R = -CH_2-$ $-CH_2OH$), 5-hydroxylysine ($R = -(CH_2)_2-\overset{*}{C}HOH-CH_2NH_2$), 4-hydroxyproline ($R = -CH_2-\overset{*}{C}HOH-CH_2-$), 3-methylhistidine ($R = -CH_2-C{=}CH-N(CH_3)-CH{=}N$, ring closed between C and N), and ornithine ($R = -CH_2)_2-CH_2NH_2$).

Each amino acid may occur in two configurations as shown in Figure 6.1, due to the four different substituents attached to the central carbon atom. These two enantiomeric forms, mirror images to each other, are usually designated by the prefix L- or D-, according to their structural relationship to the isomers of the parent glyceraldehyde. The Cahn-Prelog-Ingold designation [6] is also shown but is less commonly used. Glycine has two identical substituents (R = H attached to the α carbon and is therefore nonchiral. The amino acids with two asymmetric carbon atoms, namely isoleucine and threonine, form two enantiomeric pairs which are diastereoisomers to each other. The two pairs are denoted with the prefixes allo- and threo- in analogy to the carbohydrates allose and threose; the natural stereoisomers of both amino acids have the L-threo configuration. Inversion or epimerization of one asymmetric center leads to the corresponding diastereoisomer, not the enantiomer.

Amino acids in biological materials are present almost exclusively in the L-form. Detection of racemic amino acids is rather a sign of their nonbiotic origin [7]. On the other hand, D-amino acids are not truly unnatural, and they seem to serve special purposes mostly associated with defense mechanisms of bacteria and insects [8]. In bacterial cell walls, D-amino acids sometimes comprise more than 10% of amino acids present [9]. Their biological significance is also demonstrated by the existence of enzymes specifically accepting D-amino acids as substrates, the D-amino acid oxidases. D-Glutamic acid and other D-amino

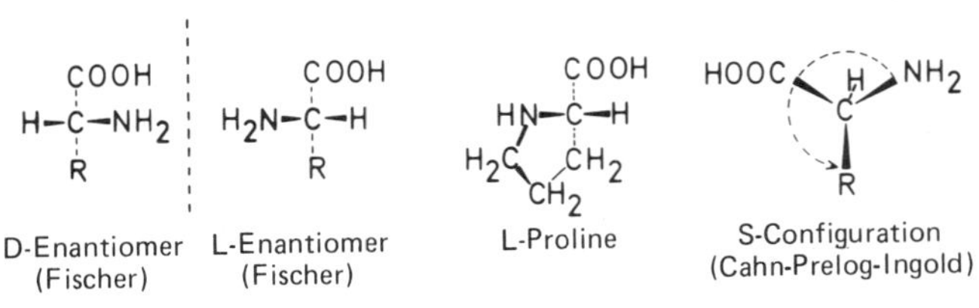

Figure 6.1 Stereostructures of D- and L-amino acids (Fischer projection).

acids were claimed to be found in tumors [10], but were later shown to be artifacts generated during isolation [11].

Amino acids play a significant role in mammalian physiology and are important intermediates of metabolism. The pool of free amino acids in the human organism is in the range of 600–700 g with a daily turnover of 400–600 g. Most important quantitatively is their role in protein biosynthesis and proteolysis, which amount to 300–500 g/day in both directions. Typical steady-state concentrations of amino acids in human blood serum range from 20 μmol (methionine) to 500 μmol (alanine, glutamine) per liter [12]. Disturbances of amino acid metabolism are usually reflected by abnormal serum concentrations [13].

It is obvious that quantitative determination of amino acids in human body fluids is important for diagnosis of metabolic disturbances, especially in pediatrics. However, not only in clinical chemistry, but also in many other fields such as synthetic and analytical protein chemistry [14], food chemistry [15], water analysis [16], geochemistry and archeometry [17], the accurate and precise determination of amino acids is of considerable interest.

Nearly three decades ago Stein, Moore, and Spackman developed an elegant and reliable method for the determination of amino acids [18], an indispensable tool for structure elucidation of peptides and proteins and for many other investigations of amino acid chemistry and biochemistry. They constructed an automatic amino acid analyzer based upon liquid-chromatographic separation of strongly acid cation exchange resins and postcolumn reaction with ninhydrin. A separation of the twenty protein amino acids took about 2 hr; of the more complex physiological mixtures, up to 6 hr. With modern amino acid analyzers the analysis time is about half as long; they are reliable instruments for multicomponent analysis. However, their purchase price and cost of operation are high, and they are dedicated instruments with little flexibility for other tasks.

Therefore, soon after the advent of gas-liquid chromatography, its application to amino acid analysis was proposed. High sensitivity, great speed of analysis, high resolving power, low cost, and great versatility of the instrument are its expedient features. Another advantage is the possibility of on-line coupling of a gas chromatograph to a mass spectrometer, enabling rapid identification of unknowns. On the other hand, gas chromatography of amino acids has a distinct disadvantage in comparison to ion exchange chromatography, namely, the need for blocking or removing ionized and polar functional groups by chemical derivatization. This is necessary to reduce intermolecular cohesion, raise the volatility, and stabilize the amino acids against thermal decomposition. Conversion of amino acids should yield derivatives which evaporate at sufficiently low temperatures and which, at the same time, retain the structural features of each individual amino acid. Another disadvantage at first sight is the low selectivity of the detectors commonly used in gas chromatography; in ion-exchange amino-acid analysis, only compounds containing primary and secondary amino acids are detected, with a few exceptions. Nevertheless, gas chromatography is a flexible technique in this respect as well, and other more selective detectors may be employed.

A method for conversion of free amino acids to volatile derivatives was first published in 1956 [19]. The chemical derivatization chosen in this report on gas chromatography of amino acid was the degradation to aldehydes. Other subtractive derivatizations which were first proposed, involved conversion to volatile amines [20], hydroxy acids [21], or amino alcohols [22]: unfortunately, the derivatives of most trifunctional amino acids are too polar and cannot be dealt with.

As an alternative, additive derivatization by blocking the functional groups present in amino acids was first proposed by E. Bayer et al. in 1957 [23]. They converted the amino acids to methyl esters, and in one of the earliest applications of capillary gas chromatography these derivatives were employed [24]. However, amino acid esters exhibit undesirable properties, especially their facile condensation to dioxopiperazines. A free amino group has strong tendency to adsorption which results in a marked tailing of the peaks (Figure 6.2). One year later, E. Bayer [25] reported that fully blocked *N*-trifluoroacetyl amino acid methyl esters (1)

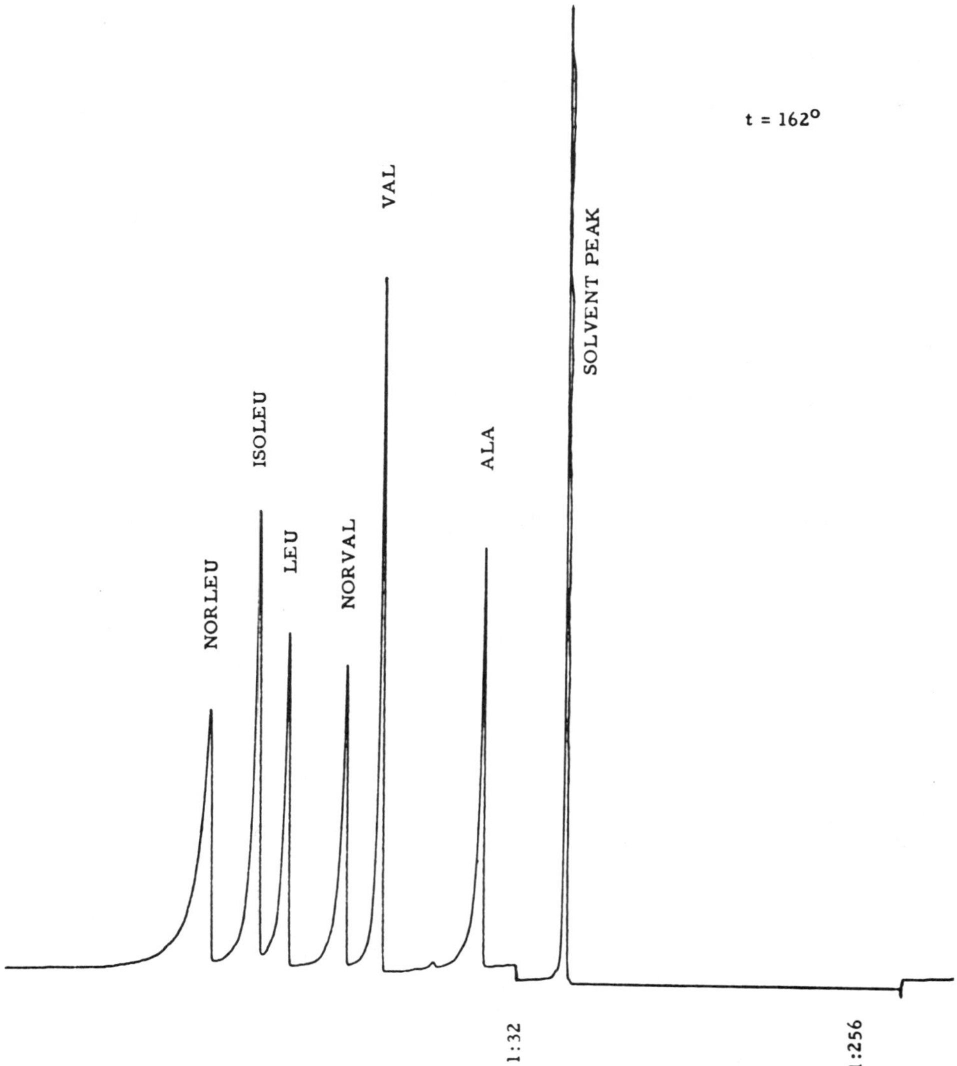

Figure 6.2 Gas chromatogram of amino acid ethyl esters on a stainless steel capillary, 150 ft × 0.01 in., coated with polypropylene glycol stationary phase (Perkin-Elmer R). (From Ref. 24.)

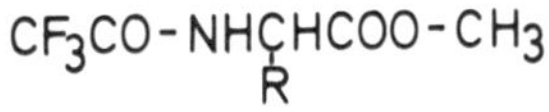

(1)

are well suited for gas chromatography. Since then, this general class of derivatives, the *N*-acyl amino acid alkyl esters (2),

(2)

have been proven to be most suitable for gas chromatography, also for quantitative determinations.

A rapid advancement ensued. The main goal at this stage was to find further efficient methods of derivatization and to optimize them with respect to yield and purity of products. The development of compatible stationary liquid phases and the preparation of sufficiently inert supports were likewise important.

As these problems became solved to a satisfactory extent, quantitative analysis by gas chromatography came into focus. Investigations were undertaken to find conditions for isolation of amino acids which ensured high recoveries from biological sources and high reproducibility and yields of derivatization. Gas chromatographic conditions for separation and detection of all protein amino acids had to be established. Most challenging were the trifunctional amino acids with their differently reactive carboxyl, nitrogen, hydroxyl, and thiol functions. In fact, they turned out to be the touchstones for any new approach. Very active proponents in this period were C. W. Gehrke and co-workers, who published many articles (e.g., Refs. 26–29) exhaustively treating all parameters of gas chromatography of amino acids. Other groups contributed alternate methods of derivatization (e.g., Refs. 30–32).

All these investigations were mainly intended for gas chromatography on packed columns, but they are equally valid for capillary chromatography. Several reviews on gas chromatography of amino acids were published during this period, one of the most comprehensive by P. Hušek and K. Macek [33].

Separation on packed columns was the prevailing method of gas chromatography until recently. Especially for quantitative work capillary gas chromatography was deemed inappropriate, due to problems associated with sample introduction. Also, the earlier capillaries were rather long and exhibited low efficiencies compared to modern standards, but capillaries were used almost exclusively right from the beginning for separation of diastereomers [34] and for direct enantiomer separation on chiral stationary phases [35], discovered by E. Gil-Av and co-workers. Highly efficient capillaries were needed to achieve complete resolution of enantiomers in spite of their small separation factors.

Most capillaries employed during the early period were made from stainless steel and had lengths of up to 150 m [36,37]. The apparent disadvantages of stainless steel are the catalytic and adsorptive properties of metal surfaces. Poor column efficiency, short lifetime or low temperature-stability of the stationary phases, decomposition of sensitive amino-acid derivatives, and tailing of trifunctional amino acids required that more suitable materials for preparation of capillaries be found. With the advent of commercially available drawing machines, glass became the material of choice. Suitable methods for deactivation of the inner surface, the support of the stationary phase, were developed.

Today, as the art of coating capillaries is becoming common knowledge to most scientists working in gas chromatography, glass capillaries enjoy increasing application for amino acid analysis in spite of their fragility. A further development is the introduction of fused silica capillaries; they are highly flexible and the danger of breakage is small. They exhibit generally lower adsorptive activity, but a typical phenomenon observed with flexible quartz capillaries is peak-splitting [38].

The development of chiral stationary phases with high thermal stability [39] enabled a further advance in gas chromatographic amino acid analysis, by which most of the difficulties previously experienced could be overcome. With these phases, complete separation of all amino acid enantiomers can be achieved. This offers the possibility of using as internal standards the D-amino acid enantiomers, chemically identical optical antipodes of the L-amino acids [40]. The method of enantiomer labeling eliminates most of the shortcomings of conventional gas chromatographic amino acid analysis, such as irreproducible yields of sample preparation and derivatization, discrimination during injection for capillary gas chromatography, or fluctuating response factors. The difficulties experienced previously with biological samples prevented the general acceptance of gas chromatography as a routine method for amino acid analysis. With enantiomer-labeling, capillary gas chromatography has become a valid alternative to ion exchange chromatography.

II. DERIVATIZATION

Amino acids, as the name indicates, contain an amino and a carboxyl group and form involatile inner salts. These groups must be blocked to abolish ion bonding and to convert amino acids into sufficiently volatile derivatives. This area of research has received considerable attention, as derivatization is decisive for success or failure of gas chromatographic amino acid analysis. Most contributions to the literature have been concerned with the development of appropriate derivatization procedures for gas chromatography on packed columns, but they apply equally to capillary gas chromatography.

An "ideal" derivatization should fulfill the following requirements:

1. Formation of only one derivative per compound
2. Formation of derivatives of high chemical and thermal stability
3. Complete and reproducible derivatization over a wide range of concentrations
4. No interference with other derivatizations
5. Derivatization of polyfunctional compounds in a minimum number of steps.
6. Reagents more volatile than derivatives
7. Reagents nontoxic
8. No interference from impurities and accompanying substances
9. Formation of derivatives with high compound-specific response in the detector
10. Formation of derivatives with maximum volatility

The main goal of derivatization is to break the strong intermolecular cohesion of the ionized amino and carboxyl groups; in principle, removal or blocking of one of these should be sufficient, an approach utilized in the early days of gas chromatography. Degradative removal of amino and carboxyl groups yields the corresponding aldehydes [20], and even amino acid esters are sufficiently volatile for gas chromatography [23] in spite of their free amino group. However, the trifunctional amino acids, because of their non-derivatized side chains, remain involatile. Current practice suggests that all polar or ionizable groups present in amino acids should be blocked for gas chromatography.

The different functional groups are of very diverse reactivity, and not all methods suggested in the past reach the goal of full derivatization, including all thirteen trifunctional amino acids. Some derivatives are only suitable for special purposes, while some seem to be reported only for the sake of their publication.

Of the trifunctional amino acids, lysine, glutamic acid, and aspartic acid present few difficulties as they contain only an additional amino or carboxyl group. However, the determination of the corresponding amides of the latter two, asparagine and glutamine, is problematic. The amide groups are cleaved under acidic conditions, forming the derivatives of the dicarboxylic acids [41]. During acid-catalyzed esterification, glutamine transiently forms pyrrolidone carboxylic acid through cyclization [42]; deamidation of glutamine to glutamic acid seems to proceed mainly via this intermediate (see Eqs. 1 and 2).

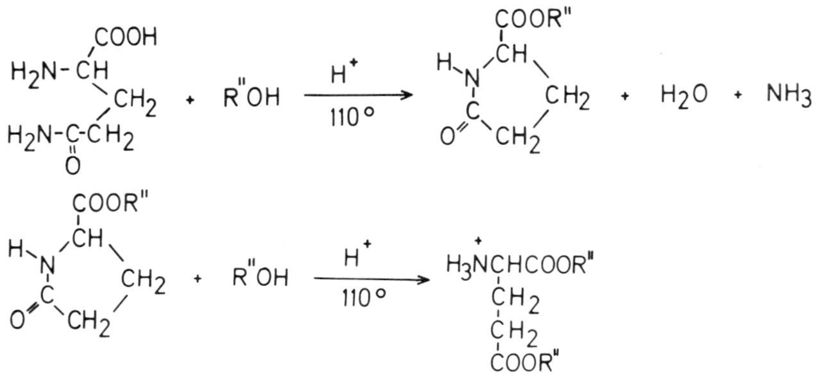

(1)

(2)

As a result, simultaneous determination of the dicarboxylic amino acids and their amides present difficulties; the concentrations of the amides in biological fluids may be of diagnostic value as they play a role in nitrogen metabolism. For protein hydrolyzates this is obviously irrelevant.

Serine and threonine contain a primary or secondary aliphatic hydroxyl function. Tyrosine has a phenolic hydroxyl group which in some cases requires special conditions or reagents for complete derivatization. Conversion of these groups to the trifluoroacetyl esters is simple and rapid, but they are not fully stable against hydrolysis and may even decompose during gas chromatography if the capillaries are not sufficiently well deactivated. Also, they are not stable enough for storage over several days.

Sulfur is present in the side chains of the three amino acids cysteine, cystine, and methionine. Cysteine has a free thiol group at the β carbon; cystine is the oxidized form, the disulfide-linked dimer. Oxidation of cysteine to cystine is rapid at high temperature in the presence of oxygen, which renders the independent determination of both difficult. A good alternative is carboxymethylation of cysteine [43] (Eq. 3).

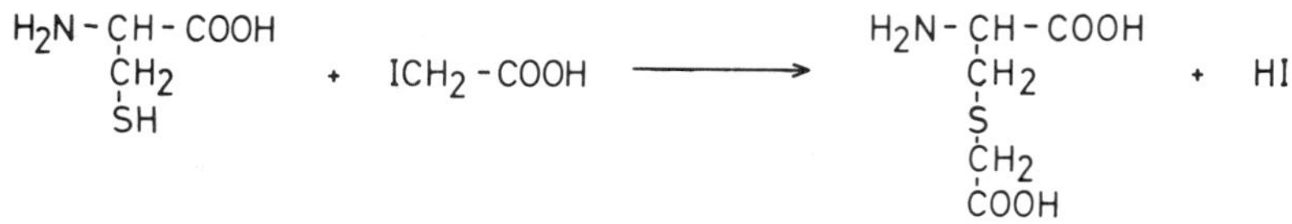

(3)

The amino acid methionine, containing a methyl thioether group, is more stable, but is prone to oxidation as well, leading to the corresponding sulfoxide and sulfone [44]. Therefore, during derivatization of samples with only small amounts of the sulfur amino acids, exclusion of oxygen and traces of metals is very important. The same applies to tryptophan; heating under acidic conditions in the presence of oxygen leads to cleavage of the indole ring and results in low yield of derivative [45].

The last two, histidine and arginine, are the quality test compounds of any gas chromatographic method for amino acid analysis. They are the most difficult to derivatize. Most of their derivatives are sensitive to water, acids, or bases. The acylated imidazole ring of histidine transfers its acyl group efficiently to weak nucleophiles, even to carboxylic acids [46]; histidine, with a free imidazole ring, is less suitable for gas chromatography. Therefore, a three-step derivatization is often preferred for histidine [47], in some cases effected by coinjection with an acylating reagent [46]; another possibility is ozonolytic degradation to aspartic acid [48], but this is time-consuming and interferes with the determination of some other amino acids. Arginine contains a strongly basic guanidino group, which is similarly difficult to derivatize. Alternatively, arginine can be degraded to ornithine [49], the lower homologue of lysine.

Naturally, complete derivatization of all functional groups in one step is most attractive. The closest approach is trimethylsilylation [31,35]. Under proper conditions, amino, carboxyl, hydroxyl, carbonyl, and thiol groups are converted to the corresponding trimethylsilyl (TMS) (Eq. 4):

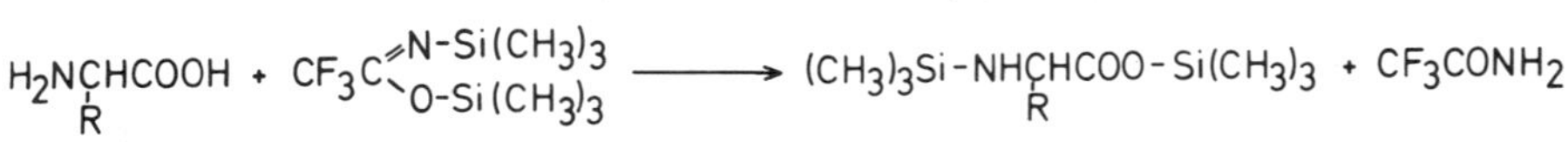

(4)

and even arginine, histidine, and tryptophan yield derivatives with acceptable gas chromatographic properties [51]. Stable derivatives are also formed with glutamine and asparagine, enabling their distinction from the free acids (Eq. 5).

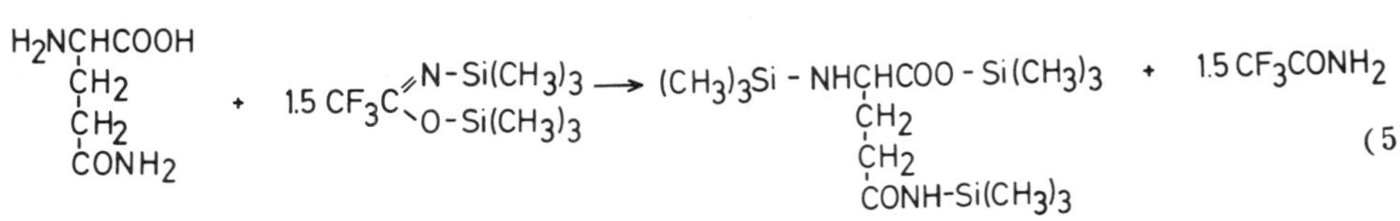

(5)

A drawback is the formation of duplicate derivatives of some amino acids by excessive trimethylsilylation: especially glycine, glutamic acid, lysine, arginine, histidine and tryptophan present difficulties. Coinjection of excess TMS reagent is required in order to block active sites of the columns. This is often necessary for protection of *N*-TMS derivatives: trimethylsilylamines and imidazoles are potent reagents themselves and transfer the TMS group easily to

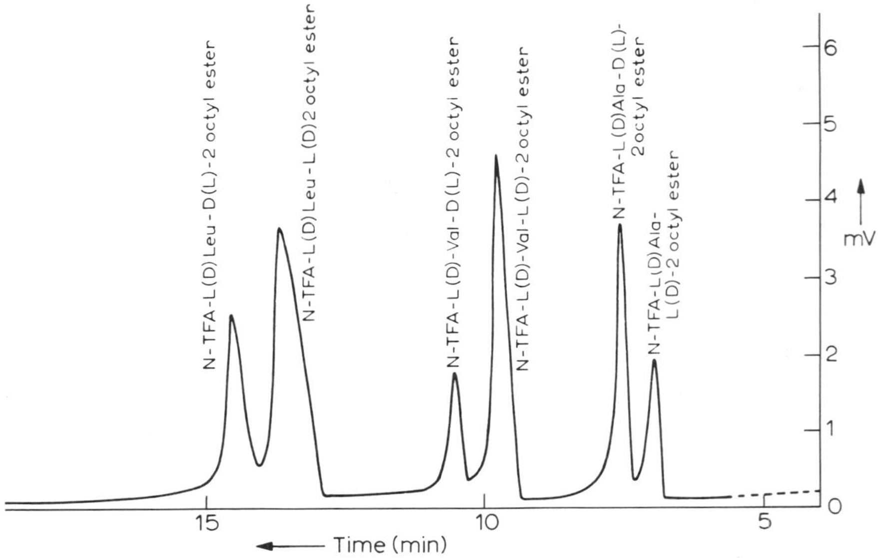

Figure 6.3 Chromatogram of diastereomers of *N*-trifluoroacetyl amino acid 2-octyl esters; stainless steel capillary FS-1265 (Perkin-Elmer), 150 ft × 0.01 in., coated with trifluoropropyl methyl polysiloxane. Temperature, 130°C isothermal; carrier gas, nitrogen; flow rate, 1.5 ml/min. (From Ref. 34.)

hydroxyl groups, including silanol groups of glass surfaces. In capillary gas chromatography this is usually more apparent since the thickness of the stationary phase covering the glass surface is small. It is therefore not surprising that they have not been employed for capillary gas chromatography, in spite of the successful separation of the pertrimethylsilyl derivatives on packed columns [52].

The most common derivatization schemes for amino acids involve two steps: esterification of carboxyl groups and acylation of the remaining amino, hydroxyl, and thiol groups and other nitrogen-containing functions [53–55]. The high volatility of the trifluoroacetyl amino acid esters was recognized relatively early [25,26], and already in the early sixties they were employed for separation of amino acid [34] (Figure 6.3) and dipeptide diastereomers [57] on capillaries (Figure 6.4). The amino acids are first esterified by heating with the corresponding alcohol in the presence of an acidic catalyst [58,59]. In the second step the amino acid esters are converted to the $N(O,S)$-acyl amino acid esters with the corresponding anhydride [60–62] (Eqs. 6 and 7).

$$H_3\overset{+}{N}\underset{R}{C}HCOOH + R''OH \xrightarrow[110^\circ]{HCl} H_3\overset{+}{N}\underset{R}{C}HCOO\text{-}R'' + H_2O \tag{6}$$

$$H_3\overset{+}{N}\underset{R}{C}HCOO\text{-}R'' + (R'CO)_2O \xrightarrow{150^\circ} R'CO\text{-}NH\underset{R}{C}HCOO\text{-}R'' + R'COOH + H^+ \tag{7}$$

The esters of all alcohols ranging from methyl to amyl have been tried in the past, but the isopropyl [63], *n*-propyl [64–66], isobutyl [59], and *n*-butyl esters [67] are the most suitable. The catalyst employed for esterification is usually hydrogen chloride, in concentrations between 1 and 8 N [59,63,64].

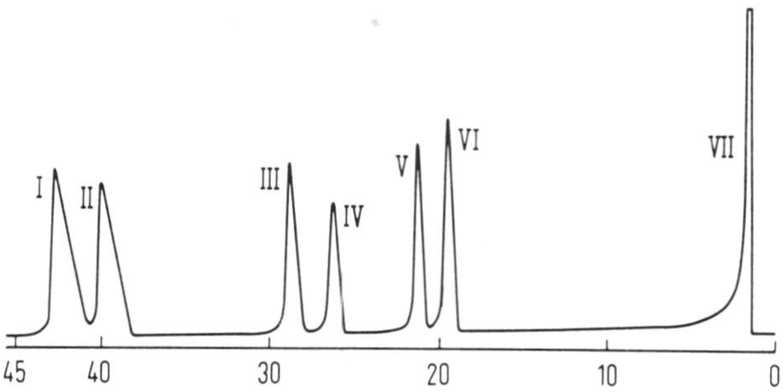

Figure 6.4 Gas chromatographic separation of diastereomeric *N*-trifluoroacetyl dipeptide methyl esters; stainless steel capillary (Perkin-Elmer R), 150 ft × 0.01 in., coated with polypropylene glycol. Temperature, 180°C isothermal; carrier gas flow rate 2.19 ml/min. I = L-Leu-D-Leu, II = L-Leu-L-Leu, III = L-Leu-D-Val, IV = L-Leu-L-Val, V = L-Val-D-Val, VI = L-Val-L-Val, VII = solvent. (From Ref. 57.)

Some amino acids are only slightly soluble in the higher alcohols [54], and even less so at higher concentrations of hydrochloric acid. Ultrasonication is therefore recommended to enhance dissolution [68]. Methyl esters are relatively volatile and are partially lost during evaporation of acylation reagent [69]. The reagents for esterification are conventionally prepared by bubbling dried hydrogen chloride into the corresponding alcohol. A convenient alternative is to prepare the reagent by mixing the appropriate amounts of acetylchloride and alcohol [61,63]. For esterification the amino acids dissolved in hydrogen chloride/ alcohol are heated for 20 min to 1 hr at temperatures between 100 and 150°C.

The second step is acylation, most frequently to the acetyl, trifluoroacetyl, pentafluoropropionyl, or heptafluorobutyryl derivatives. The latter three are preferred due to their greater volatility [70]. Conditions for N-acylation are usually chosen to block all residual functional groups; for acylation of amino groups, room temperature is sufficient, but for derivatization of other functional groups, higher temperature is required [71,72]. The trifunctional amino acids need most attention; some derivatives are very sensitive, and even silanol groups of glass surfaces and in the stationary phases may decompose them. Residual basicity of the glass surface promotes this effect. One of the most difficult amino acids is arginine, which requires high temperature for conversion to the tris-perfluoroacyl derivative [71]. Once the derivative is formed, it is sufficiently stable for capillary gas chromatography, provided it is not exposed to an alkaline glass surface [73].

Histidine is similarly problematic. The perfluoroacyl group attached to the imidazole nitrogen is very unstable. Histidine with underivatized imidazole ring is sufficiently volatile for gas chromatography, but is weakly basic and requires well-deactivated, slightly basic capillaries for tailing-free chromatography [73]; this on the other hand is detrimental to arginine. Therefore chromatography of N^{im}-acylated histidine is preferred. For packed columns this may be effected by coinjection of an acid anhydride [46], but it is inadequate for capillaries. Injection of perfluoroacyl anhydrides at high temperature definitely shortens the life-span of a capillary and depolymerizes silicone stationary phases. An alternative is the coinjection of the much less aggressive acetic acid anhydride [74], generating a stable N^{im}-acylated histidine in the injection port. A further possibility lies in an additional derivatization step for histidine,

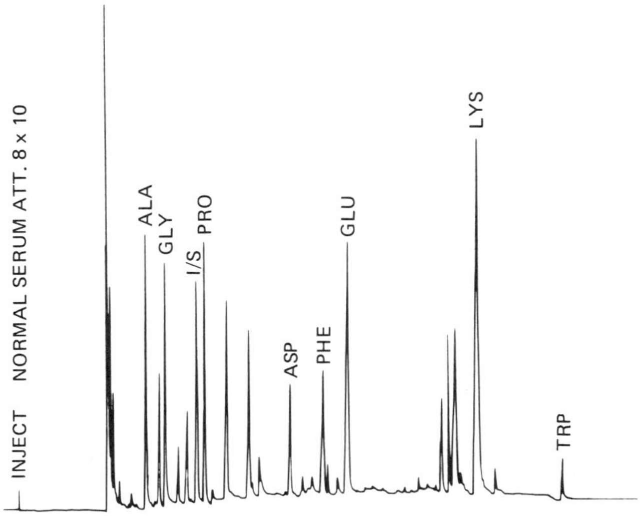

Figure 6.5 Chromatogram of *N*-acetyl amino acid *n*-propyl esters from a normal serum specimen on a glass capillary, 50 m × 0.27 mm, coated with a 1:1 mixture of polyoxyethylene (Carbowax 20 M) and a phenyl cyanopropyl silicone (Silar 5CP). Temperature program: from 110°C to 190°C at 8°/min, then ballasted to 250°C and held for 10 min; carrier gas, helium; flow rate, 0.75 ml/min; time of one chromatogram, about 20 min. The amino acids elute in the following sequence: Ala, Val, Gly, Ile, Leu, Norleu (I.S.), Pro, Thr, Ser, Asp, Phe, Glu, Tyr, Orn, His, Lys, Arg, and Trp. (Reproduced from Ref. 75, by permission of Preston Publications, Inc.)

leading to the stable N^{im}-ethoxycarbonyl histidine (Eq. 8) [47].

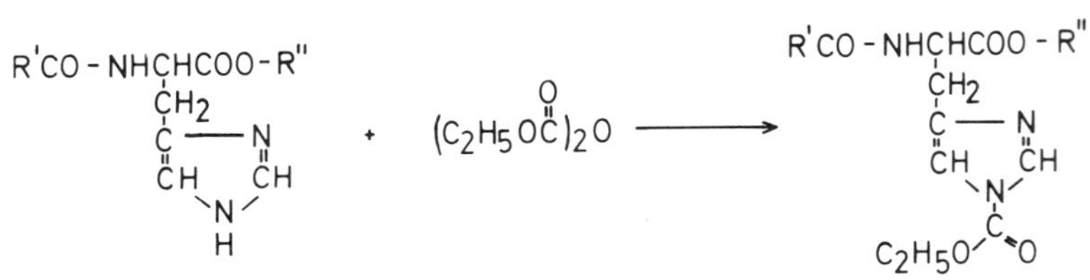

(8)

Methionine is partially degraded during this treatment.

N-Acetyl derivatives [64,65,75] have become less popular, though they are generally more stable than the perfluoroacyl derivatives. The stability of the esters of serine, threonine, cysteine, and tyrosine against hydrolysis is an especially noteworthy feature; the *N(O,S)*-acetyl amino acid propyl esters have been utilized successfully for capillary gas chromatography of amino acids in biological samples (Figure 6.5) [75].

Two interesting types of derivatives, which may be prepared in one step, are the cyclic oxazolinones (3),

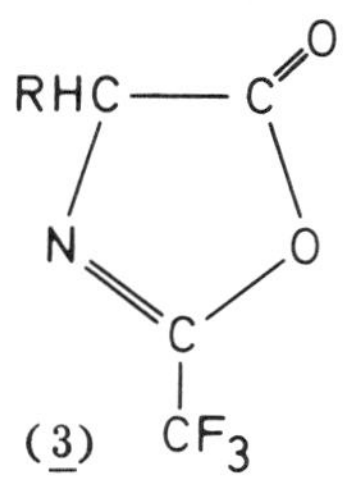

(3)

obtained by heating trifluoroacetyl amino acids to high temperature [76], and the oxazolidinones (4)

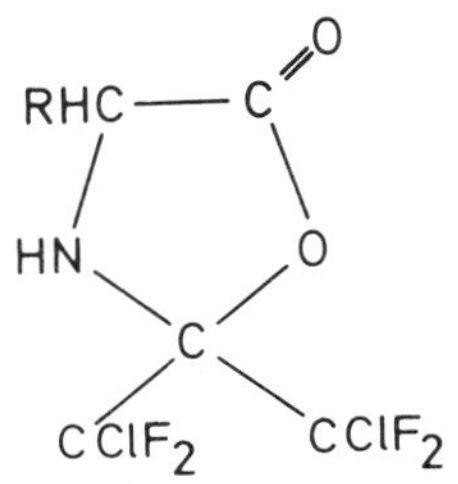

(4)

obtained by condensation with tetrafluorodichloro- or hexafluoroacetone [77]. The latter have been utilized for capillary gas chromatography for many years. These derivatives are excellently suited but trifunctional amino acids must be derivatized in a second step (Figure 6.6) [78,79]. Noteworthy is the possibility of determining the amides glutamine and asparagine. The derivatives of histidine, tryptophan, arginine, and cystine present difficulties.

Derivatives which have recently been successfully employed for quantitative determinations in serum are the *N*-alkoxycarbonyl amino acid methyl esters [80]. Asparagine and glutamine form stable and distinct derivatives, enabling their determination in the presence of the dicarboxylic acids. A disadvantage is that for determination of all amino acids, both the *N*-ethoxycarbonyl and the *N*-isobutoxycarbonyl derivatives must be prepared, and chromatography on two differently selective columns is required. It is possible to overcome this by using capillaries for gas chromatography instead of the packed columns employed thus far. Arginine must be determined in a subtractive manner from two runs, one before and one after enzymatic conversion to ornithine.

Other derivatives, such as the dinitrophenyl amino acid methyl esters [81], trifluoroacetyl amino acid trimethylsilyl esters [82], pentafluoropropionyl amino acid hexafluoroisopropyl esters [83,84], dialkyl phosphoryl amino acid esters [85,86], and some other derivatives have been used for special purposes, but are not recommended for general analysis.

For sequence determination of peptides by Edman degradation, the gas chromatography of thiohydantoins of the protein amino acids is of interest, which is mostly done on packed columns [87]. Both the trimethylsilylated hydantoins

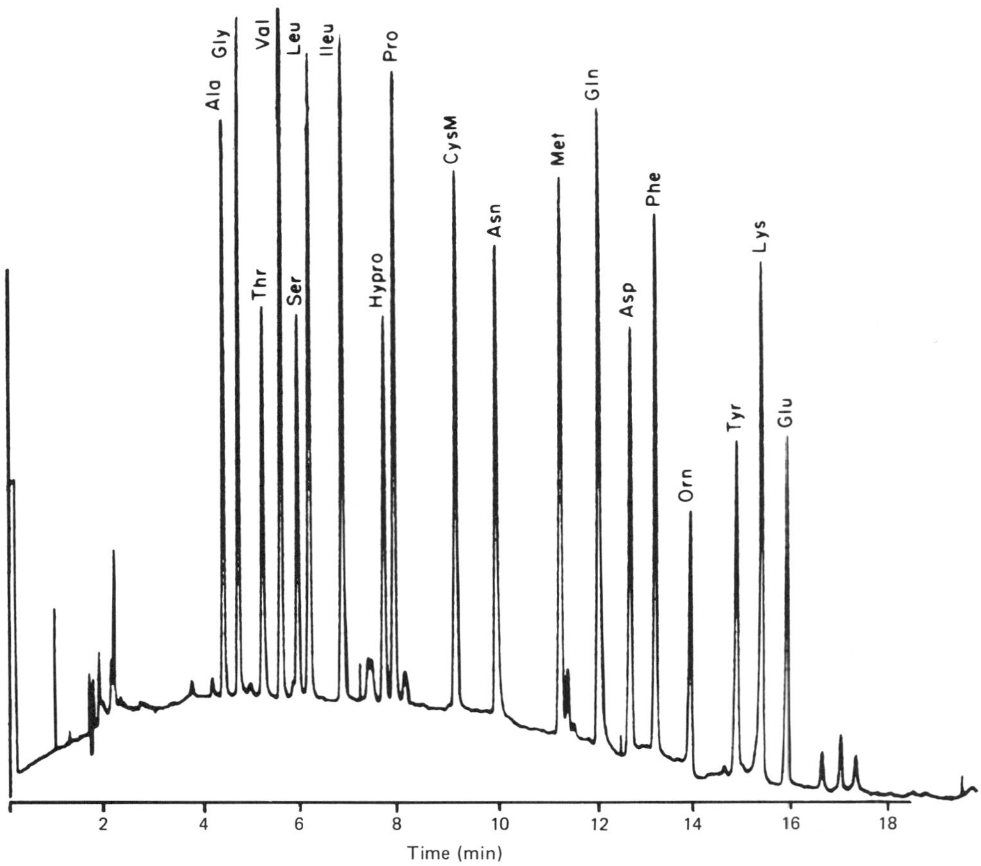

Figure 6.6 $N^{\omega}(O)$-heptafluorobutyryl amino acid oxazolidinones, standard mixture analyzed on a glass capillary, 10 m × 0.3 mm, coated with OV-17. Temperature program: from 70°C to 220°C at 8°/min; carrier gas, helium. (From Ref. 79.)

and the underivatized forms have been analyzed by gas chromatography, but complete separation was achieved only by capillary gas chromatography (Figures 6.7 and 6.8) [88] of the trimethylsilyl methylthiohydantoins.

Capillary gas chromatography has been employed relatively early for resolution of amino acid enantiomers. They can be separated either as diastereomers, formed upon derivatization with optically active reagents [34], or directly on chiral stationary phases [35]. As the separation factors of these isomers are usually small, the necessity for capillaries with high plate numbers is self-evident.

Enantiomers have been separated as diastereomeric derivatives, formed by derivatization with chiral alcohols (Figure 6.9) [89] or with chiral N-acyl groups [90]. The diastereomer approach has some severe shortcomings. Essential for correct results is high optical purity of the chiral reagents, sometimes

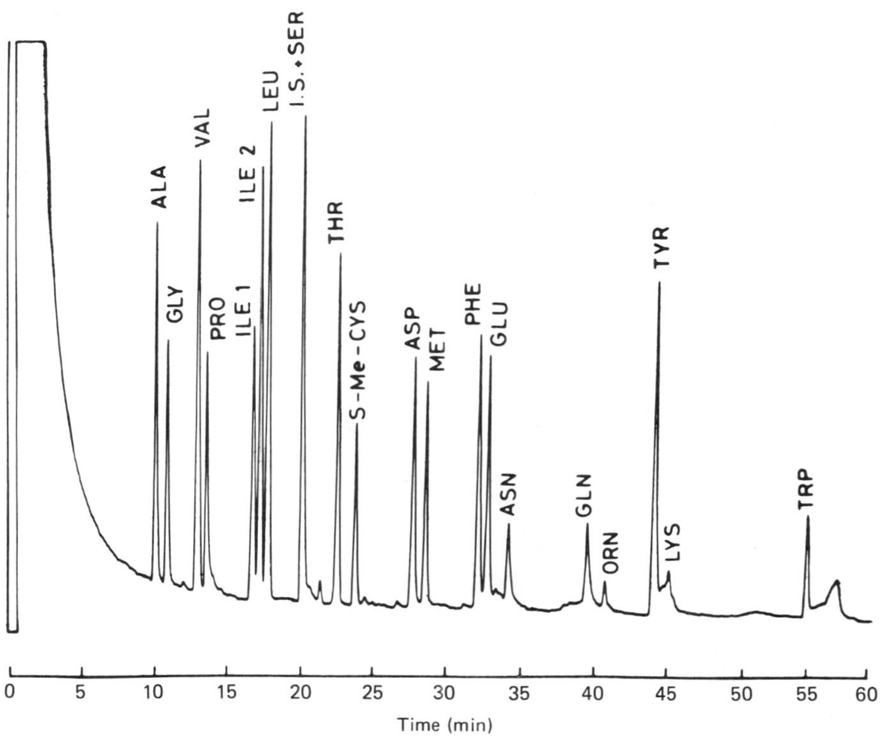

Figure 6.7 Separation of 19 trimethylsilyl methyl thiohydantoins on a glass capillary, 4.5 m × 0.13 mm, coated with a mixture of 95% methylsilicone (OV 101) and 5% phenyl cyanopropyl methyl silicone (OV 225). Temperature program: 8 min isothermal at 120°C, gradient 2.5°/min to 250°C; carrier gas, helium; flow velocity, 8 cm/sec. Ile 1 and Ile 2 are the diastereomers (From Ref. 88.)

difficult to achieve. Other facultative sources of errors are thermodynamic or kinetic differences in the reaction of two enantiomers with a chiral reagent to the respective diastereomers.

An elegant alternative is the direct separation of enantiomers derivatized with normal achiral reagents. Suitable derivatives are the *N*-trifluoroacetyl and *N*-pentafluoropropionyl amino acid methyl or propyl esters [36,37,91,92]. The trifluoroacetyl derivatives show slightly higher separation factors on some chiral phases, but the pentafluoropropionyl derivatives are somewhat more stable. Other esters can be used, but overlapping of different amino acids may occur [93].

Many other derivatives proposed in the past are mentioned in the comprehensive overview of P. Hušek and K. Macek [33].

III. COLUMNS AND STATIONARY PHASES

The suitability of a derivative for a certain analysis depends not only upon such aspects as yields or ease of formation, stability of the derivative, and availability of reagents, but also whether stationary phases for separation of

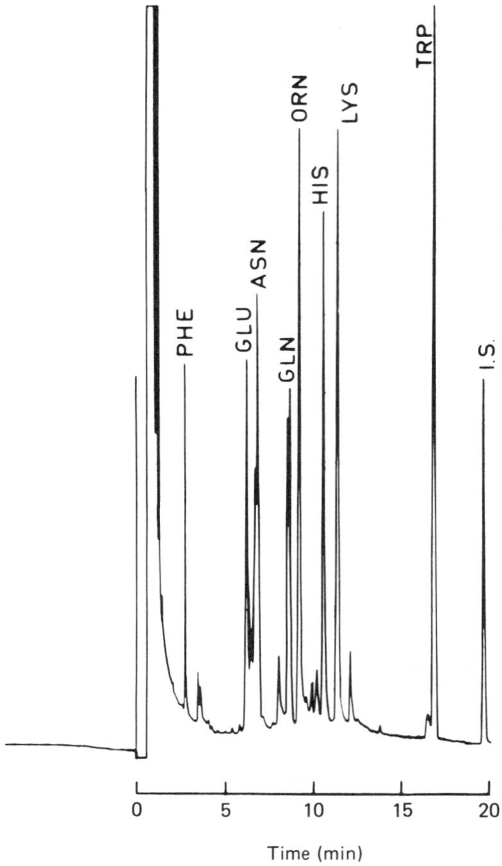

Figure 6.8 Separation of the labile trimethylsilyl methyl thiohydantoins. Conditions as in Figure 6.7, except for temperature program: 4 min isothermal at 180°C, gradient 5°/min to 260°C; carrier gas flow velocity, 14 cm/sec; internal standard is α-cholestane. (From Ref. 88.)

all derivatized constituents are available. For packed columns a stationary phase with appropriate selectivity is especially important; for capillaries, which have far higher efficiencies, this is less critical.

The search for suitable stationary phases or mixtures thereof was one of the main objectives during the past 20 years, mostly for packed columns. Separation of the *N*(*O*,*S*)-trifluoroacetyl amino acid esters was achieved on silicones and polyesters [28,68,69]. With the latter a second run on a methyl phenyl silicone is required for arginine and histidine. For the *N*(*O*,*S*)-acetyl amino acid *n*-propyl esters, a mixture of polyoxyethylene and cyanopropyl methyl silicone gave optimal results [65]. Recently the *N*(*O*,*S*)-heptafluorobutyryl amino acid esters have become the preferred derivatives due to their separability on methyl silicones like SE-30, OV-1, or OV-101 [59,74,94], stationary phases of high thermal stability. Phenyl silicones are recommended for pertrimethylsilyl derivatives [52].

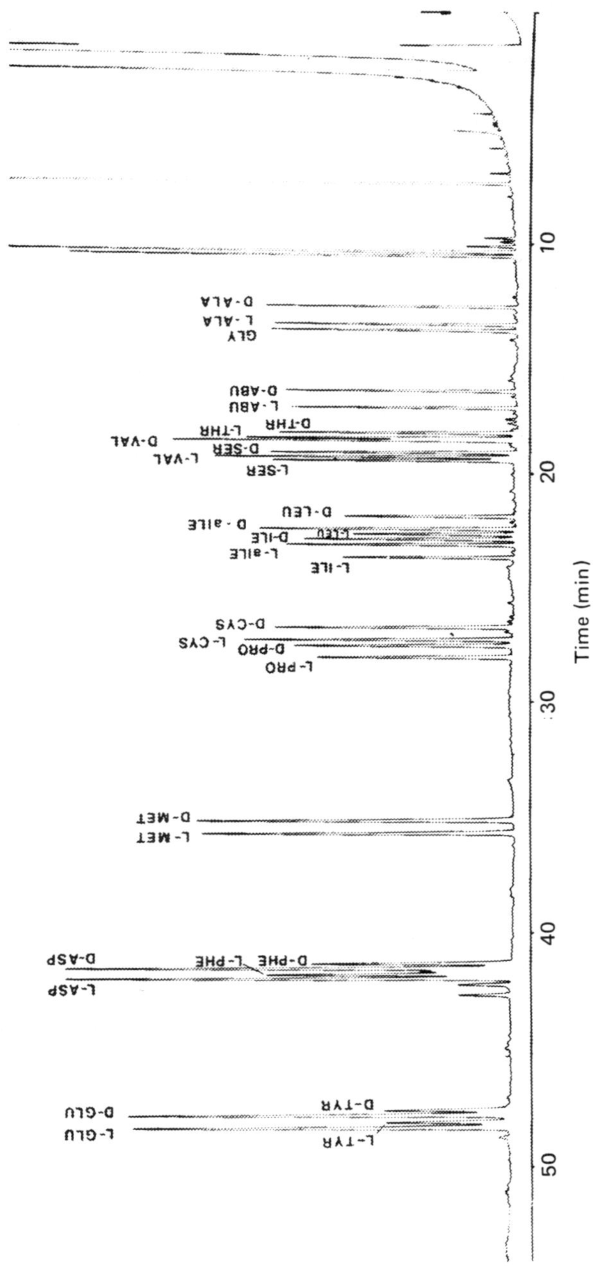

Figure 6.9 Gas chromatogram of diastereoisomeric pentafluoropropionyl amino acid (+)-3-methyl-2-butyl esters on a 25-m glass capillary, coated with methylsilicone (SE-30). Temperature program: from 85°C to 220°C at 2°/min; carrier gas, hydrogen. (From Ref. 89.)

Successful gas chromatography on packed columns stipulates proper pretreatment of the support [28,67,95], expecially for the more sensitive derivatives of threonine, serine, cysteine, tryptophan, histidine, and arginine. The material of the column wall itself is equally important. Steel tubing was early recognized as inadequate for gas chromatography of amino acids. Nickel tubing, recommended as an alternative to glass, is equally unsuitable, especially for the sulfur-containing amino acids, and it may not be used with hydrogen as carrier gas.

Capillaries were first employed by Andersson et al. [24] (Figure 6.2) for separating amino acid ethyl esters, and by Weygand et al. [57,77], who separated diastereomeric dipeptides (Figure 6.4) and oxazolidinones. These commercial capillaries were made from steel, coated with polypropylene glycol or polyphenylether, and had a length of 50 m. Nine amino acids were separated as oxazolidinones, including alanine, glycine, valine, leucine, isoleucine, proline, serine, methionine, and phenylalanine. Shortly afterwards, Halasz and Bünnig [96] reported the separation of the trifluoroacetyl amino acid methyl esters of fifteen amino acids on a capillary, 30 m × 0.25 mm, coated with a polyoxyethylene (Figure 6.10). Two years later, Pollock [70] separated eleven

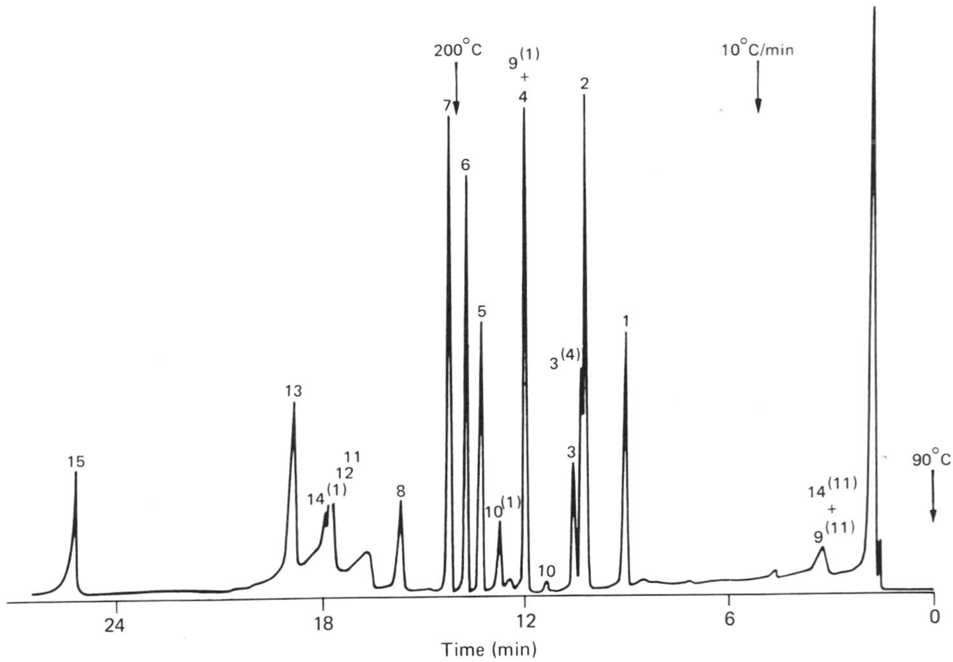

Figure 6.10 Gas chromatographic separation of *N*-trifluoroacetyl amino acid methyl esters on a capillary, 30 m × 0.25 mm, coated with polyoxyethylene (Carbowax 4000). Temperature program: 3 min isothermal at 90°C, gradient 10°/min at 200°C, 8 min isothermal; carrier gas, nitrogen; flow rate, 0.8–1.1 ml/min. 1 = valine, 2 = alanine, 3 = isoleucine, 4 = leucine, 5 = glycine, 6 = β-alanine, 7 = proline, 8 = aspartic acid, 9 = cysteine, 10 = threonine, 11 = methionine, 12 = glutamic acid, 13 = phenylalanine, 14 = serine, 15 = lysine. (From Ref. 96).

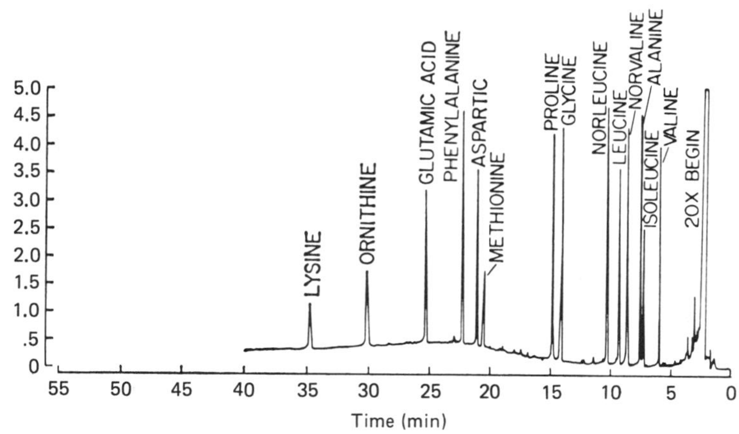

Figure 6.11 Separation of *N*-heptafluorobutyryl amino acid *n*-butyl esters of 14 amino acids on a stainless steel capillary, 150 ft × 0.02 in., coated with polyoxyethylene (Carbowax 20 M). Temperature program: 2.5 min isothermal at 100°C, gradient 4°/min to 170°C; carrier gas, helium; flow rate, 10 ml/min. (Reprinted with permission from Ref. 70. Copyright 1967 American Chemical Society.)

protein amino acids on a steel capillary, 45.7 m × 0.31 mm, coated with polyoxyethylene (Carbowax 20 M). He compared the trifluoroacetyl, pentafluoropropionyl, and heptafluorobutyryl amino acid *n*-butyl esters and found that retention times decrease in that order. The retention times of the heptafluorobutyryl derivatives of the aliphatic amino acid *n*-butyl esters (Figure 6.11) were 40–50% shorter than those of the corresponding trifluoroacetyl derivatives.

The heptafluorobutyryl amino acid *n*-propyl esters were also used by J. Jönsson (present name J. Sjödahl) et al. [66] in their important contribution to glass capillary chromatography, published in 1973 (Figure 6.12). They demonstrated the superiority of glass over steel capillaries with the complete separation of all protein amino acids except tryptophan on a glass capillary, 5.85 m × 0.2 mm, coated with a wall-bonded methyl silicone. They found that the flame ionization detector is sufficiently sensitive for analysis in the picomol range, provided derivatization of such small amounts is reliable; Zumwalt et al. [97] have shown that this is feasible if proper precautions are taken. Pearce [61] some time later reported on the superior properties of the heptafluorobutyryl derivatives, but of the isobutyl esters. His results were very similar, expectedly, since the capillary, 60 m × 0.5 mm, was coated with SE-30, also a methyl silicone. He used a faster temperature program, and the total analysis time up to cystine was slightly shorter, 29 min (Figure 6.13). In the same year Adams et al. [75] reported the separation of acetyl amino acid *n*-propyl esters on a capillary, 50 m × 0.27 mm, coated with a 1:1 mixture of polyoxyethylene (Carbowax 20 M) and a phenyl cyanopropyl silicone (Silar 5CP) (Figure 6.5). Due to lower volatility of acetyl derivatives, the starting temperature was higher than in the former cases. All protein amino acids were included, but the derivatives of the basic amino acids exhibited tailing, which probably is caused by the polyethylene fraction of the stationary phase. Shortly afterwards Poole and Verzele [98] published the use of "whiskered" glass capillaries, 30 m × 0.25 mm, coated

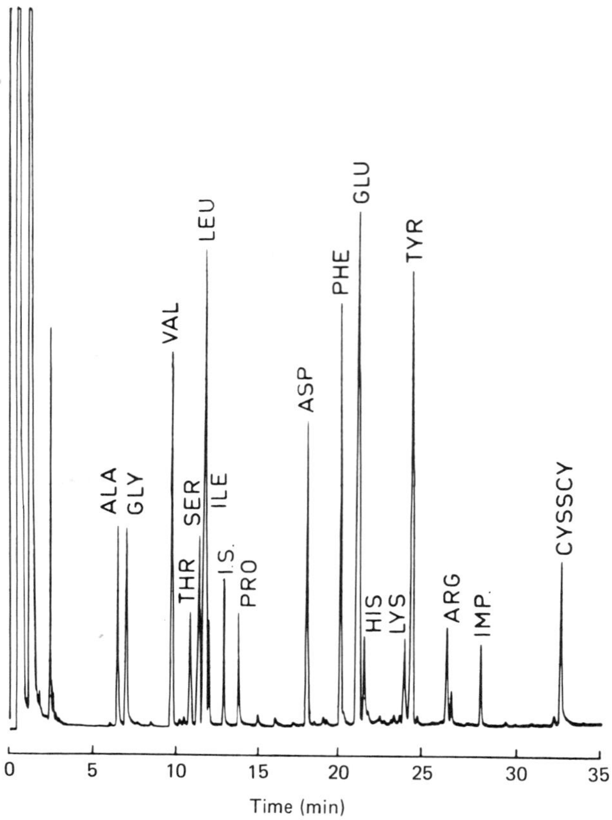

Figure 6.12 Separation of *N*-heptafluorobutyryl amino acid *n*-propyl esters from hydrolyzed insulin on a glass capillary, 5.85 m × 0.2 mm, coated with wall-bonded methylsilicone. Temperature program: 4 min isothermal at 70°C, gradient 4°/min to 230°C; carrier gas, nitrogen; flow velocity, 16 cm/sec. Methionine is not present in this case, but elutes about half a minute before aspartic acid (ASP). The impurity (IMP) is a hexosamine. (From Ref. 66.)

with SE-30, for separation of all heptafluorobutyryl amino acid *n*-butyl esters, except tryptophan. They also employed other silicones, OV-17 and OV-210, but separation on SE-30 was superior (Figure 6.14).

MacKenzie and Hogge [99] reported that amino acid derivatives, even the relatively stable heptafluorobutyryl derivatives, are sensitive indicators for the quality of the pretreatment of the glass surface. They compared two types of support-coated capillaries, one obtained by deposition of Silanox 101, the other with silanized Chromosorb R, both coated with methyl silicones (SE-30 and OV-101) as stationary phases. Only the Chromosorb capillary showed satisfactory properties and no or moderate tailing of proline, hydroxyproline, and methionine (Figure 6.15), whereas with the Silanox capillary, separation was incomplete and strong tailing was observed, even after silanization. Better results can be obtained [73] with deposition of silicic acid according to Schulte [100].

Desgres et al. [44] employed a capillary, 25 m × 0.23 mm, coated with a methyl silicone (OV-101) for determination of amino acids in biological samples

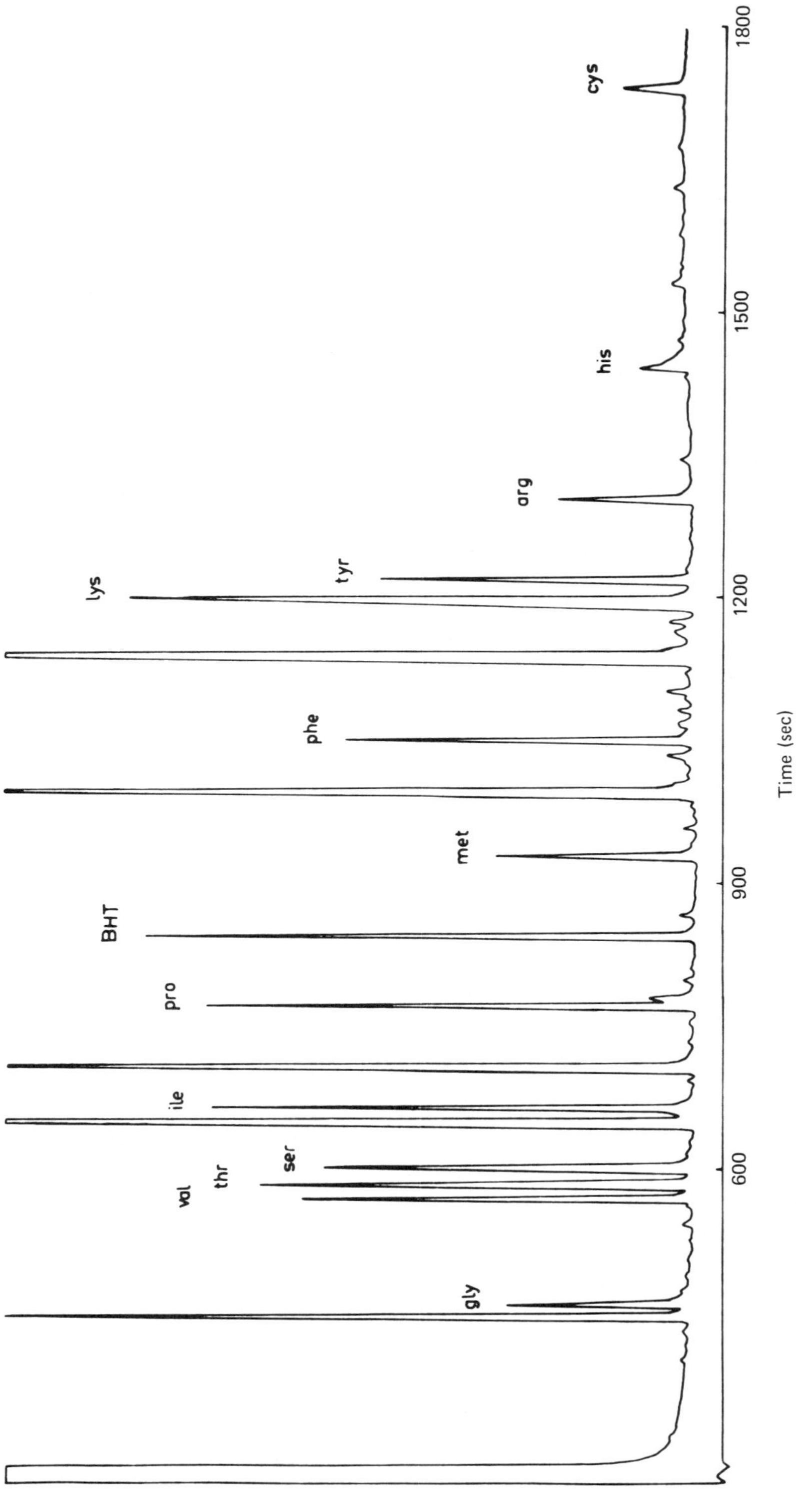

Figure 6.13 Chromatogram of *N*-heptafluorobutyryl amino acid isobutyl esters from β-lactoglobulin acid hydrolyzate (8-hr digestion) on a glass capillary, 60 m × 0.5 mm, coated with methylsilicone (SE-30). Temperature program: from 100°C to 250°C at 6°/min; carrier gas, helium; flow rate, 4 ml/min. (From Ref. 60.)

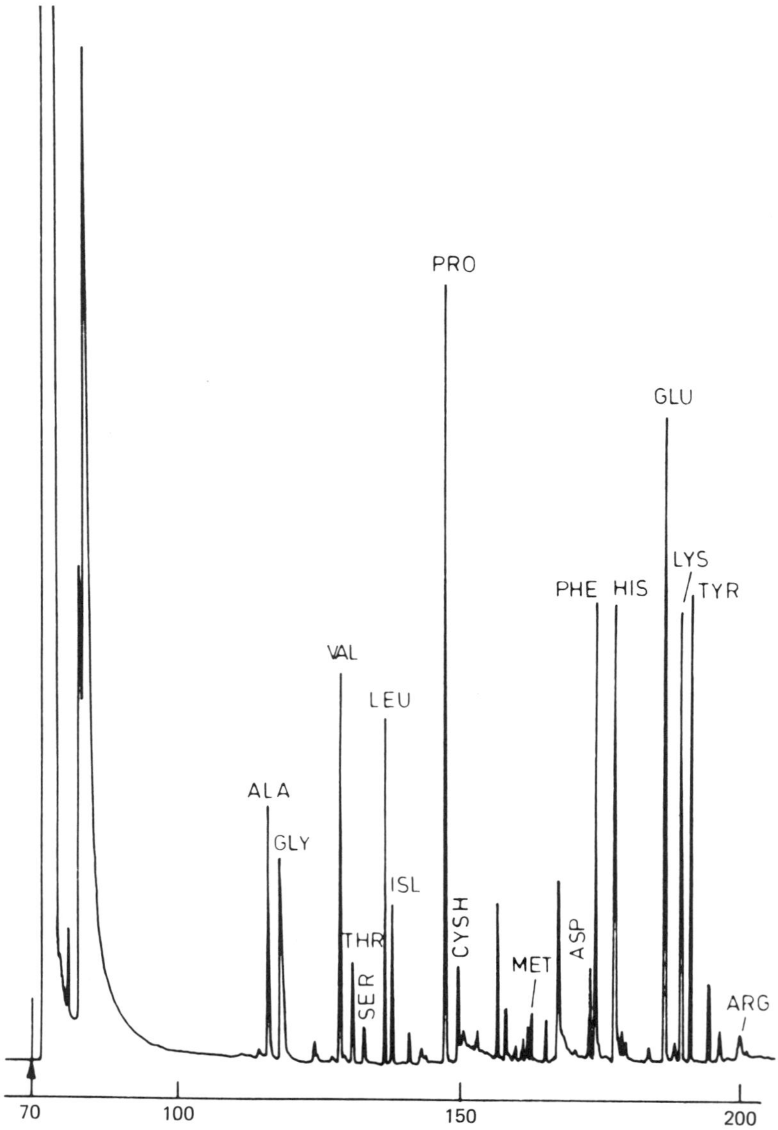

Figure 6.14 Determination of the amino acids as $N(O)$-heptafluorobutyryl n-butyl esters of uremic dialyzate after ion-exchange clean-up on a "whiskered" glass capillary, 30 m × 0.25 mm, coated with methyl silicone (SE-30). Temperature program: from 70°C to 220°C at 4°/min; carrier gas, hydrogen; flow rate, 3.3 ml/min. (From Ref. 98.)

as heptafluorobutyryl amino acid isobutyl esters (Figure 6.16). They achieved complete separation of the amino acids in slightly more than 1 hr; of the protein amino acids only tryptophan was absent, but many other physiologically important amino acids like β-alanine, β-aminobutyric acid, hydroxyproline, ornithine, lanthionine, and cystathione were included. They convincingly documented the superiority of capillary gas chromatography over packed columns for clinical amino acid analysis.

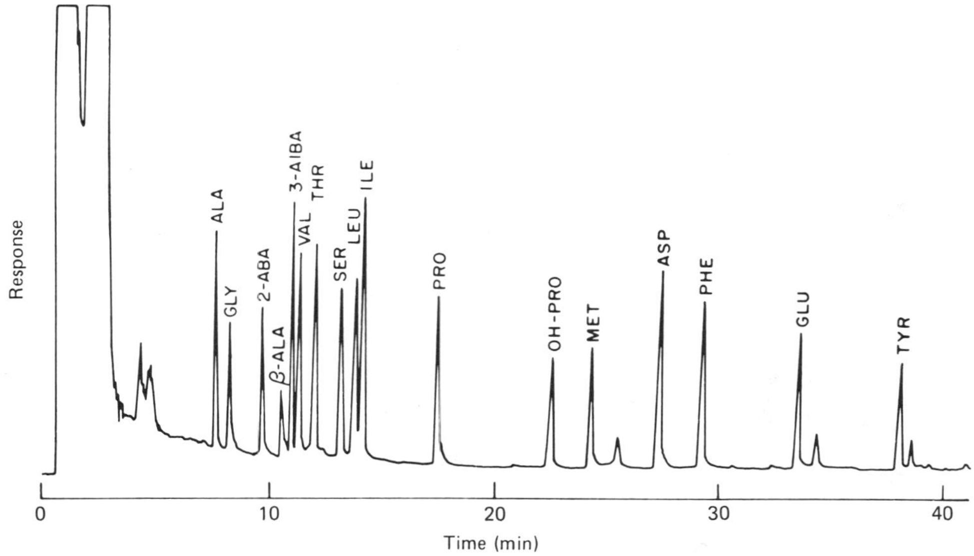

Figure 6.15 Separation of *N*(*O*)-heptafluorobutyryl amino acid isobutyl esters on a SCOT glass capillary, 20 m × 0.3 mm, coated with methyl silicone (SE-30) on Chromosorb R. Temperature program: from 90°C at 2°/min; carrier gas helium. (From Ref. 99.)

The superiority of capillary gas chromatography was also demonstrated in the publication of Eyem and Sjöquist (Figures 6.7 and 6.8) [88]. They analyzed trimethylsilyl derivatives of amino acid thiohydantoins on a capillary, 4.4 m × 0.13 mm, coated with a mixture of a methyl silicone (OV-101) and a cyanopropyl phenyl silicone (OV-225). All amino acids were separated, and only the peaks of the basic amino acids and the dicarboxylic acids showed tailing. A second injection at higher temperature and use of a steeper temperature program alleviated this. Under such conditions even histidine, the most sensitive derivative, could be eluted as a relatively sharp peak.

For separation of diastereomers and enantiomers of amino acids, capillaries have been used almost exclusively. One of the earliest publications on gas chromatography of amino acid derivatives describes the separation of 34 pairs of diastereomeric dipeptides [57]. Almost all diastereomers could be separated completely, and only those containing an N-terminal glutamyl residue showed small separation factors. The procedure was very useful for studying racemization during peptide synthesis and enabled the assessment of racemization-free, optimum conditions for peptide bond formation.

For determination of enantiomer ratios of amino acids, esterification with an optically active alcohol was also proposed [34]. These diastereomers have the advantage of higher volatility than dipeptides, enabling gas chromatographic analysis of diastereomeric derivatives of the less volatile amino acids. In 1966, the separation of the diastereomeric esters of thirteen of the nineteen optically active protein amino acids on a 50 m × 0.5 mm capillary coated with a polyoxyethylene (Carbowax 1540) was reported [101].

In the same year Gil-Av and co-workers published a momentous paper [35] dealing with the direct resolution of enantiomers on a chiral stationary phase, an amino acid derivative with low volatility. They obtained excellent results

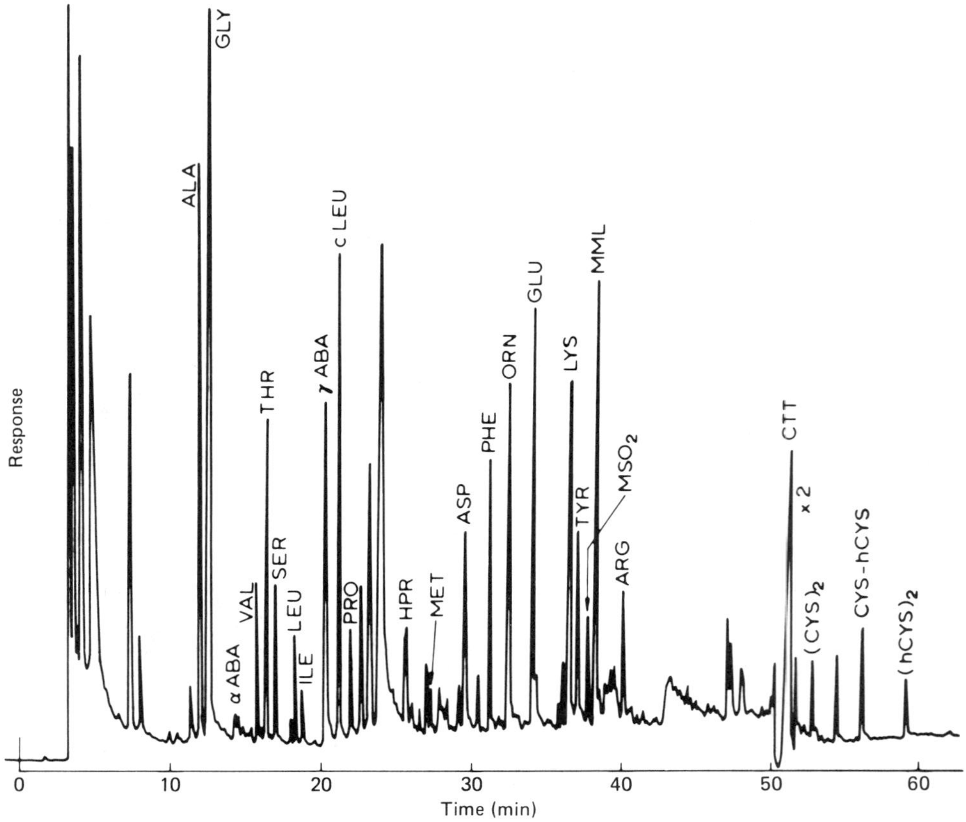

Figure 6.16 Gas chromatogram of *N(O)*-heptafluorobutyryl amino acid isobutyl esters of urine from a patient with hypercystathioninuria on a glass capillary, 25 m × 0.23 mm, coated with methylsilicone gum (OV 101). Temperature program: from 90°C to 270°C at 3°/min; carrier gas, nitrogen; flow rate, 2 ml/min. α(γ) ABA = α(γ)amino butyric acid; cLEU = cycloleucine (internal standard); HPR = 4-hydroxyproline; MSO_2 = methionine sulfone; MML = *N*ε-monomethyllysine; CTT = cystathionine; CYS-hCYS = cysteine-SS-homocysteine; $(hCYS)_2$ = homocystine. (From Ref. 44.)

with trifluoroacetyl L-isoleucine lauryl ester coated on glass capillaries, 100 m × 0.25 mm; the number of theoretical plates was determined to be between 75,000 and 140,000 m. Clearly, the large number of plates, never attainable with packed columns, was essential for the success of the experiments. All four stereoisomers of the trifluoroacetyl 2-butyl esters of alanine, valine, and leucine could be separated on this capillary. Retention times were rather long (105 min, 175 min, and 350 min, respectively) (Figure 6.17), due to the low temperature (90°C) under which these capillaries had to be operated in order to prevent the loss of the still relatively volatile stationary phase. During subsequent years, a number of similar phases of lower volatility were prepared [36,37,91, 102,102], all of them amino acid or dipeptide derivatives. Some of them exhibited sufficiently large resolution factors and could be used up to 160°C, shortening analysis time considerably and extending gas chromatographic enantiomer analysis to the less volatile amino acids glutamic acid, tyrosine, orithine, and lysine. One of the first dipeptide phases was trifluoroacetyl

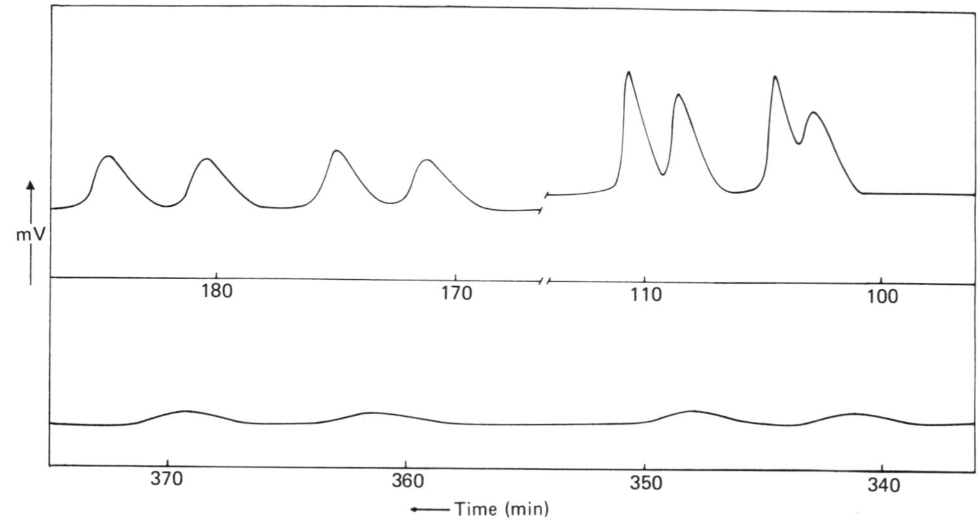

Figure 6.17 Chromatogram of the stereoisomers of *N*-trifluoroacetyl alanine (first quartet), valine (second), and leucine (third) 2-butyl esters on a glass capillary, 50 m × 0.25 mm, coated with *N*-trifluoroacetyl L-isoleucine lauryl ester. Temperature, 90°C isothermal; carrier gas, nitrogen; inlet pressure, 20 psi. (From Ref. 35.)

L-valyl-L-valine cyclohexyl ester [37], which enabled the resolution of the enantiomers of glutamic acid, tyrosine, lysine, and arginine (Figure 6.18); according to its retention time, the latter is more likely to be ornithine. Another very useful stationary phase was trifluoroacetyl L-phenylalanyl-L-aspartic acid bis(cyclohexyl) ester [104] with a melting point of 96°C and an upper temperature limit of 165°C, which allowed the separation of the two lysine enantiomers in the shortest time reported until then, i.e., 70 min (Figure 6.19). The resolution factors were somewhat lower than those on the previous phases, but still sufficiently large. This, in fact, is not always a disadvantage as it reduces the risk of overlapping of enantiomer peaks of different amino acids. Stationary phases of high enantioselectivity were reported by Feibush [102], *N*-lauroyl-L-valine *t*-butylamide, and by Charles et al. [105], *N*-docosanoyl-L-valine *t*-butylamide (5).

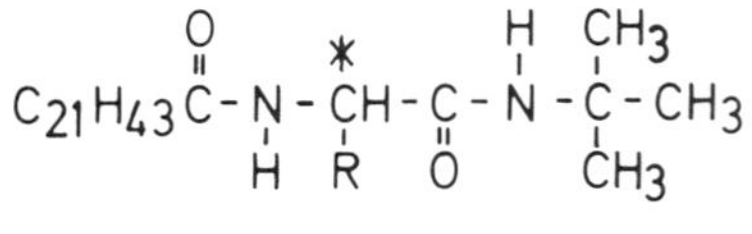

(5)

Especially the latter seems to be one of the most stable phases of this type and may be used up to 180°C. In most of these cases, stainless steel capillaries with lengths between 60 and 150 m and an internal diameter of 0.5 mm were employed, but evidently glass capillaries were considerably more efficient and column lengths could be reduced to 20–40 m [104]. Certainly the thermal stability of the stationary phases is higher when coated on glass capillaries.

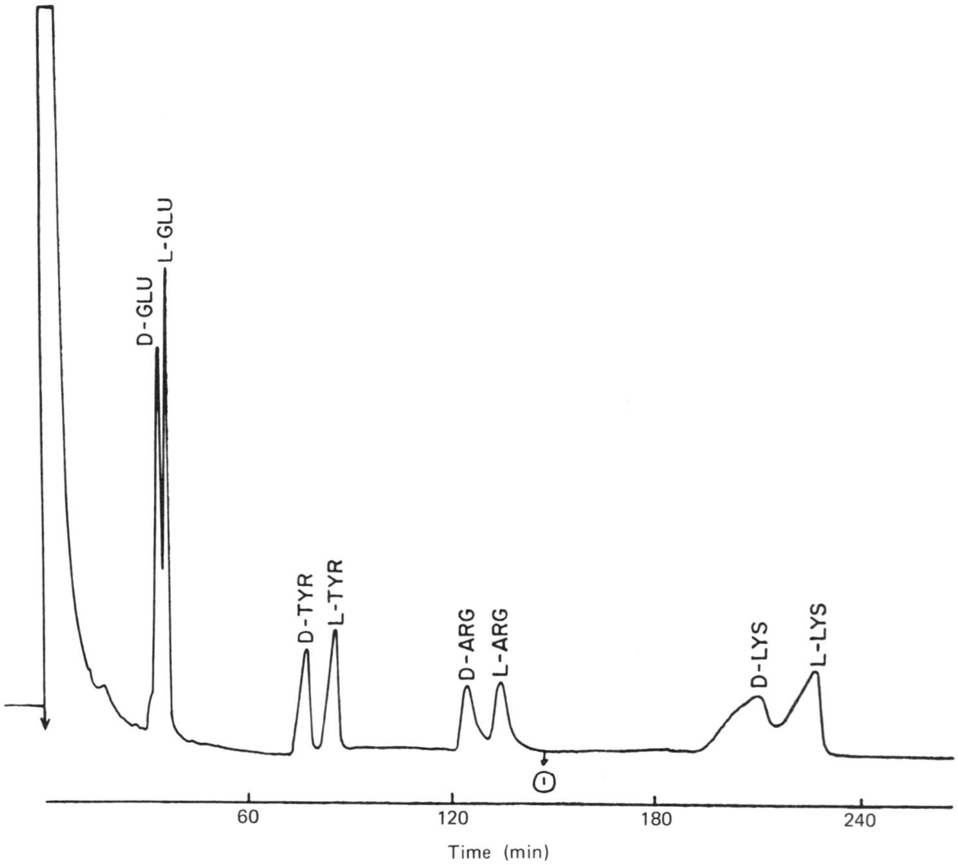

Figure 6.18 Gas chromatogram of *N*-trifluoroacetyl D,L-amino acid iso-propyl esters on a stainless steel capillary, 400 ft × 0.02 in., coated with *N*-trifluoroacetyl L-valyl-L-valine cyclohexylester. Temperature, 110°C isothermal; carrier gas, helium; inlet pressure, 10 psi. (Reprinted from Ref. 92, by permission of Preston Publications, Inc.)

A main objective in the development of enantioselective stationary phases was to increase their temperature stability; the advance from the simple amino acid derivatives of low molecular weight to docosanoyl amino acid amides and to dipeptide derivatives of the less volatile amino acids implied a considerable increase of the upper permissible temperature of operation. Another stationary phase intended for use at elevated temperature was developed by Brazell et al. [106], that is, trifluoroacetyl L-valyl-L-valine *O*-carboranyl propyl ester. They coated a mixture of the chiral phase and emulphor (1:1) on stainless steel capillaries of 165 m length and an internal diameter of 0.5 m. Separation factors of amino acid enantiomers were smaller than on most other phases, and the serine enantiomers were not resolved (Figure 6.20).

In 1977 Frank et al. [39] published the synthesis and of a chiral stationary phase of so far unequaled temperature stability. It consists of a polydimethylsiloxane with a small percentage of carboxylalkyl groups to which a chiral group is covalently attached, either through an ester or a more stable amide linkage. Silicones of different compositions and with different chiral selectors have been prepared since then [107], but the most versatile stationary phase is a silicone now known under the name Chirasil-Val. Its structure is shown in (6).

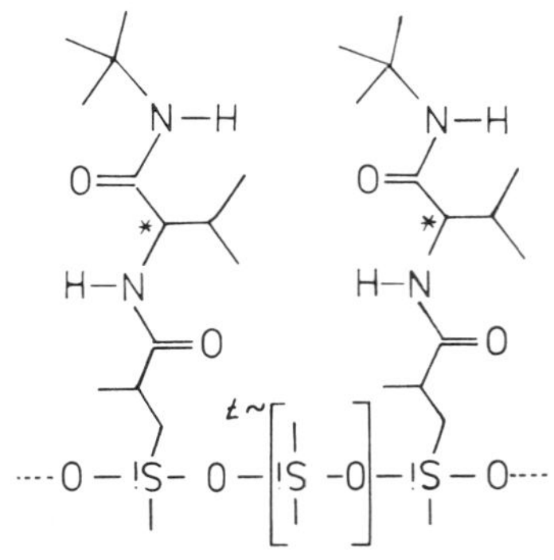

(6)

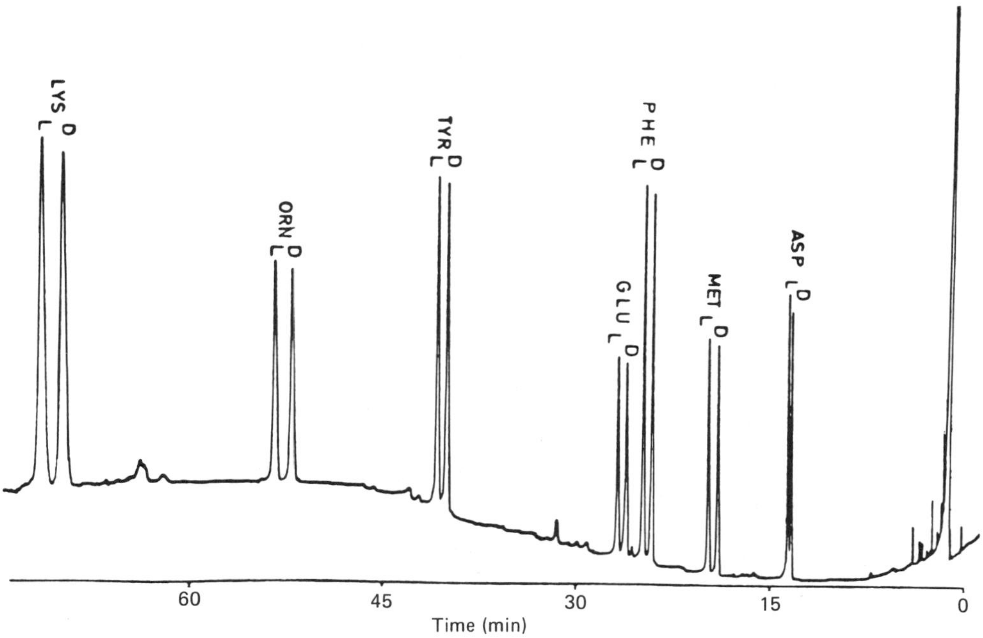

Figure 6.19 Separation of *N*-pentafluoropropionyl D,L-amino acid isopropyl esters on a glass capillary, 20 m × 0.3 mm, coated with *N*-trifluoroacetyl L-phenylalanyl-L-aspartic acid bis-cyclohexyl ester. Temperature program, from 130°C to 165°C at 1°/min; carrier gas, hydrogen. (Reprinted from Ref. 104. Copyright 1975 American Chemical Society.)

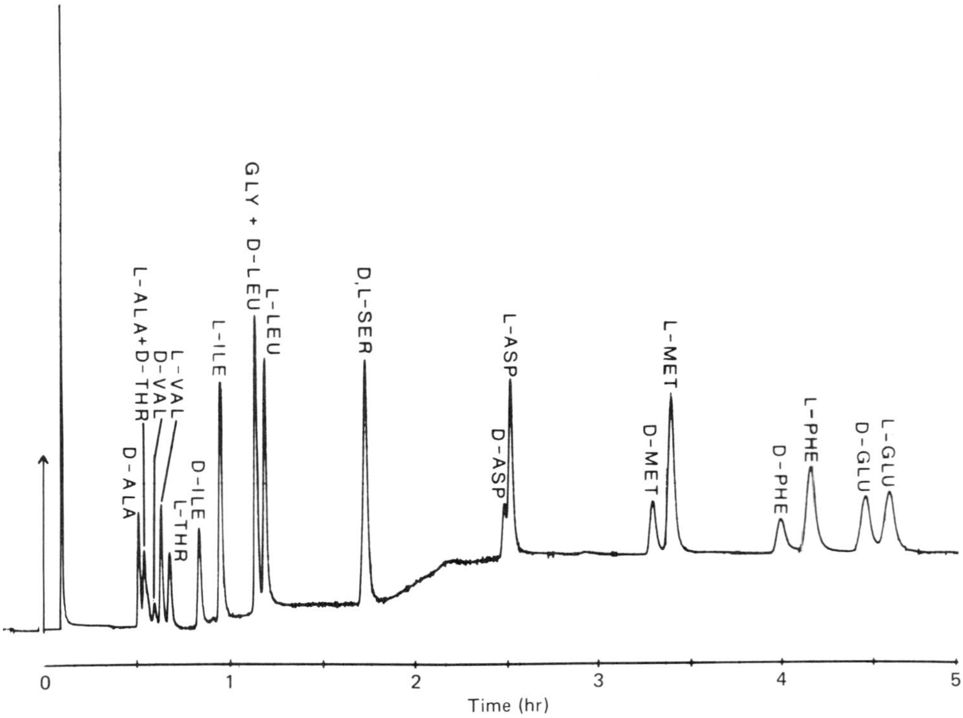

Figure 6.20 Chromatogram of *N*-trifluoroacetyl D,L-amino acid isopropyl esters on a stainless steel capillary, 165 m × 0.5 mm, coated with *N*-trifluoroacetyl L-valyl-L-valine-O-carboranyl propyl ester. Temperature program, 30 min isothermal at 75°C, gradient 0.5°C/min to 110°C; carrier gas, helium; inlet pressure, 24 psi. (From Ref. 106).

The stationary phase is fully stable up to 210°C, and for a short period may be heated to 250°C. Racemization at this high temperature is not more than a few percent per day, but depends on the pretreatment of the glass surface of the capillary [73]. Complete separation of all amino acid enantiomers is achieved in one run, including arginine, histidine, and tryptophan, in about half an hour (Figure 6.21). Glass capillaries are usually employed, but coating on fused silica has also been successful [108]. Chirsasil-Val or similar phases are applicable not only to enantiomer resolution of amino acids; enantiomers of other chiral compounds have also been resolved on these phases, namely alcohols, hydroxy acids, amino alcohols, glycols, amines, carbohydrates, and sulfoxides [109–116].

The advent of Chirasil-Val enabled the introduction of a new method for quantitative determination of amino acids in natural samples: the use of the optical antipodes of each L-amino acid as internal standards [40]. This method, called enantiomer labeling, eliminates almost all difficulties thus far encountered in quantitative amino acid analysis by gas chromatography (see below).

A number of analogous chiral stationary phases have been prepared since then [117–119]; none of them is as versatile as Chirasil-Val.

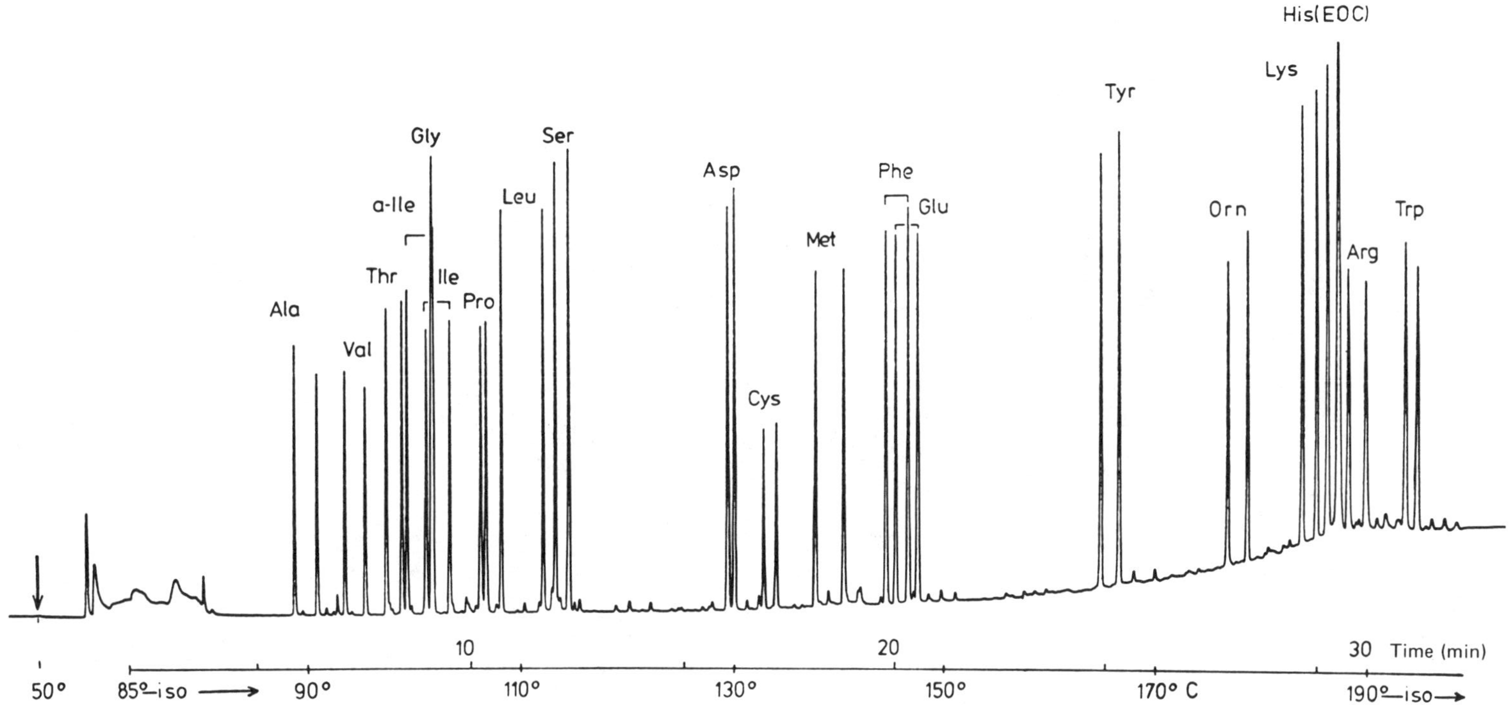

Figure 6.21 Gas chromatographic separation of the enantiomers of 19 amino acids as $N(O,S)$-pentafluoropropionyl amino acid isopropyl esters on a glass capillary, 20 m × 0.28 mm, coated with Chirasil-Val. Temperature program indicated in the figure; carrier gas hydrogen; inlet pressure, .38 KPa. (From Ref. 130.)

Interesting phases of moderate thermal stability are triazine derivatives of amino acid esters [120], as shown in (7):

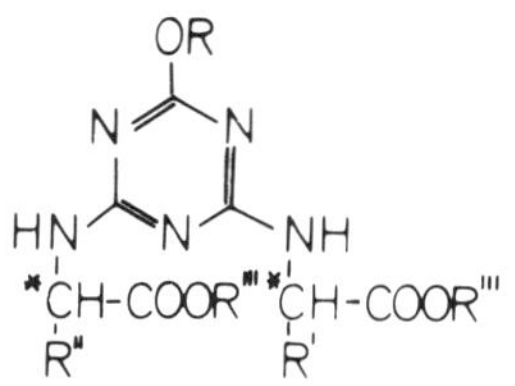

(7)

These phases exhibit relatively small resolution factors compared to the diamide phases or Chirasil-Val, but glass capillaries coated with these phases apparently tolerate temperatures up to 180°C without disintegration of the phase. Data published up to now are still insufficient for final judgement. Another dipeptide phase recently proposed has been found to be stable up to 170°C [121].

There is still some need for development of enantioselective stationary phases of high temperature stability for less volatile compounds like the thyroid hormones or β-adrenergic drugs. On the other hand, it must be kept in mind that, in general, the differences in free energy of solution between two enantiomers decrease with increasing temperature, resulting in lower resolution at higher temperatures.

Silicone phases are optimal stationary phases in many respects, especially in spreading on a glass surface and in temperature stability. The preparation of suitable capillaries coated with different, not only chiral, stationary phases can be a fortuitous matter, especially with polar silicone phases. This involves mainly the problems of deactivation of the glass surface and of proper adjustment of surface energy. These factors are interdependent, and full deactivation does not always result in an optimal surface energy, and vice versa. Droplet formation, a serious problem when polar phases are coated on silanized glass surfaces, is largely suppressed when cross-linked silicones are employed; these problems may also be alleviated with gum phases. Work has still to be done, however, in view of the different reactivities and stabilities of the amino acid derivatives on differently treated glass or quartz surfaces.

IV. QUANTITATIVE ANALYSIS

A. General Considerations

The applicability of an analytical method to quantitative analysis is the ultimate goal and an indisputable proof of its potential. For gas chromatographic amino acid analysis many different and, from the chemical point of view, interesting and elegant ways of conversion of amino acids to volatile derivatives have been published in the last twenty-five years, but only few of them have emerged as suitable for quantitative determination. Not only the gas chromatographic properties of the derivatives and the ease or reproducibility of their formation determine success or failure, but also the nature of the sample to be analyzed, the standards employed, and the instrumental parameters.

It is self-evident that the origin of the sample is often the most decisive factor for the selection of an analytical procedure and the quality of the results; quantitative determination of amino acids in pharmaceutical mixtures is clearly easier than in biological fluids.

The first step of analysis of a biological sample is enrichment and purification of the amino acids. Potentially interfering components must be removed, such as proteins and other biopolymers, lipids, and organic and inorganic salts. Every stage has its specific yields for different amino acids, or, in other words, the recovery on each stage is different for each particular amino acid. For quantitative analysis such losses must be avoided, or at least recoveries must be accurately known. Internal standards are usually added to compensate for losses.

In the next step the amino acids are converted to stable and volatile derivatives. Conditions for maximum yield of derivatization are not the same for all amino acids. At best a compromise may be reached, where countercurrent reactions of still proceeding derivative formation and already beginning derivative decomposition are approximately balanced. The next potential source of error is the loss of derivatized, now volatile amino acids during evaporation of reagents and solvents. A logical escape from this dilemma is to choose less volatile derivatives, but then higher temperatures are needed for chromatography. Gas chromatography itself involves other problems, such as decomposition of labile or adsorption of polar amino acid derivatives in the injection port or on active sites throughout the column. A specific problem of capillary gas chromatography is the so-called discrimination. The usual injection technique in capillary gas chromatography involves splitting the vaporized sample in the injector in a fixed ratio. The apparent split ratio may vary considerably for compounds of different volatility. Alternatives include splitless injection with cryogenic trapping [122], or exploitation of the "solvent-effect" [123], or injection directly "on-column" [124]. Problematic also is residual activity of the internal surface serving as the support for the stationary liquid phase.

A common problem with biological samples is deposition of involatile material in the front part of the capillary, especially with "on-column" injection. For packed columns the support may be suitably treated before coating, and if necessary, the front part of the packing may be replaced from time to time. For capillaries this is obviously not possible, and all treatments of the glass surface must be done in situ.

Subsequent to gas chromatographic separation the amino acids are detected as they emerge from the capillary. The detector most often used is the flame ionization detector (FID), but the alkali flame ionization (AFID) and the electron capture (ECD) detectors have also been employed. The amplified and recorded signal must be proportional to the amount of each derivative emerging from the column, with a constant ratio of charge produced per weight unit. Change in gas flow rates and temperature may affect the detector response. A constant ratio between the produced signal and the mass passing the detector is found for hydrocarbons; other compounds containing heteroatoms generate a smaller signal per weight ratio, depending upon the number and nature of functional groups. Therefore, molar response factors or weight response factors relative to a standard compound must be determined prior to an actual analysis. The same applies to the AFID and the ECD.

The next step is integration of the detector current over time, or, when recorded on a customary paperstrip recorder, determination of the so-called peak area. In the past, triangulation, weighing of paper-cuts, or, if chromatographic conditions and peak width are reproduced, peak-height measurements could be employed for this purpose. Most satisfactory and accurate, however, are electronic integrators, and in quantitative capillary gas chromatography they are almost indispensable.

In view of the potential sources of errors it is clear that quantitative amino acid analysis by capillary gas chromatography requires careful optimization of all chemical steps and instrumental parameters. Most of the significant publications deal with quantitative gas chromatography on packed columns, but many aspects reported there are also applicable to capillary gas chromatography.

B. Sample Preparation, Internal Standards

Protein hydrolyzates usually need not be purified, but for biological or geological samples the first step is normally isolation and enrichment of the amino acids, both for gas chromatographic analysis and ion exchange chromatography. In biological fluids or in tissues, many different compound classes may be present which can interfere with analysis, such as salts, carbohydrates, amines, lipids, proteins, nucleic acids, or hydrocarbons.

Purification of serum amino acids, not only for gas chromatography but also for ion exchange chromatography, usually starts with removal of proteins. Most commonly the serum proteins are precipitated by addition of an aqueous solution of picric acid, sulfosalicylic acid or trichloroacetic acid [125-127]. An alternative is the addition of acetone [128], which offers the advantage of high volatility. A further possibility is to add acetic acid and to isolate the amino acids by cation exchange, without previous precipitation of the proteins [129]. However, the determined values of amino acids in samples treated in this way are always too high [130].

After removal of proteins the sample is passed through a small column of a strongly acidic cation exchanger, usually Dowex 50, which retains the amino acids at a pH between 2 and 3. Excess of picric acid and other water-soluble, nonretained compounds are removed by rinsing with distilled water from the resin bed. The volume of water and its flow rate through the ion exchange bed should be small since glutamic and aspartic acids may be lost by excessive rinsing. Extraction of the sample with ether is sometimes recommended for removal of lipids [131]. For removal of the bulk of picric acid, an anion exchange step before the cation exchange has been proposed [126].

The absorbed amino acids are eluted from the resin with aqueous ammonia, usually 3 N. With lower concentrations, recoveries of the basic amino acids are low, but it has been found that concentrations higher than 1 N increase the standard deviations for phenylalanine, lysine, arginine, and cystine [132]. Finally, aqueous ammonia is evaporated and the sample is ready for derivatization.

During all these steps small losses are inevitable, and meticulous reproduction of all details of the clean-up procedure is required, such as volume of sample, flow rates, volume of eluents, and amount of ion exchanger. In principle, a more extensive clean-up results in fewer contaminant peaks in the gas chromatogram, but each manipulation of a sample potentially decreases the accuracy and precision of the analysis.

The internal standard and the stage of its addition to the sample are similarly decisive for the correctness of the results. Internal standards serve mainly to compensate for losses occurring during sample preparation and as reference for calculation of response factors. A considerable number of compounds have been suggested to this end [133], mostly amino acids not occurring naturally. Norleucine, cycloleucine, pipecolic acid and tranexamic acid have been used; others are homoserine, hydroxyproline, and ornithine, but these are less suitable since they may occur in biological samples. Carboxylic acids and aromatic hydrocarbons such as butyl stearate, decanoic acid, dimethyl dodecanedioate, fluorene, or phenanthrene have also been employed, but these are chemically very dissimilar to amino acids and serve exclusively for assessment of physical losses.

An ideal internal standard should compensate for losses of any source, but because of the large differences in chemical and physical properties of amino acids, this goal is difficult to attain. Certainly, using only one internal standard is insufficient; two or more standard compounds should be employed, each more closely related in its chemical properties to a group of chemically similar amino acids.

High chromatographic resolution is required if several internal standards are added; obviously these must not coelute with the related amino acids of the sample. On the other hand, they should have close physicochemical properties and similar gas chromatographic retention times. High efficiency of the employed column is therefore important, especially with biological samples which additionally may contain a large number of contaminant peaks. From this point of view, capillary gas chromatography with its inherently superior resolution is the method of choice.

Almost ideal internal standards are amino acids labeled with the stable isotopes deuterium [134,135], carbon-13 [136], nitrogen-15 [137], or oxygen-18 [138]. Except for a small isotope effect they are chemically and physically identical to their nonlabeled counterparts and suffer exactly the same fate during all manipulations of the sample. Recoveries need not be determined, and any losses, no matter at which stage, are fully compensated by isotope dilution methods. Isotopimers can usually not be separated by chromatographic means, and this constitutes the main drawback of stable isotope dilution methods—the need for an expensive mass spectrometer as detector. Radiolabeled amino acids could similarly serve as internal standards, but detectors for monitoring radioactivity on-line have large void volumes, cause considerable peak broadening, and are not sufficiently sensitive.

Another method for quantitative determination of natural L-amino acids involves the use of optical antipodes as internal standards, chemically identical mirror images of their L-amino acid counterparts. Obviously, the primary prerequisite is the ability to determine both isomers independently. The development of stable chiral phases was conditional for the conception of the method of enantiomer labeling [40], which has been employed for quantitative determination of amino acids in biological fluids [130] (Figure 6.22) and protein hydrolyzates. Clearly, as the number of amino acid peaks is now increased by seventeen standards, capillaries with highly resolving powers are indispensable.

Most other shortcomings of gas chromatography can be overcome with this approach. Both enantiomers suffer exactly the same fate during all physical and chemical operations in a nonchiral environment; even the losses occurring during sample preparation or incomplete derivative formation do not affect the accuracy or precision of determination. An advantage of this approach, especially attractive for routine analysis, is the possibility of simplifying considerably the clean-up of samples [139]. Low recoveries of amino acids or only fractional yields of derivatization do not affect the accuracy of analysis. Partial losses of volatile amino acid derivatives during evaporation of reagents have no influence on quantitative determinaton as long as the signal-to-noise ratio is sufficiently high. The effect of discrimination [140], a weakness of capillary gas chromatography, is now without consequences to quantitative determination. Also, relative molar response factors need no longer be determined in separate runs. In short, all the attributes of isotopic standards are also gained with enantiomeric standards without the need for an expensive mass spectrometer as detector. Any suitable detector may be employed.

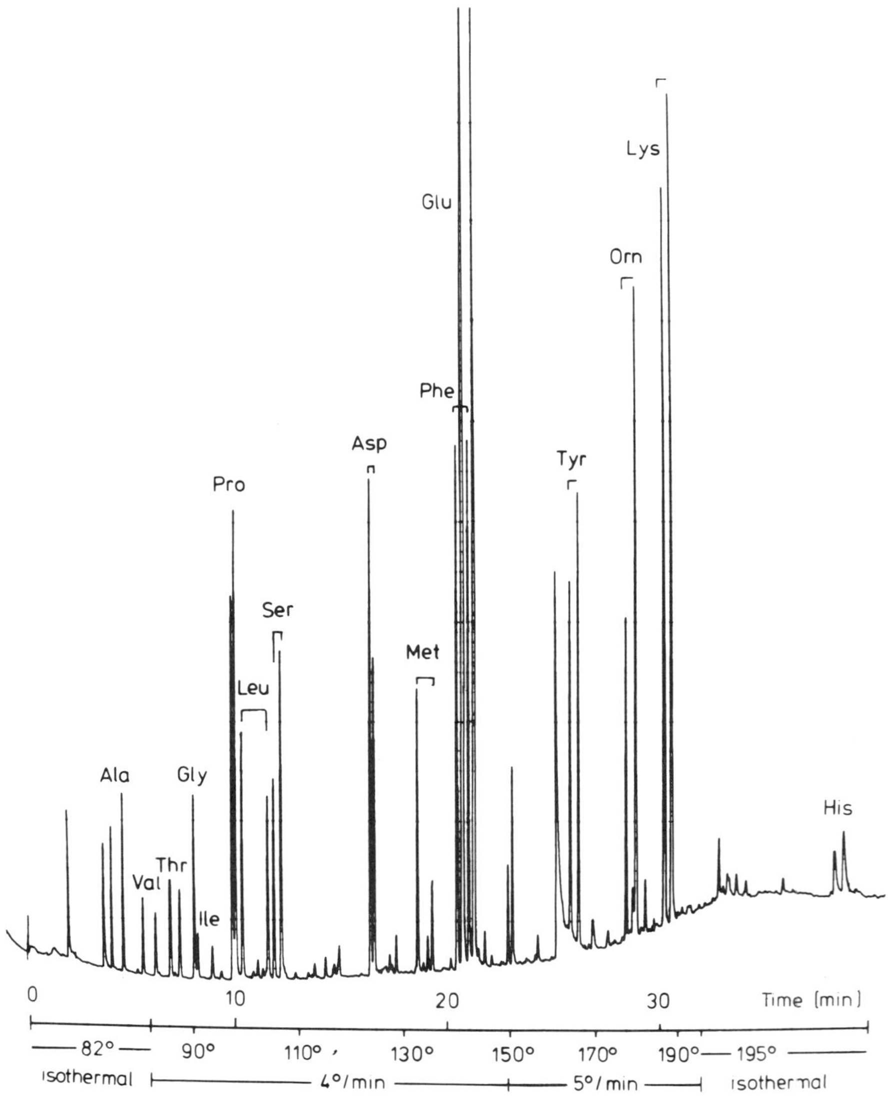

Figure 6.22 Gas chromatogram of *N*-pentafluoropropionyl amino acid isopropyl esters from a serum sample with the D-enantiomers added as internal standards on a capillary, 20 m × 0.28 mm, coated with Chirasil-Val; histidine with underivatized imidazole ring. Temperature program indicated in the figure; carrier gas, hydrogen; inlet pressure, .38 KPa. (From Ref. 130.)

Enantiomer labeling is as precise as ion-exchange chromatography and considerably faster [130]. In fact, amino acid analysis by enantiomer labeling is more accurate than ion-exchange chromatography, as in the latter method, losses due to occlusion of amino acids in the precipitated serum proteins are not accounted for.

A potential complication of enantiomer labeling arises when D-amino acids are present in the sample. This is unlikely for human body fluids; D-amino acids have not yet been found in blood serum. They could arise from digestion of microbial proteins, but resportion of D-amino acids is much slower than that of the L-enantiomers [141]. Their occurance in urine is even more unlikely. Nevertheless, the presence of endogeneous D-amino acids would not impair the accuracy of enantiomer labeling, but the ratio of both antipodes would have to be determined in a separate analysis before addition of the D-amino acid standards.

Determination of glutamine and asparagine by gas chromatogrpahy often presents difficulties. Usually the amides and free dicarboxylic acids are determined as the sum of both, but the ratio of respective amide and acid in biological fluids may be of diagnostic interest. With some derivatives this is feasible [80,85,142]. For glutamine, enantiomer labeling offers a practicable alternative. During acid-catalyzed esterification it is converted to diester via the intermediate pyrrolidone carboxylic acid ester [42]; maximum yield of the latter is achieved after a few minutes at 110°C (Eq. 1). If D-glutamine is added and an aliquot is removed after 5-10 min of heating and injected onto a capillary coated with Chirasil-Val, the separated enantiomers of pyrrolidone carboxylic acid ester are detected (Figure 6.23). The amount of endogeneous L-glutamine is directly calculated from the ratio of the two peaks [143].

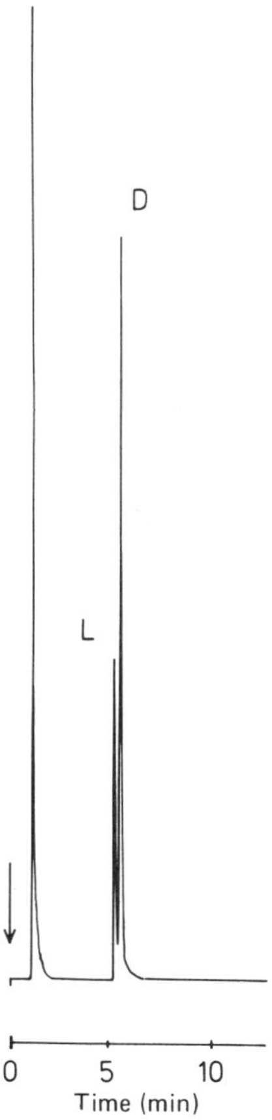

Figure 6.23 Separation of the enantiomers of pyrrolidone carboxylic acid isopropyl ester, formed from L-glutamine in a sample of cerebrospinal fluid, with D-glutamine added as internal standard, on a glass capillary, 20 m × 0.28 mm, coated with Chirasil-Val. Temperature isothermal at 160°C; carrier gas, hydrogen; gas velocity, 35 cm/min. (From Ref. 143.)

Quantitative determination of amino acids is also of interest in protein analysis and in peptide synthesis. The first important set of analytical data for a synthetic peptide or an isolated protein is the amino acid composition. The peptide is cleaved into its constituent amino acids with 6 N aqueous hydrochloric acid by heating to 110°C for 24 hr in a closed vial. For the protection of cysteine and methionine this is usually done in vacuo, sometimes in the presence of thioglycolic acid [144] or tryptamine to protect tryptophan against oxidative degradation. Cysteine and cystine are in equilibrium, depending upon the presence of oxygen and reducing agents. If these amino acids are to be determined in the presence of each other, the free thiol groups may be converted to the carboxymethyl derivative (Eq. 3) [43], and after reduction, cystine may be determined as cysteine. Glutamine and asparagine are converted to the corresponding dicarboxylic acids during hydrolysis. Serine and threonine are not fully stable; after hydrolysis for 24 hr often more than 10% has decomposed, especially when metal salts and carbohydrates are present in the sample.

Again, these effects can be compensated by adding the D-enantiomers of threonine, serine, cysteine, and tryptophan as internal standards before hydrolysis. The antipodes then serve to correct for the destruction occurring during acid hydrolysis [40]. Some attention must be directed to racemization. The configuration of the constitutent amino acids, or the extent of their epimerization, represents an important structural aspect of a peptide, and confirmation or knowledge of the stereochemistry of each amino acid is usually desired. The extent of racemization during hydrolysis is relatively small, ranging from about 3% for aspartic acid to less than 0.1% for threonine in 24 hr [145,146]. Higher percentages indicate presence of D-enantiomers in the original structure. Once the enantiomer ratios are known, the quantitative composition of the protein may be determined in a second run by enantiomer labeling.

Similar considerations apply to amino acid analyses in samples from other sources, such as geological materials, environmental samples, or food proteins. Free amino acids are isolated by ion exchange, similarly to biological fluids; in a hydrolyzed sample the sums of free and protein amino acids are determined.

Important for routine analysis is automatization. An instrument for automatic derivatization was commercially available and obviously was quite reliable [147], but the difficulties associated with gas chromatographic analysis prevented its general introduction; consequently, its production has been abandoned. With enantiomer-labeling, automatic derivatization is now again an expedient possibility, rendering gas chromatographic amino acid analysis more appealing for routine analysis.

C. Inlet Systems and Columns

A general problem of quantitative determination by capillary gas chromatography is an effect termed discrimination, observed when the sample is introduced by the common split-injection [123,124,140]. Two processes contribute to discrimination: fractionation of compounds of different volatility during injection with a hypodermic syringe [148], and inhomogeneous vaporization [124] in the injector, which is usually held at a temperature between 230°C and 260°C and equipped with a silanized glass insert. These effects lead to a virtually inconsistent splitting ratio. For gas chromatography on packed columns this is negligible or absent, since the derivatized sample is usually injected on column. Therefore, gas chromatography on packed columns has long been preferred over capillary gas chromatography for quantitative determination. Indeed, coefficients of variation determined on packed columns are often lower than those determined on capillaries (cf. Ref. 59 vs Refs. 44, 60).

Several alternative injection techniques have been proposed to overcome these difficulties of capillary gas chromatography. One possibility is to inject the sample in a splitless manner into an injection port equipped with a glass-ball packed insert [149]. One to two seconds later the splitting valve is opened; thus even for a split injection, discrimination can be largely avoided. Another possibility is splitless injection. The sample is introduced into the hot injector in a splitless mode and condensed in the front part of the capillary. Furthermore, injection of the sample can also be done directly on column, as for packed columns; this technique eliminates the problem of discrimination completely [140]. However, here biological samples present another problem. Even carefully purified samples contain involatile material which deposits in the vaporizer, upon on-column injection in the front part of the capillary. With packed columns the contaminated packing may be replaced from time to time, but capillaries must be shortened or the front coils must be replaced, an awkward procedure for routine analysis.

Another possibility is introduction of the sample with a falling needle injector [150], preventing discharge of large volumes of solvent vapors through the capillary. This mode has not been employed yet for amino acid analysis. Automatization may be difficult, more volatile derivatives may be partially lost, and the glass needle must be cleaned frequently if biological samples are analyzed.

Another commercial device allows introduction of the samples by use of capsules [151] made from aluminium or gold foil. Discrimination is minimized and the sample is injected in dry form, which avoids flooding of the capillary with solvent. Involatile or high-boiling material in the sample is withdrawn from the vaporizer together with the empty capsule after injection. Also, derivatization may be performed directly in capsules made from noble metals. Disadvantages are the cost of the capsules, the contact of amino acid derivatives with metal surfaces, and the difficulty of adapting the device to repetitive automatic injection.

A combination of the capsule injector and the falling needle is the solid sampler [152]: an open glass capillary containing the dry sample is dropped from a cooled storage container into the hot vaporizer, from which the vaporized amino acids are flushed into the column. The device is simple and inexpensive, discrimination is negligible, and the derivative vapors come into contact only with glass surfaces. Twenty-three samples can be injected automatically. A disadvantage is the potential loss of more volatile derivatives from the open capillaries during storage, but this may be minimized by closing the capillary at one end and by cooling the storage container [130].

An elegant way to deviate large solvent vapor volumes and to avoid accumulation of high-boiling compounds in the front part of the columns is to employ a precolumn, as proposed by Gehrke and co-workers [28] for packed columns with the "Sol-Vent" device. Volumes of up to 100 μl can be accommodated. A similar injector suitable for similarly large volumes has been reported [153] for capillary gas chromatography. Also, temperature-programmed injectors or column-switching techniques with packed columns or capillaries as precolumns [154] have been proposed recently to deviate solvents and high-boiling compounds from the main column. All these systems require accurate control of flow rates, temperatures, and actuation of venting and backflush valves. They are also more prone to mechanical and electronic failures than the former "basic" modes of sample injection; therefore, the potential of these modifications for routine amino acid analysis must be investigated more closely.

Many of the detriments enumerated above, such as preferential loss of the more volatile derivatives during solvent evaporation, discrimination on different stages of sample introduction, or partial breakdown of sensitive amino acid derivatives in the hot injector, may be eliminated if internal standards are added which correct for these effects on each individual amino acid. Obviously, this is not possible with only one internal standard. Similarly, as discussed in the previous section, isotopimers or enantiomers represent the optimum standards. Even if the losses of derivatives are extensive, all these effects are insignificant as long as sufficient amounts of derivatives enter the detector to generate reasonable signal-to-noise ratios.

The common carrier gases are hydrogen, helium, and nitrogen. Most appropriate for highly efficient capillary gas chromatography is hydrogen [155]. It has the lowest viscosity and highest diffusivity of all; therefore mass transfer is rapid, and high flow rates at relatively small pressure drop along the capillaries are achieved. In all, chromatography with hydrogen as carrier gas is fastest. Another advantage is that hydrogen can easily be obtained in pure form, especially free of oxygen and water traces. The only drawback is the safety hazard, but when the system is carefully checked for leaks after installation of a capillary, this should not present any risk. Also, technical devices are available which restrict the maximum flow released from a hydrogen cylinder to prevent the rapid discharge into the room atmosphere. The second best choice is helium; it is slightly less favorable, mainly due to its high price. Nitrogen may be used as well, but then chromatography is considerably slower; mass transfer is less rapid, and hence mean linear flow velocities must be slower. Also, it is often contaminated with traces of oxygen and therefore requires more meticulous purification.

Stationary phases with appropriate polarity and selectivity are important for packed columns with their relatively low efficiencies; for a long time a dual-column procedure with an apolar silicone and a polar polyester represented the best solution [28]. With highly efficient capillaries this is less important. Typically, capillaries with plate numbers between 50,000 and 80,000 and in some cases up to 100,000 have been employed. The selection of the stationary phase depends upon the type of derivative. For acetyl amino acid esters, a polar stationary phase composed of polyoxyethylene (Carbowax 20 M) and a cyanopropyl phenyl silicone gave best results [75], while the now preferred heptafluorobutyl amino acid butyl esters are resolved on nonpolar methyl silicones, such as SE-30 or OV 101 [44]. For oxazolidinone derivatives, a phenyl methyl silicone (OV 17) has been recommended [79].

Resolution of the enantiomers of all amino acids requires a special chiral stationary phase with judiciously adjusted polarity. Chirasil-Val consists of a fraction of 85 mol % nonpolar dimethylsiloxane units and a fraction of 15 mol % polar carboxypropyl methyl siloxane units carrying L-valine *t*-butylamide residues in amide linkage. The ratio of both units determines the enantioselectivity of Chirasil-Val and also its polarity and hence, the relative order of elution of the amino acids. Especially for quantitative work, complete resolution of all derivatives of interest and of contaminants must be achieved. Since siloxane copolymers with exactly defined ratio of both monomers are difficult to synthesize reproducibly, Chirasil-Val is prepared by mixing of two synthetic silicones, with a lower and a higher fraction of chiral units in the matrix.

Another important aspect of quantitative determination concerns the nature and chemical properties of the capillary-wall material. Several amino acids require well-deactivated surfaces, mainly those containing groups labile toward nucleophilic displacement, such as perfluoroacyl ester groups in serine, threonine, or tyrosine; perfluoroacyl thio esters in cysteine; acyl imidazolides in histidine; or the bis(perfluoroacyl) guanidino group in arginine [73]; but also labile trimethylsilyl groups attached to amino or carboxyl functions. Soda lime glass has an alkaline character and is difficult to deactivate to a satisfactory extent; more suitable are boron silicate glasses such as Duran, which can be deactivated relatively easily. An unfavorable characteristic of this glass type is the presence of Lewis-acid sites resulting from its boron and aluminum content. These metals can be removed by leaching with strong hydrochloric acid [156], but the glass surface may retain traces of hydrochloric acid which gradually decompose sensitive stationary phases. Another suitable method is to cover the glass surface and its active sites with silicic acid [100]. Fused silica does not contain metals, and acid leaching is therefore not necessary. Deposition of silicic acid is preferred for some more polar phases to improve the wettability of the surface; on untreated fused silica some phases may form droplets which drastically reduce the efficiency. Most appropriate is persilylation with phenyl group-containing silanes [157], which confers to glass or to fused-silica surface excellent deactivation, with a sufficiently high surface energy to enable spreading of polar phases.

Fused silica capillaries are thin walled and coated outside with a thermostable polymer. Their flexibility is a definite practical advantage, especially for novices inexperienced in handling glass capillaries. When a glass capillary breaks at the straightened front or end part, a short length of the coil must be straightened in a flame, with detrimental effects upon efficiency and deactivation of the column. With fused silica, straightening is obviously not necessary. A disadvantage of fused silica capillaries is so-called peak-splitting, an effect caused by slight temperature fluctuations in GC ovens, especially along the transition from the hot injector to the relatively cooler oven and from there to the hot detector [38]. Capillaries with thick walls and higher thermal mass do not show this effect.

The dimensions of capillaries most commonly employed for amino acid analysis are 20-30 m in length and 0.2-0.3 mm in internal diameter, but also much shorter capillaries with narrow internal diameter (4.5 m × 0.13 mm) [88] or longer capillaries with wide bore (60 m × 0.5 mm) [61] have been used. The dimensions of a capillary for a certain separation depend mainly upon the number of constituents in the derivatized sample or the number of peaks in the chromatogram, and the selectivity of the phase toward critical derivative pairs; these in turn determine the minimum required number of theoretical plates. It is difficult to give a definite value because derivative type, stationary phase, temperature program, nature and flow rate of the carrier gas, or selectivity of detection have bearings upon the number of minimum required plates. Obviously with longer capillaries analysis time is longer. This can be compensated to some extent by a steeper temperature program, but retention times of different amino acids within the whole array may change, especially of glycine, proline, hydroxyproline, threonine, and serine. Clearly, short capillaries are preferable. The plate number per meter increases with decreasing internal diameter; thus short capillaries must be correspondingly narrow for sufficiently high efficiency. On the other hand, the surface and amount of stationary phase in narrow capillaries are smaller, which entails lower capacity. Taking all these aspects into account, a capillary for quantitative determination of serum amino acids should have dimensions of about 25 m × 0.25 mm, which pertains to about 70,000–100,000 theoretical plates. The vitreous material of the capillary walls should be neutral or only slightly acidic and free of metal ions, and coated with the selected stationary phase with a film thickness of about 0.1–0.3 μm. Most suitable as carrier gas is hydrogen; otherwise, helium is the next best choice.

D. Detectors

The commonly used flame ionization detector (FID) is also preferred for amino acid analysis. Nearly all relative molar response factors reported in the literature have been determined on this detector. In principle, the flame ionization detector generates a current directly proportional to the mass of carbon passing the flame per time unit, independent of the structure of the molecule. This also holds for amino acids if appropriate corrections are made for the heteroatom-containing groups [158]. The FID response is proportional to the effective carbon number with a deviation from the average of about 7%. Due to these deviations and the sensitivity of detector response to variations of operational conditions such as flow rates or temperature, precalculation of molar response factors is inadequate for quantitative determinations. Instead, molar response factors relative to an internal standard or sample amino acids are determined in a calibration run before actual analysis.

The areas of the recorded peaks are calculated either by triangulation or with an electronic integrator. The ratio of peak area per mol of amino acid versus peak area per mole of reference compound is the relative response factor [27]. Sometimes the analogous relative weight response factor is used.

Response factors published by different groups are sometimes surprisingly diverse, even for the same internal standard and for amino acids with retention times close to the reference. This reflects a considerable inconsistency among different laboratories and indicates difficulties in transferring the exact conditions of sample work-up, derivatization, and chromatography. Comparison of published molar response factors shows that, as expected, the amino acids most similar to the standard compounds in volatility and chemical properties exhibit lowest interlaboratory variations. For the heptafluorobutyryl amino acid isobutyl esters these variations are by far the lowest, that is, between 2 and 7% from average, in spite of the relative large differences in the chromatographic conditions [59,61,62,94]. Alanine and glycine exhibit slightly larger variations, apparently a consequence of the relatively high volatility of both. The largest variations are found for arginine and histidine, which are known to present difficulties during derivatization and gas chromatography. Agreement of molar response factors relative to norleucine is acceptable for the heptafluorobutyryl amino acid *n*-propyl esters [66,159,160]. Again histidine and arginine show the highest variation.

Incompleteness of esterification of the sterically hindered carboxyl group of isoleucine causes a response factor considerably smaller than that of leucine. Both have very similar retention times and identical numbers of carbon atoms and functional groups; molar response factors of both should be fairly similar, but this is often not the case. The carboxyl group of the lower homologue, valine, is also sterically hindered. Large differences in relative response factors are often found for the amino acids cystine and tryptophan, and in some cases also for methionine; all of these are amino acids which may easily be oxidized during derivatization.

It is therefore absolutely necessary to determine relative response factors frequently when working with conventional internal standards or reference compounds. A slight or gradual change in any of the experimental conditions can result in a considerable change in response factors for one or several amino acids. Determination of response factors with a mixture of pure amino acids is not always satisfactory, as contaminants may be present in actual samples which may cause partial destruction of sensitive derivatives. Again, enantiomer labeling offers an escape from this precarious situation. Both enantiomers have close retention times and identical chemical properties and response factors.

Flame ionization detectors are sensitive enough for detection of amino acids in the range of a few nanograms, sufficient for most applications. In some cases higher sensitivity and selectivity of detection is required. Especially in the latter respect, employment of a flame ionization detector presents rather a drawback. The detector responds to any carbon-containing compound; thus not only amino acids are detected, but all other potentially interfering, nitrogen-free contaminants such as fatty acids, degradation artifacts, and monomers of the ion exchanger used for cleanup. In contrast, amino acid analysis by ion-exchange chromatography has the advantage that detection with ninhydrin is specific for amino group-containing compounds, with few exceptions. Nevertheless, for gas chromatography as well, other more selective detectors of high sensitivity are available. They respond either to certain heteroatoms or, with a mass spectrometer as detector, to selected masses.

Very useful for amino acid analysis are alkali flame ionization detectors (AFID) (Figure 6.24) [75,79,161,162]. They exhibit high selectivity and enhanced sensitivity for nitrogen-containing compounds. The response is mainly dependent upon the number of nitrogen atoms present in the molecule, but also upon a number of other factors, such as chemical structures of the compounds, technical design of the detector, and operation parameters [163].

This may explain why the AFID has received little attention for quantitative analysis of amino acids. Some exhibit large fluctuations in sensitivity upon changes in gas flows; the alkali-silicate-bead type is very responsive toward positioning and age of the bead. This demands frequent determination of response factors and is probably the reason why the sensitivity enhancement in comparison with flame ionization varies considerably; while Adams et al. [75] reported factors between 50 and 180 for amino acids, Butler and Darbre [161] found enhancements of only between 3 and 22. The hydrogen flow rate, in particular, has a strong influence upon the selective sensitivity of the AFID toward nitrogen-containing compounds; with increasing hydrogen flow, the AFID assumes more the characteristics of a normal FID. In combination with capillary gas chromatography using hydrogen as carrier gas, a specific problem arises. At constant inlet pressure the flow through the capillary decreases by more than 30% upon temperature programming from 80°C to 200°C. This causes a falling base line and modulation of detector response. In principle, there are three different possibilities to alleviate this. Helium as carrier gas may be substituted for hydrogen, but then the advantages of the latter [155] are given up as well; hydrogen needed for maintaining a "cold flame," about 1-2 ml/min, is then delivered through the usual system for the detector feed gases. Another possibility is to add such a high flow of hydrogen through the detector feed system that the small change of hydrogen flow through the capillary is insignificant [162]; as discussed above, in this mode most AFIDs behave more like an FID with slightly enhanced sensitivity for nitrogen. The most elegant solution is to provide for stabilization of hydrogen flow through the capillary. The optimum flow rate for capillaries with internal diameters of about 0.25 mm is

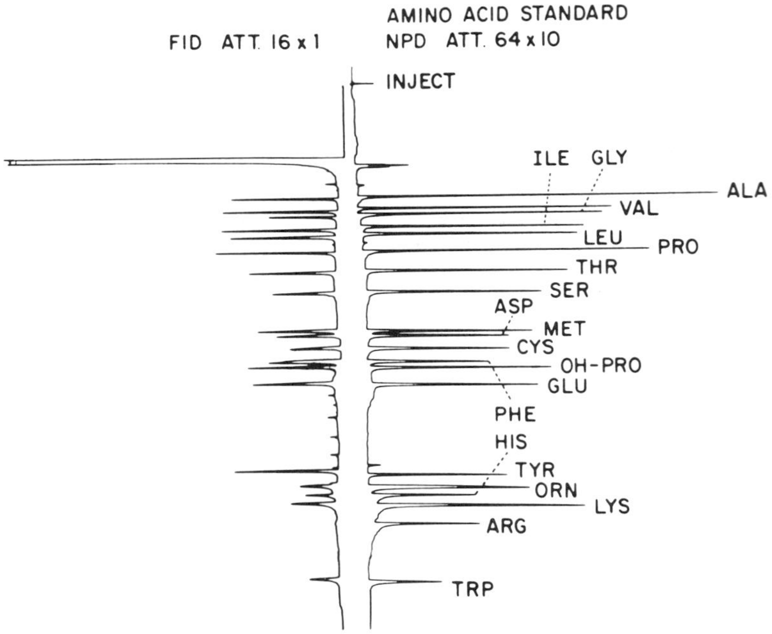

Figure 6.24 Chromatogram of an amino acid mixture consisting of 50 pmol of each *N*-acetyl amino acid *n*-propyl ester showing the different response of the alkali flame ionization detector (NPD) and the flame ionization detector (FID); gas capillary, 50 m × 0.27 mm, coated with a 1:1 mixture of polyoxyethylene (Carbowax 20 M) and phenyl cyanopropyl silicone (Silar 5CP). Temperature program, from 110°C to 190°C at 8°/min; carrier gas, helium, flow rate 0.75 ml/min. (Reproduced from Ref. 75. By permission of Preston Publications, Inc.)

1-2 ml/min; this coincides with the hydrogen flow rate affording highest selectivity. Stabilization of the carrier gas flow can be achieved by flow-regulated capillary gas chromatography instead of the conventional pressure-control, either by use of a micro-aperture or by pressure-programming simultaneously with the temperature increase. We investigated this possibility more closely [164] and obtained serum chromatograms considerably cleaner than with an FID (Figure 6.25). In combination with enantiomer labeling, this approach allows quantitative amino acid analysis with an accuracy and sensitivity not attained by most other methods.

Under modified operating conditions the AFID exhibits high selectivity for phosphorus-containing compounds. This possibility had been exploited relatively early for gas chromatography of amino acids, derivatized to the *N*-diethylphosphoryl amino acids [85]. Recently conversion to the dimethyl thiophosphinic amino acid methyl esters [86] was proposed for determination of amino acids in the upper femtogram range, but only six aliphatic amino acids were analyzed.

Another useful detector is the electron capture detector (ECD) [97,165,166]. It exhibits selective and strong response toward compounds containing electronegative moieties or atoms, especially halogens. Therefore the perfluoroacyl amino acid esters are the preferred derivatives with an ECD. Enhancement of sensitivity over an FID is small for the trifluoroacetyl derivatives but with the higher fluorinated heptafluorobutyryl derivatives, as little as 1-50 pg can be detected using packed columns with ethylene glycol adipate as stationary phase, at a signal-to-noise ratio of about three [97]. The response factors of different amino acids depend upon the chemical structure: logically, derivatives with more than one perfluoroacyl group generate a higher detector signal than the simple *N*-perfluoroacyl amino acids; an *O*-perfluoroacyl group especially affords increased detectability [161,165]. Detection with an ECD offers little advantage for biological samples; along with the amino acids, many contaminants in biological samples are converted to perfluoroacyl derivatives and hence detected. The electron capture detector proved to be useful for analysis of thyronine and thyroxine hormones (cf. Chap. 11).

Another element-specific detector is the flame photometric detector for sulfur-containing amino acids [167]. No information is available on its use in connection with capillaries. With packed columns it exhibits a somewhat lower sensitivity than the flame ionization detector.

A mass spectrometer run in the selected-ion monitoring mode represents a highly selective and sensitive detector. With respect to selectivity, it is probably the most discriminative and most versatile detector. The mass-specific detection offers great potential for quantitative determination in conjunction with isotope-labeled amino acids as internal standards. The intensities of molecular or fragment ions of a particular amino acid and its isotopimer are monitored over time. The amount of the amino acid present in the sample is directly calculated from the ratio of peak heights or peak areas, provided there is no contaminant generating ions of the same mass and eluting at the same retention time. Evidently, any contribution of the analyte compound to the ion abundance of the isotopimer and vice versa must be accounted for, such as natural abundances of heavy isotopes. In principle, four different, stable isotope labels are applicable, namely deuterium, carbon-13, nitrogen-15, and oxygen-18. Deuterium- and oxygen-18-labeled amino acids can be synthesized most easily; an inherent problem is the possibility of back exchange of both labels. For quantitation of amino acids in biological fluids this may be insignificant, if conditions are avoided which promote the exchange. However, under strongly acidic conditions, oxygen isotopes of the carboxyl group and deuterium atoms at different positions are rapidly exchanged. Carbon-13 and nitrogen-15 are therefore preferable, but the labeled amino acids are expensive due to their difficult preparation.

Properly selected ions with low baseline noise allow detection in the range of a few nanograms, in some cases down to a few picograms. Several publications report the combination of capillary gas chromatography and mass spectrometry for detection of amino acids [79,134,146,168,169]. Capillary gas chromatography/mass spectrometer offers great potential in tracing the fate of amino acids in metabolic pathways [135,170] by stable-isotope labeling, and has been employed for investigation of racemization mechanisms of amino acids [171] and determination of applicable rate constants [146,148] under different conditions. The only, but significant, disadvantages of mass spectrometers are their cost and technical complexity.

At present there are no other useful detectors for gas chromatography of amino acids; most others are too insensitive and not suited for capillary gas chromatography.

Additional instrumental requirements must be fulfilled for capillary gas chromatography, such as adaptation of the equipment to accomodate capillaries, especially of the injection system and the detector base. Important also is that the electronic processing of the detector signal is sufficiently fast. Nowadays instruments dedicated to capillary gas chromatography have electrometers with

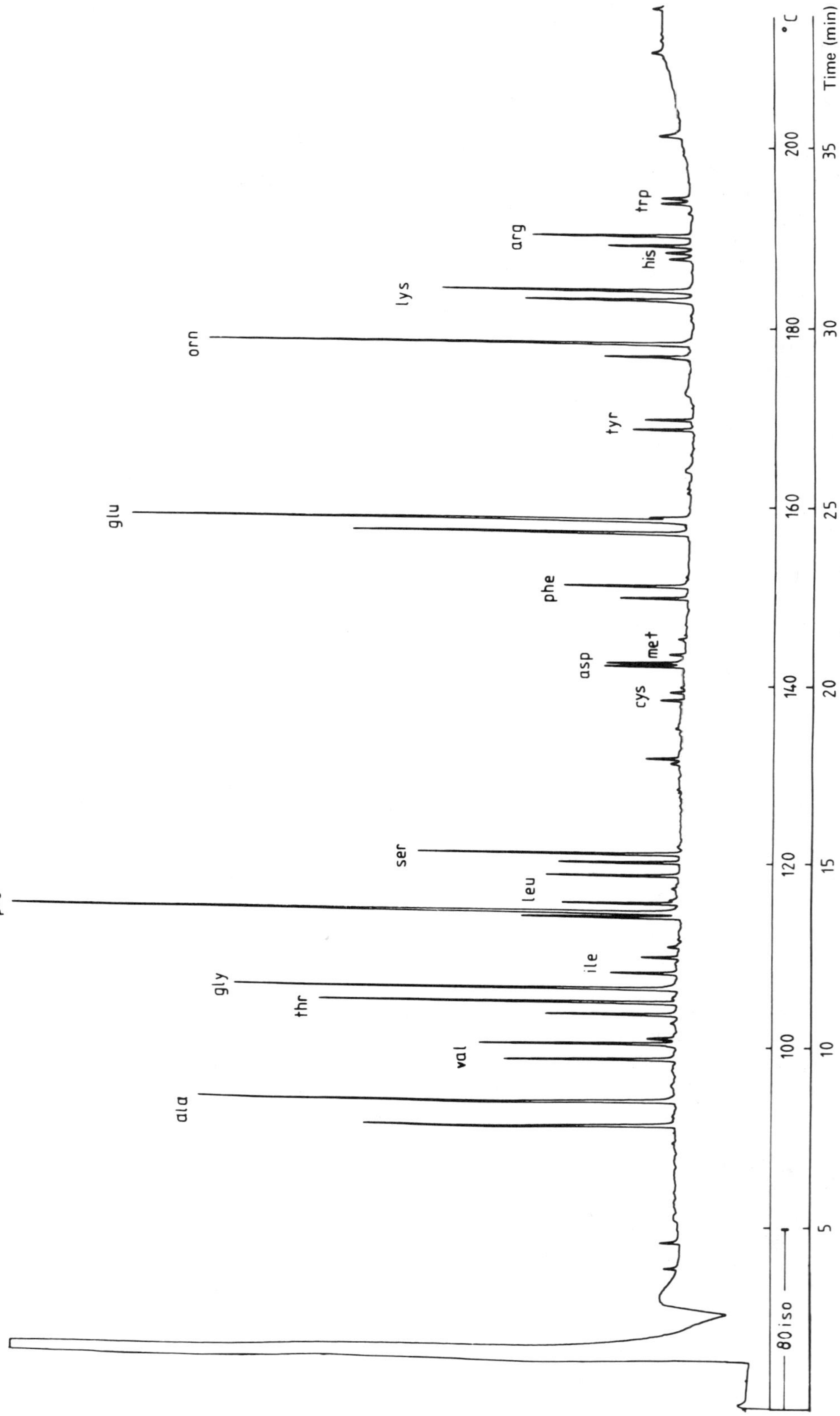
ala
val
thr
gly
ile
pro
leu
ser
cys
asp
met
phe
glu
tyr
orn
lys
his
arg
trp
80 iso
100
120
140
160
180
200
°C
5
10
15
20
25
30
35
Time (min)

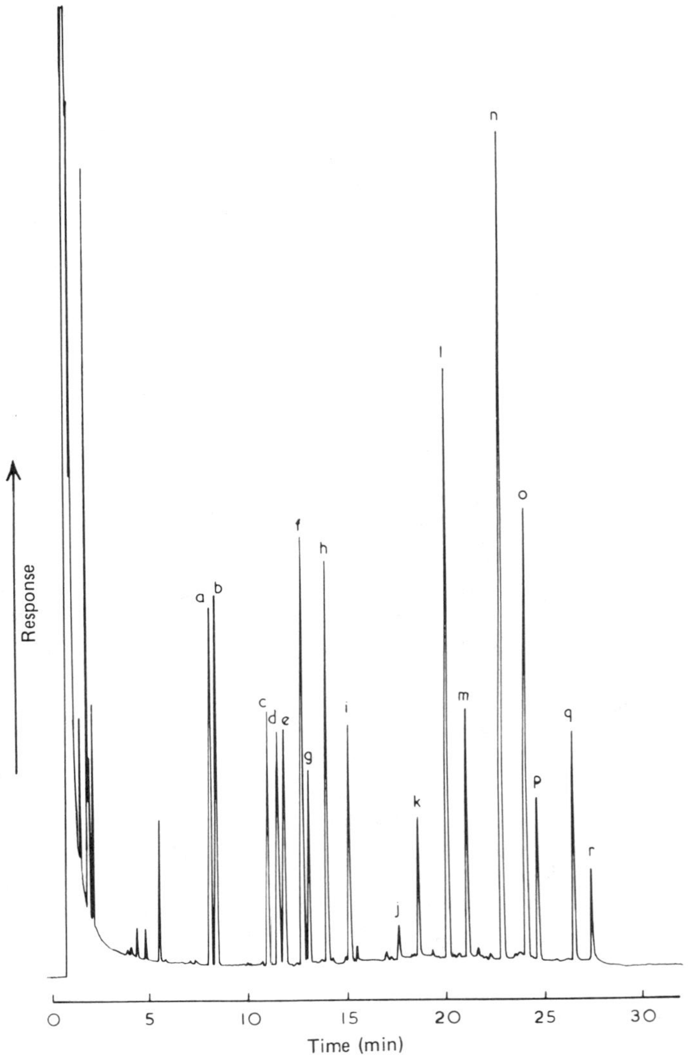

Figure 6.26 Analysis of *N*-heptafluorobutyryl amino acid isobutyl esters from an acid hydrolyzate of fish (maasbanker) muscle protein on a glass capillary, 35 m × 0.3 mm, coated with methylsilicone gum (OV 101); injection "on column." Temperature program, from 85°C to 230°C at 5°/min; carrier gas, hydrogen; flow rate, 3 ml/min. a = alanine, b = glycine, c = valine, d = threonine, e = serine, f = leucine, g = isoleucine, h = norleucine (i.s.), i = proline, j = hydroxyproline, k = methionine, l = aspartic acid, m = phenylalanine, n = glutamic acid, o = lysine, p = tyrosine, q = arginine, r = histidine. (From Ref. 140.)

Figure 6.25 Capillary gas chromatogram of serum amino acids as *N(O,S)*-pentafluoropropionyl amino acid *n*-propyl esters, histidine as N^{im}-ethoxycarbonyl derivative. Injected amounts, about 125–500 ng; split ratio, 1:50. Nitrogen detection with AFID. (From Ref. 164.)

a short time base. For further processing of the electrometer output, namely integration of peak areas, electronic integrators are commonly employed, and several different commercially available instruments are suitable. Some of them have the capacity to store raw data and complete chromatograms. Indispensable is the capability for free programming; programs for multiple internal standard methods like enantiomer labeling are not yet widely available, but they are necessary for automated routine amino acid analysis. Such programs must usually be written by the user.

V. CONCLUSIONS

Gas chromatographic amino acid analysis has found vast application in diverse areas. The obvious advantages of capillaries and the broad experience accumulated in their preparation and use will further expand their acceptance for gas chromatography of amino acids. The inherent difficulties of quantitative analysis by conventional capillary gas chromatography are overcome with enantiomer labeling. Its high accuracy and precision will boost the use of capillaries for routine analysis.

The earliest publications about capillary gas chromatography of amino acids described the separation of amino acid esters [24], of diastereomeric dipeptides for monitoring racemization during peptide synthesis [57,172], and the separation of oxazolidinone derivatives [77]. Soon afterwards the separation of trifluoroacetyl amino acid esters [96] and a study comparing the gas chromatographic properties of different perfluoroacyl amino acid esters [70] were published. In all these cases steel capillaries were employed.

For separation of amino acid enantiomers [34–37,48,56,57,89–92,103–106,172], capillaries have been used right from the beginning, either coated with conventional phases for separation of diastereomeric derivatives or coated with chiral stationary phases for direct enantiomer resolution. A large number of publications were devoted to this subject in the beginning of the seventies. Many applications involved the determination of enantiomer ratios of amino acids in fossils [173] and soils [174], and the extent of racemization during peptide synthesis [175]. Stainless steel capillaries are still used [176], as for investigations on the formation of amino acids in a primordial atmosphere [177] or the selective destruction of amino acids by ultraviolet light [177].

Considerable progress in capillary gas chromatography was achieved in 1973 with the separation of silylated thiohydantoins [88] and of heptafluorobutyryl derivatives of amino acids [66] on short glass capillaries. From 1976 onwards, the number of publications on glass capillary gas chromatography of amino acids increased; at the same time more publications were oriented toward practical applications, such as quantitative determination of amino acids in protein hydrolyzates [75,140] (Figure 6.26) or in serum and urine [44,75], assays of enzymatic activities in amino acid metabolism [179], or amino acid determination in surface waters [165].

The synthesis of Chirasil-Val enabled the complete separation of all protein amino acid enantiomers [39,40] and "started a new era for analytical studies of enantiomeric compounds, in particular amino acids" [93]. The analytical method of enantiomer labeling [40] rendered gas chromatographic amino acid analysis superior to conventional methods, and a potent alternative to ion-exchange chromatography had evolved. Several applications have been published since then, all of them, in principle, corroborating its reliability, for example, a study on the dependence of amino acid levels in human blood serum on physical excerise [130]. Others include the quantitative determination of hydroxyproline in urine and its hydrolyzate for monitoring collagen metabolism [180]; or the determination of alkylated amino acids in human hemoglobin, especially of cysteine [108], for assessing exposure to carcinogens in the workplace environment. Some animo acid analogues of pharmacological importance have also been analyzed with this general method: the optical antipode of the widely used drug D-penicillamine is toxic; accurate determination of less than 0.05% of the enantiomeric impurity L-penicillamine is now possible by reductive desulfurization to valine and enantiomer resolution on Chirasil-Val [40]. For bioavailability studies of *S*-carboxymethyl cysteine and *N*-acetyl cysteine, quantitative determination by enantiomer labeling offers accuracy, speed, and straightforwardness unattained by other methods [181].

Analysis of amino acids and their enantiomers by capillary gas chromatography currently represents the most powerful method, especially with respect to speed, sensitivity, accuracy, and cost. The exact quantitative determination of amino acids is of great importance in protein and enzyme research, in peptide synthesis, and in clinical chemistry. Elucidation of metabolism, physiology, and pathophysiology of amino acids is of prime importance for diagnosis of metabolic diseases, impossible without efficient chromatographic techniques. Although modern liquid chromatographic methods are appealing in their simplicity, the complementary method of gas chromatography is faster and more sensitive, less costly, and more versatile. Conventional gas chromatography, in spite of its long history, has not really been accepted for routine analysis of amino acids, but most difficulties have been overcome with the advent of enantiomer labeling. The commercial availability of the chiral stationary phase Chirasil-Val and the possibility of achieving automatic gas chromatographic amino acid analysis [182] will further enhance the utility of this potent analytical technique.

REFERENCES

1. J. P. Greenstein and M. Winitz, *Chemistry of the Amino Acids*, Wiley, New York, 1961.
2. P. Felig, *Ann. Rev. Biochem.* *44*:933 (1975).
3. A. Meister, *Biochemistry of the Amino Acids, Vols I, II*, Academic Press, New York, 1965.
4. P. E. Hare, in *Organic Geochemistry: Methods and Results* (G. Eglinton and M. T. J. Murphy, eds.), Springer, New York, 1969, pp. 438–463.
5. IUPAC-IUB Commission of Biochemical Nomenclature, *Biochemistry* 7: 2703 (1968).
6. W. L. Alworth, *Stereochemistry and its Application to Biochemistry*, Wiley-Intersciences, N.Y. New York, 1972.
7. E. Stephen-Sherwood and J. Oro, *Space Life Sciences* 4:5 (1973).
8. J. J. Corrigan, *Science* *164*:142 (1969).
9. J. M. Manning, *J. Biol. Chem.* *246*:2926 (1971).
10. F. Kögl, *Experientia* 5:173 (1949).
11. J. A. Miller, *Cancer Res.* *10*:65 (1950).
12. P. Soupart, in *Amino Acid Pools* (J. T. Holden, ed.), Elsevier, Amsterdam, 1962, p. 220.
13. P. Karlson, W. Gerok, and W. Groß, *Pathobiochemie*, Georg Thieme Verlag, Stutrgart, 1978, p. 66.
14. C. H. W. Hirs and S. N. Timasheff, eds., *Methods in Enzymology, Vol. XXV, Enzyme Structure*, Part B, Academic Press, New York, 1972.

15. P. M. Masters and M. Friedman, *Agric. Food Chem. 27*:507 (1979).
16. R. Feierabend, *Arch. Hydrobiol. 84*:454 (1978).
17. K. M. Williams and G. G. Smith, *Origins of Life 8*:91 (1977).
18. D. H. Spackmen, W. H. Stein, and S. Moore, *Anal. Chem. 30*:1190 (1958).
19. I. R. Hunter, K. P. Dimick, and J. W. Corse, *Chem. Industry (London) 16*:294 (1956).
20. M. Bier and P. Teitelbaum, *Ann. N.Y. Acad. Sci. 72*:641 (1959).
21. A. Liberti, in *Gas Chromatography 1958* (D. H. Desty, ed.), Butterworth Scientific Publ., London, 1958, p. 341.
22. B. Weinstein and A. H. Fenselau, *J. Chromatogr. 15*:149 (1964).
23. E. Bayer, K. H. Reuther, and F. Born, *Angew. Chem. 69*:640 (1957).
24. C. O. Andersson, E. Ryhage, S. Ställberg, and E. Stenhagen, *Ark. Kemi, 19*:405 (1962).
25. E. Bayer, in *Gas Chromatography 1958* (D. H. Desty, ed.), Butterworths Scientific Publ., London, 1958, p. 333.
26. W. A. Lamkin and C. W. Gehrke, *Anal. Chem. 37*:383 (1965).
27. C. W. Gehrke and D. L. Stalling, *Separation Sci. 2*:101 (1967).
28. C. W. Gehrke, K. Kuo, and R. Zumwalt, *J. Chromatogr. 57*:209 (1971).
29. F. Kaiser, C. W. Gehrke, R. W. Zumwalt, and K. C. Kuo, *J. Chromatogr. 94*:113 (1974).
30. C. G. Youngs, *Anal. Chem. 31*:1019 (1959).
31. K. Rühlmann and W. Giesecke, *Angew. Chem. 73*:113 (1961).
32. A. Darbre and K. Blau, *J. Chromatogr. 17*:31 (1965).
33. P. Hušek and K. Macek, *J. Chromatogr. 113*:139 (1975).
34. E. Gil-Av, R. Charles, and G. Fischer, *J. Chromatogr. 17*:408 (1965).
35. E. Gil-Av, B. Feibush, and R. Charles-Sigler, *Tetrahed. Lett., 1966*: 1009.
36. W. A. Koenig, W. Parr, H. A. Lichtenstein, E. Bayer, and J. Oro, *J. Chromatogr. Sci. 8*:183 (1970).
37. S. Nakaparksin, P. Birrell, E. Gil-Av, and J. Oro, *J. Chromatogr. Sci. 8*:177 (1970).
38. F. Munari and S. Trestianu, in *Proceedings of the Fifth International Symposium on Capillary Chromatography*, Riva del Garda 1983, Elsevier, Amsterdam, p. 327 (1983).
39. H. Frank, G. J. Nicholson, and E. Bayer, *J. Chromatogr. Sci. 15*: 174 (1977).
40. H. Frank, G. J. Nicholson, and E. Bayer, *J. Chromatogr. 167*:187 (1978).
41. H. Hediger, R. L. Stevens, H. Brandenberger, and K. Schmid, *Biochem. J. 133*:551 (1973).
42. F. S. Collins and G. K. Summer, *J. Chromatogr. 145*:456 (1978).
43. I. M. Moodie and R. D. George, *J. Chromatogr. 124*:315 (1976).
44. J. Desgres, D. Boisson, and P. Padieu, *J. Chromatogr. 162*:133 (1979).
45. M. Friedman and J. W. Finley, *J. Agric. Food Chem. 19*:627 (1971).
46. D. Roach, C. W. Gehrke, and R. W. Zumwalt, *J. Chromatogr. 43*:311 (1969).
47. I. M. Moodie, *J. Chromatogr. 99*:495 (1974).
48. W. Rahn, H. Eckstein, and W. A. Koenig, *Hoppe-Seyler's Z. Physiol. Chem. 357*:1223 (1976).
49. M. M. Shemyakin, E. J. Vinogradova, Y. A. Ovchinnikov, A. A. Kiryushkin, M. Y. Feigina, N. A. Aldanova, Y. B. Alakhov, V. M. Lipkin, B. V. Rosinov, and L. A. Fonina, *Tetrahedron 25*:5785 (1969).

50. D. L. Stalling, C. W. Gehrke, and R. W. Zumwalt, *Biochem. Biophys. Res. Commun. 31*:616 (1968).
51. C. W. Cehrke and K. Leimer, *J. Chromatogr. 53*:201 (1970).
52. C. W. Gehrke and K. Leimer, *J. Chromatogr. 57*:219 (1971).
53. C. Zomzely, G. Marco, and E. Emery, *Anal. Chem. 34*:1414 (1962).
54. D. Roach and C. W. Gehrke, *J. Chromatogr. 44*:269 (1969).
55. P. Cancalon and J. D. Klingman, *J. Chromatogr. Sci. 12*:349 (1974).
56. F. Weygand, H. Geiger, and W. Swodenk, *Angew. Chem. 68*:307 (1956).
57. F. Weygand, A. Prox. L. Schmidhammer, and W. König, *Angew. Chem. (Intern. Ed.) 2*:183 (1963).
58. S. L. MacKenzie and D. Tenaschuk, *J. Chromatogr. 111*:413 (1975).
59. S. L. MacKenzie and D. Tenaschuk, *J. Chromatogr. 171*:195 (1979).
60. P. A. Cruickshank and J. C. Sheehan, *Anal. Chem. 36*:1191 (1964).
61. R. J. Pearce, *J. Chromatogr. 136*:113 (1977).
62. S. L. MacKenzie and D. Tenaschuk, *J. Chromatogr. 173*:53 (1979).
63. H. Frank, D. Bimboes, and G. J. Nicholson, *Chromatographia 12*:168 (1979).
64. J. R. Coulter and C. S. Hann, *J. Chromatogr. 36*:42 (1968).
65. R. F. Adams, *J. Chromatogr. 95*:189 (1974).
66. J. Jönsson, J. Eyem, and J. Sjöquist, *Anal. Biochem. 51*:204 (1973).
67. C. W. Gehrke, D. R. Younker, K. O. Gerhardt, and K. C. Kuo, *J. Chromatogr. Sci. 17*:301 (1979).
68. J. Metz, W. Ebert, and H. Weicker, *Chromatographia 4*:259 (1971).
69. A. Darbre and A. Islam, *Biochem. J. 106*:923 (1968).
70. G. E. Pollock, *Anal. Chem. 39*:1194 (1967).
71. D. L. Stalling and C. W. Gehrke, *Biochem. Biophys. Res. Commun. 22*: 329 (1966).
72. I. M. Moodie, *J. Chromatogr. 208*:60 (1981).
73. G. J. Nicholson, H. Frank, and E. Bayer, *J. High Resol. Chromatogr. Commun. 2*:411 (1979).
74. C. W. Moss, M. A. Lambert, and F. J. Diaz, *J. Chromatogr. 60*:134 (1971).
75. R. F. Adams, F. L. Vandemark, and G. J. Schmidt, *J. Chromatogr. Sci. 15*:63 (1977).
76. O. Grahl-Nielsen and E. Solheim, *J. Chem. Soc., Chem. Commun. 1972*: 1092.
77. F. Weygand, *Z. Anal. Chem., 205*:406 (1964).
78. P. Hušek, V. Felt, and M. Matucha, *J. Chromatogr. 180*:53 (1979).
79. R. Liardon, U. Ott-Kuhn, and P. Hušek, *Biomed. Mass Spect. 6*:381 (1979).
80. S. Yamamoto, S. Kiyamam, Y. Watanabe, and M. Makita, *J. Chromatogr. 233*:39 (1982).
81. N. Ikekawa, O. Hosino, R. Watanuki, H. Orimo, T. Fujita, and M. Yoshikawa, *Anal. Biochem. 17*:16 (1966).
82. G. Michael, *J. Chromatogr. 196*:160 (1980).
83. L. Bertilsson and E. Costa, *J. Chromatogr. 118*:395 (1976).
84. M. Wolfensberger, U. Redweik, and H.-Ch. Curtius, *J. Chromatogr. 172*:471 (1979).
85. G. Ertingshausen, C. W. Gehrke, and W. A. Aue, *Separation Sci. 2*: 681 (1967).
86. K. Jacob, C. Falkner, and W. Vogt, *J. Chromatogr. 167*:67 (1978).
87. J. J. Pisano, T. J. Bronzert, and H. B. Brewer, *Anal. Biochem. 45*: 43 (1972).
88. J. Eyem and J. Sjöquist, *Anal. Biochem. 52*:255 (1973).

89. W. A. Koenig, W. Rahn, and J. Eyem, *J. Chromatogr. 133*:141 (1977).
90. E. A. Hoopes, E. T. Peltzer, and J. L. Bada, *J. Chromatogr. Sci. 16*:556 (1978).
91. W. Parr, C. Yang, E. Bayer, and E. Gil-Av, *J. Chromatogr. Sci. 8*:591 (1970).
92. W. Parr, J. Pleterski, C. Yang, and E. Bayer, *J. Chromatogr. Sci. 9*:141 (1971).
93. R. Liardon and S. Ledermann, *J. High Resol. Chromatogr. Chromatogr. Commun. 3*:475 (1980).
94. R. J. Siezen and T. H. Mague, *J. Chromatogr. 130*:151 (1977).
95. D. Roach and C. W. Gehrke, *J. Chromatogr. 43*:303 (1969).
96. I. Halasz and K. Bünnig, *Z. Anal. Chem. 211*:1 (1965).
97. R. W. Zumwalt, K. Kuo, and C. W. Gehrke, *J. Chromatogr. 57*:193 (1971).
98. C. A. Poole and M. Verzele, *J. Chromatogr. 150*:439 (1978).
99. S. L. MacKenzie and L. R. Hogge, *J. Chromatogr. 147*:388 (1978).
100. E. Schulte, *Chromatographia 9*:315 (1976).
101. G. E. Pollock and V. I. Oyama, *J. Gas Chromatogr. 1966*:126.
102. B. Feibush, *J. Chem. Soc. Chem. Commun. 1971*:544.
103. K. Grohmann and W.Parr, *Chromatographia* 5:18 (1972).
104. W. A. Koenig and G. J. Nicholson, *Anal. Chem. 47*:951 (1975).
105. R. Charles, U. Beitler, B. Feibush, and E. Gil-Av, *J. Chromatogr. 112*:121 (1975).
106. R. Brazell, W. Parr, F. Andrawes, and A. Zlatkis, *Chromatographia 9*:57 (1976).
107. E. Bayer and H. Frank in *Modification of Polymers* (C. E. Carraher and M. Tsuda, eds.), ACS Symposia Series No. 121, Washington, D.C., 1980, p. 341.
108. E. Bailey, P. B. Farmer, and J. H. Lamb, *J. Chromatogr. 200*:145 (1980).
109. H. Frank, G. J. Nicholson, and E. Bayer, *J. Chromatogr. 146*:197 (1978).
110. H. Frank, G. J. Nicholson, and E. Bayer, *Angew. Chem. (Intern. Ed.) 17*:363 (1978).
111. D. Thumm, Doctoral Thesis, Tübingen, 1980.
112 H. Frank, J. Gerhardt, G. J. Nicholson, and E. Bayer, *J. Chromatogr. 270*:159 (1983).
113. W. R. Sherman, A. L. Leavitt, M. P. Honchar, L. M. Hallcher, and B. E. Phillips, *J. Neurochem. 36*:1947 (1981).
114. W. A. Koenig, J. Benecke, and S. Sievers, *J. Chromatogr. 217*:201 (1983).
115. E. Bayer, E. Küsters, H. Frank, and G. J. Nicholson (Submitted for publication).
116. N. Oi, H. Kitahara, M. Horiba, and T. Doi, *J. Chromatogr. 206*:143 (1981).
117. T. Saeed, P. Sandra, and M. Verzele, *J. High Resol. Chromatogr. Chromatogr. Commun. 3*:35 (1980).
118. C. H. Lochmüller and J. V. Hinshaw, *J. Chromatogr. 202*:363 (1980).
119. W. A. Koenig, S. Sievers, and I. Benecke, in *Capillary Chromatography* (R. E. Kaiser, ed.), Bad Dürkheim, 1981, p. 703.
120. N. Oi, O. Hirvaki, and H. Schimada, *Bunseki Kagaku 27*:637 (1978).
121. I. Abe, T. Kohno, and S. Musha, *Chromatographia 11*:393 (1978).
122. J. A. Rijks, J. Drozd, and J. Novak, *J. Chromatogr. 186*:167 (1979).

123. K. Grob and K. Grob, *J. High Resol. Chromatogr. Chromatogr. Commun. 1*:57 (1978).
124. G. Schomburg, H. Behlau, R. Dielmann, F. Weeke, and H. Husmann, *J. Chromatogr. 142*:87 (1977).
125. W. D. Block, M. E. Markovs, and B. F. Steele, *Proc. Soc. Exptl. Biol. Med. 122*:1089 (1966).
126. R. W. Zumwalt, D. Roach, and C. W. Gehrke, *J. Chromatogr. 53*:171 (1970).
127. H. N. Tucker and S. V. Molinary, *Clin. Chem. 19*:1040 (1973).
128. L. Arola, E. Herrera, and M. Alemany, *Anal. Biochem. 82*:236 (1977).
129. E. D. Pellizari, C. Rising, J. H. Brown, R. W. Farmer, and L. F. Fabre, *Anal. Biochem. 44*:312 (1971).
130. H. Frank, A. Rettenmeier, H. Weicker, G. J. Nicholson, and E. Bayer, *Clin. Chim. Acta 105*:201 (1980).
131. K. L. Clay and R. C. Murphy, *J. Chromatogr. 164*:417 (1979).
132. R. J. Boila and L. P. Milligan, *J. Chromatogr. 202*:238 (1980).
133. J. F. Riordan and R. W. Giese, *Methods in Enzymology*, Vol. XLVII, *Enzyme Structure* (C. H. W. Hirs and S. N. Timasheff, eds.), Academic Press, New York, 1977, p. 31.
134. M. F. Schulman and F. P. Abramson, *Biomed. Mass Spectrom. 2*:9 (1975).
135. M. J. Zagalak, H.-Ch. Curtius, W. Leimbacher, and U. Redweik, *J. Chromatogr. 142*:523 (1977).
136. E. E. Kingston and A. M. Duffield, *Biomed. Mass Spectrom. 5*:621 (1978).
137. J. M. L. Mee, J. Korth, B. Halpern, and L. B. James, *Biomed. Mass Spectrom. 4*:178 (1977).
138. R. C. Murphy and K. L. Clay, *Biomed. Mass Spectrom 6*:309 (1979).
139. H. Frank, A. Rettenmeier, H. Weicker, G. J. Nicholson, and E. Bayer, *Anal. Chem. 54*:715 (1982).
140. I. M. Moodie and J. Burger, *J. High Resol. Chromatogr. Chromatogr. Commun. 4*:218 (1981).
141. Q. H. Gibson and G. Wiseman, *Biochem. J. 48*:426 (1951).
142. P. Hušek and V. Felt, *J. Chromatogr. 152*:546 (1978).
143. H. Frank, A. Eimiller, H. H. Kornhuber, and E. Bayer, *J. Chromatogr. 224*:177 (1981).
144. C. W. Gehrke and H. Takeda, *J. Chromatogr. 76*:77 (1973).
145. H. Frank, W. Woiwode, G. J. Nicholson, and E. Bayer, *Liebigs Ann. Chem. 1981*:354.
146. R. Liardon, S. Lederman, and U. Ott, *J. Chromatogr. 203*:385 (1981).
147. A. M. Lewis, C. Waterhouse, and L. S. Jacobs, *Clin. Chem. 26*:271 (1980).
148. K. Grob and H. P. Neukom, *High Resol. Chromatogr. Chromatogr. Commun. 2*:15 (1979).
149. E. Bayer and G. H. Liu, *J. Chromatogr. 256*:201 (1983).
150. P. M. J. van den Berg and T. P. H. Cox, *Chromatographia 5*:301 (1972).
151. B. Kolb and W. Hoser, *Chromatographia 6*:28 (1973).
152. F. Poy, P. Gagliardi and D. Bisi, Carlo Erba Strumentazione, Application Report.
153. W. Vogt, K. Jacob, A. B. Ohnesorge, and H. W. Obwexer, *J. Chromatogr. 186*:197 (1979).
154. C. Wang, H. Frank, G. Wang, L. Zhou, E. Bayer, and P. Lu, *J. Chromatogr. 262*:352 (1983).

155. K. Grob and G. Grob, *J. High Resol. Chromatogr. Chromatogr. Commun.* *2*:109 (1979).
156. K. Grob, G. Grob, and K. Grob, *Chromatographia* *10*:185 (1977).
157. K. Grob, G. Grob, W. Blum, and W. Walther, *J. Chromatogr.* *244*:197 (1982).
158. A. Islam and A. Darbre, *J. Chromatogr.* *71*:223 (1972).
159. J. F. March, *Anal. Biochem.* *69*:420 (1975).
160. T. Yoneda, *Anal. Biochem.* *104*:247 (1980).
161. M. Butler and A. Darbre, *J. Chromatogr.* *101*:51 (1974).
162. J. Chauhan, A. Darbre and R. F. Carlyle, *J. Chromatogr.* *227*:305 (1982).
163. B. Kolb, M. Auer, and P. Pospisil, *J. Chromatogr. Sci.* *15*:53 (1977).
164. H. Frank, N. Vujtovic-Ockenga, and A. Rettenmeier, *Proceedings of the Fifth International Symposium on Capillary Chromatography*, Riva del Garda, 1983, Elsevier, Amsterdam, p. 600 (1983).
165. G. Bengtsson, G. Odham, and G. Westerdahl, *Anal. Biochem.* *111*:163 (1981).
166. G. Bengtsson and G Odham, *Anal. Biochem.* *92*:426 (1979).
167. S. I. Bonvell and R. H. Monheimer, *J. Chromatogr. Sci.* *18*:18 (1980).
168. H. Frank, W. Woiwode, G. J. Nicholson, and E. Bayer, in *Stable Isotopes* (E. R. Klein and P. D. Klein, eds.), Academic Press, New York, 1979, p. 165.
169. B. A. Petersen and P. Vouros, *Anal Chem.* *49*:1304 (1977).
170. H. Wegmann, H.-Ch. Curtius, and U. Redweik, *J. Chromatogr.* *158*: 305 (1978).
171. W. Woiwode, H. Frank, G. J. Nicholson, and E. Bayer, *Chem. Ber.* *111*:3711 (1978).
172. F. Weygand and H. Fritz, *Chem. Ber* *98*:72 (1965).
173. G. Dungworth, *Chem. Geol.* *17*:135 (1976).
174. G. E. Pollock, C. N. Cheng, and S. E. Cronin, *Anal. Chem.* *49*:2 (1977).
175. E. Bayer, E. Gil-Av, W. A. König, S. Nakaparksin, J. Oro, and W. Parr, *J. Am. Chem. Soc.* *92*:1738 (1970).
176. G. G. Smith and D. M. Wonnacott, *Anal, Biochem.* *109*:414 (1980).
177. J. G. Lawless and C. D. Boynton, *Nature,* *243*:405 (1973).
178. N. Levi and J. G. Lawless, *Anal. Biochem.* *90*:796 (1978).
179. F. K. Trefz, T. Erlenmaier, D. H. Hunneman, K. Bartholome, and P. Lutz, *Clin. Chim. Acta* *99*:211 (1979).
180. W. Woiwode, D. List, and H. Weichardt, *J. Clin. Chem. Clin. Biochem.* *17*:251 (1979).
181. H. Frank, D. Thiel, and K. Langer, *J. Chromatogr.* *309*:261 (1984).
182. H. Frank, J. Gerhardt, G. J. Nicholson, and E. Bayer, *Z. Anal. Chem.* *317*:688 (1984).

7

Analysis of Sugars, Sugar Alcohols, and Related Compounds

Carl D. Pfaffenberger / *Department of Epidemiology and Public Health, University of Miami School of Medicine, Miami, Florida*

I. INTRODUCTION

High-resolution gas chromatographic separations of sugars, sugar alcohols and related compounds were first reported in 1973 by Szafranek, Pfaffenberger, and Horning [1]. Prior to this, German, Horning, and Pfaffenberger, then at Baylor College of Medicine, Houston, Texas, had been developing new types of borosilicate glass capillary columns about 60 m in length [2,3]. The columns had internal diameters of between 0.25 and 0.30 mm and possessed about 100,000–125,000 theoretical plate efficiencies. They were dynamically coated using apolar phase SE-30 which contained dispersed particles of finely divided silanized silicic acid [2,3]. Being rather heavily coated, the columns were remarkably thermally stable to 300° and, when protected by a short precolumn of SE-30 packing, were very long lived and useful for the analysis of samples acquired from biological matrices [4,5].

Since 1972, various publications in the scientific literature have demonstrated the usefulness of SE-30 borosilicate glass capillary columns, and it is with these separations that this chapter is concerned. Included herein are discussions of suitable derivatives of aldoses and alditols, obtaining and interpreting sugar/sugar alcohol metabolic profiles, separating sugar acids and lactones, and studying human cataractous lens tissue. The recently reported separations of D and L monosaccharides by Kamerling and co-workers [6,7] have also been included and illustrated.

II. ASSOCIATED CLINICAL PATHOLOGY

Whereas carbohydrates are required as energy-producing substrates for certain organs such as the central nervous system, and although they are essential components of the nucleic acids and some structural compounds such as hyaluronic acid, all necessary sugar units can be biosynthesized in the human body from precursors derived from amino acids and the glycerol of various lipids. As a result, no carbohydrate is required in the diet; and it has been shown experimentally that humans can survive in good health for months on a diet of fats and meats which provides only trace amounts of sugar. Many diets, however, provide considerable amounts of carbohydrate, both as components of vegetables and grains and as purified substances like refined sugar. Often one third to one half of an individual's caloric intake is derived from carbohydrate sources; and because it is more efficient and less expensive to use vegetables and grains for food, the proportion of carbohydrate in the diet tends to be higher in lower socioeconomic groups. Carbohydrate metabolism is thus of central importance to a large portion of the world's population.

Disturbances in carbohydrate metabolism, either congenital or developed, lead to a host of diseases which should lend themselves to investigation by the metabolic profiling techniques discussed in Secs. IV–VI.

Von Mering and Minkowski [8] first demonstrated that if a dog's pancreas is removed, serious disturbances of glucose metabolism result with sudden elevation of blood glucose concentration and development of the clinical symptoms of diabetes mellitus. Insulin, the major hormone involved in carbohydrate metabolism, affects cell membrane activity, amino acid transport, intracellular glucose metabolism, lipid metabolism, protein synthesis, and growth. Carbohydrate metabolism is also affected by epinephrine, glucagon, adrenocorticosteroids, and pituitary growth hormone.

Diabetes mellitus is a generalized, chronic metabolic disorder which manifests itself, in its fully developed form, by hyperglycemia, glycosuria, increased protein breakdown, ketosis, acidosis, and hyperlipemia. Other diseases associated with improper carbohydrate metabolism include acromegaly, hypoglycemia, fructose intolerance, congenital galactosemia, pentosuria, congenital lactose intolerance, and congenital and acquired renal glycosuria.

An array of at least seven glycogen storage diseases exist, including Type I (von Gierke's disease or hepatorenal glycogen storage disease), Type II (Pompe's disease), Type III (limit dextrinosis, debrancher glycogen storage disease, or Cori's disease), Type IV (amylopectinosis, brancher glycogen storage disease or Andersen's disease), Type V (muscle phosphorylase deficiency or McArdle's disease), Type VI (liver phosphorylase deficiency or Hers' disease) and Type VII (muscle phosphoglucomutase deficiency, another type of McArdle's disease [8].

The formation of cataroctous human lens tissue has also been linked to carbohydrate metabolism, particularly to the accumulation of *myo*-inositol within the aging lens [9].

Jellum et al. [10] published a carbohydrate profile for normal seminal fluid in 1976. Their chromatogram included trimethylsilyl and methoxime trimethylsilyl ether derivatives of glycerol, erythritol, fucose, ribose, fructose (the major component), glucose, sorbitol, a relatively large amount of *myo*inositol, and fourteen other components which were not identified until 1978 [11]. These latter compounds included: threitol, xylose, arabinose, xylitol, arabinitol, ribitol, fucitol, mannose, galactose, mannitol, galactitol, *scyllo*-inositol, 2-acetamido-2-deoxyglucose, and 2-acetamido-2-deoxygalacose.

Aldose and alditol metabolic profiles have been useful in studies of both normal and unusual carbohydrate metabolism; and many more uses should materialize, especially in studies of glycogen storage disease. The unmatched capabilities of capillary over packed columns has made it possible to diagnose maple sugar urine disease [12]. McConnell et al. [13] have published metabolic profiles of urine from both normal and diabetic patients. Koenig and Guenther [14] have analyzed an array of natural products of high polarity and low volatility such as peptides, ribose, and mannosamine. Leblanc and Ball [15] have successfully analyzed the corresponding trimethylsilyl ether derivatives of sugar and sugar phosphates.

III. DERIVATIZING ALDOSES AND ALDITOLS

Two major complications have always presented themselves in analytical procedures devised to separate complex mixtures of aldoses and related polyols. One is the well-known problem of the formation of several compounds from any one aldose, namely, an anomeric pair with pyrano and furano derivatives of each form. The other problem is associated with the fact that all the members of a family of aldoses, for example, the aldohexoses, are quite similar in structure and thus inherently difficult to separate. Invariably, any GC method employing packed columns leads to the result that at least two derivatives have identical methylene unit (MU) values [16–18].

With the advent of 60-m SE-30 open-tubular glass capillary columns having thermal stability to 300° and possessing between 100,000 and 125,000 theoretical plate efficiencies, it became feasible to attempt to separate complex mixtures of aldoses and related polyols. The procedure described here was developed by Szafranek, Pfaffenberger, and Horning [1] and is based on the first step of the classical Wohl degradation of aldoses [19–21]. Open-chain poly-*O*-acetylaldonic nitriles are formed, one from each aldose, in high yield and with retention of terminal dissymmetry [1]. The Wohl derivatives give good GC peaks; and polyols, when subjected to the Wohl conditions, lead to polyacetyl esters which also chromatograph well. Figure 7.1 is a separation

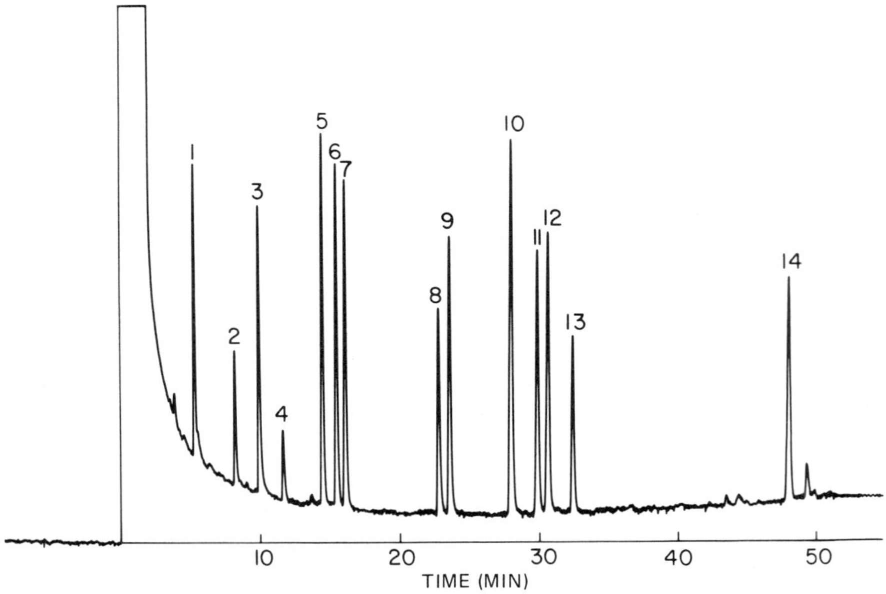

Figure 7.1 Separation of 13 aldoses as Wohl derivatives on a 64-m SE-30 glass open-tubular capillary column temperature programmed 1°/min from 155°. The components include: 1 = D-erythrose, 2 = L-threose, 3 = 2-deoxy-D-ribose, 4 = L-threitol, 5 = D-ribose, 6 = arabinose, 7 = D-xylose, 8 = 2-deoxy-D-glucose, 9 = 2-deoxy-D-galactose, 10 = 3-*O*-methyl-D-glucose, 11 = mannose, 12 = D-glucose, 13 = D-galactose and 14 = D-*glycero*-D-*gulo*-heptose.

of 13 common aldoses as poly-*O*-acetylaldonic nitriles. Methylene unit (MU) values of the derivatives are given in Table 7.1.

Complete separation of these aldoses is possible. The order of elution of the derivatives depends not only upon the molecular weight but also upon the configuration of the asymmetric carbon atoms. The only pair of compounds not separable under the experimental conditions is D-xylose and 6-deoxy-L-galactose. The latter compound was not included in the study; its MU value is 16.91 (Wohl derivative). Figure 7.2 is a separation of 22 polyols as poly-acetyl esters; Table 7.2 contains the methylene unit (MU) values for these derivatives.

Because the molecular weight of a Wohl derivative of an aldose is always lower than that of the corresponding alditol acetate, it is possible to separate complex mixtures of polyols and aldoses. From Figure 7.3 it may be observed that only two pairs of a 32-component mixture were incompletely separated, namely, the separation of D-galactose from 2-deoxy-D-glucitol and L-*glycero*-D-*gulo*-heptitol from *meso*-*glycero*-*gulo*-heptitol.

Methyl and trimethylsilyl ethers or acetyl and trifluoroacetyl esters have been used in GC analyses of aldoses, but with these derivatives it is possible for the aldoses to form anomeric pairs of cyclic furano and pyrano isomers. This situation leads to the well-appreciated multiplicity of derivatives for aldoses. A mixture containing multiple derivatives of several aldoses is often difficult to separate, even with capillary columns [22]. Although reduction of the carbonyl group of a sugar removes the possibility of forming isomers,

Table 7.1 Methylene Unit (MU) Values[a] of Wohl Derivatives of Aldoses

Aldose	MU value
D-Erythrose	14.97
L-Threose	15.10
2-Deoxy-D-ribose	15.57
D-Ribose	16.63
D-Arabinose	16.80
D-Xylose	16.94
2-Deoxy-D-glucose	18.07
2-Deoxy-D-galactose	18.15
3-*O*-Methyl-D-glucose	18.81
D-Mannose	19.06[b]
D-Glucose	19.17[b]
D-Galactose	19.40[b]
D-*Glycero*-D-*gulo*-heptose	21.37[b]

[a]Initial temperature, 150°.

[b]Initial temperature, 170°.

retention of terminal dissymmetry is likewise lost. For example, only six polyols were obtained from the eight aldohexoses. Therefore, reduction of the carbonyl group should not be applied to a mixture containing both sugars and polyols. Conversion of aldoses into aldonolactone-trimethylsilyl ethers [23] retains terminal dissymmetry, but from D-arabinose, D-xylose, D-mannose, and D-*glycero*-D-*gulo*-heptose, more than one derivative is obtained [24]. The procedure described here is based upon formation of poly-*O*-acetylaldonic nitriles (Wohl derivatives) of aldoses [19,20]. The GC analysis of mixtures of polyols and aldoses thus becomes satisfactory in that the nitriles have open-chain structures and molecular weights lower than the acetyl derivatives of the corresponding polyols.

The reactions which occur in the Wohl synthesis are included in Ref. 1 for β-D-glucose. The high temperature of acetylation used in the procedure brings about the exclusive formation [20] of poly-*O*-acetylaldonic nitriles. A cyclic by-product which is formed at low reaction temperatures is not a final product under the reaction conditions employed. Ketoses, including D-fructose, form only nonvolatile products which are not detected under the analysis conditions. Eklund et al. [25] have reported an acceptable procedure based on the formation of trifluoracetyl esters of sugars; this method should be used when ketoses are important components of the mixture being analyzed. Their method appears to be extremely sensitive, the minimum detectable amounts of monosaccharides being at the subpicogram level when an electron capture detector is being employed.

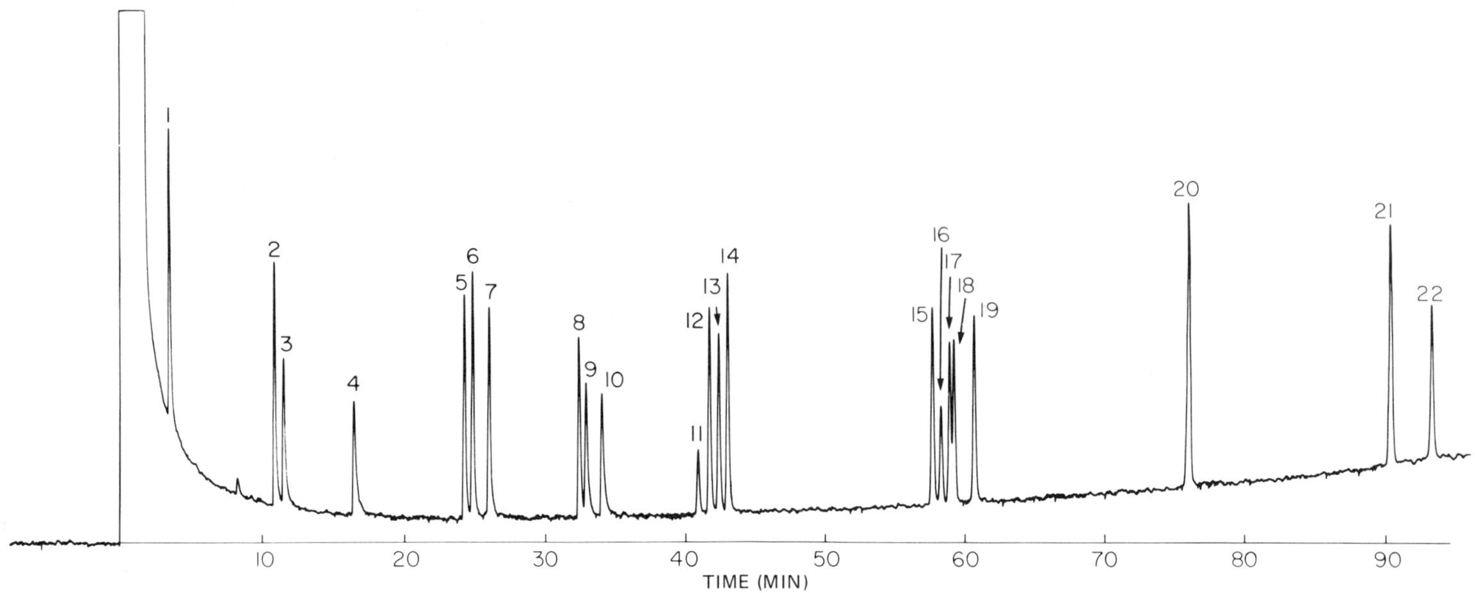

Figure 7.2 Separation of 22 polyols as polyacetyl esters; see Figure 7.1 for column description and analytical conditions. The components include: 1 = glycerol, 2 = erythritol, 3 = L-threitol, 4 = 2-deoxy-D-ribitol, 5 = ribitol, 6 = D-arabinitol, 7 = xylitol, 8 = 2-deoxy-D-glucitol, 9 = 2-deoxy-D-galactitol, 10 = 3-deoxy-D-glucitol, 11 = *myo*-inositol, 12 = D-mannitol, 13 = D-glucitol, 14 = galactitol, 15 = D-*glycero*-D-*gluco*-heptitol, 16 = D-*glycero*-D-*manno*-heptitol, 17 = L-*glycero*-D-*gulo*-heptitol, 18 = *meso*-*glycero*-*gulo*-heptitol, 19 = D-*glycero*-D-*galacto*-heptitol, 20 = D-*erthro*-L-*galacto*-octitol, 21 = D-*arabino*-D-*manno*-nonitol and 22 = D-*arabino*-L-*galacto*-nonitol.

Table 7.2 Methylene Unit (MU) Values of Polyacetyl Esters

Polyol	MU value	Polyol	MU value
Glycerol	13.98[a]	D-Mannitol	20.59
Erythritol	15.88	D-Glucitol	20.67
L-Threitol	15.98	Galactitol	20.76
2-Deoxy-D-ribitol	16.99	D-*Glycero*-D-*gluco*-heptitol	22.63
Ribitol	18.29	D-*Glycero*-D-*manno*-heptitol	22.69
D-Arabinitol	18.37	L-*Glycero*-D-*gulo*-heptitol	22.78
Xylitol	18.55	*meso-Glycero-gulo*-heptitol	22.82
2-Deoxy-D-glucitol	19.40	D-*Glycero*-D-*galacto*-heptitol	23.90
2-Deoxy-D-galactitol	19.48	D-*Erythro*-L-*galacto*-octitol	25.01
3-Deoxy-D-glucitol	19.65	D-*Arabino*-D-*manno*-nonitol	26.99
myo-Inositol	20.47[b]	D-*Arabino*-L-*galacto*-nonitol	27.40

[a]For glycerol through 3-deoxy-D-glucitol an initial temperature of 150° was employed.
[b]For *myo*-inositol through D-*arabino*-L-*galacto*-nonitol, the initial temperature was 170°.

IV. CONCERNING METABOLIC PROFILES

Kutsky [26] pointed out several years ago that with the new discoveries in molecular biology, biochemistry, physiology, and endocrinology, the emphasis in research in the life sciences is turning away from descriptive biochemistry and electron microscopy to the functional, dynamic aspects of vital metabolic processes. Moreover, particularly great interest has been generated in the actual controlling agents which accurately regulate all of the cellular enzyme systems and organelles to produce a living cell and, from that, a living multicellular organism with many highly differentiated tissues. Steroidal hormones play a key role in this area in that they are controlling agents; and because many steroids are chemically and structurally similar and exist in a constant state of physiological flux, they have been prime candidates for studies employing metabolic profiling approaches. Urinary steroid metabolic profiling has been reviewed recently by Pfaffenberger [27].

Analytical samples obtained from biological matrices often contain many components which are quite similar structurally, for example, mixtures of monosaccharides from the hydrolyzate of a complex polysaccharide. Such samples demand the use of high-resolution columns, and glass capillaries now routinely have sufficient resolving power to afford baseline separation of most components of complex mixtures. Metabolic profiling is convenient in that dietary and physiological fluctuations of various sample components can be readily determined using serial sampling over any convenient investigative period. The concept was first introduced by Horning and Horning in 1970 [28], and many biomedical

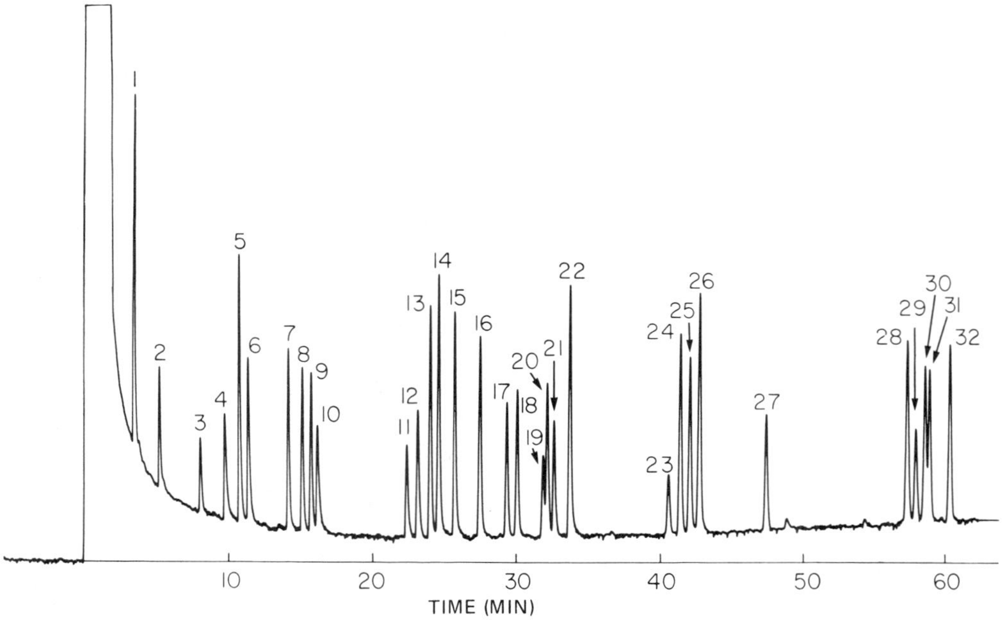

Figure 7.3 Separation of polyols and aldoses as the corresponding polyacetyl and Wohl derivatives; see Figure 7.1 for column description and analytical conditions. The components include: 1 = glycerol, 2 = D-erythrose, 3 = L-threose, 4 = 2-deoxy-D-ribose, 5 = erythritol, 6 = L-threitol, 7 = D-ribose, 8 = D-arabinose, 9 = D-xylose, 10 = 2-deoxy-D-ribitol, 11 = 2-deoxy-D-glucose, 12 = 2-deoxy-D-galactose, 13 = ribitol, 14 = D-arabinitol, 15 = xylitol, 16 = 3-*O*-methyl-D-glucose, 17 = D-mannose, 18 = D-glucose, 19 = D-galactose, 20 = 2-deoxy-D-glucitol, 21 = 2-deoxy-D-galactitol, 22 = 3-deoxy-D-glucitol, 23 = *myo*-inositol, 24 = D-mannitol, 25 = D-glucitol, 26 = galactitol, 27 = D-*glycero*-D-*gulo*-heptose, 28 = D-*glycero*-D-*gluco*-heptitol, 29 = D-*glycero*-D-*manno*-heptitol, 30 = L-*glycero*-D-*gulo*-heptitol, 31 = *meso*-*glycero*-*gulo*-heptitol and 32 = D-*glycero*-D-*galacto*-heptitol.

applications of the metabolic profiling technique have appeared in the literature since then, including analyses of human blood serum and cerebrospinal, amniotic, and seminal fluids, to name but a few. An extensive review of this area has been published by Jellum et al. [10]. The biological sample may be a single "point-in-time" collection (grab sample) or a more averaged one such as an aliquot of a 24-hr urine collection, etc. The obvious utility of the profiling approach is its multicomponent nature and, in the case of grab sampling, its inherent characteristic of indicating the physiological levels of each component at each sampling period. How to obtain some specific metabolic profiles is discussed in Sec. V, and some representative profiles of urinary aldoses and alditols appear in Sec. VI.

V. OBTAINING AND USING METABOLIC PROFILES

The recommended procedure [1] for obtaining Wohl derivatives of aldoses converts polyols to polyacetyl esters without complication. The method may therefore be applied in studies of biological samples containing both types of compounds.

A. Preparation of Derivatives

1. *Acetyl Derivatives of Reference Alditols*

One milligram of a reference alditol in 200 μl of acetic anhydride containing 2 mg of dry sodium acetate is heated to 120°C for 2 hr. It is necessary to heat longer (5 hr) only in the case of D-*arabino*-L-*galacto*-nonitol. Sodium acetate is preferable to pyridine catalyst; the reaction is considerably faster.

2. *Wohl Derivatives of Reference Aldoses*

A mixture of 1 mg of a reference aldose, 1 ml of hydroxylamine hydrochloride, and 2.5 mg of dry sodium acetate is dissolved in 200 μl of dry methanol and heated at 60° for 1 hr. The methanol is removed with a stream of dry nitrogen, and the residue is treated with 200 μl of acetic anhydride at 120°C for 1 hr. The product is the corresponding poly-*O*-acetylaldonic nitrile.

3. *Metabolic Profile Samples from Urine*

Urine samples from female and male human neonates, juveniles, and adults on unrestricted diets (excepting neonates in the nursery service) are collected in sterilized ocntainers over a 24-hr period unless only one void is obtainable [29]. No preservative is used, but the sample is kept at 4° during the collection period and stored under deep-freeze conditions as soon as possible after the volume has been measured. A creatinine determination is made on one aliquot by any standard clinical laboratory procedure.

The pH of a 20-ml aliquot of urine is adjusted to 4.5 using 10% aqueous acetic acid. This sample is poured onto a packing of DEAE-Sephadex A-25 medium (Pharmacia, Uppsala, Sweden), contained in a 10 cm × 1.3 cm I.D. glass column. Basic and neutral compounds are eluted by washing the packing with 100 ml of deionized water which has been glass-distilled from basic potassium permanganate. The eluate is collected in a 250-ml round-bottomed flask and lyophilized to constant weight. The dry residue is dissolved in 10 ml of methanol. A 1-ml aliquot of the methanolic solution is heated with 5 mg of hydroxylamine hydrochloride and 12.5 mg of sodium acetate at 60° for 30 min in a PTFE-lined screw-cap vial. The product is dried at 40° using a nitrogen stream, and derivatized in 1 hr using 300 μl of acetic anhydride at 100°. When an internal reference compound is required, 200 μl of a solution of glyceryl-1-decyl ether (6.30 mg/10.0 ml of methanol) is added prior to heating with hydroxylamine hydrochloride. Also, for samples of biological origin, 100 μl of pyridine is used as catalyst (instead of sodium acetate). The dry residue from these reactions is treated with 300 μl of acetic anhydride and 50 μl of pyridine at 100° for 1 hr.

4. *Gas Chromatography*

In the work reported here [29], separation of the urinary polyols and aldoses as polyacetyl esters and poly-*O*-acetylaldonic nitriles was carried out with a Barber Colman gas chromatograph (Model 5000 with flame ionization detection), which had been modified to accept 60-m glass capillary columns (0.3 mm I.D.) coated with SE-30 containing Silanox 101 (Cabot Corp., Boston, Mass.) prepared according to German and Horning [3]. The theoretical plate efficiencies of the capillary columns were between 100,000 and 125,000 when measured

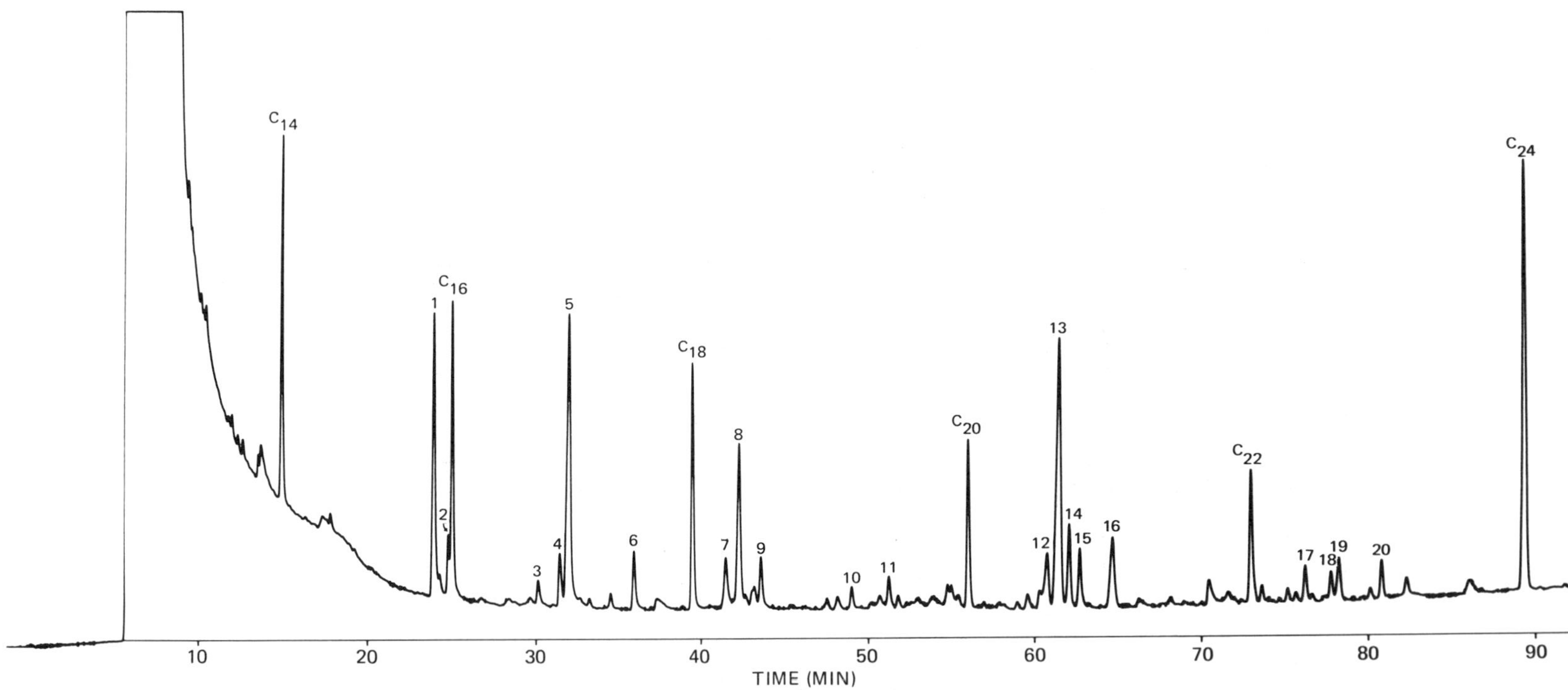

Figure 7.4 To determine methylene unit values for the urinary components, even-numbered straight-chain hydrocarbons, C_{14}–C_{24}, are coinjected with the sample under the analytical conditions given for Figure 7.1. The corresponding derivatives of tetritols (1 and 2) elute between C_{14} and C_{16}; the derivatives of pentoses (3–6), between C_{16} and C_{18}; the derivatives of pentitols and hexoses (7–11), between C_{18} and C_{20}; hexitols (12–16), between C_{20} and C_{22}. Polyacetyl derivatives of the higher alditols elute between C_{22} and C_{24}.

isothermally at 250° for *n*-tetracosane. A Keithley Model 417 picoammeter and a Texas Instruments recorder were employed. The precolumn helium [2] and flame hydrogen flow rates were controlled by Brooks Model 8744 flow controllers and measured with Matheson Model 8110-0121 mass flowmeters.

Methylene unit (MU) values [16,17] were measured through the use of *n*-alkanes having even carbon numbers, $C_{14}-C_{24}$, coinjected with the sample as shown in Figure 7.4. Temperature programming was employed at 1°/min from 150°. Other GC conditions inlcuded: sample volume, 2 μl; split ratio, 5:1; precolumn-inlet splitter, 240°; detector bath, 300°; column inlet pressure, 18 psi, resulting in a carrier gas (helium) linear velocity of 16–18 cm/sec (200°); hydrogen flow, 10 ml/min; air flow, 30 ml/min.

Table 7.3 lists methylene unit (MU) values and molar detector response factors for derivatives of the 17 polyols and aldoses identified in a study of normal

Table 7.3 Methylene Unit (MU) Values[a] for Acetyl and Wohl Derivatives of Polyols and Aldoses Identified in Human Urine

Peak no.	Assignment	MU value Column I	Column II	Molar response factor (MRF)[b]
1	Erythritol	15.78	15.89	0.62 ± 0.08
2	Threitol	15.95	16.08	0.62 ± 0.08[c]
3	Ribose	—	16.66	0.45 ± 0.02
4	Arabinose	—	16.83	0.40 ± 0.03
5	Fucose	—	16.93	0.40 ± 0.03
6	Xylose	—	16.96	0.40 ± 0.02
7	Ribitol	18.24	18.35	0.65 ± 0.03
8	Fucitol	—	18.38	—
9	Arabinitol	18.33	18.42	0.61 ± 0.08
10	Xylitol	18.50	18.61	0.62 ± 0.08
11	Mannose	—	19.12	0.52 ± 0.03
12	Glucose	19.11	19.20	0.53 ± 0.03
13	Galactose	—	19.46	0.38 ± 0.02
IS	Glyceryl-1-decyl ether	—	19.88	1.00[d]
14	*myo*-Inositol	20.54	20.63	0.66 ± 0.04
15	Mannitol	20.63	20.70	0.84 ± 0.10
16	Glucitol	20.70	20.79	0.82 ± 0.09
17	Galactitol	20.79	20.87	0.76 ± 0.02

[a]SE-30 glass open-tubular capillary column, temperature programmed at 1°/min from 150°.
[b]Determined in triplicate using column II.
[c]Assumed to be the same as for erythritol.
[d]Internal reference compound (I.S.).

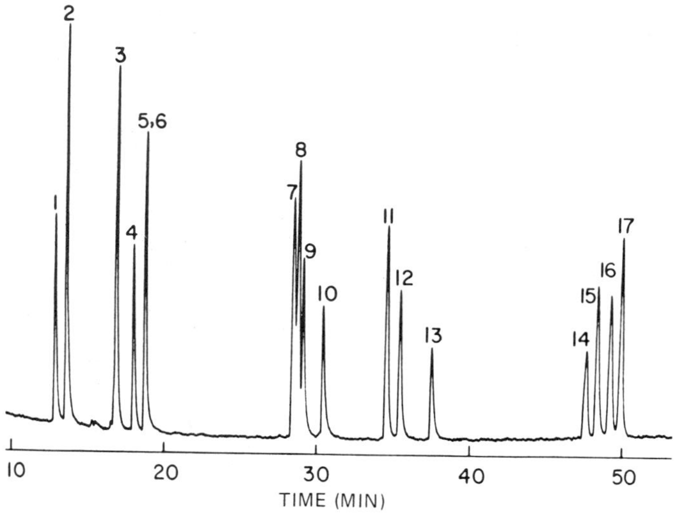

Figure 7.5 Separation of the reference mixture of 17 polyols and aldoses which were quantified as acetyl and Wohl derivatives during a study of human urine from healthy neonates, juveniles, and adults. A 60-m SE-30 open-tubular glass capillary column was temperature programmed 1°/min from 165°. The components include: 1 = erythritol, 2 = threitol, 3 = ribose, 4 = arabinose, 5 = fucose, 6 = xylose, 7 = ribitol, 8 = fucitol, 9 = arabinitol, 10 = xylitol, 11 = mannose, 12 = glucose, 13 = galactose, 14 = *myo*-inositol, 15 = mannitol, 16 = glucitol, 17 = galactitol. (From Ref. 29.)

human urine by Pfaffenberger et al. [29]. A typical GC separation of these compounds, using a 60-m glass capillary column with SE-30 liquid phase, temperature programmed 1°/min from 165°, is shown in Figure 7.5.

The methylene unit (MU) values listed in Table 7.3 are highly reproducible (±0.03 MU) for each column (Column I), but a shift of up to about 0.1 MU can occur when a different column (e.g., column II) is used. Values for all components are shifted in the same direction when this effect occurs, and no change in the order of elution or urinary components has been observed. A typical GC analysis requires about 1 hr. Molar response factors (MRF) are calculated in relation to the detector response observed for the diacetyl derivative of glyceryl-1-decyl-ether.

VI. REPRESENTATIVE METABOLIC PROFILES OF URINARY ALDOSES AND ALDITOLS

Many human pathological circumstances result in altered urinary profiles of polyols and aldoses. Common examples include diabetes mellitus [8], other glucosurias [30], galactosemia [31], pentosuria [32], and other disturbances or inborn errors in the usual carbohydrate metabolic pathways. Pfaffenberger et al. [29] chose three age groups for their initial studies of nonpathological subjects. Neonates, aged less than 6 months, were generally in the first few days of life. Juveniles, 4-10 years of age, were on unrestricted diets. Adults, also on unrestricted diets, were 20 years or more in age.

According to Figure 7.5, baseline separation of each of the components quantified was obtained except between derivatives of the fucose/xylose pair and the ribitol/fucitol/arabinitol trio. This complication proved to be unimportant because fucose and fucitol are rarely present above trace levels, and derivatives of ribitol and arabinitol are well separated. The internal reference compound (I.S.), the diacetyl ester of glyceryl-1-decyl ether, elutes in the otherwise clear region between derivatives of galactose and *myo*-inositol.

A. Excretion Profiles of Neonates

Figure 7.6 is a typical chromatogram observed for a neonate. The major components are (1) erythritol, (2) threitol, (9) arabinitol, and (14) *myo*-inositol. Erythritol and threitol are invariably present in neonatal urine, usually in a ratio of about 6:1. Average excretion values for these compounds are respectively, 120 and 21 μg/mg creatinine. Pentoses are usually not detectable, or are present in trace amounts (<4μg/mg creatinine). Fucitol is usually present in small amounts. Mannose, glucose, and galactose are often present in small amounts, with glucose predominating. Respective average excretion values for these three hexoses are 4, 24, and 8 μg/mg creatinine.

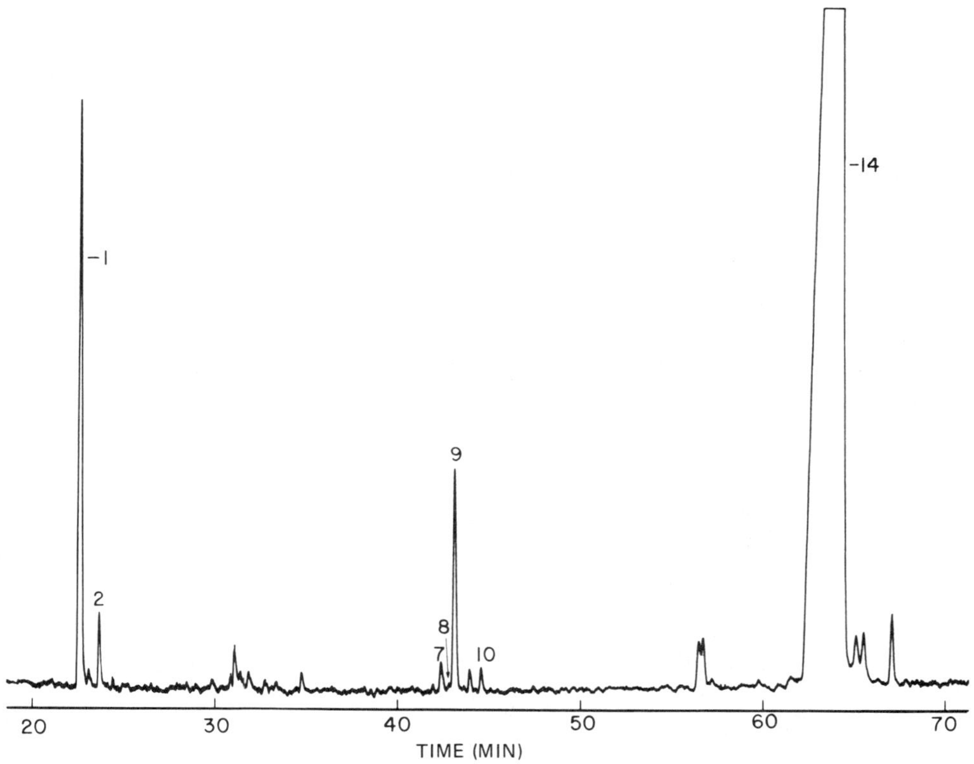

Figure 7.6 Profile of human (male neonate) urinary polyols and aldoses separated as in Figure 7.1. The identified components include: 1 = erythritol, 2 = threitol, 7 = ribitol, 8 = fucitol, 9 = arabinitol, 10 = xylitol, and 14 = *myo*-inositol, usually in relatively large amounts in neonatal urine. (From Ref. 29.)

By far the largest component of urinary samples from neonates is *myo*-inositol. The average excretion observed for *myo*-inositol is 2.2 mg/mg creatinine. The range of observed values is wide: 0.3–4.9 mg/mg creatinine for the ten neonates reported [29]. The presence of comparatively large amounts of *myo*-inositol in the urine may be related to the continuing synthesis of central nervous system tissue during the first few days of life. The samples which had the lowest *myo*-inositol levels were from the oldest infants (1 month of age).

The presence of large amounts of *myo*-inositol will mask or delay the elution of derivatives of mannitol, glucitol, and galactitol. When the amount of *myo*-inositol is low, other hexitols are readily detected.

No significant difference was noted [29] on the basis of the sex of neonates, but females appear to excrete about 2.5 times as much glucose as males. Female juveniles and female adults also usually excrete more glucose than do males; the ratio is about 10:1 for adults. (This comparison is based upon μg/mg creatinine.)

B. Excretion Profiles of Juveniles

Figure 7.7 is a typical chromatogram for a juvenile. The major components are (1) erythritol, (2) threitol, (7) ribitol, (9) arabinitol, (10) xylitol, (14)

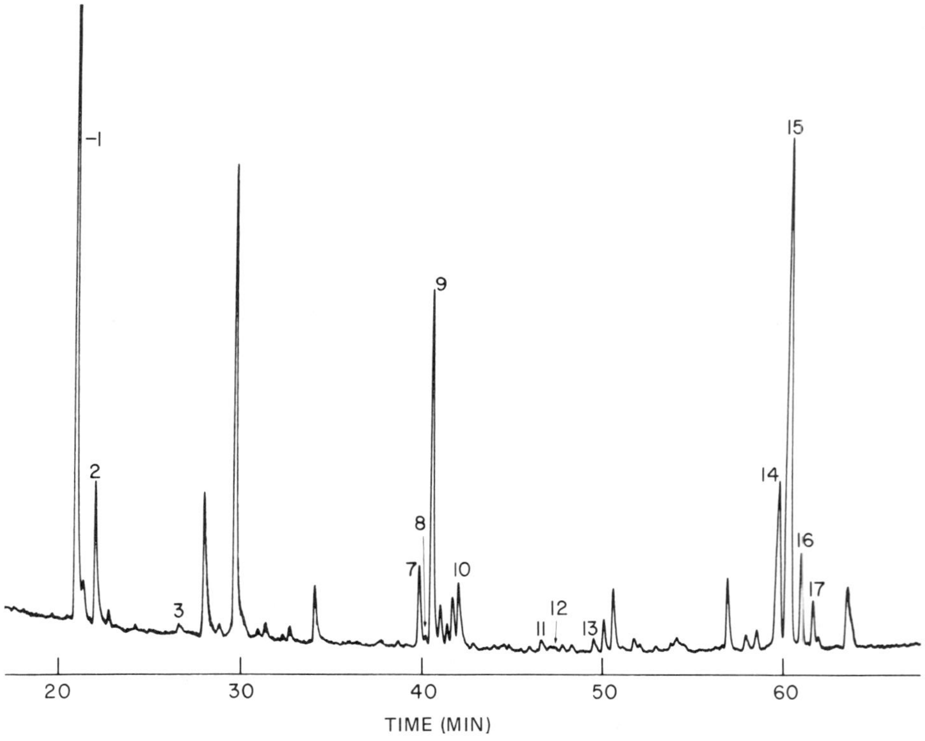

Figure 7.7 Profile of human (female juvenile) urinary polyols and aldoses separated as in Figure 7.1. The identified components include: 1 = erythritol, 2 = threitol, 3 = ribose, 7 = ribitol, 8 = fucitol, 9 = arabinitol, 10 = xylitol, 11 = mannose, 12 = glucose, 13 = galactose, 14 = *myo*-inositol, 15 = mannitol, 16 = glucitol, and 17 = galactitol. (From Ref. 29.)

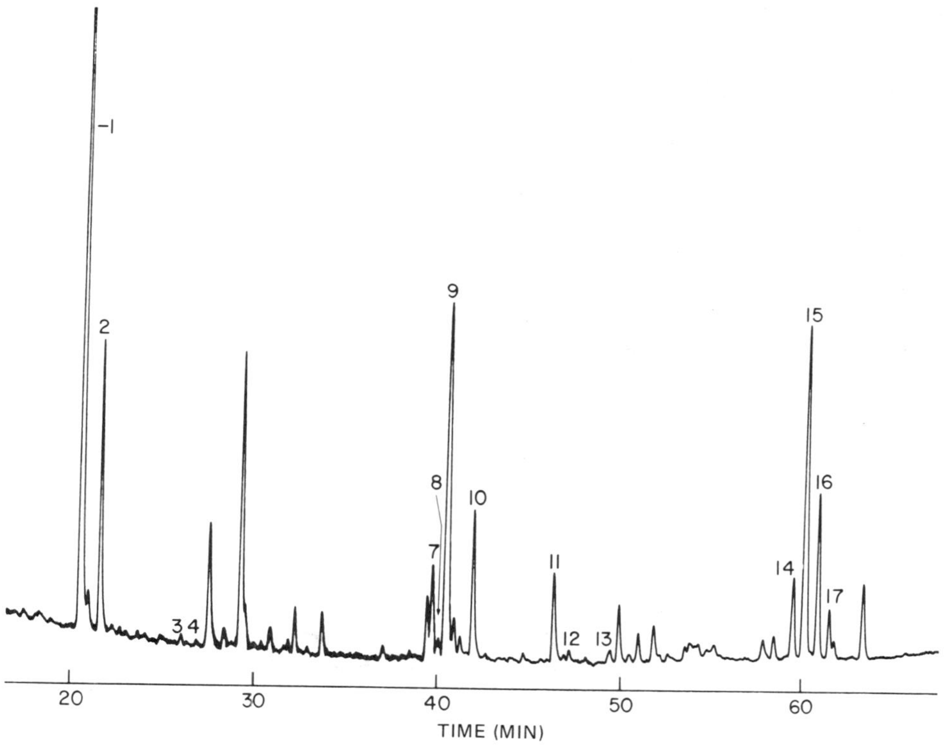

Figure 7.8 Profile of human (male adult) urinary polyols and aldoses separated as in Figure 7.1. The identified components include: 1 = erythritol, 2 = threitol, 3 = ribose, 4 = arabinose, 7 = ribitol, 8 = fucitol, 9 = arabinitol, 10 = xylitol, 11 = mannose, 12 = glucose, 13 = galactose, 14 = *myo*-inositol, 15 = mannitol, 16 = glucitol, and 17 = galactitol. (From Ref. 29.)

myo-inositol, (15) mannitol, (16) glucitol, and (17) galactitol. Erythritol and threitol are always present in a ratio of about 4:1. The average excretion values for these compounds are 110 and 25 μg/mg creatinine, respectively. Pentoses are usually present only in trace amounts. The unidentified compound which elutes around 30 min (MU - 17.04) is not a simple sugar or alditol. Ribitol, arabinitol, and xylitol are always present; the respective average excretion values are 16, 75, and 17 μg/mg creatinine. Arabinitol predominates; a trace of fucitol is usually present. Mannose, glucose, and galactose are often present in small amounts. The four hexitols—*myo*-inositol, mannitol, glucitol, and galactitol—are present in varying ratios. The average excretion values are 81, 93, 25, and 20 μg/mg creatinine, respectively. No significant differences based on sex was observed [29] except for the glucose ratios already discussed.

C. Excretion Profiles of Adults

Figure 7.8 is a typical chromatogram observed for an adult. The major components are those found in samples from juveniles, but with larger amounts of mannose. Adults excrete erythritol and threitol in a ratio of about 6:1. Average excretion values were found to be 63 and 11 μg/mg creatinine, respectively [29]. Pentoses are usually found in only trace amounts. Ribitol, arabinitol, and xylitol are always present. The respective average excretion values are 10, 53, and 22 μg/mg creatinine. A trace of fucitol is usually found. Mannose, glucose, and galactose are usually present in small amounts.

Myo-inositol, mannitol, glucitol, and galactitol were found [29], with average excretion values of 33, 26, 23, and 6 μg/mg creatinine, respectively. No sex-related differences were observed, except for the aformentioned glucose values.

D. Other Polyols and Aldoses

The region of the GC separation corresponding to MU values from 21 to 28 is usually not of interest and is often not included in the profile analysis to conserve time. No major urinary polyol has been observed in this region to date; however, the region is readily accessible. MU values for five heptitols, one octitol, and two nonitols have been reported.

Table 7.4 contains values for the maximum, minimum, and average amounts of urinary polyols and aldoses excreted according to Pfaffenberger et al. [29].

Table 7.4 Excretion Values for Urinary Polyols and Aldoses[a]

	Neonate			Juvenile			Adult		
	Min	Max	Av	Min	Max	Av	Min	Max	Av
Erythritol	50	200	120	58	200	110	33	110	63
Threitol	5	48	21	11	72	25	6	16	11
Ribose	X[b]	T[c]	X	X	5	T	X	T	T
Arabinose	X	T	X	X	T	T	X	T	T
Fucose	X	T	X	X	T	T	X	T	T
Xylose	X	T	X	X	T	T	X	T	T
Ribitol	5	24	14	8	29	16	5	40	10
Fucitol	X	T	T	X	T	T	T	T	T
Arabinitol	51	170	98	47	110	75	31	140	53
Xylitol	9	28	15	9	30	17	6	100	22
Mannose	X	20	4	X	T	T	X	T	T
Glucose	T	80	24	X	9	T	T	61	8
Galactose	X	30	8	X	17	6	X	50	7
myo-Inositol	300	4900	—[d]	11	250	81	9	72	33
Mannitol	X	60	—[e]	T	360	93	X	58	26
Glucitol	X	45	—[e]	10	44	25	T	120	23
Galactitol	7	51	—[e]	7	64	20	T	19	6

[a]In μg/mg creatinine.
[b]X = Not present in the chromatogram.
[c]T = Trace component (>4 μg/mg creatinine).
[d]Not averaged; decreased during neonatal period.
[e]Not averaged for neonatal samples.

VII. SEPARATION OF SUGAR ACIDS AND LACTONES

In 1974, Szafranek et al. [33] reported that attempts to analyze aldonic or uronic acids as the corresponding alditol acetates were successful, but terminal dissymmetry was forfeited, as evidenced by four heuronic acids yielding only two hexitols. Moreover, aldonic and uronic acids led to identical reaction products. Mixtures containing aldonic, hexuronic, and hexaric acids cannot be analyzed as the corresponding alditols because the reduction step destroys family identity [34–36]. The substances do, however, lend themselves to analysis as trimethylsilyl derivatives, the formation of which does not involve prior chemical modification. At present these compounds are not known to be of large biological significance; but this fact may change with thorough investigations of the various glycogen storage diseases. The four chemical families included in the illustrations which follow may be conveniently represented (Figure 7.9) by the linear structures for aldonic acid (I), 2-deoxyaldonic acids (II), hexuronic acids (III) and hexaric acids (IV). Members of familes I and II may exist as linear acids or as 1→4 or 1→5 lactones. Most members are found as 1→4 structures, gluconic acid being the notable exception in which the 1→5 form predominates. Hexuronic acids may form pyrano or furano hemiacetals, 6→2 or 6→3 lactones, as well as several dilactones. Figures 7.10–7.12 illustrate the utility of glass capillary gas chromatographic columns for separating these mixtures. Derivatization is accomplished using dry pyridine, bis(trimethylsilyl)-trifluoroacetamide, and trimethylchlorosilane, as described in the original reference [33]. The absolute configurations of various naturally occurring uronic acids have been reported by Kamerling et al. [7] and are discussed in Sec. IX.

VIII. STUDIES OF CATARACTOUS HUMAN LENS TISSUE

Cataract formation results in a loss of transparency of the lens of the eye. No cure or means of preventing senile cataract, the most common form of the disorder, is known. Most investigations of this circumstance have indicated that cataract formation, both in man [37–40] and in experimental animals [41–49], involves the accumulation of one of several polyols within the tissue of the lens. The osmotic balance of the lens must be carefully maintained for proper vision. Although amino acids are mainly responsible for maintaining osmotic balance [46], potassium and chloride ions make osmotic contributions [39]. Free polyols

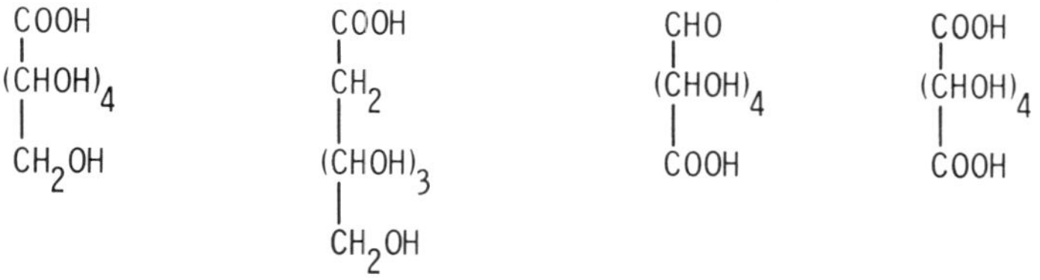

I II III IV

Figure 7.9 Four chemical families of sugar acids represented by linear structures for: aldonic acids (I), 2-deoxyaldonic acids (II), hexuronic acids (III), and hexaric acids (IV). See text for discussion of other possible forms of these compounds.

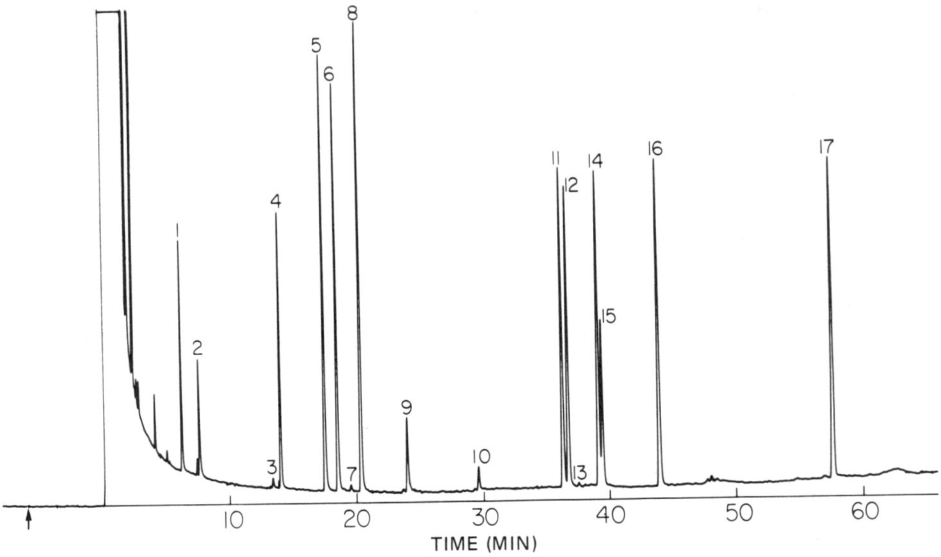

Figure 7.10 Separation of a mixture of TMS derivatives of aldonic lactones and acids on an SE-30 capillary column temperature-programmed 1°/min from 165°: 1,4 = L-threonic; 2,3 = D-erythronic; 5,7 = D-arabonic; 6 = L-xylonic; 8 = D-ribonic; 9,10 = D-lyxonic; 11,13 = D-gluconic; 12 = galactonic; 14 = D-gulonic; 15 = D-talonic; 16 = L-mannonic; 17 = D-*glycero*-D-*gulo*-heptonic. (From Ref. 33.)

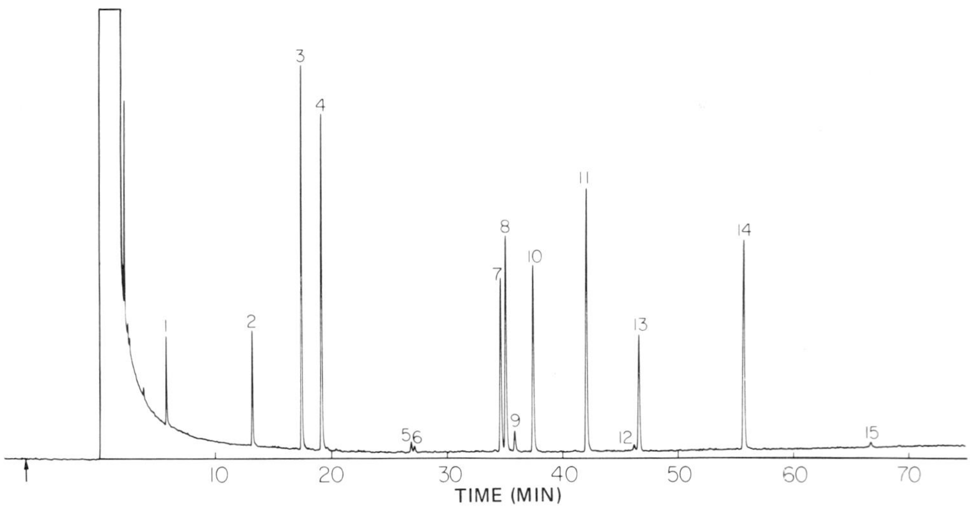

Figure 7.11 Separation of an equilibrium mixture of TMS derivatives of the aldonic lactones and acids included in Figure 7.10 under identical analysis conditions: 1,2 = L-threonic; 3,5 = L-xylonic; 4,6 = D-ribonic; 7,9,13 = D-gluconic; 8,12 = D-galactonic; 10 = D-gulonic; 11 = L-mannonic; 14,15 = D-*glycero*-D-*gulo*-heptonic. (From Ref. 33.)

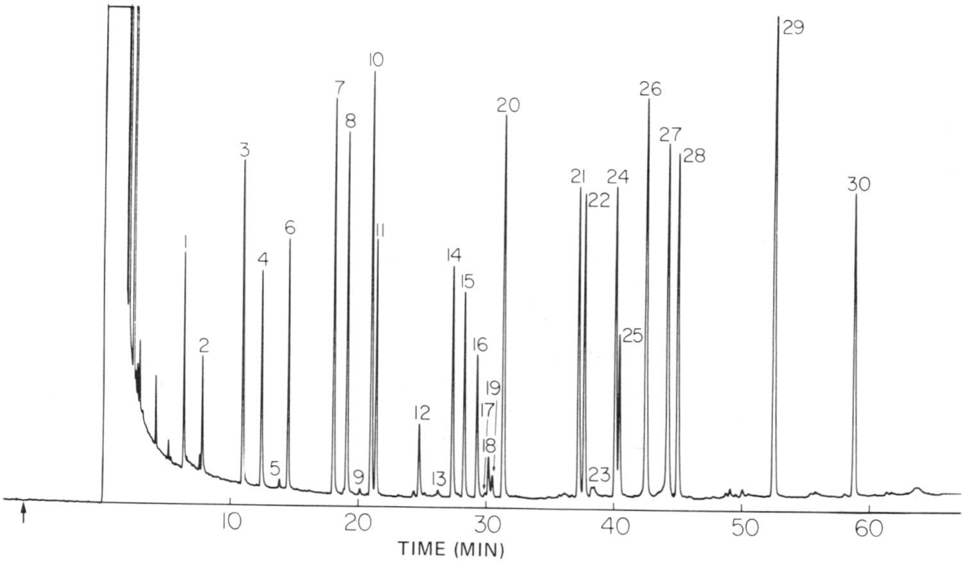

Figure 7.12 Separation of TMS derivatives of aldonic, deoxyaldonic, hexuronic and hexaric lactones and acids as in Figure 7.1. The 30 components include: 1,6 = L-threonic; 2,5 = D-erythronic; 3 = 2-deoxy-D-ribonic; 4 = 2,6-dideoxy-D-allonic; 7,9 = D-arabonic; 8 = L-xylonic; 10 = D-ribonic; 11,13 = 6-deoxy-L-galactonic; 12,19 = D-lyxonic; 14 = 3-deoxy-D-gluconic; 15 = 2-deoxy-D-galactonic; 16,17 = 2-deoxy-D-gluconic; 18,20 = D-glucuronic; 21,23 = D-gluconic; 22 = D-galactonic; 24 = D-gulonic; 25 = D-talonic; 26 = D-glucaric; 27 = D-galacturonic; 28 = L-mannonic; 29 = D-galactaric; and 30 = D-*glycero*-D-*gulo*-heptonic. (From Ref. 33.)

and aldoses also affect the osmotic balance. Accumulation of polyols within the lens is believed to modify the usual osmotic banalce [46] and to initiate loss of lens transparency.

When a sugar alcohol such as *myo*-inositol accumulates in the lens, the resulting hpertonicity is corrected by an influx of water [39]. The osmotic changes which occur initially from polyol accumulation do not seriously alter the state of viability of the lens, and in the early stages of sugar cataract formation the process is reversible. If normal osmotic balance is not restored, lens opacity develops. At this stage there is a gradual breakdown of the permeability barriers, and the electrolytes and polyols, including *myo*-inositol [50,51], become freely permeable and only the larger proteins are retained. After this stage becomes fully established, quantities of free sugar alcohol found in the cataract are insignificant, and free *myo*-inositol levels are significantly altered from normal. This agrees with the wide range of values obtained experimentally for *myo*-inositol in the study by Pfaffenberger et al. [9].

A diabetic is about five times more likely to have an initial cataract operation than is a nondiabetic of the same age and sex [52]; the polyol in these instances is believed to be glucitol.

A. Preparation of Analytical Sample From Lens Tissue

Adherent tissue (zonular ligament, ciliary muscle) is removed from surgically extracted human lenses. The intact lens is quickly weighed and frozen under an atmosphere of nitrogen. Lyophilization is effected over a 48-hr period at 0.5 torr using a Virtis Model 10-300CF lyophilizer. The dry lens is weighed (51–88 mg), pulverized to a homogeneous powder using a glass rod, and stored under nitrogen under deep-freeze conditions until analyzed.

Analytical samples are prepared by adding 10 mg of powdered lens, 10 mg of hydroxylamine hydrochloride, 0.0315 mg of glyceryl-1-decyl ether (internal reference compound, from a solution of 6.30 mg per 10.0 ml methanol) dissolved in 50 μl of methanol, 23 mg of sodium acetate, and 1 ml of methanol. The mixture is heated at 60° for 2 hr, evaporated to dryness using a nitrogen stream, and treated with 1 ml of acetic anhydride. After this mixture has been heated for 2 hr at 100°, it is allowed to stand overnight in a desiccator for complete conversion of the carbohydrates, which react more slowly in tissue preparations than they do as pure reference compounds [53]. Precipitated protein is removed by centrifugation. GC samples are removed from the clear supernatant solution and coinjected with *n*-alkenes when methylene unit (MU) [16,17] values are being determined.

B. Preparation of Analytical Sample from Urine

Urine samples from patients are collected over a 24-hr period just prior to cataract extraction. An analytical sample is prepared from a 20-ml aliquot of urine following exactly the previously described procedure.

C. Representative Lens Profile

By far the largest free polyol component of human lens tissue is *myo*-inositol [9]. This is evident in Figure 7.13. Amounts of this component range (in μg/mg lyophilized lens) from 2–19 for noncataractous lenses, 3–23 for uncomplicated senile cataractous lenses, and 7–29 for senile cataractous lenses from patients with diabetes [9].

All twenty lenses in the study by Pfaffenberger et al. [9] contained a trace of glucose, and all but two of the senile cataractous lenses from patients with diabetes contained at least a trace of glucitol.

All except two lenses (noncataractous) contained a trace of an unidentified compound (labeled 1 in Figure 7.13) with MU = 19.79, and all but four lenses contained an unidentified compound (labeled 2) with MU = 20,95. The first unknown was not one of the simple aldohexoses, which elute within an MU range of 19.02–19.56 under the analytical conditions employed [29]. The second unknown was not mannitol [1], galactitol [29], *scyllo*-inositol [54], or 1,5-anhydroglucitol [55].

D. Urinary Profile of a Controlled Diabetic

Blood-sugar levels fluctuate in diabetes, and the control achieved by therapy may vary over both short and long periods. Caird et al. [56] reported that control of diabetes is often bleow average in effectiveness in diabetics who require cataract extraction. He and his co-workers [57] later demonstrated a direct relationship between degree of control and diabetes and frequency of lens opacity in men of middle age.

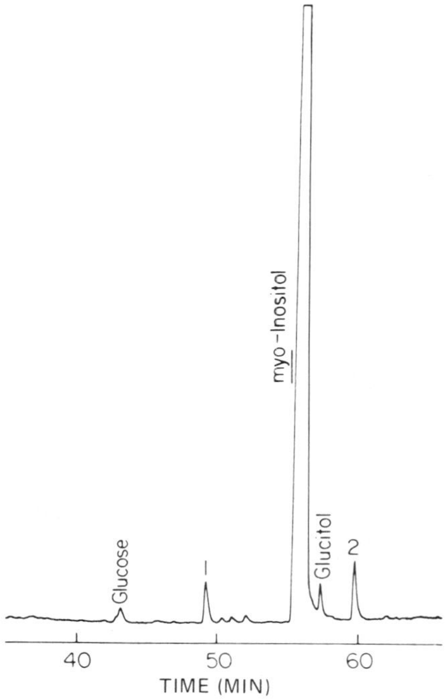

Figure 7.13 Profile of the free polyols and aldoses found in human cataractous lens tissue. Separation of the components as acetyl and Wohl derivatives as carried out on a 60-m SE-30 open-tubular glass capillary column, temperature programmed 1°/min from 150°. The largest compound is *myo*-inositol; for more information consult the text. (From Ref. 9.)

That Pfaffenberger et al. [9] found only low concentrations of glucitol in the cataractous lenses they examined was explained by the facts that diabetes is usually well controlled before surgery, and the cataractous lens (advanced) is abnormally permeable and loses glucitol to the aqueous humor by diffusion [39].

The original reference [9] contains excretion values (in mg/g creatinine) for eleven polyols and aldoses found in the urine of five of the subjects whose lenses were analyzed. The values are typical of the adult human over 50 years of age. No obvious differences were reported between noncataractous subjects and patients having senile cataracts [9].

Figure 7.14 is a urinary metabolic profile for a controlled diabetic patient with senile cataract. Excretion values for glucose, *myo*-inositol and glucitol in this sample were 56, 6, and 4 mg/g creatinine, respectively.

Figure 7.15 is a composite which includes both urinary metabolic profiles of a controlled diabetic mother and her newborn son on the day of delivery. Virtually every aldose and polyol found in the mother's urine is also present in the son's. A relatively large amount of an unknown compound appears in the infant's profile between components 9 and 10 which correspond to arabinitol and xylitol, respectively.

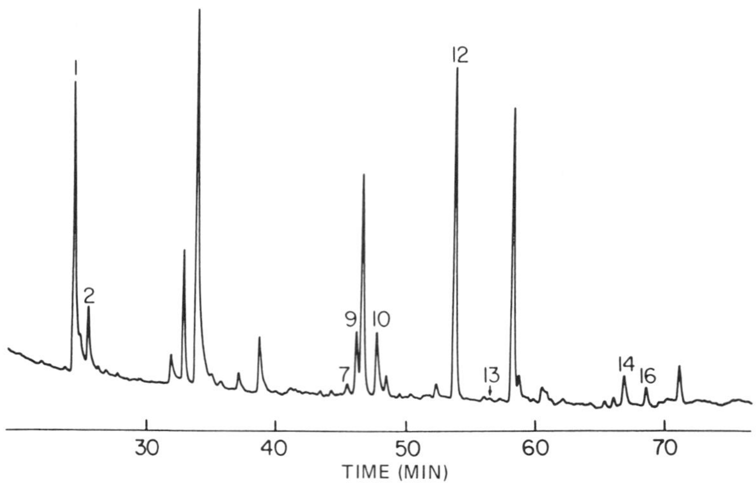

Figure 7.14 Metabolic profile of the urinary polyols and aldoses excreted by a controlled diabetic patient from whom the senile cataractous lens analyzed in Figure 7.13 was obtained. Experimental conditions are as in Figure 7.13. The identified components include: 1 = erythritol, 2 = threitol, 7 = ribitol, 9 = arabinitol, 10 = xylitol, 12 = glucose, 13 = galactose, 14 = *myo*-inositol, and 16 = glucitol. (From Ref. 9.)

IX. SEPARATING ENANTIOMERS

On the work which has been discussed thus far, absolute configurations of the various components have been given only for mixtures of reference standards. Recently, Kamerling and co-workers [6,7] have demonstrated the separation of enantiomers on SE-30 capillary columns by first converting the enantiomers into diastereomers using chiral reagents. The derivatives of choice are the trimethylsilylated (-)-2-butyl αβ-glycosides. The multiple-peak chromatographic patterns obtained are useful for the assignment of absolute configurations. (-)-2-Butyl αβ-glycosides can be prepared from either the monosaccharides or their corresponding methyl glycosides; thus hydrolysis as well as methanolysis can be applied for the analysis of oligo- or polysaccharides. Early work [6] demonstrated the applicability of the method for the (-)-2-butyl αβ-glycosides of D- and L-arabinose, L-rhamnose, D- and L-fucose, D-ribose, D- and L-xylose, D- and L-mannose, and D- and L-glucose. The pentoses and 6-deoxyhexoses were analyzed isothermally at 150° and the hexoses at 175°. Figure 7.16 shows the separation of a mixture of authentic samples of D- and L-fucose, D- and L-xylose, D- and L-mannose, D- and L-galactose, and D- and L-glucose analyzed as the corresponding silylated (-)-2-butyl αβ-glycosides. For tis analysis an SE-30 column was temperature programmed at 1°/min from 135°. The absolute configurations of the monosaccharides were easily determined.

The scope of the method was then extended by applying it to 2-acetamido-2-deoxy sugars and uronic acids and by utilizing it to analyze several biopolymers [7]. Polysaccharide from the cell-wall lipopolysaccharide of *Salmonella typhimurium* LT-2 was methanolized and the products butanolized and trimethylsilylated. The resulting chromatogram (Figure 7.17) contained the trimethylsilyl

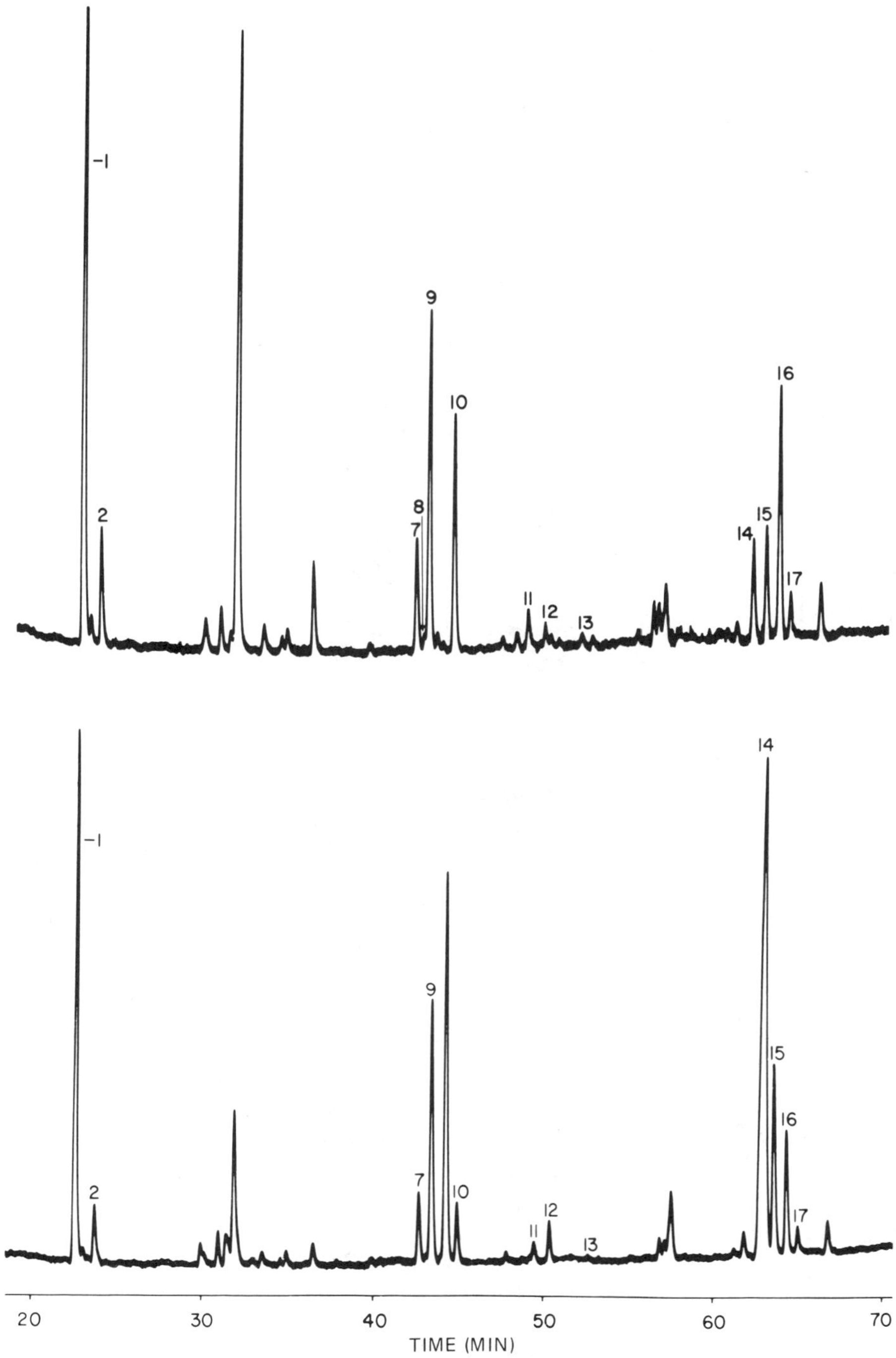

Figure 7.15 Metabolic profiles of urinary polyols and aldoses of a mother (a) and her newborn male infant (b) on the day of delivery. Experimental conditions are as in Figure 7.1. Each identified component was in both samples: 1 = erythritol, 2 = threitol, 7 = ribitol, 9 = arabinitol, 10 = xylitol, 11 = mannose, 12 = glucose, 13 = galactose, 14 = *myo*-inositol, 15 = mannitol, 16 = glucitol, and 17 = galactitol.

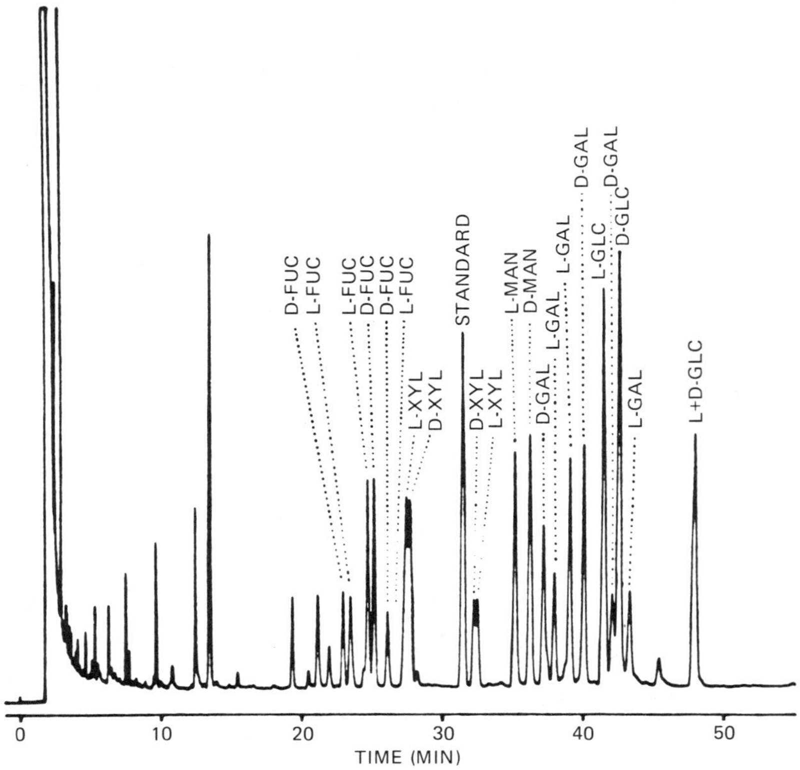

Figure 7.16 Separation of the trimethylsilylated (-)-2-butyl αβ-glycosides of D- and L-fucose, D- and L-xylose, D- and L-mannose, D- and L-galactose, and D- and L-glucose. The SE-30 glass capillary column was temperature pro-trammed 1°/min from 135°. (From Ref. 6.)

(-)-2-butyl αβ-glycosides of L-rhamnose, D-mannose, D-galactose, and D-glucose, constituents which had been previously reported [58]. When the method was applied to a capsular polysaccharide which had been isolated from *Klebsiella*, type 57, the chromatogram confirmed a literature report [59] that all of the sugar components—mannose, galactose, and galacturonic acid—belonged to the D series. α1-Acid glycoprotein was methanolized, re-N-acetylated, butanolized, re-N-acetylated, and trimethylsilylated to yield the (-)-2-butyl αβ-glycosides of L-fucose, D-mannose, D-galactose, and 2-acetamido-2-deoxy-D-glucose. The D configuration of galactose and 2-acetamido-2-deoxygalactose was confirmed in mucus glycoproteins. (It was not possible to separate the D and L enantiomers of *N*-acetylneuraminic acid 2-butyl ester under the conditions employed.)

Finally, the first example of a biopolymer which both enantiomeric forms of mannose (DL ratio, 9:1) and 6-O-methyl-mannose (DL ratio, 1:2) occur was reported [7] to be present in a complex, methylated, acidic polysaccharide isolated from the cocoliths of *Emiliania huxleyi* (Lohmann) Kamptner. The various constituent sugars included: L-galactose, D-glucose, D-mannose, L-mannose, L-rhamnose, L-arabinose, D-ribose, D-xylose, 6-*O*-methyl-D-mannose, 6-*O*-methyl-L-mannose, 2,3-di-*O*-methyl-L-rhamnose, 3-*O*-methyl-D-xylose, and D-galacturonic acid.

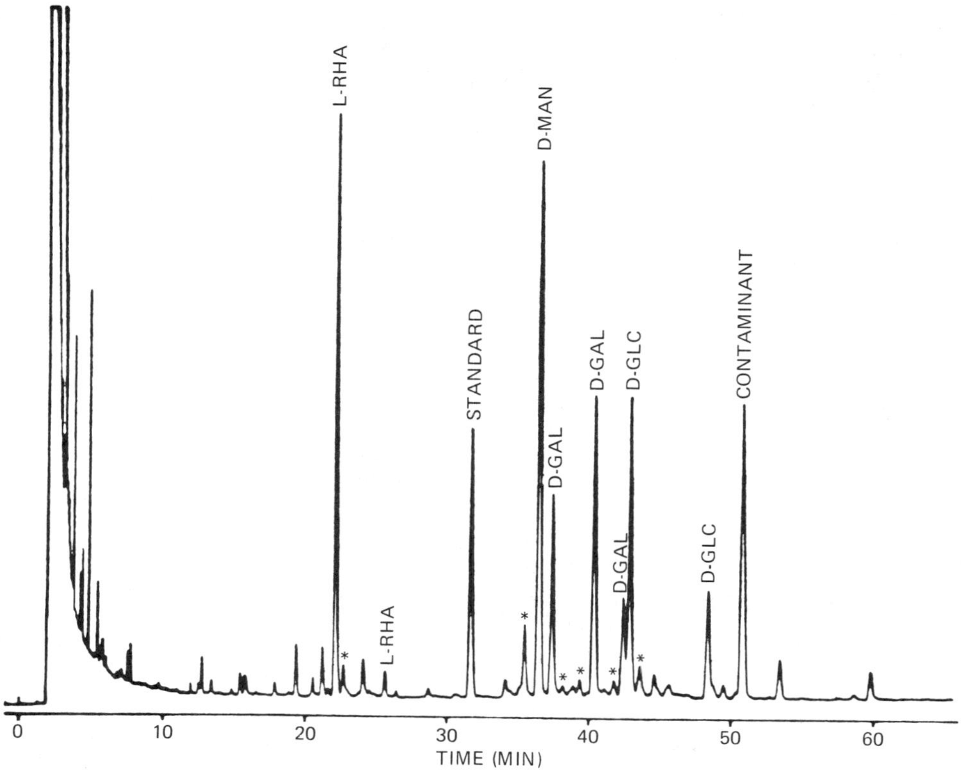

Figure 7.17 Separation of the trimethylsilylated (-)-2-butyl αβ-glycosides of the monosaccharides which occur in the cell-wall lipopolysaccharide of *Salmonella typhimurium* LT-2. The components include L-rhamnose, D-mannose, D-galactose, and D-glucose. (From Ref. 6.)

X. CONCLUDING REMARKS

The use of glass and/or fused silica capillary gas chromatography in clinical laboratories is still in its infancy. That clinically based analyses have lagged behind other areas of application such as petroleum product studies and water and air pollution analyses is probably related to the situation that almost all clinical chemistry laboratories are directed by pathologists—individuals who generally receive little (if any) exposure to gas chromatographic techniques during their formal didactic training. Moreover, most gas chromatographers obtain their degrees through analytical divisions of chemistry departments and study curricula quite different from those of pathologists. Presentations of bioanalytical studies at some of the more medically oriented symposia and inclusion of a biomedical applications section in the program of such events as the annual meeting, *Advances in Chromatography* [60-68], have served to bring these two groups of scientists with diversely differing backgrounds into closer proximity and somewhat more mutual understanding.

Although the technology and methodology for obtaining metabolic profiles from many analytical samples obtained from biological matrices have been worked out, more applications of this approach seem likely, particularly in studies related to chronic diseases and subtle metabolic disorders. The powerful separating characteristics of capillary columns have brought chromatographers to the point of almost being able to claim ability to separate any mixture of volatilizable compounds into each of its component parts. For the few instances in which compound coelution occurs, a second column of slightly different polarity usually can bring about any other separation desired. And when capillary columns are used in combination with a mass spectrometer and a totally dedicated data system, one of the most sophisticated levels of analytical capability is attained. Over the next decade, application of such analytical systems to studies of sugars, sugar alcohols, and related compounds of clinical interest will expand, and the use of capillary columns will have reached its destiny within the clinical chemistry laboratory.

ACKNOWLEDGMENTS

Sincere appreciation is expressed to many colleagues in the fields of glass capillary gas chromatography and clinical medicine. Those who have been of particular assistance to me include: C. J. W. Brooks, Ph.D., A. L. German, Ph.D., E. C. Horning, Ph.D., M. G. Horning, Ph.D., W. G. Jennings, Ph.D., S. N. Lin, Ph.D., B. S. Middleditch, Ph.D., J. Szafranek, Ph.D., J. P. Thenot, Ph.D., J. Jaeger, Ph.D., and A. Zlatkis, Ph.D.

REFERENCES

1. J. Szafranek, C. D. Pfaffenberger, and E. C. Horning, *Anal. Lett. 6*: 479 (1973).
2. A. L. German and E. C. Horning, *Anal. Lett.* *5*:619 (1972).
3. A. L. German and E. C. Horning, *J. Chromatogr. Sci. 11*:76 (1973).
4. A. L. German, C. D. Pfaffenberger, J-P. Thenot, M. G. Horning, and E. C. Horning, *Anal. Chem. 45*:930 (1973).
5. E. C. Horning, M. G. Horning, J. Szafranek, P. van Hout, A. L. German, J-P. Thenot, and C. D. Pfaffenberger, *J. Chromatogr. 91*: 367 (1974).
6. G. J. Gerwig, J. P. Kamerling, and J. F. G. Vliegenthart, *Carbohydr. Res. 62*:000 (1978).
7. Gerrit J. Gerwig, Johannis P. Kamerling, and Johannes F. G. Vliegenthart, *Carbohydr. Res., 77*:000 (1979).
8. Von Mering and Minkowski, in *Duncan's Diseases of Metabolism,* (P. K. Bondy, ed.), W. B. Saunders, Philadelphia, 1969.
9. C. D. Pfaffenberger, J. Szafranek, and E. C. Horning, *J. Chromatogr. 126*:535 (1976).
10. E. Jellum, P. Storeseth, J. Alexander, P. Helland, O. Stokke, and E. Teig, *J. Chromatogr. 126*:487 (1976).
11. R. Storseth, O. Stokke, and E. Jellum, *J. Chromatogr. 145*:351 (1978).
12. C. Jakobs, E. Solem, J. Ek., K. Halvorsen, and E. Jellum, *J. Chromatogr. 143*:31 (1977).
13. M. L. McConnell, S. Rhodes, U. Watson, and M. Novotny, *J. Chromatogr. 102*:495 (1979).
14. W. A. Koenig and W. Guenther, *J. High Resol. Chromatogr. Chromatogr. Commun. 2*:378 (1979).
15. D. J. Leblanc and A. J. S. Ball, *Anal. Biochem. 84*:574 (1978).
16. E. C. Horning, N. G. Horning, N. Ikekawa, E. Chambaz, P. I. Jaakonmaki, and C. J. W. Brooks, *J. Gas Chromatogr.* 5283 (1967).

17. W. J. A. VandenHeuvel, W. L. Gardiner, and E. C. Horning, *J. Chromatogr. 19*:263 (1965).
18. C. T. Bishop, *Adv. Carbohyd. Chem. 19*:95 (1969).
19. W. L. Wolfrom and A. Thompson, *J. Am. Chem. Soc. 53*:622 (1931).
20. V. Deulofeu, P. Cattaneo, and G. Mendivelza, *J. Chem. Soc. 147*:31 (1934).
21. B. A. Dmitriev, L. V. Bachinowsky, O. S. Chizoo, B. M. Zolotarev, and N. K. Kochelkov, *Carbohyd. Res. 19*:432 (1971).
22. K. Tesarik, *J. Chromatogr. 65*:295 (1972).
23. I. M. Morrison and M. B. Perry, *Can. J. Biochem. 44*:1115 (1966).
24. C. C. Sweeley, R. Bentley, M. Makita, and W. W. Wells, *J. Am. Chem. Soc. 85*:2497 (1963).
25. G. Eklund, B. Josefsson, and C. Roos, *J. Chromatogr. 142*:575 (1977).
26. R. J. Kutsky, in *Handbook of Vitamins and Hormones*, Van Nostrand and Reinbold Publ. Co., New York, 1973.
27. C. D. Pfaffenberger, in *Glass Capillary Gas Chromatography—The Applications* (W. G. Jennings, ed.), Marcel Dekker, Inc., New York, 1980.
28. E. C. Horning and M. G. Horning, *Methods Med. Res. 12*:369 (1970).
29. C. D. Pfaffenberger, J. Szafranek, M. G. Horning and E. C. Horning, *Anal Biochem. 63*:501 (1975).
30. F. DelGereco and L. Scapellato, *Diabetes 2*:457 (1953).
31. K. J. Isselbacher, in *The Metabolic Basis of Inherited Disease*, McGraw-Hill Publ. Co., New York, 1966.
32. H. H. Hiatt, in *The Metabolic Basis of Inherited Disease*, McGraw-Hill Publ. Col, New York, 1966.
33. J. Szafranek, C. D. Pfaffenberger, and E. C. Horning, *J. Chromatogr. 88*:149 (1974).
34. T. M. Jones and P. Albersheim, *Plant Physiol. 49*:926 (1972).
35. J. D. Blake and G. N. Richards, *Carbohyd. Res. 8*:275 (1968).
36. J. D. Blake and G. N. Richards, *Carbohyd. Res. 14*:375 (1970).
37. A. Pirie and R. VanHeyninger, *Exp. Eye Res. 3*:124 (1964).
38. R. Gitzelmann, H-C. Curtins, and I. Schneller, *Exp. Eye Res. 6*:1 (1967).
40. D. J. Heaf and D. J. Galton, *Clin. Chim. Acta 63*:41 (1975).
41. R. VanHeyninger, *Nature (London) 184*:194 (1959).
42. E. Collier, *Invest. Ophthalmol. 7*:118 (1968).
43. J. H. Kinoshita, D. Dvornik, M. Kraml, and K. H. Gabby, *Biochim. Biophys. Acta 158*:472 (1968).
44. J. H. Kinoshita, L. O. Merola, and B. Tung, *Exp. Eye Res. 7*:80 (1968).
45. L. T. Chylak and J. H. Kinoshita, *Invest. Ophthalmol. 8*:401 (1969).
46. J. H. Kinoshita, W. G. Barber, L. O. Merola, and B. Tung, *Invest. Ophthalmol. 8*:625 (1969).
47. D. Dvornik, N. Simard-Duquesne, and M. Kraml, *Science 182*:1146 (1973).
48. H. Obazawa, L. O. Merola, and J. H. Kinoshita, *Invest. Ophthalmol. 13*:204 (1974).
49. S. D. Varma and J. H. Kinoshita, *Biochim. Biophys. Acta. 338*:632 (1974).
50. R. M. Broekhuyse, *Biochim. Biophys. Acta 163*:269 (1968).
51. M. A. Steward, M. M. Kurien, W. R. Sherman, and E. V. Cotlier, *J. Neurochem. 15*:941 (1968).

52. A. Pirie, *Invest. Ophthalmol. 4*:629 (1965).
53. W. R. Sherman, M. A. Steward, M. M. Kurien, and S. L. Goodwin, *Biochim. Biophys. Acta. 158*:197 (1968).
54. W. R. Sherman, M. A. Steward, P. C. Simpson, and S. L. Goodwin, *Biochemistry 7*:819 (1968).
55. E. Pitkaenen, *Clin. Chim. Acta. 48*:159 (1973).
56. F. I. Caird, M. Hutchinson, and A. Pirie, *Br. Med. J. 2*:665 (1964).
57. F. I. Caird, A. F. Burditt, and G. J. Draper, *Diabetes 17*:121 (1968).
58. J. P. Kamerling, G. J. Gerwig, J. F. G. Vliegenthart, and J. R. Clamp, *Biochem. J. 151*:491 (1975).
59. J. P. Kamerling, B. Lindberg, J. Loenngren, and W. Nimmick, *Acta. Chem. Scand. 29*:593 (1975).
60. *Advances in Chromatography, 1973* (A. Zlatkis, ed.), University of Houston Publishers, Houston, 1973.
61. *Advances in Chromatography, 1974* (A. Zlatkis and L. S. Ettre, eds), Elsevier Publ. Co., New York, 1974.
62. *Advances in Chromatography, 1975* (A. Zlatkis, E. Bayer, L. S. Ettre, and I. Halász, eds.), Elsevier Publ. Co., New York, 1975.
63. *Advances In Chromatography, 1976* (A. Zlatkis, H. Shanfield, and L. S. Ettre, eds.), Elsevier Publ. Co., New York, 1976,
64. *Advances in Chromatography, 1977* (A. Zlatkis, L. S. Ettre, and G. Dijkstra, eds.), Elsevier Publ. Co., New York, 1977.
65. *Advances in Chromatography, 1978* (A. Zlatkis, L. S. Ettre, and H. Shanfield, eds.), Elsevier Publ. Co., New York, 1978.
66. *Advances in Chromatography, 1979* (A. Zlatkis, L. S. Ettre, and E. Kováts, eds.), Elsevier Publ. Co., New York, 1979.
67. *Advances in Chromatography, 1980* (A. Zlatkis, C. F. Poole, and L. S. Ettre, eds.), Elsevier Publ. Co., New York, 1980.
68. *Advances in Chromatography, 1981* (A. Zlatkis, R. Segura, and L. S. Ettre, eds.), Elsevier Publ. Co., New York, 1981.

8

Analysis of Steroids

Cedric H. L. Shackleton / *Children's Hospital Medical Center of Northern California, Oakland, California, and The University of California, San Francisco, California*

I. INTRODUCTION

Steroid hormones are secretion products of the adrenal glands and gonads and, during pregnancy, also of the feto-placental unit. Biologically active metabolites of vitamin D_3, a hormone produced in the skin, can also be considered steroid hormones. Originally, according to their *biological action*, human steroids were divided into five classes: androgens, esterogens, progestagens, corticosteroids, and cholecalciferols. Each of these classes have unique structural features (see Structure Group 1). *Androgens* are C_{19} steroids with a 17β-hydroxyl and 3-oxo group (testosterone and dihydrotestosterone); the *estrogens* are C_{18} steroids (e.g., estradiol) having an aromatic A-ring; the major *progestagen* is progesterone, a 21-deoxy C_{21} steroid with an A-ring containing a 3-oxo-4-ene group; the *corticosteroid hormones* are C_{21} steroids generally with 11β- and 21-hydroxyl groups (e.g., cortisol and aldosterone); and the *cholecalciferols* are hydroxylated metabolites of ring-B secosteroids with 27 carbon atoms (1,25-dihydroxyvitamin D_3).

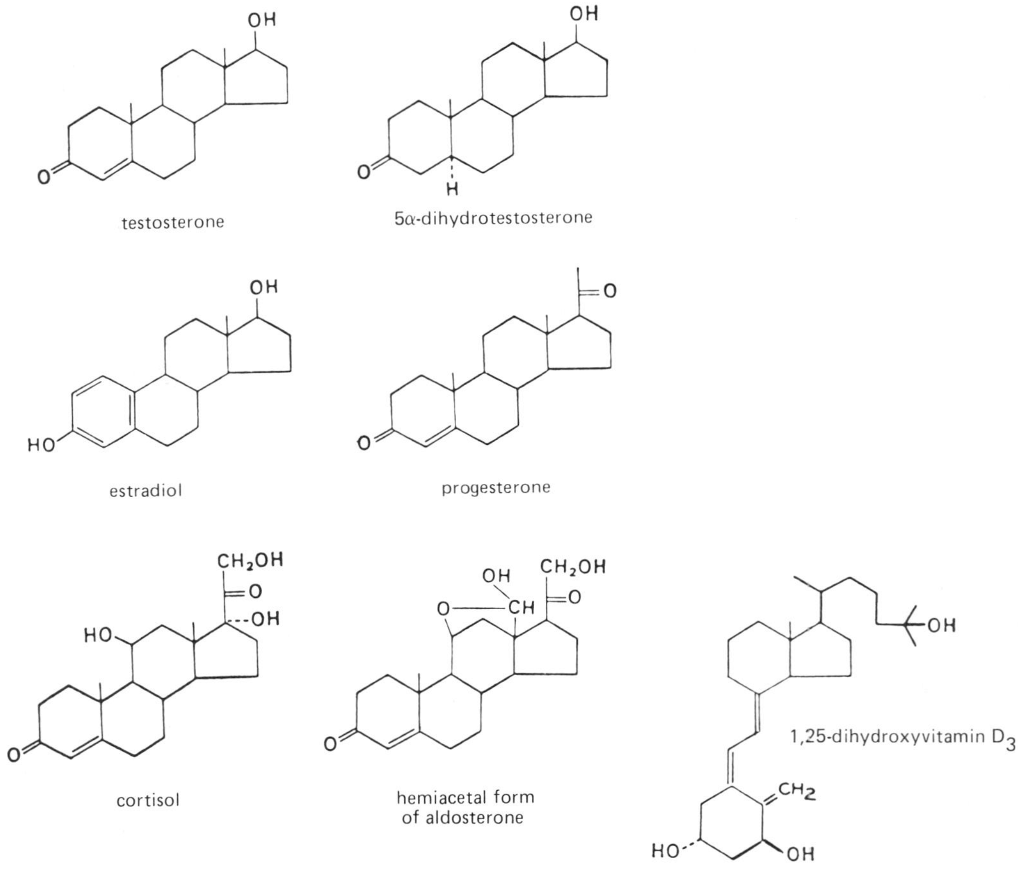

(Structure Group 1)

The generic names are often used to designate *structure* rather than a biological activity. These structural similarities may bear little or no relationship to hormonal action or origin of the compounds. For example, on structural grounds estriol is classified as an estrogen, although it has little estrogenic action and during pregnancy is principally a terminal metabolite of fetal adrenal dehydroepiandrosterone sulfate (DHA-S)* synthesis. Similarly, the urinary

*A list of steroid abbreviations appears in the glossary, just prior to References.

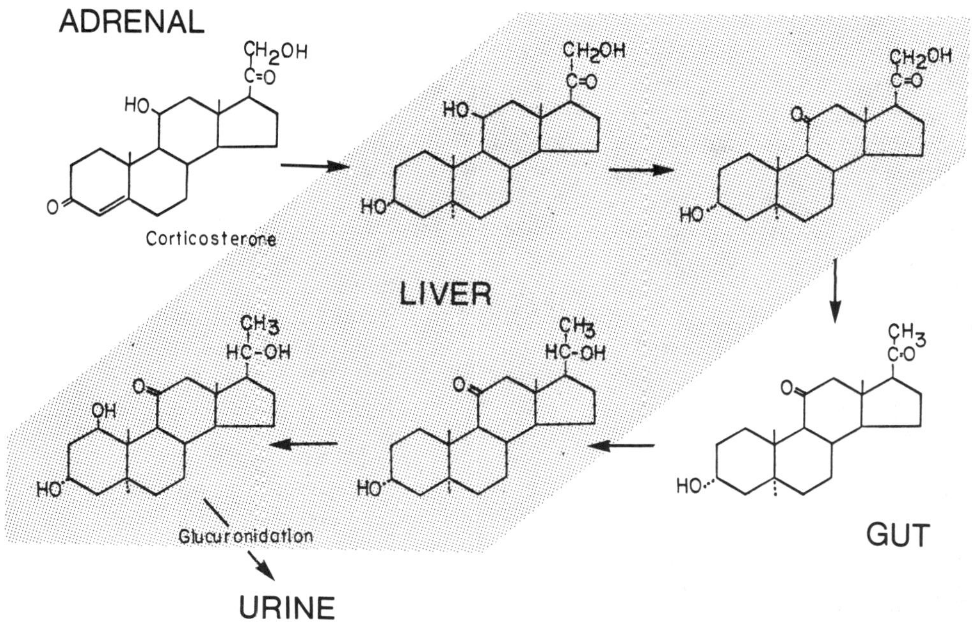

Figure 8.1 A pathway for the disposal of corticosterone in man. An example of the complexity of terminal metabolism.

C_{19} compounds 11β-hydroxyetiocholanolone are frequently referred to as androgens, although to a considerable extent they are metabolites of adrenal cortisol. Thus, although there are unique structural features present in the various forms of hormonal steroids and metabolites, these features may also be present in metabolites of steroids of different origin.

Steroid profiles of serum are relatively simple, and in many cases determination of hormonal status is fairly straightforward using appropriate liquid chromatographic and radioimmunoassay techniques. In contrast, a urinary steroid profile is exceedingly complex since the secreted hormones will have undergone extensive metabolism by a variety of routes prior to excretion, and frequently 10 or more compounds will have to be analyzed solely to cover the major metabolites of a single steroid. An example of this diversity is given in Figure 8.1.

Corticosterone is secreted by the adrenal gland and can be readily analyzed in serum where it represents the second most important glucocorticoid. Hepatic metabolism gives predominantly the 3α,5α metabolite, which in turn is oxidized at the 11-position by 11β-hydroxysteroid dehydrogenase to allotetrahydroCompound A. AllotetrahydroCompound A is partially excreted in the bile and is subject to 21-dehydroxylation by enterobacteria; the resulting "11-oxopregnanolone" is reabsorbed in the gut and is reduced to an "11-oxopregnanediol" in the liver; at some point a 1β-hydroxyl group could have been introduced, and the final excretion product would be a glucuronide of 1β-hydroxy-11-oxopregnanediol. This steroid has been shown to be an important corticosteroid metabolite in patients with 17α-hydroxylase deficiency [1]. Although in this example, six sites of the original molecule had been subject to metabolic transformation, it is still possible to have confidence that the terminal metabolite was from a single precursor produced by a single gland. There are many examples, however, of urinary metabolites which have more than one steroid precursor and more than one site of origin. These present a particularly difficult problem for multicomponent steroid analysis. The most well-known example of this is the excretion of the C_{19} steroids androsterone and etiocholanolone, which can be metabolites of gonadal hormones (testosterone), metabolites of adrenal androgens such as dehydroepiandrosterone (DHA), or metabolites of cortical steroids (Substance S).

This introduction has been designed to impress on the reader the nature and number of components present in urine which have to be measured to obtain comprehensive information on the hormonal status of the steroid endocrine glands. Gas chromatographic profile analysis on glass capillary columns is the only method currently available for analyzing compounds in such numbers and with such diversity of structure. In combination with mass spectrometry, it is at present the most powerful analytical technique for the separation and identification of compounds in complex mixtures. In addition, when in the *selected-ion-monitoring* mode, capillary GC-MS is an extremely sensitive and specific technique for measuring individual compounds of natural or drug origin circulating in serum.

We owe the original development of steroid profile analysis to Gardiner and Horning [2]. The derivatization techniques and chromatograms they produced are essentially the same that we obtain today although, of course, they used packed columns for analysis. Many of the original studies on the use of glass capillary columns for steroid profile analysis were carried out by Dr. J. A. Völlmin [3], who tragically died in the early days of the technique.

II. INSTRUMENTATION

A. Columns

1. *Preparation*

The use of glass or fused silica columns is essential for the capillary column chromatography of steroids. Through contact with metal surfaces, even derivatized steroids are readily adsorbed or decomposed, a deactivation of the surface of metal tubing presents special problems. In our laboratory we have greatest experience in the use of wall-coated glass columns. Initially, commercially available coated columns were used, but for the last few years we have prepared columns according to published methods.

Formerly, columns used in our laboratory were prepared using barium carbonate crystallization, deactivation with Carbowax, and dynamic or static coating of stationary phase [4]. This method is not ideal since active sites still remain and, the upper temperature limit for sustained operation is relatively low (about 260°), even when "stable" stationary phases are used (e.g., SE-30, OV-1). More recently, Grob et al. [5] have published a modification of their column coating method based on deactivation by HCl leaching and high-temperature silanization. This technique was shown to provide columns with few active sites, and even free acids or bases could be satisfactorily analyzed.

The columns were also stable during sustained operation at temperatures in excess of 300°. We have adopted this technique exclusively, and most of the chromatograms illustrated in this chapter were obtained following analysis on such columns.

We have coated fused silica columns (Hewlett-Packard, Inc.) using similar techniques with omission of the HCl leaching step. These columns have proven to be extremely inert and stable over long periods. Figure 8.2 illustrates the separation of urinary steroids on a fused silica column which has been in continuous use for one year with the upper limit of program always at 330°. Little deterioration in the ability of this column to separate urinary steroids has been noticed even with such continuous, high-temperature operation.

Even greater longevity will probably be achieved with the bonded fused-silica columns which have recently come on the market. Bonding of the

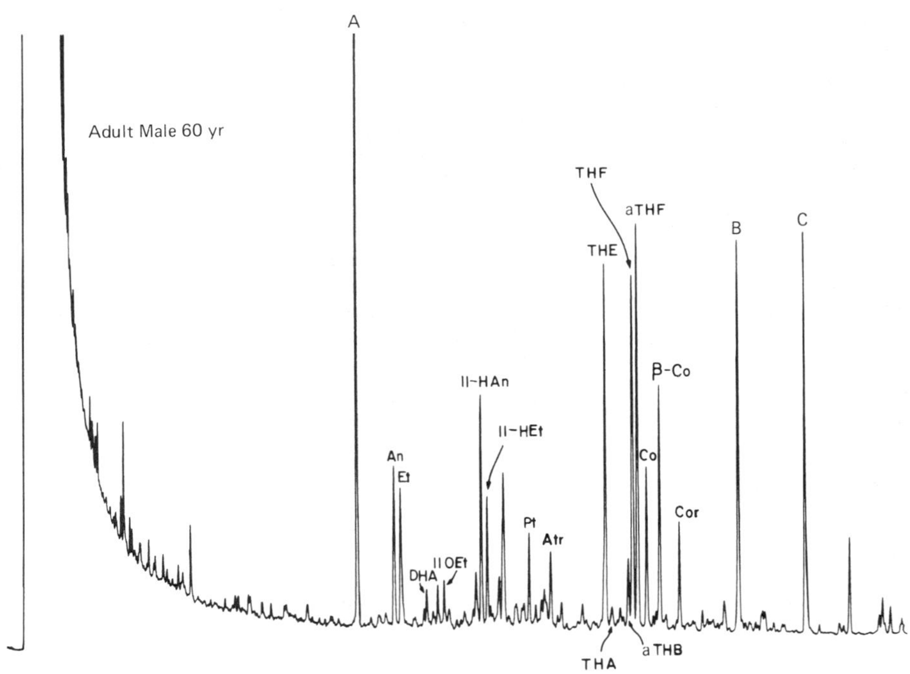

Figure 8.2 Longevity of fused silica OV-1 column. Total urinary steroid profile (methyloxime/trimethylsilyl ether derivative) obtained on an 18-m column after one year of daily use. The temperature was invariably programmed to 330° with each analysis carried out during this period. The peaks marked A, B, and C are internal standards 5α-androstane-3α,17α-diol, stigmasterol, and cholesteryl butyrate, respectively. All profiles illustrated in this chapter which originate from our laboratory contain these internal standards. The following steroid abbreviations are used: An (androsterone), Et (etiocholanolone), 11-HAn (11β-hydroxyandrosterone), 11-HEt (11β-hydroxyetiocholanolone), PD (pregnanediol), PT (pregnanetriol), Atr (5-androstene-3β,16α,17β-triol), THE (tetrahydrocortisone), THA (tetrahydroCompound A), THB (tetrahydrocorticosterone), aTHB (allotetrahydrocorticosterone), THF (tetrahydrocortisol), aTHF (allotetrahydrocortisol), CO (α-cortolone), β-Co (β-cortolone + β-cortol), and Cor (α-cortol). Conditions: 180–330° program, 3°/min, splitless injection.

stationary phase to the column wall results in extremely robust columns, and phase-stripping fron on-column or splitless injection of large solvent volumes can be eliminated. The bonded film is so tough that the columns can even be washed out with solvents.

Although most investigators employing capillary columns for steroid analysis use wall-coated, open-tubular columns, Pfaffenberger and Horning [6] have favored analysis on support-coated columns. Their method for column preparation was described by German and Horning [7]. Basically, the empty coils are silanized and are then coated with a solution of the stationary phase containing micro particles of Silanox 101. This material provides a matrix to assist stabilization of the stationary-phase coating. The steroid separations achieved using this type column do not differ materially from those produced by analysis on the published chromatogram does suggest that the resolution achieved is not as good.

2. *Testing*

The following basic information is needed for evaluating a capillary column: (1) extent of adsorption of hydroxyls, (2) separation efficiency, (3) acid-base behavior, and (4) film thickness. Grob et al. [9] introduced a simple standard mixture to enable all this information to be obtained in a single analysis. The test compounds have been carefully chosen to give reasonably distributed peaks over the whole of a temperature-programmed chromatogram.

The result obtained for analysis of a text mixture on a SE-52 column produced by the silanization method in our laboratory is illustrated in Figure 8.3a. The analysis demonstrated that the trennzahl (TZ, separation number) was 55 and that acceptable responses of the test polar compound (octanol) and of acidic and basic compounds (dimethylphenol, dimethylaniline, respectively) were obtained.

After conditioning, the stability of a column should be assessed by analyzing a sample containing components eluted over a wide temperature range. Extensive bleeding would manifest itself by a significant increase in baseline at high temperature. Figure 8.3b illustrates chromatography of n-alkanes C_{14}–C_{44} on a new fused-silica column prepared in our laboratory. This column shows little evidence of bleeding even at 350°C.

Acceptability of a column for steroid analysis finally depends on its ability to separate key components in the profile. For example, an OV-1 column for urinary steroid analysis should give baseline separation of the triplet allotetrahydrocorticosterone, tetrahydrocortisol, and allotetrahydrocortisol, and the doublet 5-pregnene-3β,17α,20α-triol and tetrahydrocortisone. Resolution of these components is often difficult on inferior columns.

3. *Attachment*

In our original studies we attached the column by heat-shrink PTFE to hand-drawn straight capillaries inserted in injector and detector [8]. By current standards this is not satisfactory because the connections tend to leak at high temperature, and peak tailing is often seen. This is now avoided by direct connection through ferrules to the detector and injector. The column ends must first be straightened by flaming, although this is not necessary with the flexible fused-silica columns now available.

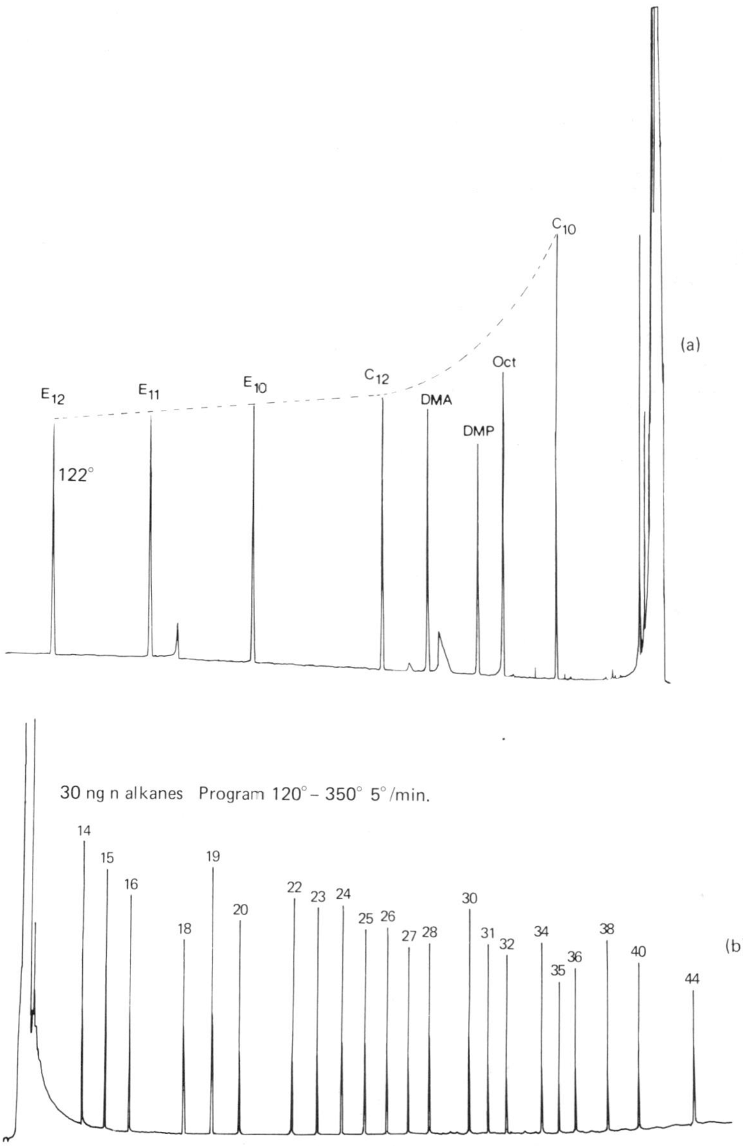

Figure 8.3 Evaluation of column performance. (a) Analysis of a test mixture on a glass SE-52 column prepared by HCl leaching, silanization, and static coating. This test mixture is composed of C_{10} and C_{12} alkanes, fatty acid methyl esters (E_{10}, E_{11}, and E_{12}), dimethylaniline (DMA), dimethyl phenol (DMP), and octanol. The temperature was programmed between 40° and 125° at 2°/min. (From Ref. 112.) (b) Testing high-temperature stability. Analysis of n-alkanes C_{14}–C_{44} on 25-m fused-silica OV-1 column coated in this laboratory by silanization and static coating method. Note the low bleed even at temperatures in excess of 300°.

B. Sampling

There are four solvent injection systems and two solid injection systems in common use: (1) on-column injection, (2) split injection, (3) splitless injection, (4) direct injection, and (5) individual solid injection (dropping needle) and automatic solid injection.

1. *On-Column Injection*

This is the most reproducible technique for quantitative analysis [9] and gives least discrimination of high-mass components eluted at high temperature. In this septumless form of injection the special long and narrow syringe needle is placed *inside* the coated capillary column and the sample is deposited in a small area. After elution of the solvent, the column temperature is raised, and individual components are volatilized and separated. This technique has not been widely used in analysis of steroids in biological materials, since there is a continual deposition of involatile materials present in the extracts onto the initial portion of the column, and these must be removed at intervals to maintain performance.

2. *Split Injection*

This is quite satisfactory in urinary steroid analysis since an excess of sample is usually available. This technique was the one initially used for capillary column steroid analysis by Völlmin [3]. Its advantage is that it can be used either with isothermal or temperature-programmed operation, since *cold-trapping* (injection into hot injector with column at low temperature) is not necessary.

3. *Splitless Injection*

This technique offers a means of injecting relatively large volumes of solvent into the instrument, so it is particularly useful in trace analysis. The sample is injected into a wide-bore glass insert, and cold-trapping is necessary to recondense solvent and sample in a small band at the top of the column. After a short interval the splitter is opened to purge the injector of residual solvent vapor. This is necessary to achieve a small "solvent front." Some sample is inevitably lost by this procedure, but this does not generally amount to more than 10%. However, this is one of the causes of discrimination against high-mass components often seen with the technique. Methods available for minimizing this discrimination have been discussed in a recent publication by Grob et al. [10]. Generally the reproducibility of splitless injection is poorer than for other sampling methods, and careful choice of internal standards for quantifications is imperative (see Table 8.1).

4. *Direct Injection*

In many respects this technique gives similar results to splitless injection, but the reproducibility is better and there is less discrimination against high-mass components. The sample is injected conventionally through a septum, and cold-trapping is employed. The major difference from splitless injection is the absence of "splitter-opening" after injection. This is made possible by greater control of carrier gas flow-rate together with the use of a narrow-bore glass insert. Solvent vapor diffusion within the injector is therefore reduced to a minimum, and an acceptable solvent front is achieved. This injection technique is almost certainly the best *liquid* sampling method for steroid analysis. Unfortunately, like splitless injection, it cannot be used with isothermal operation.

Table 8.1 Splitless Injection: Reproducibility of Two Methods of Quantification

	Height vs. C[a]		Height vs. A/B line[b]	
	Mean ($n = 6$)	C of V	Mean ($n = 6$)	C of V
Androsterone	1.52 ± .29	19.1	0.57 ± .05	9.0
Etiocholanolone	1.30 ± .27	20.7	0.49 ± .02	3.2
11β-Hydroxyandrosterone	1.45 ± .17	17.8	0.40 ± .03	7.2
Pregnanetriol	0.74 ± .11	14.8	0.35 ± .02	5.7
Tetrahydrocortisone	3.38 ± .42	12.3	1.86 ± .12	6.3
Tetrahydrocortisol	2.20 ± .28	12.7	1.32 ± .12	8.8
Allotetrahydrocortisol	1.50 ± .17	11.3	0.91 ± .07	8.2
β-Cortolone	0.87 ± .11	12.6	0.56 ± .05	8.8

[a]Peak height measured relative to the peak height of cholesteryl butyrate (C).
[b]Peak height measured relative to the height of the line joining the internal standards 5α-androstane-3α,17α-diol (A) and stigmasterol (S).

5. *Solid Sampling (Individual)*

Solid injection has been the most popular sampling technique for steroid gas chromatography. Derivatives of steroids are well suited to dry injection since they are involatile at room temperature and are generally stable, providing they are maintained in an inert gas atmosphere. Ras [11] described a technique based on a dropping-needle; this device was refined by Van der Berg and Cox [12] and is also commercially available (Packard Instrument Co., Downers Grove, Illinois). The sample is injected through a septum onto a glass needle, and the solvent is vented. The needle with dried sample is then lowered into the region of the flash heater, and volatilization and analysis can proceed. This sampling technique has also been used by Axelson and Sjövall [13], Vanluchene et al. [14], and Leunissen [15]. It has the advantage that "solvent front" components are small, and little discrimination is noticeable against the higher-mass sample analytes. It should be stressed that this technique (or the following one) is invariably used for sensitive GC-MS selected-ion-monitoring analysis of steroids (see Section IV.F). The possibility of introducing only a small portion of the total sample precludes the use of syringe injection in these cases.

6. *Solid Sampling (Multiple)*

Automatic solid samplers have been available for a number of years [8]. They consist of a revolving carousel holding 24 sample vials which are usually small, open-ended capillaries. These sample vials can be dropped at predetermined intervals into a glass insert in the region of the flash heater. The carousel is actuated by compressed air. After a number of analyses have taken place, the glass insert must be removed and the spent capsules discarded. There are several advantages to this technique: (1) Fully automatic injection and analysis result in greater reproducibility of retention time and quantification; (2) the instrument can be left unattended for long periods of time; (3) large volumes of sample can be concentrated in small vials; and (4) isothermal or temperature-programmed operation can be used. Repetitive solid sampling is an ideal sampling method for automation of GC-MS analysis (Dr. R. W. Kelly, personal communication).

In a later section (IV.B.3) the reproducibility of this technique and splitless injection will be compared.

III. SAMPLE WORK-UP PROCEDURES

A. Recovery of Steroids

Extraction of steroids from body fluids, tissue, cellular incubations, etc., is a prerequisite to gas chromatographic analysis. Great diversity can be built into extraction methods; for example, a single extract can readily be prepared which contains both neutral steroids and acidic steroids (estrogens and cortoic acids) which were present in the body fluid as free compounds, glucuronides, sulfates, and mixed conjugates. Alternatively, highly selective extraction and purification procedures can be utilized so that in the final multicomponent analysis, only one suite of compounds is represented. Examples may be fractions which contain monosulfates, glucuronides, acidic cortisol metabolites, or estrogens, exclusively. Going one stage further, extracts of steroids liberated from conjugation can be separated by chromatographic techniques into fractions based on polarity. Fractions which contain only a single steroid, therefore, may be prepared from a complex mixture by liquid chromatographic techniques.

Frequently, when gas chromatography is used for clinical urinary steroid analysis, it is preferable to prepare an extract containing virtually all the steroids excreted. Such simplicity has become possible only through the introduction of glass capillary columns of high resolving power and efficient methods for protecting the components of interest against thermal degradation. However, urine is particularly easy to work with since it contains large amounts of steroids and few other components which interfere with their analysis. Multicomponent analysis of serum hormones or tissue constituents requires more extensive sample fractionation.

1. Urine

This discussion will center first on the simplest method for deriving a urinary steroid profile; this will be followed by a description of methods of increasing complexity which allow specific and sensitive analysis of individual groups of important steroids. A total profile requires the following basic steps prior to gas chromatographic analysis: (1) hydrolysis of urinary conjugates, (2) extraction of "freed" steroids, and (3) derivatization. In addition, extraction of steroid conjugates prior to hydrolysis is frequently carried out. These procedures can be virtually nonselective: enzyme preparations can be used which *hydrolyze* almost all steroid sulfates and glucuronides; relatively polar solvents (e.g., ethyl acetate) can be used which *extract* almost all neutral and acidic steroids or, alternatively, absorptive resins can be used for extraction; and *derivatization* reagents can be used which effectively protect all groups of compounds.

Hydrolysis: The enzymes present in the digestive juice of the snail (*Helix pomatia*) are the most versatile for steroid hydrolysis. All steroid glucuronides and most sulfate esters are hydrolyzed by this preparation, although the rate of hydrolysis depends on the structure of the substrate [16]. However, sulfate esters of the 17-hydroxyl in C_{19} steroids, the 20-hydroxyl in pregnanes, and 3α-sulfates of 5α steroids are not hydrolyzed using this or other commercial enzyme preparations, but steroids of this type are quantitatively of minor importance in urine. If a homogeneous *glucuronide* fraction is required without prior conjugate fractionation, then a specific beef liver glucuronidase should be used.

Selective, nondestructive hydrolysis of steroid sulfates has traditionally been carried out by the solvolysis procedure of Burstein and Lieberman [17]. This is usually undertaken by one of two methods: (1) the urine is made 2 M with respect to sulfuric acid, and the steroid sulfates are extracted with ethyl acetate; or, (2) a dried extract of steroid sulfates is dissolved in ethyl acetate previously saturated with 2 M sulfuric acid. in each case, the ethyl acetate is kept at 40° for several hours to ensure that the hydrolysis goes to completion. The second procedure is most successful when a homogeneous sulfate fraction has been prepared using liquid chromatographic techniques (see section on Fractionation of Steroids, below). Solvolysis in acidified ethyl acetate gives clean fractions of freed steroids, but we have found that the alkali and water washes, necessary to remove acidic material at the end of the hydrolysis, tend also to remove polar neutral steroids.* Currently we are investigating a *microsolvolysis* technique in which the steroid sulfates (up to 100 μg) are hydrolyzed in 50 μl acidified ethyl acetate. The small amount of acid present can be disregarded, as this is neutralized by addition of the pyridine solution of the oximating reagent used for derivatization. The results so far produced suggest that the technique gives quantitative recovery of 3β- and 3α-monosulfates.

Another method is hydrolysis through derivatization. Methyloxime derivatives are prepared from carbonyl-containing steroids in the sulfate fraction, and the hydrolysis is carried out by the reagent used for silylation (trimethylsilylimidazole). Under the conditions used in our laboratory, silylation-hydrolysis does not go to completion and, depending on the substrate, varies between 45 and 60%. These "maximum" recoveries of free steroids are achieved by 16 hr of heating the derivative at 110°. Although the recovery is not quantitative, it is reproducible and can therefore be used routinely for hydrolysis of most serum and urinary steroid sulfates. It is probable that with suitable catalysis, recoveries could be improved. It should be stated, however, that only estrogen-3-sulfates and 3-sulfates of 3β-hydroxy-5-ene steroids are satisfactorily hydrolyzed by this technique. An example of a gas chromatographic profile obtained by this method is shown in Figure 38 in Section V.D.2.

Extraction: Ethyl acetate and diethyl ether are efficient solvents for extracting steroids freed from conjugation, and loss occurs only when the steroid molecule has more than five oxygen functions. The neutral resin Amberlite XAD-2 [18] is also highly efficient at extracting steroids over almost the whole polarity range. Recently we have shown that small C_{18} cartridges produced by Waters Associates (Sep-pak cartridges) probably offer the most versatile and rapid method for extraction of all types of steroids and steroid conjugates from water-based solution [19]. These cartridges are small columns (about 1 cm × 1 cm) containing octadecylsilane bonded-phase retained between filters. The material

---

*In a recent publication Axelson et al. [25] report the use of TEAP-LH-20 columns (in the OH^- form) for removing traces of acid present after tetrahydrofuran solvolysis.

absorbs a wide variety of polar and nonpolar organic compounds from water-based solutions, and these organic extracts are quantitatively recovered by elution of the columns with 2–3 ml of methanol.

Adlercreutz and co-workers employ a similar extraction technique for free steroids based on the use of Extralut columns (Merck, Darmstadt, Germany). The enzyme hydrolyzate (20 ml) is applied to the column and allowed to soak in for 15 min. Estrogens and neutral steroids are recovered by elution with 25 ml diethyl ether [20]. For extraction of steroid *conjugates*, these workers use Sephadex G-25 columns with sodium chloride as eluting solvent.

Unconjugated steroids of low and medium polarity can also be quantitatively extracted from aqueous solution by rapid filtration through Lipidex 1000 (Packard Instrument Co., Downers Grove, Illinois) [21]. This is a derivative of LH-20 which contains a large number of alkyl chains, giving it the properties of an organic solvent or nonpolar adsorbant while still being sufficiently polar to be wetted by water. Up to 40 ml of urine can be extracted by a column of bed volume 4 ml. To obtain quantitative extraction of steroids as polar as cortisol, pentylamine (5% v/v) has to be added to the urine prior to extraction. The extracted steroids can be recovered from the column by elution with 10 ml of methanol.

Fractionation of steroids: The previous sections have described how steroid metabolite profiles can be simplified by appropriate use of selective hydrolytic and extraction methods. This discussion will center on the use of liquid chromatography for fractionation of free steroids and conjugates prior to hydrolysis and of steroids freed by hydrolysis.

Group fractionation of free and conjugated steroids: The purification of small amounts of steroids requires the use of highly inert chromatographic media. Adsorption chromatography is therefore less suitable because of the risk of irreversible adsorption or chemical transformations. Over the last 15 years, Sjövall and co-workers have pioneered the development of substituted Sephadex derivatives for analysis of steroid metabolites. These *liquid-gel* systems for chromatography have proven to be inert, offer the choice of a wide variety of straight or reversed-phase partition systems and, by appropriate substitution, permit preparation of ion-exchange gels. The first Sephadex derivative to become commercially available and widely used in steroid analysis was LH-20. An initial application was the separation of free steroids (and glucuronides), steroid monosulfates, and steroid disulfates extracted from serum [22,23] using a solvent system of methanol: chloroform (1:1 saturated with salt).

We find a simplified version of this method useful for separating crude steroid extracts into two fractions: (1) free steroids and steroid glucuronides and (2) steroid sulfates (monosulfates and disulfates). The sample is applied to a 4-g column of Sephadex LH-20 prepared in the solvent. Thirty milliliters of this solvent elute the first fraction and 50 ml of methanol elute the sulfates.

Separation of free steroids from glucuronide-conjugates steroids by Sephadex LH-20 chromatography proved to be a probelm, and greater selectivity was achieved by preparation of ion-exchange gels. Setchell et al. [24] reported a method for the separation of neutral steroids from acidic steroids, glucuronides, monosulfates, and disulfates using diethylamino hydroxylpropyl Sephadex LH-20 (DEAP-LH-20) and eluting solvents based on a 72% ethanol with gradually increasing pH (Figure 8.4). This technique has been refined by use of gels with greater capacity, and currently the following procedure is used by Sjövall's group [25]: (1) Organic bases are removed from the sample by filtration through a column of sulfopropyl Sephadex LH-20 in H^+ form. (2) Neutral

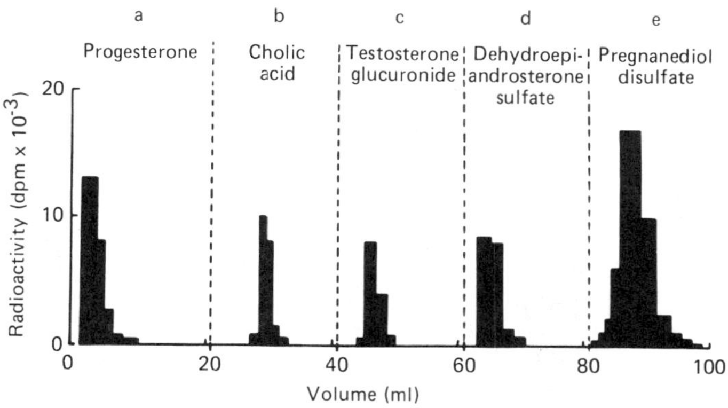

Figure 8.4 Group separation of radioactive steroids added to urine samples. Ion-exchange chromatography was performed on DEAP-LH-20 (0.6 g) in the acetate form. Stepwise elution of the steroid was achieved using the following solvents: (a) 72% ethanol, (b) 0.15 M acetic acid in 72% ethanol, (c) 0.25 M formic acid in 72% ethanol, (d) 0.3 M acetic acid-potassium acetate, pH 6.3, in 72% ethanol, and (e) 0.5 M potassium acetate, pH 10, in 72% ethanol. (From Ref. 24.)

steroids are recovered by filtration of the sample in 72% methanol through the strong anion exchanger, triethylamino hydroxypropyl LH-20 (TEAP-LH-20). (3) Estrogens are recovered by elution with 72% methanol saturated with CO_2 and are sufficiently pure to permit direct derivatization and glass capillary gas chromatographic analysis (Figure 8.5). (4) Steroid acids, steroid glucuronides, sulfates, and disulfates can then be recovered by solvent systems similar to those described in the earlier paper [24], although 72% methanol was used instead of 72% ethanol.

Adlercreutz and co-workers [26] have recently reported two similar methods designed primarily for the selective extraction of urinary estrogens. In the first method, all urinary conjugates are extracted and hydrolyzed prior to removal of neutral steroids. The method recovers the labile 2- and 4-hydroxyestrogens which were formerly lost completely because of their tendency to undergo autoxidative decomposition. By the utilization of ascorbic acid for their protection during almost all the isolation steps, it has now been demonstrated that the 2-hydroxyestrogens are among the most important metabolites in human subjects. The procedure is as follows: (1) gel filtration on Sephadex G-25 to prepare a fraction containing all estrogen conjugates, (2) hydrolysis of conjugates with *Helix pomatia* enzyme preparation; (3) recovery of freed estrogens by extraction on Extralut columns using diethyl ether as eluant; (4) removal of organic acids by filtration on column of DEAE-Sephadex A-25 anion-exchange resin in acetate form; (6) chromatography on Sephadex LH-20 in 3% methanol in benzene and elution of three fractions of varying polarity with (a) 3% methanol in benzene, (b) 6% methanol in benzene, and (c) 25% methanol in benzene (7) The two less polar fractions are purified on a column of DEAE-Sephadex A-25 anion exchanger in the OH^- form. This procedure effectively removes the neutral compounds. The most polar fraction from Sephadex LH-20 chromatography can be derivatized directly without further purification.

The alternative method described by Adlercreutz's group depends on the fractionation of conjugate types prior to hydrolysis [17]. In contrast to the

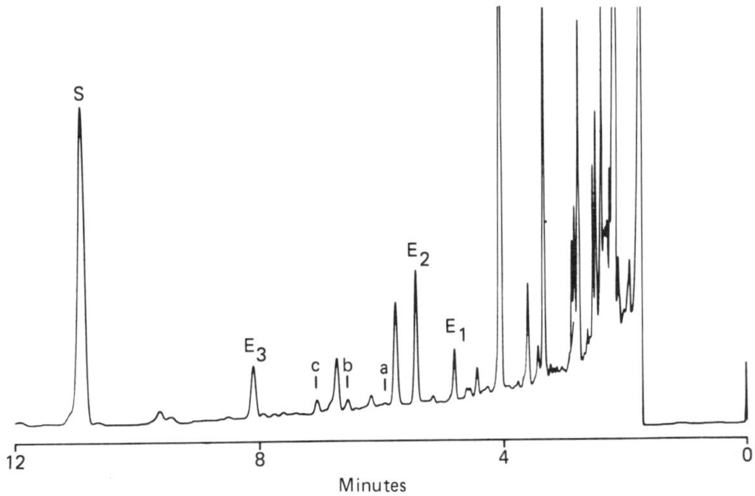

Figure 8.5 Gas chromatographic analysis of trimethylsilyl ethers of unconjugated estrogens in plasma from a pregnant woman (fortieth week). This fraction was prepared by ion-exchange chromatography on TEAP-Lipidex. Peaks E_1, E_2, and E_3 indicate the derivatives of estrone, estradiol, and estriol, respectively, and the letters a, b, and c, the retention times of the derivatives of 2-methoxyestrone, 16α-hydroxyestrone, and 16-oxoestradiol. Peak S is the internal standard. Conditions: 20-m SE-30 column, isothermal operation at 260°. Solid injection. (From Ref. 96.)

previous method, constituents of the final fractions are dependent on original conjugation rather than polarity. This technique uses a standard Sephadex ion-exchanger resin rather than the lipophilic variety generally favored by Sjövall's group. DEAE-Sephadex is packed into a 0.5 × 13 cm column. The column is impregnated with 0.005 mol/l ascorbic acid in methanol, and the sample is applied in 0.5 ml of the same solvent. The fraction collection is summarized in Figure 8.6 which illustrates the separation achieved for a mixture of authentic, labeled steroids. These fractions can then be hydrolyzed in the normal fashion and analyzed by capillary gas chromatography. An illustration of the separation achieved is given in a later section (V.B).

Tetsuo et al. [28] report a method for the specific analysis of ethinyl steroids based on the TEAP-Lipidex method described above [24]. The fractions of neutral and phenolic steroids from the columns of TEAP-Lipidex are passed through a small columns of SP-Sephadex in Ag^+ form. This column is packed in 72% methanol on top of a 30-mm layer of SP-Sephadex in H^+ form and has a high capacity to bind ethinyl steroids. The latter steroids are displaced from the column by elution with hexyne in methanol, and the ion exchanger in the H^+ form serves as a filter to remove Ag^+ ions which may form insoluble salts with the steroids.

Purification and fractionation of free steroids: It would be impossible here to review all methods for chromatographic fractionation of neutral steroids prior to gas chromatographic analysis. A decade ago, small silicic acid columns were particularly popular for group fractionation of steroids [29], but lipophilic Sephadex derivatives have since been favored because of their inert nature. Convenient methods for separation of neutral steroids on Sephadex LH-20 have

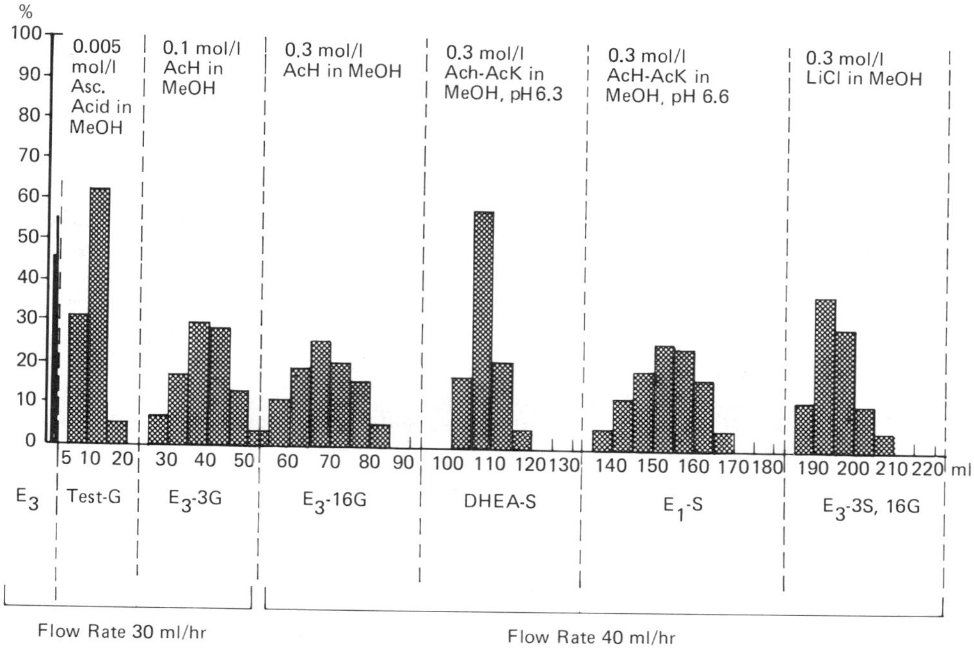

Figure 8.6 Stepwise fractionation of reference steroids and steroid conjugates on DEAE Sephadex. This system has been primarily designed for analysis of estrogens. (From Ref. 27.)

been described by Nyström and Sjövall [30], Setchell and Shackleton [31], and Murphy [32]. Straight-phase chromatography of free steroids on Lipidex 5000 has also been reported [33]. In addition, high-pressure liquid chromatography (HPLC) is increasingly used for steroid separation.

2. *Serum*

Free serum steroids can be recovered almost quantitatively by extraction with methylene chloride or ethyl acetate. By dilution of the serum with saline, Amberlite XAD-2 columns can be used for free and conjugate extraction, but the columns must be heated during the process to minimize protein binding [34].

A total sterol and steroid extract can be prepared by slow addition of serum to 10 vol. acetone:ethanol (1:1 v/v) or methanol:chloroform (1:1, v/v). This procedure precipitates the plasma proteins and should ideally be carried out with continuous agitation, commonly achieved by holding the flask in a sonic bath. This procedure is only suitable if the extract is subjected to extensive fractionation (e.g., LH-20, DEAP-LH-20, or TEAP-LH-20) prior to gas chromatographic analysis, and it has been generally used for the analysis of plasma steroid sulfates [22].

A method has been published by Axelson and Sjövall [35] which appears valuable for the selective isolation of 3-oxo-4-ene steroids. Oximes of these compounds are positively charged in methanol and are retained by the lipophilic strong cation exchanger, sulfohydroxypropyl Sephadex LH-20. The method involves four steps: (1) fractionation of the biological sample through the cation exchanger in H^+ form; (2) formation of oximes and filtration of the reaction mixture through an anion exchanger to remove reagent; (3) filtration of the oximes through the cation exchanger in H^+ form; and (4) derivatization and purification through Lipidex 5000 (see next section).

B. Derivatization

Many underivatized steroids cannot be analyzed by glass capillary gas chromatography. Although it should now be possible with the sophisticated column deactivation methods available to obtain good peak shape for "free" *nonpolar* steroids (e.g., 17-oxosteroids, pregnanediol, etc.), derivatization of all polar steroid hormones or metabolites will always be necessary. The preparation of volatile, thermostable derivatives of corticosteroid metabolites is essential for the analysis of all extracts of biological materials. For a comprehensive listing of the numerous methods reported, the reader should consult the book by Knapp [36].

The formation of trimethylsilyl ethers and methyloxime-trimethylsilyl ethers reported in the original profile analyses by Gardiner and Horning [2] remains the preeminent technique.

1. Oxime Formation

The condensation of steroid carbonyl fractions with methoxyamine hydrochloride resulting in the formation of methyloxime derivatives was first described by Fales and Luukkainen [37]. This serves two functions: (1) The products are more thermally stable and less polar than the underivatized steroids; and (2) during the following trimethylsilylation, enol-ether formation in is prevented.

All steroid oxo-groups with the exception of the sterically hindered 11-oxo-group can be quantitatively converted into the oxime derivatives using 1–10% of methoxyamine hydrochloride in pyridine. The reaction can be carried out at room temperature overnight or at 60°C for 15–60 min. The wide range of conditions described relates to the diversity of methods reported in the literature from different laboratories; the end results, however, are very similar.

The presence of a trivalent nitrogen atom in the oxime derivative results in formation of two configurations. In gas chromatographic literature, these configurations are termed *syn* and anti forms of the derivative, but such prefixes do not allow the unambiguous designation of the absolute configurations of both isomers. *E* and *Z* labeling [15,38] eliminates this problem, and Figure 8.7 shows the absolute configuration of the isomers of 3-, 17-, and 20-methoxime groups. For most urinary steroids, either one of the configurations is formed almost exclusively, or the two forms are not resolved since single peaks are obtained. In 3-oxo-4-ene steroids, 16-hydroxylated 17-oxosteroids and 17-deoxycorticosteroid metabolites, two isomers are invariably formed and are frequently separated by high-resolution gas chromatography (Figure 8.8). It was considered that the lack of derivatization of the 11-carbonyl might lead to enol-ether formation during the subsequent silylation reaction, but Thenot and Horning [39] have shown that less than 2% of the C-11 carbonyl groups of tetrahydrocortisone are enolized.

Particular care has to be taken for the derivatization of 18-oxogenated C_{21} steroids. 18-Hydroxylated metabolites of corticosteroids (e.g., 18-hydroxytetrahydroDOC) exist predominantly as C18–C20 cyclic hemiketals. The cyclic structure is broken during oximation and the C-20 carbonyl can be derivatized, but this derivatization requires much harsher conditions than those required

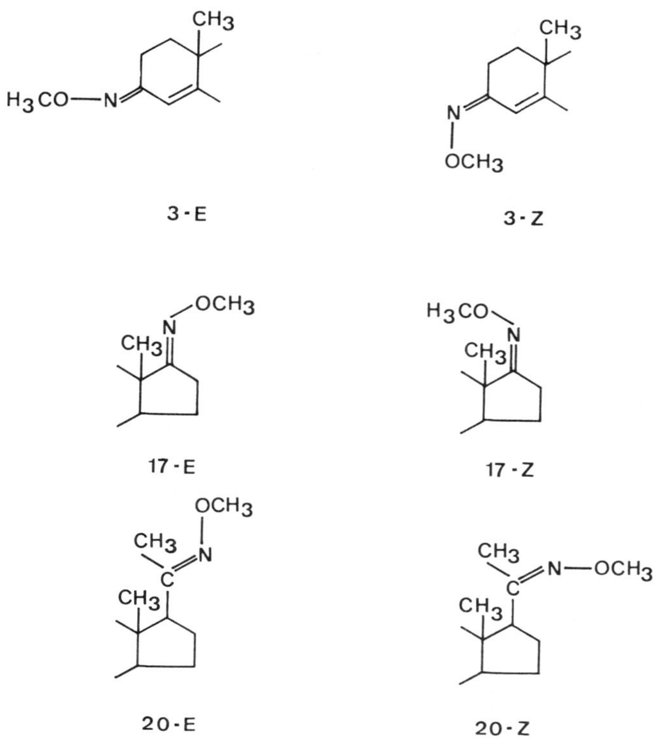

Figure 8.7 E and Z isomers of steroidal 3-, 17-, and 20-methoxime groups. (From Ref. 15.)

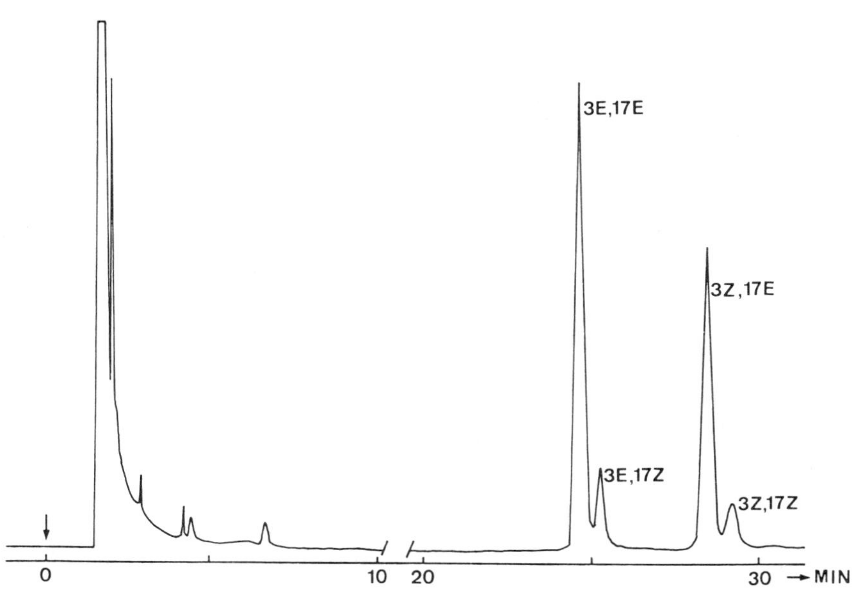

Figure 8.8 Chromatogram of the four isomers of androstenedione bismethoxime separated on Silar 5-CP column operated isothermally. (From Ref. 15.)

for other conventional carbonyl functions. We have found that optimum conversion to complete derivatives is obtained by a 72-hr reaction at room temperature or by heating the derivative mixture at 60° for 3 hr, but this does give rise to some degradation [40].

Aldosterone and its metabolites also require particular conditions for successful analysis [41,42]. We investigated the derivatization of tetrahydroaldosterone, the major urinary metabolite of aldosterone. The natural form of this metabolite contains a hemiacetal ring which is cleaved during derivatization, allowing condensation of both 18- and 20-carbonyls. This cleavage is extremely slow at room temperature, and complete oximation of both C-18 and C-20 carbonyls was achieved only after 72 hr. At shorter reaction times, the monomethoxime of the hemiacetal was found as well as the bismethoxime. The rate of formation of the bismethoxime is illustrated in Figure 8.9. Short, hot derivatization (8 hr, 60°C) resulted in dehydration of tetrahydroaldosterone and could not be recommended.

2. Trimethylsilylation

Silylation is invariably carried out following methoxime formation and, until the early 1970s relatively mild reagents were used for preparing derivatives for profile analysis. Hexamethyldisilazane added to the methoxyamine reagent enabled all primary and secondary hydroxyl groups to be derivatized. The reagent used for oxime preparation acts as a catalyst since hexamethyldisilazane will not normally derivatize 11β-hydroxyl groups unless trimethylchlorosilane is added. The popularity of this form of derivatization arose from the convenience of using volatile reagents. Following derivatization, the reagents

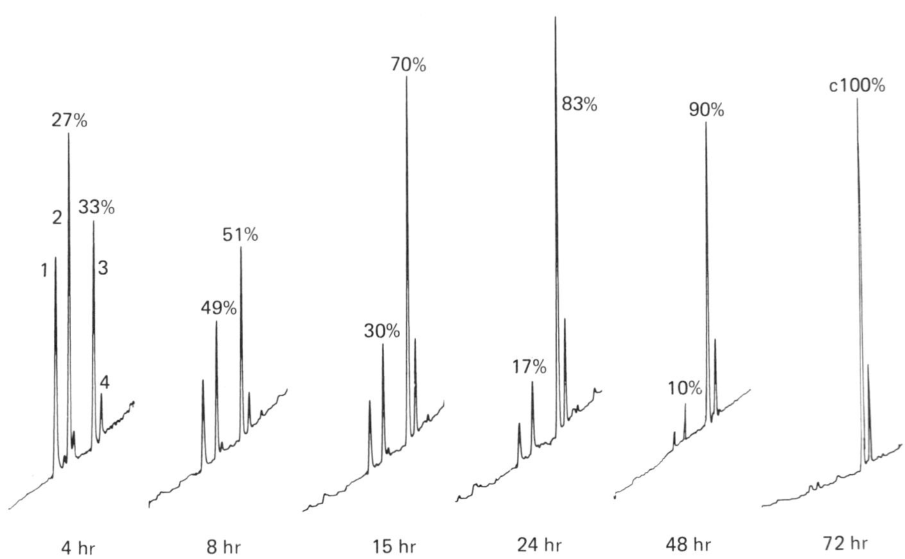

Figure 8.9 Part of the gas chormatograms (OV-101) of methyloxime-trimethylsilyl ether derivatives of tetrahydroaldosterone. The time course of methoxime formation prior to an 8-hr silylation is illustrated. The percentage figures represent the sum of peak heights of 1 plus 2 and 3 plus 4. (From Ref. 42.)

could be evaporated with nitrogen, and the steroid oxime-silyl ethers could be taken up in a nonpolar solvent such as hexane.

During the period when packed columns were used for steroid profile analysis, this form of derivatization was adequate, since good peak shape was achieved for corticosteroid metabolites, even though a 17α-hydroxyl group was underivatized. Capillary columns are less forgiving, and the free hydroxyl frequently gives rise to poor peak shape due to absorption on active sites on the glass surface or in the injector and detector. Whereas slight absorption of a "large" sample analyzed on a packed column could be tolerated, significant absorption of a much smaller sample analyzed on a glass capillary was extremely significant and could not.

Complete derivatization of all hydroxyl groups in corticosteroids is therefore recommended for profile analysis. Thenot and Horning [39] introduced the powerful silyl donor TSIM for silylation and, although the reagent permits quantitative derivatization of most steroids, it does suffer from the disadvantage of being extremely involatile so that excess reagent cannot be removed by evaporation under nitrogen. Small amounts of the reagent can be injected directly onto packed columns or even onto capillary columns, providing split injection is employed. However, for solid injection or splitless injection, the reagent must be removed prior to gas chromatographic analysis because poor volatilization of sample, rapid fouling of injector, and premature destruction of the wall-coated column result if this is not carried out. This problem was resolved in 1974 with the introduction by Axelson and Sjövall of a liquid chromatographic technique suitable for removal of excess reagent [34]. The support used is a lipophilic Sephadex derivative (Lipidex 5000), and small columns of this material are prepared in the solvent system hexane:hexamethyldisilazane:pyridine (98:1:1). Following evaporation of the pyridine, the crude derivative is taken up in the solvent system and is applied to the column. A total eluant volume of 3 ml is collected. The extremely apolar steroid derivatives are eluted immediately, while the polar derivatization reagents remain on top of the Lipidex column.

Even with TSIM, steroids with hydroxyl groups in both C-17 and C-20 positions, such as cortolones and pregnanetriol, are fairly resistant to silylation because of the immediate silylation of the C-20 hydroxyl which further hinders reaction of the tertiary 17α-hydroxyl. We have found that complete derivatization of the 17-hydroxyl in α-cortolone by addition of TSIM directly to the methoxyamine/pyridine solution requires heating at 100° for more than 16 hr (Figure 8.10). Purified oxime-trimethylsilyl derivatives can be kept in hexane or similar solvent for several months without sample deterioration. Use of such nonpolar solvents for sample storage permits rapid vaporization when split or splitless injection is employed and allows complete "drying" of sample on the needle in sample vials used for solid injection.

This sample derivatization method is now most widely used for steroid profile analysis. The basic method has been adapted slightly in several investigations. For example, Axelson and Sjövall [35] found that substitution of hydroxyamine hydrochloride for methoxamine hydrochloride in the condensation reaction prior to trimethylsilylation resulted in derivatives suitable for the analysis of plasma steroid hormones. Similarly, specific analytical requirements may demand the use of other "oximation" reagents producing, for example, O-benzyloximes [52].

Silylation without prior oxime condensation has also remained popular since the problem of *syn*- and *anti*-oxime formation is avoided, although others certainly exist. Chambaz et al. [44] prepared "enol-TMS" derivatives for

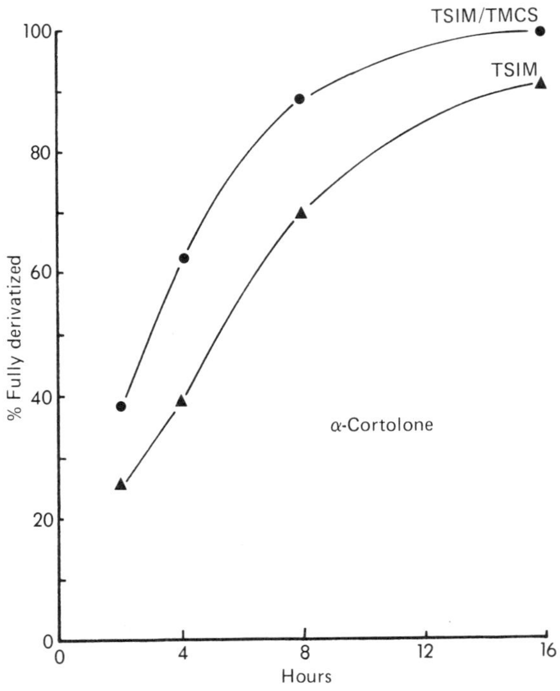

Figure 8.10 Silylation of α-cortolone. Plot of percentage of fully derivatized compound at increasing time intervals determined by relating the peak height of "full" derivatives to the total of "full" and "partial" derivatives. The tertiary hydroxyl of α-cortolone (or α-cortol) is the most difficult group to derivatize. Derivatization with TSIM approaches completeness after 16 hr at 100° in the presence of TMCS.

protection of the dihydroxyacetone side-chain of corticosteroids and showed that these derivatives were suitable for glass capillary gas chromatography. They were prepared by addition of bis(trimethylsilyl)trifluoroacetamide, (BSTFA) (100 μl) and crystalline potassium acetate (0.5 mg) to the sample which was heated at 60° for 40 min. The reagents were removed by evaporation under nitrogen. Most steroids gave one peak under these conditions, exceptions being 11-oxoandrosterone and 11-oxoetiocholanolone. The C-20 carbonyl reacted completely (except when it was the only function in the side-chain) but the C-17 carbonyls in 17-oxosteroids were not derivatized. Neither 11β- nor 17α-hydroxyls formed derivatives. This technique did not gain wide acceptance since multiple peaks were still formed from single compounds in certain instances, and losses by interactions of underivatized groups with active sites on glass surfaces occurred.

Spiteller's group continue to use silylation without prior oxime formation, and mixed derivatives (trimethylsilyl ethers and enol-ethers) have been reported in their communications. Also they report significant degradation of polar corticosteroids [45]. In more recent work from their laboratory, some of these probelms have been overcome, but a remaining disadvantage of the method is the inability to separately determine metabolites of corticosterone. Trimethylsilyl-enol-TMS derivatives were also used in the method for urinary steroid analysis published by Fantl and Gray [46].

VI. QUALITATIVE AND QUANTITATIVE ANALYSIS

A. Resolution of Steroids

1. *Choice of Column Phase*

Although the introduction of glass capillary columns has greatly reduced the problems associated with resolution of biologically important steroids, there are still many examples of compounds which coelute. Nonpolar columns coated with methylsilicone gums (SE-30, OV-1, OV-101) are the most versatile for analysis of methyloxime-trimethylsilyl derivatives of steroids. Generally the "overlaps" are unimportant for most aspects of clinical investigation since they represent metabolites of the same parent steroid; for example, the 11-oxo derivatives of etiocholanolone and androsterone coelute, as do β-cortol and β-cortolone and two of the metabolites of corticosterone (allotetrahydrocorticosterone and allotetrahydroCompound A). One example of a more unfortunate "overlap" is 11β-hydroxyandrosterone and 17α-hydroxypregnanolone since these represent metabolites of very different precursors.

A significant change in elution pattern can be made by slight variation of stationary phase. The introduction of phenyl groups (5%) to a methylsilicone gum (SE-52 column) will permit the resolution of β-cortolone (Figure 8.11) but at the cost of loss of resolution of allotetrahydroCompound A, allotetrahydrocorticosterone, and tetrahydrocortisol. Clearly this is a less satisfactory solution since under these conditions cortisol and corticosterone metabolites will be

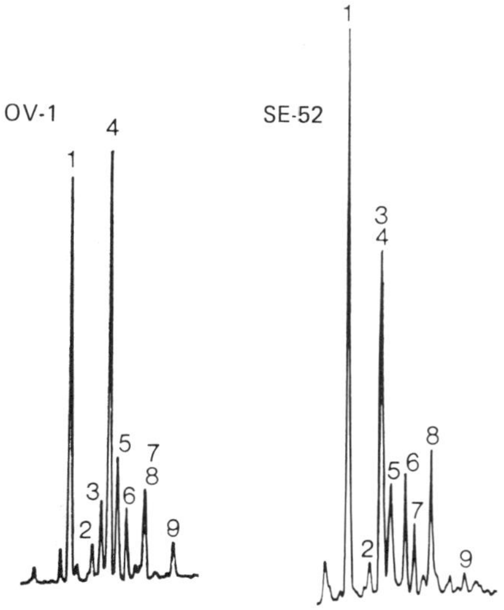

Figure 8.11 Achievement of required steroid separation by a slight change in stationary phase. This figure illustrates portions of chromatograms containing the following corticosteroid metabolites: 1, tetrahydrocortisone; 2, tetrahydrocorticosterone; 3, allotetrahydrocorticosterone; 4, tetrahydrocortisol; 5, allotetrahydrocortisol; 6, α-cortolone; 7, β-cortol; 8, β-cortolone; and 9, α-cortol. β-Cortol and β-cortolone are not resolved by analysis on OV-1 or SE-30 methylsilicone stationary phases. Introduction of 5% phenyl groups (SE-52 phase) allows all cortols and cortolones to be resolved but at the expense of the allotetrahydrocorticosterone/tetrahydrocortisol separation.

mixed. A useful compromise would entail the ready availability of instruments with both types of columns.

Certain separations do require use of polar columns. Leunissen [15] used OV-225 for the resolution of different "pairs" of steroids. An example from his work is illustrated in Figure 8.12 which shows the separation of 11-oxo-androsterone from 11-oxoetiocholanolone and allotetrahydroSubstance S from allotetrahydroDOC, resolution of which is currently impossible to achieve by separation of methyloxime-trimethylsilyl ethers on nonpolar columns such as OV-1.

In our studies, we experienced great difficulty in the separation of the major acidic cortisol metabolites (as methyl ester, trimethylsilyl ethers) on OV-1 or SE-52 capillary columns (Figure 8.13, upper two chromatograms). However, separation of α-cortolic, β-cortolic, α-cortolonic, and β-cortolonic acids was readily achieved using a column of Carbowax 20 M. Unfortunately, this latter phase is subject to high bleed at the temperatures required for volatilization and separation of the high-molecular-weight components, and columns have to be periodically recoated. Since Carbowax columns are prepared by dynamic coating, rejuvenation is a simple procedure and takes only a few minutes. As this manuscript was going to press, we prepared a very polar silicone SP-2330 column (68% cyano propyl) by the "barium carbonate" method and dynamic coating [4]. Our initial experience with this column has demonstrated its superior resolving power for the cortoic acid derivatives, but information on longevity is unavailable as yet. Table 8.7 gives retention time data for the cortoic acids on OV-1, SE-52, Carbowax 20 M, and SP-2330 columns.

An ideal solution to problems of coelution of important metabolites is two-dimensional chromatography, and this facility is now commercially available on

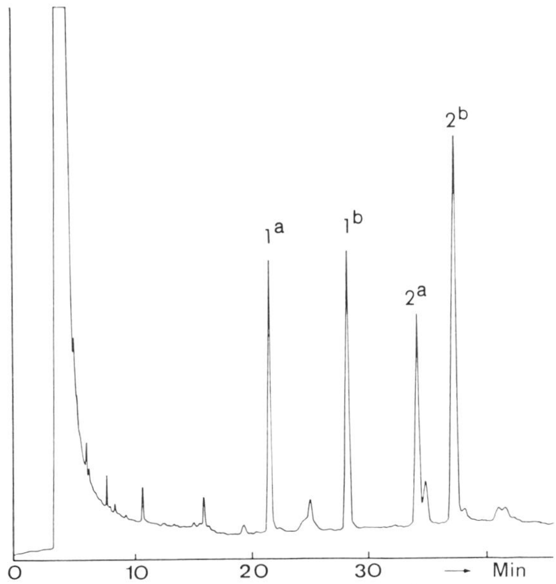

Figure 8.12 Separation of steroids unresolved on nonpolar column. The polar column OV-225 can be used for resolution of alloTHS (1a), alloTHDOC (1b), 11-oxoandrosterone (2a) and 11-oxoetiocholanolone (2b). (From Ref. 15.)

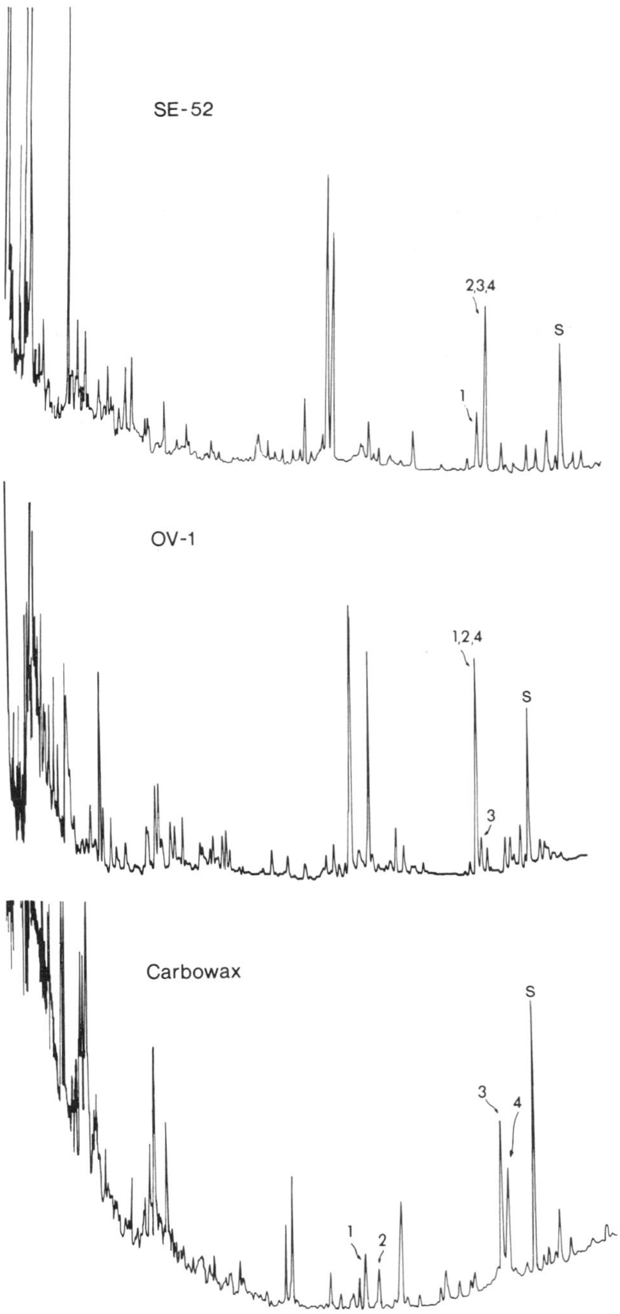

Figure 8.13 Separation of urinary cortoic acids as methylester-trimethylsilyl ethers on glass capillary columns coated with SE-52, OV-1, and Carbowax stationary phases. Peak 1 is α-cortolic acid; 2, β-cortolic acid; 3, α-cortolonic acid; and 4, β-cortolonic acid. (From Ref. 93.)

Packard instruments. This consists of two columns of different selectivity connected in series. By diverting to the first detector part or all of the flow exiting from the first column, it is possible to selectively transfer a portion of a chromatographic run from the first column to the second. Using this system, transference of only one or a few peaks from one column to the other can be achieved—a technique known as "heart-cutting." Both columns need not be open-tubular; an initial crude separation could be carried out on a packed column, and the final separation carried out on a capillary analytical column. It has been found that optimum use of this technique is achieved if compounds of interest are "trapped" before they are transferred to the second column. The trap is cooled by circulating cold nitrogen gas and is heated electrically by applying a large current to a surrounding graphite jacket. Volatilization takes place in about 30 sec, and the selected material is transferred to the second column (Figure 8.14). One of the great advantages of this system is the possibility of extending the dynamic range achievable by gas chromatography. Small and contaminated peaks eluting from the primary column can be resolved on the secondary column, providing care has been taken in the choice of stationary phase. Obviously this system is utilized to its fullest extent by coupling of a nonpolar (SE-30, OV-1) column to a polar (Carbowax or OV-225) column.

Finally, in addition to careful choice of column phase for a particularly difficult separation, change of derivative frequently can also be utilized. Either the reagent used for oxime condensation can be varied or, alternatively, different silylating reagents can be used. It is usually preferable to change the oxime since this generally leads to a smaller increase in derivative weight than a change to a "heavier" silylating reagent. O-Benzyloximes have frequently been employed for steroid profile analyses since, when combined oxime-silyl ethers are prepared, a group of carbonyl-containing steroids can frequently be completely separated from compounds with only hydroxyl groups (or nonderivatizable carbonyls) because the retention index differential between methyl- and benzyloximes is so great. An example of this technique is illustrated in Figure 8.15. The sample was obtained from a patient with

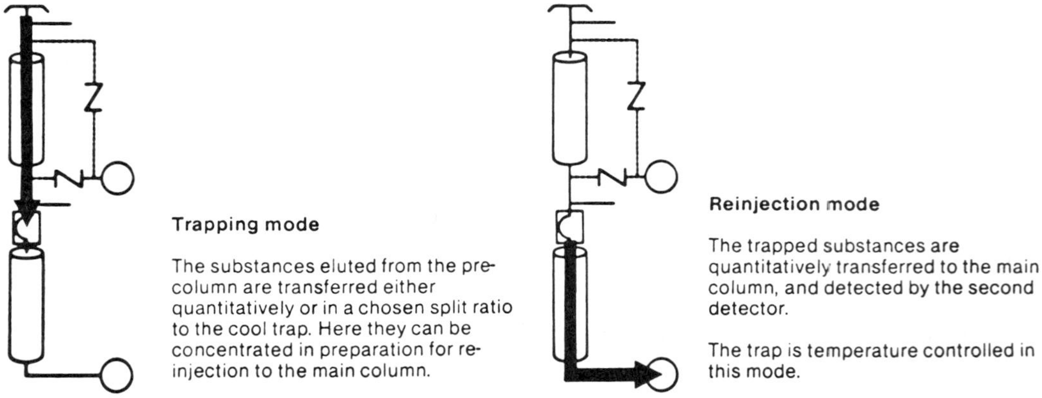

Figure 8.14 "Heart-cutting" transference of individual peaks or groups of peaks frone one column to a second column of different stationary phase. (Courtesy of Packard-Becker, Delft, The Netherlands.)

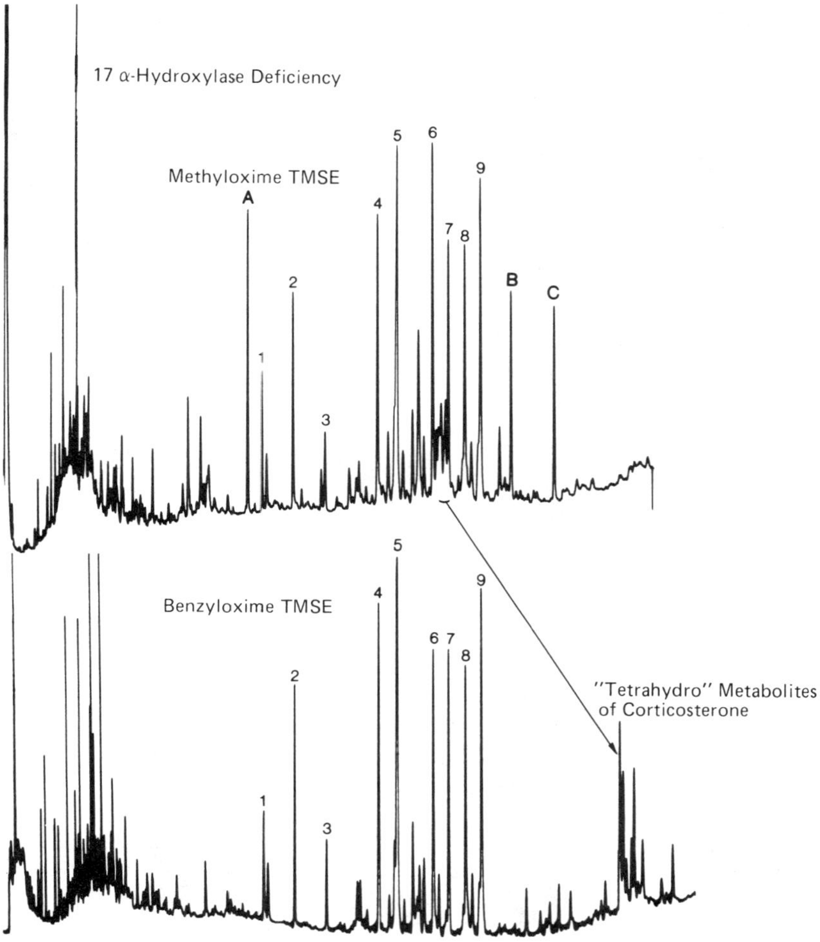

Figure 8.15 Use of benzloxime derivatives to clearly separate steroids with and without derivatizable carbonyl groups. In the example shown from a patient with 17α-hydroxylase deficiency, the major urinary steroids are 21-deoxymetabolites of corticosterone. The profile is highly complex, and accurate independent measurement of the classical corticosterone metabolites (e.g., alloTHB) is impossible. Substitution of a benzyloxime for the methyloxime permits simplification of the profile by increasing the retention time of these steroids (e.g., THA, THB, alloTHB, etc.) which contain derivatizable carbonyl groups so they are no longer eluted in the same region as steroids with only trimethylsilyl groups. The following steroids are numbered. 1, unidentified; 2, unidentified pregnenediol; 3, pregnanediol; 4, 5-pregnene-3β,20α-diol; 5, 11-oxopregnanediol (5α and 5β epimers); 6, 5-pregnene-3β,-16α,20β-triol; 7, 1,3,20-trihydroypregnan-11-one; 8, hexahydro A (5α and 5β epimers); 9, hexahydro B (5α and 5β epimers). Only the upper chromatogram includes the internal standards A, B, and C. Conditions: 20-m OV-1 glass column programmed 180–330°C at 3°/min; solid injection.

17α-hydroxylase deficiency, and benzyloximes were prepared because the carbonyl-containing compounds were obscured by the quantitatively more important components containing C-20 hydroxyls.

A particularly important application of benzyloxime derivatives in steroid profiling is the separation of 11β-hydroxyandrosterone and 17α-hydroxypregnanolone, wich coelute as methyloxime/trimethylsilyl derivatives on OV-1, SE-52, and SE-30 columns. This was particularly significant in our studies of 5α-reductase deficiency, where accurate measurement of the 11β-hydroxyandrosterone/11β-hydroxyetiocholanolone ratio was required.

In summary, by a careful choice of column coatings and derivatization agents, all steroids of importance can be separated. The increasing availability of two-dimensional gas chromatography will have a strong influence on multicomponent steroid assay in the coming years and, in particular, will permit the accurate analysis of clinically important, but quantitatively minor, steroid metabolites.

Other steroid silylating reagents have been studies recently and in certain cases offer particular advantages since they are more stable or give more useful mass spectrometer fragmentation patterns [47,48]. Most of the alternative derivatives have much higher mass than methyloxime-trimethylsilyl ethers and will prove useful only in combination with columns designed for high temperature operation.

2. *Recording of Steroid Retention Times*

Gas chromatographic retention behavior should be described through comparison with retention time values for two reference substances. This was first suggested by Kovats [49], who proposed the use of saturated, unbranched aliphatic hydrocarbons as reference substances. Two hydrocarbons were used for reference purposes in an isothermal separation, and the "Retention Index" was calculated as a four-digit value. Methylene unit vales are equivalent to the Kovats retention index values divided by 100, and these units have been in common use for reporting and comparing steroid retention data since the earliest days of profile analysis [50]. Reference hydrocarbons are available for almost all useful values in the *odd* and *even* series, but for practical purposes a reference mixture for steroid analysis need contain only those alkanes between C_{24} and C_{32}. Methylene unit values are determined by coinjection of this reference mixture with the steroid derivatives. A methylene unit value of 28.25 for a particular compound would indicate that it was eluted at a time 0.25 of the interval between the reference hydrocarbons with 28 and 29 carbon atoms in a linear temperature-programmed separation. In contrast to Kovats retention indices, methylene unit values are normally determined during temperature-programmed operation. In practice, for most methyloxime-trimethylsily derivatives, values secured by measurements of Kovats retention indices obtained isothermally differ little from methylene unit values determined during temperature-programmed operation. Methylene unit values obtained in our laboratory for analysis on OV-1 columns are listed in Table 8.2. Similar relative retention time values should be obtained by all investigators using similar stationary phases (e.g., OV-101, SE-30).

B. Quantification

1. *Internal Standards*

Internal standards are incorporated into derivatives primarily to reduce the necessity for accurate sampling. They offer a reference against which other peaks in the chromatogram can be measured. In the analysis of one or two purified components, internal standards can be chosen which are extremely similar in structure to the analytes. It could then be predicted that these standards would behave similarly during the derivatization procedure and that they would have similar stability to the analyte and would have similar stability to the analyte and would have similar FID response factors. The only requirement would be that they had different retention indices from the analytes. However, in multicomponent urinary analysis this is really not possible since it would be impractical to incorporate appropriate internal standards for each of the analytes. It is the opinion of the author that there are two essential functions of internal standards in glass capillary chromatography. The first is that they should give accurate representation of the variable discrimination which occurs towards high-mass components during chromatography. When splitless injection is used, small variations in injection technique can give rise to marked differences in the height (or area) of late peaks *relative* to height (or area) of early peaks, even for consecutive analyses. A measure of this variation could only be achieved providing standards are incorporated which are eluted *before* and *after* the series of compounds of interest. Fortunately, on-column and solid injection are less susceptible to this variable, high-mass discrimination. For this function, simple standard compounds are adequate, and several authors use the C_{24} and C_{32} *n*-alkanes as standards for urinary steroid analysis [15].

The second important function relates to the stability of the derivatives. Methoxime-trimethylsilyl ethers of steroids are stable in the absence of acid or water. However, all chromatographers have the experience of occasional derivatives which degrade for unidentifiable reasons. Natural contaminants in the biological extracts may well be a cause of some of these problems. The occurrence of inadequate derivatization or the presence of unstable derivatives is readily observed if internal standards are incorporated which are subject to derivatization in the same manner as the analytes. It is not necessary to include standard components which form oxime derivatives since these derivatives invariably form quantitatively and are not generally susceptible to degradation. However, the inclusion of standards which can be silylated is obviously useful.

The internal standard that has been most commonly used over the years is cholesteryl butyrate. This was used in the initial studies of Gardiner and Horning [2] for steroid profile analysis on packed columns since it was conveniently elutes after the last of the corticosteroid metabolites. Use of this standard alone was adequate in many cases when packed columns were employed since there was no discrimination against high-mass components. However, use of this compound alone in capillary column analysis is inadvisable because of the problem of high-mass discrimination and also because cholesteryl butyrate is more subject to "tailing" on older columns than are derivatives of urinary steroids.

We have settled on the use of three internal standards and these are incorporated into all derivatives: (1) 5α-androstane-3α,17α-diol, A; (2) stigmasterol, B; and (3) cholesteryl butyrate, C. Most of the standards in the chromatograms illustrated in this review which were obtained in our laboratory are labeled A, B, and C. respectively. The androstanediol (as trimethylsilyl ether, is conveniently eluted before the first major urinary steroid (androsterone), and stigmasterol is eluted after the last major steroid (α-cortol). The cholesteryl butyrate is incorporated because it does *not* derivatize and therefore allows a judgment to be made on the stability of the adjacent stigmasterol. A

Table 8.2 OV-1 Retention Times (Methylene Units) of TMS Ether, Methyloxime-TMS, and Benzyloxime-TMS Ethers of Major Urinary Steroids[a]

Steroid	MU value		Steroid	MU value	
	MO	BO		MO	BO
Androsterone	25.11	32.25	16α-Hydroxypregnenolone	29.38	
Etiocholanolone	25.28	32.15	3β,16α-Dihydroxy-5α-pregnan-20-one	29.42	
DHA	25.75	32.90	5-Pregnene-3β,17α,20α-triol	29.46	
5-Androstene-3β-17β-diol	25.85		HexahydroSubstance S	29.54	
11-Oxoandrosterone	25.95		5-Androstene-3β,15β,16α,17β-tetrol	29.64	
11-Oxoetiocholanolone	25.95	32.90	Tetrahydrocortisone	29.65	35.48
11β-Hydroxyandrosterone	27.00	33.48	5-Androstene-3β,16,17β,18-tetrol	29.72	
17α-Hydroxypregnanolone	27.00	33.60	18-HydroxytetrahydroDOC	29.73/30.50	
Androstenetrione	27.15/27.22		TetrahydroCompound A	29.77	36.08
11β-Hydroxyetiocholanolone	27.18	33.40	Tetrahydrocorticosterone	30.00/30.16	35.90
16α-HydroxyDHA	27.38/27.42		1,3,16-Trihydroxypregnan-20-one	30.10	
Pregnanediol	27.64		*allo*-TetrahydroCompound A	30.16/30.65	
16β-HydroxyDHA	27.74		*allo*-Tetrahydrocorticosterone	30.16/30.80	36.16
Pregnanetriol	28.00		Tetrahydrocortisol	30.24	35.73
5-Pregnene-3β,20α-diol	28.16		21-Hydroxypregnenolone	30.30/30.60	
16-Oxoandrostenediol	28.19		*allo*-Tetrahydrocortisol	30.39	36.00

15β, 16α-DihydroxyDHA	28.20	
Androstenetriol	28.46	
3α, 16α-Dihydroxy-5β-pregnan-20-one	28.52	
3α, 6α-Dihydroxy-5β-pregnan-20-one	28.52	
11β-Hydroxyandrostenedione	28.55/28.62	
TetrahydroSubstance S	28.62	
16α, 18-DihydroxyDHA	28.65/28.81	
3α, 20α-Dihydroxy-5β-pregnan-11-one	28.66	
Estriol	28.75	
3β, 15α-Dihydroxy-5α-pregnan-20-one	28.75	
3α, 20α-Dihydroxy-5α-pregnan-11-0ne	28.75	
5β-Pregnane-3α, 16α, 20α-triol	28.75	
3α, 21-Dihydroxy-5β-pregnan-20-one	28.75/29.30	35.05
allo-TetrahydroSubstance S	29.05	
11-Oxopregnanetriol	29.10	
5β-Pregnane-3α, 11β, 20α-triol	29.19	
5-Pregnene-3β, 16α, 20α-triol	29.20	
5α-Pregnane-3α, 11β, 20α-triol	29.31	
α-Cortolone	30.51	
18-HydroxytetrahydroCompound A	30.64/31.44	
β-Cortolone, β-Cortol	30.73	
HexahydroCompound A	30.76	
6α-Hydroxytetrahydrocortisone	30.80	
Tetrahydroaldosterone	30.86/30.91	
6α-HydroxytetrahydroCompound A	30.86/31.52	
5-Pregnene-3β, 16α, 20α, 21-tetrol	31.00	
18-Hydroxytetrahydrocorticosterone	31.00/31.45	
5β-Pregnane-3α, 11β, 20α, 21-tetrol	31.15	
α-Cortol	31.20	
5α-Pregnane-3α, 11β, 20α, 21-tetrol	31.23	
6α-Hydroxytetrahydrocorticosterone	31.82	
6α-Hydroxycortolone	31.82	
1β-Hydroxycortolone	32.00	
Cortisol	32.54/32.63	
6β-Hydroxycortisol	32.73	
20α-Dihydrocortisol	32.45/32.93	

Source: Ref. 66.

greatly reduced peak height of stigmasterol relative to cholesteryl butyrate would immediately put preparation or stability of the derivatives in question. Such derivatives should be remade. An example of this is illustrated in Figure 8.16. Both chromatograms represent the same urinary steroid extract, but the derivative giving rise to the upper chromatogram was unsatisfactory. This was determined by the inappropriately low peak given by stigmasterol relative

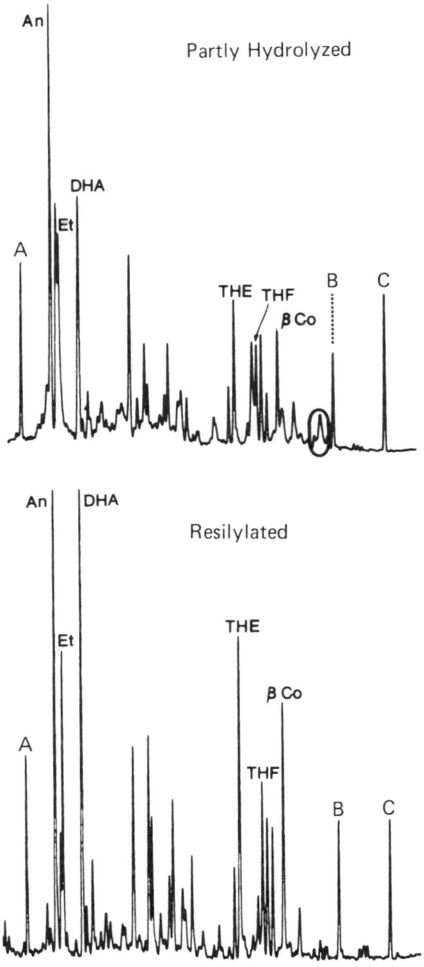

Figure 8.16 Unstable trimethylsilyl ethers. Using the derivatization procedure described in the text, a defective derivative is readily detected by a decrease in the peak height of stigmasterol (B) relative to cholestryl butyrate (C). Underivatized stigmasterol is seen as a broad, small peak immediately preceding the derivatized compound (circled). The top chromatogram in this illustration shows a partly hydrolyzed urinary steroid extract, and the lower chromatogram shows the same extract resilylated and purified by Lipidex 5000 chromatography. The following steroids are named by abbreviation: An (androsterone), Et (etiocholanolone); DHA (dehydroepiandrosterone), THE (tetrahydrocortisone), THF (tetrahydrocortisol), and βCo (β-cortol and β-cortolone).

to cholesteryl butyrate and the presence of a "hump" representing underivatized stigmasterol immediately preceding the trimethylsilyl ether. Re-silylation and Lipidex purification gave a good derivative, the analysis of which is illustrated in the lower chromatogram. Stigmasterol is not new as an internal standard for steroid gas chromatographic analysis, but it could only be used for urinary steroid profiling with the introduction of the persilylation reaction. When derivatives of corticosteroid metabolites were prepared which retained a free 17α-hydroxyl, they eluted in the same area of the chromatograms as stigmasterol.

A demonstration of the inaccuracies of splitless injection when a single *late* standard is used for quantification is illustrated in Table 8.1. In this experiment the same sample was analyzed six times and the peak heights of selected steroids were determined either relative to the androstanediol/stigmasterol line or relative to the height of cholesteryl butyrate. The coefficient of variation for the latter method is approximately double that obtained with the androstanediol/stigmasterol method.

The standards currently used in steroid profile analysis by various groups are summarized in Table 8.3.

2. *Response Factors*

The response of the flame ionization detector to equal mass of individual steroids decreases with increasing number of oxygen and silicone atoms [51]. In addition, steroids are subject to absorption throughout the gas chromatographic system, and the resulting FID response reflects this.

Until very recently, relative response factors (RRF) varied considerably when measurements were carried out on different columns and instruments. The response given by individual components depended heavily on the presence of active sites in the column, but this variation has been greatly reduced by improvement in column preparation technology. In addition, unsatisfactory methods were frequently used to attach the column to the detector and injector;

Table 8.3 Standards Used in Glass Capillary Chromatography of Steroids

1. Standards eluted before androsterone:		
C_{24} alkane	Spiteller, Leunissen	[15,76]
5α-Androstan-17β-ol	Pfaffenberger and Horning	[69]
Estratetratrienol	Teller and Homoki	[54]
5α-Androstane-3α,17α-diol	Shackleton et al.	[66]
2. Standards eluted after cortisol metabolites:		
C_{32} alkane	Spiteller, Leunissen	[15,76]
Cholesteryl butyrate	Gardiner and Horning	[2]
Cholesteryl butyl ether	Pfaffenberger and Horning	[69]
5α-Cholestane-3β-ol	Vanluchene et al.	[14]
Stigmasterol	Shackleton et al.	[66]
3β-Hydroxy-5-cholesten-7-one	Axelson et al.	[78a]

for some years our group and others connected the columns via heat-shrink PTFE tubing to straight capillary gas detector and injector inserts. These unsatisfactory connections also contributed to the great variations in RRFs between instruments and columns.

Improvement of column deactivation techniques carried out prior to stationary phase coating has greatly improved the response of many important steroid metabolites; for example, until two years ago we routinely obtained a RRF of 0.6 for steroids with underivatized 11-carbonyl(tetrahydrocortisone, 11-oxo-etiocholanolone), but now these RRFs rarely drop below 0.85 (measured according to the androstanediol/stigmasterol line technique previously described). Also, it is now normal practice to straighten the ends of glass columns so that the end may be placed within a millimeter of the FID jet, and the front can be centrally placed in the injector. The recent availability of flexible fused-silica columns offers an even better solution since end-straightening is no longer required, and by their nature they have fewer active sites than most glass columns.

Our recent studies have shown that response factor measurements can be remarkably consistent between columns and instruments. For example, we have analyzed a single standard mixture containing tetrahydrocortisone and tetrahydrocortisol several times over a one-year period. These analyses were carried out on three instruments (Carol Erba 2150, Varian 3700, and Packard 430), with two types of column (glass and fused silica), two types of stationary phase (OV-1 and SE-52), and sampling either by solid or splitless injection. The mean RRF for tetrahydrocortisone was 0.90 with a coefficient of variation of 7.9% ($n=16$). For tetrahydrocortisol the mean RRF was 1.10 with coefficient of variation of 5.3%. Two of the chromatograms obtained during this study are illustrated in Figure 8.17. It is of great interest that the discrepancy between individual splitless injections carried out under the same conditions was just as great, so the influences of column, instrument, and injection system are minor. When repetitive solid injection was used, reproducibility was found to be considerably better. Frequent determination of response factors is therefore generally unnecessary in our laboratory.

However, our methods of preparation of columns, end-straightening of glass columns, and attachment in the instruments are consistent and these may be more significant areas where variations can arise.

It would not be useful here to list the RRFs we obtain since the method we use for quantification is unique to our laboratory. These factors must be determined by individual laboratories according to the method they use for peak measurement. According to our method of analysis, the RRFs are between 0.85 and 1.30 for most steroids. However, where a steroid gives two separable peaks for oximes, the height of the larger peak may give a RRF as low as 0.55. Examples of this are tetrahydrocorticosterone, 16α-hydroxyDHA, and cortisol. Cortols and cortolones give RRFs around 1.5. If a reference material is not available for RRF determination, then a factor of unity is assumed.

Finally, in the determination of RRFs the results are only as accurate as the reference materials are pure. During our studies no additional purification of authentic steroids has been carried out so the response factors used may be slight underestimates.

3. *Reproducibility*

In an earlier discussion relating to the choice of internal standards (Section IV.B.1), data were presented which illustrated the reproducibility of individual urinary steroid estimation using splitless injection of samples. Table 8.1

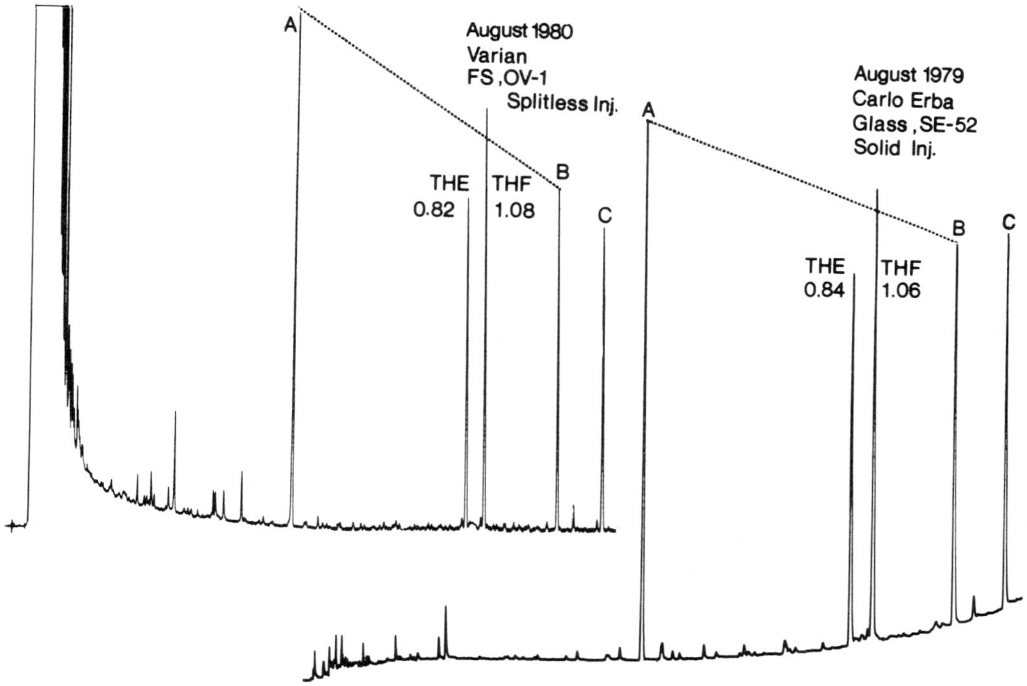

Figure 8.17 Consistency of response factor measurement. A derivative containing the standards tetrahydrocortisone (THE) and tetrahydrocortisol (THF) was analyzed occasionally over a 1-year period. These chromatograms were obtained under very different conditions. Right chromatogram: original analysis carried out by Carlo Erba 2150 chromatograph housing glass SE-52 column with solid injection. Left chromatogram: recent analysis carried out on Varian 3700 chromatograph, housing fused-silica OV-1 column with splitless injection. In each case the relative response factors were very similar. Additionally, the figure shows that steroid methyloxime/trimethylsilyl derivatives are stable for long periods providing they are kept in dry, nonpolar solvents.

showed that the coefficient of variation was between 4.0 and 9.0 for measurement of eight steroids in six consecutive analyses.

For automatic solid sampling, the reproducibility of relative peak heights for 12 consecutive 1-hr analyses on an OV-1 column housed in a Packard 430GC is shown in Table 8.4. In general, the variation was greatest for the smallest peaks determined, suggesting that slow pen response and inaccuracies in hand measurement contributed greatly to this error. Automatic programmed peak-height or area determination would probably improve reproducibility.

Thus, auto-solid injection is preferable to splitless injection with regard to reproducibility, and derivatized urinary steroids are stable in the carousel of auto-solid injectors for at least 11 hr. The latter consideration is important since concern has often been expressed regarding stability of derivatives in the dry state. Stability is assisted by the sample vials being kept in an inert gas atmosphere and also by the complete removal of excess derivatization agents by the Lipidex procedure of Axelson and Sjövall [34].

Leunissen [15] has undertaken a similar investigation of the reproducibility of urinary steroid estimation. Using a "dropping-needle" solid injection system,

Table 8.4 Reproducibility of Repetitive Solid Injection Using Automated Carousel[a]

	ng in peak ± SD (n = 12)	C of V
Androsterone	68.0 ± 3.1	4.2
Etiocholanolone	78.0 ± 2.9	3.7
DHA	21.6 ± 0.66	3.0
11-Oxoetiocholanolone	22.8 ± 0.83	3.6
11β-Hydroxyandrosterone[b]	90.8 ± 2.0	2.2
11β-Hydroxyetiocholanolone	39.7 ± 1.3	3.3
Tetrahydrocortisone	224.9 ± 6.2	2.8
TetrahydroCompound A	14.5 ± 0.96	6.6
Allotetrahydrocorticosterone	53.4 ± 1.9	3.6
Tetrahydrocortisol	148.9 ± 4.1	2.8
Allotetrahydrocortisol	149.5 ± 3.5	2.4
α-Cortolone	29.2 ± 1.1	3.8
β-Cortolone + β-Cortol	49.2 ± 1.9	3.8
α-Cortol	19.0 ± 1.5	8.1

[a]Quantification achieved by measuring peak heights relative to line joining top of peaks of the internal standards, androstanediol, and stigmasterol (100 ng).
[b]Also contains 17α-hydroxypregnanolone.

with C_{26} and C_{32} alkanes as internal standards and automatic peak integration, he obtained a coefficient of variation for several steroids in 10 consecutive analyses of 1.9 and 8.2%; individual measurements were not reported. Few other papers on capillary column chromatography of steroids include data on reproducibility of peak height or area measurement in consecutive analyses.

C. Identification

1. *Combined GC-MS*

The introduction of glass capillary columns with superior resolving power has increased the confidence with which component identity can be assigned by retention time alone. Also, in many of the disorders of steroid biosynthesis and metabolism which are investigated using gas chromatographic methods, the overall pattern is frequently highly specific for a given condition. Therefore, combined GC-MS is less important today than it was a decade ago for identification of steroid metabolites in clinical samples. However, there are always unidentified components in complex mixtures which can be characterized only by mass spectrometry. In addition, GC-MS with selected ion monitoring probably offers the only technique for specific quantification of minor components since gas chromatography does not generally have the specificity required for such analysis. Common examples of biologically important steroids which cannot be quantified in normal urine are tetrahydrodeoxycorticosterone, 18-hydrotetrahydroCompound A, and tetrahydroaldosterone. Selected-ion monitoring is also particularly important in the analysis of circulating levels of hormonal steroids and steroid drugs.

In some respects capillary columns are more suited to a combination with mass spectrometers than are packed columns. The low carrier-gas flow can often be passed directly into the ion source without requiring molecular separation. Currently in our Hitachi M-52 we have a direct connection (heat-shrink Teflon) between the column and a glass restrictor, surrounded by a brass jacket through which helium is flushed to protect against air leaking into the mass spectrometer. The carrier gas flow-rate through the column is about 0.8 ml/min. In repetitive scanning mode, the mass spectrometer scans the full mass range (10–800 amu) every 3.6 sec. Axelson and Sjövall favor retention of a molecular separator [13] since they maintain that carrier gas flow is easier to control. Certainly, under their operating conditions this does not appear to result in loss of sensitivity as they work with analyte concentrations of 10–100 pg.

Effective deactivation of GC-MS systems is imperative if reliable mass spectral data are to be obtained. When packed chromatography columns were used, few problems of steroid degradation outside the ion source were noted, but this was because relatively large samples were being analyzed. Decomposition of a few percent of the total for a particular analyte did not give rise to markedly altered spectra or total-ion-current chromatograms. However, the small samples analyzed by capillary column GC-MS can be completely degraded by contact with bare metal surfaces or active areas in injector, column, or interface glassware. Axelson and Sjövall [13] found that when using methyloxime derivatives, compounds with a carbonyl group at position 3 (e.g., progesterone) were particularly susceptible to absorption and degradation when small amounts (10–100 pg) were being analyzed. These workers effectively deactivated glass surfaces by silanization with dimethyldichlorosilane; steel surfaces inside the molecular separator were made inert by coating with water glass followed by treatment with benzyltriphenylphosphorium chloride.

In our initial experience with capillary column/mass spectrometer linkage, we found that although the total-ion-current chromatograms for analysis of certain methyloxime/trimethylsilyl ethers were indistinguishable from equivalent GC/FID chromatograms, the mass spectra obtained from individual components were incorrect. An example of this anomaly is shown in Figure 8.18. The illustration is that part of the GC-MS total-ion-current chromatogram of urinary steroids which contains the corticosteroid metabolites. The elution pattern is very similar to the corresponding area on a GC/FID recording (e.g., Figure 8.11) so there was not doubt as to the identity of the major peaks. However, the mass spectra were inappropriate. For example, whereas the normal spectrum of allotetrahydrocorticosterone shows a strong molecular ion (*m/z* 595) and fragment resulting from loss of methoxy group (*m/z* 564, *M*-31), the spectrum produced during this analysis lacked these ions, and the major high-mass ion was at *m/z* 446 (*M*-149). See Structure 2.

Accurate mass measurement by high-resolution mass spectrometry showed that these steroid components had undergone Beckman fission prior to electron-impact ionization [52]. This probelm was rectified by vigorous deactivation of the glass restrictor, by slight lowering of the interface temperature, and by positioning the interface restrictor closer to the poles of the ion source. These problems should emphasize the need for vigilance in the connection of capillary columns to mass spectrometers—in particular, the effective removal of all active sites in the injector and interface.

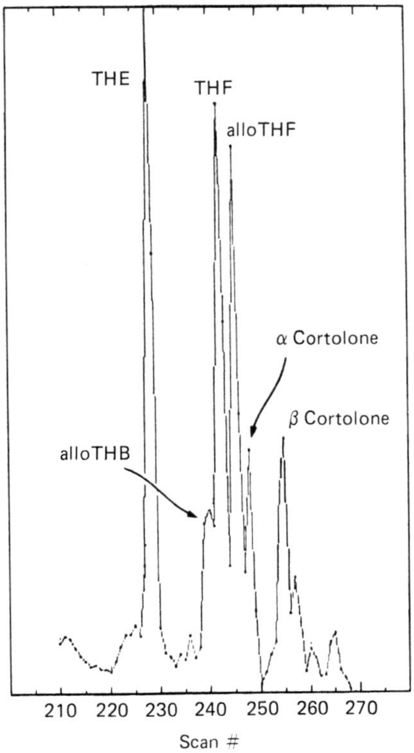

Figure 8.18 Part of reconstructed total-ion current record for GC-MS analysis of urinary steroids. The peaks in this section are metabolites of cortisol and corticosterone. Conditions: Hitachi M-52 GC-MS, 50-m glass OV-1 SCOT column; programmed 200–280°, 5°/min; splitless injection.

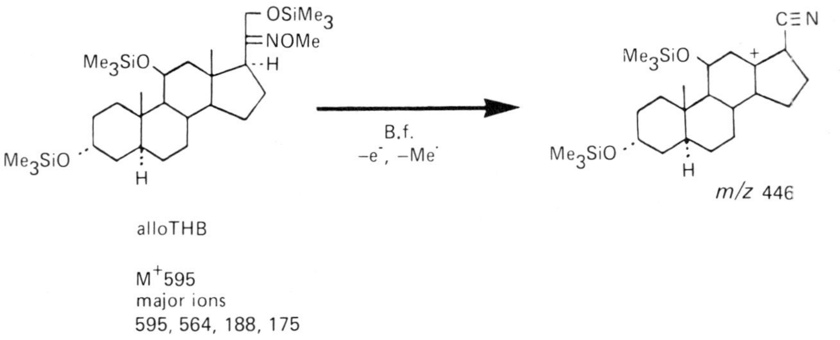

(Structure Group 2)

V. APPLICATIONS

A. Analysis of Urinary C_{19} and C_{21} Steroids

1. *Disorders of Steroid Synthesis*

General: Many of the publications on capillary column chromatography of steroids relate to investigation of patients with disorders of steroid biosynthesis. Extensive studies have been made on the following forms of congenital adrenal hyperplasia: 21-hydrolase deficiency [15,53,54], 3β-hydroxysteroid dehydrogenase deficiency [53,55], 11β-hydroxylase deficiency [15,57,58], and 17α-hydroxylase deficiency [1,59,60]. In addition, studies have been carried out on the excretion of steroids by patients with abnormal aldosterone secretions, e.g., pseudohypoaldosteronism [61,62], hypoaldosteronism [63,64], and certain preterm infants [65]. Many representative capillary GC steroid profiles from these disorders have been published collectively in a monograph by the author and his collaborators [66].

A comprehensive description of the role of capillary column chromatography in investigations of all these disorders cannot be undertaken in this chapter, and discussion of cortisol biosynthetic defects will be limited to the two forms of *hypertensive* congential adrenal hyperplasia: the 11β-hydroxylase and 17α-hydroxylase defects.

11β-Hydroxylase deficiency: Deficiency of this enzyme prevents cortisol synthesis so there is little feedback control of adrenal steroid synthesis, and adrenal hyperplasia results. There is hypersecretion of DOC and Substance S since these are the steroids immediately before the metabolic block in the aldosterone and cortisol biosynthetic pathways, respectively. The hypersecretion of DOC gives rise to the hypertension seen in patients with this disorder.

Steroid profiles of patients with 11β-hydroxylase deficiency have been published [15,57,58,66] and these show elevated excretion of tetrahydroSubstance S, hexahydroSubstance S and adrenal androgen metabolites (androsterone and etiocholanolone, (see Structure Group 3).

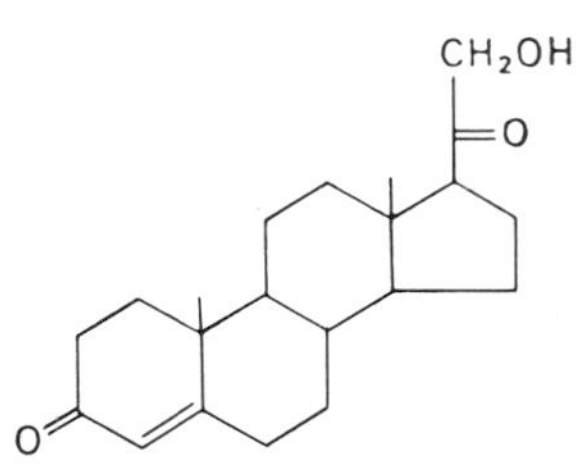

DOC

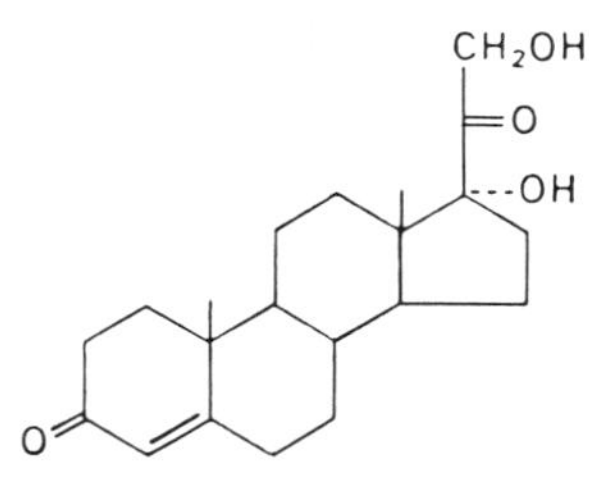

Substance S

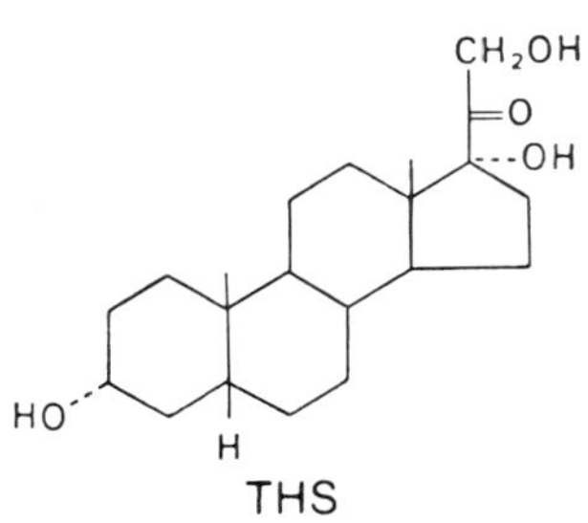

THS

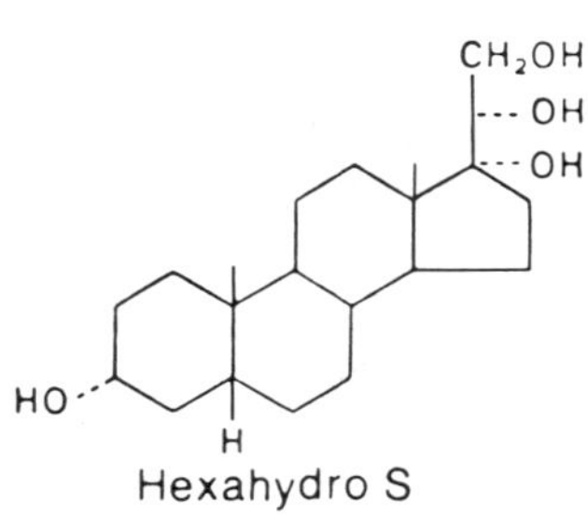

Hexahydro S

(Structure Group 3)

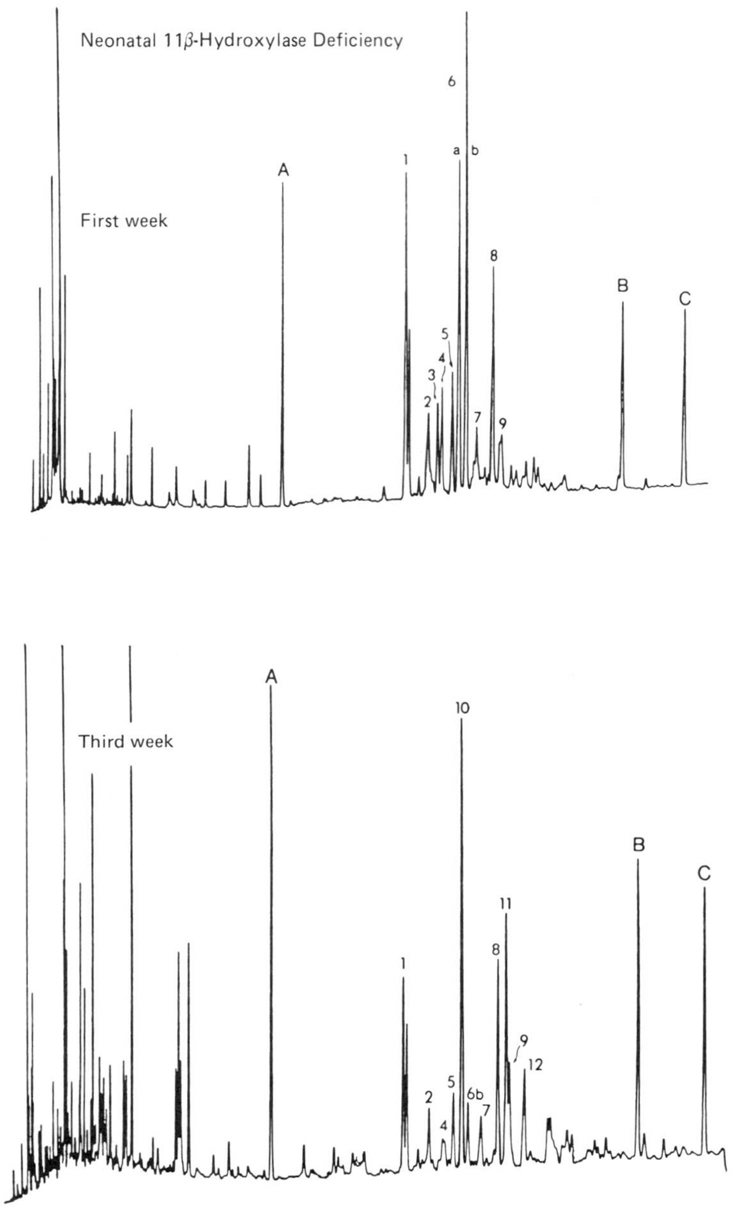

Figure 8.19 Diagnosis of 11β-hydroxylase deficiency in the neonatal period. The upper and lower chromatograms illustrate profiles obtained during the first and third weeks of life, respectively. The following steroids are numbered: 1, 16α-hydroxyDHA (double peak); 2, 16β-hydroxyDHA; 3, 16-oxo-androstenediol; 4, 15β,16α-dihydroxyDHA + 5-androstene-3β,16β,17α-triol; 5, 5-androstene-3β,16α,17β-triol; 6, 16α,18-dihydroxyDHA (double peak, a and b); 7, androstenetetrol; 8, 16α-hydroxypregnenelone; 9, androstenetetrol; 10, tetrahydroSubstance S; 11, hexahydroSubstance S; and 12,6α-hydroxy-tetrahydroSubstance S. The chromatogram obtained during the first few days of life is totally normal and shows no evidence for biosynthetic abnormality. The diagnostic Substance S (17-deoxycortisol) metabolites appear in the urine during the second week. Conditions: 20-m, OV-1, WCOT column programmed 180–280°, 3°/min; solid injection.

Even though tetrahydroDOC excretion is highly elevated in the disorder, the quantitative excretion of this compound is small compared to the aforementioned compounds, and accurate quantification by capillary GC is difficult. We have recently analyzed urine from a neonate with this condition [58]. Like the 21-hydroxylase deficiency [53], this disorder is extremely difficult to diagnose in the first days of life because of the predominance of the fetal 3β-hydroxy-5-ene steroid metabolites. In fact, the urinary steroid profile obtained during the first three days of independent life in one patient with 11β-hydroxylase deficiency does not appear to differ from normal. During this period the neonate is probably relying on the remnants of maternal cortisol to some extent so excessive stimulation of the cortisol biosynthetic pathway is not taking place. However, by the second week of life, Substance S metabolites are clearly seen in the urine. This change in the profile during the neonatal period is evident in the profiles illustrated in Figure 8.19. In addition to the normal metabolites of Substance S, neonates also produce a *hydroxy*tetrahydroSubstance S which on the basis of corresponding data for cortisol metabolism [67] is probably a 6α-hydroxy metabolite.*

17α-Hydroxylase deficiency: Congenital adrenal hyperplasia caused by deficiency of steroid 17α-hydroxylase is the most recent form of the disorder to be characterized and, since the first report, only just over 20 cases have been described in the literature. Steroid 17α-hydroxylation is essential not only for adrenal biosynthesis of cortisol but also for the formation of androgens and estrogens by both gonads and adrenals. Due to this enzymatic block, the major secretory products of the adrenal glands are DOC and corticosterone; it is the former compound which is largely responsible for the hypertension associated with the disorder.

The urinary steroid profile in this condition was originally studied by Honour et al. [1] and Shackleton et al. [59]. These results were verified by Lisboa et al. [60] who also employed gas chromatography on glass capillary columns and GC-MS for steroid identification.

As expected, the urine from affected individuals contains substantial amounts of tetrahydro metabolites of corticosterone (e.g., tetrahydroCompound A, allotetrahydrocorticosterone). However, the profiles also contain steroids which, in addition to having undergone hepatic metabolism, have been considerably modified by enterobacteria. The explanation for that is that, unlike cortisol and many other steroids, extensive biliary excretion of corticosterone metabolites occurs, rendering them susceptible to bacterial transformation in the gut. Most of the metabolites are reabsorbed and appear in the urine. (See Structure Group 4.)

A steroid profile of a patient with 17α-hydroxylase deficiency is illustrated in Figure 8.20. The major differences from normal are: (1) the absence of C_{19} steroids and cortisol metabolites; (2) the presence of large amounts of conventional corticosterone metabolites (marked by a white star); (3) the presence of enterobacterial corticosterone metabolites (white star and black star); and (4) the presence of a series of pregnane and pregnene steroids which may or may not be metabolites of DOC or corticosterone (black star). The short retention times of the latter compounds suggest they have a 17α side chain and therefore must have been transformed by enterobacteria. The major enterobacterial metabolites of corticosterone are 3α,20α-dihydroxy-5α(and β)-pregnan-11-one, 5α(and β)pregnane-3α,11β,20α-triol, and a compound provisionally identified as 1,3,20-trihydroxypregnan-11-one (see Fig 8.1). These metabolites

*Confirmation of the identity of this compound as 6α-hydroxytetrahydroSubstance S has now been obtained [116].

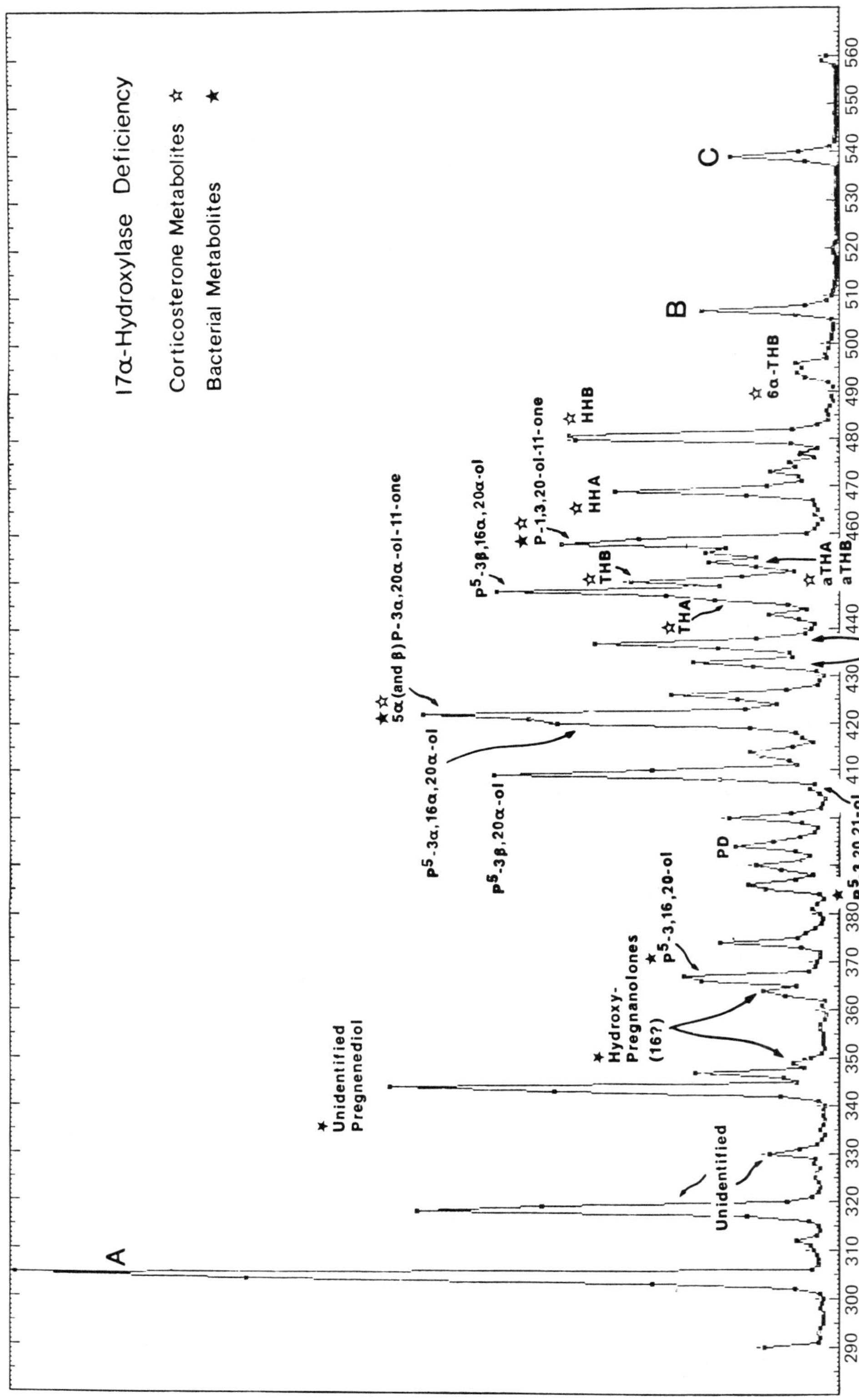

17α-Hydroxylase Deficiency
Corticosterone Metabolites ☆
Bacterial Metabolites ★
A
B
C
★ Unidentified Pregnenediol
Unidentified
★ Hydroxy-Pregnanolones (16?)
★ P5-3,16,20-ol
PD
★ P5-3,20,21-ol
P5-3β,20α-ol
P5-3α,16α,20α-ol
★☆ 5α(and β)P-3α,20α-ol-11-one
P5-3β,16α,20α-ol
☆ THA
☆ THB
★☆ P-1,3,20-ol-11-one
☆ HHA
☆ HHB
☆ 6α-THB
☆ aTHA
aTHB
★☆ 5β(and 5α)-3α,11β,20α-ol
290 300 310 320 330 340 350 360 370 380 390 400 410 420 430 440 450 460 470 480 490 500 510 520 530 540 550 560

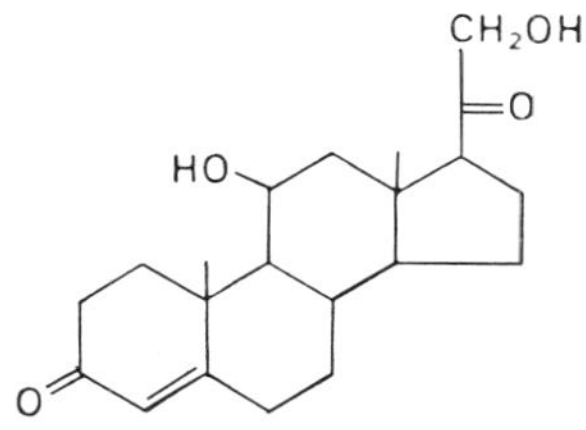

Corticosterone(B)

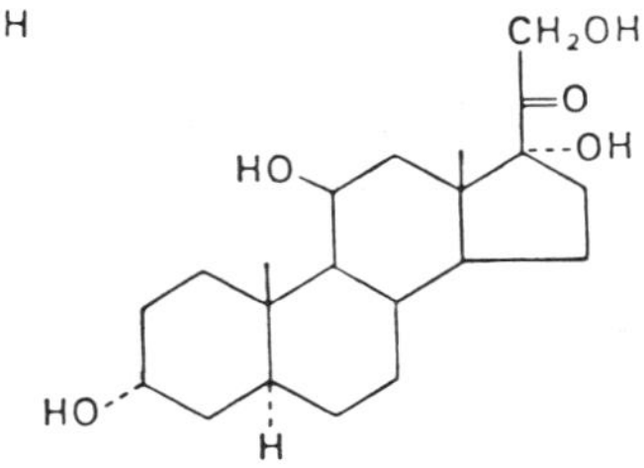

alloTHB

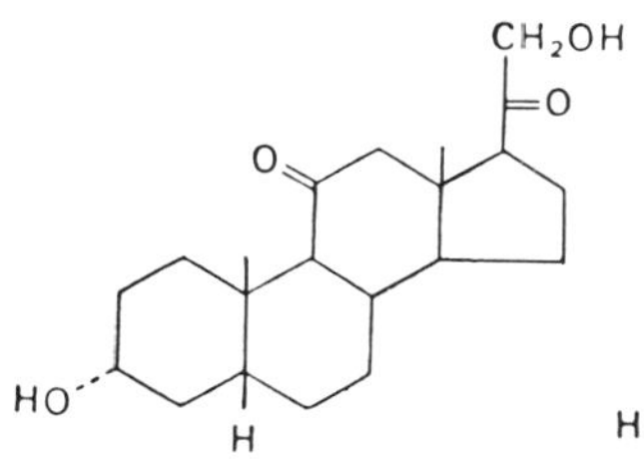

THA

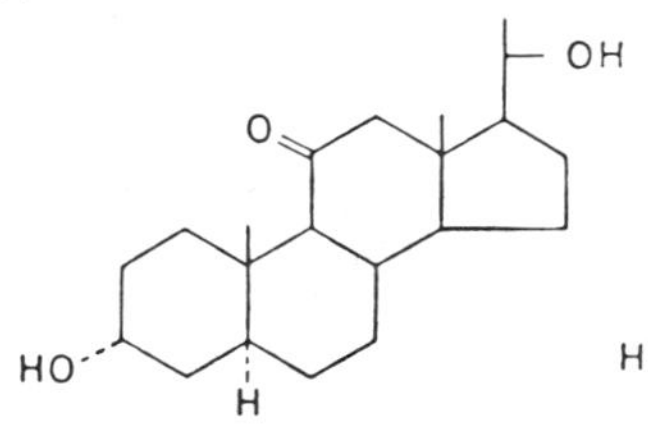

11-Oxo-Pregnanediol
5α and β

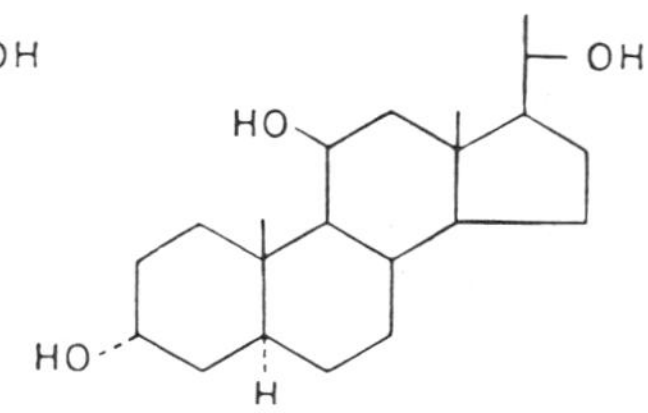

11β-OH-Pregnanediol
5α and β

(Structure Group 4)

are formed by 21-dehydroxylation undertaken by the obligate anaerobe *Eubacterium lentum*. Between 9 and 25% of such metabolites have been found in the neutral steroid profiles of 15 patients with 17α-hydroxylase deficiency (Shackleton, Biglieri, and Honour, unpublished data). None of the urine samples from the affected individuals contained significant amounts of any C_{19} steroids or metabolites of cortisol.

The profile of urine from a neonate with the disorder [68] appeared similar to that of 21-hydroxylase deficiency (high 16α-hydroxypregnenolone/16α-hydroxyDHA ratio), and differentiation of these two conditions by GC alone would be virtually impossible at this period of life [66]. The presence of small amounts of 6-hydroxytetrahydroCompound A was confirmed by GC-MS analysis. At this age the infant is host to few gut microorganisms so significant bacterial metabolism can be discounted. Within a few months, this infant gave a urine profile similar to other patients with 17α-hydroxylase deficiency.

Figure 8.20 Total steroid profile (GC-MS reconstructed total-ion-current recording) in 17α-hydroxylase deficiency. This example is more complex than is usually seen in this disorder and can be considered to contain four populations of steroids: 1, classical hepatic metabolites of corticosterone (e.g., THA, alloTHB, etc.); 2, hepatic/enterobacterial metabolites of corticosterone (e.g., 5β(and α)pregnane-3α,11β,20α-triol and pregnane-1,3,20-triol-11-one); 3, conventional 3β-hydroxy-5-ene steroids (e.g., 5-pregnene-3β,20α-diol and 5-pregnene-3β,16α,20α-triol); and 4, pregnane and pregnene (5-ene or other unsaturation) steroids with shorter retention time than appropriate for *conventional* structures. Some of these metabolites probably have a 17α side-chain which was introduced during bacterial metabolism. Conditions: Hitachi M-52 GC-MS, 20-m OV-1 WCOT column programmed 180–270° at 5°/min; splitless injection. This analysis was carried out in collaboration with Dr. Paul Fennessey of the University of Colorado Health Sciences Center.

2. Disorders of Steroid Metabolism

The advent of glass capillary chromatography has permitted the detection of minor abnormalities in steroid biosynthesis and metabolism. In this section I shall discuss, primarily, variations in the catabolism of steroid hormones between normal individuals and patients with disease conditions associated with steroid hormone imbalance. The conditions that will be discussed are diverse, namely, breast cancer, anorexia nervosa, a metabolic defect associated with hypertension, and pseudohermaphroditism. However, they all have in common a divergence from normal in their metabolism of steroids. The enzymes of "abnormal" activity are responsible for 5α reduction, 5β reduction, 3α-hydroxysteroid dehydrogenation, and 11β-hydroxysteroid dehydrogenation.

Metabolite ratios in normal men and women: Pfaffenberger and Horning [69] made a study of the excretion of two pairs of 5α and 5β steroids, androsterone/etiocholanolone, and tetrahydrocortisol/allotetrahydrocortisol, (see Structure Group 5), by normal postpubertal men and by women between puberty and the menopause. An example of one of their published chromatograms is illustrated in Figure 8.21, and it is similar to those published by us (see Figure 8.3) and other workers for normal individuals.

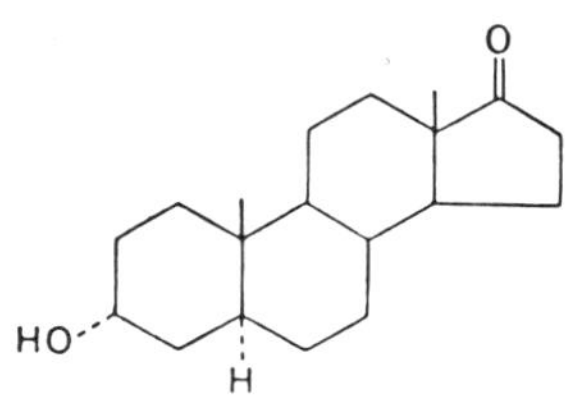

androsterone

5α

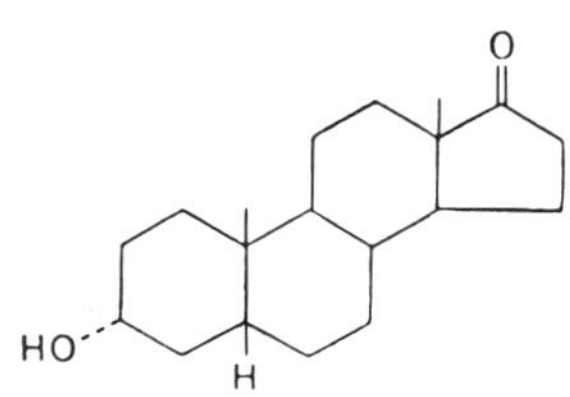

aetiocholanolone

5β

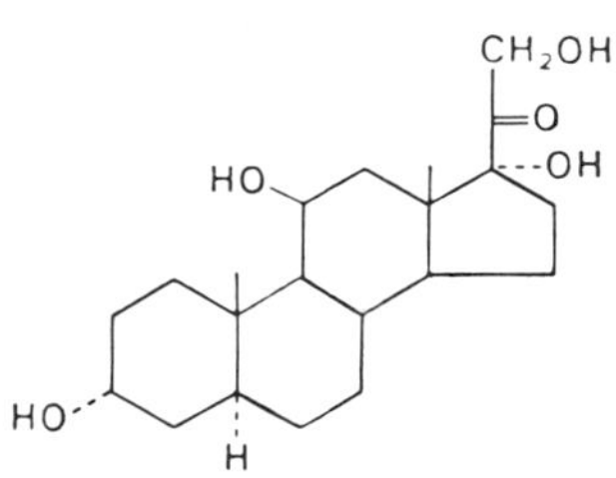

allo-THF

5α

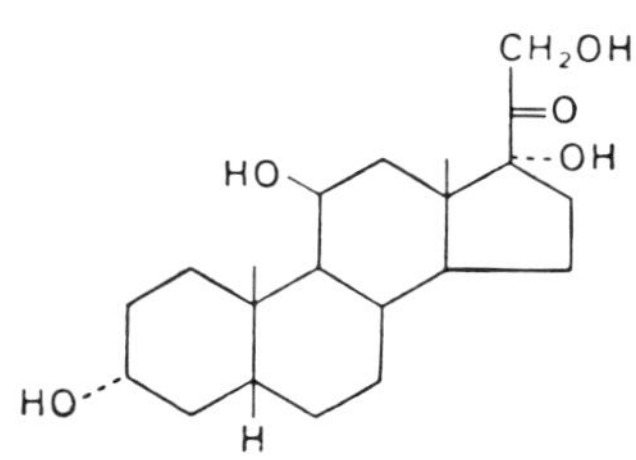

THF

5β

(Structure Group 5)

The quantitative results for ratio measurement are shown diagrammatically in Figure 8.22. In normal males the ratio of etiocholanolone to androsterone is always less than unity, while in females it can range from 0.2 to 1.8, and Pfaffenberger and Horning believe there are two discernable groups: one in which the ratio of etiocholanolone to androsterone is greater than 1 and the ratio of tetrahydrocortisol to allotetrahydrocortisol is between 2 and 4.5 (Pattern A females), and a second group in which the etiocholanolone/androsterone ratio is less than 1 and the tetrahydrocortisol/allotetrahydrocortisol ratio is less than 1.6 (Pattern B females). The Pattern B females therefore have a steroid metabolic profile which does not differ from that of males.

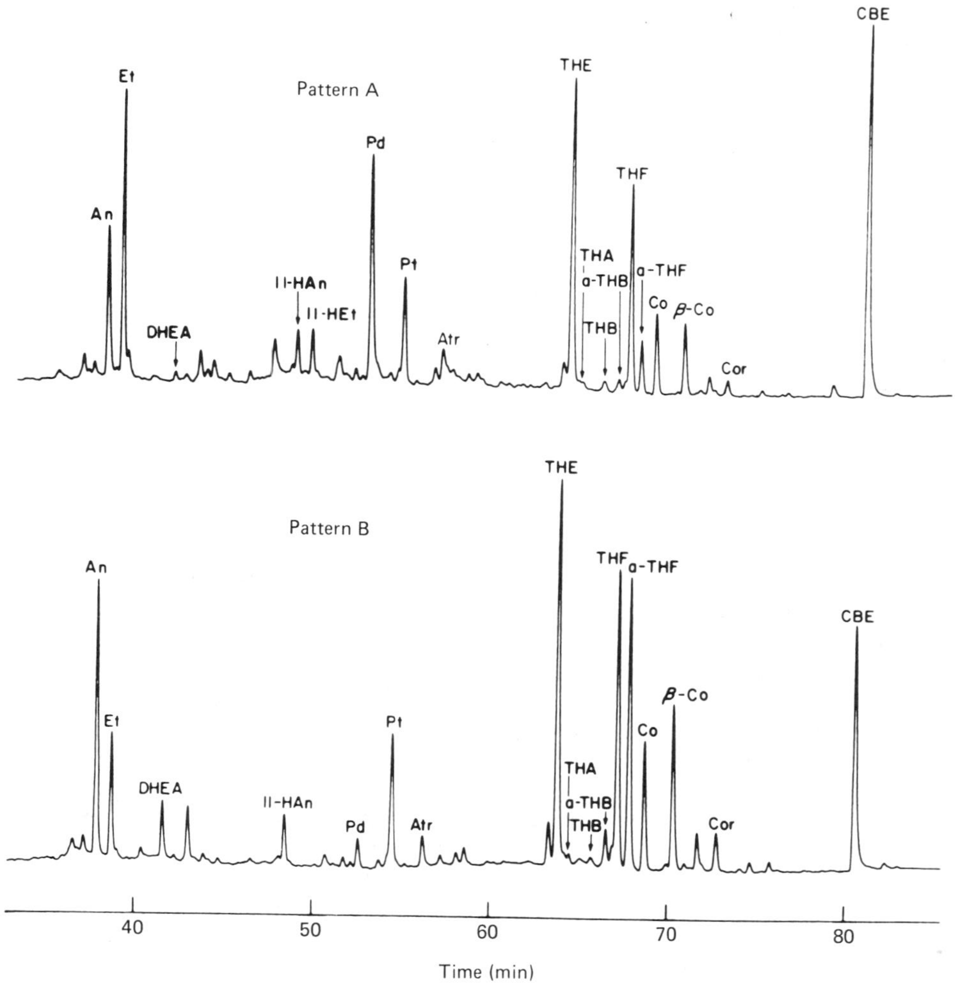

Figure 8.21 Steroid metabolic profiles of two premenopausal females. Cholesteryl butyl ether (CBE) was used as internal standard. The upper profile shows pattern A ratios of Et to An (2:1) and THF to aTHF (3:1) while the lower chromatogram shows pattern B ratios (Et/An, 1:2, THF/aTHF, 1:1). Conditions: 60-m SE-30 Silanox column, programmed 200–290° at 1°/min; split injection. Figure reproduced from Pfaffenberger and Horning [69] with permission of authors and publisher.

Although it is quite clear from these studies that there are considerable differences between males and females in the activities of 5α- and 5β-reductases, no evidence is presented that suggests there are really two *distinct* groups of females. A greater range of "normality" could equally well be the interpretation put on these results. The results are interesting, however, since they provide evidence for the difference between male and female hepatic metabolism which may be the result of neonatal *imprinting* [70]. This aspect is discussed in some detail in Ref. 69, but their conclusions are tenuous. Nevertheless,

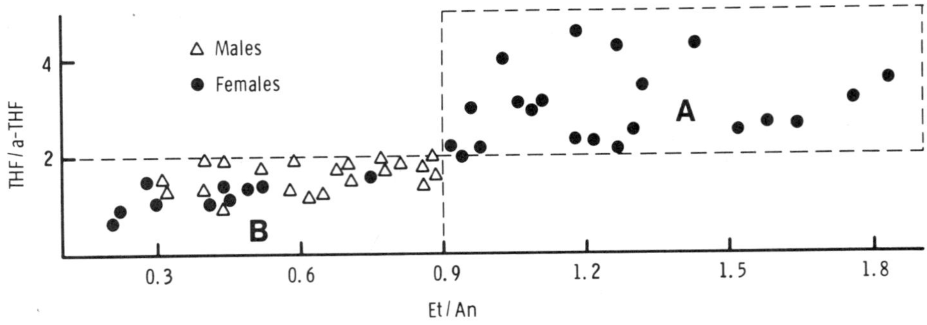

Figure 8.22 Ratio values (Et/An and THF/aTHF) for premenopausal females and male subjects. Two female groupings, A and B, are evident; one group has ratio values essentially identical to those of males. Figure reproduced from Pfaffenberger and Horming [69] with permission of authors and publisher.

the study provided new data on steroid excretion in normals and illustrated the power of capillary column analysis for this type of investigation.

Steroid metabolism in patients with breast lesions: The studies of Pfaffenberger and Horning were expanded by the examination of urinary steroid profiles in patients with breast lesions [71]. All premenopausal women with or without benign breast lesions had metabolite ratios indistinguishable from type A and B normals, but postmenopausal patients showed a higher ratio of 5β to 5α steroids than was obtained by analysis of type A, and this was referred to as type C. No normal premenopausal patients studied gave a type C profile. These results, although perliminary, are of interest as they reinforce the evidence provided almost two decades ago that patients with breast cancer have altered metabolism of steroids hallmarked by altered ratio of urinary metabolites [72]. However, this remains a very controversial issue, and multicomponent steroid analysis has really not contributed to clarification. Leunissen [15] measured many androgen and corticosteroid metabolites in urine from 17 breast cancer patients but could find no significant difference from normal. In fact, the results obtained suggested that cancer patients had slightly higher mean excretions of 11-deoxy-17-oxosteroids than normal women, which contrasts with earlier studies [72]. Part of the problem may rest in the difficulty of measuring absolute amounts of steroids excreted daily. In the studies of Leunissen [15], no attempt was made to accurately collect complete samples, and values were expressed relative to creatinine excretion. However, even "24-hr" excretions can seldom be relied upon when undertaken outside a strictly controlled environment. To detect statistically significant abnormalities in quantitative excretions of steroids, particular care would also have to be taken over age-matching of the controls and the stage of the menstrual cycle in premenopausal females.

Altered metabolism in *anorexia nervosa*: Vanleuchene et al. [14] have made a detailed study of 5α and 5β ratio and 11β-hydroxy to 11-oxo ratios in anorexia nervosa. Following analysis on OV-1 capillary columns, they measured androsterone, etiocholanolone, tetrahydrocortisone, tetrahydrocortisol, and allotetrahydrocortisol. Their results were similar to those published earlier by Bradlow et al. [73] and demonstrate a "hypothyroid-like state" whereby 5α reduction and 11-hydroxysteroid dehydrogenation are decreased. These results are illustrated in diagrammatic form in Figure 8.23 which shows the ratios of tetrahydrocortisone to tetrahydrocortisol and androsterone to etiocholanolone. It is probable that in this condition the hepatic activity of 5α-reductase and 11β-hydroxysteroid dehydrogenase are directed by triiodothyronine concentrations, and it is this control mechanism which affects the altered steroid metabolite excretion.

There is some evidence that cortoic acid (Section V.C.) metabolism is altered in anorexia nervosa. Excretion of 20β-cortolic acid exceeds that of the 20α epimer while in normals the reverse is true.

5α-Reductase deficiency: Another condition which gives rise to markedly altered ratios of 5α and 5β urinary steroids is 5α-reductase deficiency. This enzyme is essential for the conversion of testosterone to 5α-dihydrotestosterone, the hormone jointly responsible with testosterone for male differentiation. Since in this case this enzyme is involved in hormone synthesis rather than metabolism, it is questionable whether the condition should be discussed as a defect of metabolism.

Affected males frequently have genitalia of female appearance and may have been brought up as females until puberty when the testosterone surge results in muscular development and masculinization [74]. Diagnosis of this disorder has been achieved by the measurement of the ratio of the major urinary 5α-reduced steroid androsterone to the major 5β-reduced steroid etiocholanolone. In affected individuals this ratio is markedly reduced compared to normal. Other diagnostic tests include monitoring the metabolism of radioactive steroids administered to the affected individuals and investigation of the metabolism of steroid precursors in fibroblasts grown in organ culture. These tests frequently give ambiguous results.

Although the clinical problem arises from deficiency in production of an important androgen, it has been found that lack of 5α-reductase affects the

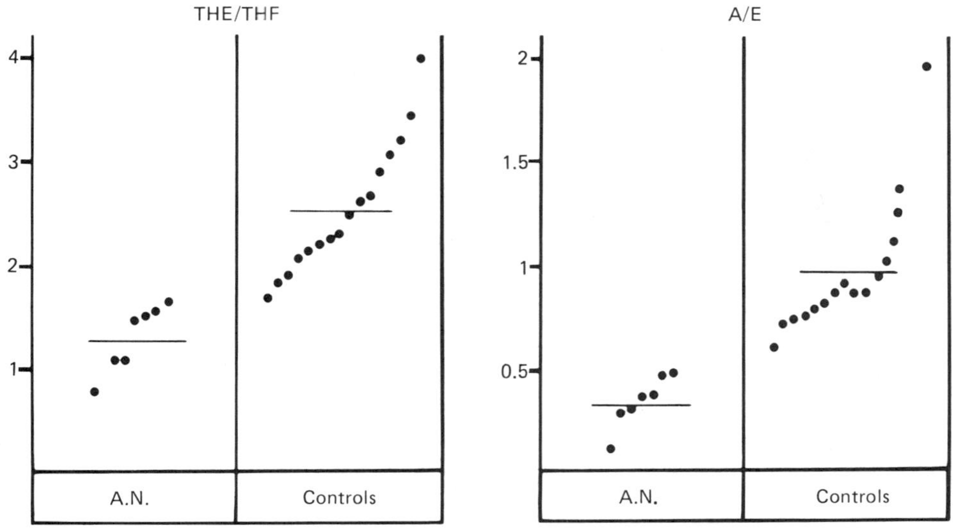

Figure 8.23 Ratios of tetrahydrocortisone (THE) to tetrahydrocortisol (THF) and of androsterone (A) to etiocholanolone (E) excretions for patients with anorexia nervosa (A.N.) and for a control group. Figure reproduced from Vanluchene et al. [14] with permission of authors and publisher.

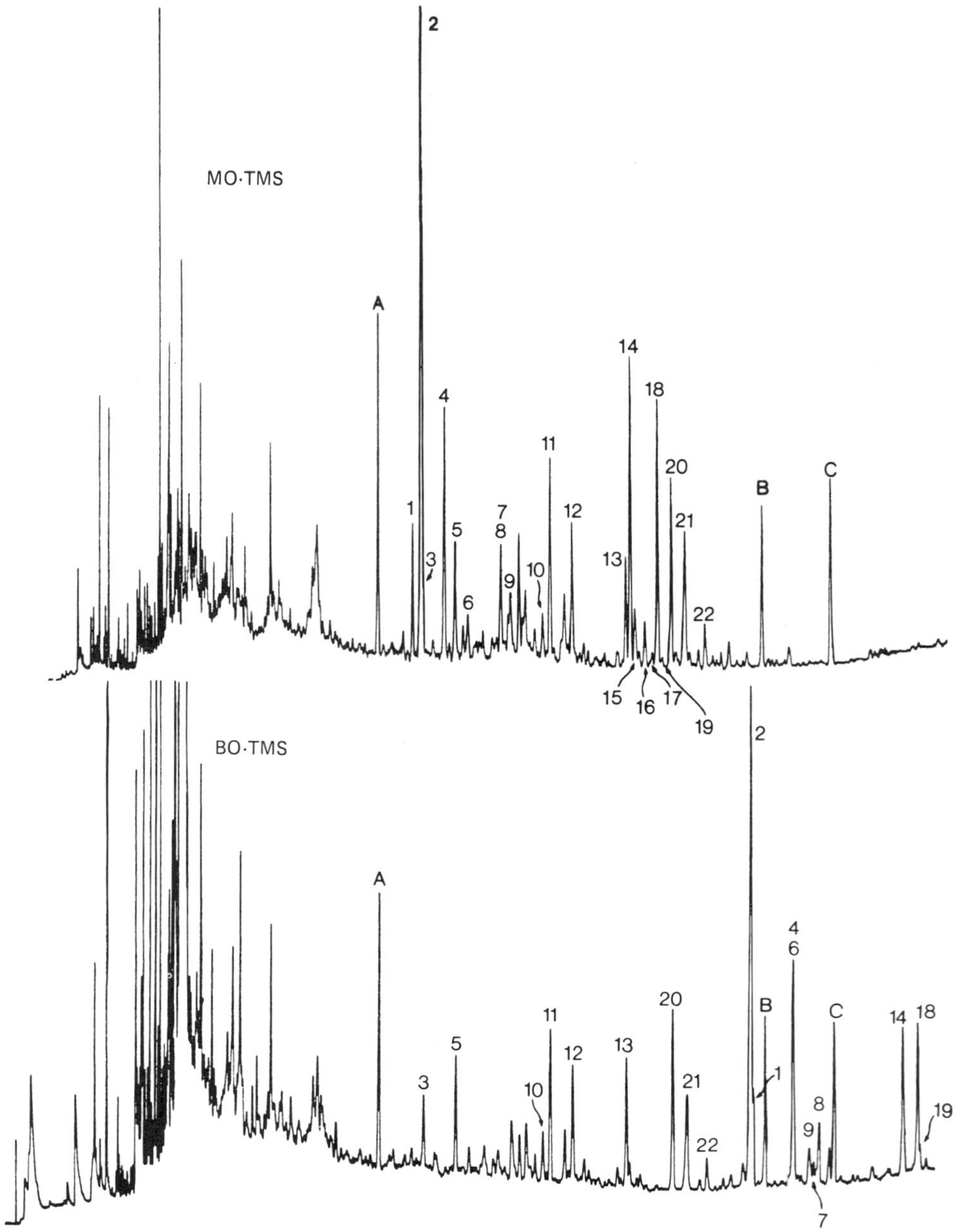

Figure 8.24 5α-Reductase deficiency. Separation of urinary steroids as methyloxime or benzyloxime trimethylsilyl ethers. The following steroids are numbered: 1, androsterone; 2, etiocholanolone; 3, 5β-androstane-3α,17β-diol; 4, dehydroepiandrosterone; 5, 5-androstene-3β,17β-diol; 6, 11-oxoetiocholanolone; 7, 11β-hydroxyandrosterone; 8, 17α-hydroxypregnanolone; 9, 11β-hydroxyetiocholanolone; 10; pregnanediol; 11, pregnanetriol; 12, 5-androstene-3β,16α,17β-triol; 13, 5-pregnene-3β,17α,20α-triol; 14, tetrahydrocortisone; 15, tetrahydroCompound A; 16, tetrahydrocorticosterone; 17, allotetrahydrocorticosterone; 18, tetrahydrocortisol; 19, allotetrahydrocortisol; 20, α-cortolone; 21, β-cortolone, β-cortol; and 22, β-cortol. The steroids underlined are all 5α-reduced steroids and have very low excretions in this disorder. Conditions: 25-m OV-1 fused-silica columns, programmed 180-230° at 3°/min. Solid injection.

metabolism of all steroids. In addition to the androsterone deficiency, there is reduced excretion of 11-oxoandrosterone, 11β-hydroxyandrosterone, allotetrahydrocorticosterone, and allotetrahydrocortisol. Simultaneous measurement of several pairs of 5α- and 5β-reduced steroid metabolites therefore should increase the certainty of diagnosis.

This is a problem ideally suited to capillary column chromatographic analysis. The retention index differential between 5α and 5β metabolites is so slight that their separate measurement on packed columns is impractical or impossible. Even on glass capillary columns, complete resolution of all components of interest is not possible on a single-column phase. For example, 11-oxoandrosterone is not resolved from 11-oxoetiocholanolone on nonpolar stationary phases (SE-52, OV-1), and accurate measurement of 11β-hydroxyandrosterone is not possible because of coelution with 17α-hydroxypregnanolone. However, by utilizing the benzyloxime (BO) derivative, 11β-hydroxyandrosterone can be resolved from 17α-hydroxypregnanolone and 11β-hydroxyetiocholanolone. We have analyzed urinary steroids from about 35 patients with 5α-reductase deficiency (C. H. L. Shackleton, J. Imperato-McGinley, and R. E. Peterson, unpublished), and typical profiles of the MO-TMS and BO-TMS derivatives are illustrated in Figure 8.24. Note the particularly low levels of the 5α steroids: androsterone (1), allotetrahydrocorticosterone (17), and allotetrahydrocortisol (19) in the chromatogram of the MO-TMS derivative. The profile of the BO derivative is included to demonstrate the resolution of 11β-hydroxyandrosterone (7) from 11β-hydroxyetiocholanolone (9) and 17α-hydroxypregnanolone (8).

5α/5β Ratio measurements have been made for the patients studied and these are illustrated in Figure 8.25. It can be seen that this form of analysis gives

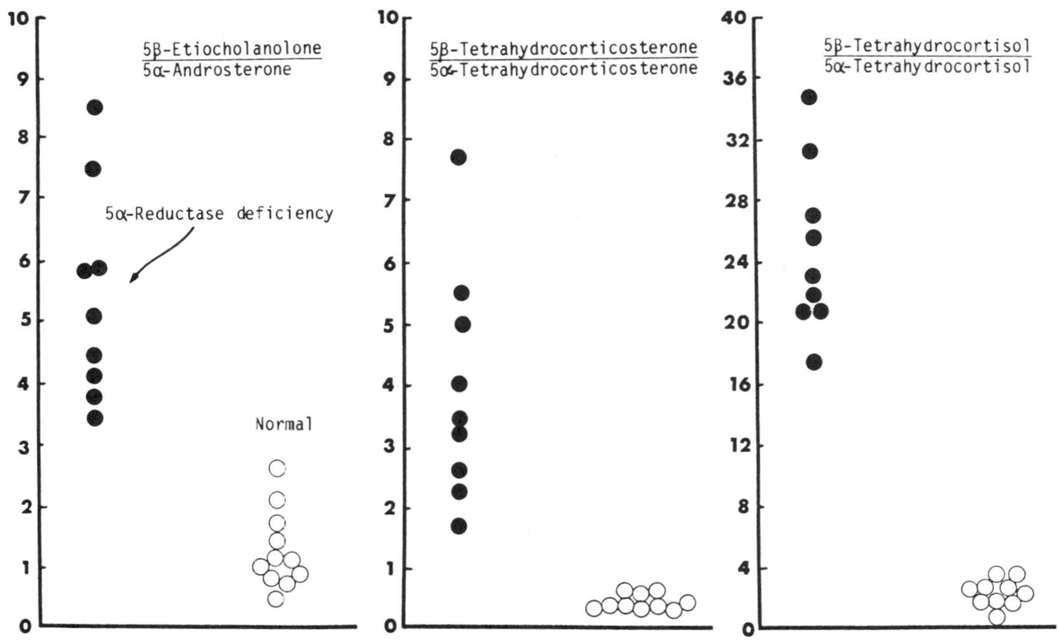

Figure 8.25 Ratio of 5β to 5α steroids in normal individuals and nine patients with 5α-reductase deficiency. Data obtained by capillary GC quantification. (From C. H. L. Shackleton, J. Imperato-McGinley, and R. E. Peterson, unpublished.)

clear delineation of the condition and that, although the individuals are affected clinically by androgen deficiency, the most reliable discriminant is the ratio of tetrahydrocortisol to allotetrahydrocortisol.

Altered steroid catabolism in hypertension: Hypertension frequently results from excessive mineralocorticoid secretion due to a variety of causes, for example: (1) excessive production of aldosterone, (2) excessive production of other mineralocorticoids (e.g., deoxycorticosterone) often associated with abnormal adrenal steroid biosynthesis (e.g., 17α-hydroxylase or 11β-hydroxylase defects), and (3) unidentified hypermineralocorticoid conditions.

The latter case represents a heterogeneous condition in which patients frequently have normal or low aldosterone secretion associated with inappropriately low plasma renin activity. The etiology of this type of disorder (frequently termed low-renin hypertension, LRHT) continues to puzzle endocrinologists. Results of provocative therapeutic tests in many of these individuals suggest the overproduction of a steroid hormone with mineralocorticoid properties, but to date no conclusive identification of an unknown steroid with these properties has been obtained.

This diverse condition would seem ideally suited to study by multicomponent gas chromatography since the excretion of new steroids or inappropriate amounts of known steroids should be readily detected. We have examined urine from several adult patients with the disorder but could find no significant abnormality in steroid production or excretion in most cases. However, differences may exist, and examination of a much larger sample of a well-defined group may be required to uncover them.

There is one subgroup of children with low-renin hypertension who do have metabolic profiles different from normal. It is significant that these abnormalities were initially diagnosed by glass capillary gas chromatography. Other techniques could have been used but they would have been more laborious. In 1974, we analyzed urine from a child with hypertension (Shackleton and Zachman, unpublished) and found that there were marked abnormalities in the relative amounts of individual corticosteroids present. In particular, the relative excretion of tetrahydrocortisone was greatly diminished, and the excretion of tetrahydrocortisol and allotetrahydrocortisol was relatively high. The predominant 20-dihydro metabolites were also *cortols* rather than the normally favored *cortolones*, and significant amounts of cortisol metabolites retaining the 3-oxo-4-ene group were present.

Since that time, three other young patients have been investigated who give similar evidence of altered steroid metabolism. Two patients studied by Ulick, New, and co-workers [75] also showed evidence for elevated excretion of free cortisol and metabolites of cortisol (e.g., 5- and 20-dihydrometabolites) and reduced excretion of steroids with the 11-carbonyl group (e.g., tetrahydrocortisone) relative to steroids with the 11β-hydroxyl group (tetrahydrocortisol). Results of capillary column GC-MS analysis carried out on their two patients are illustrated in Figure 8.26. The main points to note are the low excretion of tetrahydrocortisone relative to the tetrahydrocortisols and the presence of significant amounts of tetrahydroSubstance S (THS).

A fourth case of this "disorder" has also recently been investigated in our laboratory [76]. This patient was a four-year-old with hypertension, hypokalemia, low renin, and low aldosterone. Figure 8.27 shows a gas chromatogram of the total steroids obtained from analysis of a urine specimen obtained while the child was being administered ACTH. The analyses show the now familar pattern of high excretion of unmetabolized or partially metabolized cortisol and low conversion to metabolites with the 11-carbonyl function. In addition, there is a high excretion of 5α-tetrahydrocortisol relative to the 5β metabolite.

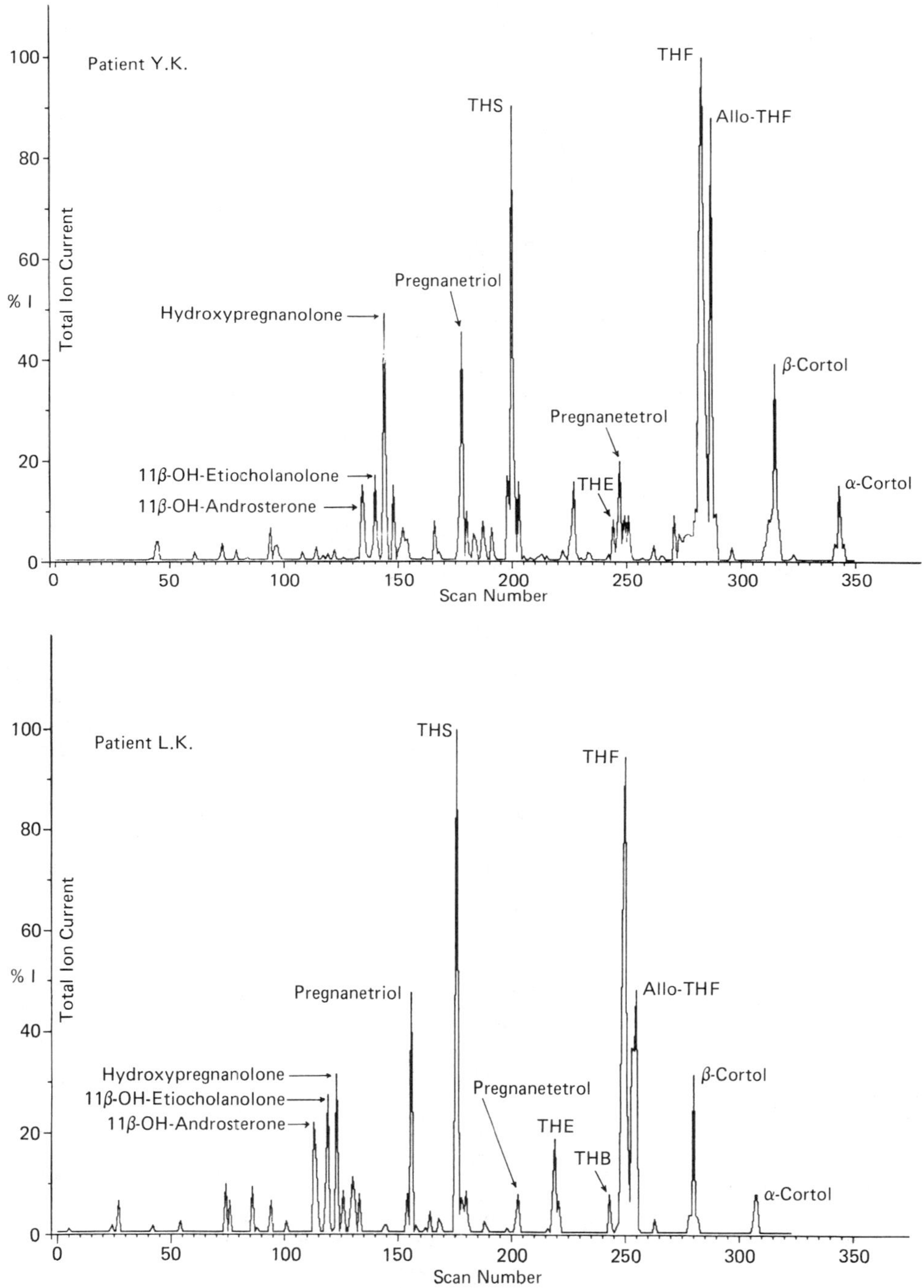

Figure 8.26 Computer-generated plots of mass spectrometer ion current vs scan number during capillary gas chromatography of the urinary steroid glucuronide fraction of two young hypertensive patients with altered corticosteroid metabolism. The major abnormality is the reduced amount of tetrahydrocortisone (THE) relative to the tetrahydrocortisols, (THF, alloTHF). Conditions: VG Micromass instrument, 50-m glass OV-101 SCOT column. (From Ref. 75.)

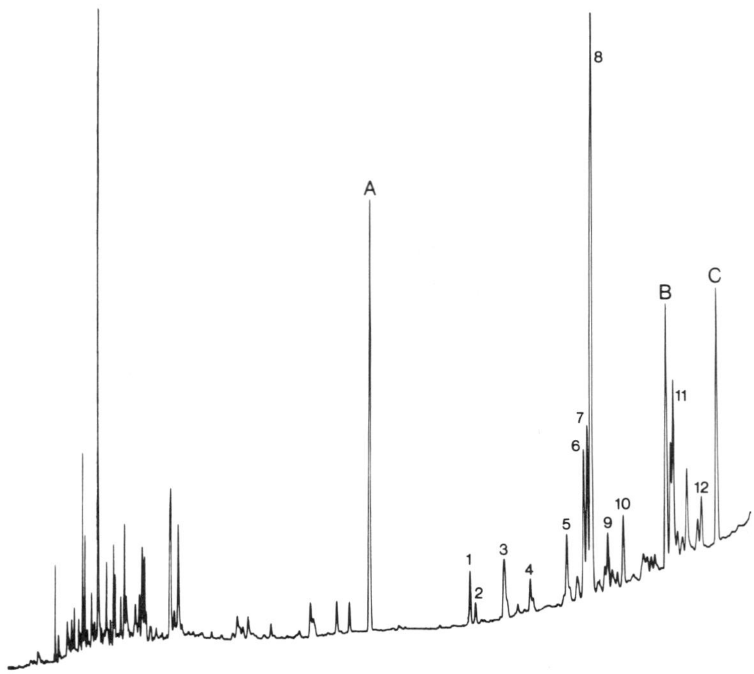

Figure 8.27 Capillary gas chromatographic profile of hypertensive child with abnormal corticosteroid metabolism. Peak 1, 11β-hydroxyandrosterone; 2, 11β-hydroxyetiocholanolone; 3, pregnanetriol; 4, tetrahydroSubstance S; 5, tetrahydrocortisone; 6, allotetrahydrocorticosterone; 7, tetrahydrocortisol; 8, allotetrahydrocortisol; 9, β-cortol; 10, α-cortol; 11, cortisol (double peak), and 12, 20α-dihydrocortisol. The major abnormalities in the profile are the reduced tetrahydrocortisone excretion relative to the excretion of the tetrahydrocortisols and the high excretion of free cortisol. Conditions: 20-m OV-1 column programmed 180–280° at 3°/min; solid injection. (From Ref. 76.)

Clearly, juvenile hypertension of the form described is a fascinating disorder, but many questions remain unanswered. In particular, how does the deficiency in hepatic steroid metabolism relate to the apparent hypermineralocorticoid condition? Also, since no adult cases of this type have been described, is the metabolic disorder corrected as the children get older? It is possible that other patients with low-renin hypertension have a similar form of altered hepatic metabolism, and a systematic study of urine from these patients should be undertaken. It should be evident that glass capillary gas chromatography is an appropriate methodology for studying this condition.

3. *Miscellaneous Conditions*

Emotional stress: Spiteller's group has examined the steroid profiles of 10 healthy adults who had just undertaken academic examinations and reports that the excretion of DHA is increased 10- to 20-fold by the stress of this occasion [77]. The excretions of most other steroids were not affected, although accurate measurement of the corticosteroids was not achieved. Typical chromatograms obtained for urine from one adult collected just after the examination and several days later (control specimen) are illustrated in Figure 8.28.

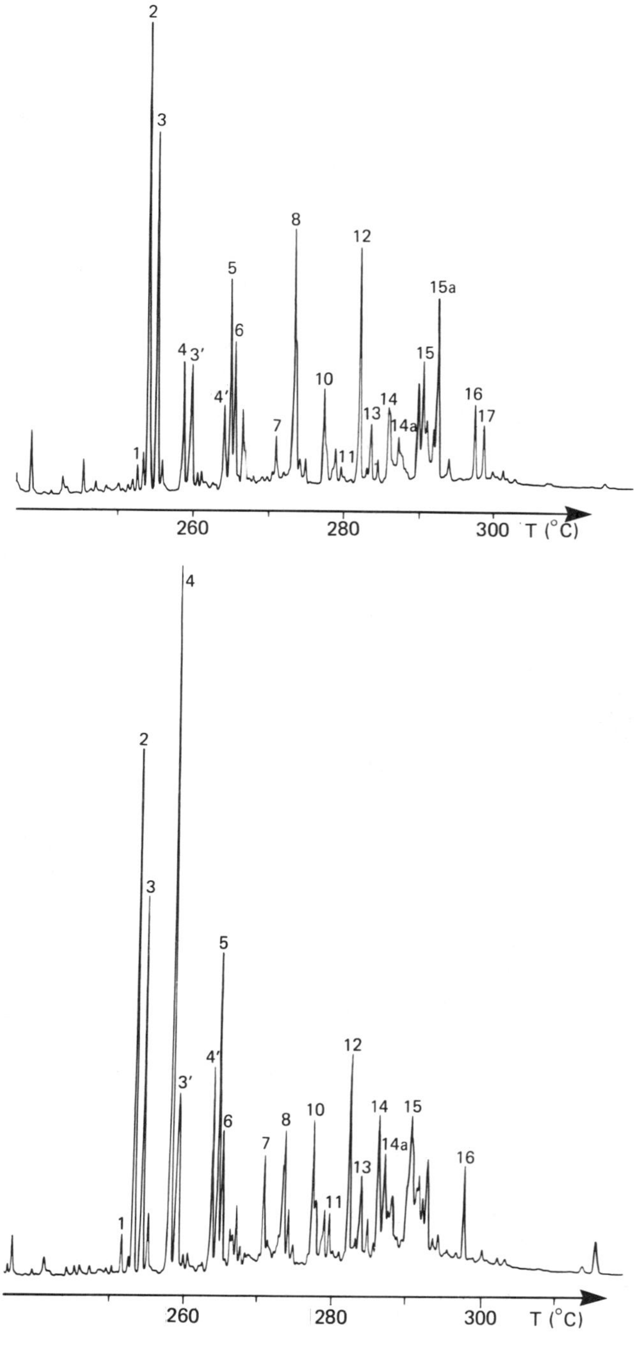

Figure 8.28 Apparent effect of emotional stress on DHA excretion. Steroid profiles of man (a) before and (b) after academic examination. The following steroids are numbered: 1, 3α-hydroxy-5-androsten-17-one; 2, androsterone; 2a, enol of 2; 3, etiocholanolone; 3a, enol of 3; 4, DHA; 4a, enol of 4; 5, androstenediol; 6, 11β-hydroxyandrosterone; 7, 16α-hydroxyDHA; 8, pregnanediol; 10, androstenetriol; 11, 5-pregnen-3,16,20-triol; 12, pregnanetriol; 13, pregnenetriol; 14, 11-oxopregnanetriol; 14a, a pregnanetetrol; 15, tetrahydrocortisone; 15a, a pregnanetetrol; 16, β-cortolone; and 17, α-cortolone. The steroids were analyzed as trimethylsilyl ether derivatives, and some destruction of corticoids occurred. (From Ref. 77.)

Idiopathic hirsutism: In idiopathic hirsutism, the excretion of 17-oxosteroids in the urine may be above or in the upper part of the normal range. Administration of dexamethasone results in suppression of the 17-oxosteroid excretion to low levels. This suggests that they are of adrenal origin. Egger and co-workers [78] have determined the urinary steroids of hirsute women by profile analysis and have reported that in 50% of the cases there is a 10- to 100-fold increase in DHA production compared to normal values. Dexamethasone administration for one week reduced the DHA excretion to almost zero, providing further evidence for the adrenal origin of this steroid.

Effect of alcohol ingestion on urinary steroid profile: An interesting application of steroid profile analysis has recently been reported by Axelson et al. [78a]. Previous studies by this group had shown that the ratio of plasma concentration of 17β-hydroxy- and 17-oxo-steroid monosulfates increased during ethanol metabolism due to the changed redox state of the hepatic pyridine nucleotides. In the latest study, they investigated urinary excretion of several redox couples (e.g., 17β-hydroxy/17-oxo; 11β-hydroxy/11-oxo, and 20α-hydroxy/20-oxo) prior to and following alcohol ingestion. This study was only possible by use of a combination of a method for separation of free steroids, steroid glucuronides, and sulfates (TEAP-LH-20 method, Ref. 25) and high-resolution gas chromatographic analysis. It was found that the ratio between the excretion rates of 17β-hydroxy- and corresponding 17-oxo-steroid monosulfates increased 2–5 times, but neither did 20α(and β)-hydroxy/20-oxo-steroid couples form from cortisol, nor did various 11β-hydroxy/11-oxo-steroid couples in the glucuronide fraction change toward a more reduced state during ethanol metabolism. Thus, these latter ratios do not appear to be dependent on ethanol oxidation. The excretion of glucuronides of 11-oxygenated C_{19} steroids relative to that of cortisol metabolites of the C_{21} series decreased slightly, indicating an inhibition of side chain cleavage.

4. *The Perinatal Period*

Infancy urine: Gas chromatography and GC-MS have played a dominant role in the analysis of steroids produced by the human fetus and newborn. It was really only in the late 1950s that the unique steroid biosynthetic and metabolic activity of the newborn was recognized, generating a requirement for identification and measurement that coincided with the introduction of steroid gas chromatography. Also it was early recognized that the classical group steroid assays which were used for assessing steroid production could not be used on urine obtained neonatally.

During the 1960s, 16α-hydroxyDHA, 5-androstene-3β,16α,17β-triol, 21-hydroxypregnenolene, and 16α-hydroxypregneneolone were identified (Structure Group 6) and were shown to be important constituents of neonatal urine. These steroids were secreted as sulfate conjugates. Shackleton and co-workers first studied the excretion of steroids by neonates and infants using gas chromatography with glass capillary columns, and they reported the changing pattern of metabolites excreted during the first year of life [79].

For a number of years it had been known that some of the most important metabolites present in neonatal urine were uncharacterized androstenetriolones and androstenetetrols [80]; these were analyzed by glass capillary gas chromatography and identified by Shackleton and Taylor [81] although, for some of the components, doubt remains regarding the stereochemistry of the hydroxyl

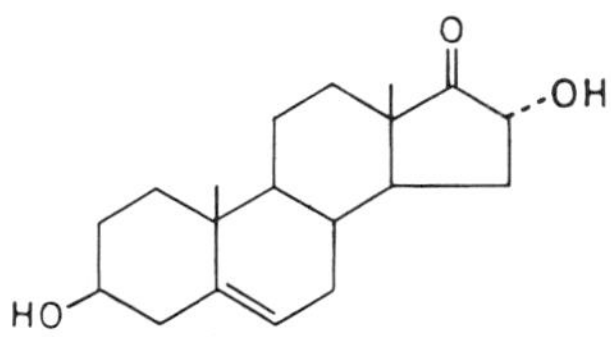

16α-OH-DHA

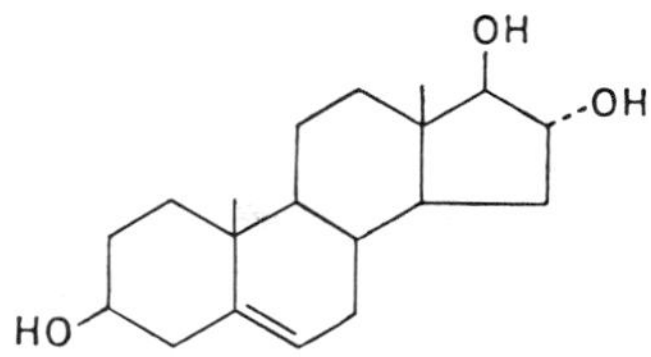

Androstenetriol

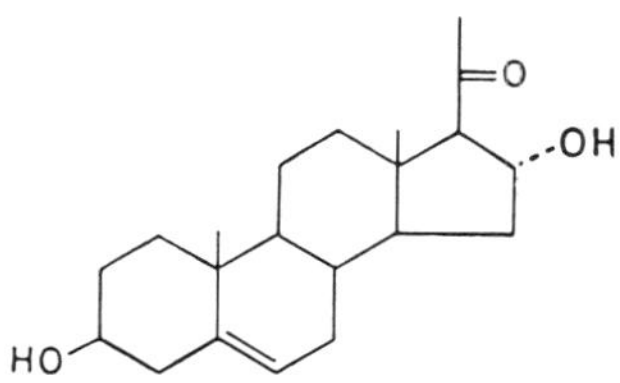

16α-OH-Pregnenolone

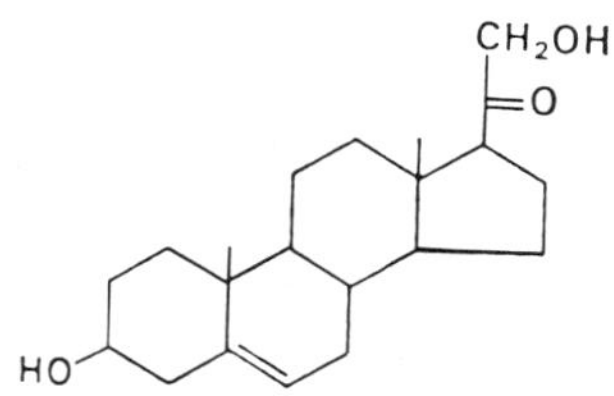

21-OH-Pregnenolone

(Structure Group 6)

groups. These steroids (Structure Group 7) were identified as 16α,18-dihydroxy-DHA, 15β,16α-dihydroxyDHA, 3,15,16,17-tetrols, and 3,16,17,18-tetrols, and the first of these compounds is generally quantitatively the most important steroid excreted by human newborns.

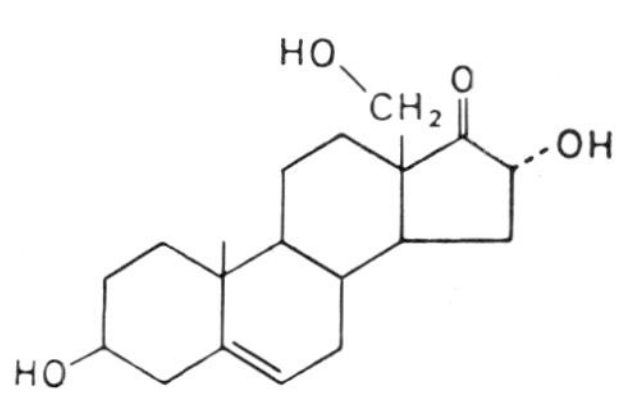

16α,18-DiOH-DHA

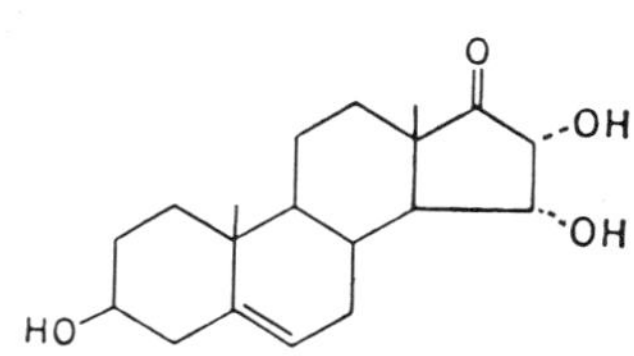

15β,16α-DiOH-DHA

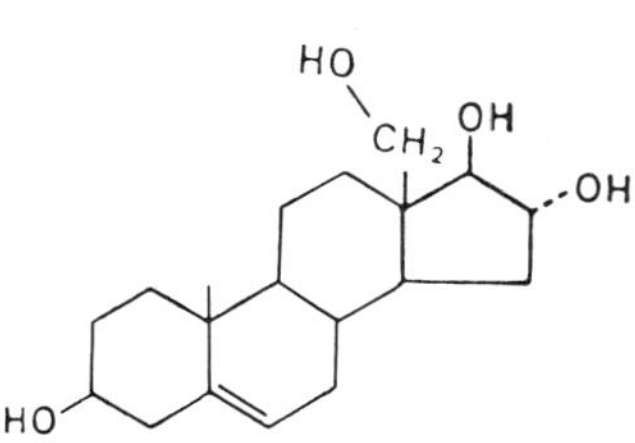

5-Androstene-
3β,16α,17β,18-tetrol

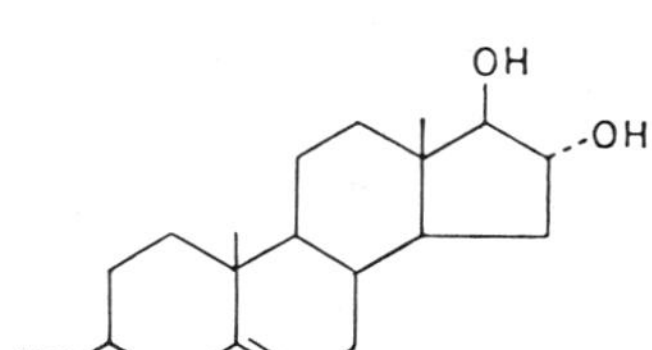

5-Androstene-
3β,15β,16α,17β-tetrol

(Structure Group 7)

Studies were also carried out in an attempt to identify metabolites of cortisol in newborn infants since it had long been recognized that these were in many cases different from those excreted in urine of adults. Radioactive tracer

Table 8.5 Excretion of Cortisol Metabolites by Newborn Infants (μg/24)[a]

	A, Mean (n = 13)	B, Mean (n = 6)
Tetrahydrocortisone (THE)	114.2	115.9
Tetrahydrocortisol (THF)	NA[b]	2.4
6β-Hydroxycortisone	NA	6.1
6β-Hydroxycortisol	NA	19.6
α-Cortolone	NA	7.7
β-Cortolone	22.7	13.8
6α-HydroxyTHE	110.8	90.2
1β-HydroxyTHE	22.2	NA
6α-Hydroxy-α-cortolone	15.1	35.0
6α-Hydroxy-β-cortolone	56.2	93.1
1β-Hydroxy-β-cortolone	27.8	NA

[a]Results obtained in studies of (A) Shackleton et al. [84], and (B) Derks and Drayer [67].
[b]NA = not assayed in this study.

experiments had indicated that less than 50% of the metabolites excreted were conventional metabolites. Shackleton and Snodgrass [61] identified 1β-hydroxycortolone as a prominent metabolite in 1973, and later studies showed that 6-hydroxytetrahydrocortisone and 6-hydroxycortolone were also quantitatively important [82] (see Structure Group 8). Coincidentally, Derks and Drayer, by elegant synthesis of the appropriate reference steroids [83], demonstrated that these metabolites were 6α-hydroxylated. Derks and Drayer [67] and Shackleton et al. [84] have quantified these important metabolites in infancy urine by glass capillary column analysis (Table 8.5). There is also evidence that 6-hydroxytetrahydro-11-dehydrocorticosterone (6-hydroxyTHA) is the principle metabolite of corticosterone during the first weeks of life [84].

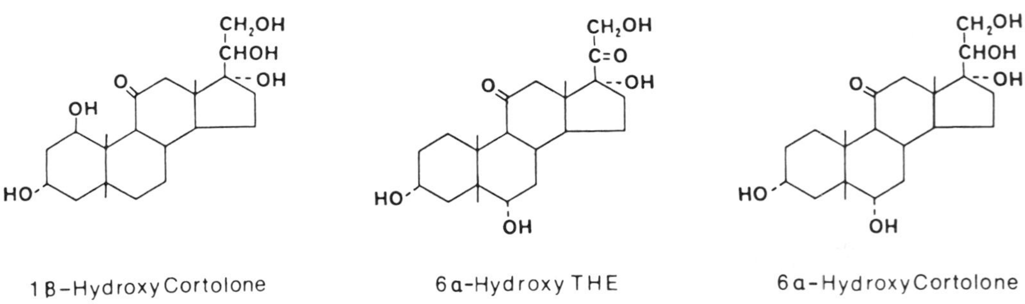

(Structure Group 8)

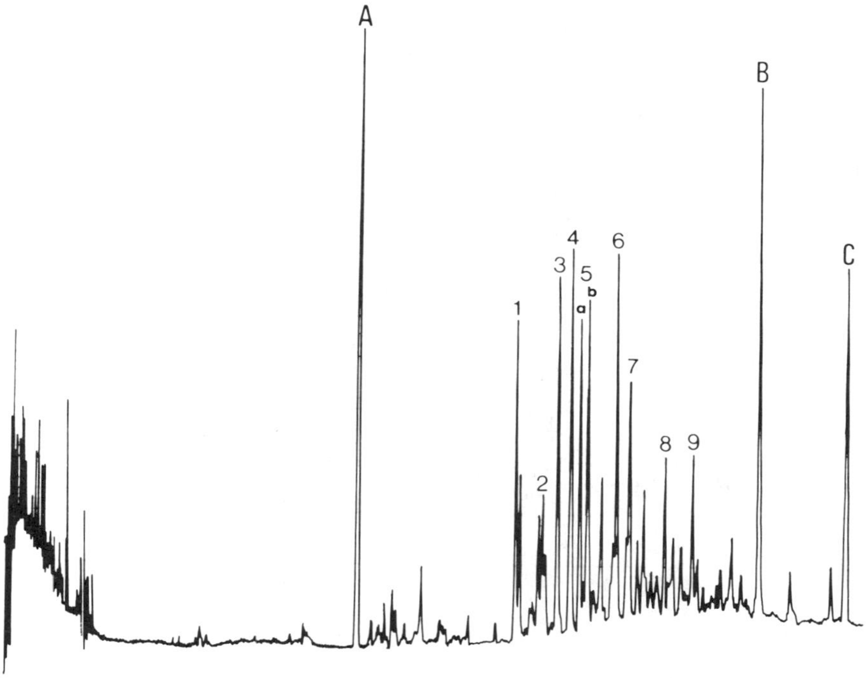

Figure 8.29 Total profile of hydrolyzed neonatal urinary steroids as methyloxime trimethylsilyl ethers. The following steroids are numbered: 1, 16α-hydroxyDHA (double peak); 2, 16β-hydroxyDHA; 3, 16-oxoandrostenediol; 4, androstenetriol; 5, 16α,18-dihydroxyDHA (double peak); 6, 16α-hydroxypregnenolone; 7, androstenetetrol + tetrahydrocortisone; 8, 6α-hydroxytetrahydrocortisone; and 9, 5-pregnene-3β,16α,20α-tetrol. This technique is unsatisfactory for analysis of neonatal *corticosteroid* metabolites since the peaks are frequently small and may be masked by the more predominant 3β-hydroxy-5-ene steroids. Conditions: 20-m OV-101 column programmed between 180 and 260° at 2.5°/min; solid injection. (From Ref. 66.)

It is probable that almost all the major neonatal neutral steroid metabolites have now been identified; a typical glass capillary chromatogram of a urine extract is illustrated in Figure 8.29. Quantitatively, 3β-hydroxy-5-ene steroids predominate. This group of steroids is excreted as sulfate esters, and so the enzyme preparation used for hydrolysis must contain sulfatase. Also it should be noted that two of the most important steroids, 16α-hydroxyDHA and 16α,18-dihydroxyDHA, give double peaks when methyloxime derivatives are prepared.

The complexity of a neonatal profile renders accurate quantification of the important corticosteroid metabolites difficult, even when capillary columns of high resolving power are used. In the latter part of the chromatogram there are many peaks of hydroxylated 3β-hydroxy-5-ene steroids as well as the metabolites of cortisol and corticosterone. We have found that for steroid analysis of infancy urine (and amniotic fluid) it is advantageous to separate

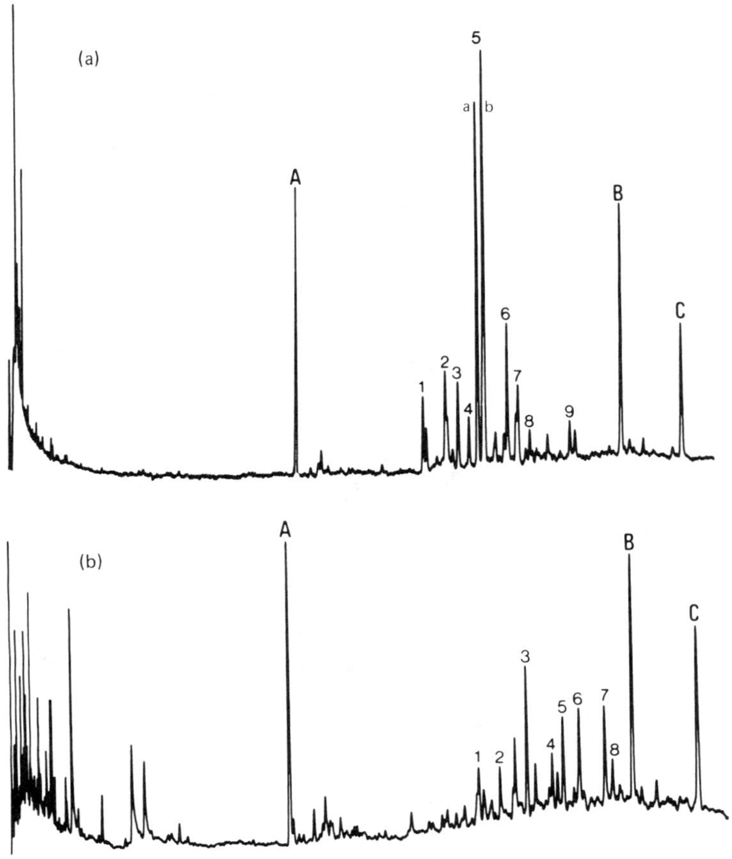

Figure 8.30 Separate analysis of methyloxime/trimethylsilyl ethers of the sulfate and glucuronide fractions of neonatal urine. These fractions were prepared by the method of Jänne et al. [23] employing Sephadex LH-20 chromatography. In the sulfate fraction (a), the following steroids are numbered: 1, 16α-hydroxy-DHA (double peak); 2, 16β-hydroxyDHA; 3, 16-oxoandrostenediol; 4, androstenetriol; 5, 16α,18-dihydroxyDHA (double peak, a and b); 6, 16α-hydroxypregnenolone; 7, two androstenetetrols; 8, 21-hydroxypregnenolone; and 9, 5-pregnene-3β,16α,20α,21-tetrol. In the free + glucuronide fraction (b), the following steroids are numbered: 1, 6α-hydroxypregnanolone; 2, 16α-hydroxypregnanolone + 5β-pregnane-3α,16α,20α-triol; 3, tetrahydrocortisone, 4, 1,16-dihydroxypregnanolone; 5, 6α-hydroxytetrahydrocortisone; 6, cholesterol; 7, 6α-hydroxycortolone; and 8, 1β-hydroxycortolone. Conditions: 20 m OV-101 WCOT column. Temperature programmed from 180-260° at 2.5°/min. Solid injection. (From Ref. 66.)

the steroid glucuronides (and free steroids) from the steroid sulfates by Sephadex LH-20 chromatograph prior to hydrolysis (see Section III.A.1, Fractionation of Steroids). An example of a urine extract which has been analyzed by this method is illustrated in Figure 8.30. The success of this method is based on a convenient biological coincidence, that is, that 3β-hydroxy-5-ene steroids are primarily sulfate conjugated, and saturated steroids are primarily excreted free or as glucuronide conjugates. Thus, the glucuronide and free fraction are largely devoid of interfering 3β-hydroxy-5-ene steroids, and the cortisol metabolites can be accurately determined.

Placental sulfatase deficiency (PSD): It is particularly convenient to consider this disorder immediately following a discussion of neonatal steroids, since the analytical problem is remarkably similar. Deficiency of the enzyme for cleaving the sulfate group of steroid sulfates prevents placental metabolism of fetal 3β-hydroxy-5-ene steroid sulfates during pregnancy. Under normal conditions, fetal 16α-hydroxyDHA sulfate, for example, is transferred to the placenta where the sulfate is removed and the free steroid is converted to estriol. The estriol is conjugated by maternal hepatic glucuronidation and is eliminated in the urine. In PSD, aromatization of 3β-hydroxy-5-ene steroid sulfates is not possible, and they are excreted almost unchanged in maternal urine.

The condition of PSD can be clearly diagnosed by detecting the elevated excretion of 3β-hydroxy-5-ene steroid sulfates. However, it was found in the studies of Taylor and Shackleton [86] that these could not be readily measured in the presence of the exceedingly large amounts of saturated neutral metabolites normally found in pregnancy urine, so it was necessary to prepare a fraction which contained only sulfates. This was most easily obtained by the method of Setchell et al. [24], although the method of Jänne et al. [23] using Sephadex LH-20 chromatography, or selective hydrolysis of steroids by solvolysis [17] would achieve the same end.

Capillary gas chromatography of a urinary sulfate fraction from a normal pregnant woman and a woman with PSD is illustrated in Figure 8.31. It can be seen that the affected patient has high levels of 3β-hydroxy-5-ene steroids compared to the normal. Several urinary 3β-hydroxy-5-ene steroids have been measured in nine patients with this disorder [86], and Figure 8.32 illustrates diagrammatically the elevated excretion of one of the steroids, 16α-hydroxyDHA, by these patients.

The affected male infants retain the congenital deficiency of sulfatase throughout life, and it has recently been recognized that the disorder is identical to a skin condition, recessive X-linked ichthyosis (RXLI). Our studies have shown that patients with RXLI have extremely elevated plasma concentrations of cholesteryl sulfate (see Section V.D.2).

Maternal corticosteroid metabolites: The newborn infant is capable of producing substantial amounts of cortisol, and yet it has proved difficult to determine what contribution the fetus makes to maternal corticosteroid excretion. If the 1β- and 6α-hydroxylated metabolites of cortisol found in neonatal urine (see previous section, Infancy urine) were also present in pregnancy urine, measurement of these might provide a valuable indication of fetal cortisol synthesis and metabolism. Taylor and Shackleton (to be published) have analyzed polar corticosteroids in urine from several third-trimester pregnant women. Since these metabolites are present in much smaller amounts than the classical cortisol metabolites, it was necessary to prepare a fraction of the steroid extract which contains only metabolites of cortisol with an additional hydroxyl group (six oxygen functions). This is conveniently carried out by the method of Setchell and Shackleton [31] employing Sephadex LH-20 and a solvent system cyclohexane:ethanol (4:1). When a 6-g column is used, the polar corticosteroid metabolites are all eluted with between 120 and 250 ml.

Quantitative results were obtained for four pregnant women and the results are reported in Table 8.6. These should not be considered normal since the women had pregnancy complications, but the results are likely to approximate values obtained during normal pregnancy. In contrast to the neonatal polar corticosteroid metabolites, several metabolites are present which retain the

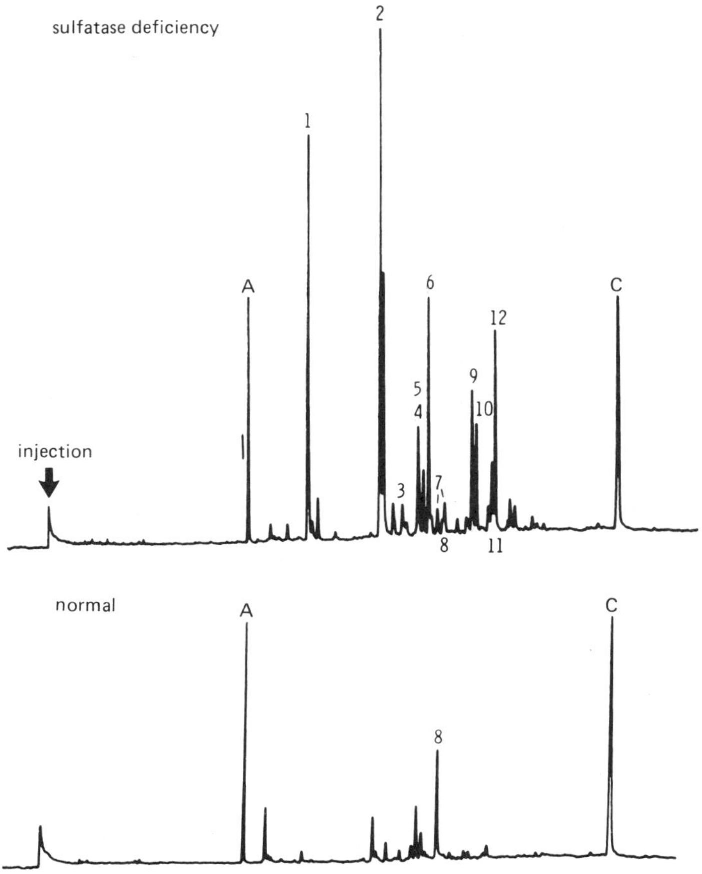

Figure 8.31 Profiles of sulfate fractions from normal pregnancy urine and from urine of a patient with placental sulfatase deficiency. The steroids were analyzed as methyloxime/trimethylsilyl ether derivative. The following steroids are numbered: 1, DHA; 2, 16α-hydroxyDHA; 3, 16β-hydroxyDHA; 4, 16-oxo-androstenediol; 5, 15β,16α-dihydroxyDHA (4 and 5 not resolved by this column), 6, androstenetriol; 7, 16α,18-dihydroxyDHA (two peaks); 8, estriol; 9, 16α-hydroxypregnenolone; 10, 16α-hydroxypregnanolone; 11, 5-pregnene-3β,16α,-20α-triol; 12, 5α-pregnane-3β,16α,20α-triol. Conditions: 20-m OV-101 WCOT column; temperature programmed from 180–260° at 2.5°/min; solid injection. (From Ref. 86.)

11β-hydroxyl group, e.g., 6α-tetrahydrocortisol. In agreement with earlier studies, 6β-hydroxycortisol is the major component of this group, but significant amounts of the new metabolites are excreted. In one normal pregnant woman, the polar metabolites were measured monthly from the tenth week of gestation, and it was found that almost all the metabolites showed a dramatic increase in excretion from the thirtieth week of pregnancy. It is premature to suggest that this increase is caused by increased production and metabolism of fetal cortisol, but more comprehensive studies on a large group of "normals" and patients with known disorders of pregnancy should surely be carried out in an attempt to answer this question. Without doubt, capillary gas chromatography and GC-MS are the only techniques currently available for carrying out

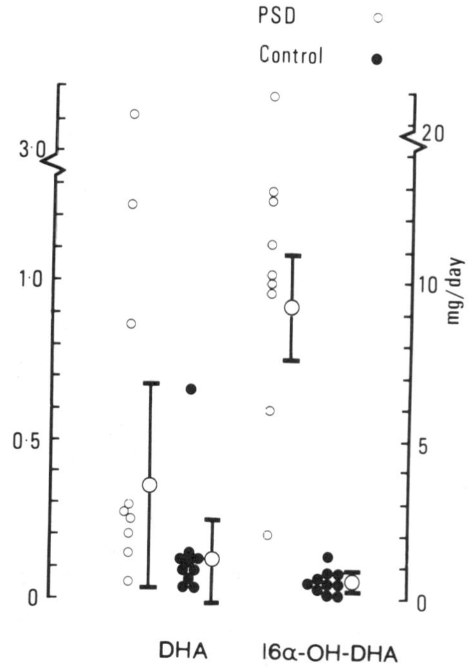

Figure 8.32 Urinary excretion (mg/24 hr) of DHA and 16α-hydroxyDHA by normal pregnant women (control) and pregnant women with placental sulfatase deficiency. Vertical lines represent the mean ± SE. (From Ref. 86.)

Table 8.6 Excretion of Polar Cortisol Metabolites in the Third Trimester of Pregnancy (μg/24 hr) Determined by Capillary GC[a]

	Subject				
	1	2	3	4	Mean
6α-HydroxyTHE	150	102	NA[b]	45	101
1β-HydroxyTHE	127	NA	NA	NA	∿30
6α-HydroxyTHF	183	103	137	47	117
6α-Hydroxy-α-cortolone	454	65	223	69	203
1β-Hydroxy-α-cortolone	738	27	261	37	266
6α-Hydroxy-alloTHF	96	25	18	33	43
6α-Hydroxy-β-cortol	43	14	88	NA	36
6α-Hydroxy-β-cortolone	101	57	164	33	89
1β-Hydroxy-β-cortolone	657	36	503	58	311
6β-Hydroxycortisol	3279	1900	2857	245	2003
6β-Hydroxy-20β-dihydrocortisol	546	630	488	119	446

[a]Only one of the four subjects had a completely normal pregnancy; the quantitative results should therefore be considered preliminary. For 6β-hydroxycortisol the results were corrected by the relative response factor; all other steroids were considered to have response factors of unity.
[b]NA = detected, but present in too small a concentration for accurate assay.

this type of study. It has also been proposed that corticosteroid sulfates (e.g., cortisol sulfate, corticosterone sulfate) may be specific products of fetal adrenal synthesis which appear in the maternal urine without prior metabolism by the placenta or mother [87]. Possibly these steroids should also be measured as a monitor of fetal adrenal synthesis.

B. Analysis of Estrogens

The clinical application of estrogen analysis by capillary gas chromatography is still in its infancy. Generally, classical colorimetric methods and radioimmunoassays are quite satisfactory for determination of estrogen status, so there has been little call for development of multicomponent analysis.

However, recently there has been increasing enthusiasm for the analysis of individual estrogens, and to some extent this can be attributed to the identification of new metabolites, some of which may have biological importance. In particular there is growing interest in the 2- and 4-hydroxyestrogens which are collectively referred to as the catechol estrogens [Structure Group 9) [88]. These estrogens (particularly the 4-hydroxy compounds) may have important biological functions because of their receptor affinity, their interference with biosynthesis and methylation of catecholamines, and their possible role in the regulation of gonadotrophin secretion. One of these estrogens, 2-hydroxyestrone, is the principal metabolite of estradiol in the human, and it has also been shown to give rise to suppression of serum prolactin [89].

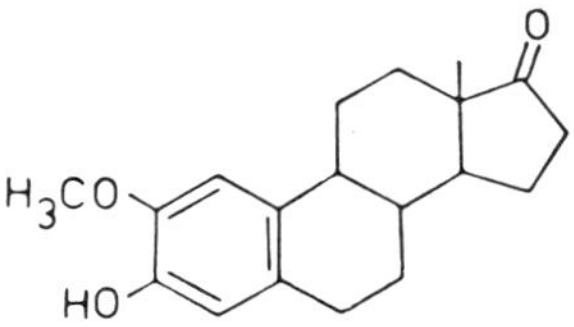

2-Methoxy Estrone

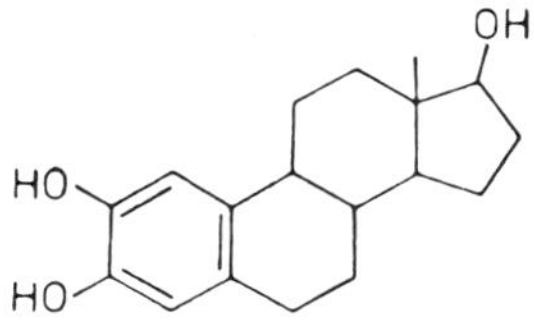

2-Hydroxy Estradiol

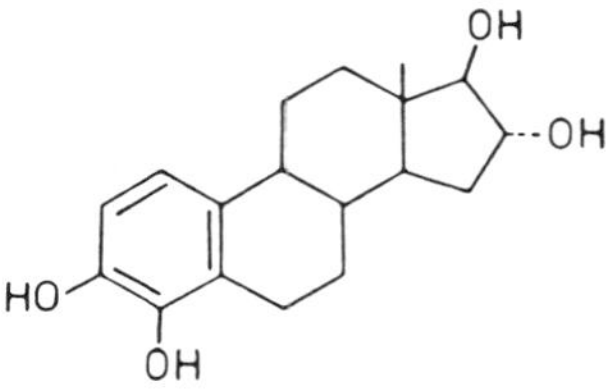

4-Hydroxy Estriol

(Structure Group 9)

Unfortunately, catechol estrogens are subject to rapid autoxidative decomposition, and only recently have methods become available for their satisfactory recovery. The secret of successful analysis by gas chromatography lies in the continued presence of an antioxidant such as ascorbate during the work-up procedure. Figure 8.33 illustrates capillary column separation and quantification of urinary estrogens excreted by a pregnant woman. These conjugate fractions were prepared by the method of Fotsis and co-workers (unpublished) which allows separation of the various forms of steroid conjugates by DEAE-Sephadex prior to enzymatic hydrolysis (see SectionIII.A.1, Fractionation of steroids). The presence in the chromatograms of the catechol estrogens should be noted. Development of method for the direct analysis of conjugates (see Section VI) may have particular importance in estrogen analyses since any techniques which reduce procedural losses will surely give more satisfactory recoveries of the more labile steroids. Retention of the glucuronide in urinary catechol estrogens should prevent destruction of these important molecules by autoxidation.

Capillary gas chromatography has also been used for analysis of estrogens fractionated by lipophilic ion-exchange chromatography [25,90], and Figure

8.5 illustrates separation of a free estrogen fraction prepared by the method of Axelson et al. [25]. An important application of their technique has recently been described [90a]. Glass capillary GC-MS was used for measurement of estradiol in nuclear and cytoplasmic components of rat uterine target cells. High sensitivity was required since nuclear estradiol concentration in adult rat uteri was in the range of 1 pg/mg tissue.

At present there are only two methods which are potentially useful for direct estimation of steroid hormones in target tissues: immunoassay and GC-MS. However, when the steroid levels are low, the specificity of radioimmunoassay is difficult or impossible to establish without the use of GC-MS. Low-resolution mass spectrometry can only be used for estimation of minute amounts of steroids, providing good purification methods are used to reduce sample contamination. It is through the use of ion exchange and partition chromatographic methods that Axelson and co-workers have found it possible to analyze trace amounts of estrogen in target cell nuclei. The use of glass capillary columns for these studies is essential for several reasons. The columns have a high separating ability which increases the specificity of the analyses. They reduce the peak width and increase the height of peaks and give less bleed of the stationary phase, thereby enhancing the signal/noise ratio and sensitivity. Less steroid is absorbed and decomposed on these columns, and possible losses are further reduced by use of a deactivated GC-MS system.

Following sample extraction, the estrogen fraction was isolated by TEAP-LH-20 chromatography. 5β-Cholestan-3α-ol was added as internal standard and trimethylsilyl ether derivatives were prepared. The samples were introduced into the SE-30 capillary column by solid injection, and the fragment ions *m/z* 416 (estradiol, M^+) and *m/z* 370 (internal standard) were monitored by the mass spectrometer. Quantitative analyses were made by comparison of peak heights in the selected-ion-current chromatogram with those given by known amounts of reference steroids.

Using this method it was found that the levels of estradiol in nuclei from target (uterus) and nontarget (spleen, lung, and heart) organ in the immature rat were 0.30 pg/mg and 0.06 pg/mg tissue, respectively. The sensitivity of the method allowed quantification to be made on subcellular extracts of a single uterus.

C. Analysis of Steroid Acids

For many years it was apparent that a significant fraction of the cortisol metabolites in urine either remained in the residual urine after hydrolysis with glucuronidase and extraction with ether, or was recovered in the alkali wash of the organic extract. From their solubility properties it seemed likely that the latter fraction contained acidic metabolites. The introduction of Amberlite XAD-2 by Bradlow [18] offered a means of recovering these acids in high yield from water-based solution, and following esterification and chromatography, several components were identified [91]. The four major constituents of this group comprised 5–25% of the identifiable metabolites of cortisol and had structures similar to α- and β-cortol and α- and β-cortolone, the 21-hydroxyl having been oxidized to a carboxylic acid. These steroids were given the generic name *cortoic acids*, and the four principal examples were termed α- and β-cortolic acid and α- and β-cortolonic acid (Structure Group 10).

Further research on this major gorup of urinary metabolites has been hampered by the unavailability of specific methods for their quantitative analysis. A method was reported by Farhi and Monder [92] which permits the complete

(a)

		µg/24 h
1.	11-Dehydroestradiol-17α	374.2
2.	Unknown dehydroestradiol	229.9
3.	Unknown compound (not estrogen)	
4.	Estradiol-17β	245.2
5.	Cholestane (I.S.)	
6.	2-Hydroxyestradiol-17β	149.8
7.	Unknown tri-OH-estrogen	156.0
8.	Unknown tri-OH-estrogen	110.1
9.	Unknown tri-OH-estrogen	112.2
10.	15α-Hydroxyestradiol-17β	141.3
11.	Estriol	6880.3
12.	16-Epiestriol	788.1
13.	Estrone	2069.6
14.	16α-Hydroxyestrone	2486.6
15.	2-Methoxyestriol	60.9
16.	16β-Hydroxyestrone	297.4
17.	2-Hydroxyestrone	522.6
18.	16-Oxoestradiol-17β	822.5
19.	Unknown tri-OH-mono-oxo-estrogen	353.3
20.	2-Methoxyestrone	549.1
21.	15α-Hydroxyestrone	344.1

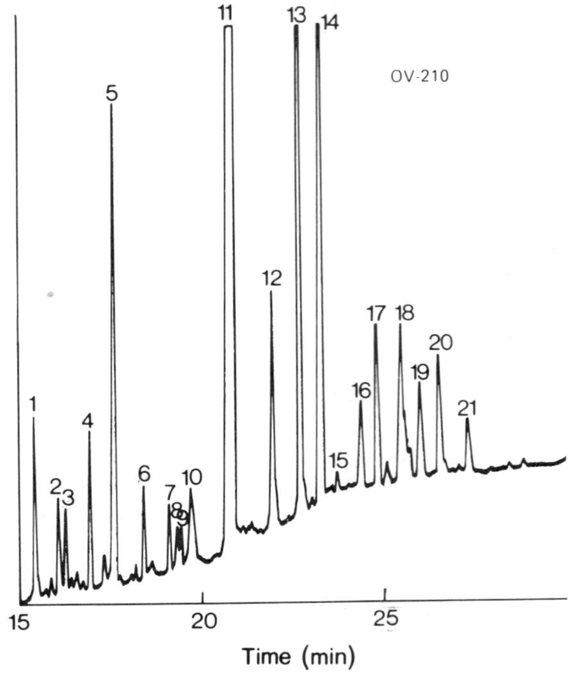

(b)

		µg/24 h
1.	Estradiol-17β	52.0
2.	Cholestane (I.S.)	
3.	Estriol	20847.0
4.	Unknown tetra-OH-estrogen	570.6
5.	2-Hydroxyestriol	209.3
6.	4-Hydroxyestriol	253.0
7.	16α-Hydroxyestrone + unknown tetra-OH-estrogen	740.9
8.	15α-Hydroxyestriol	946.6
9.	Unknown tri-OH-mono-oxo-estrogen	313.8

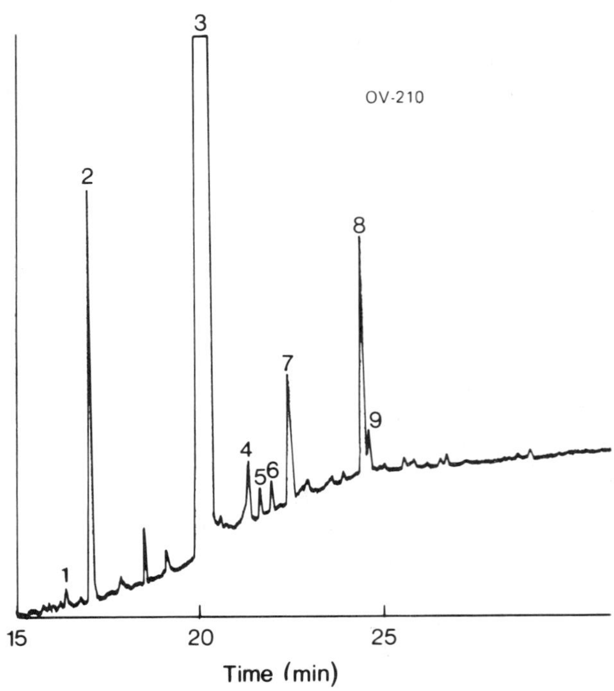

(c)

	µg/24 h
1. Estradiol-17β	51.3
2. Cholestane (I.S.)	
3. Unknown tri-OH-estrogen	328.5
4. Unknown tri-OH-estrogen	32.3
5. 2-Methoxyestradiol-17β	46.8
6. Estriol	696.8
7. 16-Epiestriol	40.5
8. Estrone	494.8
9. 16α-Hydroxyestrone	348.1
10. 2-Hydroxyestrone	72.5
11. 16-Oxoestradiol-17β	1000.5
12. 2-Methoxyestrone	186.1

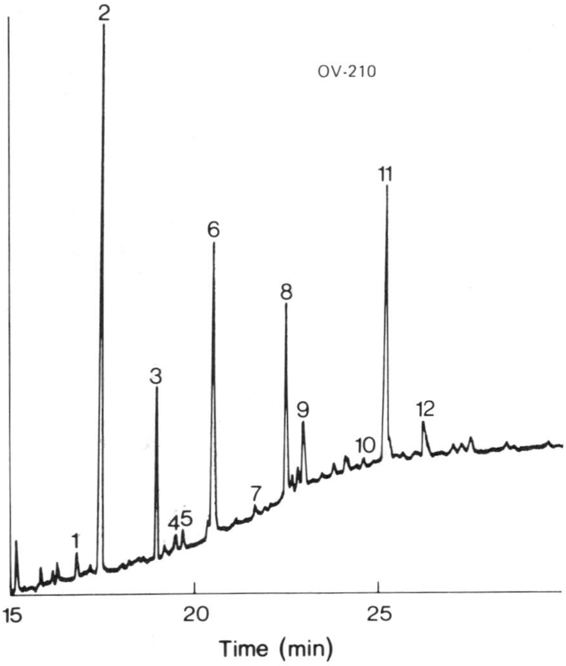

Figure 8.33 Capillary chromatography of separated estrogen conjugate fractions from a sample of pregnancy urine. The conjugates were hydrolyzed prior to derivatization and gas chromatographic analysis on a glass OV-201 column. The quantitative excretions of all the metabolites are listed. (a) A-ring glucuronides. (b) D-ring glucuronides. (c) Monosulfates. (From unpublished analyses carried out by Fotsis and coworkers.)

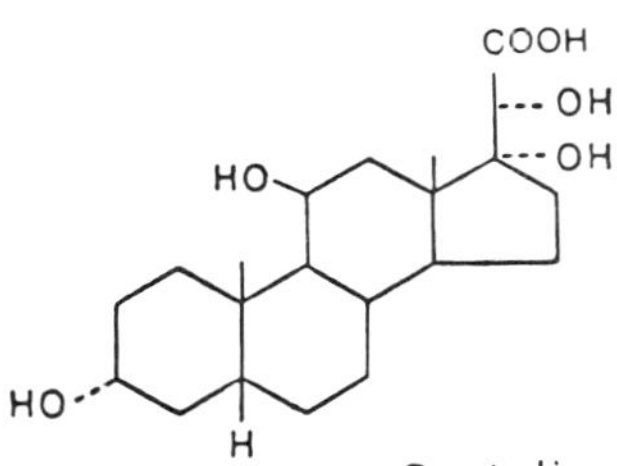

α Cortolic

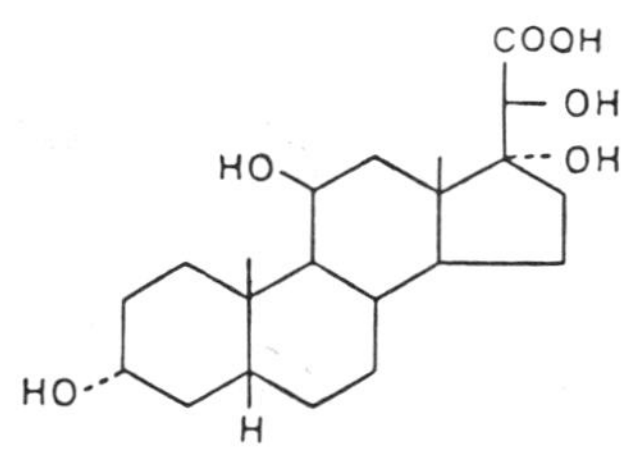

β Cortolic

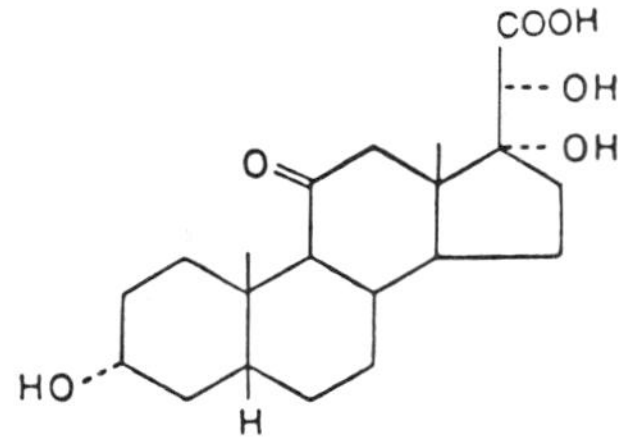

α Cortolonic

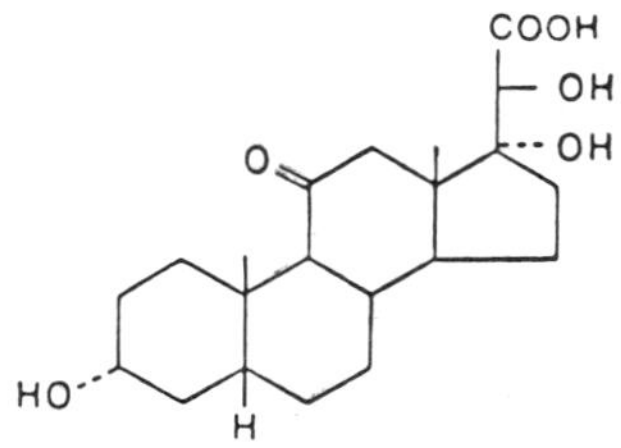

β Cortolonic

(Structure Group 10)

resolution of all four acids by HPLC, but in practice it was found that the method was unsuitable for measuring these steroids in urine due to the presence in the extracts of nonspecific chromogens. Provided that a suitable stationary phase could be found which would permit the resolution of all four of the acids, gas chromatography seemed an ideal method for their analysis. Consequently, we have recently developed an extraction method which may be summarized as follows: (1) hydrolysis of urinary steroid conjugates; (2) extraction with ether; (3) alkali extraction of the ether; (4) extraction of the alkali wash with Amberlite XAD-2; (5) methylation and trimethylsilylation; and (6) derivative purification on Lipidex 5000 [34].

The methyl ester/trimethylsilyl ether derivatives of the cortoic acids were formed quantitatively and had good gas chromatographic properties. Separation was unsuccessful on the nonpolar gas chromatography phases favored for neutral steroid analysis. Complete resolution of the four major components was readily achieved using Carbowax 20 M (Figure 8.13 lower chromatogram). Reference cortoic acids were subjected to combined GC-MS analysis on a Carbowax 20 M column (Figure 8.34), and the identity of the cortoic acid peaks in urinary samples was verified using this technique. Preliminary quantitative

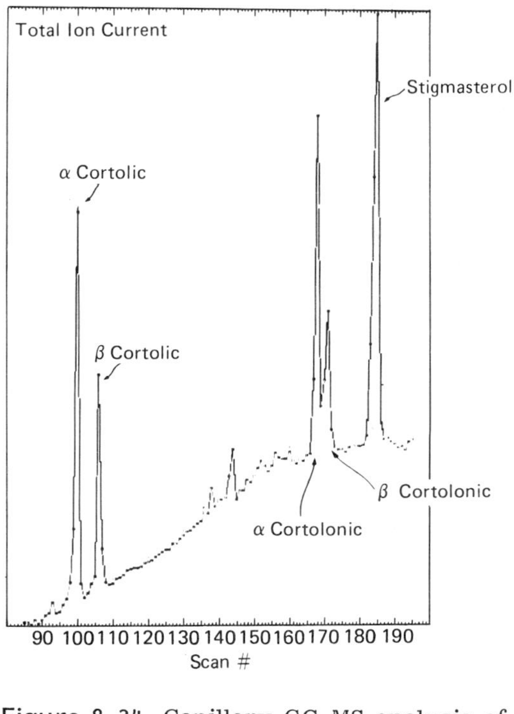

Figure 8.34 Capillary GC-MS analysis of methyl ester/trimethylsilyl ether derivatives of reference cortoic acids with stigmasterol as internal standard. The illustration shows part of the total-ion-current recording obtained from repetitive scanning data. Conditions: The analysis was carried out on a Carbowax 20 M column in a Hitachi M-52 GC-MS instrument. The sample was introduced by splitless injection, and the temperature was programmed from 180–250° at 3°/min. Scans (0–800 amu) were taken every 3–6 sec. (From Ref. 93.)

Table 8.7 Urinary Cortoic Acids: Retention Times on Capillary Columns (Methylene Units) with Three Stationary Phases, and Daily Excretion (μg/24 hr) by Six Cancer Patients

	Retention time				Excretion	
Steroids	OV-1	SE-52	Carbowax	SP-2330	Mean ($n = 6$)	Range
α-Cortolic	30.58	30.54	31.12	34.40	247 ± 114	106-438
β-Cortolic	30.55	30.76	31.43	34.60	106 ± 17	80-131
α-Cortolonic	30.38	30.76	34.05	38.43	518 ± 104	399-649
β-Cortolonic	30.38	30.76	34.16	38.00	402 ± 126	219-575

data have been obtained for the analysis of individual cortoic acids in six cancer patients and the results are reported in Table 8.7. We have found that the gas chromatographic method is suitable for the analysis of these important metabolites, even in relatively small volumes of urine [93].

Little is known yet of the changes brought about in cortoic acid excretion by diseases affecting steroid biosynthesis and metabolism. This has been principally due to the unavailability of specific methods for analysis of these steroids. However, as mentioned in a previous section, there is evidence for increased excretion of α-cortolic acid in anorexia nervosa.

Many more steroid acid metabolites probably remain to be identified. Presumably, appropriate 21-oic acids will be found in urine of patients with the classical enzyme deficiencies, e.g., acid metabolites of Substance S or corticosterone.

D. Analysis of Steroids in Serum and Tissue

1. *Free Steroids*

Little attention has been paid to the capillary column analysis of plasma unconjugated steroids. The principal reason is the relatively small number of these steroids and the availability of specific sensitive radioimmunoassays for each of the major hormones. Multicomponent capillary column analysis is neither quick enough nor sensitive enough for clinical diagnosis.

Methods have been developed which allow efficient recovery of unconjugated steroids from serum [21,35], and quantitative results have been published. For example, Axelson and Sjövall [94] have quantified 13 steroids in plasma from pregnant women by glass capillary GC-MS (Table 8.8). For research purposes this technique is obviously of great importance since alternative analytical methods are not readily available for most of the steroids listed. An example of their chromatography is shown in Figure 8.35 in the form of fragment-ion chromatograms. The method allows quantification in the concentration range of 1–300 ng/ml.

Conventional GC-MS with selected ion detection is frequently of inadequate specificity for quantitative analysis of steroid hormones owing to the presence of many interfering compounds in the serum extracts. Gaskell and Millington [95] have shown that considerable improvement in specificity can be achieved by monitoring metastable peaks corresponding to fragmentations occurring in

Table 8.8 Concentrations (ng/ml) of Some C_{21} Steroids in Plasma from a Pregnant Woman

	Week of pregnancy					
Steroids	20	24	28	32	37	38
3α-Hydroxy-5β-pregnan-20-one	4.2	5.3	6.0	10.3	11.4	15.3
3α-Hydroxy-5β-pregnan-20-one	2.3	3.1	5.0	7.5	8.9	12.5
3β-Hydroxy-5α-pregnan-20-one	3.6	9.2	7.8	13.8	16.4	20.6
3β-Hydroxy-5-pregnan-20-one	7.3	5.6	5.6	9.6	7.2	8.5
5α-Pregnane-3α, 20α-diol	0.8[a]	2.1	2.0	2.6	3.1	2.5
5β-Pregnane-3α, 20α-diol	1.0[a]	2.4	2.2	4.0	3.7	5.3
5α-Pregnane-3β, 20α-diol	0.9[a]	6.4	4.0	7.9	5.3	6.1
5α-Pregnane-3, 20-dione	9.6[a]	5.2[a]	7.4[a]	11.2	14.4	25.6
Progesterone	25.4	39.4	51.5	78.1	92.8	98.9
20α-Dihydroprogesterone	11.2	12.0	17.2	22.6	30.3	37.7
17α-Hydroxyprogesterone	—	4.6	9.6	4.7	6.2	7.0
Cortisone	—	18.5	29.8	24.3	36.0	44.9
Cortisol	124	97	212	224	244	295

[a]Value uncertain due to noisy baseline.
Source: Axelson and Sjövall [94].

the first field-free region of a double focusing mass spectrometer. They developed a method capable of measuring 20 pg of 5α-dihydrotestosterone (5αDHT) in plasma. The major ion in the mass spectrum of the *tert*-butyl dimethylsilyl ether (TBDMS) of 5αDHT is M-57 (m/z 347), and this ion further fragments primarily to a daughter ion at m/z 271 in the first field-free region. For capillary GC-MS analysis with metastable peak monitoring, the metastable peak corresponding to the transition m/z 347 → m/z 271 was focused using the linked-scan facility of a VG Micromass instrument. Results obtained are illustrated in Figure 36B. Four major peaks were observed, and the peak corresponding to a 5αDHT-TBDMS is indicated. An indication of the specificity of the metastable peak monitoring is provided by a comparison of the selected-ion-detection profiles obtained by monitoring ions of mass m/z 271 formed in the ion source (Figure 8.36A).

The gas chromatographic analysis of 3-oxo-4-ene steroids in tissue or cellular extracts has been carried out using glass capillary columns. Maume et al. [96] have been studying the secretion of steroids by mouse adrenocortical cells grown in tissue culture. Following deproteinization and delipidation (70% methanol at -20°), the steroids were recovered by ethyl acetate and dichloromethane extraction. After purification by TLC, methyloxime/trimethylsilyl derivatives were prepared. These were analyzed by capillary gas chromatography using SE-30 columns and solid injection. A typical chromatogram from

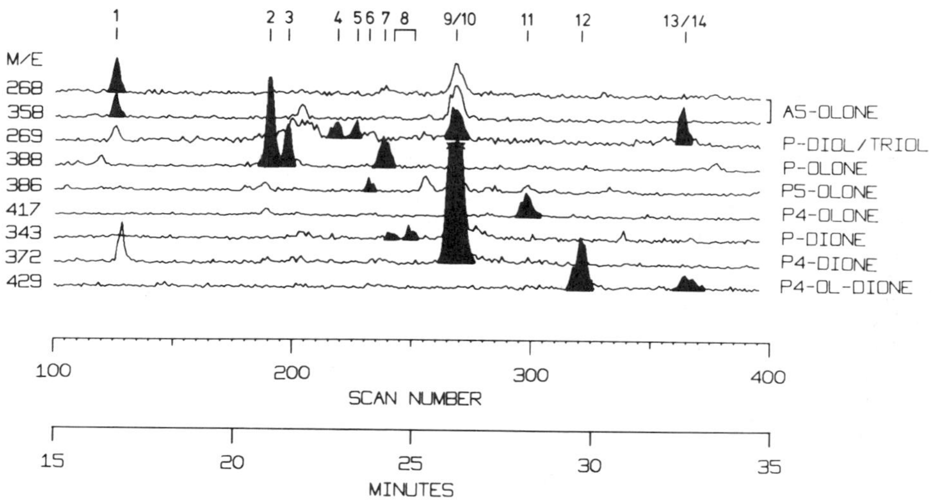

Figure 8.35 Capillary GC-MS analysis of unconjugated plasma steroids from a woman in the thirty-seventh week of pregnancy. The recordings are fragment-ion-current chromatograms constructed by the computer from repetitive scanning data. The numbers mark the peaks produced by the following components: 1, DHA; 2, 3α-hydroxy-5α-pregnan-20-one; 3, 3α-hydroxy-5β-pregnan-20-one; 4, 5α-pregnane-3α,20α-diol; 5, 5β-pregnane-3α,20α-diol; 6, 3β-hydroxy-5-pregnen-20-one; 7, 3β-hydroxy-5α-pregnan-20-one; 8, 5α-pregnane-3,20-dione; 9, 4-pregnene-3,20-dione; 10, 5α-pregnane-3β,20α-diol; 11, 20α-hydroxy-pregnen-3-one; 12, 17α-hydroxy-4-pregnene-3,20-dione; 13, 16α-hydroxy-4-pregnene-3,20-dione; 14, 5α-pregnane-3α,20α,21-triol. Conditions: 25-m WCOT OV-1 column attached via single-stage separator to LKB 9000 GC-MS instrument. The temperature of the column was programmed between 220 and 275° at 1.2°/min. (From Ref. 94.)

analyses of one of these TLC fractions is illustrated in Figure 8.37. The steroids were quantified by relating the peak heights to the peak height of the internal standard cholesteryl butyrate. Corrections were made for FID response factors and recovery of radioactive steroids carried through the procedure.

2. *Conjugated Steroids*

Serum 3β-hydroxy-5-ene steroid sulfates can be conveniently analyzed by a simple procedure described in Section III.A entailing simultaneous hydrolysis and derivatization. A profile obtained by this procedure is illustrated in Figure 8.38a.

However, although this method gives reproducible results, more quantitative recovery of saturated as well as 3β-hydroxy steroids is obtained by solvolysis. The profile obtained by the direct derivatization procedure illustrated in Figure 8.38a represents the circulating steroid sulfates in an adult patient with the dermatological condition, recessive X-linked ichthyosis (RXLI), a condition associated with deficiency of steroid sulfatase. This enzyme defect is identical to the *placental sulfatase deficiency* previously described (see Section V.A.4).

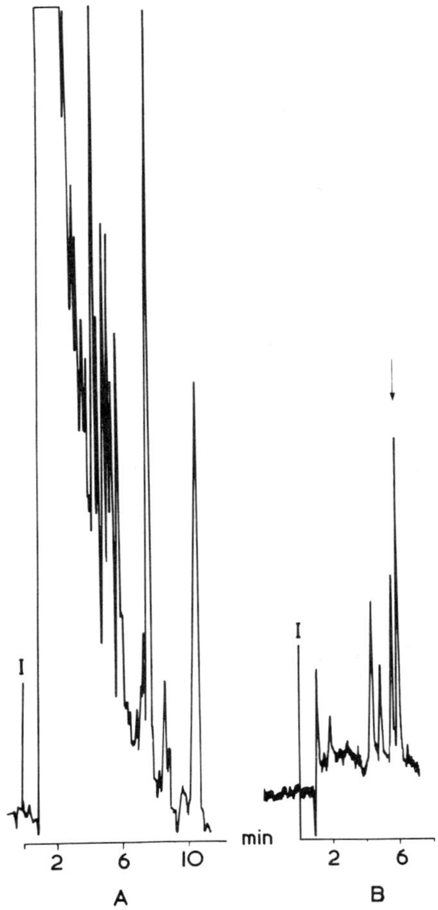

Figure 8.36 Analysis of plasma 5α-dihydrotestosterone (TBDMS derivative) by metastable peak monitoring. A, selected detection ions of m/z 271 formed in the ion source. B, selected detection of ions of m/z 271 derived by fragmentation of ions of m/z 347 in the first field-free region of the double-focusing mass spectrometer (VG-micromass 70-70F instrument with linked scanning capability). Conditions 20-m OV-1 WCOT column programmed between 225 and 260° at 6°/min. (From Ref. 95.)

When the male infants reach a few months of age, the skin starts scaling but there are no other major problems associated with the condition. In six patients with this condition, we found that the levels of DHA-sulfate and other steroid sulfates were slightly higher than normal [117], but the major abnormality was a 15- to 20-fold increase in the plasma concentration of cholesteryl sulfate above normal levels (Figure 8.38b). How the increased cholesteryl sulfate concentration relates to the skin condition is not known at this time. With the exception of cholesteryl sulfate, the 3β-hydroxy steroid profile is essentially identical to that obtained for normal individuals.

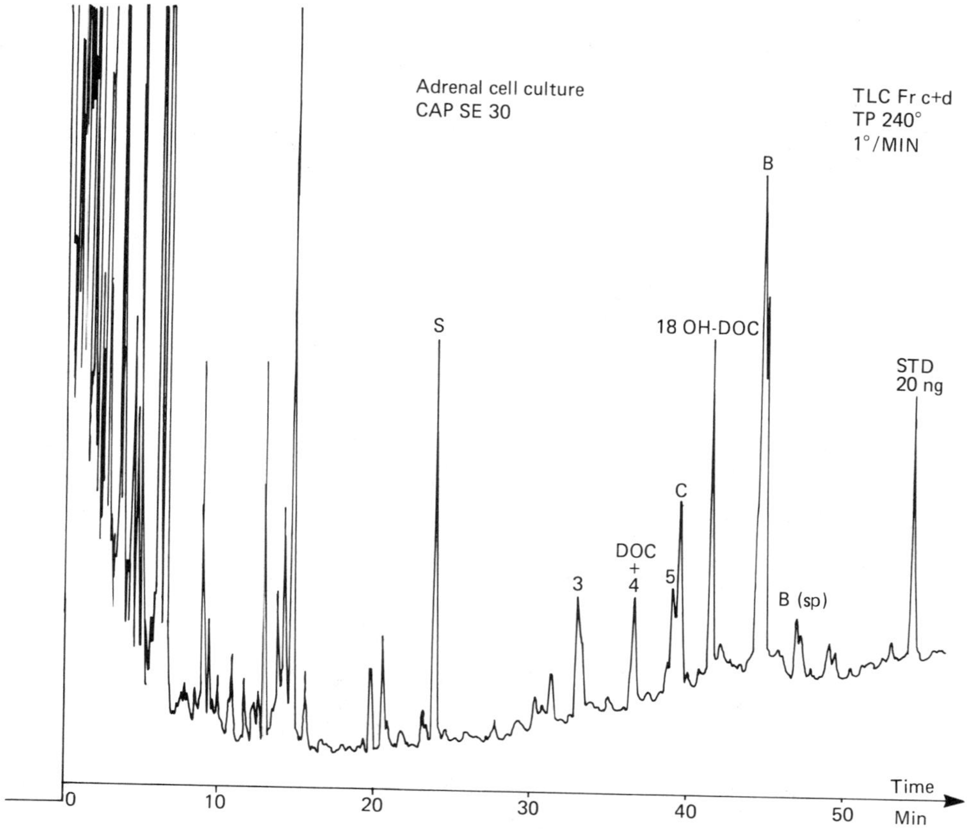

Figure 8.37 Capillary gas chromatographic separation of steroids synthesized by mouse adrenal cell culture. This analysis represents one of the three fractions purified by TLC. The main steroids are 18-hydroxyDOC (18-OH-DOC) and corticosterone (B). The steroids were analyzed as methyloxime/trimethylsilyl ethers. Cholesteryl butyrate was the internal standard. Conditions: 25-m SE-30 column programmed from 240° at 1°/min. Solid injection of sample. (From Ref. 96.)

Ludwig et al. [97] have examined the steroid profiles of hemofiltrates by capillary gas chromatography and have compared these profiles of serum steroids in normal individuals. The profiles obtained from hemofiltrates differ from the normal steroid pattern by the presence of much larger amounts of sulfates of saturated steroids, particularly 11-oxygenated C_{19} steroids. The major 3β-hydroxy-5-ene steroid was identified as 5-androstene-3β,17β-diol rather than DHA. In this condition the steroids have passed through the liver so often that they have undergone a much greater number of oxidation and reduction reactions.

E. Analysis of Vitamin D_3 Metabolites

Only during the last decade has the complexity of the vitamin D_3 hormonal system been properly appreciated [98]. It is now known that vitamin D_3 is

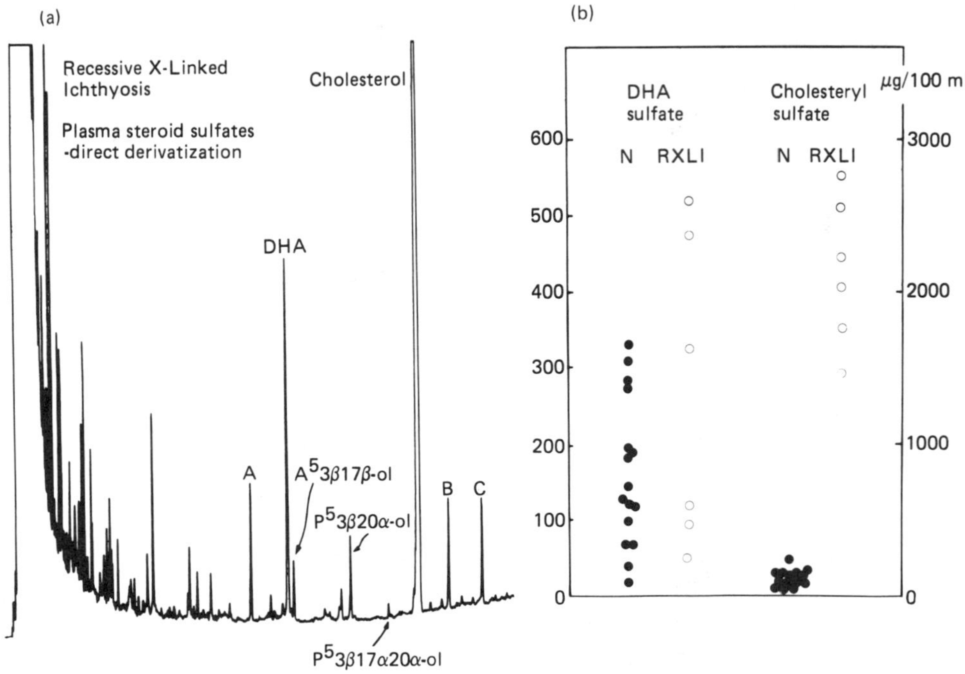

Figure 8.38 Plasma steroid sulfates. (a) This example shows the analysis of a steroid sulfate fraction of plasma prepared by the method of Jänne et al. [23]. The sample was analyzed following direct derivatization whereby the hydrolysis was carried out by the silylation reagent. The plasma was obtained from a patient with X-linked ichthyosis, a skin condition associated with congenital absence of steroid sulfatase. The major feature of the chromatogram is the large cholesterol peak (derived from cholesteryl sulfate). Conditions: 15-m, OV-1 fused silica column, splitless injection. (b) Concentration of DHA sulfate and cholesteryl sulfate in normal controls (N) and six pateints with recessive X-linked ichthyosis (RXLI). Whereas there was considerable overlap in the concentration of DHA sulfate, there was a markedly elevated plasma concentration of cholesteryl sulfate.

merely the precursor of a series of hydroxylated metabolites, generally of greater activity with respect to maintenance of calcium homeostatsis. Vitamin D_3 is first hydroxylated in the liver to form 25-hydroxyvitamin D_3, which in turn is converted renally to the primary "hormone" 1,25-dihydroxyvitamin D_3 (Structure Group 11). As part of the control system for regulating vitamin D_3 synthesis and breakdown, 24,25-dihydroxyvitamin D_3, 25,26-dihydroxyvitamin D_3, and 1,24,25-trihydroxyvitamin D_3 are also formed [99]. Corresponding metabolites are also formed of the major dietary supplement, vitamin D_2, which is distinguished from vitamin D_3 by a methyl group at C-24 and unsaturation at C22–C23.

1. *Serum*

The importance of this group of hormones rests in their absorption and metabolism so there is a requirement for sensitive and specific assay procedures. The

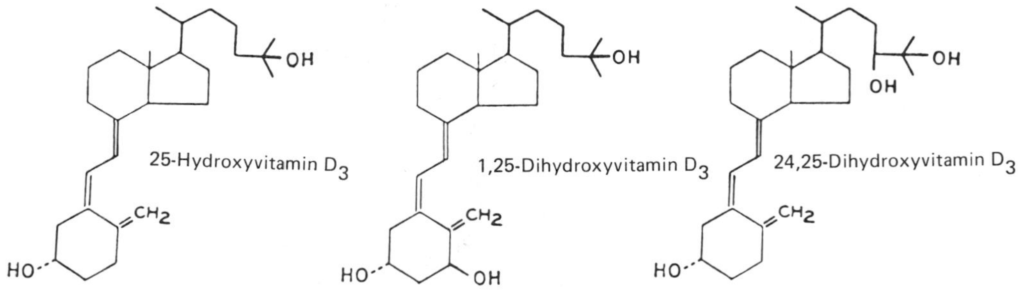

(Structure Group 11)

concentrations of these hormones in serum vary from 5–25 ng/ml (25-hydroxyvitamin D_3) to 5–25 pg/ml (1,25-dihydroxyvitamin D_3), so naturally most interest has centered on radioimmunoassay. However, packed column GC-MS SIM methods have been published by Björkem et al. [99,100] for 25-hydroxyvitamin D_3 and 1,25-dihydroxyvitamin D_3, although these are operated at the absolute limit of sensitivity, at least for the latter hormone. Methods for synthesis of trideuterated internal standards for use in the SIM method were described in these communications and in a recent paper from our laboratory [101]. It is probable that the introduction of glass capillary columns may permit improvement in sensitivity.

The metabolites of vitamin D_3 are readily persilylated by trimethylsilylation, and mild reagents (hexamethyldisilazane/trimethylchlorosilane) may be used even though most of the metabolites have a tertiary 25-hydroxyl. The derivatives chromatograph well on nonpolar WCOT columns, but cyclization takes place at the high temperature required for volatilization, and two peaks are formed for each compound dependent on the way the B-ring fuses (Figure 8.39, Table 8.9). These are referred to as *pyro*- ($9\alpha,10\alpha$) and *isopyro* ($9\beta,10\beta$) forms and are inevitably produced both for underivatized and derivatized vitamin D_3 metabolites [101,102]. The electron-impact mass spectra of trimethylsilyl ethers of all vitamin D_3 metabolites with a 25-hydroxyl group are dominated by an ion at m/z 131 formed by fragmentation of the side chain [101,102]. This ion is of lower mass than is generally desirable for selected-ion-monitoring GC-MS, so Lisboa and Halket [102] have examined the capillary GC and mass spectrometric properties of other derivatives; for example, acetates, *n*-butylboronate/trimethylsilyl ethers, and *tert*-butyldimethylsilyl/trimethylsilyl mixed derivatives. However, they concluded that ordinary trimethylsilyl ethers are still probably best for identification and quantification of all the metabolites.

2. *Urine*

The terminal metabolism of vitamin D_3 is virtually unknown, in contrast to that of other steroid hormones, and confirmatory evidence for the structure of a single urinary or fecal metabolite has not been obtained. However, the ability to analyze unique urinary metabolites of vitamin D_3 would offer a secondary test of the production and turnover of this hormone and would be especially useful in the investigation of young infants and children where frequent blood sampling is undesirable. Measurement in 24-hr urine would give valuable information of the turnover rate—information difficult to obtain by serum assay. In preliminary studies it has been shown that 25-hydroxyvitamin D_3 is principally excreted as a glucuronide, and tentative identification by capillary GC-MS of urinary metabolites have been obtained in children with vitamin D

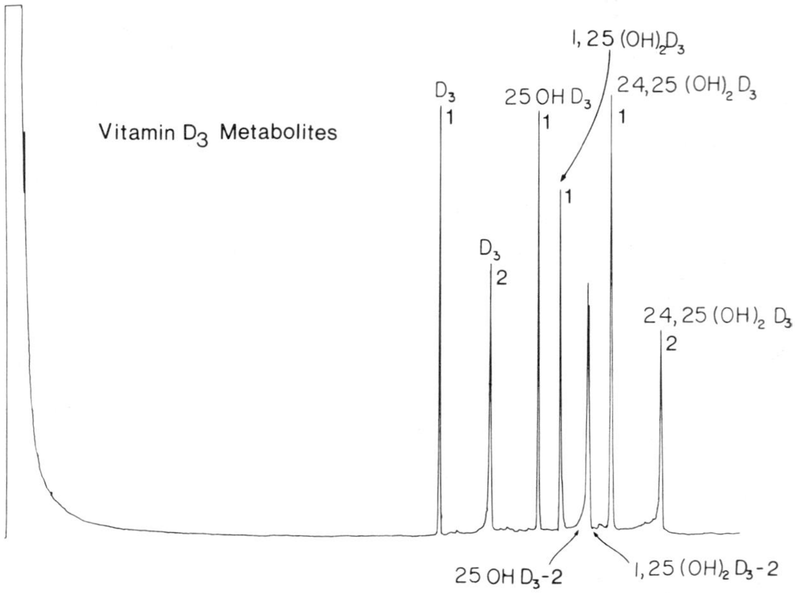

Figure 8.39 Analysis of vitamin D_3 metabolites as trimethylsilyl ethers. Cyclization of the seco-steroids occurs during vaporization, and two peaks are given for each compound. These are formed by 9α, 10α ring fusion (pyro form) and 9β, 10β ring fusion (isopyro form). Conditions: 18-m OV-1 fused-silica column, programmed 180–300° at 3°/min, splitless injection.

resistant rickets [103]. Recent evidence from the rat has suggested that about 13% of 1,25-dihydroxyvitamin D_3 is excreted in a bile as 1α-hydroxy-23-carboxytetranorvitamin D_3 [104], and it will be interesting to discover whether side chain-oxidation is important in man.

Table 8.9 OV-1 Retention Times (Methylene Units, MU) of Vitamin D_3 Metabolites as Trimethylsilyl Ethers

Steroid	MU	MU
Vitamin D_3	29.58	31.05[a]
25-Hydroxyvitamin D_3	32.45	33.95[a]
1,25-Dihydroxyvitamin D_3	33.16	34.00[a]
24,25-Dihydroxyvitamin D_3	34.67	36.25[a]

[a]Vitamin D_3 metabolites thermally cyclize during vaporization to give pyro and isopyro forms which are well separated by gas chromatography.

F. Analysis of Steroid Drugs

Compared to the study of endogenous steroid hormones and their metabolites, capillary gas chromatography has played a relatively small part in the analysis of steroid drugs. One of the reasons for this is that many of these compounds are phenolic or have α, β unsaturation which renders them particularly suitable for UV detection following HPLC analysis.

1. Quantification in Serum

Steroid contraceptive drugs with an ethinyl group are noted for their associated side effects, but the mechanisms by which these drugs produce these adverse reactions are not fully understood. In spite of extensive metabolic studies, little is known about relationships between metabolic conversions, concentrations in target organs, and appearance of side effects. This is due mainly to the lack of method having the sensitivity and specificity required for the determination of the concentrations of ethinyl steroids (Structure Group 12) in tissues and body fluids.

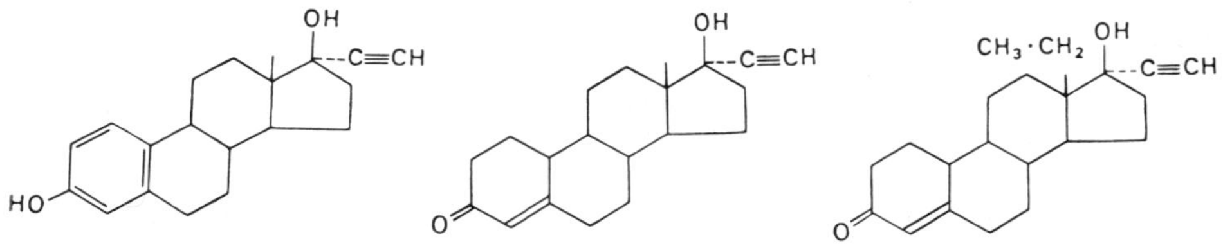

Ethinylestradiol Norethisterone Norgesterol

(Structure Group 12)

Siekman and co-workers [105] have described a method for the quantification of 17α-ethinyestradiol-17β in serum of women taking oral contraceptive drugs containing this steroid. They developed a SIM method using [$^3H_2$]ethinylestradiol as internal standard. Separation of the drug (as heptafluorobutyric ester) was achieved on an OV-101 glass capillary column coupled to a mass spectrometer by a molecular separator. When ethinylestradiol is reacted with heptafluorobutyric anhydride, the C-17 hydroxyl is removed as water and replaced by a double bond. Simultaneously the phenolic hydroxyl reacts with the reagent, yielding the final product 19-nor-1,3,5(10)-pregnatetraen-20-yn-3-ol-3-heptafluorobutyric ester. The molecular ion at m/z 474 was selected for monitoring, and the corresponding ion for the labeled material was at m/z 478. Typical results for analysis of a serum sample are illustrated in Figure 8.40. It has been found that the sensitivity limit of this method was 25 pg/ml, and a coefficient of variation of 2.5% could be achieved for analysis of duplicates.

Tetsuo et al. [28] also developed a capillary column GC-MS method for the analysis of ethinyl steroids and metabolites in plasma and urine. They employed a novel Ag^+ Sephadex ion-exchange chromatographic system for specific recovery of these steroids, and this was described in a previous section (see Section III.A). Trimethylsilyl ethers and methyloxime trimethylsilyl ether derivatives were prepared from the phenolic and neutral fractions, respectively, and individual steroids were determined by single-ion-monitoring GC-MS

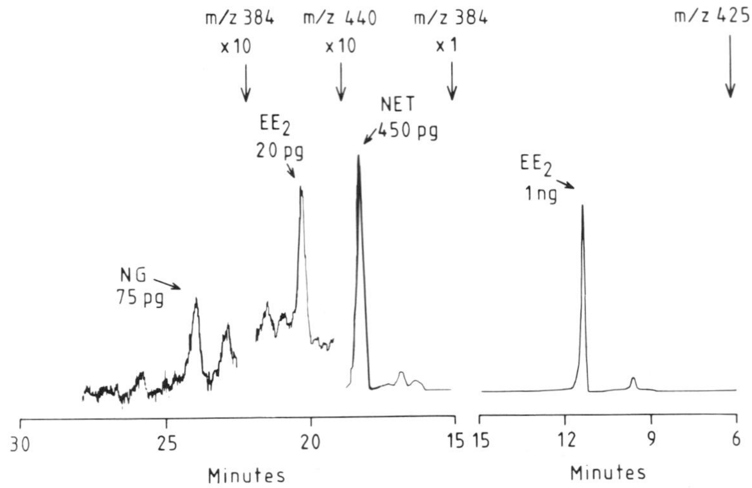

Figure 8.40 GC-MS analysis of the heptafluorobutyric ester derivative of 17α-ethinylestradiol-17β (*m/z* 474, lower trace) and the corresponding tritiated internal standard (*m/z* 478, upper trace). This steroid sample was recovered from plasma. Conditions: 10-m, OV-101 WCOT column attached to mass spectrometer via molecular separator. Solid injection of sample. (From Ref. 105.)

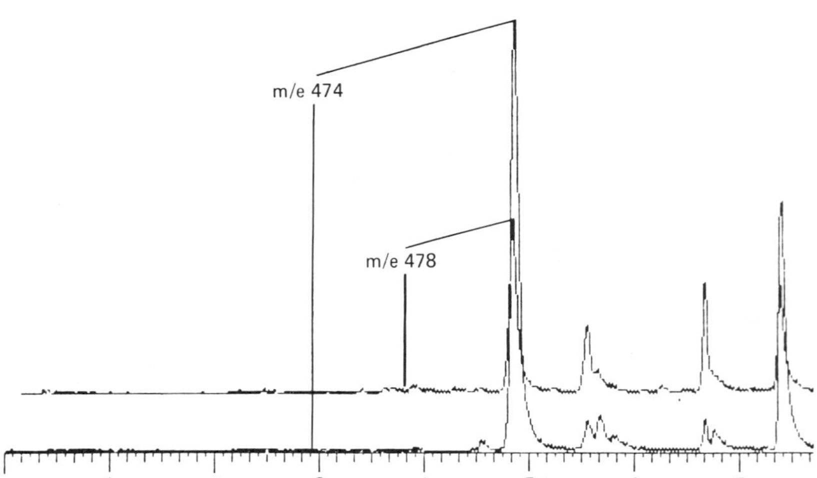

Figure 8.41 GC-MS analysis of unconjugated ethinyl steroids (left chromatogram) and total ethinylestradiol (after hydrolysis, right chromatogram) purified from urine of a woman taking 250 μg (-)-norgesterol (NG) and 50 μg ethinylestradiol (EE_2) daily. Norethisterone (NET) was added to urine as internal standard, and ethinyl steroids were isolated using the procedure of Tetsuo et al. [28]. Methyloxime/trimethylsilyl ethers were prepared, and ions at *m/z* 384 (NET,NG) and *m/z* 425 or 440 (EE_2) were monitored. Conditions: 25-m SE-30 WCOT column attached via single-stage separator to LKB 9000 instrument. Isothermal operation (250°) and solid injection. (From Ref. 28.)

employing OV-1 or SE-30 capillary columns. Ethinyl estradiol was determined by analysis of the molecular ion (m/z 440) or M-15 (m/z 425) fragments and norgesterol by analysis of the ion at m/z 384. Figure 8.41 illustrates GC-MS analyses of unconjugated ethinyl steroids (left chromatogram) and total ethinyl-estradiol after enzymatic hydrolysis, (right chromatogram) purified from urine of a woman taking 250 μg norgesterol and 50 μg ethinylestradiol daily. The method described permitted analysis of levels of ethinyl steroids excreted during normal use of the contraceptive steroids.

Gaskell and co-workers have developed a method for analysis of the experimental drug 16α-cyano-3β-cyclopentylpregn-5-en-20-one in which open-tubular columns are used [106]. The drug was extracted with ethyl acetate and purified by chromatography on Lipidex 5000. Oximes were formed with hydroxylamine hydrochloride, and silylation was carried out with t-butyldimethylchlorosilane. Quantification was achieved by SIM using the [$^2H_9$]cyclopentyloxy analogue (Structure 13) as internal standard after chromatography on a Silanox OV-1 capillary column. Figure 8.42 illustrates a representative selected-ion chromatogram from their studies.

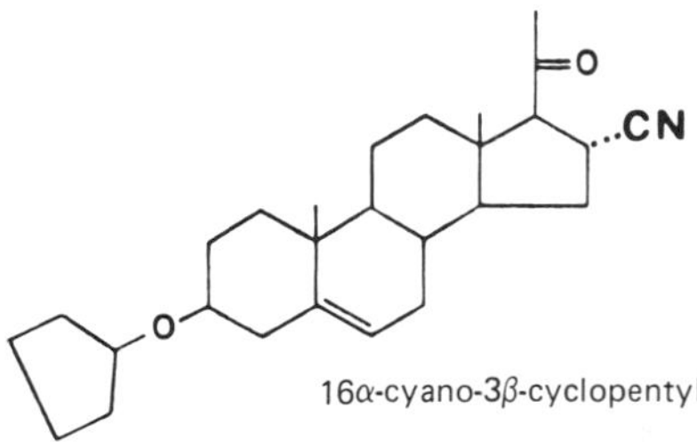

16α-cyano-3β-cyclopentyloxypregn-5-en-20-one

(Structure 13)

2. *Detection of Anabolic Steroids*

In recent years there has been widespread misuse of anabolic steroids by athletes attempting to improve performance through increase in muscular mass. The governing bodies of international sport, including the International Olympic Committee, have imposed a ban upon their use, and it is for this reason that sensitive and specific tests are required.

GC-MS is the only acceptable method for confirming the presence of these drug metabolites, although radioimmunoassay is particularly useful for prior screening of samples. The drug most communly used by athletes is Dianabol (Structure Group 14),

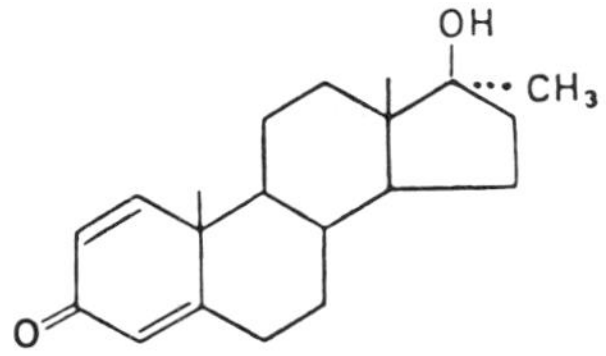

Dianabol

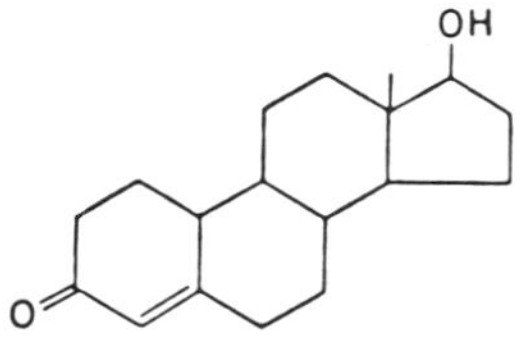

19-nor-Testosterone

(Structure Group 14)

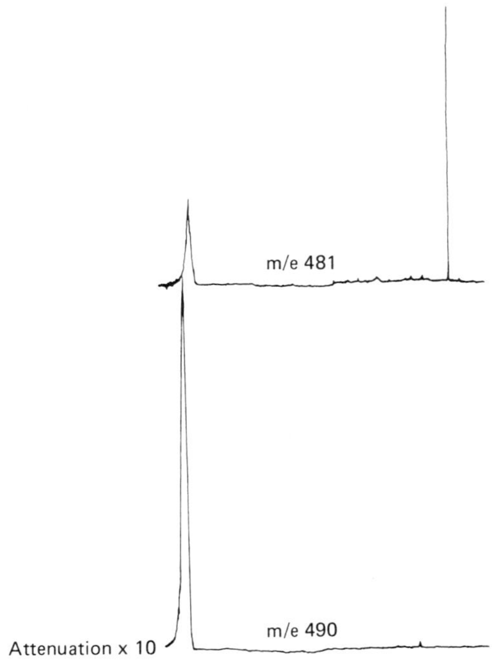

Figure 8.42 GC-MS analysis of plasma 16α-cyano-3β-cyclopentyloxypregn-5-en-20-one (500 pg) as TBDMS-oxime (in the presence of 25 ng of the [$^{2}H_{9}$] analogue. Conditions: 30-m OV-1 Silanox column. Isothermal operation at 300° with solid injection. (From Ref. 106.)

and it is excreted in urine principally as the 6β-hydroxy derivative, although to a certain extent there is inversion of the constituents at position C-17 [107]. The metabolites of Dianabol can be conveniently extracted by solvent without prior hydrolysis since they are primarily excreted unconjugated, but if a general method is required which would include other anabolic steroid drug metabolites, then the urine samples should be hydrolyzed first. Dianabol metabolites are analyzed primarily as trimethylsilyl ethers, and an example of a separation achieved by Dürbec et al. [107] is illustrated in Figure 8.43.

Next to Dianabol, the anabolic steroid drugs most favored by athletes are derivatives of 19-nortestosterone, such as Durabolin (19-nortestosterone phenylpropionate). Durabolin readily loses phenylpropionate in the body, and the major metabolites excreted in urine are the 19-nor homologues of the testosterone metabolites, androsterone and etiocholanolone. The simplest method for analysis utilizes β-glucuronidase hydrolysis, chloroform extraction, and formation of methyloxime-trimethylsilyl ethers [108]. Results obtained using this technique are shown in Figure 8.44. This illustrates: (1) a conventional OV-101 capillary column analysis with FID detection; (2) selected-ion chromatograms from repetitive scanning of GC-MS data of ions characteristic of 19-norandrosterone; and (3) the mass spectrum of the component identified as 19-norandrosterone methyloxime-trimethylsilyl ether. This technique therefore, permits a full mass spectrum to be obtained, and little doubt need remain regarding the identification, particularly since accurate retention time data are also available. However, when such a simple analytical procedure is used for sample preparation, analysis on glass capillary columns is virtually obligatory.

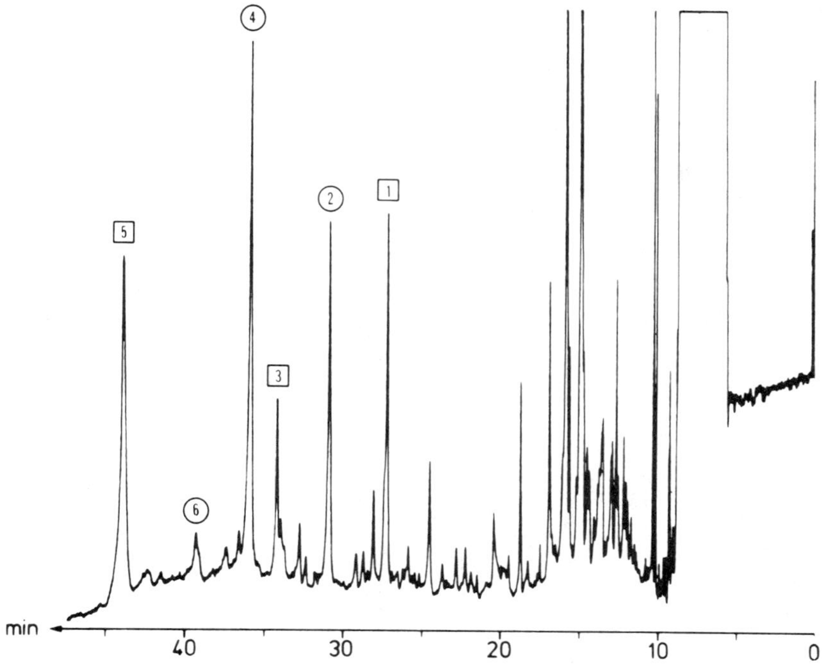

Figure 8.43 Gas chromatography of a urinary extract containing unconjugated metabolites of ingested Dianabol. The following peaks are marked: 1, a phthalate; 2, 17-epi-Dianabol; 3, squalene; 4, 6β-hydroxy-17-epiDianabol; 5, cholesterol; and 6, 6β-hydroxy-Dianabol. The steroids were analyzed as trimethylsilyl ethers. Conditions: 40-m SE-54 WCOT column programmed from 200–260° at 2.5°/min. Splitless injection. (From Ref. 107.)

Glass capillary column chromatography has also been used to a considerable extent in studies of the metabolism of other anabolic steroid drugs, e.g., orabolin and nilevar [109].

3. *Synthetic Corticosteroids*

Data on the use of glass capillary columns for the gas chromatographic separation of synthetic corticosteroids are almost nonexistent. HPLC is now virtually exclusively the technique of choice for investigation of the purity of pharmaceutical preparations, and most studies on the metabolism of these compounds were carried out before the introduction of glass capillaries. However, there would be few problems in the gas chromatography of steroids of this type.

VI. CURRENT AND FUTURE DEVELOPMENTS—THE ANALYSIS OF INTACT CONJUGATES

A major drawback of the established methods for urinary steroid analysis which have been described in this chapter is the necessity of hydroylzing the steroid conjugates. It is rarely possible to be completely confident that all major steroids have been quantitatively released from conjugation, and the

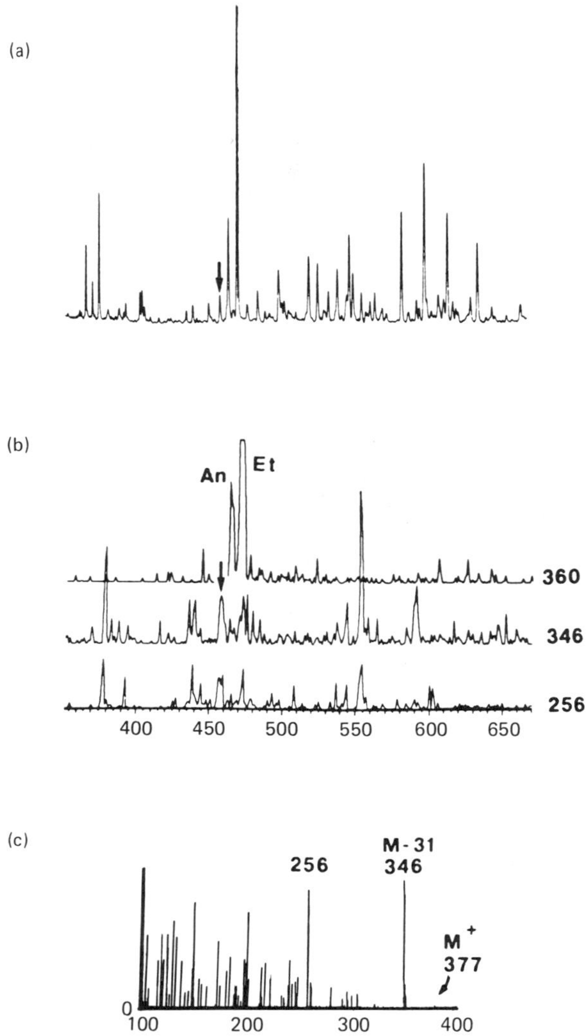

Figure 8.44 Detection of misuse of anabolic steroid drugs. Identification of a trace amount of 19-nortestosterone metabolite in urine by GC-MS. A mixture of all urinary steroids (hydrolyzed, ethyloxime trimethylsilyl ether derivatives) were analyzed by repetitive scanning GC-MS. The upper part of the figure (a) shows the capillary gas chromatogram. A minor peak eluting just ahead of androsterone (arrowed) was finally identified as 3α-hydroxy-5ξ-estrone-17-one from selected-ion profiles (b) and the complete mass spectrum (c). The broadness of the peak suggests that it may be a mixture of two isomers (5α and 5β). The selected-ion profile of m/z 360 is given by the endogenous androgen metabolites androsterone (An) and etiocholanolone (Et) and represents the M-31 fragment. Conditions: 20-m OV-1 column programmed between 180 and 260° at 3°/min. Splitless injection.

hydrolytic step introduces further contamination and is often time-consuming.* The possibility of direct derivatization and analysis of steroid conjugates is therefore extremely attractive. Gas chromatography of steroid glucuronides is not a new concept. Jaakomaki et al. [110] demonstrated the separation of androsterone, etiocholanolone, and pregnanediol glucuronides (as methyl ester, trimethylsilyl derivatives) more than 10 years ago. However, it is possible that only now have work-up procedures gained sufficient sophistication to allow conjugate methods to be developed further. Advances required for the development of such a method for capillary column analysis were: (1) the ability to quantitatively recover microgram quantities of conjugates from urine, (2) the ability to purify steroid trimethylsilyl ether derivatives, and (3) the preparation of capillary columns of high stability, since high temperatures are required for analysis. These developments have occurred, and now there is little which should inhibit development of steroid conjugate analysis. For example, the use of Waters Sep-pak cartridges permits the rapid concentration of urinary constituents [19]; the method of Axelson and Sjövall [34] permits the purification of steroid trimethylsilyl derivatives prepared by the trimethylsilylimidazole method; and current methods for column preparation by initial silanization [5] allow preparation of nonpolar columns of high stability.

Steroid glucuronides are readily protected for gas chromatography by formation of methyl ester, methyloxime, and trimethylsilyl ether derivatives. As in derivatization of free steroids, a potent silylation reagent (TSIM) is required to fully protect all classes of steroids and steroid conjugates. Sulfates of 3β-hydroxy-5-ene steroids and estrogens readily lose their sulfate group during trimethylsilylation, so steroids of this type are effectively analyzed as free compounds. We have not yet examined the behavior of other forms of sulfated steroids, such as, 17β-sulfates, 20α-sulfates, or sulfates of 3α-hydroxy-5α steroids, but there is evidence from other studies that elimination of the sulfate group and the ester hydroxyl takes place [111].

The method that we are currently using for conjugate analysis is as follows: (1) extraction of steroid conjugates by Sep-pak, (2) methylation of carboxylic acid by ethereal diazomethane, (3) methyloxime formation, and (4) 8-hr trimethylsilylation with TSIM. The derivatives are purified by Lipidex 5000 chromatography according to the method previously described.

An example of direct analysis of steroid sulfates and glucuronides is illustrated in Figure 8.45. The urine used for this analysis was obtained from a patient with 3β-hydroxy-5-ene dehydrogenase deficiency and contained principally 3β-hydroxy-5-ene steroid sulfates, although significant amounts of saturated steroid glucuronides were present. The upper chromatogram shows steroids analyzed following conventional enzymatic hydrolysis and derivatization, and the lower chromatogram shows results obtained by direct derivatization. The retention time of the derivatized glucuronide conjugates is very much

*Although hydrolytic periods of one or more days are used by many workers, hydrolysis with the *Helix pomatia* preparation can be completed in 1 hr at 62° [25]. It should therefore be possible to have a free steroid extract ready for derivatization within 2 hr.

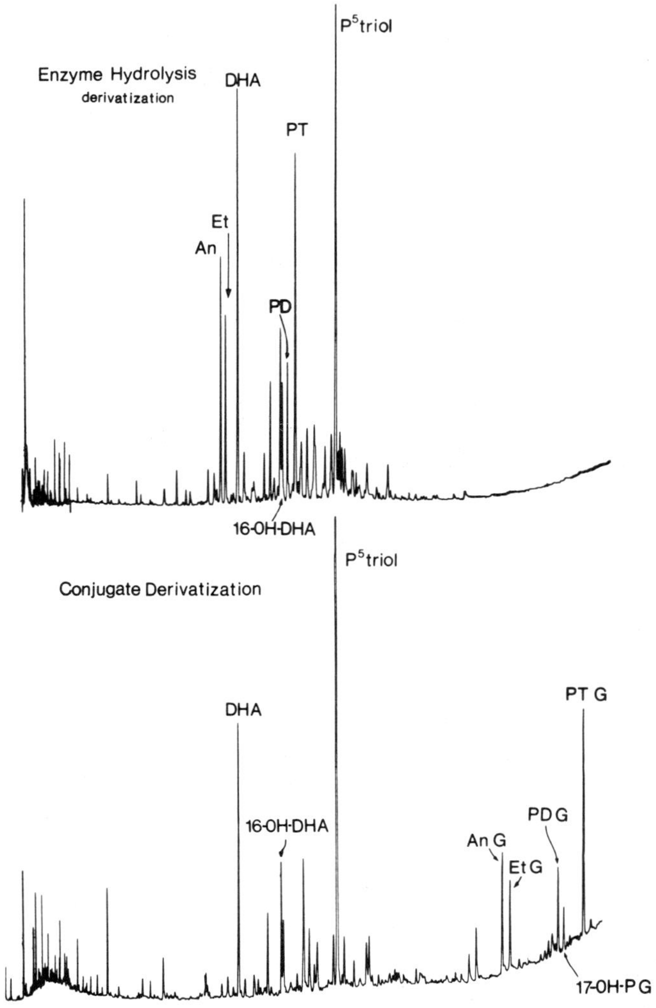

Figure 8.45 Conjugate analysis of steroids excreted by patient with 3β-hydroxy steroid dehydrogenase (3β-OH-SD) deficiency. The upper chromatogram represents derivatives of steroids analyzed following enzymatic hydrolysis of the urine, while the lower chromatogram was obtained by direct derivatization and analysis of steroid sulfates and glucuronides. The following abbreviations are used for glucuronide (G) conjugated steroids: An (androsterone); Et (etiocholanolone); PD (pregnanediol); 17-OH-P (17α-hydroxypregnanolone); and PT (pregnanetriol). Steroids present in urine as sulfate esters are abbreviated as follows: DHA (dehydroepiandrosterone), 16-OH-DHA (16α-hydroxydehydroepiandrosterone), and P^5-triol (5-pregnane-3β,17α,20α-triol). Conditions: 20-m high-temperature OV-1 WCOT column, temperature-programmed 180–330° at 3°/min. Solid injection of samples. (From Ref. 113, with permission of publishers).

greater when direct derivatization is carried out, while the retention time of the steroid "sulfates" is unchanged since they undergo hydrolysis during silylation.

This example only contained significant amounts of *simple* glucuronides, yet a conjugate method would have generally utility only if even the most complex molecules could be analyzed. The analysis of corticosteroid glucuronides would cause the greatest difficulty, and we have begun investigations in the analysis of such compounds using reference materials synthesized by Dr. Vernon Mattox of the Mayo Clinic. C_{36} and C_{44} alkanes have been used as internal standards since they are eluted before androsterone glucuronide and after β-cortol glucuronide, respectively. Response factors (peak height) are generally lower than those obtained for free steroids. For example, the glucuronides of tetrahydrocortisone, tetrahydrocortisol, and tetrahydroSubstance S each give a response of 25% relative to the alkane internal standards (Figure 8.46), even though they elute as single peaks. However, the glucuronides of β-cortol and β-cortolone give response factors of approximately unity.

A mixture of reference estrogen conjugates has also been analyzed according to this method. The results are illustrated in Figure 8.47. The first two steroid peaks represent estrogen sulfates which have been hydrolyzed during silylation, while the latter peaks are estrogen glucuronides. It can be seen that there is good separation between estriol-3-glucuronide and estriol-16-glucuronide. Development of this methodology may prove particularly useful for the analysis of the catechol estrogens since retention of the glucuronide during analysis should stabilize them.

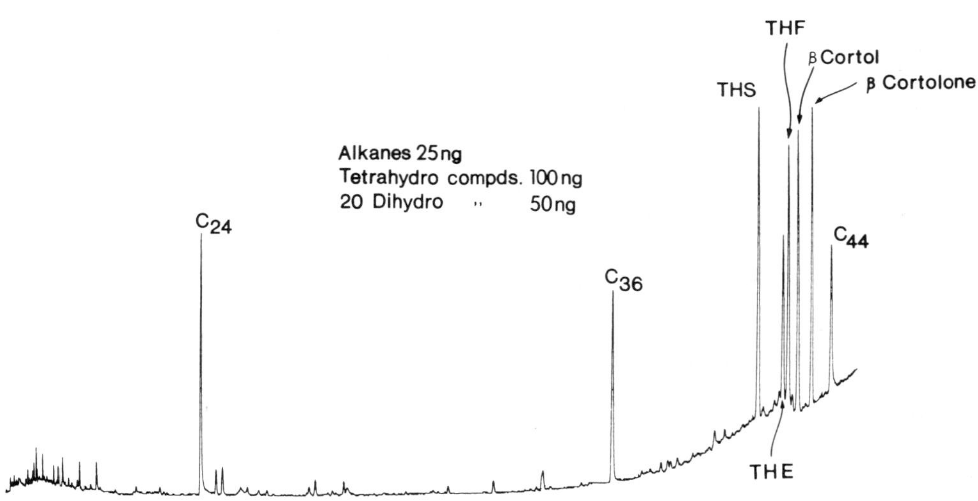

Figure 8.46 Corticosteroid metabolite metabolite glucuronides as methyl ester, methyloxime, trimethylsilyl ethers. Abbreviations used: THS (tetrahydro-Substance S), THE (tetrahydrocortisone), and THF (tetrahydrocortisol). For conditions of analysis, see legend of Figure 8.45. (From Ref. 113.)

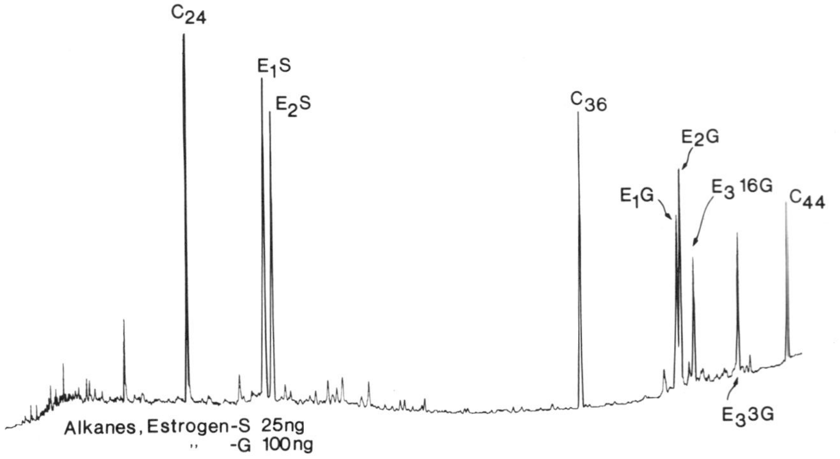

Figure 8.47 Derivatization and analysis of reference estrogen conjugates. The following abbreviations are used: E_1S (estrone sulfate, as free compound), E_2S (estradiol 3-sulfate, as free compound), E_1G (estrogen glucuronide), E_2G (estradiol 3-glucuronide), $E_3$16G (estriol 16-glucuronide), and $E_3$3G (estriol 3-glucuronide). For conditions of analysis, see legend of Figure 8.45. (From Ref. 114, with permission of publishers.)

Retention indices of many of the glucuronides of common urinary steroids have now been determined on OV-1 open-tubular columns, and these are listed in Table 10. Some of these figures were determined by analysis of authentic steroid conjugates while others were deduced from analysis of urine from patients with known hypersecretion of particular compounds.

The current methodology is already useful for the analysis of urinary steroids in a variety of clinical conditions. Figure 8.48, for example, illustrates the separation of steroids from a patient with an ovarian tumor producing large amounts of 17α-hydroxyprogesterone. In the upper chromatogram the steroids have been analyzed following enzymatic hydrolysis, and in the lower chromatogram they have been analyzed by direct derivatization. The major components are metabolites of progesterone and 17α-hydroxyprogesterone, and in most respects the profile can be considered to be identical to that given by the 21-hydroxylase defect form of congenital adrenal hyperplasia. The only difference is the lack of a peak due to 11-oxopregnanetriol. Using this methodology, diagnosis of 21-hydroxylase deficiency could be made within a few hours of receipt of sample, no longer than it would take to analyze serum 17α-hydroxyprogesterone by radioimmunoassay.

Placental sulfatase deficiency (see Section V.A.2) is another disorder which is quickly diagnosed by analysis of urinary steroid conjugates. Rapid diagnosis is desirable since, if the patient has been shown to have the disorder, other more serious conditions affecting fetal well-being can be excluded. A chromatogram of urinary steroids from a patient with PSD is illustrated in Figure 8.49. The diagnostic steroids are 16α-hydroxyDHA sulfate and adrostenetriol sulfate, which are of course hydrolyzed during the derivatization process and are effectively analyzed as free steroids.

Table 8.10 Relative Retention Times (Methylene Units, MU) for Derivatives of Unconjugated and Glucuronide Conjugates Steroids[a]

Steroid	MU unconjugated	MU conjugated	ΔMU[b]
Androsterone	25.20	36.84	11.64
Etiocholanolone	25.42	37.16	11.74
11-Oxoetiocholanolone	26.00	37.89	11.89
11β-Hydroxyandrosterone	27.00	37.61	10.61
11β-Hydroxyetiocholanolone	27.20	37.89	10.69
Pregnanediol	27.64	39.50	11.86
17α-Hydroxypregnanolone	27.00	39.80	12.80
Pregnanetriol	28.00	40.87	12.87
TetrahydroDOC	28.70	40.70	12.00
TetrahydroSubstance S	28.62	41.29	12.67
TetrahydroCompound A	29.77	41.47	11.70
Tetrahydrocorticosterone	29.80	41.29	11.49
Allotetrahydrocortiocosterone	30.00	40.46	10.46
Tetrahydrocortisone	29.61	42.14	12.98
Tetrahydrocortisol	30.20	42.33	12.13
Allotetrahydrocortisol	30.29	41.25	10.90
α-Cortolone	30.51	43.00	12.49
α-Cortol	31.20	43.00	11.80
β-Cortol	30.73	42.70	11.97
β-Cortolone	30.73	43.26	12.53
Estrone	26.60	39.59	12.99
Estradiol	26.60	39.76	13.16
Estriol	28.75	40.88[c]	12.13
Estriol	28.75	40.33[d]	11.58

[a] Unconjugated steroids, methyloxime/trimethylsilyl ethers; glucuronides, methyl ester/methyloxime trimethylsilyl ethers.
[b] ΔMU represents the increase in methylene unit value by addition of a glucuronide.
[c] Estriol 3-glucuronide.
[d] Estriol 16-glucuronide.

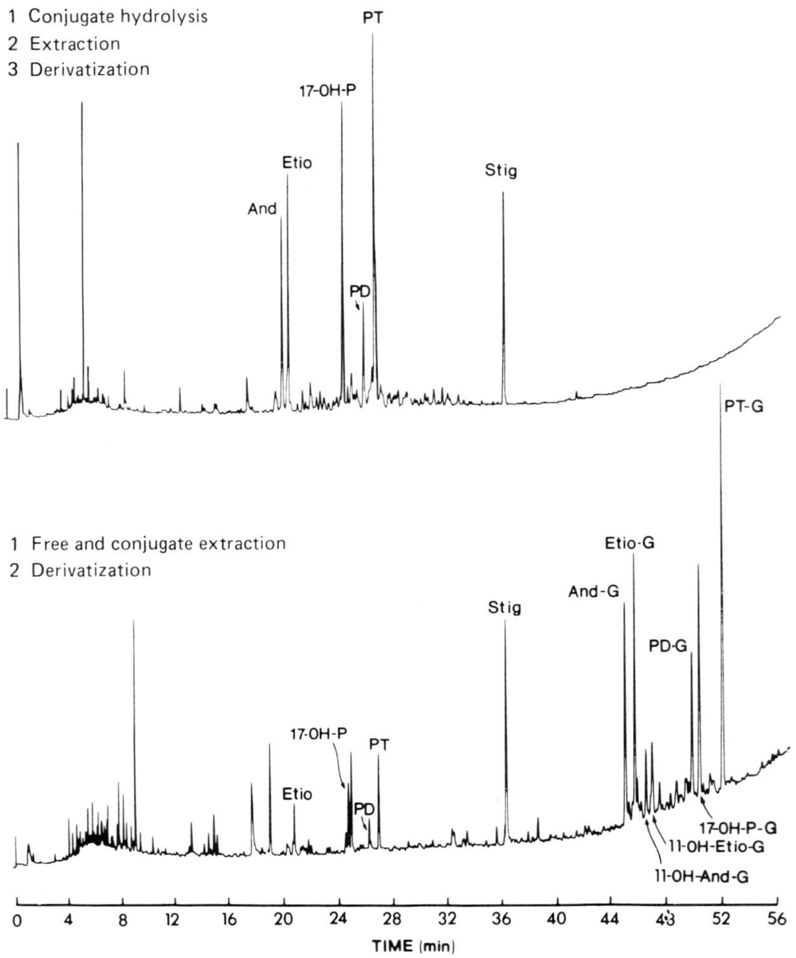

Figure 8.48 Analysis of steroids excreted by patients with 17α-hydroxyprogesterone-secreting ovarian tumor. The upper chromatogram is of steroids obtained following hydrolysis of urine, and the lower chromatogram shows the same steroids directly derivatized. The following abbreviations are used: And (androsterone), Etio (etiocholanolone), 17α-OH-P (17α-hydroxypregnanolone), PD (pregnanediol), PT (pregnanetriol), and Stig (stigmasterol, internal standard). Suffix G represents glucuronide. The urinary profile of steroids excreted by patients with the 21-hydroxylase-deficiency from a congenital adrenal hyperplasia would be very similar to those shown. For conditions of analysis, see legend from Figure 8.45. (From Ref. 115.)

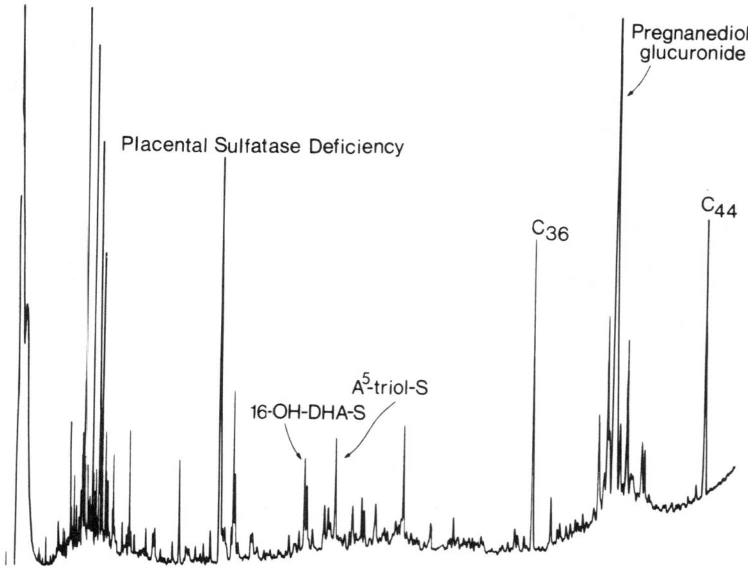

Figure 8.49 Diagnosis of placental sulfatase deficiency by conjugate analysis of maternal urine. Hypersecretion of 16α-hydroxyDHA sulfate (16-OH-DHA-S) and androstenetriol sulfate (A^5-triol-S) is readily seen. These sulfates are hydrolyzed during silylation so the compounds are effectively analyzed as free compounds. For conditions of analysis, see legend of Figure 8.45.

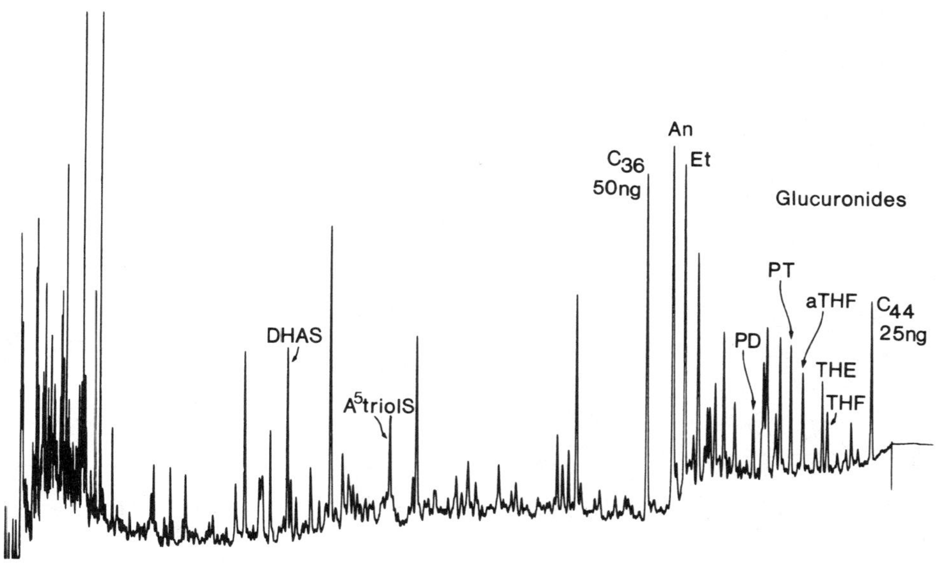

Figure 8.50 Direct conjugate analysis of urinary steroids of normal male adult. For conditions of analysis and steroid abbreviations, see legends for Figures 8.45–8.49. (From Ref. 113.)

Finally, Figure 8.50 illustrates the separation of steroids from a normal adult male. The identities of many of the major peaks are now known by comparison of retention times to authentic standards; others are yet to be identified.

GLOSSARY

Steroid Nomenclature and Abbreviations

Androsterone (An)	3α-Hydroxy-5α-androstan-17-one
Etiocholanolone (Et)	3α-Hydroxy-5β-androstan-17-one
11-Oxoandrosterone (11-O-An)	3α-Hydroxy-5α-androstane-11,17-dione
11-Oxoetiocholanolone (11-O-Et)	3α-Hydroxy-5β-androstane-11,17-dione
11β-Hydroxyandrosterone (11-OH-An)	3α,11β-Dihydroxy-5α-androstan-17-one
11β-Hydroxyetiocholanolone (11-OH-Et)	3α,11β-Dihydroxy-5β-androstan-17-one
Dehydroepiandrosterone (DHA)	3β-Hydroxy-5-androsten-17-one
16α-HydroxyDHA (16α-OH-DHA)	3β,16α-Dihydroxy-5-androstene-17-one
Androstenetriol (AT)	5-Androstene-3β,16α,17β-triol
Pregnanolone	3α-Hydroxy-5β-pregnan-20-one
Pregnanediol (PD)	5β-Pregnane-3α,20α-diol
Pregnanetriol (PT)	5β-Pregnane-3α,17α,20α-triol
Pregnenolone	3β-Hydroxy-5-pregnen-20-one
Pregnenediol (Δ^5PD)	5-Pregnene-3β,20α-diol
Pregnenetriol (Δ^5PT)	5-Pregnene-3β,17α,20α-triol
11-Oxopregnanediol (11-O-PD)	3α,20α-Dihydroxy-5α(and β)-pregnan-11-one
11-Oxopregnanetriol (11-O-PT)	3α,17α,20α-Trihydroxy-5β-pregnan-11-one
11β-Hydroxypregnanediol (11-OH-PD)	5α(and β)-Pregnane-3α,11α,20α-triol
TetrahydroDOC (THDOC)	3α,21-Dihydroxy-5β-pregnan-20-one
TetrahydroCompound A (THA) and allo- (aTHA)	3α,21-Dihydroxy-5β(α)-pregnane-11,20-dione
Tetrahydrcorticosterone (THB) and allo- (aTHB)	3α,11β,21-Trihydroxy-5β(α)-pregnan-20-one
TetrahydroSubstance S (THS)	3α,17α,21-Trihydroxy-5β-pregnan-20-one
Tetrahydrocortisone (THE)	3α,17α,21-Trihydroxy-5β-pregnan-11-one

Tetrahydrocortisol (THF) and allo- (aTHF)	3α,11β,17α,21-Tetrahydroxy-5β(α)-pregnan-20-one
α(β)-Cortolone	3α,17α,20α(β),21-Tetrahydroxy-5β-pregnan-11-one
α(β)-Cortol	5β-Pregnane-3α,11β,17α,20α(β),21-pentol
α(β)-Cortolic acid	3α,11β,17α,20α(β)-Tetrahydroxy-5β-pregnan-21-oic acid
α(β)-Cortolonic acid	3α,17α,20α(β)-Trihydroxy-11-oxo-5β-pregnan-21-oic acid

ACKNOWLEDGMENTS

I wish to thank my colleagues John Honour, Norman Taylor, and Esther Roitman who have contributed greatly to the studies described which were carried out in my laboratory. I am indebted to the Medical Research Council Steroid Reference Collection (Curator, Prof. D. N. Kirk) for supplying most of the authentic compounds used in our investigations. Dr. V. Mattox of the Mayo Clinic contributed the steroid glucuronides, and I am indebted to him for this generosity. I am grateful to Virtinia Schutz for undertaking the typing of the manuscript and to all the investigators who supplied data and illustrations used in this review. Some of these data were generously supplied prior to publication. Financial support for many of the investigations described came from the National Institutes of Health Division of Research Resources Grant RR01614 (A. L. Burlingame, Principal Investigator).

REFERENCES

1. J. W. Honour, J. Tourniaire, E. G. Biglieri, and C. H. L. Shackleton, *J. Steroid Biochem. 9*:495 (1978).
2. W. L. Gardiner and E. C. Horning, *Biochem. Biophys. Acta 115*:524 (1966).
3. J. A. Völlmin, *Clin. Chim. Acta 34*:207 (1971).
4. K. Grob and K. Grob, Jr., *J. High. Res. Chromatogr. and CC 1*:57 (1978).
5. K. Grob, G. Grob, and K. Grob, Jr., *J. High Res. Chromatogr. and CC 2*:31 (1979).
6. C. D. Pfaffenberger and E. C. Horning, *J. Chromatogr. 112*:581 (1975).
7. A. L. German and E. C. Horning, *J. Chromatogr. Sci. 11*:76 (1973).
8. C. H. L. Shackleton and J. W. Honour, *Clin. Chim. Acta 67*:287 (1976).
9. K. Grob, Jr., G. Grob, and K. Grob, *J. Chromatogr. 156*:1 (1978).
10. K. Grob, Jr., and H. P. Neukom, *J. High Resol. Chromatogr. and CC 1*:15 (1979).
11. A. Ras, *J. Chromatogr. Sci. 3*:252 (1965).
12. P. M. J. Van den Berg and Th. P. H. Cox, *Chromatographia* 5:301 (1972).
13. M. Axelson and J. Sjövall, *Res. on Steroids 8*:119 (1979).
14. E. Vanluchene, W. Aertsens, and D. Vandekerckhove, *Acta Endocrinol. 90*:133 (1979).

15. J. J. Leunissen, Quantitative aspects of the determination of steroid profiles from urine by capillary gas chromatography, Thesis: Technical Highschool, Eindhoven, The Netherlands (1979).
16. P. Vestergaard, J. F. Sayegh, J. H. Mowat, and L. Hemmingsen, *Acta Endocrinol. Suppl. 217*:(1978).
17. S. Burstein and S. Lieberman, *J. Biol. Chem. 233*:331 (1958).
18. H. L. Bradlow, *Steroids 11*:265 (1968).
19. C. H. L. Shackleton and J. O. Whitney, *Clin. Chim. Acta 107*:231 (1980).
20. T. Fotsis, P. Järvenpää, and H. Aldercreutz, *J. Steroid Biochem. 12*: 503 (1980).
21. A. Dyfverman and J. Sjövall, *Anal. Lett. B11*:485 (1978).
22. J. Sjövall and R. Vihko, *Acta Endocrinol 57*:247 (1968).
23. O. Jänne, R. Vihko, J. Sjövall, and K. Sjövall, *Clin. Chim. Acta. 23*: 405 (1969).
24. K. D. R. Setchell, B. Almé, M. Axelson, and J. Sjövall, *J. Steroid Biochem. 7*:615 (1976).
25. M. Axelson, B.-L. Sahlberg, and J. Sjövall, *J. Chromatogr. 224*:335 (1981).
26. T. Fotsis, P. Järvenpää, and H. Adlercreutz, *J. Steroid Biochem. 12*: 503 (1980).
27. T. Fotsis, H. Adlercreutz, P. Järvenpää, K. Setchell, M. Axelson, and J. Sjövall, *J. Steroid Biochem. 14*:457 (1981).
28. M. Tetsuo, M. Axelson, and J. Sjövall, *J. Steroid Biochem. 13*:847 (1980).
29. C. H. L. Shackleton, J.-Å. Gustafsson, and J. Sjövall, *Steroids 17*:265 (1971).
30. E. Nyström and J. Sjövall, *Archiv. Kemi. 29*:107 (1968).
31. K. D. R. Setchell and C. H. L Shackleton, *Clin. Chim. Acta 47*:381 (1973).
32. B. E. P. Murphy, *Nature (New Biol.) 232*:21 (1971).
33. R. A. Anderson, G. Defay, C. Madani, E. M. Chambaz, and C. J. W. Brooks, *J. Chromatogr. 99*:485 (1974).
34. M. Axelson and J. Sjövall, *J. Steroid Biochem. 5*:733 (1974).
35. M. Axelson and J. Sjövall, *J. Chromatogr. 126*:705 (1976).
36. D. R. Knapp, *Handbook of Analytical Derivatization Reactions*, Wiley interscience, New York, 1979.
37. H. M. Fales and T. Luukkainen, *Anal. Chem. 37*:955 (1965).
38. J. E. Blackwood, C. L. Gladys, K. L. Loening, A. E. Petrarca, and J. E. Rush, *J. Am. Chem. Soc. 90*:509 (1968).
39. J.-P. Thenot and E. C. Horning, *Anal. Lett. 5*:905 (1972).
40. C. H. L. Shackleton and J. W. Honour, *J. Steroid Biochem. 8*:199 (1977).
41. E. C. Horning and B. F. Maume, *J. Chromatogr. Sci. 7*:411 (1969).
42. J. W. Honour and C. H. L. Shackleton, *J. Steroid Biochem. 8*:299 (1977).
43. E. C. Horning, P. G. Devaux, A. C. Moffat, C. D. Pfaffenberger, N. Sakauchi, and M. G. Horning, *Clin. Chim. Acta 34*:135 (1971).
44. E. M. Chambaz, G. Defaye, and C. Madani, *Anal. Chem. 45*:1090 (1973).
45. G. Spiteller, *Pure and Appl. Chem. 50*:205 (1978).
46. V. Fantl and D. H. Gray, *Clin. Chim. Acta 79*:237 (1977).
47. G. Phillipou, D. A. Bigham, and R. F. Seamark, *Steroids 26*:516 (1975).

48. M. A. Quilliam, J. F. Templeton, and J. B. Westmore, *Steroids 29*: 613 (1977).
49. E. Kovats, *Helv. Chim. Acta. 41*:1915 (1958).
50. E. C. Horning, in *Gas Phase Chromatography of Steroids* (N. B. Eik-Nes and E. C. Horning, eds.), Springer-Verlag, New York, 1968, p. 32.
51. R. W. H. Edwards, *J. Chromatogr. 153*:1 (1978).
52. C. J. W. Brooks, *Phil. Trans. R. Soc. Lond. A. 293*:53 (1979).
53. C. H. L. Shackleton, *Clin. Chim. Acta. 67*:287 (1976).
54. W. M. Teller and J. Homoki, *Pädiatrie und Pädologie Suppl. 5*:29 (1977).
55. M. Zachman, J. A. Völlmin, G. Mürset, H.-Ch. Curtius, and A. Bader, *J. Clin. Endocrinol. Metab. 30*:719 (1970).
56. R. L. Rosenfield, B. H. Rich, J. I. Wolfsdorf, F. Cassorla, J. S. Parks, A. M. Bongiovanni, C. H. Wu, and C. H. L. Shackleton, *J. Clin. Endocrinol. Metab. 51*:345 (1980).
57. M. Zachmann, J. A. Völlmin, M. I. New, and H.-Ch. Curtius, *J. Clin. Endocrinol. Metab. 33*:501 (1971).
58. J. W. Honour and C. H. L. Shackleton, *Acta Endocrinol. 103*:101 (1983).
59. C. H. L. Shackleton, E. G. Biglieri, E. Roitman, and J. W. Honour, *J. Clin. Endocrinol. Metab. 48*:976 (1979).
60. B. P. Lisboa, J. M. Halket, I. Ganschow, and M. del C. Ruiz-Gonzalez, *Acta Medica Portugesa 1*:433 (1979).
61. C. H. L. Shackleton and G. H. A. I. Snodgrass, *Ann. Clin. Biochem. 11*:91 (1974).
62. M. J. Dillon , J. W. Leonard, J. M. Buckler, D. Ogilvie, D. Lillystone, J. W. Honour, and C. H. L. Shackleton, *Arch. Dis. Childh. 55*:427 (1980).
63. C. H. L. Shackleton, J. W. Honour, M. Dillon, and P. Milla, *Acta Endocrinol. 81*:762 (1976).
64. P. J. Milla, R. Trompeter, M. J. Dillon, D. Robbins, and C. H. L. Shackleton, *Arch. Dis. Childh. 52*:580 (1977).
65. J. W. Honour, H. B. Valman, and C. H. L. Shackleton, *Acta. Paed. Scand. 66*:103 (1977).
66. C. H. L. Shackleton, N. F. Taylor, and J. W. Honour, *An Atlas of Gas Chromatographic Profiles of Neutral Urinary Steroids in Health and Disease*, Packard-Becker, Delft, The Netherlands, 1980.
67. H. J. G. M. Derks and N. M. Drayer, *Steroids 31*:289 (1978).
68. H. Dean, J. S. D. Winter, and C. H. L. Shackleton, *J. Clin. Endocrinol. Metab.* in press.
69. C. D. Pfaffenberger and E. C. Horning, *Anal. Biochem. 80*:329 (1977).
70. J.-Å. Gustafsson and A. Stenberg, *Endocrinology, 96*:501 (1975).
71. C. D. Pfaffenberger, L. R. Malinak, and E. C. Horning, *J. Chromatogr. 158*:313 (1978).
72. R. D. Bulbrook, J. L. Hayward, C. C. Spicer, and B. S. Thomas, *Lancet ii*:1239 (1962).
73. H. L. Bradlow, R. M. Boyar, J. O'Connor, B. Zumoff, and L. Hellman, *J. Clin. Endocrinol. Metab. 43*:571 (1976).
74. R. E. Peterson, J. Imperato-McGinley, T. Gautier, and E. Sturla, *Am. J. Med. 62*:170 (1977).
75. S. Ulick, L. S. Levine, P. Gunczler, G. Zanconato, L. C. Ramirez, W. Rauh, A. Rosler, H. L. Bradlow, and M. I. New, *J. Clin. Endocrinol. Metab. 49*:757 (1979).

76. C. H. L. Shackleton, J. W. Honour, M. J. Dillon, C. Chantler, and R. W. A. Jones, *J. Clin. Endocrinol. Metab. 50*:786 (1980).
77. H. Ludwig, M. Spiteller, H. J. Egger, and G. Spiteller, *Israel J. Chem. 16*:7 (1977).
78. H. J. Egger, J. Reiner, G. Spiteller, and R. Haffele, *J. Chromatogr. 145*:359 (1978).
78a. M. Axelson, T. Cronholm, B.-L. Sahlberg, and J. Sjövall, *J. Steroid Biochem. 14*:155 (1981).
79. C. H. L. Shackleton, J.-Å. Gustafsson, and F. L. Mitchell, *Acta Endocrinol. 74*:157 (1973).
80. E. M. Chambaz, C. J. W. Brooks, M. G. Horning, E. C. Horning, and R. M. Hill, *Compt. Rendu, 268*:2817 (1969).
81. C. H. L. Shackleton and N. F. Taylor, *J. Steroid Biochem. 6*:1393 (1975).
82. N. F. Taylor, D. H. Curnow, and C. H. L. Shackleton, *Clin. Chim. Acta 85*:219 (1978).
83. H. J. G. M. Derks and N. M. Drayer, *Steroids, 31*:9 (1978).
84. C. H. L. Shackleton, J. W. Honour, and N. F. Taylor, *J. Steroid Biochem. 11*:523 (1979).
85. C. H. L. Shackleton, P. G. F. Swift, D. C. L. Savage, and J. W. Honour, *J. Clin. Endocrinol. Metab. 49*:247 (1979).
86. N. F. Taylor and C. H. L. Shackleton, *J. Clin. Endocrinol. Metab. 49*:78 (1979).
87. M. deM. Fench, R. J. Stillman, J. Cohen, and D. Tulchinksy, *Nature 287*:225 (1980).
88. P. Ball and R. Knuppen, *Acta Endocrinol. 93*:Suppl. 232 (1980).
89. J. Fishman and D. Tulchinsky, *Science 210*:73 (1980).
90. K. D. R. Setchell, N. F. Taylor, H. Adlercreutz, M. Axelson, and J. Sjövall, *Res. on Steroids 8*:131 (1979).
90a. M. Axelson, J. H. Clark, H. A. Eriksson and J. Sjövall, *J. Steroid Biochem.*, in press.
91. H. L. Bradlow, B. Zumoff, C. Monder, H. J. Lee and L. Hellman, *J. Clin. Endocrinol. Metab. 37*:811 (1973).
92. R. L. Farhi and C. Monder, *Anal. Biochem. 90*:58 (1978).
93. C. H. L. Shackleton, E. Roitman, C. Monder, and H. L. Bradlow, *Steroids 36*:289 (1980).
94. M. Axelson and J. Sjövall, *J. Steroid Biochem. 8*:683 (1977).
95. S. J. Gaskell and D. S. Millington, in *Quantitative Mass Spectrometry in Life Sciences*, Vol. 2 (A. P. de Leenheer, R. R. Roncucci, and C. Van Peteghem, eds.), Elsevier, Amsterdam, 1978, p. 135.
96. B. F. Maume, C. Millot, D. Mesnier, D. Patouraux, J. Doumas, and E. Tomori, *J. Chromatogr. 186*:581 (1979).
97. H. Ludwig and G. Spiteller, *J. Chromatogr. 146*:381 (1978).
98. H. F. De Luca, *Vitamin D Metabolism and Function*, Springer-Verlag, Berlin, 1979.
99. I. Björkhem and I. Holmberg, *Clin. Chim. Acta 68*:215 (1976).
100. I. Björkhem and I. Holmberg, *Clin. Chem. 25*:584 (1979).
101. J. O. Whitney, C. H. L. Shackleton, C. G. Edmonds, A. L. Burlingame, and C. F. Piel, in *Stable Isotopes: Proceedings of the Third International Conference* (E. R. Klein and P. D. Klein, eds.), Academic Press, New York, 1979, p. 37.

102. B. P. Lisboa and J. M. Halket, in *Recent Developments in Chromatography and Electrophoresis* (A. Frigerio and L. Renoz, eds.), Elsevier, Amsterdam, 1979, p. 144.
103. C. H. L. Shackleton, E. Roitman, and J. O. Whitney, *J. Steroid Biochem.* *12*:521 (1980).
104. B. L. Onisko, R. P. Esvelt, H. K. Schnoes, and H. F. De Luca, *Biochemistry* *19*:4124 (1980).
105. L. Siekman, A. Siekman, and H. Breuer, *Fresenius Z. Anal. Chem.* *290*:158 (1978).
106. S. J. Gaskell, C. J. W. Brooks, and S. B. Matin, *Biomed. Mass Spectrom,* *5*:460 (1978).
107. H. W. Dürbeck, I. Buker, B. Scheulen, and B. Telin, *J. Chromatogr.* *167*:117 (1978).
108. C. H. L. Shackleton, R. J. Ward, and A. M. Lawson, unpublished.
109. R. J. Ward, A. M. Lawson, and C. H. L. Shackleton, *J. Steroid Biochem.* *8*:1057 (1977).
110. P. I. Jaakomaki, K. A. Yarger, and E. C. Horning, *Biochim. Biophys. Acta* *137*:216 (1967).
111. S. Murray and T. A. Baillie, *Biomed. Mass Spectrom.* *6*:82 (1979).
112. C. H. L. Shackleton, "Multicomponent analysis by capillary gas chromatography," in *Current Developments in Clinical Applications of HPLC, GC, and MS* (A. M. Lawson, C. K. Lim, and H. Richmond, eds.), Academic Press, London, England, 1980, p. 247.
113. C. H. L. Shackleton, V. R. Mattox, and J. W. Honour, *J. Steroid Biochem.* *19*:209 (1983).
114. C. H. L. Shackleton, *Steroids* *39*:485 (1982).
115. C. H. L. Shackleton, *Clin. Chem.* *27*:509 (1981).
116. J. W. Honour, J. M. Anderson, and C. H. L. Shackleton, *Acta Endocrinologica* *103*:101 (1983).
117. E. H. Epstein, R. M. Krauss, and C. H. L. Shackleton, *Science* *214*:659 (1981).

9

Analysis of Thyroid Hormones

Jeffrey A. Corkill,* Simon H. Kuttab, and Roger W. Giese / *College of Pharmacy and Allied Health Professions, and Barnett Institute of Chemical Analysis, Northeastern University, Boston, Massachusetts*

I. INTRODUCTION

There are two thyroid hormones, triiodothyronine and tetraiodothyronine, which are abbreviated T_3 and T_4, respectively. The structures of these hormones are shown in Figure 9.1.

The iodothyronine hormones are found in the blood in both free and protein-bound forms. Although only about 0.3% of T_3 and 0.02% of T_4 are normally

---

**Current affiliation:* Eastern Washington University, Cheney, Washington

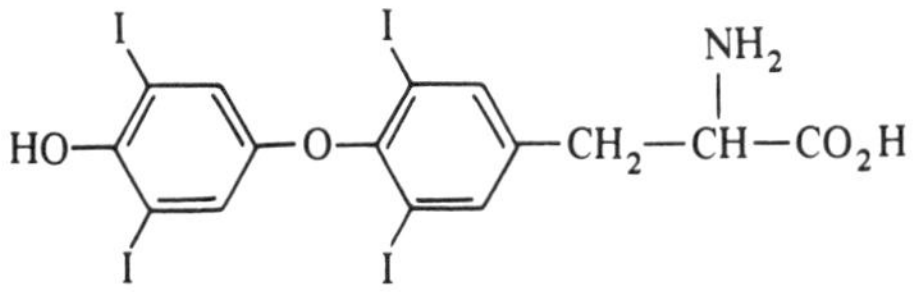

Tetraiodothyronine (Thyroxine)
(T4)

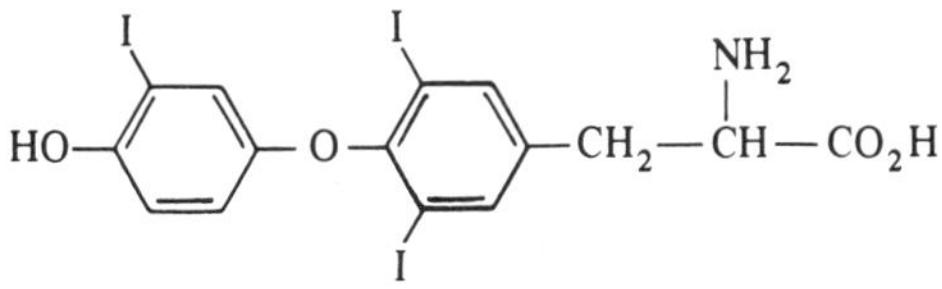

Triiodothyronine (Liothyronine)
(T3)

Figure 9.1 Structures of the thyroid hormones.

present in the free form, corresponding to about 4 and 18 pg/ml of serum, respectively, it is the concentration of free hormone that best correlates with the functional thyroid status.

Other iodothyronines occur in the body as well, but apparently arise only as inactive metabolites of T_3 and T_4. These compounds are reverse T_3 (3,3',-5'-T_3 or RT_3, a metabolite arising from removal of an iodine atom from the inner ring of T_4), and other iodothyronines containing one or two iodine atoms at one or two of the positions substituted by iodine in T_4. Monoiodotryosine and diiodotyrosine also are found to a minor extent in certain physiological fluids as precursor side products from the biosynthesis of T_3 and T_4.

For routine clinical purposes, only the measurement of T_3 and T_4 in serum or plasma is currently of interest, involving measurements of both the total (free plus protein-bound) and free forms of these substances. In contrast, for research purposes, where the objective is to achieve a better understanding of thyronine physiology as well as pathophysiology, measurements of some of the iodothyronines other than T_3 and T_4, such as RT_3 and 3,3'-T_2, are of interest particularly in tissue samples.

The practical need for capillary gas chromatography-electron capture detection (GC-ECD) analysis of iodothyronines depends largely on whether routine or research assays are involved. The routine assays involve measurements of the thyroid hormones mostly in serum or plasma samples. For these analyses the radioimmunoassay methods are quite convenient, limiting the need for analysis by GC-ECD. Nevertheless, capillary GC-ECD can be useful in this context as a reference method for the analysis of total, and especially free, levels of T_3 and T_4 in serum, to help to establish and maintain the accuracy of the results by radioimmunoassay (RIA), or by some kinetic methods. These kinetic methods have been introduced recently in the form of commercial kits for estimating free T_3 and free T_4 levels in serum.

There is a major need for capillary GC-ECD to measure iodothyronines in research studies attempting to advance our understanding of thyroid hormone physiology, because increasingly, measurements are required of extremely small trace amounts of the iodothyronines in tissues. Radioimmunoassay can be inadequate for some of these measurements because of its limited sensitivity, and ready susceptibility to interferences. Capillary GC-ECD potentially can make a significant contribution to these studies because of its extraordinary sensitivity for suitably derivatized iodothyronines, its compatability with the sample cleanup procedures needed to overcome interferences (these cleanup procedures tend to substitute one interference for another in analyses of trace amounts of iodothyronines by radioimmunoassay), and its ability to analyze several iodothyronines at the same time. This latter property is particularly helpful to assess the equilibrium ratios of iodothyronines in tissue samples.

Because of their polar composition and high molecular weights, the iodothyronines require derivatization prior to analysis by GC-ECD. The molecular weights of the resultant derivatives typically range from 900 to 1200, which is extremely high for polar trace solutes by GC-ECD. This feature places severe demands on the performance of the GC-ECD instrument for reliable results to be obtained. At the same time, this feature probably facilitates the analysis of iodothyronines in complex biological samples such as serum, since it affords high-capacity factors for the derivatized iodothyronines by GC-ECD, thereby facilitating the separation of these substances from an extensive elution of less retained compounds (e.g., Figure 9.2). Thus, this feature, in combination with the special sensitivity of these compounds by ECD, due to their content of iodine atoms, has allowed moderately successful analyses of T_3 and T_4 in serum to be achieved with only a limited sample cleanup. (These cleanup techniques will be discussed in more detail later.) Nevertheless, the chromatograms obtained could benefit from more sample cleanup, not only for a cleaner baseline in the region of T_3 and T_4, but to allow resolution of the mono- and diiodothyronines from the broad and complex pattern of early eluting background material. Also, it remains to be seen whether the sample cleanup procedures developed thus far are adequate for the analysis of iodothyronines in tissue samples.

II. BACKGROUND

A. Previous Iodothyronine and Iodotyrosine Analysis by GC (prior to 1981)

Only a single article appeared prior to 1981, in which capillary gas chromatography was used for iodoamine acid analysis [1]. The major focus of that article was to conduct a mass spectral study of derivatized iodothyronines. This included both an extensive analysis of the fragmentation patterns of the *N,O*-diheptafluorobutyryl methyl ester derivatives of the thyronines, and also a confirmation of the identity of a GC peak from a dialyzed serum sample as free reverse T_3. A value of 2 pg/ml was obtained, consistent with an estimated level of 1 pg/ml by immunoassay and radiolabel dialysis techniques.

The primary purpose of the capillary GC column in this work was to minimize the flow of carrier gas into the mass spectrometer, which is somewhat of a peripheral role for the capillary column. A modified injection system was developed for a Varian 2700 gas chromatograph in order to allow direct capillary injection without splitting. The schematic for this device is shown in Figure 9.3. The GC column used in this work was a wall-coated glass capillary (20 mm × 0.15 mm ID) coated with OV-101. The column temperature was 290°C and the flow rate was 2 ml/min, giving retention times for derivatized T_3 and RT_3 of about 11 and 16 min, respectively.

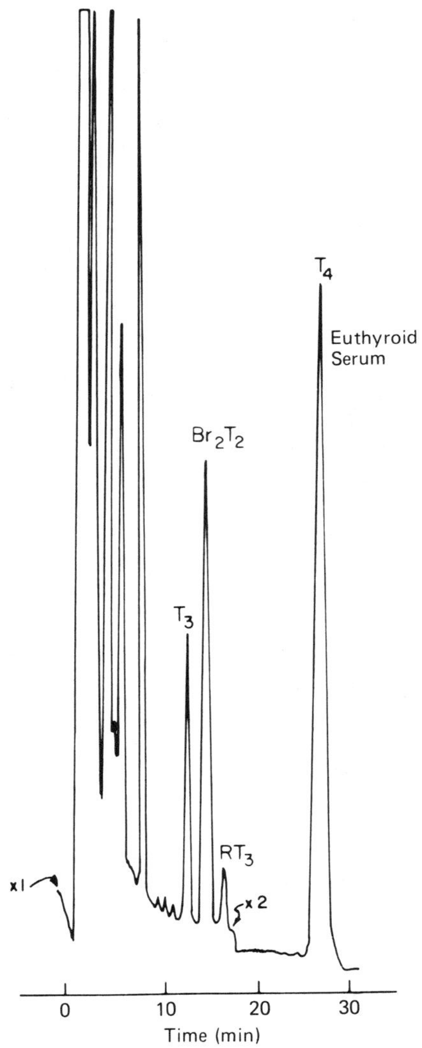

Figure 9.2 Free iodothyronines in euthyroid serum. Reprinted with permission from [20].

In contrast, several papers appeared before 1981 in which iodothyronines were analyzed on conventional packed GC columns. Some of these articles have been dicussed in sections of books or chapters [2,3]. We are unaware of any iodothyronine analysis on micropacked GC columns.

B. Derivatization of Iodothyronines for Analysis by GC

Perhaps the aspects of the previous GC work on the iodoamino acids employing conventional packed columns that are most relevant to this report are the derivatization schemes employed and the detection limits achieved. Table 9.1 provides a summary of these features along with the detectors and samples involved. As seen, the derivatives can be organized into the five categories

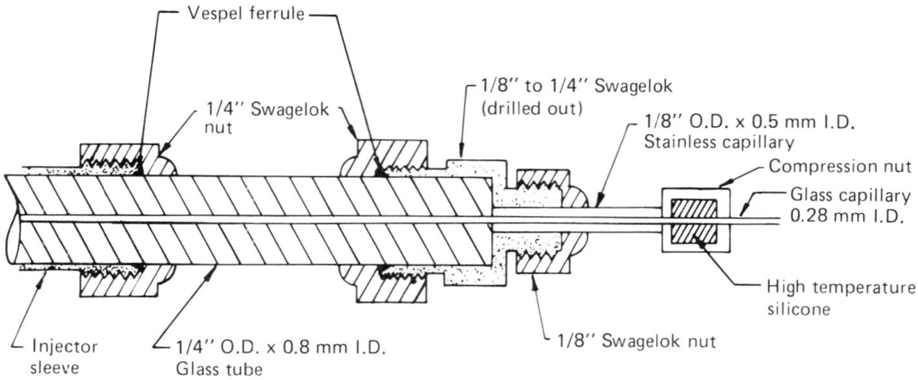

Figure 9.3 Schematic of capillary column connection to injection port of GC-MS system. (From Ref. 1.)

of trimethylsilyl, acyl esters, perhalogen-acyl esters, oxazolidinones, and permethyl esters. We shall discuss each of these in turn.

It seemed as though the availability of *N,O*-bis(trimethylsilyl) acetamide (BSA) was instrumental in stimulating the trimethylsilyl work on the iodoamino acids, since it was used in all of the papers cited in part A of Table 9.1 (sometimes as part of a mixture with other silylating agents), and nearly all of these articles appeared about the same time. Either the authors simply employed BSA without comment, or compared it relative to other silylation agents and found it to be advantageous. This was somewhat expected based on prior success of BSA with amino acids. Seven of the nine papers appeared in 1968–1971, and the other two were published in 1974 and 1978. The typical analytical technique was to react the iodoamino acids in a single step with excess BSA and then inject the reaction mixture (perhaps after dilution, but still containing unreacted BSA) directly into the gas chromatograph.

Only one of these papers showed a chromatogram from the analysis of a "real sample." This particular paper [10] presented the results of an analysis of iodothyronines in pharmaceutical tablets. A second paper was concerned with this type of sample, but only the chromatographic results from standards were shown [9]. Most of the other authors commented on a future interest in extending their methods to iodoamino acids in biological samples, but were not heard from further in this regard. Probably a combination of factors was involved, including the difficulty of these samples, the more favorable properties of some of the other iodothyronine derivatives in this table, and the success of competing methodology such as radioimmunoassay.

As will be discussed subsequently, for the iodothyronines it was possible to generate either di- or tri-TMS derivatives. The latter was apparently involved in all of the articles except one, and several of the authors concerned with the tri-TMS derivatives noted the lability of these products, especially when reaction mixtures containing these substances were diluted to lower levels (e.g., ng/ml). Several of the authors blamed moisture for this instability and assumed that it was the moisture in air which accounted for the air sensitivity.

These tri-TMS derivatives were quite sensitive to the conditions of the GC column. For example, Manius et al. [12] found that lag times of more than

Table 9.1 GC Analysis of Iodothyronines and Iodotyrosines

Reported derivative[a]							
OH (phenolic)	NH_2	CO_2H	Compounds analyzed[b]	Detector	Detection limit	Samples	Reference
A. Trimethylsilyl							
TMS	TMS	TMS	tyr, MIT, DIT, T_0, T_3, T_4	FID	—	Standards	4
Depends[c]	TMS	TMS	tyr, MIT, DIT, T_2, T_3, T_4[d]	FID ECD	∿20 ng —	Standards Standards	5
—	—	—	MIT, DIT, T_3, T_4	FID ECD	0.1 μg 0.3 ng	Standards	6
TMS	TMS	TMS	tyr, DIT, T_2, T_3, T_3', T_4[d]	FID ECD	— —	Standards Standards	7
TMS	TMS	TMS	tyr, MIT, DIT, T_0, T_1, T_2, T_2', T_3, T_3', T_4[d]	FID ECD MS	— 0.2 ng —	Standards Standards Standards	8
—	—	—	tyr, MIT, DIT, T_0, T_2, T_3, T_4	FID FID	— —	Standards Dried thyroid extract	9
—	—	—	T_0, T_1, T_1', T_2, T_2', T_2'', T_3, T_3', T_4	FID FID	5.0 ng —	Standards Tablets	10
TMS	TMS	TMS	tyr, MIT, DIT, T_0, T_2, T_3, T_4[d,e]	MS MS	— 0.10 ng	Standards Standards	11

TMS TMS	TMS TMS	Me TMS	T_0, T_2, T_3, T_4[d]	ECD ECD	 0.2 ng	Standards Standards	12
B. Acyl ester							
Pivaloyl	Pivaloyl	Me	tyr, MIT, DIT, T_3, T_4	FID	—	Standards	13
Pivaloyl	Pivaloyl	Me	MIT, DIT, T_3, T_4	ECD	0.5 ng	Standards Serum	14
Pivaloyl	Pivaloyl	Me	T_3, T_4	ECD	1.0 ng	Standards Serum	15
Pivaloyl	Pivaloyl	Me	MIT, DIT, T_2, T_3, T_4, Br_2tyr	FID	—	Standards	16
Pivaloyl	Pivaloyl	Me	T_3, T_4	ECD	50 pg	Standards Serum	17
Pivaloyl	Pivaloyl	Me	T_3, T_4	ECD	0.25 ng	Standards Serum	18
Pivaloyl	Pivaloyl	Me	T_2, T_3, T_4	ECD	1.5 ng	Standards	19
Pivaloyl	Pivaloyl	Me	tyr, MIT, DIT, T_0, T_2, T_3, T_4[e]	MS	—	Standards	11
Pivaloyl	Pivaloyl	Me	T_3, T_3', T_4, Br_2T_2	ECD	—	Standards	20
C. Perfluoroacyl ester							
TFA	TFA	Me	MIT, DIT, T_2, T_3, T_4	FID	—	Standards	21
TFA	TFA	Me	T_2, T_3, T_4	ECD	—	Standards	19

Table 9.1 (Continued)

Reported derivative[a]							
OH (phenolic)	NH_2	CO_2H	Compounds analyzed[b]	Detector	Detection limit	Samples	Reference
(Perfluoroacyl ester)							
TFA	TFA	*n*-Bu	tyr, MIT, DIT, T_2, T_3, T_4	—	—	Standards	5
HFB	HFB	Me	T_3, T_3', Cl-T_3, T_4,Br_2T_2[d]	ECD	0.2 pg 0.2 pg	Standards Serum	20,22
HFB	HFB	Me	T_2, T_3, T_3', T_4,Br_2T_2	MS	0.5 pg 0.5 pg	Standards Serum	1
HFB	HFB	Me	tyr, MIT, DIT	ECD	70 pg	Standards	2,25
D. Oxazolidinones							
X[f]	Oxazo	Oxazo	tyr, MIT, DIT	FID,ECD	—	Standards	25

E. Permethyl esters

Me	Me	Me	T_4	FID	—	Standards	23
ME	—[g]	Me	tyr, MIT, DIT, T_0, T_2, T_3, T_4[d,e]	MS	—	Standards	11

[a]In some cases, the authors didn't comment on the particular sites derivatized.
[b]Abbreviations: tyr - tyrosine; Br_2tyr = 3,5-dibromotyrosine; MIT = monoiodotyrosine; DIT = diiodotyrosine, T_0 = thyronine; T_1 = 3-monoiodothyronine; T_1' = 3'-monoiodothyronine; T_2 = 3,5-diiodothyronine; T_2' = 3,3'-diiodothyronine; T_2'' = 3',5'-triiodothyronine; T_3 = 3,5,3'-triiodothyronine; T_3' = 3,3',5'-triiodothyronine; T_4 = 3,3',5,5'-tetraiodothyronine; Cl-T_3 = 3'-chloro-3,5,5'-triiodothyronine; and Br_2T_2 = 3',5'-dibromo-3,5-diiodothyronine.
[c]See the text. The authors reported tri-TMS derivatives for iodotyrosines, and di-TMS derivatives for iodothyronines.
[d]Structural evidence for the derivatives was provided by mass spectrometry, infrared analysis, or elemental analysis.
[e]Deamino acids were also analyzed.
[f]X = acetyl, TFA, HFB, pivaloyl, and TMS.
[g]For permethylation, this author worked only with deaminothyronines or deaminothyronine analogues.

30 min between injections (where the injection solutions contained excess BSA and thereby were constantly resilanizing the column) led to a changed response. Similarly, Funakoshi and Cahnmann [8] reported partial destruction of the TMS derivatives on their GC column when lower nanogram amounts were injected. Although these TMS derivatives are interesting for the iodoamino acids because only a single heating step is required for their formation, it is not very appealing that the products are so unstable.

The issue of di- and tri-TMS derivatives of the iodoamino acids was mentioned previously. There is complete agreement among all the authors who addressed this issue that the iodo*tyrosines* are tri-substituted with TMS (i.e., TMS on the phenolic oxygen, the amino nitrogen, and one of the carboxyl oxygens). However, Shahrokhi and Gehrke [5] presented convincing evidence (IR and elemental analysis), at least for the preparative samples, that their procedure produced di-substituted TMS derivatives (TMS labeling at only the amino and carboxyl sites) of the iodo*thyronines*. The other workers either assumed, supported (by IR data; [7]), or proved (by mass spectral data; [8,11,12] that tri-substituted products are obtained. Thorough work was reported by both groups here (e.g., Funakoshi et al. [8] analyzed the derivatives by mass spectroscopy both before and after analysis by GC); thus procedural differences must have led to the two types of products. Perhaps the more severe BSA reaction conditions employed by Shahrokhi and Gehrke (30 min at 150°C vs much shorter times of lower temperatures by the other authors), choice of reaction solvent, distillation step, or a combination of these factors contributed to these differences. It would be interesting to know the moisture sensitivity and overall stability of the di-trimethylsilylated iodothyronines relative to the corresponding tri-trimethylsilylated products. At least the former derivatives could be isolated in preparative amounts for elemental analyses. The authors reported that their products were faintly yellowish, viscous oily liquids when maintained under anhydrous conditions, and that these compounds solidified to a yellowish wax on exposure to air [5].

The dipivaloyl methyl esters of the iodoamino acids are more difficult to prepare than the TMS derivatives, since these esters involve a multistep derivatization sequence. However, these esters are much more stable; Volpert [16] prepared them on a preparative scale, obtained solids in all cases, and was able to weigh these products without any special precautions. All of the authors prepared the methyl esters first and then employed pivalic anhydride in conjuction with various amounts of triethylamine, with typical acylation conditions of 70° for 30 min. As shown in Table 9.1, part B, analyses of iodothyronines in serum were achieved in a few of the cases, although the chromatograms suggested that significant T_4-to-T_3 conversion was a persistent problem. Even a few percent of conversion is significant because of the large ratio of T_4 to T_3 in serum. The mean level of T_4 in euthyroid serum is known to be about 50-fold greater than the mean value for T_3, but falsely lower ratios than these were observed in certain cases.

Two kinds of perfluoroacyl ester derivatives have been investigated, trifluoroacetyl (TFA) and heptafluorobutyryl (HFB), as cited in part C of Table 9.1. Although Richards and Mason [21] in a brief report commented that the *N,O*-bis(TFA) methyl esters of the iodoamino acids gave satisfactory results by GC analysis, Docter and Hennemann abondoned these derivatives after encountering severe reproducibility probelms in the analysis of these compounds by GC. They attributed this problem to substantial breakdown of the products on the columns and reported that it was not helpful to change the stationary phase in the column [19]. Consistent with these latter results, it was reported that peaks could not be obtained from *N,O*-bis(TFA)-*n*-butyl esters of iodothyronines by GC analysis [5].

Analyses by GC of the corresponding *N,O*-bis(HFB) methyl ester derivatives of the iodothyronines have been more successful [1,2,10,22]. These products are sufficiently stable to be stored as solids, recrystallized, and give correct elemental and mass spectral analyses. Their volatility is much higher than that of the corresponding dipivaloyl derivatives; for instance, they elute after 8 min under the same conditions eluting the dipivaloyl derivatives after 2 hr [20]. They have been utilized for the determination of free iodothyronines in serum, giving values in agreement with independent methods, and affording a detection limit of 0.2 pg [20]. Thus, this is the derivative that we are pursuing, at least initially, in our capillary GC work to be discussed subsequently.

The formation of oxazolidinone derivatives of iodinated tyrosines and thyronines has been reported [2,25]. Although the amino and carboxyl groups of these iodoamino acids are derivatized in a single step, a second step is still required to derivatize the phenolic OH. These derivatives are reported to be stable and have been analyzed by both GC-flame ionization detection (GC-FID) and GC-ECD, as indicated in part D of Table 9.1.

The last type of derivative which has been investigated for iodoamino acid analysis by GC are permethyl esters. Two articles have reported this type of derivative, as summarized in part E of Table 9.1. Diazomethane was used in one case on various iodothyronines [23], and dimethylsulfoxide anion with methyl iodide quenching was employed by other workers, but only on deamino iodoamino acids [11]. The results were not encouraging in either case, because multiple GC peaks were obtained from individual iodoamino acids.

C. GC Conditions and Sample Cleanup Techniques

We will now complete our discussion of Table 9.1 by reviewing the general GC conditions employed, and the cleanup techniques used for the analysis of iodothyronines in serum samples.

The stationary phases employed in the columns were usually OV or SE materials except for use of a polysulfone [14] and Dexsil 300 [18] phase. The isothermal or programmed upper temperature limit ranged from 225 to 305°C. The most typical temperature, irrespective of the type of derivative, was about 280°C, consistent with the high molecular weights of derivatized iodothyronines (in the range of 800–1200).

Two types of approaches were employed for cleanup of serum samples: organic extraction and ion-exchange chromatography. The former was carried out by Jaakonmaki and Stouffer [14], and involved extraction of acidified serum with several volumes of *n*-butanol, which released the iodothyronines from their protein binding sites and also precipitated the proteins. The aqueous/butanol extract was reduced to a small volume, filtered to remove salts, and washed with chloroform, probably to remove lipids. The iodothyronines were then partitioned into aqueous ammonium hydroxide. After evaporation of this phase, these analytes were converted into methyl esters with methanol/thionyl chloride or methanol/hydrochloric acid, acylated with pivalic anhydride in the presence of triethylamine, diluted with methanol, and injected into a gas chromatograph equipped with an ECD. Peaks for T_3 and T_4 were seen, with T_3 eluting earlier than T_4, as expected from its lower molecular weight. Both peaks were significantly located on the tail of a broad, early eluting peak containing numerous, early eluting impurities in addition to the injection solvent. Nevertheless, few discrete peaks other than those for T_3 and T_4 were seen. It is noteworthy, and the reasons were discussed earlier, that such little sample cleanup allows reasonable peaks for T_3 and T_4 to be obtained from a serum sample.

The second approach to iodoamino acids extraction from serum involved cation-exchange chromatography. An initial procedure by Nihei et al. [17] seemed to have been passed on first to Tajudden and Elfbaum [18], and then to Petersen [20], with some modifications involved in each case. In the original procedure [17], the serum was applied to a small column of AG 50W-X2, which bound the iodothyronines. After many of the interferences were washed out with 0.15 M ammonium acetate, pH 8.5, the iodothyronines were eluted with 8.4 M ammonium hydroxide. The thyroid hormones were then converted to dipivaloyl methyl esters, and these derivatives were purified by anion-exchange chromatography on Amberlite IR-45 prior to analysis by GC. Although numerous background peaks were seen in the chromatogram, the peaks for T_3 and T_4 eluted after these other peaks.

The overall procedure of Tajuddin andElfbaum employed essentially the same steps and reagents, except that a wash with methanol was added on the cation-exchange column, to provide a cleaner final extract for the GC analysis. The thyronines were eluted from the column with a dilute solution of ammonium hydroxide in methanol. The final GC chromatogram showed peaks for T_3 and T_4 and may have involved less backgound elution that that from Nihei et al., but this is hard to determine, since a higher temperature was employed for the GC column, which placed the T_3 and T_4 peaks more onto the tail of early eluting material. The relatively large size of the T_3 peak indicated that some conversion of T_4 to T_3 had occurred, consistent with the reported mean value of 1.9 μg/l for T_3 in this paper, which is higher than the currently accepted value of about 1.4–1.5 μg/l.

Petersen and co-workers reported the analysis of the free rather than total thyronines in serum [20.22]. The overall procedure involved dialysis, internal standard addition (3,5-diiodo-3',5'-dibromothyronine), cation-exchange chromatography, derivatization (to *N,O*-di-HFB-methyl esters), and gas chromatography with electron capture detection. The cation-exchange chromatographic conditions were essentially identical to those developed by Tajuddin and Elfbaum [18]. A notable lack of conversion of T_4 to T_3 was observed by these workers [20,22], perhaps due to careful cleaning and silanization of the glassware, and the use of highly purified reagents. Euthyroid, hypothyroid, and hyperthyroid sera were analyzed [20], and a typical chromatogram from a euthyroid serum is shown in Figure 9.2. As seen, in addition to peaks for free T_3 and free T_4, a peak for free reverse T_3 is observed. An important feature of this paper is the use of heptafluorobutyric anhydride in the derivatization sequence. The resulting di-HFB-methyl esters of the thyronines were reasonably stable and highly volatile relative to derivatives reported previously of the iodothyronines for analysis by GC. The acylation step with this anhydride involved heating the thyronine methyl esters with heptafluorobutyric anhydride and acetonitrile for 1 hr at 60°C, and then evaporation with a stream of nitrogen.

III. CAPILLARY GC

In the remainder of this article, we will summarize intial results from our laboratory on the analysis of thyronines by capillary GC-ECD. This is basically a continuation of our previous work utilizing conventional packed column

GC-ECD, as summarized in the previous section. We are ultimately interested in the analysis of thyronines in subcellular tissue fractions, and thus have turned to capillary GC particularly for its potential for providing more sensitivity and less susceptibility to the interferences arising in physiological samples.

A. Materials

The iodothyronines (T_2, T_3, and T_4) were obtained as free amino acids from Sigma (St. Louis, Missouri, USA) and stored in a desiccator over Drierite at -4°C.

For all solvent distillations, a middle fraction (about 60–70%) was taken. The methanol (Burdick and Jackson, Muskegan, Michigan, USA) was redistilled over magnesium/I_2 (e.g., 5 g Mg and 1 g I_2 for 250 ml of methanol) and stored in the dark under a nitrogen atmosphere at room temperature in a Pyrex flask fitted with a glass stopper.

Hydrogen chloride gas (electronic grade) and a stainless steel valve were obtained from Matheson (Gloucester, Massachusetts, USA).

The heptafluorobutyric anhydride (aldrich, Milwaukee, Wiscosin, USA), was refluxed over a phosphorus pentoxide (approximately 0.5 g P_2O_5 for 5 ml HFBA; [24]) and stored under a nitrogen atmosphere in a glass vial with a Teflon-lined cap at -4°C. On the day of use, a small quantity was distilled employing a Wheaton/Hickman flask (Wheaton Scientific Co., Millville, New Jersey, USA) on a rotary evaporator (Buchi, Basel, Switzerland) at 60°C (bath temperature) and a vacuum of 15 mm Hg.

Acetonitrile (Budick and Jackson) was distilled from calcium hydride (e.g., 10 g CaH_2 for 250 ml of acetonitrile) and stored under a nitrogen atmosphere. The toluene (Burdick and Jackson) was used as received.

The glassware, including the 0.1, 1.0, and 5.0 cm^3 Reacti-Vials, matching caps (open-topped), Mininert valve caps, "Tuf-bond" Teflon-silicone discs, and "Silli-vap" nitrogen evaporator were acquired from Pierce (Rockford, Illinois, USA).

Stock solutions of derivatized iodothyronines were stored in 2-ml screw cap borosilicate septum vials with open-topped caps (Varian) but refitted with the "Tuf-bond" Teflon-silicone discs (Pierce) rather than the discs supplied by Varian.

Microliter syringes (size/type 10/701N, 50/705, and 250/725) were from Hamilton Co. (Reno, Nevada, USA) and were used in all microtransfers and additions.

The silanizing compounds, trimethylchlorosilane (TMCS) and hexamethyldisilazane (HMDS) were obtained from Pierce and were "Specially Purified Grade."

The water was always doubly distilled.

B. Instrumentation

A Varian Model 3740 gas chromatograph (Walnut Creek, California, USA) equipped with a constant-current, pulse-modulated ^{63}Ni (8 mCi) electron capture detector was utilized. The detector geometry is two displaced coaxial cylinders with a volume of 300 μl^3. Integration of the peak areas and measurements of the retention times were carried out by a Varian CDS-111A data system. The electrometer response was monitored with a Varian model 9176 recorder (full-scale response time <0.5 sec).

Ultrahigh-purity nitrogen (Matheson) was used as the carrier gas. A high-purity brass all-metal pressure regulator (Matheson Model 3104) provided a constant delivery pressure. The carrier gas was filtered twice before reaching the injection port. The first filter was a "large carrier gas or inlet filter" containing activated charcoal and 13× molecular sieve beads (for the removal of moisture and hydrocarbon contamination). The second filter was an Oxyclear Disposable Purifier (Oxyclear, Oakland, California, USA), to remove oxygen. The Oxyclear cartridge filter was changed after every second tank of nitrogen.

The samples were introduced onto the capillary columns with a "direct injection insert" supplied by Varian. This glass insert has a capillary bore of 0.9 mm, an outer diameter of 6.3 mm, and a length of 155 cm.

Before the glass direct insert was installed into the injector assembly, it was treated as follows. First, it was heated at 80°C in concentrated nitric acid for 12 hr, flushed with 20 volumes of methanol and dried for 24 hr at 110°C. Then it was silanized by soaking it in a solution of 5% (v/v) TMCS in toluene for 15 min, washed with 20 volumes of methanol and dried for 24 hr at 110°C, and stored wrapped in aluminum foil in an oven at 60° until installation. Typically, a treated insert was installed every 15 days of use, or whenever a decrease in detector response and an increase in peak tailing with standard solutions of the derivatized thyronines were observed. When a treated glass insert was installed in the injector, the capillary column was disconnected from the brass union at the base of the insert. A 15-cm length of capillary tubing was cut off this end of the column before reassembly.

A Teflon seal in the injector assembly separated the carrier gas from the septum purge flow. The latter flow was set by a fixed restrictor inside a 1/8-inch Swagelok union. The septum purge flow was 2 cm^3 min^{-1} per 5 psi column head pressure (standard on this instrument).

The capillary columns were connected to the injector and detector assembly units with graphite ferrules (0.4 and 0.8 mm for the fused silica and borosilicate columns, respectively; Supelco Inc., Bellefonte, Pennsylvania, USA).

High-temperature blue septa (Alltech Associated, Arlington Heights, Illinois, USA) and Thermogreen LB-1 (Supelco) septa (both are silicone rubber) were utilized in the injector. The blue septa could tolerate more injections than the green ones but required significant conditioning. The Thermogreen septa needed only about 1 hr of conditioning (including one or two injections at the end of this hour) to stabilize, and thus were installed during the day if a septum had to be replaced, whereas a blue one was installed whenever 12 hr or more were available for the necessary conditioning (e.g., overnight) at 290°C and flow rate of 1 $cm^3 min^{-1}$.

The gas chromatographic conditions used in the analyses of the *N,O*-diheptafluorobutyryl methyl esters of the thyronines are given in Table 9.2.

The flow rates through the gas chromatography column reported in this work are all operating flow rates, calculated from the observed flow rate measured at room temperature. The make-up flow rate settings were 20 $cm^3 min^{-1}$ at the capillary insert base and 10 $cm^3 min^{-1}$ at the detector base, following the suggestions of Varian. These flow rates were measured at room temperature and are uncorrected.

C. Derivatization of T_2, T_3, Br_2T_2, and T_4

For the preparation of the *N,O*-diheptafluorobutyryl methyl esters of the thyronines, the procedure of Petersen et al. [22] was followed: The derivatives obtained were purified by recrystallization twice from toluene. They

Table 9.2 Conditions for Capillary Gas Chromatography

Injection:	300°C	
Detection:	310°C	
Columns:	(1)	Fused silica capillary column (J and W Scientific, Inc., Orangevale, California, USA), 15 m × 0.25 mm ID; stationary phase SE-52 (manufacturer-observed coating efficiency, 106%)
		Temperature: 265°C
		Flow rate: 8.1 $cm^3 min^{-1}$
	(2)	Borosilicate glass capillary column (J and W Scientific 30 m × .25 mm ID, stationary phase SP 2100 (manufacturer-observed coating efficiency, 69%)
		Temperature: 275°C
		Flow rate: 6.4 $cm^3 min^{-1}$

gave single peaks when analyzed by GC under the conditions listed in Table 9.2. The crystalline compounds were stored in air in the dark over Drierite at -4°C. No decomposition of these derivatives was detected by GC-ECD over a period of 9 months. For the calibration curves, the more dilute solutions (pico-, femto-, and attomolar concentrations) were unstable at room temperature. Some loss in peak intensity could be seen as early as 2–6 hr under these conditions, even when the solutions were stored in the dark in silanized glass vials.

D. Cleaning Procedure and Silanization of Glassware

All glassware used in this work was initially cleaned in a solution of ethanol and 2 N hydrochloric acid (1:1 v/v) in an ultrasonic bath (Bransonic, Inc., Danbury, Connecticut, USA) for 15 min. This was followed by heating in concentrated nitric acid at 85°C overnight. The glassware was then rinsed five times with water and heated in boiling water for 1 hr. After rinsing with water and methanol (Burdick and Jackson), it was dried at 115°C overnight. The interior surfaces were silanized by filling the vessels with a solution of 4% (v/v) TMCS in toluene and left to stand at ambient temperatures for 15 min. (This solution was prepared in a 100-ml quantity and used repeatedly for two weeks.) The glassware was then rinsed with 1 volume of methanol and dried in an oven at 110°C for at least 4 hr. It was stored at room temperature in a desiccator over Drierite. All the reagents used in this cleaning and silanizing procedure were stored either in glass bottles with Teflon-lined caps or in glass/Teflon Oxford Pipettors (VWR Scientific, Boston, Massachusetts, USA).

The Teflon-coated silicone discs of the open-topped screw caps were rinsed twice with water, twice with acetone, allowed to stand 30 min in methanol, dried overnight at 60°C, and stored in a desiccator over Drierite.

E. Cleaning and Silanizing the Syringes

Careful cleaning of the syringes was important. These syringes were used both for the injections into the gas chromatograph and the stepwise dilutions.

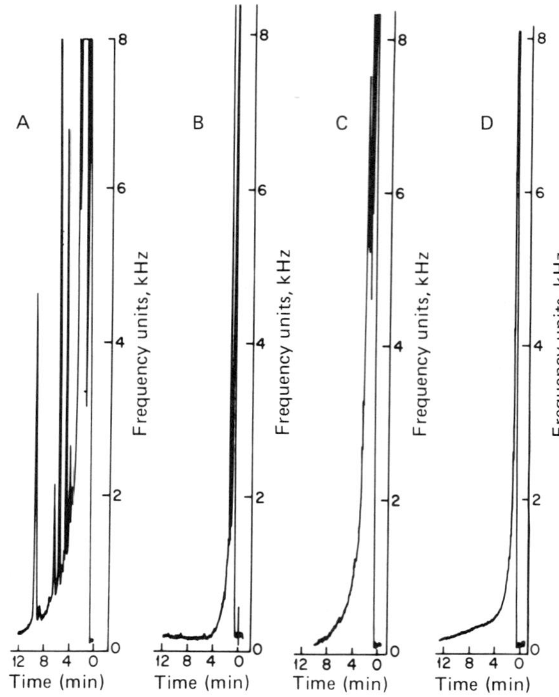

Figure 9.4 GC-ECD chromatograms of 0.5-μl aliquots of 1:9 (v/v) acetonitrile: toluene injected with 10-μl syringes after these syringes were first contaminated with a 0.25 μM solution of derivatized T_2, T_3, BrT_2, and T_4), and then cleaned as follows (with further details defined in the text): A, by drawing through 1:1 (v/v) acetonitrile:toluene; B, by drawing through hot concentrated nitric acid, water, and methanol through the syringe used in A; C, by further cleaning a syringe, initially cleaned as in A in an ultrasonic bath containing 1:1 (v/v) ethanol: 2 N HCl, and then drawing up water and methanol through this syringe; D, by subsequently cleaning the syringe used in B with the conditions described in C. The chromatographic conditions were as described in Table 9.2 utilizing the fused silica column.

Four unsilanized 10-μl syringes (arbitrarily designated A, B, C, D, for purposes of clarity here) were filled with a 0.25 μM solution of derivatized T_2, T_3, Br_2T_2, and T_4 and allowed to stand at 25°C for 5 min. Forty volumes of 1:9 (v/v) acetonitrile-toluene solution were then drawn through the syringes. The syringe plunger was washed with a 1:1 (v/v) acetonitrile-toluene solution, and the syringes were then dried with a syringe cleaner/dryer (Hamilton Inc., Reno, Nevada, USA). This device consists of a stainless steel tube at 195°C with a septum at one end and a vacuum outlet at the other. Figure 9.4A represents the chromatogram of 0.5 μl of a 1:9 (v/v) acetonitrile-toluene solution injected into the gas chromatograph using syringe A. The conditions were as noted in Table 9.2 using the fused silica column. The peaks observed for the derivatives indicated that the solvent was contaminated with approximately 300 pg of each of the compounds (or 0.25% of that in the original solution). Syringe B gave values double this level. Thus, this cleaning procedure was both incomplete and irreproducible.

Syringes A and B were then flushed with 20 volumes of concentrated nitric acid at 85°C, followed by 100 volumes each of water at 85°C and methanol at 50°C through the needle and barrel of the syringes. The plungers were immersed in the 85°C nitric acid for 5 sec, then allowed to stand in 85°C water for 5 min and finally rinsed in a methanol bath at 50°C for 30 sec. The syringes and plungers were heated in an oven at 60°C for 2 hr.

These two syringes (A and B) were used once again to inject 0.5 μl of the 1:9 acetonitrile-toluene solution. The resultant chromatograms were identical but revealed that the syringes were still not clean. One of these chromatograms is shown in Figure 9.4B. The small but discernible peaks represent less than 0.5 pg of the derivatives.

The other two "contaminated" syringes (C and D, which had only been rinsed with acetonitrile/toluene and dried at this stage) were placed in an ultrasonic bath containing ethanolic hydrochloric acid (1:1 (v/v) ethanol/2 N HCl) for 10 min. Water (100 volumes) was drawn through the syringes followed by 100 volumes of methanol both at 25°C. The plungers were placed in water at 25°C for 5 min and then washed with methanol for 30 sec. The syringes and plungers were heated for 12 hr in an oven at 60°C. Figure 9.4C illustrates the chromatogram of an injection of 0.5 μl of the 1:9 acetonitrile-toluene solution using one of these syringes. Although the amounts of the derivatized iodothyronines were quite small again (comparable to those seen in Figure 9.4B), the interference from early eluting materials was greater. The chromatogram from the other syringe was essentially the same.

Further removal of these iodothyronine derivatives to nondetectable levels was achieved by continuing to clean the first pair of the syringes (A and B) in the ultrasound bath by the treatment used for the second pair of syringes. An injection of 0.5 μl of the 1:9 acetonitrile-toluene solution with these syringes then finally showed no peaks for the derivatized thyronines, as shown for one of the identical chromatograms in Figure 9.4D.

This chromatogram was repeated with both syringes after they were "recontaminated" with the original solution of derivatized thyronines and carried through this overall cleanup procedure (omitting the intermediate GC injections). It also was successful when the syringes were silanized (see below) prior to filling with the solution of derivatized thyronines.

The cleaned syringes were stored sealed at room temperature, and each syringe was utilized only for solutions in (1) micromolar, (2) nano- to picomolar, or (3) femto- to attomolar concentration ranges.

For silanized syringes used to transfer or inject solutions containing picomolar (or lower) concentrations of the derivatized thyronines in 1:9 acetonitrile-toluene, a less tedious cleaning procedure was found to give clean chromatograms from injection of a solvent blank. In this procedure, the syringe was immediately cleaned by drawing 25 volumes of the 1:9 acetonitrile-toluene mixture through the needle and barrel with the use of a water aspirator. The plunger, after wiping dry, was immersed in the same solvent mixture for 30 sec and again wiped dry. The silanizing solution (8 μl) was drawn up into the syringe with the plunger and expelled. A further 8 μl of this solution was drawn up and allowed to stand for 2 min at room temperature. The plunger was removed, and the syringe was emptied and dried utilizing the syringe cleaner/dryer. The plunger was reintroduced into the barrel and pumped twice before the syringe was removed from the cleaner/dryer. This cleaned and silanized syringe was then stored sealed with the plunger fully inserted until its use for another injection within 1 hr.

Table 9.3 Effect of the Silanization of 10-μl Syringes on the Gas Chromatography Recovery and Precision of Derivatized Thyronines[a]

Syringes	Peak heights (mm) or peak ratios				
	T_3	Br_2T_2	T_4	T_3/Br_2T_2	T_4/Br_2T_2
Nonsilanized					
1.	13	16	10	0.83	0.62
2.	46	47	26	0.98	0.55
3.	58	58	34	1.0	0.58
Mean	39	40	23	0.94	0.58
Range	45[b]	42	24	0.17	0.07
Silanized					
4.	59	63	36	0.94	0.57
5.	68	73	41	0.93	0.56
6.	77	88	49	0.89	0.56
Mean	68	75	42	0.92	0.56
Range	18	25	13	0.05	.01

[a] 23.5 pg of derivatized T_3, 22.0 pg of derivatized Br_2T_2, and 22.5 pg of derivatized T_4 in 0.5 μl 1:9 acetonitrile-toluene were injected; attenuation ×8. Gas chromatographic conditions were as in Table 9.2, with the fused silica column.
[b] 45 = 58 − 13.

F. Effect of Silanization of Syringes on Recovery and Precision

It was of interest to examine the effect of silanization of the syringes on the recovery of the derivatized thyronines at trace levels. The syringes (cleaned according to the method used in Figure 9.4D) on the day of use were silanized by drawing up (e.g., to the 8μl mark for a 10-μl syringe) a 5% (v/v) solution of HMDS in toluene. The solutions were pumped out after 5 min with a Hamilton syringe cleaner/dryer, and the syringes and plungers (unassembled) were dried in an oven at 60°C for 30 min. The syringes were then stored sealed at room temperature until use.

Six syringes were employed for the evaluation of the silanization, three silanized and three nonsilanized. For each syringe, 0.5 μl of a 1:9 (v/v) acetonitrile-toluene solution was first injected into the gas chromatograph to confirm the absence of any interfering peaks. Then 0.5 μl of a solution of derivatized T_3, Br_2T_2, and T_4 (47, 44, and 45 pg/μl, respectively, in 1:9 acetonitrile-toluene) was injected. Peak heights were measured and compared as given in Table 9.3. Both recovery and precision were better with the silanized syringes. The absolute peak heights with the silanized syringes were always greater, the ranges of these values were always lower, and the values were more consistent for the peak ratios of T_3 and T_4 relative to Br_2T_2, the internal standard, with the silanized syringes.

G. Preparation of Standard Solutions of Derivatized Thyronines

Amounts of the solids of the order of 1–2 mg were weighed in a silanized vial and dissolved with typically 4 ml of 1:9 (v/v) acetonitrile-toluene). Dilutions to lower concentrations were made with silanized 10-μl syringes which were first checked for possible contamination by injection of a 0.5-μl solvent blank of 1:9 acetonitrile-toluene into the gas chromatograph. One syringe was dedicated to each dilution step, and the successive transfers were made after first flushing the new syringe with the standard being diluted in order to equilibrate the syringe and remove the air. In establishing the calibration curve, the most dilute solution under study was injected first, gradually increasing the concentration of the derivatized thryonines in subsequent injections, periodically injecting solvent blanks between injections to rule out carry over peaks and syringe contamination.

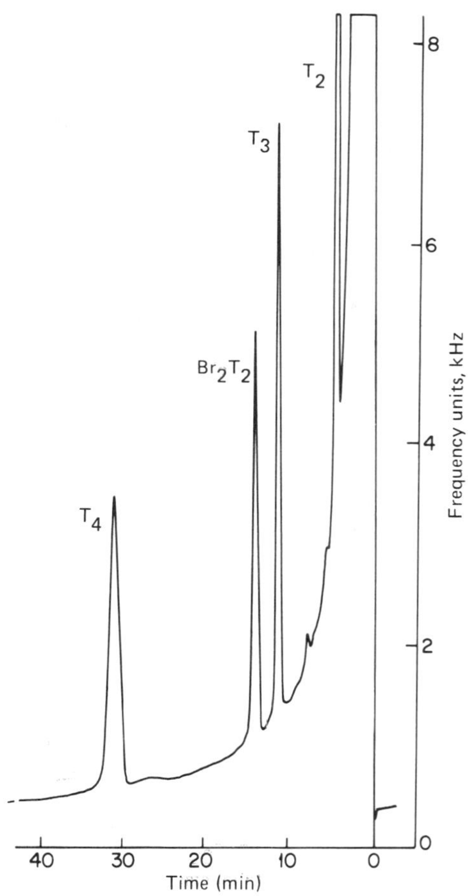

Figure 9.5 Separation of the standards of the thyroid hormone derivatives by gas chromatography utilizing a packed column with electron capture detection. Each peak represents about 400–800 pg in 1 μl injected onto a 200 cm × 6 mm ID glass column packed with 3% SP2250 on 100/120 Supelcoport. The flow rate was 55 $cm^3 min^{-1}$. The injector, column, and detector temperatures were 300, 275, and 300°C, respectively.

H. Comparison of GC Columns

Figures 9.5, 9.6, and 9.7 show the chromatograms from injection of standard solutions of the derivatized thyroid hormones (400–800 pg of each) in 1:9 (v/v) acetonitrile-toluene on a packed glass column, narrow-bore borosilicate glass capillary column, and narrow-bore fused-silica capillary column, respectively. In each case, the injection volume was 1 μl and the column temperature was 275°C. A major contrast is seen especially between the chromatograms in Figures 9.6 and 9.8. The retention times and degree of interference of the solvent peak are much less in the latter chromatogram involving fused silica capillary column. The performance of the borosilicate capillary column, shown in Figure 9.6, was intermediate in these respects. Although the length

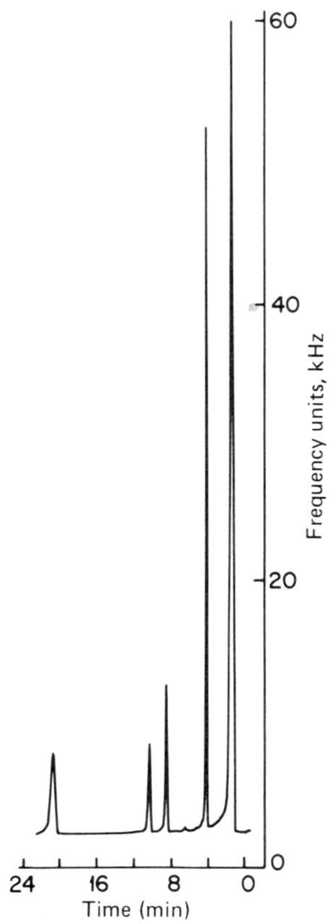

Figure 9.6 Separation of the standards of the thyroid hormone derivatives by gas chromatography utilizing a borosilicate glass capillary column with electron capture detection. Each peak represents 400–800 pg in 1 μl injected onto a 30 m × 0.25 mm ID borosilicate glass capillary column coated with SP2100. The flow rate through the column was 6.4 cm^3min^{-1} and through the make-up was 48.6 cm^3min^{-1}. The injector, column, and detector temperatures were 300, 275, and 300°C, respectively.

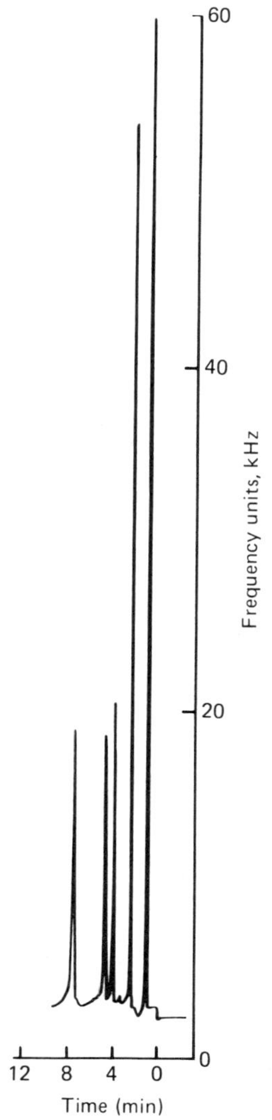

Figure 9.7 Separation of the standards of the thyroid hormone derivatives by gas chromatography utilizing a fused silica capillary column with electron capture detection. Each peak represents 400–800 pg in 1 μl injected onto a 15 m × .25 mm ID fused silica capillary column coated with SE-52. The flow rate through the column was 3.8 cm^3min^{-1} and through the make-up was 51.2 cm^3min^{-1}. The injector, column, and detector temperature were 300, 275, and 310°C, respectively.

of the borosilicate capillary column was twofold that of the fused silica column, the flow rate through the latter was nearly twofold less than that on the borosilicate column, and the internal diameters of the two columns were the same. Thus, the differences in the chromatograms in Figures 9.6 and 9.7 can be

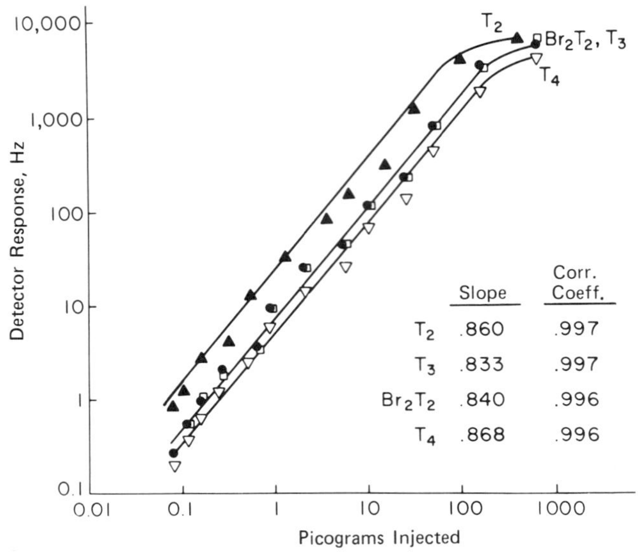

Figure 9.8 Calibration plot of the diHFB methyl esters of the thyronines. A stock solution containing derivatized T_2, T_3, Br_2T_2, and T_4 (156, 261, 241, and 240 μg/ml, respectively) in 1:9 (v/v) acetonitrile-toluene was diluted sequentially with toluene, and then 0.5 μl volumes were injected into the gas chromatograph. T_2 = ▲, T_3 = □, Br_2T_2 = ●, T_4 = ▽. The chromatographic conditions were as in Table 9.2 for the fused silica column.

attributed primarily to the different surface and/or coating characteristics of the fused silica column.

I. Calibration and Detection Limit

A calibration plot for the derivatized thyronines is shown in Figure 9.8. As seen, the linear range is approximately 10^4. Even though each point represents a single injection, high values are obtained for the correlation coefficients, and the slopes (from a least squares calculation) are nearly the same. For this experiment, a typical detection limit of 0.08 pg is observed at a signal-to-noise ratio of 2. When a freshly cleaned and silanized injection glass insert is used (as was not the case here), a slightly improved detection limit of 0.04 pg is seen. One of these chromatograms is shown in Figure 9.9.

J. Temperature Programming

An analysis of the derivatized iodothyronines at the 30- to 60-pg level by capillary GC-ECD under temperature programming conditions is shown in Figure 9.10. The use of temperature programming will facilitate the analysis of the iodothyronines in physiological samples by increasing the separation of the anlayte peaks from earlier eluting sample interferences.

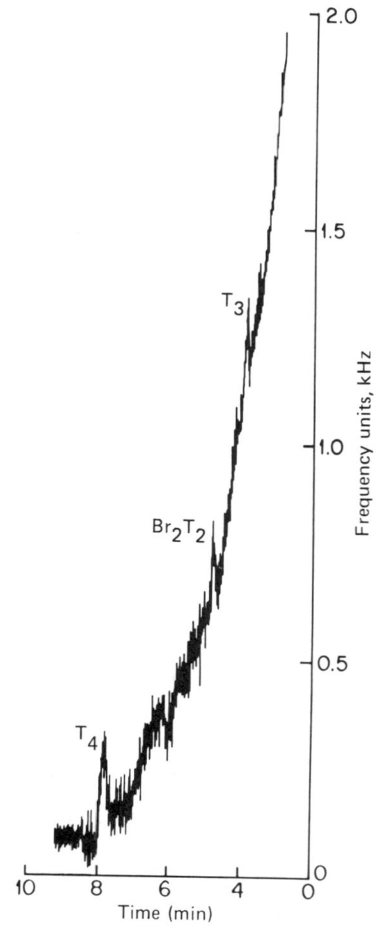

Figure 9.9 GC-ECD chromatogram of the standards of the thyroid hormone derivatives at the detection limit. The peak for the diHFB derivative of T_4 methyl ester represents 40 fg (3.1×10^{-17} mol) in a 0.5-μl injection volume. The conditions were as in Table 9.2 for the fused silica column.

Apparently this is the first reported example of temperature programming by GC-ECD at the picogram level. This type of analysis has not been possible previously because of excessive baseline drift and noise, especially due to column bleed. The inert surface offered by the deactivated fused-silica column used in this analysis provides a lower bleed rate presumably due to less degradation of the stationary liquid phase [27]. The detection limit reported for derivatized T_4 by this approach was 30 fg [26]. This article also reported alternate cleaning procedures for the syringes, the use of fused silica capillary columns as short as 7.5 m, the importance of a high flow-rate of carrier gas (above 6 cm^3min^{-1}) in the injector and column to maximize the recovery, and a

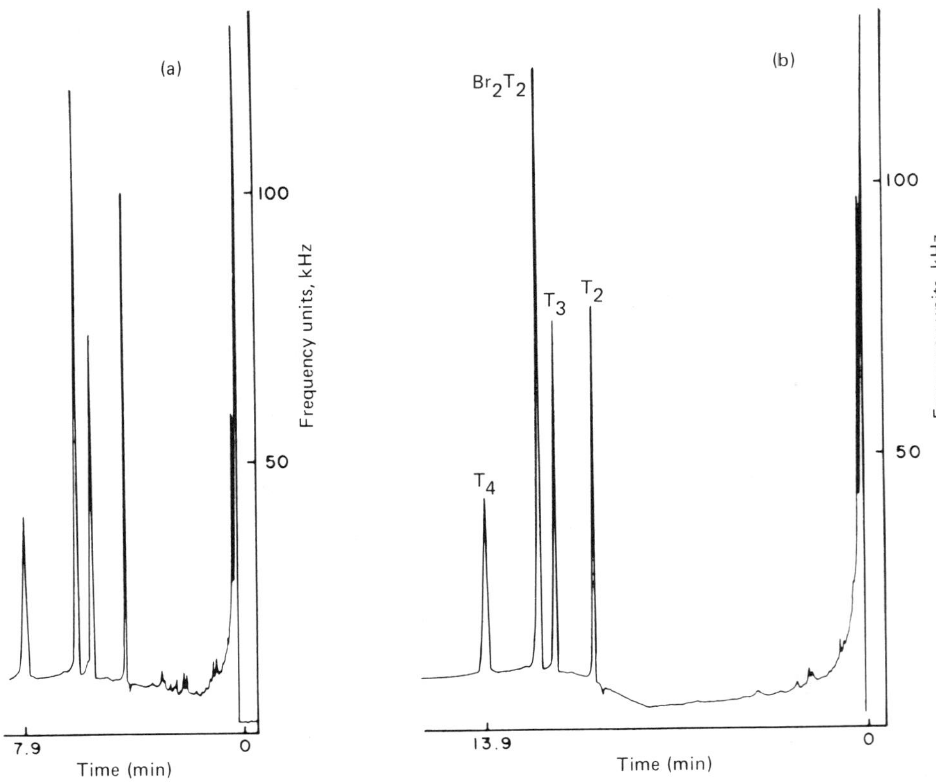

Figure 9.10 Chromatograms with temperature programming from injection of 0.5 μl of a solution 32, 30, 38, and 55 pg of T_2, T_3, T_4, and Br_2T_2, respectively, with an initial column temperature of 200°C, and a hold for 2 min (A), and 8 min (B). The fused silica column was then programmed at 60°/min to a final hold of 275°C. The flow rate was 7.1 cm^3min^{-1} at 275°C. (From Ref. 27.)

linear increase in peak height with increasing injection volumes at least up to 6 μl under temperature-programming conditions.

ACKNOWLEDGMENTS

We wish to acknowledge NIH for support of this work under grant AM21797-03, and also grant RR07143, a Biomedical Research Support Grant given to Northeastern University by the Department of Health, Education, and Welfare, Contribution Number 81 for the Barnett Institute of Chemical Analysis.

REFERENCES

1. B. A. Petersen and P. Vouros, *Anal. Chem.* *49*:1304 (1977).
2. P. Husek and K. Macek, *J. Chromatogr.* *113*:139 (1975).
3. A. Niederwieser, in *Chromatography* (E. Heffmann, ed.), Van Nostrand Reinhold Co., New York, 1975, p. 393.

4. N. M. Alexander and R. Scheig, *Anal. Biochem.* *22*:187 (1968).
5. F. Shahrokhi and C. W. Gehrke, *Anal. Biochem.* *24*:281 (1968).
6. E. T. Backer and V. J. Pileggi, *J. Chromatogr.* *36*:351 (1968).
7. L. B. Hansen, *Anal. Chem.* *40*:1587 (1968).
8. K. Funakoshi and H. J. Cahnmann, *Anal. Biochem.* *27*:150 (1969).
9. R. Bilious, E. Koehler, and J. J. Windheuser, *J. Pharm. Sci.* *60*:1270 (1971).
10. B. M. R. Heinl, H. M. Ortner and H. Spitzy, *J. Chromatogr.* *60*:51 (1971).
11. R. Hoffenberg, A. M. Lawson, D. B. Ramsden, and P. J. Ray, in *Mass Spectrometry in Biochemistry and Medicine*, (A. Frigerio and N. Castagnoli, eds.), Raven Press, New York, 1974.
12. G. J. Manius, P. Fallon, and R. Tscherne, *Anal. Biochem.* *87*:496 (1978).
13. J. E. Stouffer, P. I. Jaakonmaki, and T. J. Wenger, *Biochim. Biophys. Acta* *127*:261 (1966).
14. P. I. Jaakonmaki and J. E. Stouffer, *J. Gas Chromatogr.* *6*:149 (1967).
15. C. S. Hollander, *Trans Assoc. Amer. Physicians*, *81*:76 (1968).
16. E. M. Volpert, N. Kundu, and J. B. Dawiclzik, *J. Chromatogr.* *50*: 507 (1970).
17. N. N. Nihei, M. C. Gershengorn, T. Mitsuma, L. R. Stringham, A. Cordy, B. Kuchmy, and C. S. Hollander, *Anal. Biochem* *43*:433 (1971).
18. M. Tajuddin and S. G. Elfbaum, *Clin. Chem.* *19*:109 (1973).
19. R. Docter and G. Hennemann, *Clin. Chim. Acta.* *34*:297 (1971).
20. B. A. Petersen, R. W. Giese, P. R. Larsen, and B. L. Karger, *Clin. Chem.* *23*:1389 (1977).
21. A. H. Richards and W. B. Mason, *Anal. Chem.* *38*:1751 (1966).
22. B. A. Petersen, R. N. Hanson, R. W. Giese, and B. L. Karger, *J. Chromatogr.* *126*:503 (1976).
23. R. O. Zimmerer, Jr., and L. T. Grady, *J. Pharm. Sci.* *60*:493 (1971).
24. T. Walle and H. Ehrsson, *Acta Pharm. Suecica* *7*:389 (1970).
25. P. Husek, *J. Chromatogr.* *91*:475 and 483 (1974).
26. J. A. Corkill and R. W. Giese, *Anal. Chem.* *53*:1667 (1981).
27. S. R. Lipsky, W. J. McMurray, M. Hernandez, J. E. Purcell, and K. A. Billeb, *J. Chromatogr. Sci.* *18*:1 (1980).

10

Glass Capillary GLC of Fatty Acids

Halvor Jaeger / *L.A.B. Gesellshaft für pharmakologische Untersuchungen m.b.H. & Co. (Laboratory for Pharmacological Research), Neu-Ulm, Federal Republic of Germany*

Hans U. Klör and Hans Ditschuneit / *University of Ulm, Ulm, Federal Republic of Germany*

Hartmut Frank / *Eberhard-Karls-Universität, Tübingen, Federal Republic of Germany*

I. INTRODUCTION

A. Structure and Nomenclature

Fatty acids are an alkyl chain attached to a carboxylic acid group. The total number of carbons in almost all naturally occurring fatty acids is even. Most of the 300 naturally occurring fatty acids can be grouped into two general families, linear and branched. The linear group is found in most animal tissues and fluids while the branched are more abundant in plant tissues. Each of these families can be further subdivided into saturated and unsaturated fatty acids. The saturated acids occupy more space in a lipid structure because there is less coiling than in the unsaturated analogues. The polyunsaturated fatty acids (PUFA) play an important role in membrane fluidity [1] because they fit hydrophobic areas in cell membranes.

Fatty acids are generally classified by chain length and degree of unsaturation since the physiological role they play depends most often on these two factors. Those acids with 2–8 carbons are short-chain, those with 8–14 are medium-chain, and those with 14–24 are long-chain fatty acids. The number of double bonds in a fatty acid can vary from a single one, as in oleic acid, to many, as in a polyunsaturated fatty acid (PUFA), where the arrangement of the bonds can vary from isolated to fully conjugated. The most important saturated acids are palmitic, which has 16 carbon atoms and zero double bonds, and stearic acid, with 18 carbon atoms and zero double bonds. Often a shorthand notation is used, C16:0 for palmitic and C18:0 for stearic. These trivial names often refer to the original source from which the fatty acids were first isolated.

The biologically important PUFAs contain mostly isolated double bonds arranged in the so-called divinyl-methane rhythm, which means that a methyl group is interposed between double bonds. The systematic chemical nomenclature numbers the position of the double bond counting from the carboxyl group; however, this name does not make clear the biochemical relationship between the various PUFAs. As a result, a biochemical nomenclature based on the position of the double bond relative to the methyl end has become more widely used. As an example, in oleic acid the double bond is nine carbon atoms away from the methyl end so that all acids related metabolically to oleic acid also have their first double bond nine carbons from the methyl end, thus, they belong to the ω9 family. A fatty acid with 20 carbon atoms such as 5,8,11-eicosatrienoic acid is not by its name obviously related to oleic acid, but the biochemical nomenclature (20:3ω9) makes this relationship more obvious. This acid is an important indicator of essential fatty acid deficiency and is chemically, but not biochemically, related to 8,11,14-eicosatrienoic acid, which belongs to the linoleic acid family (or the ω6 family) and is not related to fatty acid deficiency. The biochemical designation for this acids would be 20:3ω6. This biochemical nomenclature serves as a useful way to classify the families of the many isomers of PUFA, both for analytical purposes and for metabolic studies.

B. Fate and Function

Fatty acids are found free (free fatty acids, FFA) in both intracellular and vascular space. In addition they comprise a vital part of larger molecules and molecular complexes such as triglycerides and phospholipids. The represent in the free form a transport system and a source of energy for tissues, and as such they are rapidly turned over (up to 40%/min) with an average half-life in man of 2–3 min [2]. As parts of other compounds they represent structural building blocks for membranes and vital hydrophobic areas for maintaining the structure of complex lipids. In plasma, about 50% of the FFAs are bound to albumin, while the rest are associated with various lipoproteins [3].

Free fatty acids are generated from hydrolysis of triglycerides which are the storage form. During the fasting state the source is usually adipose tissue while postprandial FFAs are generated by lipoprotein lipase, which is found on the surface of endothelial cells in capillaries of adipose and muscle tissue [4]. This enzyme hydrolyzes the triglyceride core of the triglyceride-rich lipoproteins (chylomicrons) or very low-density lipoproteins (VLDL), converting them into low-density lipoproteins (LDL) or high-density lipoproteins (HDL). The FFA thus released can remain in the plasma bound to protein or can cross the capillary wall for cellular utilization. The plasma levels of FFA which can be measured by gas-liquid chromatography (GLC) may reveal the source of the general metabolic condition of the subject [5].

Once the FFA has entered the cell, it can be used as an energy source or it can be incorporated into lipids for the cell's internal needs, such as phospholipids for formation of internal and external cellular membranes. The liver and gut have cells that synthesize cholesterol esters and triglycerides which are then released into the plasma together with free cholesterol and protein in the form of complex lipoproteins.

C. Analysis of Fatty Acids—Biochemical Applications

The analysis of fatty acids both in the free form and as parts of more complex molecules has been very important in solving problems related to form and function of lipids in plants and animals. A good example is the elegant method for determining which fatty acids are bound to the three carbons of glycerol in naturally occurring triglycerides. The method involves hydrolysis of the 1- and 3-position ester linkages of the triglyceride with pancreatic lipase, followed by isomerization, synthesis of artificial phospholipids as an intermediate product, and determination of the fatty acid species located at the three different positions of the triglyceride molecule [6]. Using this method, swine adipose tissue triglycerides differ from those of other animals in the fatty acid groups at the 1-, 2-, and 3-positions in the glycerol group [7]. Phospholipids can be analyzed in a similar manner using phospholipase A from snake venom (Naja naja) which selectively splits off the fatty acids at the 2-position of lecithin or other phosphoglycerides. The fatty acids thus released are then compared to those remaining in the 1-lysophosphatide. Use of this method has shown that naturally occurring lecithins in erythrocytes, plasma, etc., generally have an unsaturated fatty acid in the 2-position and a saturated one at the 1-position [8,9]. Use of radiolabeled fatty acids and their precursors has been applied to metabolic studies. In these cases, after GLC analysis of specific activities of the single fatty acids are determined [10].

A "cold labeling" technique has also been used to study the metabolic fate of exogenous fatty acids without using radiolabeled compounds [11]. In this method linolenic acid (18:3ω3), which is present only in trace amounts in body fluids and is not degraded by the body, is given and its incorporation along with other PUFA into lipids and cellular components is followed [12]. Other cold markers have been tried [13,14], but the differences in fatty acid composition of the lipids in lipoprotein classes were not great enough for these to be effective. For the study of the metabolic conversions of fatty acids in body fluids and in tissues without radiolabeled compounds, the determination of positional isomers is required. This kind of analysis has shown that normal liver cells incorporate the isomeric C18:1 fatty acids oleic acid (ω9) and vaccenic acid (ω11) in a fixed ratio in different lipids, while hepatoma cells have lost this specificity and incorporate these acids in a random fashion [15]. Hepatoma cells have other subtle differences in the metabolism of the C18:1 isomers which require very sophisticated assay techniques to find [16].

Tumor cells (Ehrlich ascites) were used as a model to study the incorporation and intracellular metabolism of cholesterol esters of fatty acids in vitro. By comparison of the medium levels and cellular levels, metabolic changes and the selectivity of incorporation of the mono-enoic fatty acids into tissue cholesterol esters could be determined. This selectivity is very important in atherogenesis [17].

The fatty acid composition of aortal lipids was compared to that of plasma lipoproteins in swine given different lipid-elevating and lipid-lowering diets. Because of the differences in fatty acid pattern between tissue and plasma, it was concluded that the atherogenic proteins were not retained unchanged [18]. Because PUFA has been reported to lower lipid levels, the fate of dietary linoleate and its incorporation into various plasma lipids have been studied extensively [19,20]. A nearly selective accumulation of dietary linoleate in cholesterol ester and phospholipids was observed in hypercholesterolemic patients when they were given linoleate-enriched lecithin (EPL substance) [21].

The induction of cardiac arrythmias by elevated FFA levels in myocardial infarction has been investigated to see how the spectrum of fatty acids varies [22,23]. The effect of alcoholism on fatty acid metabolism has also been investigated to determine the relative contributions of endogeneous and exogenous fatty acids. Because of the stimulation of lipolysis, the fatty acid spectrum in both body fluids and tissues was derived mainly from adipose tissue fatty acids or their derivatives [24,25]. In fact, the spectrum of an alcoholic in samples from liver and adipose tissue was so unique compared to nonalcoholics that this analysis might be applicable as a postmortem diagnostic technique for alcoholism. Cirrhosis of the liver, both compensated and decompensated, could be diagnosed by an analysis of the short-chain fatty acids in the breath of patients.

One disease that can be diagnosed only by an analysis of the fatty acid is Refsum's syndrome, where there is an abnormal accumulation of a branched C20 fatty acid, phytanic acid. This acid cannot be metabolized; thus it can accumulate to the point where it represents 5–30% of the total fatty acids [26–28].

D. Fatty Acid Analysis in Nutrition

The field of nutrition has used analysis of fatty acids extensively. Human milk has been found to contain a higher percentage of PUFA than cow's milk, explaining in part why human milk is best [29]. Fatty acid analysis of stool samples has shown that substituting medium-chain triglycerides for normal dietary fat gives children with fat malabsorption sufficient calories for normal growth [30]. The analysis of hydrogenated fats and margarines has become increasingly important since it was shown that *trans*-PUFA, which can be formed under certain conditions during hydrogenation, are atherogenic whereas the naturally occurring cis isomers act antiatherogenic [31].

The most important application of fatty acid analysis is the diagnosis of essential fatty acid (EFA) deficiency, a condition that was first observed in animals which were on a fat-free diet. This condition is due to a lack of linoleic acid (C18:2ω6) and its derivatives [32]. In man the lack of the ω6 family leads to EFA deficiency, whereas in simians linolenic acid (C18:3ω3) is the key acid [33]. Essential fatty acid deficiency cannot easily be detected in adults without fatty acid analysis, although clinical symptoms appear readily in newborns because they do not at that time have stores of PUFA [34]. In adults who have been on an extensive fat-free diet, such as intravenous nutrition, skin rashes and sores develop. The first biochemical signs appear much earlier. As little as two weeks after the start of a fat-free diet there is a significant decrease of linoleic and arachidonic acid (C20:4ω6) in normal subjects [35]. As a result, patients on long-term parenteral nutrition, or with absorption problems due to surgery or inflammatory bowel disease, or with liver disease should have routine fatty acid analysis to signal the development of this condition so that appropriate intervention can be initiated [41].

II. SAMPLE PREPARATION AND DERIVATIZATION

A. Extraction Procedures

As noted above, in addition to the free fatty acids there are a large variety of lipids in serum and tissues. The most abundant of these lipids are the glycerides in which from one to three fatty acids are bound to glycerol. The phospholipids are a large polar group of lipids bound to protein. This group can be generally subdivided into glycerophospholipids and sphingolipids. The first subgroup contains several important fractions such as phosphatidylcholine, phosphatidylserine, lysophosphatides, and phosphatidylinositol. The sphingolipids group contains sphingomyelins, cerebrosides, and gangliosides.

The procedures used to extract this wide array of samples do not vary much with the source of the sample. In most cases these methods can be applied for serum, stool, bile, tissues, oils, and butterfat samples. Neutral lipids can be easily extracted, but with the more polar phospholipids some problems arise. McFarlane [36] described the first successful methods, which involved successive extractions with ether at -20° to -25°C. This method did not completely remove all lipid, leaving about 35% with the protein material. The method was first modified by Scanu and co-workers [37,38] who used ethanol-ether (3:1, v/v) instead of ether and repeated the procedure to make it more applicable for serum samples [39,40]. The disadvantages of the long extraction periods, the low temperature requirements, and the large volumes of solvents required still remained and severely limited the widespread application of this method.

The method reported by Folch et al. [41] is still used extensively today. In this method the plasma sample is diluted 20-fold with chloroform-methanol (2:1, v/v) which precipitates the protein. After removal of the protein by centrifugation, the filtrate is washed with 4 ml of a 0.02% solution of calcium chloride. The upper phase is discarded, and the organic phase is evaporated to dryness at 40°C under a stream of nitrogen. In our laboratory, using radiolabeled compounds we found that the recovery rates with this procedure were very low, being about 60% for phosphatidylcholine after removal of the protein, and 41% after the calcium chloride wash. For free fatty acids the corresponding recoveries were 82% and 62% and for triglycerides, 83% and 82%. Cholesterol esters gave the best recoveries—86% and 84%—after removal of protein and washing, respectively. These results limit the application of this method to the nonpolar lipid fractions.

Cham and Knowles [42] overcame the need to resolubilize the protein by using a mixture of butanol and diisopropyl ether (DIPE). In this method, the serum or plasma (5 ml) containing 0.5 mg of ethylenediamine tetraacetate (EDTA) was combined with 10 ml of butanol-DIPE. After mixing by rotation for varying periods the two phases were separated by centrifugation at 2000 rpm. The aqueous phase was removed and subjected to analysis for lipids. Using a ratio of butanol-DIPE of 25:75 (v/v) and mixing for 3 hr, the authors reported 100% delipidization of triglycerides, cholesterol, phospholipids, and unesterified fatty acids. We had the same results with the nonlipid-containing aqueous phase, but the recovery as measured with radiolabeled compounds in the organic phase was 41–47% for phospholipids, 63% for free fatty acids, and 56% for triglycerides, indicating that much of the lipids remain bound to the protein.

Our best results have been obtained using a modification of the Folch method developed by Nelson [43,44], which we further modified. In this procedure, 0.2 ml of serum is added dropwise to 3 ml of methanol with stirring. Subsequently 8 ml of chloroform is added in two equal portions with additional stirring and, finally, shaking. After filtration, the nonlipid material is removed on addition of 0.2 g of Sephadex G-25 and filtration. The solvent is removed by evaporation on a water bath. An additional 4 ml of chloroform is added to dissolve the residue and is evaporated. Finally the residue is taken up in 4 ml of chloroform, and after addition of a few drops of sulfuric acid, the sample is transferred to a small column. The lipids elute with 26 ml of chloroform. The recoveries were as follows: for triglycerides, 99%; for cholesterol esters, 98%; for phosphatidylcholine, 97%; for phosphatidylserine, 95%; and for phosphatidylinositol, 87%. Wells and Dittmer also used Sephadex G-25 for purification of lipid extracts [45]. No other method to our knowledge gives equivalent recoveries for all classes of lipids including phospholipids [46].

Wren and co-workers [47] have noted the importance of using an antioxidant when working with lipids and especially with phospholipids [47]. We have obtained very good results with BHT (butylated hydroxytoluene).

Once separated from the primary sample, the lipids are then separated into general classes, usually by adsorption chromatography. The details for these separations can be found elsewhere [48–52].

B. Derivatization

1. *Esters*

Free fatty acids are usually esterified while fatty acids bound in lipids have to be saponified before a simpler, more volatile ester can be prepared. The saponification step can be avoided by transesterification. Most of the reported methods now in practice used this transesterification so that all lipids and free fatty acids in the sample are converted into their esters. The method reported by Thenot and co-workers [51] is a good example of this type of procedure. Based on the finding of Brechbühler et al. [54], which demonstrated the usefulness of amide acetals for esterification reactions, this method requires the addition of the fatty acid sample dissolved in pyridine to 0.2 ml of *N,N*-dimethylformamide-dimethylacetal (DMF-DMA). An aliquot of this solution is injected into the gas chromatograph. One could as well use a "sandwich" injection where the sample is an organic solvent is taken up in the syringe followed by a few microliters of DMF-DMA.

Diazomethane has also been used to prepare methyl esters [55], but its use is hazardous because of the danger of explosions and because it is a possible carcinogen [56]. Even though a precolumn reactor technique has been developed [57] which decreases the risks somewhat, we still are of the opinion that it is not the method of choice.

The methyl esters of the higher-molecular-weight acids have suitable volatility for gas chromatographic analysis, but those of the smaller acids are so volatile that some losses occur during derivatization.

Although both acids and bases can be used to prepare esters, the acid catalysts are preferred in order to avoid isomerization of double bonds [58,59]. The most commonly used acids are hydrochloric and sulfuric. In a typical procedure as described by Stoffel and co-workers [60], the fatty acid sample is boiled under reflux with a 20-fold excess of 5% hydrochloric acid in methanol for 60–90 min. The methyl esters are isolated from the reaction mixture by sublimation in vacuo. Using the sublimator redesigned by Sheppard and Ford

[61], excellent recoveries have been obtained. Neither the concentration of hydrochloric acid nor the temperature of the reaction appears to be critical since methods using 2.5–5% (w/v) hydrochloric acid and reaction temperatures of 60–100°C for intervals of 1–17 hr have been described [62–68]. This same procedure has also been applied to lipids and free fatty acids in the same sample [69].

Sulfuric acid has been used extensively for this reaction and has given quantitative esterification of both cis and trans forms of acid [70–72]. As for hydrochloric acid, the concentration of the acid and the reaction conditions do not seem to be critical. The use of acid concentrations from 1–20% (w/v) and temperatures from 65 to 100°C for 1–16 hr has been reported [73–80]. Mason et al. [81–83] have found that by adding 2,2-dimethoxypropane to the reaction mixture, one can avoid losses during refluxing and formation of polymers. This material has been used with both hydrochloric acid catalyst [83] and with sulfuric acid catalyst [84]. It also effectively removes the water formed during the reaction.

Boron trifluoride was first developed as a catalyst for esterification of organic acids by Metcalfe and Schmitz in 1931 [85–87]. They used a 12.5% solution of boron trifluoride in methanol, which gave the same results as diazomethane and HCl-methanol. When 50% solutions were used by Lough [88], methoxy-substituted fatty acids were formed in considerable quantities. As the percentage of boron trifluoride was reduced, this problem was circumvented [89]. The Association of Official Analytical Chemists (A. O. A. C.) replaced the sulfuric acid-methanol procedure [93] with the Metcalfe procedure [90], which incorporated the saponification procedure of Ast [59,91,92].

The best reference on transesterification of fatty acids using boron trifluoride has been published by Morrison and Smith [94]. The method used by these authors gives excellent results for fatty acids derived from various lipids of human and animal origin. This reference is especially useful in treating the more uncommon lipid samples. The procedure they report adds boron trifluoride-methanol reagent to the lipid sample after it had been evaporated to dryness. The reagent is added under nitrogen in the ratio of 1 ml of reagent to 4–16 mg of lipid. The reaction tube with a Teflon-lined screw cap is heated at 100°C for a suitable time. On cooling, the esters are partitioned between 2 volumes of pentane and 1 volume of water. A modification of this procedure was required for triglycerides and sterol esters, which required heating at 100°C for 45 min in a mixture of 35% boron trifluoride-methanol, 30% benzene, and 35% methanol. Other lipids required neat boron trifluoride-methanol reagent (14% w/v). Reaction times varied from 2 min for free fatty acids to 90 min for sphingomyelins.

Sodium methoxide has also been used as a catalyst for transesterifications [95].

Ethyl esters have been used to decrease the volatility of the lower acid esters [96], as have butyl esters. Both sulfuric acid (2%) [97] and hydrochloric acid [98–100] have been employed as catalysts. The very short-chain fatty acids have been derivatized as their benzyl esters [101,102] and as para-substituted benzyl esters [103]. In principle, these acids can be chromatographed without derivatization, but these techniques have not as yet been perfected for routine analysis [106].

In fact, some lipids such as triglycerides can be chromatographed without additional derivatization and separated by their carbon content. This technique developed by Kukis and co-workers [107] requires very high temperatures, which has limited its use. It is hoped that new columns designed for high-temperature work [108] and new on-column injection systems [109] will facilitate this work.

If halogenated alcohols are used to prepare the esters of fatty acids, one can use an electron capture detector and achieve ultratrace analysis. An example of this is the pentafluorobenzyl esters prepared by Kawahara [110]. He reacted α-bromo-2,3,4,5,6-pentafluorotoluene and potassium carbonate in acetone at reflux for 3 hr. The esters were prepared in 81–88% yield.

2. Silyl Ethers

Silyl ethers are especially useful when fatty acids are to be analyzed in the presence of other components. Pierce [111] has shown that silylating reagents react with the carboxylic acid group as well as other groups in the molecule to give esters and ethers. Tallent and Kleinman [112] have applied this derivatization procedure to lipid hydrolyzates to determine not only the fatty acids but glycerol as well. In all cases the yield was quantitative. Others [107, 113–115] have applied this reaction for fatty acid analysis and have used these derivatives to determine the "actual activity" of columns—an expression of the liquid phase and support quality of the column.

3. Miscellaneous

The other procedures in the literature are of minor interest and are usually directed to a particular problem. The tetramethylammonium salts can be directly injected into a gas chromatograph to give the methyl esters [116]. Direct hydrogenation with palladium as catalyst on column has been used [117]. The microscale analysis of polyenoic fatty acids has also been reported by Pohl and co-workers [118]. The mercury adducts of these polyenoic acids were separated by thin-layer chromatography. They were then oxidatively cleaved with potassium permanganate in glacial acetic acid. The cleavage products were esterified and identified by programmed gas-liquid chromatography.

4. Derivatization—Summary

The derivatization of fatty acids has been summarized by many authors over the years, beginning with Fontell et al. in 1960 [119] and then by Kuksis [120], Sheppard and Iverson [121], Drozd [122], Ahuja [123], and Ackman [124]. As early as 1959, Craig and Murty [125] compared the diazomethane procedure with the use of sodium methoxide as a catalyst for ester formation. They found that both methods gave comparable results.

In most of the following comparison studies, essentially comparable results were obtained. In 1960, Luddy and co-workers [126] found that esterification by diazomethane or HCl-methanol or sulfuric acid-methanol gave quantitative yields but that the use of sodium or potassium methoxide was somewhat less quantitative. Unfortunately, no yield data are available for the latter reaction.

When a procedure using diazomethane [127] was compared with HCl-methanol combined with sublimation as prepurification [60], with HCl-methanol using ion-exchange resins to purify the extract [128], and with boron trifluoride-methanol [87], it was found that the standard deviation for the diazomethane method when applied to higher-molecular-weight acids was slightly better (±0.25%) than the boron trifluoride-methanol (±0.46%) the HCl-methanol with sublimation (±0.52%), or the HCl-methanol coupled with resin cleanup (0.77%). For shorter-chain fatty acids there were somewhat greater losses during sublimation with the HCL-methanol method. For short-chain acids the best results were obtained using diazomethane, which gave a standard deviation of ±0.32%. The standard deviations of the boron trifluoride and the HCl-methanol methods were much larger—±13.8% and 126.5%, respectively.

Four years later, Jamieson and Reid [128a] reported the comparison of six different methods: (1) sodium methoxide, using a reaction time of 40 min [129]; (2) potassium hydroxide; (3) sodium methoxide, using a reaction time of 4 hr [130]; (4) saponification with 7.8% KOH in ethanol; (5) saponification with 0.5 N KOH in ethanol [131]; and (6) cold saponification with 50% aqueous KOH [132]. The results were essentially the same as those described above.

In 1967, de Man [133] compared the HCl-methanol coupled with the use of ion-exchange resins as described by Hornstein et al. [128], the boron trifluoride-methanol method as described by Morrison and Smith [94], and the "sealed-tube method" described by de Man [134]. de Man found that, as noted above, there were some losses of lower-molecular-weight fatty acids but that for the higher-molecular-weight fatty acids the methods were comparable. In general, most of these methods give comparable results, but they should be carefully checked on each problem to make sure they are applicable to that type of sample. In our laboratory, we have found the boron trifluoride-methanol method [94] to have the widest range of applicability for samples from natural sources.

III. INSTRUMENTATION

In general, the fatty acid samples can be analyzed on the usual columns with the state-of-the-art inlets and detectors. However, because of the decreased sample capacity of capillary columns, sample introduction is much more complicated. Normally, only nanograms of sample are injected onto a capillary column whereas a packed column sample is usually 200 times larger. As a result, injection splitters were developed so that the bulk of the sample injected is vented to the atmosphere and only a small portion actually enters the column [135,136]. With normal split ratios of 1:20 and 1:1000, one difficulty with splitters has been maintaining the exact ratio during the long periods of time required for routine analysis. Because of this difficulty, Grob and Grob [137] developed a method for injecting the complete sample without splitting. This method uses the principle of cold trapping whereby the sample is injected at a temperature at least 30–40°C below the separation temperature. The sample is thereby retained on the top of the column, and the solvent is flushed through the column ahead of the sample. The temperature is then raised to effect the separation of the sample into small, well-defined peaks.

In some cases, the septum will absorb some of the sample and bleed this part out during the analysis leading to extensive tailing and poor peak shape. In general, this absorption does not occur if the carrier gas flow is high enough. Capillary columns, because of their normal low flow, are more susceptible to this problem than packed columns. The problem can be overcome by using a technique of septum flush which is carried out as follows [137,138]. The sample is injected on column 30°C below the normal separation temperature with the splitter closed. After 0.5–3.0 min, the splitter is reopened to flush the septum of any retained sample and to vent it. The "cold-trapped" sample is then analyzed by heating the column to the desired initial temperature and commencing the temperature program.

Although our experience with this technique has been quite good, others [139] have criticized it because the sample is first vaporized, then condensed, and then vaporized again. To avoid this problem they use a double-column technique. All of these problems can probably best be overcome by the use of an on-column injector which can inject up to 5 μl. By injecting directly on column, septum uptake and size differences are eliminated. Of the several types reported [105,109,109a], one is commercially available from Brechbühler A. G., Urdorf, Switzerland.

Most of the methods of injection reported here can be adapted to automatic injection. In addition to these types of sample introduction based on syringe injection, there are now available solid samplers, such as the one from Carlo Erba S.p.A., Milan, Italy. In this method the sample is taken up in a capillary previously coated with a silicone liquid phase such as SE-30. Prior to coating the tube with a 0.05% solution of SE-30 in methylene chloride (w/v), the walls of the tube are treated with a 4% solution of trimethylchlorosilane in chloroform (v/v). This pretreatment gives a uniform coating by decreasing the contact angle between liquid and surface [140].

After evaporation of the solvent from the capillaries with a stream of nitrogen at 4°C, they are placed into the automatic injector. The use of this sampling system is restricted to liquids and solids and is limited by the volatility of the sample. However, a good reproducibility could only be obtained for fatty acids with more then 14 carbon atoms. This sampling system does combine the advantages of injection of the entire sample, absence of a solvent peak which increases column life, lack of septum effects, low cost, and good stability.

A great deal of work to develop good columns for fatty acid analysis has been carried out. Orr and Callen [141] introduced the use of polyesters of short-chain dicarboxylic acids and low-molecular-weight dialcohols, while Murty [142] and James [69] reported the use of polyesters of ethylene glycol, diethylene glycol, and 1,4 butanediol. Packed columns for the analysis of fatty acids have been used by many authors [143–146], but the wide range of samples and the wide range of volatility have limited the use of such columns. The work has been reviewed by Haken [146a]. Because of these limitations, capillary columns have become the method of choice for characterizing fatty acid patterns. The method for preparing these columns has been adequately described by many authors [136, 147–158]. The principal liquid phases tested and compared are summarized in Table 10.1.

We have obtained our best results to date on FFAP (free fatty acid phase) columns. This liquid phase described by Goretti and Liberti [164] is a reaction product of Carbowax 20 M and 2-nitroterephthalic acid. Grob and coworkers [167] compared FFAP columns to Silar-10C and found that in their hands a systematic low result was obtained for the trans acids on the Silar phase and for the cis acids on the FFAP column. It is possible that the efficiency of separation on these two columns was not optimal since we find that the ω9, ω7, and ω5 isomers are easily analyzed on a FFAP column [168]. In addition, the FFAP column was stable for 2 years while undergoing automated injections 24 hr per day. Although Silar columns should separate better than FFAP columns, none of the results reported to date shows this. There have recently appeared on the market commercial columns which report to give extraoridinarily good chromatograms [166]. To date we have not been able to obtain one and verify these results.

Flame ionization detectors (FID) are generally quite sufficient for most fatty acid analysis. Electron capture detectors (ECD) have a large detector volume which decreases the separation factor. The newer micro-ECDs, which are especially designed for capillary columns, overcome this problem [169]. One

Table 10.1 Comparison of Liquid Phases for the Analysis of Fatty Acids

Special features of fatty acid sample	Type of liquid phase	[Ref.]	Remarks
Methyl esters	DEGS	[159]	
	BDS	[159]	
	Polyester-coated SCOT	[159]	
Complex mixtures	BDS	[160]	Apiezon-L is good for mixtures. Silar-5CP gave a shorter elution time. Stability was good
	Silar-5CP	[160]	
	Apiezon-L	[160]	
Methyl esters	SE-30	[161]	Had wide polarity range, decreased overlap, and helped with identification.
	OV-225	[161]	
	Silar-5CP	[161]	
	Silar-10C	[161]	
Unsaturated FA	Carbowax 20 M	[162]	Did not separate cis/trans isomers.
	PZ-176	[163]	Comparison showed that Silar gave better results but column stability was poor. (See also [108].
	HI-EFF-8BP	[163]	
	Silar-5CP	[163]	
	OV-17	[163]	
	SP-2401	[163]	
	Silar-10C	[165]	These cyanopropyl polysiloxanes gave comparable results
	SP-2340	[165]	
	OV-275	[165]	
	Polyethylene glycol	[177]	

which we have used is commerically available from Carlo Erba, S.p.A., Milan Italy. Radio-gas chromatography [170–173], a radioactivity monitoring system is used in place of a conventional detector, has been successfully applied to the biochemical problems of the fatty acids. This procedure is not superior in accuracy but is more specific when specifically labeled precursors are used.

Mass spectrometry has been used as a detection method, but it has not been proved effective in determination of structure because unsaturated fatty acids give similar spectra regardless of the position of the double bond. This is due to the fact that the double bonds tend to migrate up and down the chain under electron impact prior to fragmentation. Several groups are working on ways to overcome this problem [174–177].

The routine analysis of fatty acids generates a large number of peaks. The large amount of data is best handled by automatic integration. Even punch tapes and magnetic tape transfer to computers is too time-consuming for this volume. The best methods use computers as dedicated integrators [178].

IV. APPLICATIONS

The need to study the wide and complicated spectrum of fatty acids found in lipid samples has favored the use of capillary columns with larger separation factors over the use of packed columns. Packed columns are also reported to have a high interassay variability, precluding an accurate measurement [179–180).

Gas chromatography of free fatty acids on capillary column was carried out by Verzele and co-workers using on-column injection [182]. They preferred to use columns with an I.D. of less than 0.5 mm. To avoid breakage, they used gold capillaries as an interface. Gold is an excellent choice because it is catalytically inert with respect to fatty acids. Thus the advantages of their method which allowed good separation of fatty acids included the low dead-volume of the injection port and the lack of catalytically active surfaces. Using hydrogen as a carrier gas at high flow-rates, they were able to obtain HETPs of less than 1 mm [183]. They obtained separations on FFAP, WCOT, and SCOT columns up to linolenic acid (C18:3). Figure 10.1 shows a chromatogram of free fatty acids on a 10-m WCOT column with a temperature programming

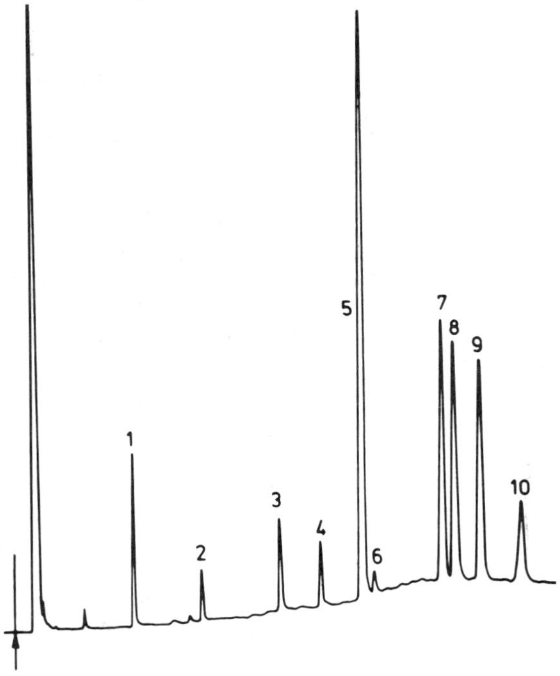

Figure 10.1 Analysis of long-chain free fatty acids. Column: 10 m × 0.5 mm; FFAP; WCOT; temperature programmed 140–204°C at 2°C min^{-1}; detector and injector, 250°C; carrier gas, hydrogen; flow-rate, 4 ml min^{-1}; elution time C18:3, 75 min. Peaks: 1, C10:0; 2, C12:0; 3, C14:0; 4, C15:0; 5, C16:0; 6, C16:1; 7, C18:0; 8, C18:1; 9, C18:2; 10, C18:3. (Reproduced by permission of Prof. M. Verzele, University of Ghent, Belgium.)

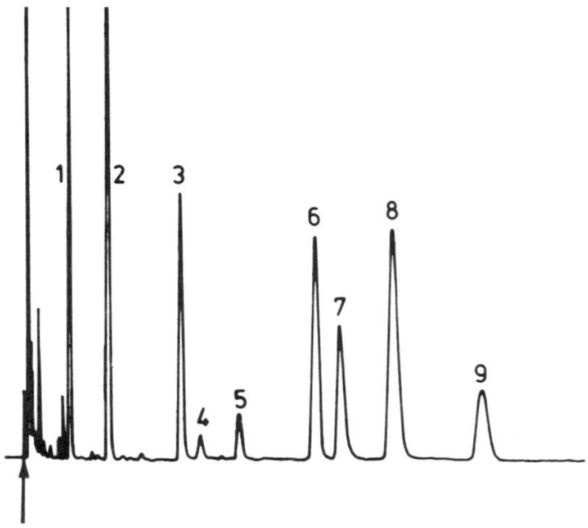

Figure 10.2 Analysis of long-chain free fatty acids. Column: 28 m × 0.6 mm ID; FFAP; SCOT; temperature, 200°C; detector and injector, 300°C; carrier gas, hydrogen; flow-rate, ±10 ml min^{-1}; elution time C18:3; 60 min. Peaks: 1, C12:0; 2, C14:0; 3, C16:0; 4, C16:1; 5, C17:0; 6, C18:0; 7, C18:1; 8, C18:2; 9, C18:3. (Reproduced by permission of Prof. M. Verzele, University of Ghent, Belgium.)

from 140 to 204°C. The elution times are somewhat long but still very good, considering that the acids were analyzed without prior derivatization. Figure 10.2 shows a similar mixture up to (C10:0) using isothermal analysis at 200°C. Overall, the quantitative analysis was better with on-column injection than with any of the splitters used. Others [181] have tried to use Dexsil-300GC but there was considerable tailing even using a short, 12-m glass column.

The low-molecular-weight fatty acids C_2–C_6 were intensively studied by Hirvňák and co-workers [104]. They were able to obtain baseline separation of 16 saturated fatty acids in this range using a stainless steel capillary column 45 m long and with an I.D. of 0.25 mm. Several liquid phases were tested, but the best separation was obtained on Ucon LB-550-X with phosphoric acid added, using isothermal conditions at 125°C. At this temperature the number of theoretical plates was less than 100,000 for all pairs of acids when calculated using the equation $m = 5.54[t_R(w_{0.5})^2]$. The method was easily applicable for analysis of biological samples and of food samples.

In 1974, Ettre and March [184] reported on a study where packed and open-tubular columns were compared for fatty acid analysis. Their results are summarized in Table 10.2. The liquid phase on all columns was the same. The phase ratio compares the volume of the gas phase to the volume of the liquid phase, while the partition ratio is a measure of the time spent in the liquid phase relative to the gaseous phase. From these two figures the partition coefficient for methyloleate was calculated. Figure 10.3 shows a reconstructed chromatogram for each column calculated from the number of theoretical plates, the retention times, and assuming a Gaussian distribution for the peak.

The true performance of a column can be assessed by comparing retention times at constant resolution, by comparing resolution at constant retention

Table 10.2 Comparison of Packed and Capillary Columns for the Analysis of Fatty Acids

Column	Phase ratio (β)[a]	Partition ratio (k)[b]	Partition coefficient[c]	Resolution[d]	HETP (mm)
Packed 8 ft × 2.16 mm chromosorb W-80-100 with 10% DEGS	10.3	58.6	597.4	1.22	1.61
Stainless steel WCOT; 150 ft × 0.25 mm with 10% DEGS	120	5.1	612.0	5.85	1.46
Stainless steel SCOT; 50 ft × 0.5 mm with 10% DEGS on Aerosil + Minusil	54	11.2	604.8	3.85	1.13

[a] $\beta = V_G/V_L$,
[b] $k = t'_R/t_M$
[c] For methyl oleate at 180°C.
[d] For methyl stearate/methyl oleate peaks at optimal gas flow constant retention time.

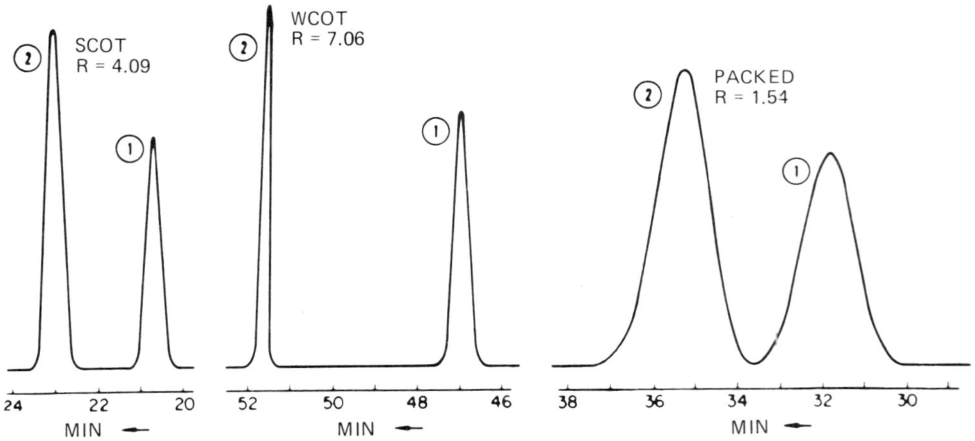

Figure 10.3 Reconstructed chromatograms showing the methylstearate/methyloleate peak pair on the three columns, at optimum average linear gas velocity. Peaks: 1, methylstearate; 2, methyloleate. (From Ref. 196. Reproduced by permission of Elsevier Scientific Publishing Company, Amsterdam.)

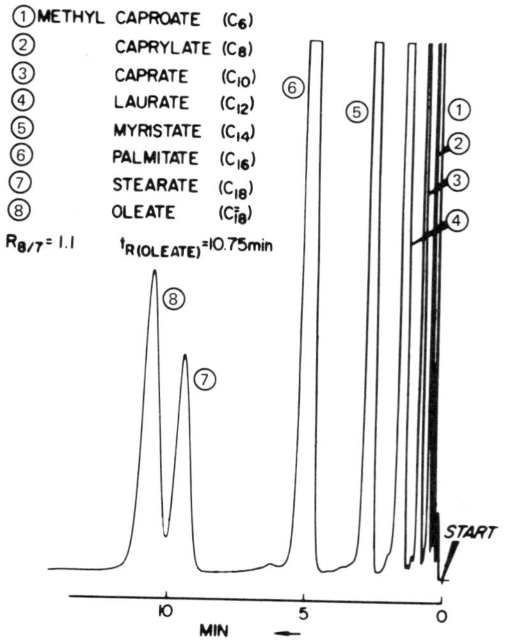

Figure 10.4 Chromatogram of a fatty acid methyl ester mixture obtained on a 8-ft × 1/8-in. I.D. packed column coated with DEGS stationary phase. Temperature, 180°C; average linear carrier gas (helium) velocity, 34 cm sec^{-1}; resolution of the methylstearate/oleate peaks 1.1. (From Ref. 197. Reproduced by permission of Friedr. Vieweg und Sohn Verlagsgesellschaft, Wiesbaden.)

times, or by comparing plots of resolution versus retention time. The results of the second comparison are found in Table 10.2. Figures 10.4, 10.5, and 10.6 compare the performance at constant retention times for oleate on the three columns [185]. In general, these open columns must be considered less than optimal. The same group [186] reported on the analysis of fatty acid methyl esters on a 100-m WCOT glass capillary column with an I.D. of 0.27 mm coated with DEGS. With this column the resolution of methyl oleate/methyl stearate was 12.8. The chromatogram is seen in Figure 10.7.

The problem of cis and trans isomers in the same sample was studied by Ackman and Hooper [187]. They reported that as far as cis/trans mixtures are concerned, there was no "loading effect," i.e., where large amounts of earlier eluting peaks interfere with the analysis of tailing peaks [188]. Figure 10.8 shows that as early as 1969, quite good separations of cis/trans acid mixtures were obtained on stainless steel columns. In this case the column was 150 ft × 0.25 mm I.D. and coated with BDS (butanediol succinate). Pollock [189] tried to separate unsaturated, saturated, and cis/trans mixtures on Dexsil-400GC phases, in which the columns were open and where the ends were

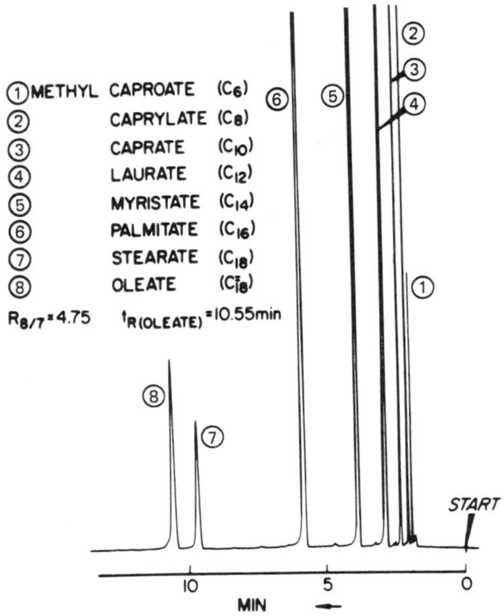

Figure 10.5 Chromatogram of a fatty acid methyl ester mixture obtained on a 150-ft × 0.010-in. I.D. WCOT column coated with DEGS stationary phase. Temperature, 180°C; average linear carrier gas (helium) velocity, 42.3 cm sec^{-1}; resolution of the methylstearate/oleate peaks, 4.75. (From Ref. 197. Reproduced by permission of Friedr. Vieweg und Sohn Verlagsgellschaft, Wiesbaden, and Elsevier Scientific Publishing Company, Amsterdam.)

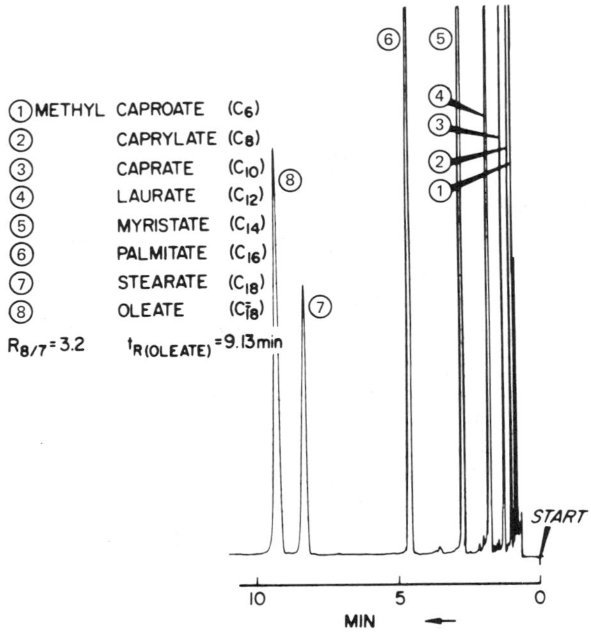

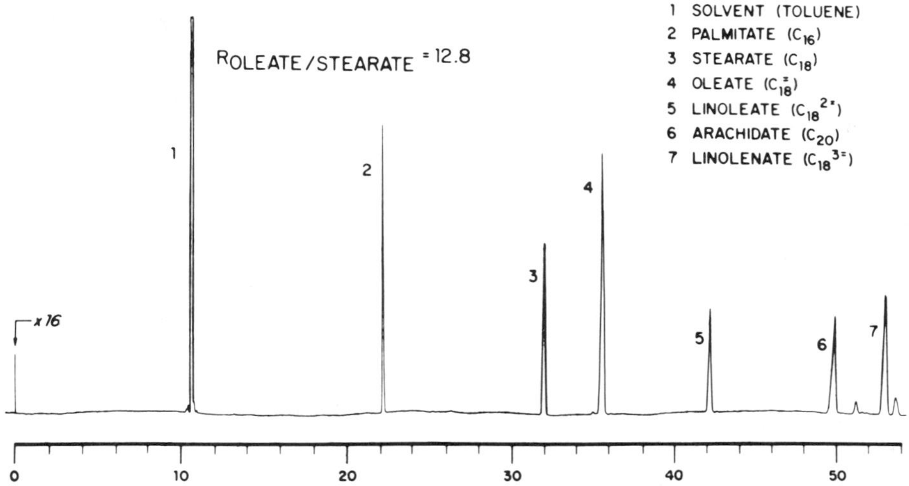

Figure 10.7 Analysis of fatty acid methyl ester sample. Column: 100 m × 0.27 mm I.D. open-tubular coated with DEGS liquid phase; glass tubing. Column temperature: isothermal at 185°C. Resolution (R) between methylstearate and methyloleate peaks is 12.8. (From Ref. 198. Reproduced by permission of Perkin Elmer, Inc.)

capped. The columns were 150 ft × 0.25 mm I.D. The tailing of the esters of the C_6–C_{11} acids was more severe when the caps were on, as was the bleeding. The uncapped columns were also more stable. As can be seen in Figure 10.9, they were able to separate saturated from unsaturated acids, but the cis/trans isomers could not be separated.

Apiezon-L and BDS have been used to separate cis/trans isomers [193]. Although BDS gave generally poorer separations, as can be seen in Figure 10.10, this liquid phase did allow the separation of positional isomers. The column used to obtain the chromatogram in Figure 10.10 had 100,000 theoretical plates.

The minor constituents of human milk were investigated by Egge and coworkers [190]. The branched-chain acids, as shown in Figure 10.11, were concentrated 100- to 500-fold by urea fractionation and then analyzed as their methyl esters after catalytic hydrogenation of the unsaturated constituents.

Figure 10.6 Chromatogram of a fatty acid methyl ester mixture obtained on a 50-ft × 0.020-in. I.D. SCOT colum prepared with DEGS stationary phase. Temperature, 180°C; average linear carrier gas (helium) velocity, 34.0 cm sec^{-1}; resolution of the methylstearate/oleate peaks, 3.2. For peak identification, see the legend to Figure 10.5. (From Ref. 197. Reproduced by permission of Friedr, Vieweg und Sohn Verlagsgesellschaft, Wiesbaden, and Elsevier Scientific Publishing Company, Amsterdam.)

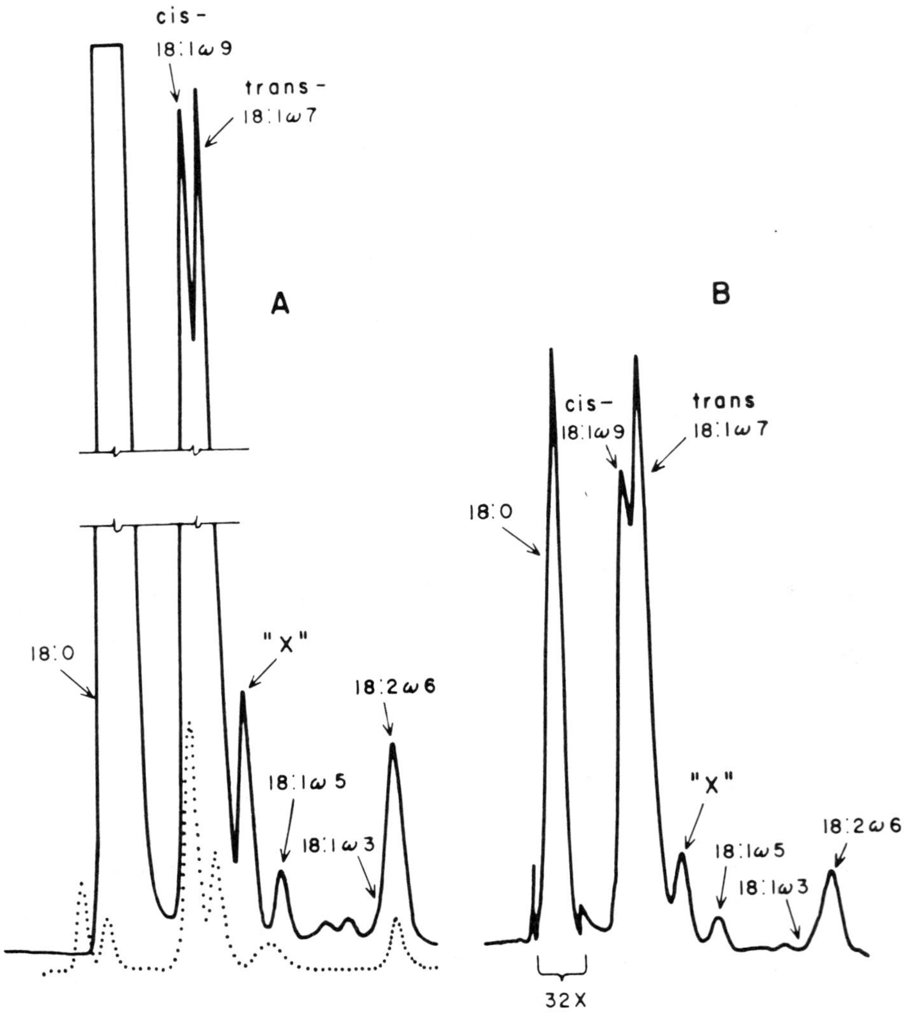

Figure 10.8 Parts of the chromatograms of C-18 monoethylenic and adjacent esters from two ruminant fats. In A the normal separation of *cis*-18:1ω9, ω7, and ω5 in nonruminant fat (fish oil) is shown by the dotted curve at bottom. Attenuation of B was 8× except for 32× for duration of 18:0 component. Attenuation of A was 16× for ruminant material, 16× for fish material. (From Ref. 199. Reproduced by permission of Preston Publications, Inc.)

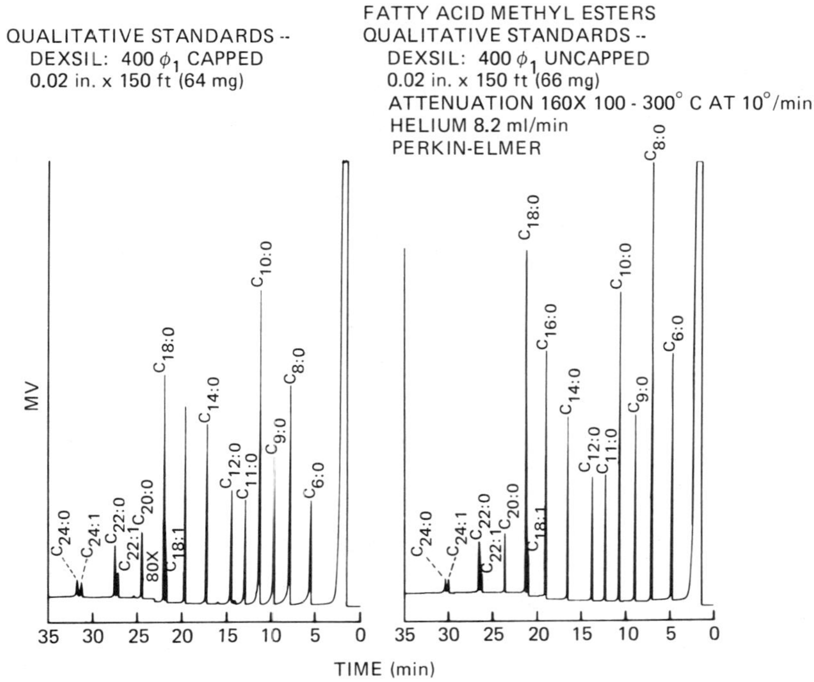

Figure 10.9 Separation of saturated and unsaturated fatty acid methyl esters on capped and uncapped Dexsil 400GC stationary phase. (From Ref. 201. Reproduced by permission of the American Chemical Society.)

The column was a 100 ft × 25 mm I.D. open-tubular column coated with DEGA (diethylene glycoladipate polyester).

The analysis of the fatty acids in butterfat has been extensively studies. When BDS was initially used, it was found that a good separation of the *cis* 16:1 from the *trans* 16:1 was obtained and that the methyl esters of the saturated fatty acids eluted before those of the unsaturated fatty acids, due primarily to the polar nature of the polyester liquid phase. A sample chromatogram is shown in Figure 10.12 [191]. The column was 100 ft × 0.25 mm I.D. and had initially 190,000 theoretical plates. Unfortunately the polarity of the stationary phase was poorly defined and changed as the column aged. Later the same authors [194] reported that the optimal separation of the fatty acids in butterfat could be obtained using combinations of several columns of differing polarity. They used columns prepared from Apiezon-L, DEGS, and BDS (See Figures 10,13, 10.14, and 10.15 for sample chromatograms). In all three cases, an FID detector was used. The column dimensions were 45 m × 0.25 mm I.D. 45 m × 25 mm I.D., and 100 m × 0.25 mm I.D., which gave 40,000, 60,000, and 190,000 theoretical plates for the Apiezon-L, DEGS, and BDS columns, respectively.

Shen-Nan Lin and co-workers [195] developed a method for preparing a glass capillary column coated with PZ-176 which did not break up on repeated heating and cooling. This was accomplished by incorporating very fine particles

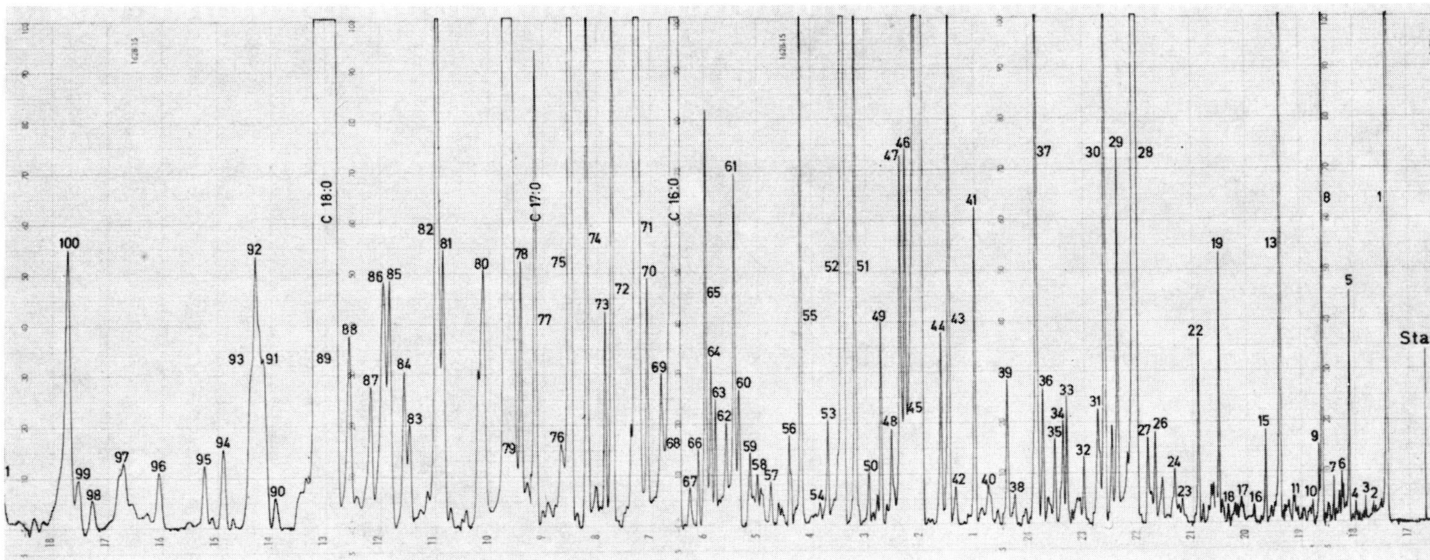

Figure 10.10 Gas chromatogram of methyl esters of branched-chain fatty acids derived from the lipids of human milk. Open-tubular column 0.25 mm × 50 m coated with polyphenylether (OS-138, Perkin Elmer); temperature program, 160–220°C, 1°C min^{-1}; carrier gas He, 1 ml min^{-1}; injection block, 280°C; molecule separator; 200°C; samplc size, 20 nl injected without splitting; chart speed 0.5 cm min^{-1}. Peaks: 1, methyl-*n*-decanoate; 2, methyl-3 and 4-methyldecanoate; 3, ethyldecanoate and methyl-7- and 5-methyldecanoate; 4, methyl-9-methyldecanoate; 5, methyl-8-methyldecanoate; 6, methyl-*n*-undecanoate; 7, methyl-*n*-undecanaote; 8, methyl-3-methylundecanoate; 9, methyl-4- and 5-methylundecanoate; 10, methyl-10-methylundecanoate; 11, methyl-*n*-dodecanoate; 12, methyl-3-methyldodecanoate; 13, methyl-5- and 6-methyldodecanoate; 14, methyl-4-methyldodecanoate; 15, methyl-2-, 7-, and 8-methyldodecanaote; 16, methyl-11-methyldodecanoate; 17, methyl-

10-methyldodecanoate; 18, 2,6-di-*tert*-butyl-*p*-cresol; 19, methyl-*n*-tridecanoate; 20, methyl-*n*-tridecenoate; 21, methyl-3,7,11-trimethyldodecanoate; 22, methyl-5-methyltridecanoate; 23, methyl-4-methyltridecanoate; 24, methyl-12-methyltridecanoate, 25, methyl-11-methyltridecanoate; 26 and 27, isomers of peak 18; 28, methyl-*n*-tetradecanoate; 29, methyl-7- and 8-methyltetradecanoate; 30 and 31, isomers of peak 18; 32, methyl-5- and 6-methyltetradecanoate; 33, methyl-4,8,12-trimethyltridecanoate; 34, methyl-4- and 2-methyltetradecanoate; 35, methyl-13-methyltetradecanoate; 36, methyl-12-methyltetradecanoate; 37, solvent impurity; 38, methyl-*n*-pentadecanoate; 39, methyl-*n*-pentadecanoate; 40, methylpentadecanoate; 41, methyl-5,9,13-trimethyltetradecanoate; 42, methyl-4-methylpentadecoanate; 43, methyl-14-methylpentadecanoate; 44, methyl-13-methylpentadecanoate; 45, solvent impurity; 46, methyl-*n*-hexadecanoate; 47, methyl-9-cyclohexylnonanoate; 48, solvent impurity; 49, *o*-dimethylphthalate, 50, methyl-2,6,10,14-tetramethylpentadecanoate; 51, methyl-7- and 8-methylhexadecanoate; 52, methyl-4-methylhexadecanoate; 53, methyl-12-methylhexadecanoate; 54, methyl-15-methylhexadecanoate; 55, methyl-14-methylhexadeconate; 56, methyl-*n*-heptadecanoate; 57, 58, and 59, solvent impurities; 60, methyl-9-,5-, and 10-methylheptadecanoate; 61, methyl-3,7,11,15-tetramethylhexadecanoate, phthanic acid methyl ester; 62, methyl-11- and 12-methylheptadecanoate; 63, methyl-16-methylheptadecanoate; 64, solvent impurity; 65, methyl-*n*-octadecanoate; 66, methyl-11-cyclohexylundecanoate; 67, oleic acid methyl ester; 68, linoleic acid methyl ester; 69, methyl-10- and 11-methyloctadecanoate; 70, methyl-*n*-nonadecanoate; 71, di-isobutylphthalate; 72 , not identified; 73, *n*,*i*-butylphthalate. (From Ref. 202. Reproduced by permission of North Holland Publishing Company, Amsterdam.)

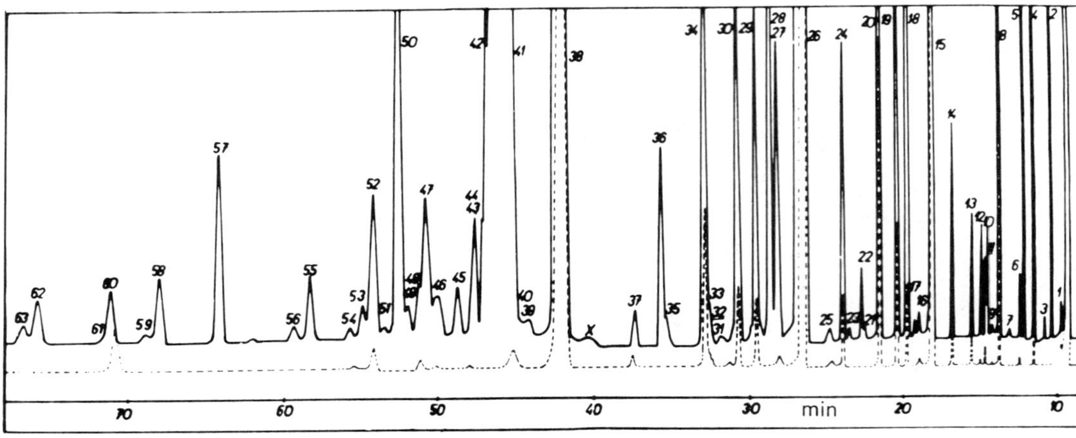

Figure 10.11 Separation of the methyl esters of butterfat acids on BDS at 180°C. Solid line, original sample; dotted line, sample after hydrogenation. Peaks: 1, 7:0; 2, 8:0; 3, 9:0, 4:10:0; 5, 10:1; 6, 11:0; 7, *iso*-12:0; 8, 12:0; 10, 12:1; 11, 12:1 + *iso*-13:0; 12, 12:1; 13, 13:0; 14, *iso*-14:0; 15, 14:0; 18, 14:0 + *iso*-15:0; 19, *anteiso*-15:0; 20, 15:0; 22, 15:1; 24, *iso*-16:0; 26, 16:0; 27, *trans*-16:1; 28, *cis*-16:1; 29, *iso*-17:0; 31, *anteiso*-17:0; 34, 17:0; 36, 17:1; 37, *iso*-18:0; 38, 18:0; 41, *cis*-18:1 (9); 50, *cis,cis*-18:2; 52, 19:0; 57, *cis,cis,cis*-18:3 (9,12,15); 60, 20:0; 62, 20:1. (From Ref. 203. Reproduced by permission of Elsevier Scientific Publishing Company, Amsterdam.)

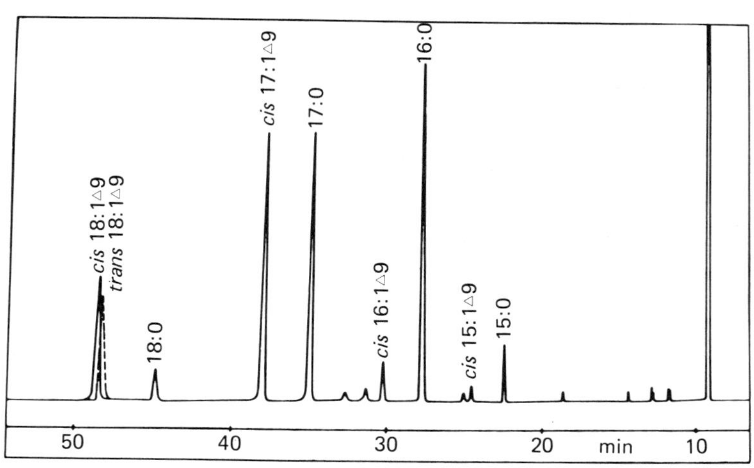

Figure 10.12 Separation of the methyl esters of fatty acids isolated from biological material by capillary gas chromatography on BDS at 180°C. Methylelaidate (*trans*-18:1Δ9) (dotted peak) was added to the mixture. (From Ref. 205. Reproduced by permission of the American Oil Chemists Society.)

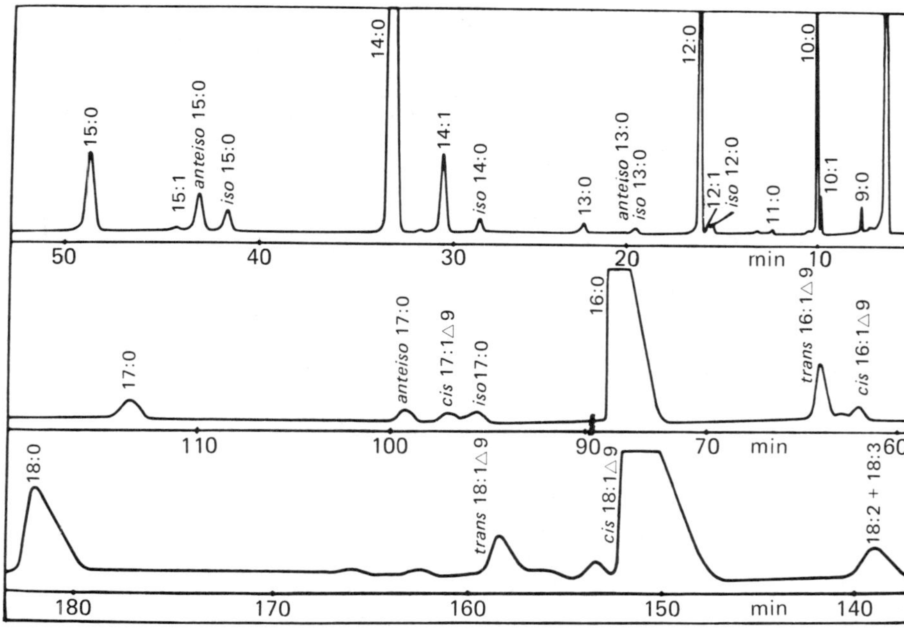

Figure 10.13 Chromatogram of fatty acid methyl esters from butterfat analyzed on Apiezon-L. For further details see text. (From Ref. 206. Reproduced by permission of Volkswirtschaftlicher Verlag GmbH, Kempten, West Germany.)

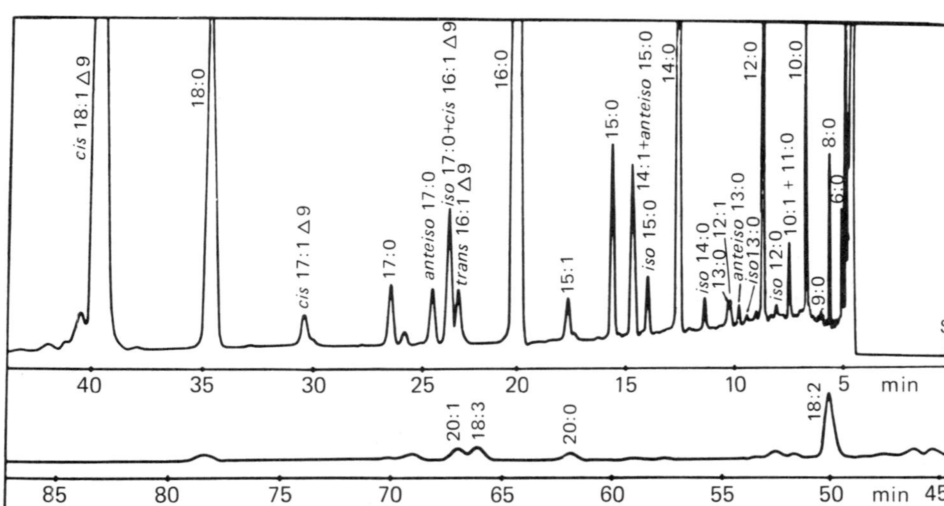

Figure 10.14 Chromatogram of the same sample as in Figure 10.13 analyzed on DEGS. (From Ref. 206. Reproduced by permission of Volkswirtschaftlicher Verlag GmbH, Kempten, West Germany.)

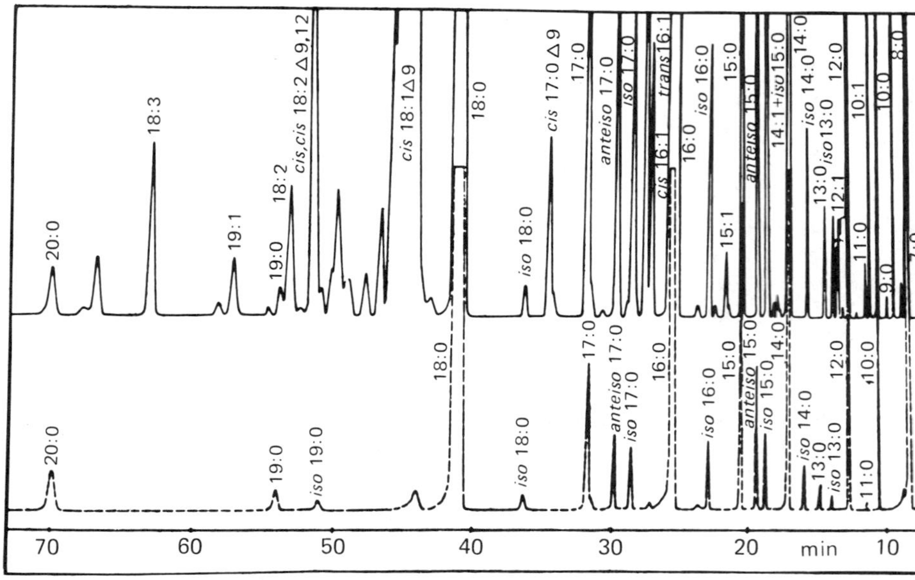

Figure 10.15 Chromatogram of the same sample as in Figure 10.13 analyzed on BDS. (From Ref. 206. Reproduced by permission of Volkswirtschaftlicher Verlag CmbH, Kempten, West Germany.)

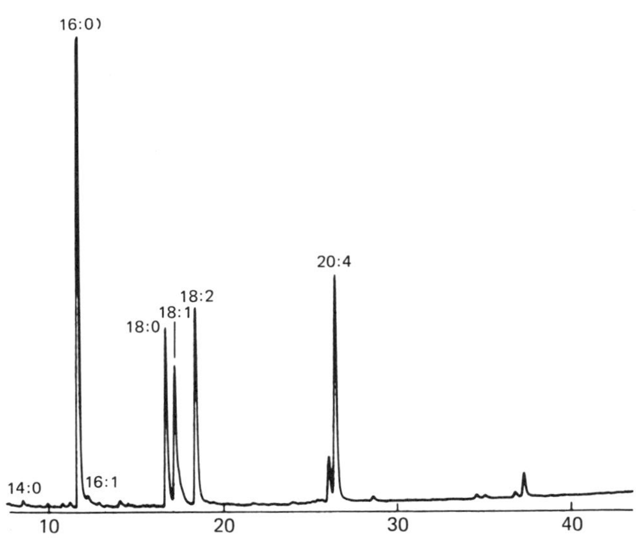

Figure 10.16 GC analysis of long-chain methyl esters from plasma phosphatidylcholines of a stroke patient. (From Ref. 210. Reproduced by permission of Elsevier Scientific Publishing Company, Amsterdam.)

of silanized silicic acid (Silanox 101) in the liquid phase [196] and by using the two-step coating procedure developed by German and Horning [197]. PZ-176 is a phenyl ether-phenyl sulfone polymer which can be used in the temperature range 180–320°C. The average column had an efficiency of 1700-1800 theoretical plates per meter.

These columns were used to analyze the fatty acids in human plasma phosphatidylcholines as their methyl esters and esterified cholesterol [198]. The lipids were extracted as described by Folch et al. [41], separated by TLC, transesterified with methanol-sulfuric acid at 60° for 3 hr, and extracted with isooctane. A typical chromatogram showing the fatty acid methyl esters derived from the phosphatidylcholines of a stroke patient is shown in Figure 10.16. The column was a 32 m × 0.25 mm I.D. glass capillary column coated with PZ-176; the injector temperature, 250°C; detector temperature, 280°C; temperature program, 2°/min.

An attempt to separate the C_{14}–C_{18} fatty acid methyl esters on Carbowax 20 M and the more polar DEGS was made by Schomburg and Husmann [147].

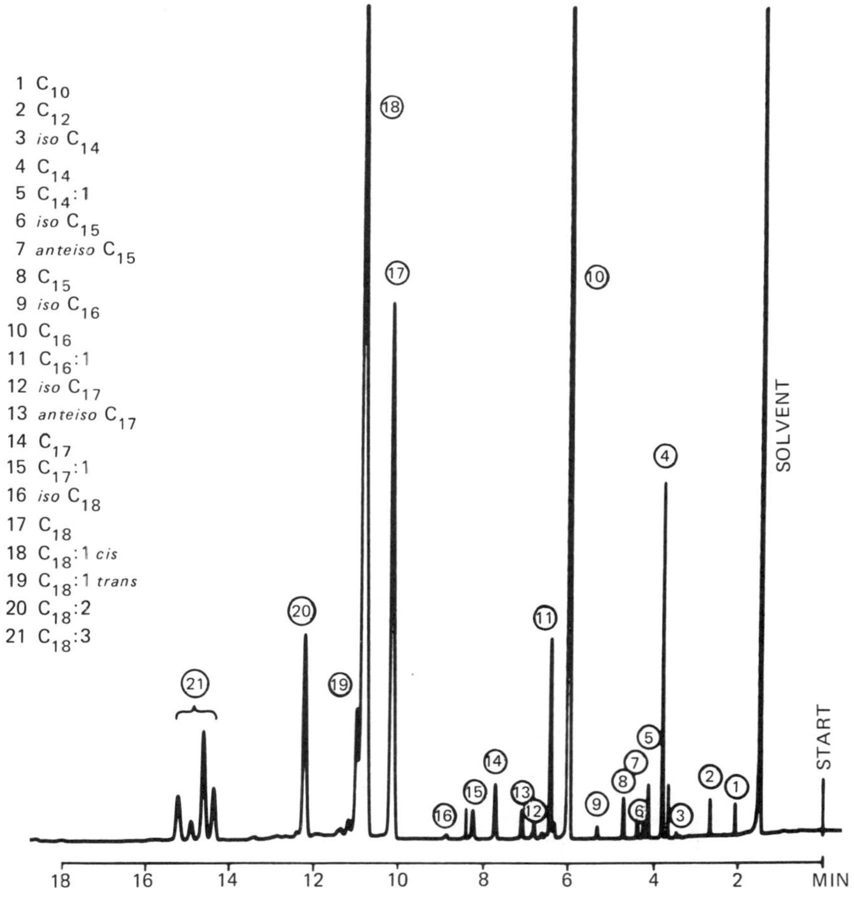

Figure 10.17 Fatty acid methyl esters (sample courtesy of Henkel and Cie, Düsseldorf). Column: 50 m × 0.25 mm CW 20 M; glass, 200°C, 2.0 bar H_2. (From Ref. 159. Reproduced by permission.)

They found that although Carbowax was more stable, it could not adequately separate oleic and elaidic acids (see Figure 10.17). DEGS could effect this separation but was considerably less stable over 260°C (Figure 10.18).

Carbowax 20 M has been used by Flanzy and co-workers [199] to analyze nutritionally important fats and oils. Figures 10.19 and 10.20 show chromatograms of human milk fat and sheep peritoneal fat obtained using this type of column. The column was 100 m × 0.5 mm I.D. and made of stainless steel. The total number of theoretical plates was about 100,000. The method could be used routinely for the analysis of the usually well-defined raw materials. The Carbowax 20 M used here had the advantages of being as efficient as the nonpolar Apiezon-L and having the resolving capabilities of BDS plus greater stability.

In an effort to improve coating efficiency of columns in general, Sandra and co-workers [200–202] developed a technique for roughening the surface of a

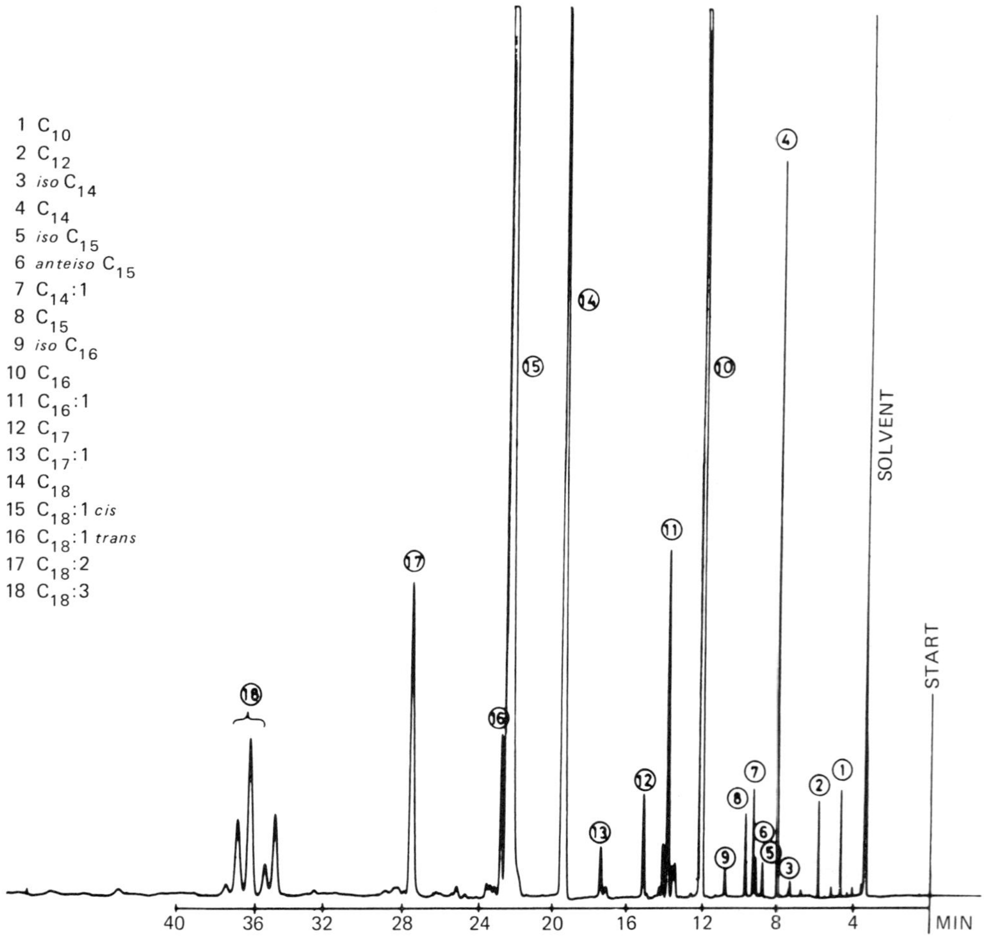

Figure 10.18 Fatty acid methyl esters (sample courtesy of Henkel and Cie., Düsseldorf). Column: 50 m × 0.25 mm DEGS; glass, 180°C, 0.5 bar N_2. (From Ref. 159. Reproduced by permission.)

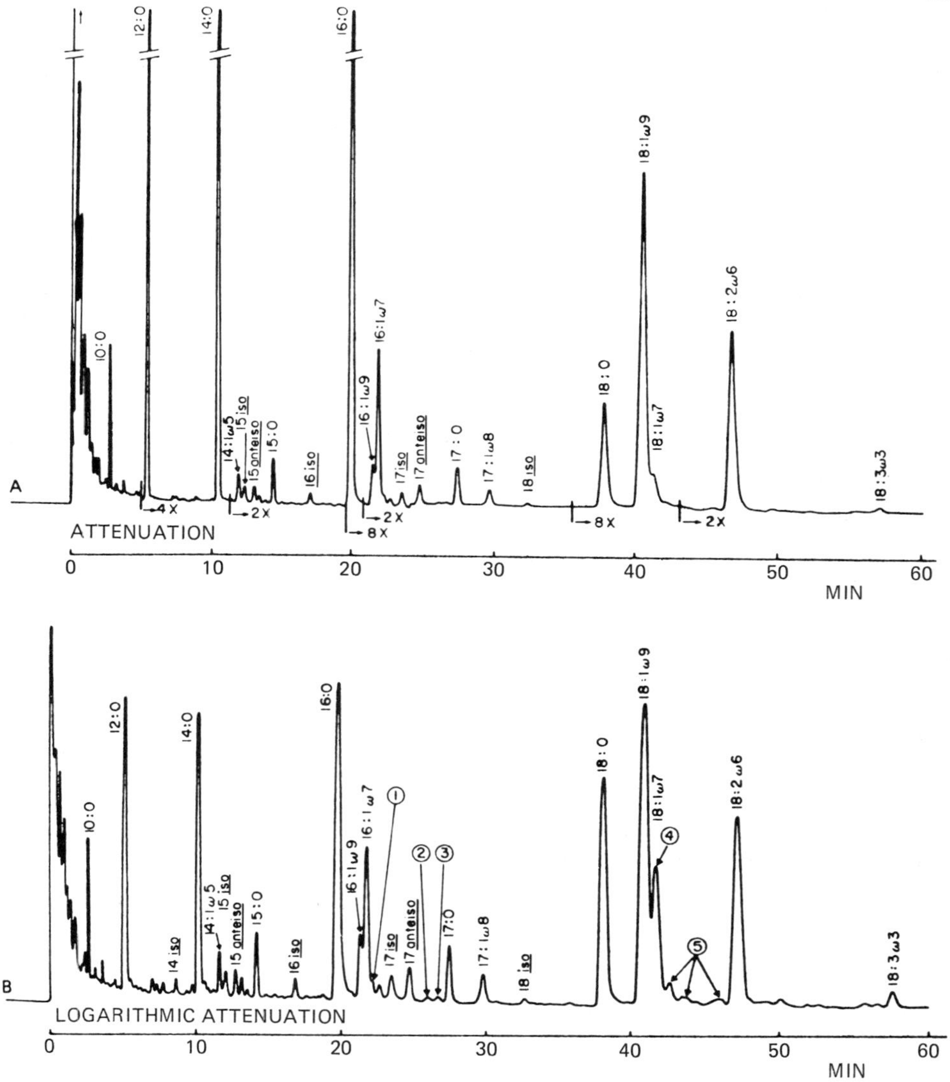

Figure 10.19 Comparison of standard linear recorder response (A) and logarithmic recorder response (B) for same sample of methyl esters of fatty acids from human milk. Note manual attenuations in linear recording mode and the fact that all peaks are on scale in the logarithmic mode of recording. (From Ref. 211. Reproduced by permission of Preston Publications, Inc.)

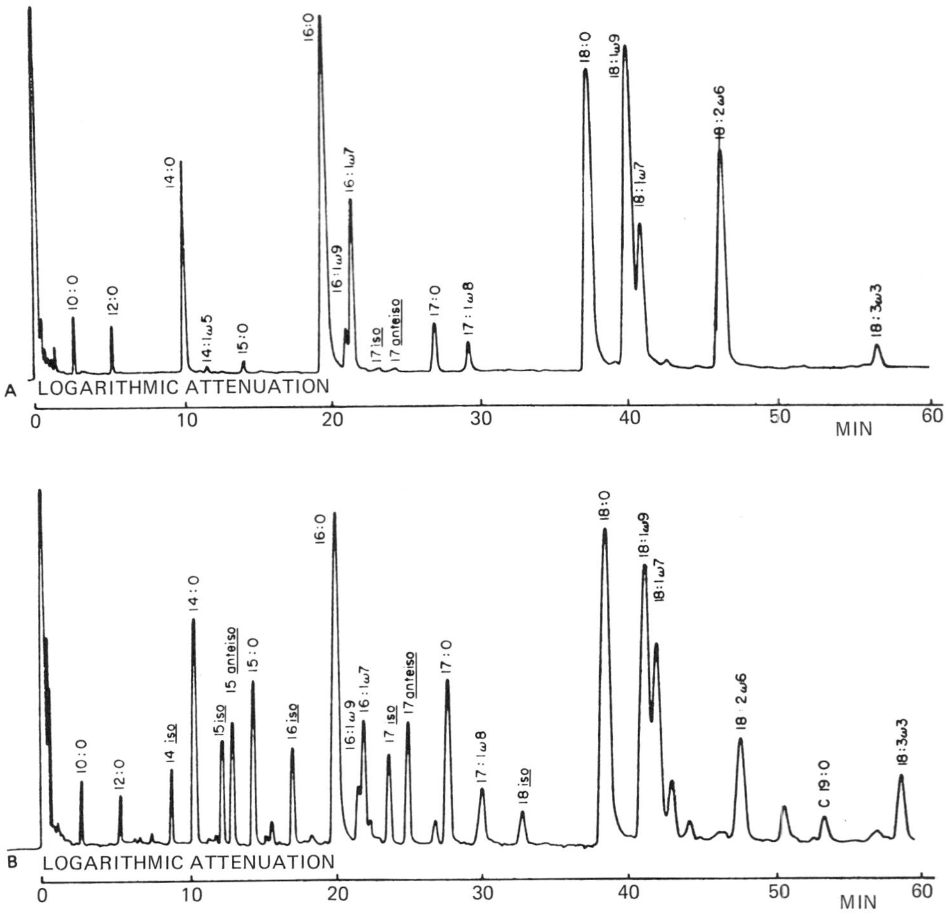

Figure 10.20 Comparison of methyl esters of fatty acid of pig leaf fat (A) and sheep peritoneal fat (B), both recorded in logarithmic mode. The additional complexity of the ruminant (sheep) fat is apparent. (From Ref. 211. Reproduced by permission of Preston Publications, Inc.)

glass capillary column to form so-called "whiskers." The method as originally described has been modified by Sandra and Verzele [203]. The use of a column with a whisker surface coated with Silar-5CP to analyze a standard mixture of fatty acid methyl esters is shown in Figure 10.21. The very high polar Silar-5CP as well as Silar-10CP is particularly useful in separating cis/trans isomers as well as positional isomers. The effectiveness of Silar-5CP can be seen in the good separation of the cis and trans C18:1 acids, oleic and elaidic acids, in Figure 10.21.

Sandra and Verzele [204] have also reported on Superox-4, a polyethylene glycol liquid phase with a molecular weight of 4 million. This liquid phase, which has high-temperature stability and good adherence to glass, has a separation efficiency comparable to Carbowax 20 M as seen in Figure 10.22. The separation of fatty acid methyl esters up to linolenic acid (C18:3), completed in 138 sec, represents a shortening of the time for analysis to a minimum.

However, it can be applied only when separation of cis/trans or positional isomers is not required.

SP-2340 has been used to separate the fatty acids associated with the phospholipids of red cell ghosts [205,206]. The glass capillary column was 30 m × 0.3 mm I.D. and was coated with SP-2340, a cyanopropylsiloxane having an average molecular weight of 2800. The temperature program was as follows: starting at 100°, increasing the temperature at 5°/min for 5 min, then 0.5°/min for 30.5 min and then 1°/min to 190°, and holding isothermal at this temperature

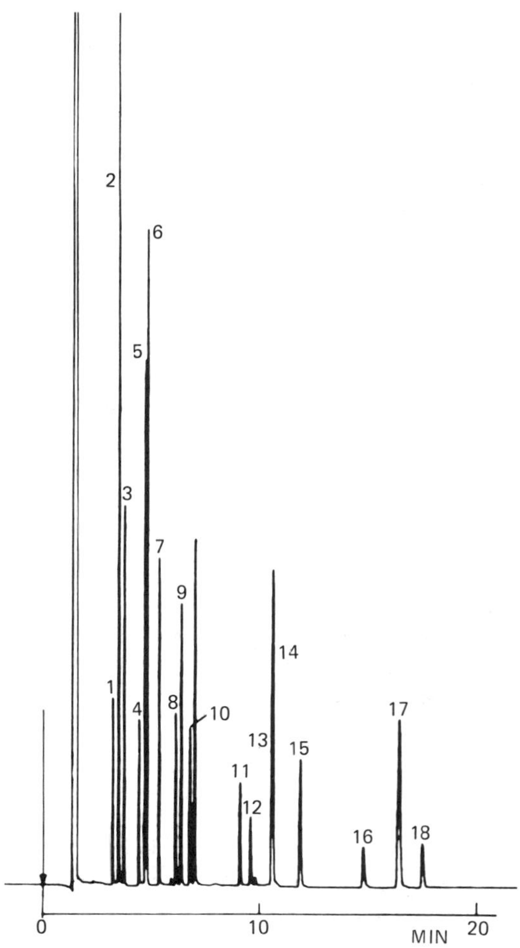

Figure 10.21 Gas chromatogram of a fatty acid methyl ester standard mixture. Column: 50 m × 0.25 mm I.D. deactivated with *N*-cyclohexyl-3-azetidinal (CHAZ) and coated with Silar-5CP. Column temperature 240°C, carrier gas flow ∿2 ml min^{-1} hydrogen. Peaks: 1, C16:0; 2, C16:1; 3, C17:0; 4, C18:0; 5, C18:1 (*trans*); 6, C18:1 (*cis*); 7, C18:2; 8, C18:3; 9, C20:0; 10, C20:1; 11, C20:4; 12, C22:0; 13, C20:5; 14, C22:1; 15, C23:0; 16, C24:0; 17, C24:1; 18, C22:6. (From Ref. 212. Reproduced by permission of Dr. Alfred Hüthig Verlag GmbH, Heidelberg, West Germany.)

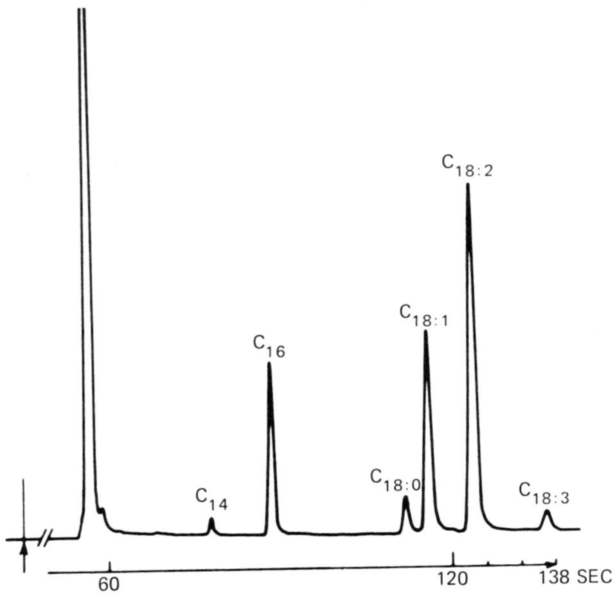

Figure 10.22 FAME analysis on a 25-m × 0.5-mm I.D. NaCl dendrite column coated with Superox-4. Temperatures: column, injector, and detector, 250°C. Carrier gas (hydrogen) flow-rate: 5 ml min^{-1}. (From Ref. 216. Reproduced by permission of Elsevier Scientific Publishing Company, Amsterdam.)

for 10 min. The total analysis time was 95 min. The carrier gas was nitrogen, flowing at 2.5 ml/min with a split ratio of 1:70. The injection port temperature and the detector temperature were 240°C. The separation number (TZ) is a measure of column efficiency and is based on the degree of separation of two members of a homologous series, taking into account their relative retention times and relative peak shapes. The following equation is used:

$$TZ = \frac{t_{R(B)} - t_{R(A)}}{W_{0.5(A)} + W_{0.5(B)}} - 1$$

where t_R is the retention times for A and B, respectively, and $W_{0.5}$ is the width of peaks A and B at half height, respectively. For this column TZ was calculated to be 7.76 or, normalized for column length, TZ/m = 0.26. The number of theoretical plates was 7674 or, per meter, 256 TP/m. The loadability for decanoic acid methyl ester was 1.80×10^{-6}g.

In spite of the low separation number and the relatively low number of theoretical plates of this SP-2340 column, the separation of geometric and positional isomers was very good. As can be seen in Figures 10.23 and 10.24, all five pairs of geometric isomers of monoethylenic fatty acids, the *cis,cis*- versus *trans,trans*-octadecadiethylenic and the *cis,cis,cis*- versus the *trans, trans,trans*-octadecatriethylenic acids are separated by more than 4.6σ. Although all technical problems with application of these coatings have not yet been solved, the cyanopropylsiloxane phases like SP-2340 and Silar-10C appear to be the liquid phases of choice for the future in fatty acid analysis. Heckers

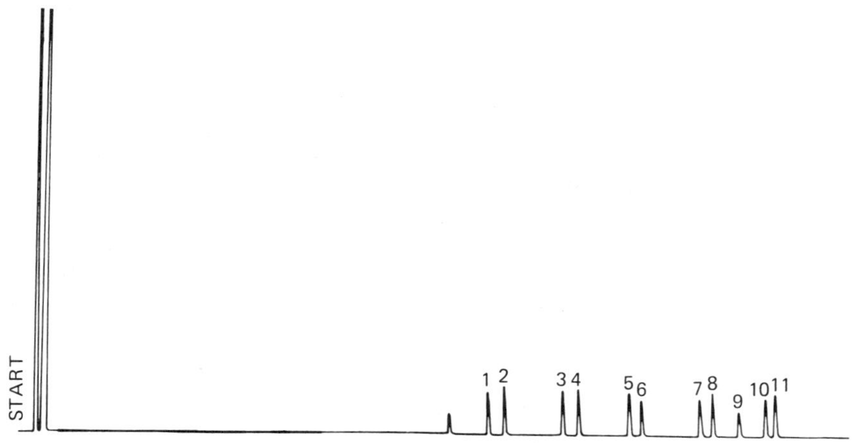

Figure 10.23 Chromatogram of a standard mixture of cis/trans fatty acid methyl esters on SP-2340. For chromatographic conditions, see text. Peaks: 1, C14:1 *trans*; 2, C14:1 *cis*; 3, C15:1 *trans*; 4, C15:1 *cis*; 5, C16:1 *trans*; 6, C16:1 *cis*; 7, C17:1 *trans*; 8, C17:1 *cis*; 9, C18:0; 10, C18:1 *trans*; 11, C18:1 *cis*. (Reproduced courtesy of Dr. Heckers, University of Giessen, West Germany.)

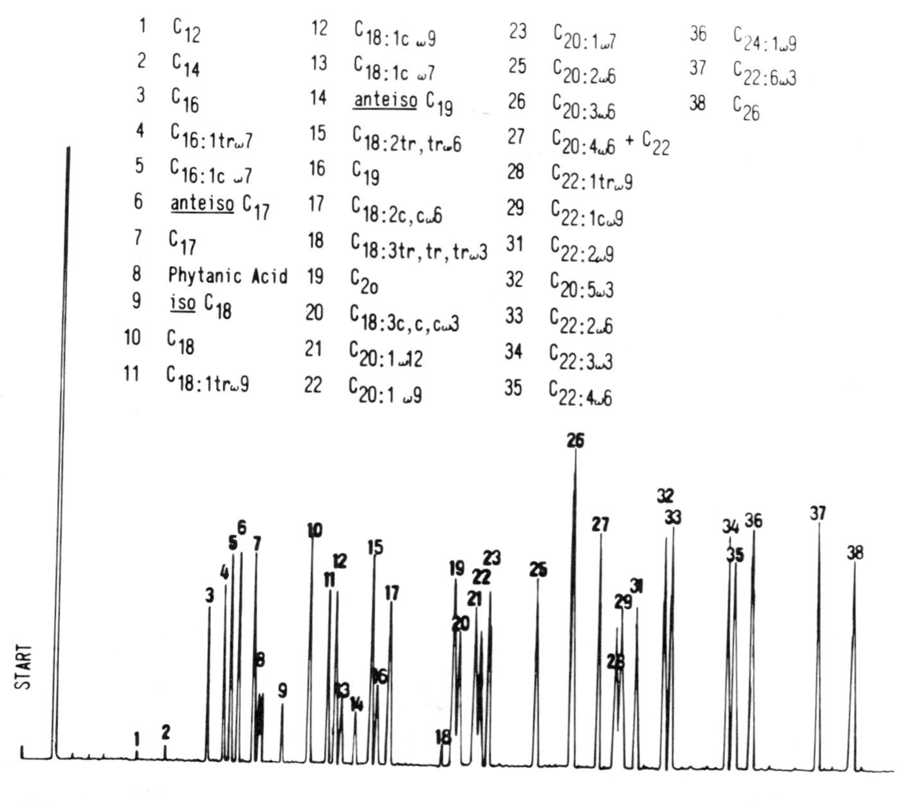

Figure 10.24 Chromatogram of a standard mixture of fatty acid methyl esters including positional and geometrical isomers on SP-2340. (Reproduced courtesy of Dr. Heckers, University of Giessen, West Germany.)

and co-workers successfully applied their technique to the analysis of the trans isomeric fatty acids in the human myocardium, jejunum, and aorta and their role in the development of atherosclerosis [207]. The various margarines, shortenings, frying oils, and cooking fats on the German market have also been analyzed for the presence of the trans isomeric acids using this technique [208].

The first experiences with glass capillary chromatography of fatty acids from our laboratory were reported in 1975. We extracted the lipids from human serum using the method described by Folch et al. [41] and transesterified them to their methyl esters using the method described by Morisson and Smith [94], which was described earlier. Purified hydrogen was used as a carrier gas at a flow-rate of 4 ml/min at room temperature (0.9 atm pressure). The flow of hydrogen and air to the FID were 12 ml/min and 150 ml/min, respectively. The injection port temperature was 250° and injections were carried out using a solid sampler as described above at a temperature of 40°C. The samples were "cold trapped" using the method described by Grob [138] with subsequent heating to 95°C. The temperature of the column was increased to 198°C at 1.2°/min and then held isothermal at 198° for 7 min before cooling to the initial temperature.

The glass capillary column was 50 m × 0.33 mm I.D. (purchased from H. & G. Jaeggi, Trogen, Switzerland). The coating was FFAP (free fatty acid phase), a modified polyether material. The separation number (TZ) was 47.65 or, normalized for column length, 0.93/m. Although good nonpolar columns easily reach values of TZ/m greater than 1.0, 0.93/m represents a very good value for such a polar column. The number of theoretical plates (n) was 227,052 or 4541/m, when normalized for column length. The gas holdup time (t_M) was calculated to be 16.04 by the following expression:

$$t_M = \frac{[t_{M+R}(a) \times t_{M+R}(c)] - [t_{M+R}(b)]^2}{t_{M+R}(a) + t_{M+R}(c) - 2t_{M+R}(b)}$$

where (a), (b), and (c) denoate octanoic, nonanoic, and decanoic acid, respectively, and where t_{M+R} is the total retention time for the peaks due to (a), (b), and (c). The values used for this calculation were obtained from an isothermal analysis at 90°C. From t_M one can calculate the partition ratio (k) from the relationship:

$$k = \frac{t_R}{t_M}$$

where t_R is the adjusted retention time. Finally, one can then calculate the capacity of the column for any component by using k and the following equation,

$$C = 0.05\ M\ d^3\ (1 + k)\ 10^{-6}\,g$$

where M is the molecular weight, and d is the inner diameter of the column. For decanoic acid the capacity was calculated to be 5.28×10^6g.

Using this column, the standard deviation from 10 chromatograms of the relative retention times did not exceed 0.26%, while those of the weight percent values did not exceed 1.53%. The correction factors for saturated esters varied from 965 to 992 and for the unsaturated esters from 1010 to 1100. There was less than a 10% variation in the factors from column to column and with the same column, with time up to 1 year.

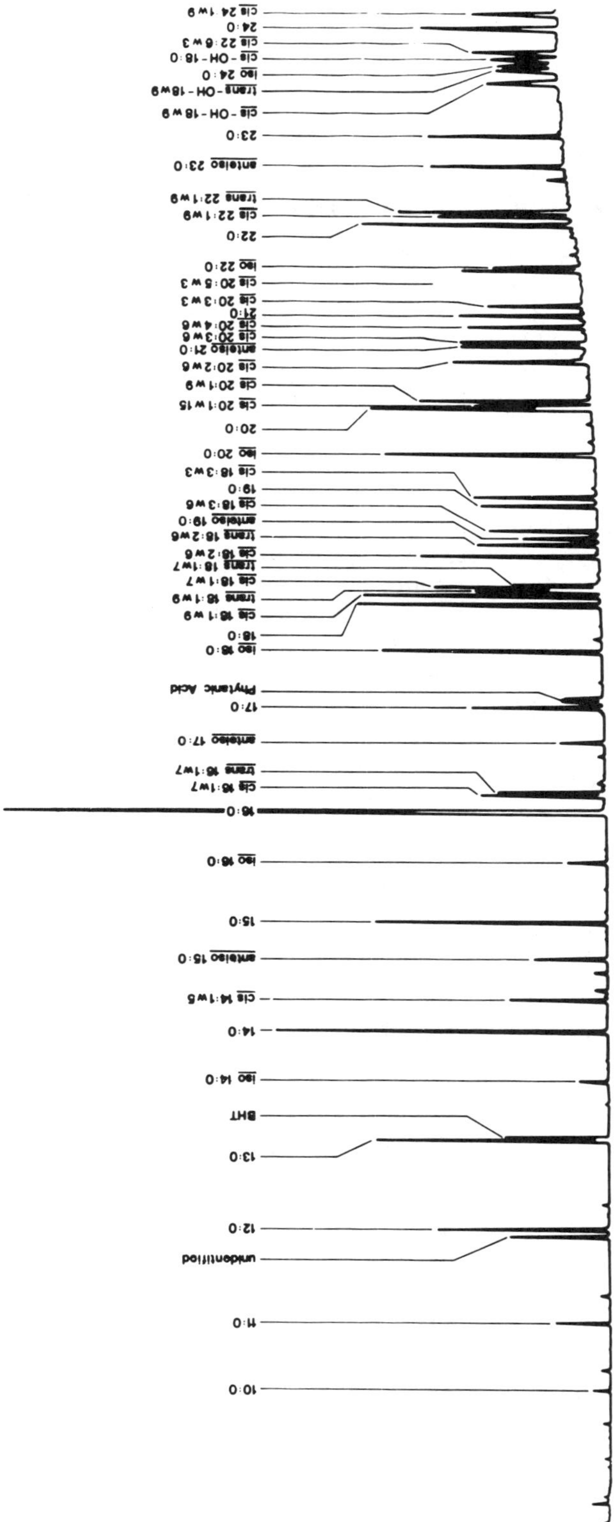

Figure 10.25 Chromatogram of a standard mixture of fatty acid methyl esters including positional and geometrical isomers on FFAP. Chromatographic conditions are given in the text.

The separation of fatty acid methyl esters shown in Figure 10.25 is typical of what can be achieved with such a column. The separation of cis and trans isomers is particularly noteworthy. *Cis*- and *trans*-palmitoleic acid (C16:1ω7) are separated with less than 1–2% overlap (4.6σ) allowing the use of electronic integrators with a precision of about ±0.3%. In Figure 10.26, which is an enlargement of part of Figure 10.25, one can see that a separation was obtained for the following important pairs of cis/trans or positional isomers:

cis- and *trans*-linoleic acids (C18:2ω6)
α, and γ-linolenic acids (*cis*-C18:3ω3, *cis*-18:3ω6)
8,11,14- and 11,14,17-eicosatrienoic acids (*cis*-20:3ω6, *cis*-20:3ω3)
5- and 11-eicosaenoic acids (*cis*-20:1ω15, *cis*-20:1ω9)
four C18:1 isomers: oleic (*cis*-18:1ω9), elaidic (*trans*-18:1ω9), *cis*-vaccenic and *trans*-vaccenic (*trans*-18:1ω7) acids
cis- and *trans*-C18:1ω6 acids
cis- and *trans*-13-docosenoic acids (erucic and brassidic acids, 22:1ω9)
cis- and *trans*-ricinoleic acids (HO, 18:1ω9)

In most cases a baseline separation, or a 4.6σ separation, was obtained. In general, positional isomers were separated better than cis/trans isomers.

Analysis of human plasma phospholipids fatty acids (Figure 10.27) showed that almost all of the fatty acids in the standard mixture were also present in that sample. This type of analysis can be applied to all lipid fractions obtained from man. Typical chromatograms of normal plasma triglycerides (Figure 10.28) and of normal cholesterol esters (Figure 10.29) are shown.

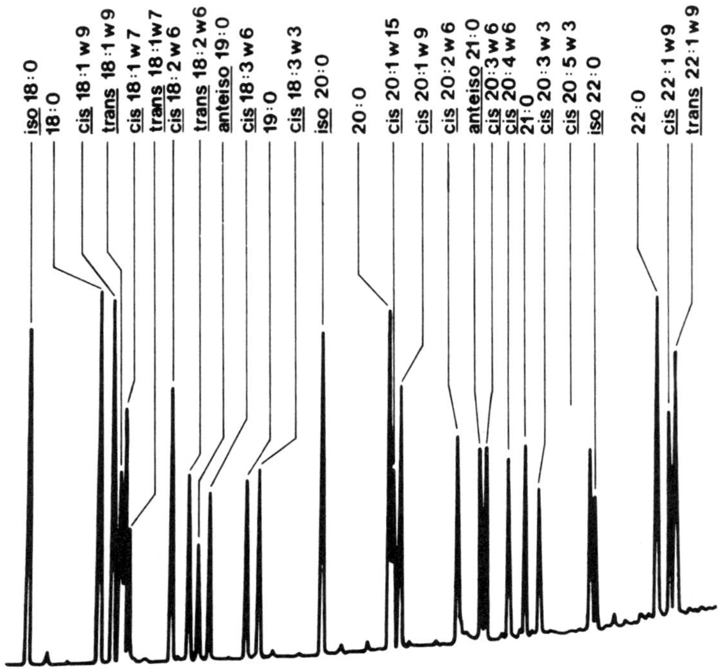

Figure 10.26 Part of Figure 10.25 in higher magnification.

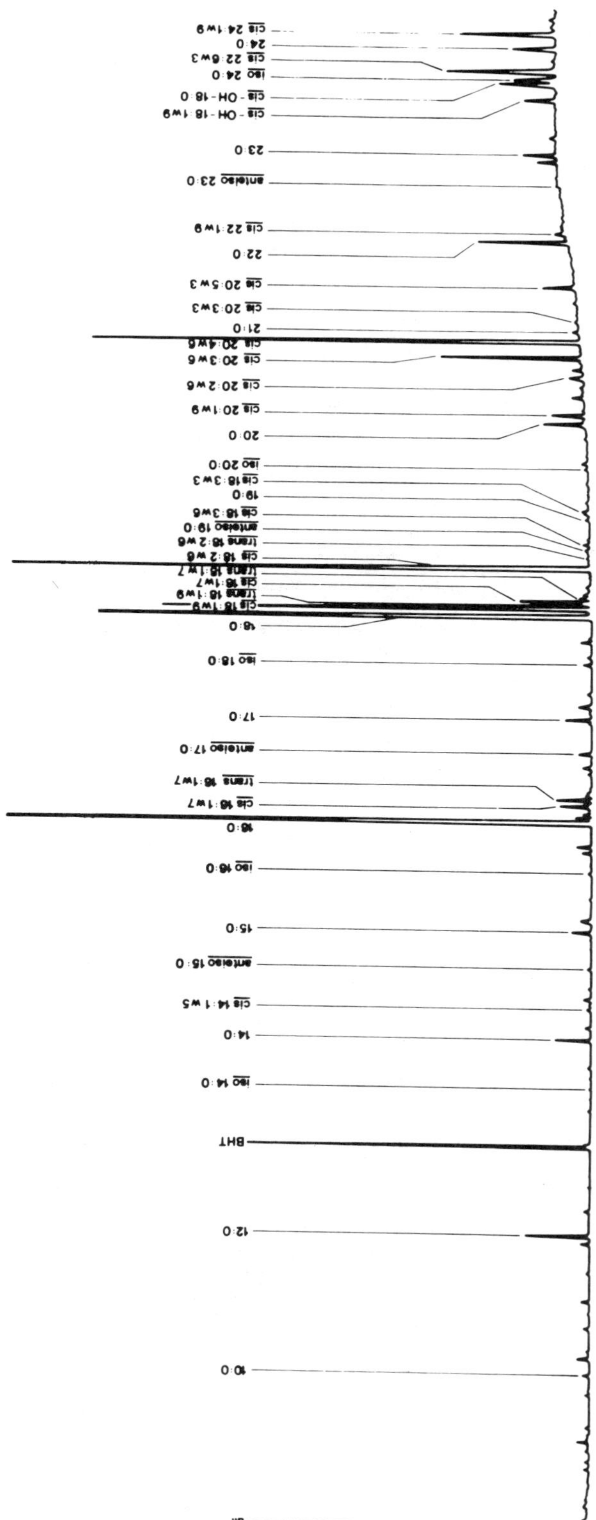

Figure 10.27 Chromatogram of fatty acid methyl esters from normal human phospholipids on FFAP. Chromatographic conditions are given in the text.

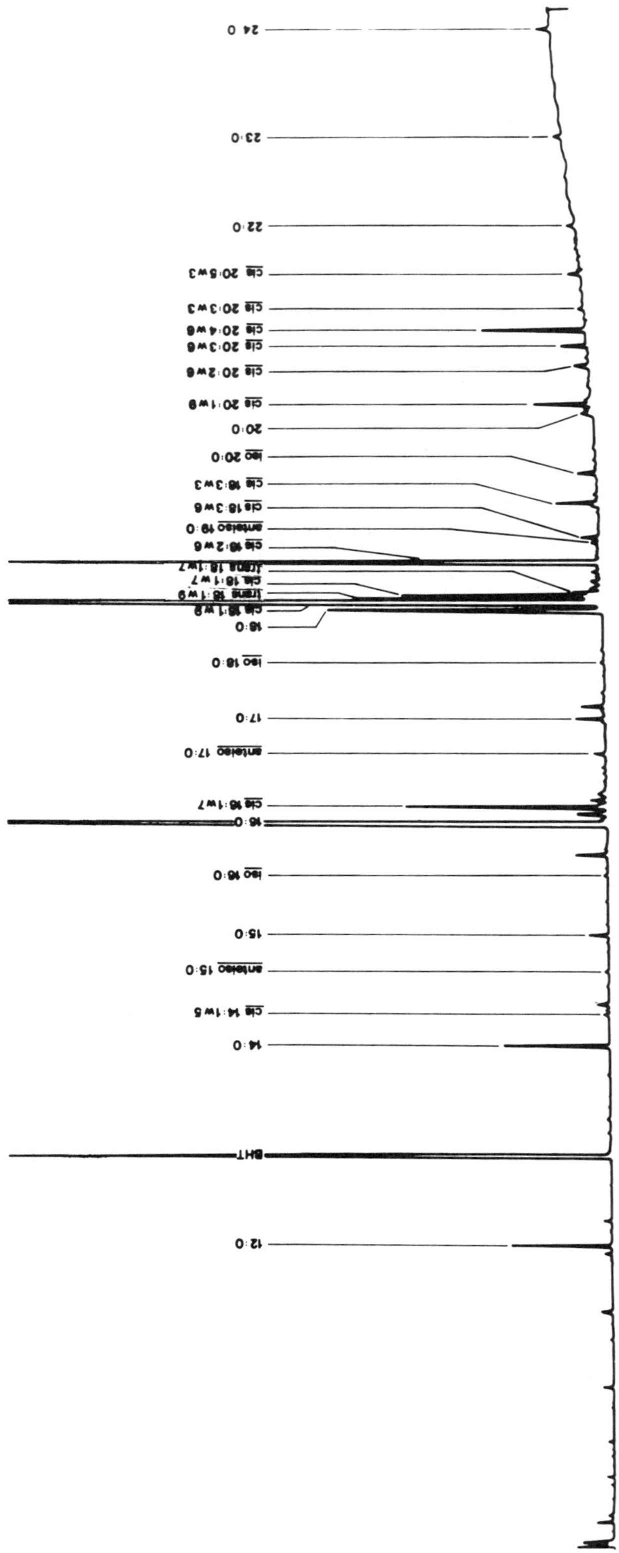

Figure 10.28 Chromatogram of fatty acid methyl esters from normal human triglycerides on FFAP. Chromatographic conditions are given in the text.

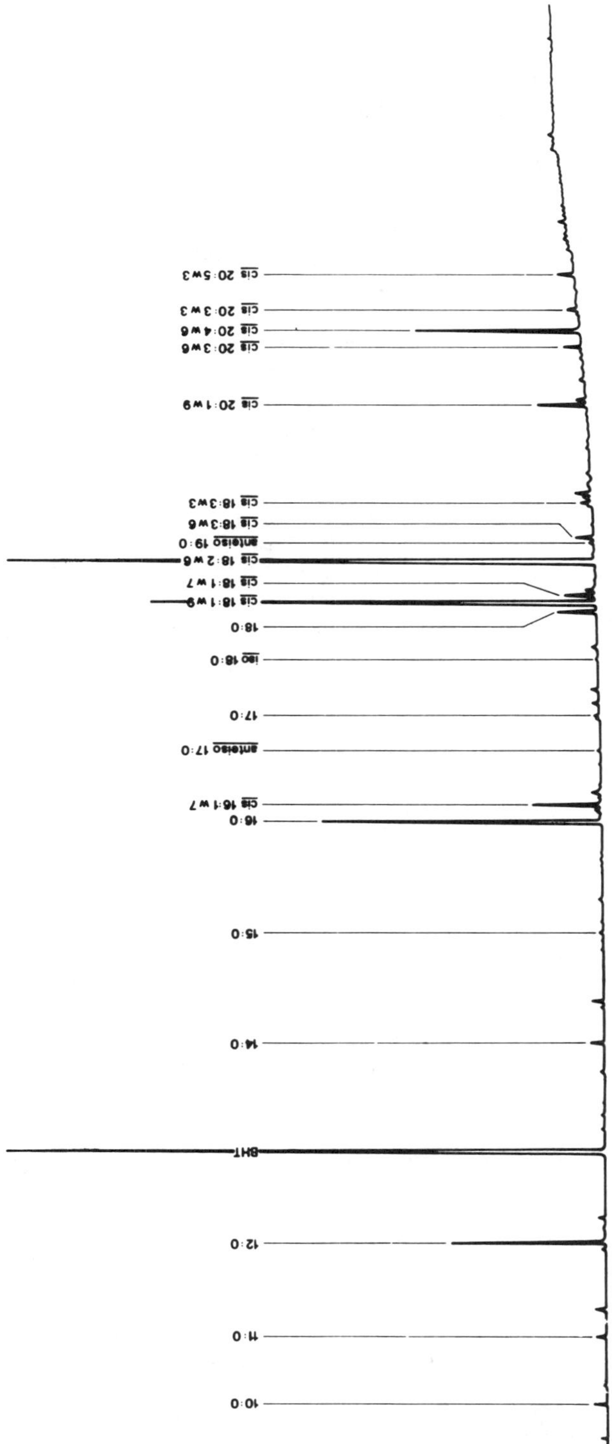

Figure 10.29 Chromatogram of fatty acid methyl esters for normal human cholesterol esters on FFAP. Chromatographic conditions are given in the text.

V. SUMMARY

The extraction and derivatization of samples has been well developed over the last 10 years. The catalysts of choice for preparation of the fatty acid methyl esters appear to be the acidic ones: methanolic HCl or sulfuric acid, boron trifluoride-methanol. The alkaline catalysts do not seem to give reproducible results.

Injection techniques have also advanced, but much work remains to be done to improve on-column injection and to automate the chromatography through use of autosamplers and computer control overall, including calculation of final results.

For routine analysis where separation of positional or cis/trans isomers is not needed, one can use one of the new columns with very short analysis times, When analysis of isomers becomes important, one of the polyether phases such as FFAP, which gives equal or superior separations compared to the cyanopropylsiloxane phases, appears to be the column of choice. Once some technical problems with these cyanopropylsiloxane phases have been overcome, one can expect separations far better than those presently available.

In general, the analysis of fatty acids either as such or derivatized as methyl esters has improved considerably in the last 5 years. Detection limits have been lowered and a wide range of new fatty acids can be analyzed. Much of this development is due to the impact of the use of glass capillary chromatography. With the discovery of more and more different fatty acids in nature and with the need to know what role each plays in physiology or in the pathophysiology of such diseases as atherosclerosis, the development of more sophisticated analytical techniques will be necessary. Glass capillary chromatography will play a vital role in this development.

REFERENCES

1. L. L. M. van Deenen, *Pure Appl. Chem. 25*:25 (1971).
2. V. P. Dole, and M. A. Rizack, *J. Lipid Res. 2*:90 (1961).
3. V. P. Skipski, M. Barclay, R. K. Barclay, V. A. Fetzer, J. J. Good, and F. M. Archivald, *Biochem. J. 104*:340 (1967).
4. T. Olivecrona, G. Bengtsson, M. Höck, and U. Lindahl, in *Lipoprotein Metabolism* (H. Greten, ed.), Springer, Berlin, 1976, p. 13.
5. D. G. Cornwell, F. A. Kruger, G. J. Hanwi, and J. B. Brown, *Metabolism 11*:840 (1962).
6. W. W. Christie and J. H. Moore, *Biochim. Biophys. Acta 176*:445 (1969).
7. W. W. Christie and J. H. Moore, *Biochim. Biophys. Acta 210*:46 (1970).
8. L. M. G. van Golde, V. Thomasi, and L. L. M. van Deenen, *Chem. Phys. Lipids 1*:282 (1967).
9. L. Marai and A. Kuksis, *J. Lipid Res. 10*:141 (1969).
10. L. Hagenfeld and J. Wahren, *Metabolism 24*:1299 (1975).
11. K. Oette, W. Frese, and R. Phlippen, *z. Ges. Exp. Med. 154*:208 (1971).
12. P. J. Nestel and E. A. Couzens, *J. Lipid Res. 7*:487 (1966).
13. F. T. Lindgren and A. V. Nichols, *Ann. N.Y. Acad. Sci. 94*:55 (1961).
14. D. S. Goodman and T. Shiratori, *J. Lipid Res. 5*:30 (1964).
15. R. Wood and R. D. Wiegand, *Lipids 10*:746 (1975).

16. D. E. Wamerstrom and H. M. Jenkin, *Biochim. Biophys. Acta 431*:469 (1976).
17. D. E. Brenneman, R. McGee, V. A. Liepkalnis, and A. A. Spector, *Biochim. Biophys. Acta 388*:301 (1975).
18. D. P. J. Goldsmith and H. P. Jacobi, *Lipids 13*:174 (1978).
19. A. Chait, A. Onitiri, A. Nicole, E. Rabaya, J. Davies, and B. Lewis, *Atherosclerosis 20*:347 (1974).
20. P. J. Nestel, N. Havenstein, T. W. Scott, and L. J. Cook, *Aust. N.Z. J. Med. 4*:497 (1974).
21. J. Peeters, V. Bloton, F. Soetewey, B. Declercq, and D. Vandamme, *Münch. Med. Wochenschr. 115*:1358 (1973).
22. Grünert, A. M. Olbermann, and M. Grünert-Fuchs, *Med. Welt 27*:1304 (1976).
23. L. Hagenfeldt and P. O. Wester, *Acta Med. Scand. 194*:357 (1973).
24. C. Alling, S. J. Dencker, L. Svennerholm, and J. Tichy, *Acta Med. Scand. 185*:99 (1969).
25. J. Jaillard, G. Sezille, P. Scherpereel, and J. C. Fruchert, *Nutr. Metabol. 13*:114 (1971).
26. E. Klenk and W. Kahlke, *Hoppe-Seylers Z. Physiol. Chem. 333*:133 (1963).
27. J. A. Herndon, D. Steinberg, B. W. Uhlendorf, and H. M. Fales, *J. Clin. Invest. 48*:1010 (1969).
28. W. Kahlke, *Klin Wochenschr. 42*:1011 (1964).
29. W. Droese, E. Pape and H. Stolley, *Europ J. Pediat. 122*:57 (1976).
30. F. C. Leyland, A. S. Fosbrooke, J. K. Lloyd, M. M. Segale, J. Tamir, R. Tomkins and O. H. Wolff, *Arch Dis. Child. 44*:170 (1969).
31. D. Sgoutas and F. A. Kummerow, *Am. J. Clin. Nutr. 23*:1111 (1970).
32. R. T. Holman, in *Progress in the Chemistry of Fats and other Lipids*, Vol. 9, Pergamon Press, Oxford, 1968, pp. 279ff.
33. R. N. Fiennes, A. J. Sinclair, and M. A. Crawford, *J. Med. Prim. 2*:155 (1973).
34. J. R. Paulsrud, C. F. Witten, S. E. Steward, and R. T. Holman, *Am. J. Clin. Nutr. 25*:897 (1972).
35. J. D. Wene, W. E. Connar and L. den Besten, *J. Clin. Invest. 56*:127 (1975).
36. A. S. McFarlane, *Nature 149*:439 (1942).
37. A. Scanu, L. A. Lewis, and F. M. Bumpus, *Arch. Biochem. Biophys. 74*:390 (1958).
38. A. Scanu, *J. Lipid Res. 7*:295 (1966).
39. A. Scanu and N. Schiano, *Riv. 1st. Sieroter., Italy 29*:276 (1954).
40. A. Scanu and N. Schiano, *Riv. 1st. Sieroter., Italy 29*:457 (1954).
41. J. Folch, M. Lees, and G. H. Sloane Stanley, *J. Biol. Chem. 226*:497 (1957).
42. B. E. Cham and B. R. Knowles, *J. Lipid Res. 17*:176 (1976).
43. G. J. Nelson, *J. Lipid Res. 3*:71 (1962).
44. G. J. Nelson, *J. Lipid Res. 8*:374 (1967).
45. M. A. Wells and J. C. Dittmer, *Biochemistry 2*:1259 (1963).
46. J. Sigrist, K. Sigrist-Nelson, and C. Gitler, *Biochem. Biophys. Res. Commun. 74*:178 (1977).
47. J. J. Wren and . . Szezepanowska, *J. Chromatogr. 14*:405 (1964).
48. B. Borgström, *Acta Physiol. Scand. 25*:111 (1952).
49. J. Hirsch and E. J. Ahrens, Jr., *J. Biol. Chem. 233*:311 (1958).
50. J. J. Wren, *J. Chromatogr. 4*:173 (1960).
51. G. V. Marinetti, ed., *Lipid Chromatographic Analysis*, Vol. 1, Marcel Dekker, New York, 1967.

52. G. J. Nelson, *J. Amer. Oil Chem. Soc. 44*:86 (1967).
53. J.-P Thenot, E. C. Horning, M. Stafford and M. G. Horning, *Anal. Lett. 5*(4):217 (1972).
54. Brechbühler, H, H. Büchi, E. Hatz, J. Schreiber, and A. Eschenmoser, *Angew. Chem. (Internat. Edit.) 2*:211 (1963).
55. H. Schlenk and J. L. Gellerman, *Anal. Chem. 32*:1412 (1960).
56. B. S. Middleditch and D. M. Desiderio, *Anal. Lett. 5*:605 (1972).
57. D. B. de Oliveira and W. E. Harris, *Anal. Lett. 6*(12):1107 (1973).
58. J. W. Farquhar, W. Insull, Jr., P. Rosen, W. Stoffel, and E. H. Ahrens, Jr., *Nutr. Rev. Suppl. 17*:1 (1959).
59. H. T. Ast, *Anal. Chem. 35*:1539 (1963).
60. W. Stoffel, F. Chu, and E. H. Ahrens, Jr., *Anal. Chem. 31*:307 (1959).
61. A. J. Sheppard and L. A. Ford, *J. Assoc. Offic. Anal. Chem. 46*:947 (1963).
62. G. J. Nelson and N. K. Freeman, *J. Biol. Chem. 235*:578 (1960).
63. T. J. Heath, E. P. Adams, and B. Morris, *Biochem. J. 92*:511 (1964).
64. W. A. Spencer and R. Schaffrin, *Can. J. Biochem. 42*:1659 (1964).
65. M. Kates, *J. Lipid Res. 5*:132 (1964).
66. J. L. Toote, R. J. Allan, and B. W. Agranoff, *J. Lipid Res. 6*:518 (1965).
67. Y. Kishimoto and N. S. Radin, *J. Lipid Res. 6*:435 (1965).
68. J. S. O'Brien, D. L. Fillerup, and J. F. Mead, *J. Lipid Res. 5*:329 (1964).
69. A. T. James, in *Methods of Biochemical Analysis*, Vol. 8 (D. Glick, ed.), Interscience, New York and London, 1960, pp. 18–20.
70. American Oil Chemists Society, *Official and Tentative Methods of the American Oil Chemists Society*, 3rd, ed.; Method Ce 2-66, Champaign, Ill., 1966.
71. American Oil Chemists Society, *Official Methods of Analysis*, 10th ed., Washington, D.C., 1965, Sec. 26.052, p. 429.
72. American Oil Chemists Society, *J. Amer. Oil Chem. Soc. 41*:158 (1964).
73. F. Davidoff and E. D. Korn, *J. Biol. Chem. 238*:3199 (1963).
74. P. Ways, C. F. Reed, and D. J. Hanahan, *J. Clin. Invest. 42*:1248 (1963).
75. D. E. Bowyer, W. M. F. Leat, A. N. Howard, and G. A. Gresham, *Biochim. Biophys. Acta. 70*:423 (1963).
76. A. M. Chalvardjian, *Biochem. J. 90*:518 (1964).
77. G. L. Feldman and G. Ronser, *J. Amer. Oil Chemists Soc. 42*:290 (1965).
78. A. S. Fosbrooke and I. Tamir, *Clin. Chim. Acta 20*:517 (1968).
79. M. Rogozinski, *J. Gas Chromatogr. 2*:136 (1964).
80. M. Rogozinski, *J. Gas Chromatogr. 2*:328 (1964).
81. M. E. Mason and G. R. Waller, *Anal. Chem. 36*:583 (1964).
82. M. E. Mason, M. E. Eager, and G. R. Waller, *Anal. Chem. 36*:587 (1964).
83. M. E. Mason, B. Johnson, and M. C. Hamming, *Anal. Chem. 37*:769 (1965).
84. I. L. Shapiro and D. Kritschevsky, *J. Chromatogr. 18*:599 (1965).
85. H. Bowlers and J. A. Nieuwland, *J. Amer. Chem. Soc. 53*:3835 (1931).
86. J. D. Hinton and J. A. Nieuwland, *J. Amer. Chem. Soc. 54*:2017 (1932).

87. L. D. Metcalfe and A. A. Schmitz, *Anal. Chem. 32*:363 (1961).
88. A. K. Lough, *Biochem. J. 90*:40 (1964).
89. A. K. Lough, *Nature 202*:795 (1964).
90. L. D. Metcalfe, A. A. Schmitz, and J. R. Pelka, *Anal. Chem. 38*: 514 (1966).
91. D. Firestone, *J. Assoc. Off. Anal. Chem. 52*:254 (1969).
92. R.T. O'Conner, R. R. Allen, J. R. Chipault, S. F. Herb, and C. W. Hoerr, *J. Amer. Oil. Chem. Soc. 45*:103 (1968).
93. American Oil Chemists Society, *Official and Tentative Methods of the American Oil Chemists Society*, Vols. 1 and 2, 3rd ed., Champaign, Ill., 1964.
94. W. R. Morrison and L. M. Smith, *J. Lipid Res. 5*:600 (1964).
95. J. E. Storry and J. A. F. Rook, *Biochem. J. 98*:210 (1965).
96. R. J. Dummel and E. Kun, *J. Chromatogr. 54*:130 (1971).
97. J. Bezard and M. Bugant, *J. Chromatogr. Sci. 7*:639 (1969).
98. W. A. Aue, C. R. Hastings, K. O. Gerhardt, J. O. Pierce, H. H. Hill, and R. F. Moseman, *J. Chromatogr. 72*:259 (1972).
99. K. D. Gerhardt and W. A. Aue, *J. Chromatogr. 82*:382 (1973).
100. G. E. McCone and R. J. Hance, *J. Chromatogr. 69*:204 (1972).
101. D. L. Corina and P. M. Dunstan, *Anal. Biochem. 53*:571 (1973).
102. H. P. Klemm, U. Hintze, and G. Gercken, *J. Chromatogr. 75*:19 (1973).
103. J. R. Watson and P. Crescuolo, *J. Chromatogr. 52*:63 (1970).
104. N. Hirvnák, L. Soják, B. Beska and J. Janak, *J. Chromatogr. 68:* 55 (1972).
105. M. Verzele, M. Verstappe, P. Sandra, E. Van Luchene, and A. Vuge, *J. Chromarogr. Sci. 10*:688 (1972).
106. G. C. Cochrane, *J. Chromatogr. Sci. 13*:440 (1975).
107. A. Kuksis, O. Stachnyk, and B. J. Holub, *J. Lipid Res. 10*:660 (1969).
108. J. Jaeggi, personal communication.
109. K. Grob and K. Grob, Jr., *J. Chromatogr. 151*:311 (1978).
109a. F. W. Karasek, *Anal. Chem. 43*:1982 (1971).
110. F. K. Kawahara, *Anal. Chem. 40*:2073 (1968).
111. A. E. Pierce, *Silylation of Organic Compounds*, Pierce Chemical Co., Rockford, Ill., 1968.
112. W. H. Tallent and R. Kleiman, *J. Lipid Res. 9*:146 (1968).
113. A. Kuksis, L. Marai, and D. A. Gornall, *J. Lipid Res. 8*:352 (1967).
114. C. E. Dalgliesh, C. E. Horning, M. G. Horning, K. L. Knox, and K. Yarger, *Biochem. J. 101*:792 (1966).
115. L. Birkofer and M. Donike, *J. Chromatogr. 26*:270 (1967).
116. D. T. Downing and R. S. Greene, *Anal. Chem. 40*:827 (1968).
117. V. Koman, *J. Chromatogr. 45*:311 (1969).
118. P. Pohl, H. Glasl, and H. Wagner, *J. Chromatogr. 42*:75 (1969).
119. K. Fontell, R. T. Holman, and G. Lambertsen, *J. Lipid Res. 1*:391 (1960).
120. A. Kuksis, *Fette, Seifen, Anstrichm. 73*(2):130 (1971).
121. A. J. Sheppard and J. L. Iverson, *J. Chromatogr. Sci. 13*:448 (1975).
122. J. Drozd, *J. Chromatogr. 113*:303 (1975).
123. S. Ahuja, *J. Pharm. Sci. 65*:163 (1976).
124. R. G. Ackman, in *Methods in Enzymology*, Vol. 14 (J. M. Lowenstein, ed.), Academic Press, New York and London, 1969, pp. 354–357.

125. B. M. Craig and N. L. Murty, *J. Amer. Oil Chem. Soc. 36*:549 (1959).
126. F. E. Luddy, R. A. Barford, and F. W. Riemenschneider, *J. Amer. Oil Chem. Soc. 37*:447 (1960).
127. D. M. roper and T. S. Ma, *Microchem. J. 1*:245 (1957).
128. I. Hornstein, J. A. Alford, L. E. Elliot, and P. F. Crowe, *Anal. Chem. 32*:540 (1960).
128a. G. R. Jamieson and E. H. Reid, *J. Chromatogr. 17*:230 (1965).
129. S. F. Hert, P. Magidman, F. E. Luddy, and R. W. Tiemenschneider, *J. Amer. Oil. Chem. Soc. 39*:142 (1962).
130. S. Anselmi, L. Boniforti, and R. Monacelli, *Boll. Lab. Chim. Prov. 2*:257 (1960).
131. H. P. Burchfield and E. E. Storrs, *Biochemical Applications of Gas Chromatography*, Academic Press, New York, 1962, p. 528.
132. K. Paeck and M. W. Tracey, eds., *Modern Methods of Plant Analysis*, Vol. 2, Springer, Berlin, 1955, p. 626.
133. J. M. de Man, *Lab. Pract. 16*:150 (1967).
134. J. M. de Man, *J. Dairy Sci. 47*:546 (1964).
135. I. Halasz and W. Schneider, *Anal. Chem. 33*:978 (1961).
136. G. Schomburg, H. Husmann, and F. Weecke, *J. Chromatogr. 99*:63 (1974).
137. K. Grob and G. Grob, *Chromatographia 5*:1 (1972).
138. U. Grob and K. Grob, Jr., *J. Chromatogr. 94*:53 (1974).
139. G.Schomburg, R. Dickmann, H. Husmann, and F. Weeke, *J. Chromatogr. 122*:55 (1976).
140. H. Jaeger, H. U. Klör, and H. Ditschuneit, *J. Lipid Res. 17*:185 (1976).
141. C. H. Orr and J. E. Callen, *J. Amer. Chem. Soc. 80*:249 (1958).
142. B. M. Craig and N. L. Murty, *J. Amer. Oil Chem. Soc. 36*:549 (1959).
143. S. R. Lipsky, R. A. Landowne, and M. R. Godet, *Biochim. Biophys. Acta 31*:336 (1959).
144. I. Hornstein, L. E. Elliott, and P. F. Crowe, *Nature 184*:1710 (1959).
145. C. J. F. Böttcher, F. P. Woodford, E. Boelsma-van Houte, and E. M. van Glut, *Rec. Trav. Chim. Pays-Bas. 78*:794 (1959).
146. B. M. Craig, *Acta Chem. Scand. 14*:1442 (1960).
146a. J. K. Haken, *J. Chromatogr. Sci. 13*:430 (1965).
147. G. Schomburg and H. Husmann, *Chromatographia 8*:517 (1975).
148. G. Alexander and G. A. F. M. Rutten, *J. Chromatogr. 99*:81 (1974).
149. D. H. Desty, I. M. Haresmape, and B. H. F. Whyman, *Anal. Chem. 32*:302 (1960).
150. M. Novotny and Tesarik, *Chromatographia 1*:332 (1968).
151. M. Novotny and U. D. Bartle, *Chromatographia 7*:122 (1974).
152. M. H. J. van Rijswick and K. Tesarik, *Chromatographia 7*:135 (1974).
153. J. Szafranek, C. D. Pfaffenberger, and E. C. Horning, *J. Chromatogr. 88*:149 (1974).
154. K. Grob, *Helv. Chim. Acta. 48*:1362 (1965).
155. K. Grob, *Helv. Chim. Acta. 51*:718 (1968).
156. K. Grob and H. J. Jaeggi, *Chromatographia 5*:382 (1972).
157. K. Grob and G. Grob, *J. Chromatogr. 125*:471 (1976).
158. K. Grob, G. Grob, and K. Grob, Jr., *Chromatographia 10*:181 (1977).

159. G. J. Nelson, *Lipids 9*:254 (1974).
160. R. G. Ackman and S. N. Hooper, *J. Chromatogr. Sci. 12*:131 (1974).
161. R. V. Golovnya, V. P. Kralets, and T. E. Kuzmenko, *J. Chromatogr. 121*:118 (1976).
162. A. F. Prevot and R. X. Mordret, *Rev. Franc. Corps Gras 23*:409 (1976).
163. P. Van Hout, J. Szafranek, C. D. Pfaffenberger, and E. C. Horning, *J. Chromatogr. 99*:103 (1974).
164. G. Goretti and A. Liberti, *J. Chromatogr. 61*:334 (1971).
165. E. S. van Vleet and J. G. Quinn, *J. Chromatogr. 151*:396 (1978).
166. *Gas-Chromatogr. Newsletter* (Applied Science Lab., State College, Pa.), 19 (1978).
167. K. Grob, Jr., H. P. Neukom, D. Fröhlich, and R. Battaglia, *HRC & CC 1*:94 (1978).
168. H. Jaeger, H. U. Klör, G. Blos, and H. Ditschuneit, *Chromatographia 8*:507 (1975).
169. B. Brechbühler, L. Gay, and H. Jaeger, *Chromatographia 10*: (1977).
170. A. Karmen, in *Methods in Enzymology*, Vol. 14 (J. M. Lowenstein, ed.), Academic Press, New York, 1969, p. 465.
171. L. Swell, *Proc. Soc. Exp. Biol. Med. 121*:1290 (1966).
172. W. C. Breckenridge and A. Kuksis, *Lipids 5*:342 (1970).
173. B. Kolb and E. Wiedeking, *Z. Anal. Chem. 243*:129 (1968).
174. P. Capella and C. M. Zorzut, *Anal. Chem. 40*:1458 (1968).
175. G. Eglinton and D. H. Hunnemann, *Phytochemistry 7*:313 (1968).
176. D. J. Argoudelis and E. G. Perkins, *Lipids 3*:379 (1968).
177. E. G. Perkins and D. J. Argoudelis, *Lipids 4*:619 (1969).
178. G. Schomburg, F. Weeke, B. Weimann, and E. Ziegler, *Angew Chemi(internat. Edit.) 11*:366 (1972).
179. E. C. Horning, E. H. Ahrens, Jr., S. R. Lipsky, F. H. Mattson, J. F. Mead, D. A. Turner, and W. H. Goldwater, *J. Lipid Res. 5*:20 (1964).
180. F. H. De la Court, N. J. P. Van Cassel and J. A. M. van der Falk, *J. Chromatogr. 72*:249 (1972).
181. M. Novotny, R. Segura, and A. Zlatkis, *Anal. Chem. 44*:9 (1972).
182. M. Verzele, K. Van Cauwenberghe, and J. Bouche, *J. Gas Chromatogr. 5*:114 (1967).
183. R. E. Kaiser, *Chromatographie in der Gasphase*, Bibliographisches Institut, Mannheim, 1961, p. 20.
184. L. S. Ettre and E. W. March, *J. Chromatogr. 91*:5 (1974).
185. L. S. Ettre, *Kapillarsäuren in der Gaschromatographie: Eine Einführung, Vieweg, Braunschweig, 1976.*
186. W. Averill and E. W. March, *Chromatography Newsletter 4*:20 (1976).
187. R. G. Ackman and S. N. Hooper, *J. Chromatogr. Sci. 7*:549 (1969).
188. R. G. Ackman and J. D. Castell, *J. Gas Chromatogr. 5*:489 (1967).
189. G. E. Pollock, *Anal. Chem. 44*:634 (1972).
190. H. Egge, U. Murawski, R. Ryhage, R. György, W. Chatronanon, and F. Zilliken, *Chem. Phys. Lipids 8*:42 (1972).
191. J. Krupcik, J. Hrivnák, L. Barnoky, and J. Janak, *J. Chromatogr. 65*:323 (1972).
192. A. Prevot, in *Chromatography and Methods of Immediate Separation*, (G. Parissatiis, ed.), Union of Greek Chemists, Athens, 1966, p. 141.

193. J. Hrivnák, L. Soják, J. Krupcik, and Y. P. Duchesne, *J. Amer Oil Chem. Soc. 49*:68 (1972).
194. J. Hrivnák, V. Palo, and J. Krupcik, *Milchwissenschaft 28*:11 (1973).
195. S.-N. Lin, C. D. Pfaffenberger, and E. C. Horning, *J. Chromatogr. 104*:319 (1975).
196. J. Merle d'Aubigne, C. Landault, G. Guichon, *Chromatographia 4*:309 (1971).
197. A. L. German and E. C. Horning, *J. Chromatogr. Sci. 11*:76 (1973).
198. S.-N Lin and E. C. Horning, *J. Chromatogr. 112*:483 (1975).
199. J. Flanzy, M. Boudon, C. Leger, and J. Pihet, *J. Chromatogr. Sci. 14*:17 (1976).
200. P. Sandra, M. Verstappe, and M. Verzele, *J. HRC & CC 1*:28 (1978).
201. J. D. Schieke, N. R. Comins, and V. Pretorius, *J. Chromatogr. 112*:97 (1975).
202. J. D. Schieke, N. R. Comins, and V. Pretorius, *Chromatographia 8*:354 (1975).
203. P. Sandra and M. Verzele, *Chromatographia 10*:419 (1977).
204. M. Verzele and P. Sandra, *J. Chromatogr. 158*:111 (1978).
205. H. Heckers, K. Dittmar, F. W. Melcher, and H. O. Kalinowski, *J. Chromatogr. 135*:93 (1977).
206. H. Heckers, F. W. Melcher and U. Schloeder, *J. Chromatogr. 136*:311 (1977).
207. J. Heckers, M. Korner, T. W. L. Tuschen, and F. W. Melcher, *Atherosclerosis, 28*:389 (1977).
208. H. Heckers and F. W. Melcher, *Am. J. Clin. Nutr. 31*:1041 (1978).

11

Analysis of Lower Fatty Acids, Phenols, and Indoles

Yasuyuki Hoshika* / *Aichi Environmental Research Center, Aichi, Japan*

I. INTRODUCTION

Simultaneous analysis of lower fatty acids, phenols, and indoles is a commonly encountered problem in organic analysis of human body fluids, foods, cigarette smoke, drugs, and environmental pollution, because of the toxic effects including skin and eye irritation [1]. The acids are also present in breath and serum of some patients with hepatic coma [2], in the breath of some patients with cirrhosis of the liver [3], and are known as intermediates and end products in bacterial fermentation and as metabolites from alcohols via aldehydes [4–7]. It is well known that the phenols are metabolites of alkylbenzenes

*Present address: Department of Hygiene, Shinshu University School of Medicine, Nagano, Japan

[8–10]. It has been reported that the free and conjugated phenols are present in blood, organs, urine, saliva, sweat, and feces [11]. Indoles are of interest as metabolites of tryptophan in cerebrospinal fluid, brain, plasma, and urine [12], and are present also in ovine plasma and rumen fluid [13].

The co-occurrence of the lower fatty acids, phenols, and indoles has been determined in cigarette smoke and its condensate [14–17], and in accumulated liquid poultry manure [18] and liquid swine manure [19]. Furthermore, these compounds have low odor-threshold values at levels below 1 ppb (10^{-9}) in air. Therefore the simultaneous analysis of these compounds is of considerable interest. The constituents of feces depend on gastrointestinal motility, nature of diet, and degree of digestion and fermentation of foods. Identification of steroids, organic acids (such as acetic, lactic, and butyric), hydrogen sulfide, carbon dioxide, methane, hydrogen, ammonia, phenols, mercaptans, and indoles in fecal matter has been reported [20,21].

Direct gas chromatographic (GC) analysis of these compounds in the free form at low concentrations has been limited by adsorption and decomposition in the column, ghosting phenomena, the tailing of eluting peaks, and the difficulty of separating many isomeric compounds. In general, when packed columns are employed, the simultaneous GC separation of the mixture of lower fatty acids, phenols, and indoles is difficult because the fatty acids exhibit relatively low boiling points, whereas the phenols and indoles have higher boiling points and include many isomers.

Recently, Hrivňák and co-worker [22,23], and Barber et al. [24] reported the separation of lower fatty acids and phenols by using open-tubular columns containing tricresylphosphates; capillary columns containing Emulphor O [15, 16], Squalane [15], Ucon [25,26], SF-96 [27], and SE-52 [28] have also been used for the analysis of lower fatty acids, phenols, and indoles. Two reviews on recent developments in methodology and application of the GC to the analyses of free fatty acids in the underivatized form have been reported [29,30]. Various types of liquid phase have been proposed for the GC of phenolic isomers [31].

The analysis of organic components in human breath and urine with chromatographic and mass spectrometric techniques was reviewed by Pierce and Gearhart [32]. Jellum [33] reviewed the profiling of human body fluid in healthy and diseased states using GC and mass spectrometry. The application of capillary columns to biological and biomedical samples is of interest because of their high resolving power which makes them suitable for the profile analysis of the trace components in body fluid samples.

This author has previously reported the simultaneous GC analysis of low concentrations of the lower fatty acids, phenols, and indoles by using glass capillary columns (high-resolution columns containing free fatty acid phase (FFAP) and PEG 20 M [34]. Few other references exist for this technique. This paper describes the use of three kinds of flexible fused-silica capillary columns (50 m × 0.2 mm I.D.) of high resolution to achieve the simultaneous GC analysis of the low concentrations of the lower fatty acids, phenols, and indoles. The method has been applied to the analysis of these substances in a fresh normal urine of a Japanese male and in feces of a Persian cat. The peaks of the typical phenols were also identified as their bromophenols which were prepared by the reaction with bromine [35].

II. EXPERIMENTAL

A. Reagents

Fourteen lower fatty acids, 11 phenols, 7 indoles, ethanol, diethyl ether, and bromine were obtained from Poly Science (Niles Illinois, USA), Aldrich (Milwaukee, Wisconsin, USA), Katayama Chemical Industries, Ltd. (Osaka, Japan), Tokyo Kasei Kogyo, Ltd. (Tokyo, Japan), Wako Pure Chemical Industries, Ltd. (Osaka, Japan). All reagents used were guaranteed or reagent grade chemicals without further purification and were proven sufficiently pure for GC analysis.

B. Apparatus

The gas chromatograph used was a Shimadzu Model GC5AP$_5$ (dual-column system) equipped with on-column injection, flame ionization detector (FID), and a digital integrator (Shimadzu Model ITG-2A) for the determination of retention time and quantitative analysis. Three flexible fused-silica capillary columns (WCOT type, 50 m × 0.2 mm I.D., made by Hewlett-Packard Avondale Division, Pennsylvania, USA) containing VFA (Volatile Fatty Acid, Carbowax 20 M non-deactivated fused silica), Carbowax 20 M (Carbowax 20 M deactivated), and Methyl Silicone (SP-2100, Carbowax 20 M deactivated) were obtained from Yokogawa-Hewlett-Packard, Ltd., Tokyo.

C. Chromatographic Conditions

The three capillary columns were preconditioned at 200°C for 1 hr for VFA and Carbowax 20 M and at 280°C for 1 hr for Methyl Silicone with a constant flow of nitrogen (about 1 ml/min) before being connected to the FID. The chromatographic conditions were as follows:

1. For stationary phases VFA and Carbowax 20 M; column temperature, held for 1 min at 100°C; column oven, heated at 4°C/min for 100–220°C, maintained at 220°C for 9–30 min, and then cooled to 100°C; injection port and detector temperatures, 250°C; carrier gas (nitrogen) flow-rates, 0.85 ml/min for VFA and 0.97 ml/min for Carbowax 20 M; purge gas (nitrogen) flow-rate, 40 ml/min; splitting ratios, 1:10 for VFA and 1:7 for Carbowax 20 M; air and hydrogen flow-rates for the FID, 1.0 liter/min and 50 ml/min, respectively.

2. For stationary phase Methyl Silicone: column temperature, held for 1 min at 100°C; column oven-heated at 4°C/min from 100°C to 255 or 260°C, and then cooled to 100°C; injection port and detector temperatures, 250°C; carrier gas (nitrogen) flow-rate, 0.94 ml/min; purge gas (nitrogen) flow-rate, 40 ml/min; splitting ratio; 1:9; air and hydrogen flow-rates for the FID, 1.0 liter/min and 50 ml/min, respectively.

D. Sample Preparation Methods

1. Acidic diethyl ether extraction: 20 ml of conc. sulfuric acid was added to a flask containing 130 ml of fresh normal urine from a Japanese male (35 years old). Then the mixture was transferred into a 300-ml flask fitted with a condenser and cooled with a continuous flow of water (18°C). The flask was

heated in boiling water for 1 hr for deproteination, then cooled to room temperature, and the contents were filtered. The filtrate was transferred into a 200-ml separate funnel; then 40 ml (two times 20 ml) of diethyl ether was added. After shaking vigorously for 1 min, the diethyl ether solution layer was separated. The extract was concentrated to 8 ml by bubbling through nitrogen gas for 40 min at a constant flow-rate of 200 ml/min.

2. Preparation of bromophenols: A 20-ml aliquot of 1% (w/v) aqueous bromine solution was added to a 1-ml aliquot of the concentrated diethyl ether solution, and an additional 4 ml of pure diethyl ether was added; then the mixture was kept at room temperature for 4 hr. After the bromination reaction, the organic layer was separated, and concentrated to 1 ml by bubbling through nitrogen for 10 min at a constant flow-rate of 200 ml/min. For the bromophenol derivative formation, the eight phenols (i.e., phenol, *o*-, *m*-, and *p*-cresols and 2,3-, 2,5-, 3.4-, and 3,5-xylenols) with concentrations of 10^{-8}mol/ml of ethanol and mol ratios (bromine/phenols) greater than 400 were allowed to stand at room temperature for 4 hr; however, 2,4-, 2,6-xylenols, *o*- and *p*-ethylphenols did not give the corresponding bromophenols in the reaction period [35].

3. Persian cat feces sample: The feces (2 g) and ethanol (100 ml) were shaken vigorously and the solution obtained was then filtered. 25 ml of the filtrate obtained was concentrated to 2 ml by bubbling with nitrogen constant carrier gas 200 ml/min at 25°C. 5 μl of the concentrated sample was injected with a 10 μl-Hamilton microsyringe (Reno, Nevada 701-N) into the GC.

The standard reagents used in this investigation were dissolved in water, ethanol, and diethyl ether to give about 0.25×10^{-4} mol/10 ml solvents. The volume injected of the standard solution was usually 1–5 μl and that of urine samples was usually 1–6 μl.

III. RESULTS AND DISCUSSION

A. Typical Gas Chromatograms

Typical gas chromatograms of a mixture of 14 lower fatty acids, 11 phenols, and 7 indoles obtained by use of the VFA and Carbowax 20 M flexible fused-silica capillary columns are shown in Figures 11.1 and 11.2. Figure 11.1 shows a typical gas chromatogram of a mixture of 32 compounds obtained using the VFA capillary column. The time required for analysis was about 45 min. The separation of the 14 lower fatty acids was complete with no tailing; however, that of the 11 phenols was incomplete, especially for pairs of compounds such as phenol and *o*-cresol, and *o*-ethylphenol and 3,5-xylenol, whose peaks overlapped. The separation of the 7 indoles was complete except for 2- and 3-methylindoles whose peaks were poorly separated. However, the separation of the peaks of *p*-ethylphenol, 2,5-xylenol, *p*- and *m*-cresols was more favorable than that obtained with the FFAP G-SCOT glass capillary column (20 m × 0.28 mm I.D.) described in a previous report [34], while the separation of the peaks of 2-methylindole and *n*-lauric acid was poorer than that obtained with the FFAP G-SCOT glass capillary column.

Figure 11.2 shows a typical gas chromatogram of a mixture of the 32 compounds obtained using the Carbowax 20 M capillary column. The analysis time of the chromatogram was about 40 min. The separation of the 14 lower fatty acids was complete with no tailing; however, that of the 11 phenols was incomplete, especially for pairs or groups of compounds such as phenol and *o*-cresol; and *o*-ethylphenol and 3,5-xylenol and *n*-pelargonic acid; and

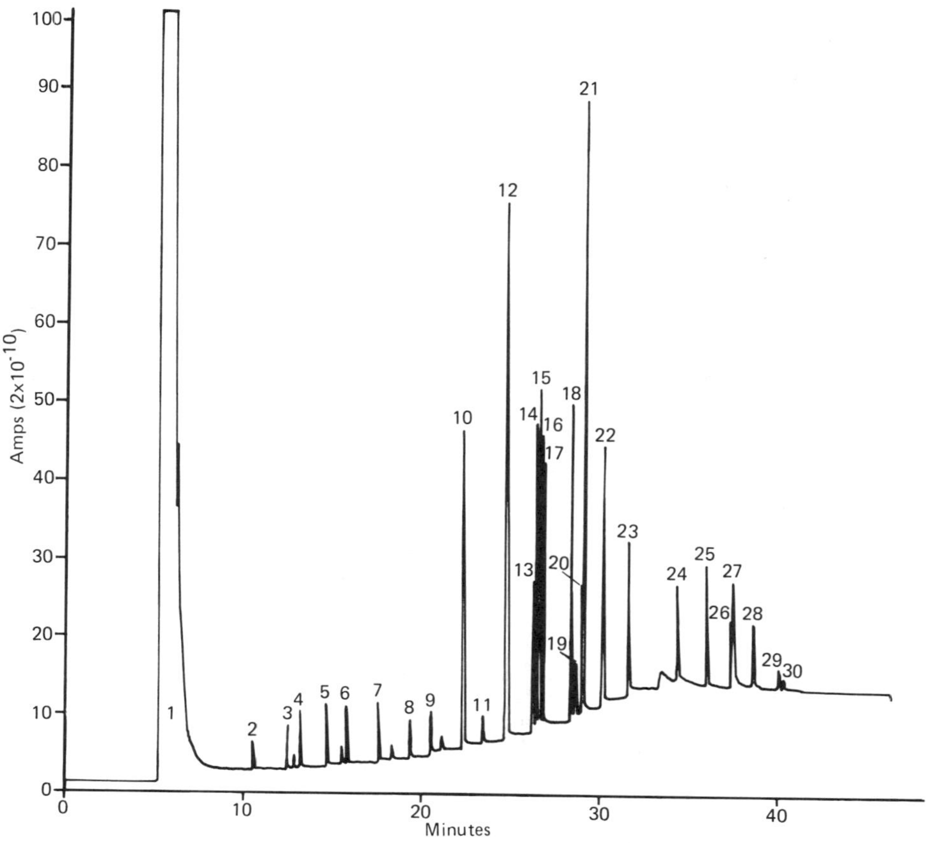

Figure 11.1 Typical gas chromatogram of a standard mixture of 14 lower fatty acids, 11 phenols, and 7 indoles. Flexible fused-silica capillary column (containing VFA, 50 m × 0.2 mm I.D.); FID sensitivity, 2×10^2. Peaks: 1, solvent (water-ethanol); 2, acetic acid (0.1 μg); 3, propionic acid (0.2 μg); 4, isobutyric acid (0.2 μg); 5, *n*-butyric acid (0.2 μg); 6, isovaleric acid (0.2 μg); 7, *n*-valeric acid (0.2 μg); 8, isocaproic acid (0.3 μg); 9, *n*-caproic acid (0.3 μg); 10, 2,6-xylenol (0.6 μg); 11, *n*-heptanoic acid (0.3 μg); 12, phenol (0.6 μg) + *o*-cresol (0.6 μg); 13, *n*-caprylic acid (0.6 μg); 14, *p*-ethylphenol (0.6μg); 15, 2,5-xylenol (0.6 μg); 16, *p*-cresol (0.6 μg); 17, *m*-cresol (0.6 μg); 18, 2,3-xylenol (0.6 μg); 19, 1,2-dimethylindole (0.2 μg); 20, *n*-pelargonic acid (0.5 μg); 21, *o*-ethylphenol (0.6 μg) + 3,5-xylenol (0.6 μg); 22, 3,4-xylenol (0.6 μg); 23, *n*-capric acid (0.5 μg); 24, *n*-undecanoic acid (0.6 μg); 25, indole (0.3 μg); 26, 3-methylindole (0.3 μg); 27, 2-methylindole (0.2 μg) + *n*-lauric acid (0.6 μg); 28, 5-methylindole (0.3 μg); 29, 2,3-dimethylindole (0.2 μg); 30, 2,5-dimethylindole (0.2 μg).

2,3-xylenol and 1,2-dimethylindole, whose peaks overlapped. The separation of the 7 indoles was complete, except for 2- and 3-methylindole, whose peaks were poorly separated. However, the separation of the peaks of *p*-ethylphenol, 2,5-xylenol, and *p*- and *m*-cresols was more favorable than that obtained with the PEG-20 M G-SCOT glass capillary column (20 m × 0.28 mm I.D.) described in a previous report [34]. The separation of the peaks of

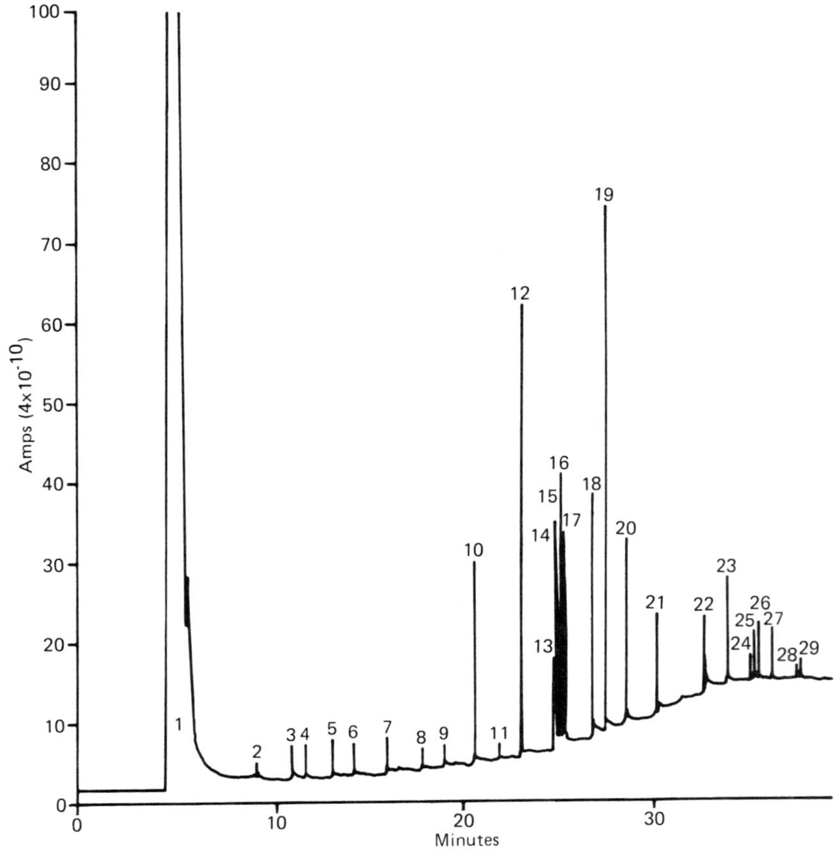

Figure 11.2 Typical gas chromatogram of a standard mixture of 14 lower fatty acids, 11 phenols, and 7 indoles. Flexible fused-silica capillary column (containing Carbowax 20 M, 50 m × 0.2 mm I.D.); FID sensitivity, 4×10^2. Peaks: 1 through 17, as in Figure 11.1; 18, 2,3-xylenol + 1,2-dimethylindole; 19, *n*-pelargonic acid + *o*-ethylphenol + 3,5-xylenol; 20, 3,4-xylenol; 21, *n*-capric acid; 22, *n*-undecanoic acid; 23, indole; 24, 3-methylindole; 25, 2-methylindole, 26, *n*-lauric acid; 27, 5-methylindole; 28, 2,3-dimethylindole; 29, 2,5-dimethylindole. The amounts of each peak are the same as in Figure 11.1.

2,3-xylenol and 1,2-dimethylindole, and of *o*-ethylphenol and 3,5-xylenol and *n*-pelargonic acid was poorer than that obtained with the PEG-20 M G-SCOT glass capillary column.

Figure 11.3 shows the typical gas chromatograms of a mixture of the 11 phenols in the free form (A) and 8 bromophenols (B) obtained using the Methyl Silicone capillary column. The analysis times of the chromatograms were about 15 min for the 11 phenols in the free form and about 35 min for the 8 bromophenols. The separation of these free phenols was complete, except for pairs of the compounds such as 2,3- and 3,5-xylenols and *p*-ethylphenol, whose peaks overlapped. However, the separation of phenol and *o*-,*m*- and *p*-cresols was completed within 11 min. The separation of the 8 bromophenol was complete, that is, the bromophenols derived from phenol, *o*-, *m*-, and *p*-cresols were also sufficiently separated. However, the peak of the tribromo-3,5-xylenol was obtained as double peak. The elution order of the peaks of the 8 bromophenols in the present GC conditions was as follows: dibromo-*o*-cresol, dibromo-*p*-cresol, tribromo-2,5-xylenol, bromo-2,3-xylenol, tribromo-3,4-xylenol, tribromophenol, tribromo-*m*-cresol, and tribromo-3,5-xylenol. This order was compared with that of the column of (DEGS + H_3PO_4) [35]. The elution order obtained with the (DEGS + H_3PO_4) column was as follows: dibromo-*o*-cresol, tribromo-2,5-xylenol,

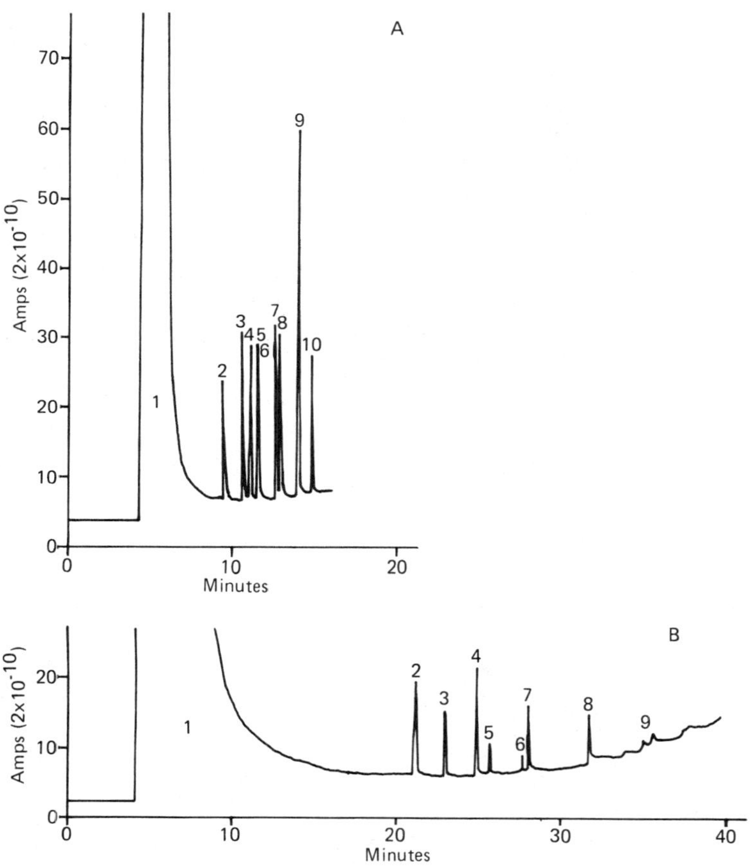

Figure 11.3 Typical gas chromatograms of a standard mixture of 11 phenols (A) and 8 bromophenols (B). Flexible fused-silica capillary column (containing Methyl Silicone, SP-2100, 50 m × 0.2 mm I.D.); FID sensitivity, 2×10^2. Chromatogram (A) peaks: 1, solvent (ethanol); 2, phenol (0.5 μg); 3, *o*-cresol (0.5 μg); 4, *m*-cresol (0.5 μg); 5, *p*-cresol (0.5 μg); 6, 2,6-xylenol (0.5 μg); 7, *o*-ethylphenol (0.5 μg); 8, 2,5-xylenol (0.5 μg); 9, 2,3-xylenol (0.5 μg) + 3,5-xylenol (0.5 μg) + *p*-ethylphenol (0.5 μg); 10, 3,4-xylenol (0.5 μg). Chromatogram (B) peaks: 1, solvent (ethanol); 2, dibromo-*o*-cresol (1.1 μg); 3, dibromo-*p*-cresol (1.2 μg); 4, tribromo-2,5-xylenol (0.6 μg); 5, bromo-2,3-xylenol (0.7 μg); 6, tribromo-3,4-xylenol (0.7 μg); 7, tribromophenol (1.0 μg); 8, tribromo-*m*-cresol (0.5 μg); 9, tribromo-3,5-xylenol (0.5 μg).

bromo-2,3-xylenol, dibromo-*p*-cresol, tribromo-3,4-xylenol, tribromophenol, tribromo-*m*-cresol and tribromo-3,5-xylenol. As has been previously reported [35], the bromo derivatives of phenols has strong electron-capturing properties. The sensitivity of the electron capture detector (ECD) to the bromophenols was about 100 times greater than that of the FID to the free phenols.

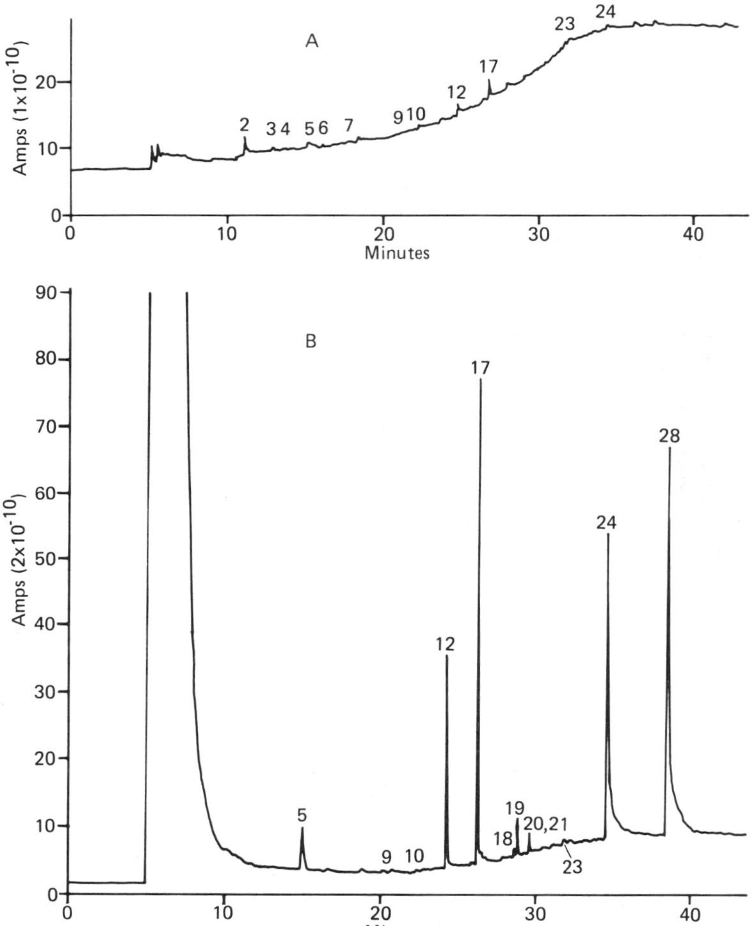

Figure 11.4 Typical gas chromatograms of a fresh normal urine sample of a Japanese male (35 years old) (A) and an acidic diethyl ether extraction of the urine (B). Flexible fused-silica capillary column (containing VFA, 50 m × 0.2 mm I.D.). Chromatogram (A): FID sensitivity, 1×10^2; the volume injected of the urine sample was 2.0 μl. Peaks: 2, acetic acid (about 0.05 μg/2.0 μl of the urine). The other peak numbers are the same as in Figure 11.1 (A). Chromatogram (B): FID sensitivity, 2×10^2; the volume injected of the concentrate of the acidic diethyl ether extraction of the urine sample was 5.0 μl. The peak numbers are the same as in Figure 11.1(B).

B. Urine Samples

Figure 11.4 shows a typical gas chromatogram of the lower fatty acids, phenols, and indoles in the fresh normal urine of a Japanese male (35 years old) using the VFA capillary column. As shown in chromatogram (A) of Figure 11.4, at least 16 peaks were recognized, and 19 compounds were identified as 12 lower fatty acids, 7 phenols, and 3 indoles, that is, acetic, propionic, *iso*- and *n*-butyric, *iso*- and *n*-valeric, *n*-caproic, *n*-heptanoic, *n*-pelargonic, *n*-capric, *n*-undecanoic, and *n*-lauric acids; and phenol, *o*-, and *p*-cresols, 2,5-, 2,6-, and 3,5-xylenols, *o*-ethylphenol; and indole, and 2- and 3-methylindoles. The identification of several peaks in the urine sample was carried out by the standard addition method. The quantitative results for the acetic acid expressed in microgram per injected sample volume was about 0.05 μg/2.0 μl of urine. The variation of the retention times resulting from the use of a different sample solvent in the Carbowax 20 M capillary column was larger than that of the VFA capillary column.

Chromatogram (B) of Figure 11.4 shows a typical gas chromatogram of the concentrate of the acidic diethyl ether extraction of the urine. The peaks of *n*- butyric, *n*-caproic, and *n*-capric acids were identified; however, the peaks of the acetic and propionic acids did not appear. It is possible that these compounds were evaporated during the sample concentrating procedure.

The recovery test of the 32 compounds using the overall pretreatment with the acidic diethyl ether extraction and concentration was measured by using 0.1 ml of the standard solutions and 1 ml of the urine and heating for 1 hr in boiling water for deproteination, then adding 1 ml of diethyl ether. The detected concentration values were corrected with percentage recovery during the concentrating procedure of these compounds in the sample, as follows. The results of this recovery test show that the lower fatty acids of C_2–C_5 were 0–40%, and of C_6–C_{12} were 40–70%; that of the 11 phenols ranged 40–100%, and that of the 7 indoles ranged 75–100%. In conclusion, the pretreatment of the acidic diethyl ether extraction and the concentration of the lower fatty acids and some of phenols in the urine sample were insufficient.

Figure 11.5 shows the typical gas chromatograms of the concentrate of the acidic diethyl ether extraction of the fresh normal urine of a Japanese male (35 years old) (A), and after bromination of the concentrate (B), using the Methyl Silicone capillary column. As shown in chromatogram (A) of the Figure 11.5, phenol, *p*-cresol, 2,3-, 2,6-, and 3,5-xylenols, and *p*-ethylphenol were identified. From chromatogram (B) of Figure 11.5, the presence of the phenols and of the *o*-, *m*-, and *p*-cresols was confirmed as bromophenols. Chromatogram (C) shows the blank of the bromination using diethyl ether solvent.

Figure 11.6 shows the typical gas chromatograms of the lower fatty acids, phenols, and indoles in the fresh normal urine sample of a nonsmoking Japanese male (35 years old). Chromatogram (A) of Figure 11.6 represents a pattern of urine 0.5 hr before smoking; chromatogram (B) represents a pattern of urine 0.5 hr after smoking; chromatogram (C) represents a pattern of urine 1 hr after smoking; chromatogram (D) represents a pattern of urine 2 hr after smoking. The cigarette used for the smoking was a Japanese filterless cigarette. As shown in Figure 11.6, when the pH values were decreased from 7.1 to 6.0, all of the peaks 1 through 12 increased; however, 2 hr after smoking this behavior disappeared, especially, for the peaks of acetic, propionic, *n*-butyric, and isovaleric acids which had increased to about twice that of the peaks 0.5 hr before smoking.

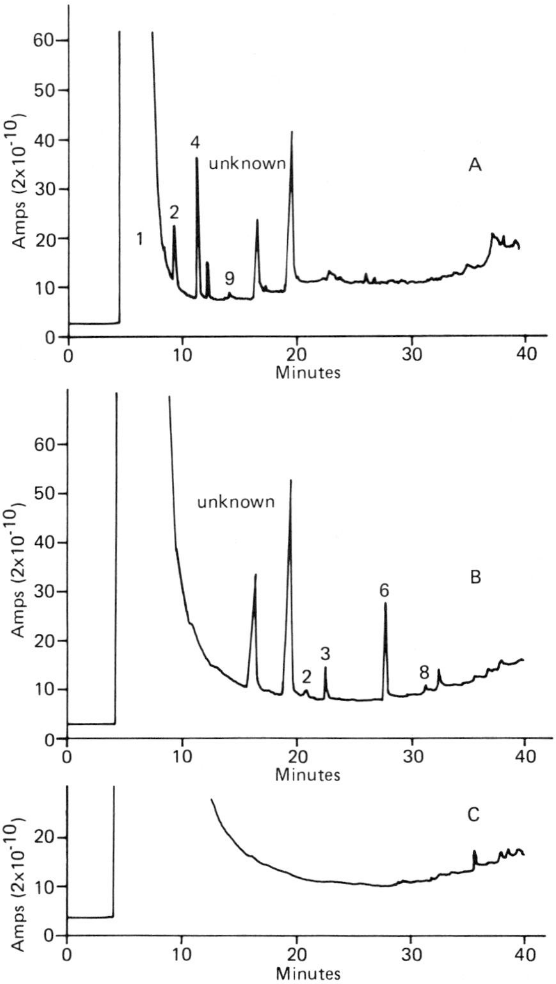

Figure 11.5 Typical gas chromatograms of the concentrate of the acidic diethyl ether extraction of the fresh normal urine of the Japanese male (35 years old) (A), and the subsequent bromination reaction of the concentrate (B), and the blank of the bromination reaction (C). Flexible fused-silica capillary column (containing Methyl Silicone, SP-2100, 50 m × 0.2 mm I.D.); FID sensitivity, 2×10^2. Chromatogram (A): The volume of the diethyl ether concentrate solution injected was 5.6 μl. The peak numbers are the same as in Figure 11.1(A). Chromatogram (B): The volume of the brominated diethyl ether concentrate injected was 5.6 μl. The peak numbers are the same as in Figure 11.1(B). Chromatogram (C): The volume of the brominated diethyl ether injected was 5.6 μl. The peak numbers are the same as in Figure 11.1(C).

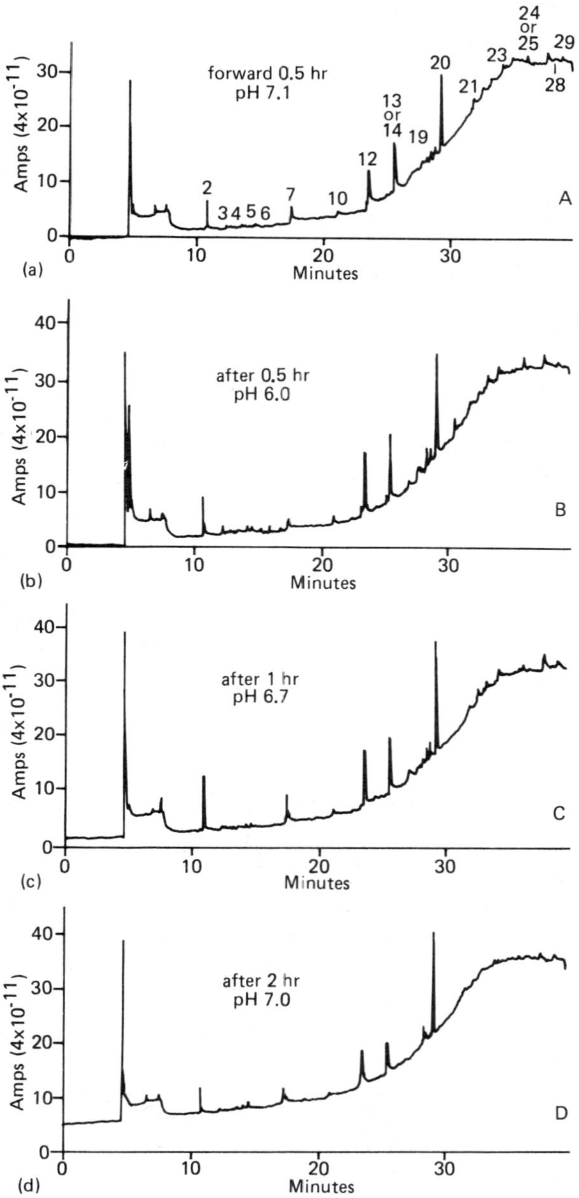

Figure 11.6 Typical gas chromatograms of fresh normal urine from a Japanese male (35 years old, nonsmoker). Flexible fused-silica capillary column (containing Carbowax 20 M, 50 m × 0.2 mm I.D.), FID sensitivity, 4×10^3. Chromatogram (A): urine sample from 0.5 hr before his smoking. Chromatogram (B): urine sample from 0.5 hr after his smoking. Chromatogram (C): urine sample from 1 hr after his smoking. Chromatogram (D): urine sample from 2 hr after his smoking. All of the chromatograms were obtained by the injection of 5.0 μl of the urine sample. Peaks: 2, acetic acid (0.1–0.2 μg/5.0 μl of the urine); 20, 3,4-xylenol (about 0.2 μg/5.0 μl of the urine). The other peak numbers are the same as in Figure 11.2.

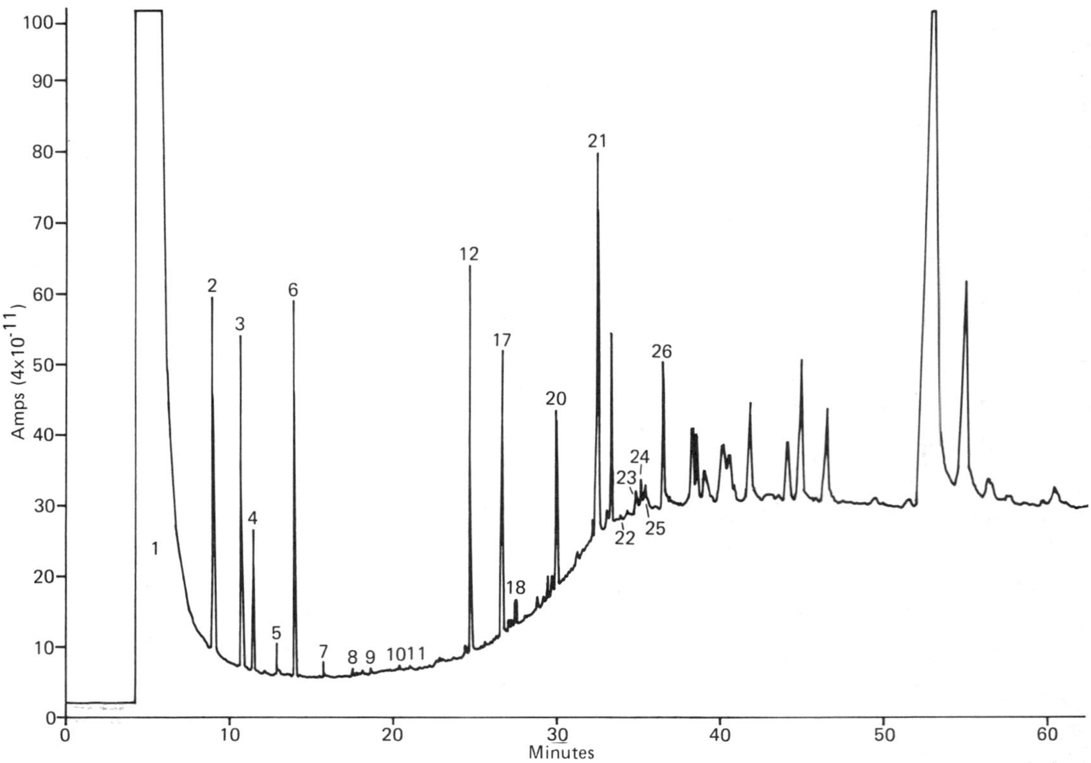

Figure 11.7 Typical gas chromatogram of lower fatty acids, phenols, and indoles isolated from the ethanol concentrate of feces of a Persian cat (4 years old). Flexible fused-silica capillary column (containing Carbowax 20 M, 50 m × 0.2 mm I.D.); FID sensitivity, 4 × 10^3. The peak numbers are the same as in Figure 11.2. The volume injected was 0.5 μl.

C. Feces Sample of Persian Cat

Figure 11.7 shows a typical gas chromatogram of the lower fatty acids, phenols, and indoles isolated from the ethanol concentrate of the feces of a 4-year-old Persian cat. More than 90 peaks were recognized. The main group in the feces sample consisted of lower fatty acids.

IV. SUMMARY

The simultaneous gas chromatographic separation of a mixture in the free form of the lower fatty acids, phenols, and indoles has been carried out by using a flexible fused-silica capillary column. Complete separation of the mixture has been obtained with a VFA capillary column, except for phenol and *o*-cresol, *o*-ethylphenol and 3,5-xylenol, 2-methylindole and *n*-lauric acid, whose peaks overlapped; and with a Carbowax 20 M capillary column, except for phenol and *o*-cresol, *o*-ethylphenol and 3,5-xylenol, and *n*-pelargonic acid, 2,3-xylenol and 1,2-dimethylindole, whose peaks overlapped; and 2- and 3-methylindoles, which were poorly separated.

The optimum conditions are as follows: stationary phase, VFA (Volatile Fatty Acid, Carbowax 20 M nondeactivated) or Carbowax 20 M (Carbowax 20 M deactivated); column size, 50 m × 0.2 mm I.D.; column temperature held for 1 min at 100°C, heated at 4°C/min from 100 to 220°C, maintained at 220°C for 9–30 min; carrier gas (nitrogen) flow-rates, 0.85 ml/min for the VFA capillary column, 0.97 ml/min for the Carbowax 20 M capillary column; detector, FID. The method has been applied to the analysis of the lower fatty acids, phenols, and indoles in the urine sample of a normal Japanese male (35 years old) and in the feces of a Persian cat (4 years old). The peaks of the phenols were also identified as their bromophenols, which were prepared by the reaction with bromine using a capillary column (stationary phase: Methyl Silicone SP-2100, Carbowax 20 M deactivated). The detection sensitivity of the analytical instrument in the present investigation was sufficient to measure 1–5 μl of the concentrated urine sample. The time required for analysis was about 40 min for standard compounds, about 60 min for complex sample solutions. In the fresh normal urine sample, at least 19 compounds were identified as 12 lower fatty acids, 7 phenols, and 3 indoles. In the feces sample of a Persian cat, more than 90 peaks were recognized.

ACKNOWLEDGMENTS

The author is thankful to Dr. K. Yoshimoto, Director of Aichi Environmental Research Center, and Professor Dr. N. Murayama, Shunshu University School of Medicine, for useful suggestions.

REFERENCES

1. D. W. Fassett, in *Industrial Hygiene and Toxicology*, Vol. II, *Toxicology* (F. A. Patty, ed.), 2nd ed., Interscience, New York, 1962, p. 1771.
2. E. A. Annison, K. J. Hill, and D. Lewis, *Biochem. J. 66*:592 (1957).
3. J. W. Gardner and G. E. Thompson, *Analyst(London) 99*:326 (1974).
4. R. D. Hawkins and H. Kalant, *Pharmacol. Rev. 24*:67 (1972).
5. F. Lundquist, in *Alcohols and Derivatives*, Vol. I (J. Trĕmoliĕres, ed.), Pergamon Press, Ltd., Oxford, 1970, p. 95.
6. V. M. Sardesai, ed., *Biochemical and Clinical Aspects of Alcohols Metabolism*, Charles C. Thomas, Springfield, Ill., 1969, p. 319.
7. M. Kŏivusalo, in *Alcohols and Derivatives*, Vol. II (J. Trĕmoliĕres, ed.), Pergamon Press, Ltd., Oxford, 1970, p. 465.
8. O. M. Bakke and R. R. Scheline, *Toxicol. Appl. Pharmacol. 16*:691 (1970).
9. R. T. Williams, in *Detoxication Mechanisms: The Metabolism and Detoxication of Drugs, Toxic Substances, and Other Organic Compounds*, 2nd ed., John Wiley & Sons, New York, 1959, p. 188.
10. K. Nomiyama, *Ind. Health 3*:53 (1965).
11. W. Deichmann and L. Schafer, *Am. J. Clin. Pathol. 12*:129 (1942).
12. G. M. Anderson and W. C. Purdy, *Anal. Chem. 51*:283 (1979).
13. G. Norheim and J. Rygge, *J. Chromatogr. 154*:291 (1978).
14. D. Hoffmann and G. Rathkamp, *Anal. Chem. 42*:366 (1970).
15. K. Grob, *Helv. Chim. Acta, 48*:1362 (1965).
16. K. Grob and J. A. Voellmin, *J. Chromatogr. Sci. 8*:218 (1970).
17. D. A. Buyske, P. Wilder, Jr., and M. E. Hobbs, *Anal. Chem. 29*:105 (1957).

18. W. E. Burnett, *Environ. Sci. Technol. 3*:744 (1969).
19. A. Yasuhara and K. Fuwa, *Bull. Chem. Soc. Japan 50*:731, 3029 (1977).
20. E. S. West and W. R. Todd, *Textbook of Biochemistry*, MacMillan, New York, 1951, p. 522.
21. T. C. Ruth and J. F. Fulton, eds., *Medical Physiology and Biophysics (Howell's Textbook of Physiology*, 18th ed.), W. B. Saunders, Philadelphia and London, 1960, p. 952.
22. J. Hrivňák and E. Beske, *J. Chromatogr. 54*:277 (1971).
23. J. Hrivňák, *J. Chromatogr. Sci. 8*:602 (1970).
24. E. D. Barber, E. Sawicki, and S. P. McPherson, *Anal. Chem. 36*:2442 (1964).
25. J. Krupcík, J. Hrivňák and J. Janák, *J. Chromatogr. Sci. 14*:7 (1976).
26. R. Gloor and H. Leidner, *Chromatographia 9*:618 (1976).
27. K. Grob and G. Grob, *J. Chromatogr. Sci. 8*:635 (1970).
28. M. L. Lee, D. L. Vassilaros, C. M. White, and M. Novotny, *Anal. Chem. 51*:768 (1979).
29. S. Preston, R. G. Ackman, L. Metcalfe, *Analysis of Fatty Acids and Their Esters by Chromatographic Methods*, Preston Publications, Niles, Ill., 1977, p. 547.
30. A. Kuksis, *Sep. Purific. Methods 6*:353 (1977).
31. L. Fishbein, *Chromatography of Environmental Hazards*, Vol. II, *Metals, Gaseous and Industrial Pollutants*, Elsevier, Amsterdam, 1973, p. 491.
32. S. K. Pierce and H. L. Gearhart, *Talanta 24*:473 (1977).
33. E. Jellum, *J. Chromatogr. 143*:427 (1977).
34. Y. Hoshika, *J. Chromatogr. 144*:181 (1977).
35. Y. Hoshika and G. Muto, *J. Chromatogr. 179*:105 (1979).

12

Analysis of Phospholipids After Enzymatic Hydrolysis

Simon J. Gaskell / *Tenovus Institute for Cancer Research, Welsh National School of Medicine, Cardiff, Wales*

I. INTRODUCTION

The phospholipids are an important class of natural compounds with structures based on, or related to, phosphatidic acid [1]. The structures of major phospholipid classes are shown in Figure 12.1.

Analyses of phospholipids most commonly involve the characterization of compound class alone. In certain clinical situations, such analyses are of proven value. Thus, for example, determination of the lecithin/sphingomyelin ratio in amniotic fluid is an import indicator of fetal lung maturity [1], particularly when such analyses are complemented by determinations of minor phospholipid classes (e.g., phosphatidyl glycerol) [2].

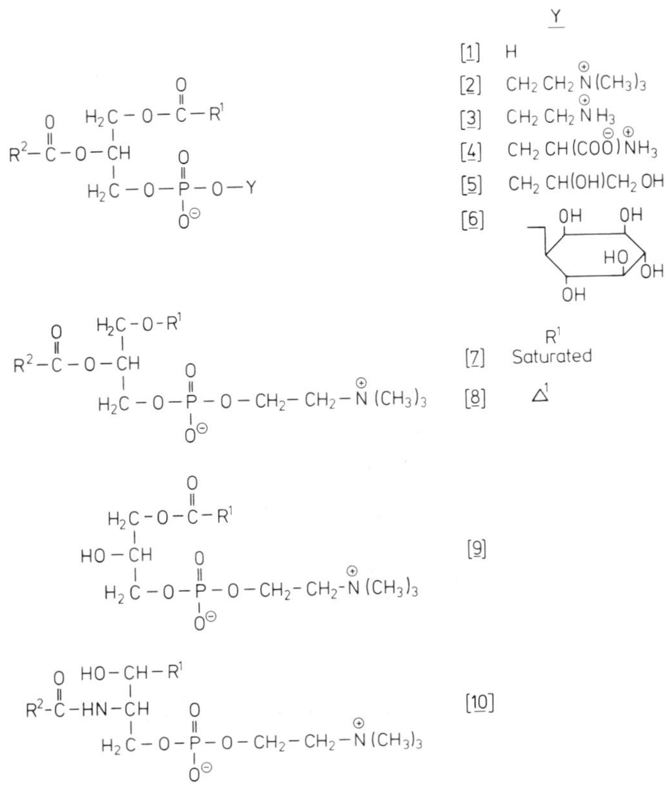

Figure 12.1 Structures of major classes of phospholipid. R represents a long alkyl chain. (1) phosphatidic acid; (2) phosphatidylcholine (lecithin); (3) phosphatidylethanolamine; (4) phosphatidylserine; (5) phosphatidylglycerol; (6) phosphatidylinositol; (7) glyceryl ether phospholipids; (8) choline plasmalogen; (9) lysolecithin; (10) spingomyelin; R^1, R^2 = alkyl. For the glycerophospholipids, R^1COO and R^2COO groups are shown at positions sn-1 and sn-2, respectively.

A more detailed approach to phospholipid analysis which is commonly employed involves degradation by enzymatic (see below) or chemical means to afford the constituent fatty acids and, in the case of sphingomyelin, long-chain bases. Simple procedures for the analysis of molecular species of phospholipids were devised by the Hornings and co-workers [3–5]. Dephosphorylation of glycerophospholipids was achieved by high-temperature reaction with phenyl ether and water or by direct displacement by a trimethylsilyl group. Higher yields were obtained by the former procedure which was also applicable to the analysis of sphingomyelin [5]. The trimethysilylated products were analyzed by GLC and GC-MS. The principal disadvantage of the approach was the observed isomerization of 1,2-diglycerides (derived from glycerophospholipids) to 1,3-diglycerides and the dehydration of ceramides (derived from sphingomyelin).

The use of the phospholipase enzymes provides procedures for the degradation of phospholipids under mild conditions. The sites of hydrolytic action of the phospholipases are shown in Figure 12.2 for the examples of lecithin and

[2] 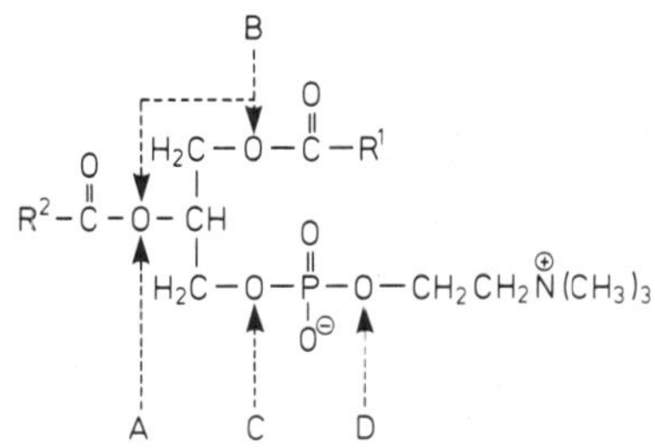

[10] 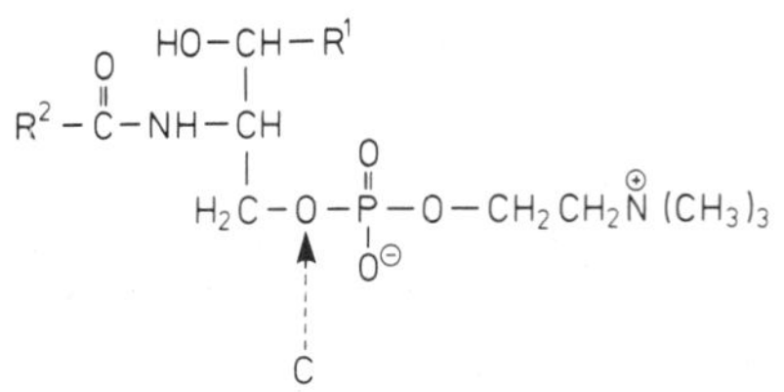

Figure 12.2 Sites of hydrolytic action of phospholipases.

sphingomyelin. Phospholipase A is of value in providing a means of distinguishing between acyl substituents of positions *sn*-1 and *sn*-2 of glycerophospholipids. Its use has established, for example, that unsaturated acids are preferentially esterified at position *sn*-2. For nonselective hydrolysis of acyl ester linkages, phospholipase B may be employed. Clearly, the mixture of fatty acids derived from phospholipase A or B hydrolysis may be most effectively analyzed, after suitable derivatization, by glass capillary gas chromatography (see Chap. 11). Phospholipase C provides a means of enzymatic degradation of phospholipids of particular value since major structural features are retained in the lipid product, which is amenable, after suitable derivatization, to gas-phase analysis. Thus, initial separation of phospholipids into compound class followed by phospholipase C hydrolysis will, in many cases, permit the subsequent determination of individual molecular species.

While the relationship between acyl substituents of phospholipids and biochemical properties is well known, the significance of individual molecular species as such is less well established. This may be attributed, or least in part, to the difficulties associated with detailed analysis. In one of the few published examples of such studies, Breimer [6] reported on the sphingomyelin composition of different parts of the digestive tract. Following phospholipase C hydrolysis, the resulting ceramides were analyzed by packed column GC-MS as the TMS ethers. The distribution of molecular species varied, presumably reflecting different membrane characteristics. The importance, or otherwise, of the determination of individual molecular structures will be more firmly established when readily applied procedures are developed. In any case, the availability of procedures for detailed analysis will facilitate biosynthetic and metabolic studies.

This review examines one approach to the study of molecular species of phospholipid and evaluates its advantages and shortcomings. The procedure is based on phospholipase C hydrolysis, derivatization, and glass capillary gas chromatography in combination with mass spectrometry.

II. PROCEDURE FOR PHOSPHOLIPASE C HYDROLYSIS [7–9]

Enzymatic hydrolysis of phospholipid (1–15 mg) dissolved in diethyl ether (1 ml) may be achieved by mixing with 1 ml of 0.05 M Tris buffer (pH 7.3) containing calcium chloride (0.03 M) and approximately 50 μg of phospholipase C. The solution of enzyme in buffer is extracted with diethyl ether before use to minimize contamination of hydrolysis products. Hydrolysis is allowed to proceed for 3–15 hr at room temperature, and products are extracted with diethyl ether. Hydrolysis of isolated phosphatidylethanolomine does not take place under these conditions but is reported to be effective in the presence of other classes of phospholipid [7].

III. DERIVATIVES FOR GAS CHROMATOGRAPHIC-MASS SPECTROMETRIC ANALYSES

A. Hydrolysis Products of Glycerophospholipids: Diglycerides and Monoglycerides

Diglycerides are generally converted to bis-trimethylsilyl (TMS) ethers for gas chromatographic and mass spectrometric analyses. The derivatives have good GLC properties, and their mass spectra have been studied in detail [4,8,10]. Molecular ions are not observed but $[M\text{-}CH_3]^+$ ions are present. Intense ions associated with the glycerol moiety are observed, together with fragments (e.g., $[RCO]^+$, $[RCO+74]^+$) indicating the acyl substituents. The ions $[M\text{-}RCOOH]^{+\cdot}$ provide further evidence for acyl substitution; studies with mixed diglycerides (i.e., those containing two different acyl substituents) indicate that loss of an unsaturated acyloxy group from *sn*-2 generally occurs without concomitant loss of H [8].

1,3-Diglycerides may be formed as artifacts during chemical degradation of glycerophospholipids or during derivatization of 1,2-diglycerides [4]. The TMS ethers of the isomeric compounds are readily separated by GLC and are further differentiated by the presence in the mass spectra of the 1,3-diglyceride derivatives of intense ions attributable to fragmentation of the C1-C2 or C2-C3 bonds of the glycerol moiety [4].

In the work described here, TMS ethers were prepared by dissolving the sample in a mixture of pyridine:hexamethyldisilazane:trimethylchlorosilane (5:4:3 v/v/v; 100 μl) and allowing to stand at 20° for 30 min [8]. Reagents were removed under a stream of nitrogen, and the products were dissolved in ethyl acetate prior to GLC and GC-MS. No isomerization of 1,2- to 1,3-diglycerides was observed under these derivatization conditions. Thus, for example, GC-MS (30-m OV-1/Silanox glass capillary column) of the TMS ether of the phospholipase C hydrolysis product of synthetic 1-stearoyl-2-oleoyl-lecithin revealed a single product with a retention index of 4113 and a mass spectrum consistent with the structure 1-stearoyl-2-oleoylglycerol TMS ether.

TMS derivatives have also been shown to be suitable for gas chromatographic and mass spectrometric analyses of 1-monoglycerides [11]. Mass spectrometric studies of the acetates have also been reported [12]. Molecular ions are not generally observed at significant intensity in the spectra of the TMS ethers, but the molecular weight is clearly indicated by the presence of $[M\text{-}CH_3]^+$ and $[M\text{-}CH_2OTMS]^+$ ions. $[RCO]^+$ ions further indicate the acyl substituent. Methaneboronates are suitable alternative derivatives for GC-MS analyses of monoglycerides [13,14]; retention times are substantially shorter than for the corresponding TMS derivatives. Molecular ions are observed, though at low

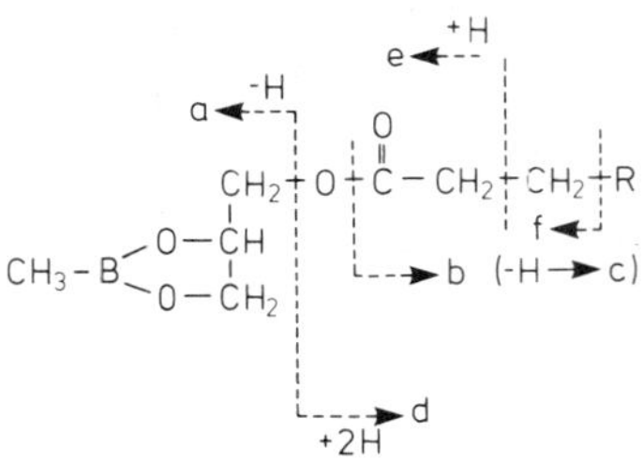

Figure 12.3 Fragmentation of 1-monoglyceride methaneboronates under electron impact. R = alkyl. (Redrawn from Ref. 23.)

intensity; the acyl substituent is again indicated by $[RCO]^+$ ions. Further diagnostically useful fragmentations are indicated in Figure 12.3.

The methaneboronates of 1-monoglycerides (the expected products of phospholipase C hydrolysis of lysolecithin) are readily distinguished from the 2-monoglyceride isomers by the intensity of m/z 99 (corresponding to fragmentation *a*, Figure 12.3, without associated loss of H) in the spectra of the latter.

Methaneboronate derivatives of monoglycerides (and of ceramides, discussed below) are readily prepared by treatment with a slight molar excess of methaneboronic acid in ethyl acetate. An aliquot of the reaction mixture may be analyzed after about 30 min at 20°C.

B. Hydrolysis Products of Sphingomyelins: Ceramides

Table 12.1 lists derivatives of ceramides used in analyses by gas chromatography or mass spectrometry. Some form of derivatization is required to permit GC analyses and to promote mass spectrometric fragmentations of diagnostic value. Trimethylsilyl ethers have attracted most interest since they are readily prepared and provide useful mass spectra. Major modes of fragmentation under electron impact are shown in Figure 12.4.

Table 12.1 Derivatives for the Gas Chromatographic and Mass Spectrometric Analyses of Ceramides

		Published work	
Derivative	Δm^a	Gas Chromatography	Mass spectrometry
None	—	—	15,16
Permethyl	42 (56)	17	15,16
O-acetyl	84 (126)	—	15
O-TMS	144 (216)	3,5,18,19,20,21	
Me boronate	24 (48)	13,22,23	

[a]Mass increment accompanying derivatization. Figures in parentheses refer to ceramides with 2-hydroxyacyl substituents.

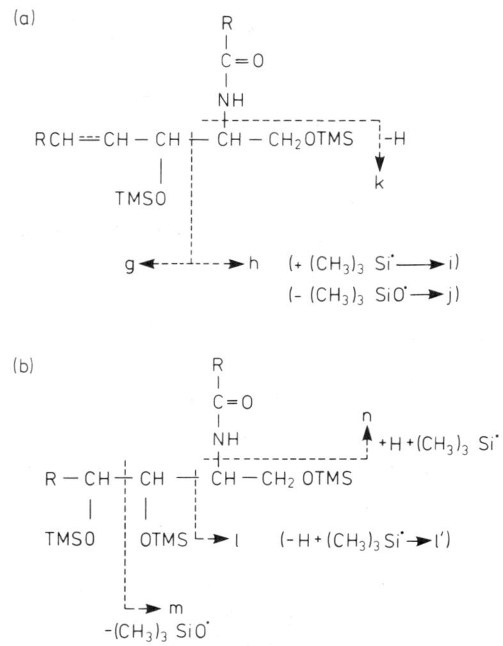

Figure 12.4 Fragmentation of ceramide trimethylsilyl ethers under electron impact. (a) Ceramides derived from dihydroxy bases. (b) Ceramides derived from trihydroxy bases. R = alkyl.

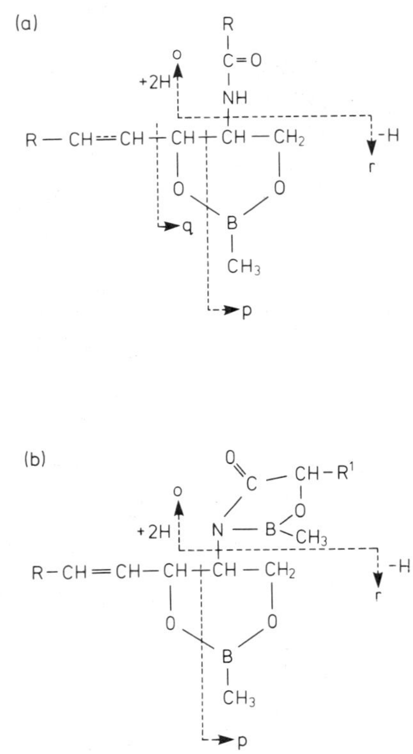

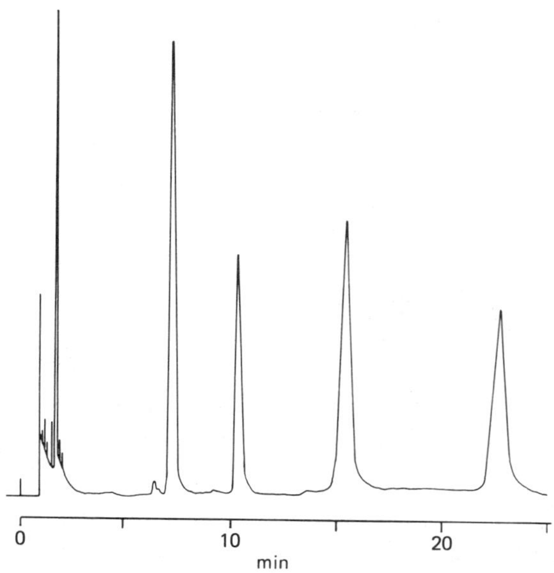

Figure 12.6 Gas chromatographic analysis of ceramide methaneboronates derived from the reaction of sphinganine methaneboronate with a mixture of C_{12}, C_{14}, C_{16}, and C_{18} acyl chlorides. Glass capillary column (30 m × 0.55 mm, I.D.) of OV-1/Silanox at 300°, with a helium flow-rate, measured at 20°, of 5 ml/min.

The utility of the methaneboronate derivative is suggested by the selectivity of formation of the cyclic structure and by the low mass increment accompanying derivatization. The mass spectrometric fragmentation is generally simple and affords ions permitting the determination of acyl substituents and parent long-chain bases. Major fragmentations are summarized in Figure 12.5. A preliminary study (discussed below) of ceramides containing 2-hydroxyacyl substituents indicated the preparation of bis-methaneboronate derivatives with the postulated structure shown in Figure 12.5b.

For the accumulation of reference data, standard ceramides as methaneboronate derivatives may be conveniently prepared, either singly or as mixtures, by reaction of the appropriate long-chain base methaneboronate with the acyl chloride(s) [22]. Products may generally be analyzed without further treatment.

Figure 12.6 shows the GLC analysis of ceramide methaneboronates prepared from sphinganine methaneboronate and a mixture of acyl chlorides.

The value of electron impact mass spectrometry in structural studies of ceramide derivatives is generally complemented by chemical ionization (CI) mass spectrometry in that the latter technique generally affords intense protonated molecular ions (MH^+) or ions derived from MH^+ by simple neutral

---

Figure 12.5 Fragmentation of ceramide methaneboronates under electron impact. R = alkyl. (a) Ceramides with fatty acyl substituents. (b) Ceramides with 2-hydroxyacyl substituents.

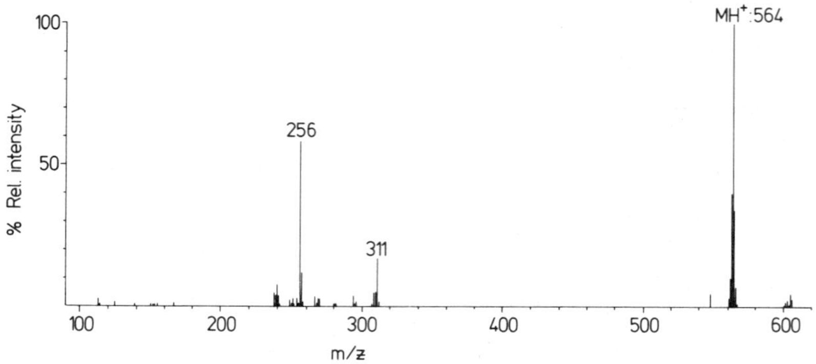

Figure 12.7 Isobutane chemical ionization mass spectrum of *N*-palmitoyl sphinganine methaneboronate.

losses. Figure 7 shows the isobutane CI mass spectrum of *N*-palmitoyl sphinganine methaneboronate; MH^+ is the base peak, and an intense ion of m/z 256 indicates the palmitoyl substituent.

Comparative GC retention data are not available for all the derivatives listed in Table 12.1. A comparison of TMS and methaneboronate derivatives, however, indicates that, for ceramides containing fatty acyl substituents, the retention indices of the two derivatives are similar [13].

Table 12.2 records retention indices of selected synthetic ceramide methaneboronates, analyzed using a glass open-tubular column coated with OV-1 on Silanox [22]. Complete resolution of individual structures is clearly difficult using a nonselective phase such as OV-1. The problem of mixture analysis, however, is substantially reduced using the GC-MS combination (see below).

Table 12.2 Retention Indices for Synthetic Ceramide Methaneboronates (OV-1 Liquid Phase)[a]

Ceramide	Long-chain base	Acyl substituent	Retention index
N-palmitoyl sphinganine	18:0	16:0	3896
N-palmitoyl 4-sphingenine	18:1	16:0	3912
N-stearoyl sphinganine	18:0	18:0	4097
N-oleoyl sphinganine	18:0	18:1	4073
N-linoleoyl sphinganine	18:0	18:2	4074

[a] Analyses were performed using a glass capillary column (40 m × 0.55 mm, I.D.) of OV-1 on Silanox, at 300°, with a helium carrier gas flow-rate of 15 ml/min. Samples were introduced via a "falling-needle" type dry injection system. Kovats' retention indices were determined by coinjection of the standard ceramides with n-C_{38} and n-C_{42} alkanes.

Source: Data from Ref. 22.

IV. GLASS CAPILLARY GC FOR THE ANALYSIS OF PRODUCTS OF PHOSPHOLIPASE C HYDROLYSIS

In each application described below, the products of phospholipase C hydrolysis of phospholipid fractions were suitably derivatized and analyzed using glass capillary columns of the OV-1/Silanox type [24]. These columns have the particular merit, for this application, of tolerating high sample loads and therefore providing a wide concentration range for the detection of mixture constituents. A further advantage is the very low level of liquid phase bleed observed at high temperatures; this consideration is particularly important when minor components are to be detected during GC-MS.

In general, the use of capillary GC columns provides a much shorter analysis time, for a given chromatographic resolution, than is achieved with packed columns. For the analysis of compounds of low volatility, such as derivatized diglycerides and ceramides, the ability to obtain good separations at moderate temperatures and with reasonably short retention times is clearly important. Versatility in the use of glass capillary columns for such analyses may be achieved, where necessary, by operating at a carrier gas flow rate well above the optimum indicated by the van Deempter plot. This means of achieving shorter retention times (with the consequent sacrifice in chromatographic resolution) is preferred to the alternative of using a shorter column since the analytical conditions are readily adjusted to suit the current application. Thus the analyses, reported below, of the relatively volatile monoglyceride methaneboronates were performed at the near-optimal flow rate of 5 ml/min, whereas analyses of ceramide methaneboronates on the same column were carried out at 10 ml/min.

A dry injection system, of the "falling-needle" type [25], was used throughout the analyses cited below. This method of sample introduction is particularly suitable for GC-MS application since the absence of a pressure surge associated with the solvent helps preserve the stability of ion source focusing conditions and obviates the need for an isolation valve at the GC-MS interface.

V. APPLICATIONS

A. Phospholipase C Hydrolysis of Lysolecithin [23]

Figure 12.8 shows the total-ion-current trace obtained during a rapid GC-MS analysis of the methaneboronate derivatives of products of phospholipase C hydrolysis of egg lysolecithin. No solvent peak is observed following sample introduction via the "falling-needle" device. The high sample capacity and the low level of background bleed permit the reliable identification of components present at widely different concentrations. Table 12.3 records GC-MS data and proposed assignments.

Only monoglycerides containing saturated acyl substituents were available as reference standards, but mono-unsaturated acyl glyceride methaneboronates were identified by interpretation of mass spectra. The retention indices are about 20 units less than those of the unsaturated acyl analogues.

B. Phospholipase C Hydrolysis of Sphingomyelin [22,23]

A ceramide fraction was obtained by preparative TLC following phospholipase C hydrolysis of phospholipids from atherosclerotic human arterial tissue (Figure 12.9). The gas chromatogram obtained by analysis of the ceramides, as

Table 12.3 GC-MS for Methaneboronate Derivative of the Products of Phospholipase C Hydrolysis of Egg Lysolecithin

Constituent[a]	Acyl substituent	Retention index	Mass spectrum[b]						
			$M^{+\bullet}$	*a*	*b*	*c*	*d*	*e*	*f*
A	16:1	2355	352 (3)	98 (100)	237 (16)	236 (28)	— —	158 (10)	171 (5)
B	16:0	2373	354 (2)	98 (100)	239 (24)	238 (7)	257 (12)	158 (36)	171 (10)
C	17:0	2478	— —	98 (100)	253 (13)	252 (8)	271 (9)	158 (28)	171 (10)
D	18:1	2552	380 (9)	98 (100)	265 (23)	264 (57)	— —	158 (15)	171 (15)
E	18:0	2572	382 (2)	98 (100)	267 (13)	266 (7)	285 (6)	158 (33)	171 (10)

[a]See Figure 12.7.
[b]Electron impact (20 eV). Ions are listed as m/z (relative intensity).

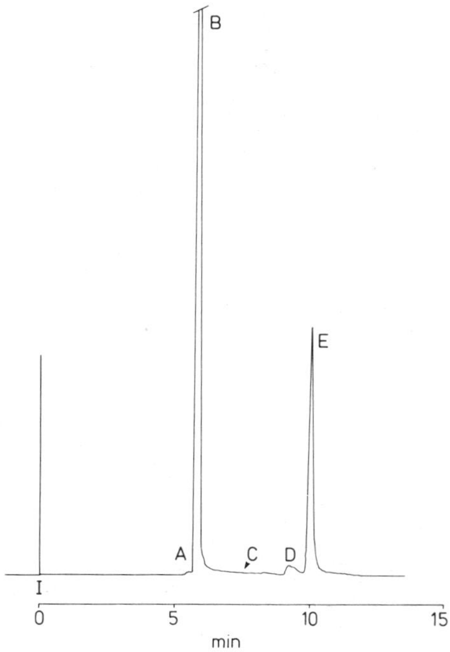

Figure 12.8 Total-ion-current trace recorded during GC-MS of the methaneboronate derivatives of the products of phospholipase C hydrolysis of egg lysolecithin. Glass capillary column (30 m × 0.55 mm, I.D.) of OV-1/Silanox at 230°, with a helium flow-rate, measured at 20°, of 5 ml/min. (Redrawn from Ref. 23.)

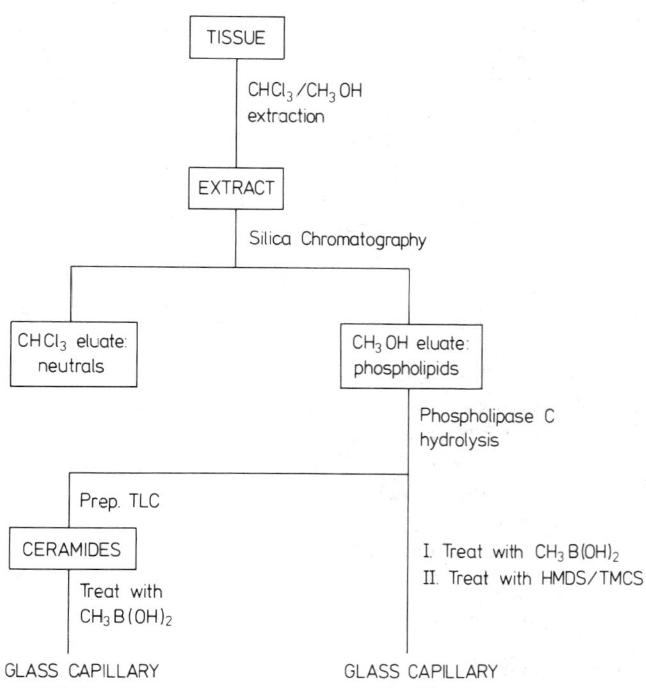

Figure 12.9 Procedure for the analysis of phospholipids in human arterial tissue. HMDS, hexamethyldisilazane. TMCS, trimethylchlorosilane.

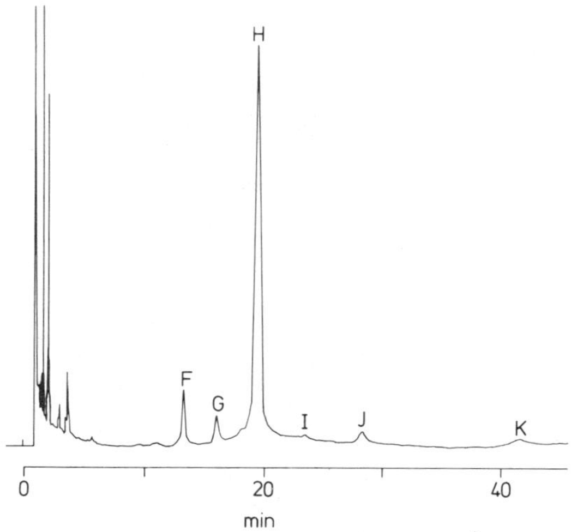

Figure 12.10 Gas chromatographic analysis of methaneboronate derivatives of ceramides derived from sphingomyelin of human arterial tissue. Glass capillary column (40 m × 0.55 mm, I.D.) of OV-1/Silanox at 300°C, with a helium flow-rate, measured at 20°, of 15 ml/min. Flame ionization detector. (Redrawn from Ref. 22.)

methaneboronate derivatives, is shown in Figure 12.10. Each of the major constituents was identified by Kovats' retention index and mass spectrum (Table 12.4). Peak F was attributed to a mixture of the methanoboronates of *N*-palmitoyl-hexadeca-4-sphingenine and *N*-myristoyl-4-sphingenine. It is difficult to envisage a gas chromatographic resolution of these ceramides so that their separate identification by GC-MS is dependent on the simple, informative fragmentation pattern afforded by the methaneboronate derivatives. Ceramides containing very long-chain ($>C_{20}$) acyl substituents were not detected. An increasing proportion of shorter-chain ($<C_{20}$) acyl substituents in sphingomyelin from arterial tissue with increasing atherosclerosis has been suggested [26], but an assessment of the significance of the relative proportions of molecular species requires further detailed analyses.

Figure 12.11 shows the GLC analyses of the methaneboronate derivative of a commercially obtained mixture of ceramides derived by enzymatic hydrolysis of sphingomyelin from bovine brain. The mixture was coinjected with n-C_{38}, -C_{42}, and -C_{44} alkanes to permit the determination of retention indices. (Since higher n-alkanes were not available, retention indices of later-eluting components were obtained by extrapolation.) GC-MS analysis indicated that the major constituents were bis-methaneboronates of 2-hydroxyacyl ceramides. The spectrum of component L is shown in Figure 12.12, together with the proposed origins of the major fragment ions (Figure 12.5b).

C. Hydrolysis Products of a Phospholipid Mixture [23]

The applications of GC-MS of the products of phospholipase C hydrolysis given above have been concerned with separated phospholipid fractions (lysolecithin, sphingomyelin). To explore the possibility of rapid analysis of phospholipid mixtures, phospholipase C hydrolysis followed by derivatization and GC-MS was

Table 12.4 GC-MS Data for Methaneboronate Derivatives of Ceramides Derived from Phospholipase C Hydrolysis of Phospholipids from Human Arterial Tissue

Constituent[a]	LCB[c]	Acyl[c]	Retention index	Mass spectrum[b]					
				$M^{+\cdot}$	$[M\text{-}CH_3]^+$	o	p	q	r
F_1	16:1	16:0	3710	533 (6)	518 (10)	256 (100)	323 (33)	—	278 (34)
F_2	18:1	14:0				228 (64)	295 (22)	324 (6)	306 (29)
G	17:1	16:0	3815	547 (2)	532 (5)	256 (100)	323 (32)	352 (1)	292 (29)
H	18:1	16:0	3913	561 (2)	547 (7)	256 (100)	323 (34)	—	306 (35)
I	18:1	17:0	4012	575 (2)	560 (5)	270 (100)	337 (31)	366 (4)	306 (30)
J	18:1	18:0	4109	589 (3)	574 (3)	284 (100)	351 (29)	380 (2)	306 (44)
K	18:2	20:0	4307	615 (2)	600 (5)	312 (100)	379 (35)	408 (1)	304 (25)

[a]See Figure 12.9.
[b]Electron impact (20 eV). Ions are recorded as m/z (relative intensity)
[c]Assigned structures are indicated by long-chain base (LCB) and acyl substituent.

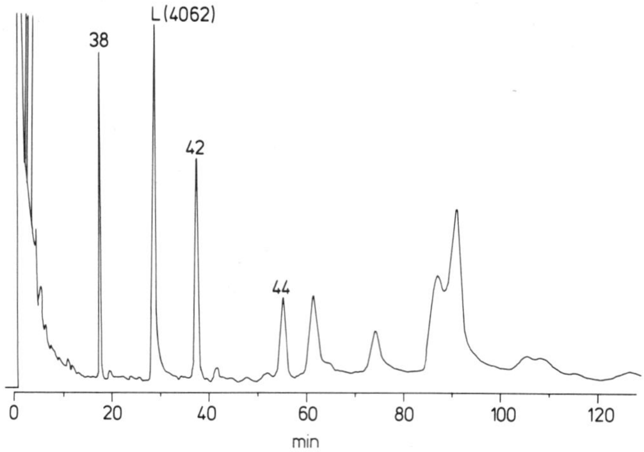

Figure 12.11 Gas chromatographic analyses of methaneboronate derivatives of ceramides derived from sphingomyelin of bovine brain. Analytical conditions as for Figure 12.10. (Redrawn from Ref. 22.)

applied to the analysis of total phospholipids derived from atherosclerotic human aorta. The experimental procedure is outlined in Figure 12.9. Following enzymatic hydrolysis, ceramides were selectively converted to methaneboronates, and diglycerides were subsequently derivatized to form the TMS ethers. Under the chosen conditions for trimethylsilylation (described above), ceramide boronates were unchanged.

Micro-TLC of an aliquot of the products of enzymatic hydrolysis indicated ceramides (derived from sphingomyelin) and diglycerides (derived from glycerophospholipids) to be the major constituents. GC-MS analysis of the total derivatized products confirmed this and allowed identification of individual constituents. Figure 12.13 shows the total-ionization-current trace. Each of the major

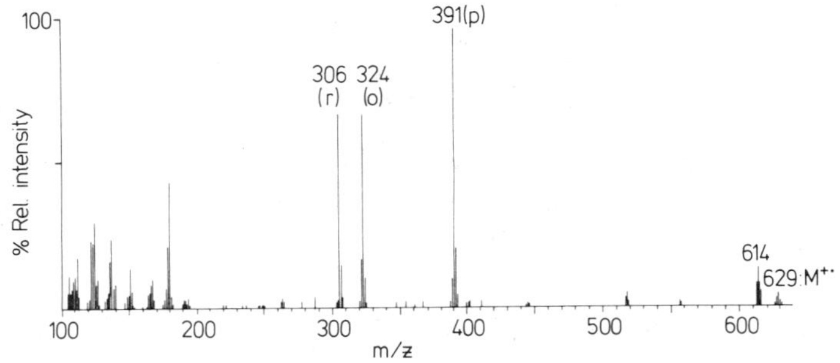

Figure 12.12 Mass spectrum of component L (Figure 12.11) of ceramides derived from sphingomyelin of bovine brain. The component was identified as the bis-methaneboronate of 2-hydroxystearoyl-4-sphingenine. (Redrawn from Ref. 22.)

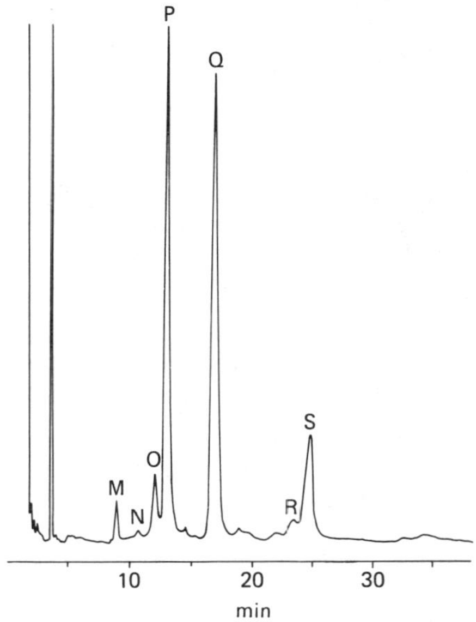

Figure 12.13 Total-ion-current trace recorded during GC-MS of the derivatized (methaneboronates, TMS ethers) products of phospholipase C hydrolysis of phospholipids from human arterial tissue. Glass capillary column (30 m × 0.55 mm, I.D.) of OV-1/Silanox at 300°, with a helium flow-rate, measured at 20°, of 5 ml/min. (Redrawn from Ref. 23.)

peaks indicated was shown by mass spectrometry to be attributable to one or two ceramides or diglycerides. Monoglycerides were not detected.

Preliminary GC-MS analysis was performed with repetitive scanning and data acquisition by computer. Mass chromatograms were subsequently selected to indicate the compound class to which the major mixture constituents belonged. Thus, for examples, ions of *m/z* 129 are attributable to the structure shown in Figure 12.14, which is derived from the glycerol moiety of diglyceride TMS ethers. A mass chromatogram of *m/z* 129 indicated peaks O, Q, R, and S (Figure 12.13) to be diglyceride TMS ethers. Ceramide methaneboronates incorporating an *N*-palmitoyl constituent were identified in the mass chromatogram of *m/z* 256 (Figure 12.5, fragmentation *o*). Peaks M, N, and P were so identified.

A detailed analysis was performed by examination of full mass spectra obtained for each component; data are summarized in Table 12.5. The major ceramide constituents are those previously identified in sphingomyelin from human arterial tissue. It is noteworthy, however, that the minor, late-eluting ceramides

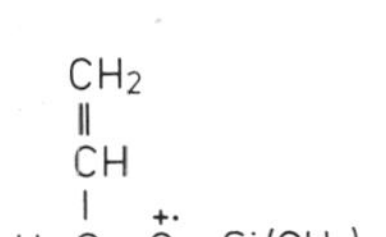

m/z 129

Figure 12.14 Ion of *m/z* 129 derived from diglyceride TMS ethers.

Table 12.5 GC-MS Data for Derivatives (Methaneboronates, TMS Ethers) of Products of Phospholipase C Hydrolysis of Phospholipids from Human Arterial Tissues

I. Ceramide methaneboronates							
			Mass spectrum[b]				
Constituent[a]	LCB[c]	Acyl[c]	$M^{+\cdot}$	$[M\text{-}CH_3]^+$	*o*	*p*	*r*
M1	16:1	16:0			256 (100)	323 (53)	278 (47)
			533 (4)	518 (5)			
M2	18:1	14:0			228 (100)	295 (23)	306 (28)
			547 (2)	532 (4)			
N	17:1	16:0			256 (100)	323 (48)	292 (47)
			561 (3)	546 (7)			
P	18:1	16:0			256 (100)	323 (34)	306 (45)

II. Diglyceride TMS ethers

Mass spectrum[b]

Constituent[a]	Acyl substituent[c]	$[M\text{-}CH_3]^+$	$[M\text{-}RCOO]^+$		$[M\text{-}RCOOH]^+$	
O	16:0, 16:1	625 (33)	385 (40)		384 (24)	
Q1	16:0, 18:1	651 (16)	385 (100)	409 (10)	410 (20)	
Q2	16:0, 18:2	649 (14)		—	408 (21)	
R	18:1, 18:2?	675 (13)	411 (58)	409 (11)	410 (10)	408 (15)
S1	18:0, 18:2	677 (12)	413 (100)	409 (13)	408 (25)	
S2	18:0, 18:1	679 (10)		411 (5)	410 (17)	

[a]See Figure 12.11.

[b]Electron impact (20 eV). Ions are recorded as m/z (relative intensity).

[c]Assigned structures are indicated by long-chain base (LCB) and acyl substituent (for ceramides) and by acyl substituents (for diglycerides).

previously detected (Figure 12.10) are obscured in the analysis of the more complex mixture. For both ceramide and diglyceride constituents, complete resolution is not achieved (Table 12.5), but the mass spectra of the binary mixtures involved are readily interpreted. The spectrum recorded for peak Q, for example, indicated the presence of two diglyceride TMS ethers, differing in the degree of unsaturation of one of the acyl substituents. Two $[M\text{-}CH_3]^+$ ions are observed; prominent $[M\text{-}(CH_3)_3SiOH]^{+\bullet}$ ions confirm the assignments of molecular weights. The base peak of *m/z* 385 $[M\text{-}RCOO]^+$) suggests a common palmitoyl substituent; the other acyl substituents (presumably at positions *sn*-2 are indicated as octadecadienoyl and octadecenoyl by the doublet at *m/z* 408, 410 $[M\text{-}RCOOH]^{+\bullet}$).

VI. DISCUSSION

The results of this and other work indicate that cleavage of the phosphate ester bond of phospholipids and subsequent GLC analysis of derivatized products provides a powerful means of determination of molecular species. The use of glass capillary GLC columns is advantageous in affording good resolution of mixtures, reasonably short retention times at moderate temperatures, and low levels of liquid phase bleed. Complete resolution of phospholipid components has not been achieved in the analyses reported here (which employed a nonselective silicone phase), but the use of GC-MS with appropriate choice of derivative permits the identification of individual constituents by the interpretation of mass spectra of binary mixtures.

The analysis of a total phospholipid fraction by phospholipase C hydrolysis, derivatization, and glass capillary GC-MS provides a rapid means of preliminary analysis, the results of which may suggest the need for more detailed investigation. An elegant example of such a detailed analysis is the study of phosphatidylcholines by Curstedt and Sjövall [8]. Phosphatidylcholines were separated from other classes of phospholipid by gel chromatography using an anion exchange derivative of lipophilic Sephadex (dibutylamino-hydroxypropyl Sephadex LH-20 in the acetate form); the method provides a convenient alternative to the preparative TLC procedure generally employed. Following phospholipase C hydrolysis, TMS ethers of the resulting diglycerides were separated by reverse-phase chromatography on hydroxyalkoxypropyl Sephadex. The diglyceride derivatives were eluted in order of increasing chain length and decreasing degree of unsaturation. Consequently, in the example studied, the subsequent GC-MS analysis required only a separation according to carbon number to effect complete resolution of molecular species. Little advantage, in terms of the requisite resolution, was therefore achieved through the use of glass capillary columns [27], reflecting the sophistication of the preliminary separation step. Further detailed structural studies were performed by GC-MS following lipase hydrolysis of the 1-acyl linkage in 1,2-diglyceride TMS ethers. Subsequent studies by the same group [27] have evaluated the utility of glass capillary columns for the analysis of synthetic mixtures of diglyceride TMS ethers. The separation of diglycerides which differ by the presence of a mono-unsaturated acyl, in place of a saturated acyl, residue may be achieved using nonselective stationary phases (OV-1, OV-73), as indicated by the data reported above (Table 12.5). A selective phase (SP-2340) is, however, required to effect separation of diglycerides containing saturated and mono-unsaturated acyl residues from those containing saturated and di-unsaturated acyl groups.

The products of phospholipase C hydrolysis of phospholipids, after derivatization, remain of low volatility, and high temperatures are required for gas phase analysis. It is likely that the problem of sample adsorption on the GC column (though minimized by modern glass capillary column technology) and GC-MS interface will prove the limiting factor in highly sensitive analyses (using, for example, selected ion monitoring during GC-MS). Recent improvements in the resolution achieved by high-pressure liquid chromatography (HPLC) suggest an alternative procedure for the separation of molecular species of phospholipid, before or after enzymatic hydrolysis. HPLC has been applied, for example, to the partial separation of sphingomyelin derived from bovine brain and sheep and pig erythrocytes [28]. The difficulties presently associated with reverse-phase HPLC-MS suggest that this tandem technique does not yet parallel GC-MS in its power of combined separation and identification. The off-line combination of HPLC and field desorption MS [29], however, is likely to be valuable for phospholipid analyses.

In summary, a variety of separation and identification methods now available for the analysis of molecular species of phospholipids. The combination of gas chromatography (particularly incorporating glass capillary columns) and mass spectrometry is especially valuable for rapid and simple analyses of derivatized products of phospholipase C hydrolysis.

ACKNOWLEDGMENTS

The work reported in this review was carried out during the tenure of a postdoctoral Fellowship at the Department of Chemistry, University of Glasgow. The guidance of Professor C. J. W. Brooks and the collaboration of Dr. C. G. Emonds are gratefully acknowledged. The author's present work is generously supported by the Tenovus Organization.

REFERENCES

1. L. Gluck, M. V. Kulovich, R. C. Borer, P. H. Brenner, G. G. Anderson, and W. N. Spellacy, *Am. J. Obstet. Gynecol. 109*:440 (1970).
2. M. Hallman, B. H. Feldman, E. Kirkpatrick, and L. Gluck, *Pediat. Res. 11*:714 (1977).
3. G. Casparrini, M. G. Horning, and E. C. Horning, *Anal. Lett. 1*:481 (1968).
4. M. G. Horning, G. Casparrini, and E. C. Horning, *J. Chromatogr. Sci. 7*:267 (1969).
5. M. G. Horning, S. Murakami, and E. C. Horning, *Am. J. Clin. Nutr. 24*:1086 (1971).
6. M. E. Breimer, *J. Lipid Res. 16*:189 (1975).
7. O. Renkonen, *J. Am. Oil Chem. Soc. 42*:298 (1965).
8. T. Curstedt and J. Sjövall, *Biochim. Biophys. Acta. 360*:24 (1974).
9. K.-A. Karlsson, *Acta Chem. Scand. 22*:3050 (1968).
10. T. Curstedt, *Biochim. Biophys. Acta 360*:12 (1974).
11. J. J. Myher, L. Marai, and A. Kuksis, *J. Lipid Res. 15*:586 (1974).
12. C. B. Johnson and R. T. Holman, *Lipids 1*:371 (1966).
13. S. J. Gaskell, C. G. Edmonds, and C. J. W. Brooks, *Anal. Lett. 9*: 325 (1976).
14. C. J. Brooks, C. G. Edmonds, and S. J. Gaskell, in *Advances in Mass Spectrometry*, Vol. 7 (N. R. Daly, ed.), Heyden/Institute of Petroleum, London, 1978.

15. S. P. Markey and D. A. Wenger, *Chem. Phys. Lipids 12*:182 (1974).
16. G. Aranda, H. E. Audier, and M. Fetizon, *C. R. Acad. Sci. (Ser. C), 280*:1505 (1975).
17. R. T. C. Huang, *Hoppe-Seyler's Z. Physiol. Chem. 352*:1306 (1971).
18. B. Samuelsson and K. Samuelsson, *J. Lipid Res. 10*:41 (1969).
19. S. Hammarstrom, B. Samuelsson, and K. Samuelsson, *J. Lipid. Res. 11*:150 (1970).
20. S. Hammarstrom, *J. Lipid Res. 11*:175 (1970).
21. K. Samuelsson and B. Samuelsson, *Chem. Phys. Lipids.* 5:44 (1970).
22. S. J. Gaskell, C. G. Edmonds, and C. J. W. Brooks, *J. Chromatogr. 126*:591 (1976).
23. S. J. Gaskell and C. J. W. Brooks, *J. Chromatogr. 142*:469 (1977).
24. A. L. German and E. C. Horning, *J. Chromatogr. Sci. 11*:76 (1973).
25. P. M. J. van den Berg and T. P. H. Cox, *Chromatographia 5*:301 (1972).
26. J. C. Dousset, G. Soula, A. M. Fourcade, L. Douste-Blazy, M. Jullian, M. T. Pieraggi, and H. Bouisson, *Paroi Arterielle 2*:123 (1974).
27. J. Sjövall, personal communication.
28. F. B. Jungalwala, V. Hayssen, J. M. Pasquini, and R. H. McCluer, *J. Lipid Res. 20*:579 (1979).
29. G. W. Wood, P.-Y. Lan, and G. M. S. Rao, *Biomed. Mass Spectrom. 3*:172 (1976).

13

Glass Capillary GLC of Bile Acids

Halvor Jaeger / *L.A.B. Gesellshaft für pharmakologische Untersuchungen m.b.H. & Co. (Laboratory for Pharmacological Research), Neu-Ulm, Federal Republic of Germany*

Wilhelm Nebelung, Hans U. Klör, and Hans Ditschuneit / *University of Ulm, Ulm, Federal Republic of Germany*

Hartmut Frank / *Eberhard-Karls-Universität, Tübingen, Federal Republic of Germany*

I. INTRODUCTION

Bile acids have long been recognized as essential for maintenance of body function and health [1]. A resurgence of interest in their role in physiology and pathophysiology in the last 20 years has largely paralleled to development of more and more sophisticated techniques for elucidation of their structures and for quantitative analysis. This paper will primarily deal with the analytical aspects of the problem.

A. General Structure and Function

Bile acids are mainly surfactants or detergents with the primary responsibility of dispersing lipids such as lecithin, glycerides, and cholesterol. The structure of bile acids which are derived from cholesterol are ideally suited to perform this function. They consist of a hydrophobic side composed of a sterol backbone with modifications due to metabolism primarily by the liver (dehydroxylation at one or more positions or oxidation of a hydroxyl to a carbonyl) coupled with other metabolic changes such as oxidation to a carboxylic acid or conjugation to give a hydrophilic portion of the molecule. The dual functionality allows bile acids to coat lipids to form droplets called micelles [2] wherein the hydrophilic side faces the aqueous polar phase and the hydrophobic phase is in the lipid core. This arrangement serves to disperse dietary lipids in the gut lumen for hydrolysis and subsequent absorption. Bile acids also play an important role in the transport of lecithin and cholesterol plus other bile components from the liver to the gut. The concentration of bile acids in the bile and the ratio of bile acid to cholesterol are critical. If the concentration of cholesterol relative to bile acids becomes too high, the cholesterol can crystallize to form gallstones [3].

B. Formation and Metabolism

The modifications of cholesterol leading to the formation of bile acids are carried out primarily by enzymes in the mitochondria or in the smooth endoplasmic reticulum. Side-chain oxidation is carried out by the former while ring metabolism is performed by the latter. The overall rate-limiting step seems to be the reaction catalyzed by cholesterol 7α-hydroxylase [4] which leads to the synthesis of the primary bile acids such as cholic (3α, 7α, 12α-trihydroxy-5-cholanic acid), chenodeoxycholic acid (3α, 7α-dihydroxy-5β-cholanic acid), and perhaps ursodeoxycholic acid (3α, 7β-dihydroxy-5β-cholanic acid) [5]. The formation of β-muricholic acid (3α, 6β, 7β,-trihydroxy-5β-cholanic acid) has been observed in rat liver when bile flow was impaired [6]. Both ursodeoxycholic acid and β-muricholic acid are formed in the liver from chenodeoxycholic acid.

The primary bile acids are conjugated via an amide linkage to either glycine or taurine prior to leaving the liver. In this form they exhibit less cellular toxicity [7]. Although the majority of the primary bile acids are reabsorbed from the gut unchanged, some undergo modification by the gut flora to give secondary bile acids such as deoxycholic acid (3α, 12α-dihydroxy-5β-cholanic acid) from cholic acid and lithocholic acid (3α-hydroxy-5β-cholanic acid) from chenodeoxycholic acid. As can be seen, the primary modification results in removal of the 7α-hydroxyl group. It is not yet clear if ursodeoxycholate is also metabolized by the intestinal bacteria.

The conversion of ursodeoxycholate to lithocholic acid in man is suspected since infusion of the former leads to increased amounts of the latter in the feces [9]. A similar result was not obtained when rhesus monkeys were infused perhaps due to a higher specificity of the bacterial dehydroxylase enzyme present in the monkey for the 7α position. Intestinal bacteria are also capable of deconjugation by hydrolysis of the amide linkage in both primary and secondary bile acids. The reactions occur primarily in the mid and terminal ileum and in the colon, where bacterial growth is most pronounced [10]. During inflammation of the ileum or in "blind-loop syndrome" after gastric resection, the degree of deconjugation may greatly increase due to bacterial overgrowth in the proximal portions of the intestine [10].

Both free and conjugated primary and secondary bile acids are reabsorbed and transported back to the liver by the portal vein where they are essentially removed from the blood during the first pass. This enterohepatic circulation of bile acids prevents the need for continuous supplementation of fecal loss by de novo synthesis by the liver.

C. Occurrence and Distribution

Over 90% of all bile acids in bile, serum, and urine are represented by glycyl conjugates of cholate, chenodeoxycholate, and deoxycholate [5]. Ursodeoxycholate and lithocholate conjugates appear to represent less than 5% of the bile acids in human bile [10]. They have been found in small amounts in serum and urine of patients with cholostasis, along with hyocholic and hydeoxycholic (3α,6α-dihydroxy-5β-cholanic) acids [11]. Epilithocholic acid has been reported in trace amounts in normal human urine [12].

D. Specific Properties

While all bile salts function as detergents and lead to micelle formation, they differ somewhat in their overall effect. Cholic and β-muricholic acid appear to be relatively weak detergents when tested as inhibitors of liver microsomal enzymes [6]. The dihydroxylated bile acids, especially chenodeoxycholate, appear to be much more potent in that they interfere with the microsomal enzymes important for the metabolism of bile acids, lipids, and drugs [13]. At higher concentrations they are toxic in the rhesus monkey [8]. Deoxycholate has been observed to induce severe parenchymal liver damage in rats due to its detergent effect on the membranes of subcellular organelles [14].

The excellent detergent properties of chenodeoxycholate have prompted its use to remove or dissolve cholesterol gallstones [15]. At low, nontoxic doses, the lithogenic index of the bile is reduced, resulting in the dissolution of the cholesterol gallstone. Recent reports favor ursodeoxycholate for this purpose in that it appears to have no toxicity in monkey or man at doses sufficient to dissolve the biliary calculi [8,9]. Both chenodeoxycholate and ursodeoxycholate are potent inhibitors of hydroxymethylglutaryl-CoA reductase, which is important for cholesterol synthesis [8]. The trihydroxy bile acids are less potent in this respect. Deoxycholate and chenodeoxycholate inhibit the reabsorption of water and sodium ions from the colon, giving rise to diarrhea when present in large concentrations [16].

Lithocholate appears to exhibit the most liver toxicity in most animal species including man [17], by disruption of microsomal enzymes leading eventually to parenchymal necrosis. The principal means of detoxification appears to be conjugation as a sulfate even in cases of obstructive jaundice [18]. Once sulfated, lithocholate is only poorly reabsorbed from the gut. The little that is reabsorbed is excreted in the urine.

Although lithocholate is formed from chenodeoxycholate—and one would expect large amounts to be formed when the latter is used to treat gallstones—the amount of free lithocholate is increased only slightly because the sulfation pathway is greatly induced [19].

E. Determination—Applications

The differentiation of bile acids by various analytical procedures is becoming increasingly important in biochemical and clinical studies because of their somewhat different properties. Several sophisticated techniques have been used to study the contribution of different precursor pools, such as newly synthesized cholesterol [20–22], lipoprotein-cholesterol [23], and cholesterol ester [24], to the synthesis of different bile acids. The clearance of the individual bile acids by the liver has been studied by comparing blood levels in portal and systemic blood to learn something about the mechanism [25]. The various bile acids have been studied in bile, duodenal fluid, and feces to learn more about their pool size and the turnover rates both in normals [26] and in patients with liver cirrhosis, where bile acid metabolism has changed [27,28]. The binding properties of various antacids for bile salts have been studied to see if the potential damage to the stomach can be reduced [29].

Because an increased incidence of gallstones appears to be a hazard of some lipid-lowering regimens, an attempt was made to clarify the mechanism of this side effect by measuring the bile acid concentrations in the bile, serum, and feces of patients with hyperlipidemia [30–32]. Much work has been done to analyze bile acids in patients with liver disease since this condition gives rise to the most marked alterations in bile acid metabolism. Significant changes in bile acids in bile and serum have been found in patients with cirrhosis. In these patients, the amount of chenodeoxycholate was increased while cholate was decreased [33–36]. In many cases the serum concentration of deoxycholate was reduced, often to the point where it could not be detected [37–39].

With more sophisticated techniques such as gas-liquid chromatography coupled with mass spectroscopy, it was possible to detect traces of several abnormal bile acids in serum and urine of cholostatic patients [11]. Other studies in this type of patient demonstrated that the pathways leading to conjugation were not the normal pathways, in that large amounts of sulfated monohydroxyl bile acids were excreted into the urine [40]. In addition, large amounts of bile acid glucuronides were excreted into the urine after synthesis by the cholostatic liver [41].

II. PREPARATION AND DERIVATIZATION OF SAMPLES

VandenHeuvel and co-workers [42] were the first to analyze bile acids by gas-liquid chormatography as their methyl esters. This method suffered from the disadvantage that the high boiling points of these esters lead to extensive decomposition. Later it was shown that the methyl esters could be made more volatile by derivatizing them as their trifluoroacetyl esters. This allowed their analysis at lower temperatures with improved recoveries. Improved columns and the use of capillary GLC have solved most of these problems in a relatively short time. The analysis of bile acids in physiological fluids posed additional problems. The only relatively easy analysis was that of bile acids in the gallbladder bile because the concentration was very high. Fecal matter and gallstones were also high in many bile acids and were easier to analyze than urine and serum, in which only small amounts of bile acids were present. The extraction from urine and serum required more sophisticated techniques especially since most of the bile acids were in the conjugated forms such as the glycine and taurine conjugates. These conjugates required hyrolysis prior to analysis. The small amounts of sample required the optimal in standardization, and necissitated that high-purity solvents be used for extraction, hydrolysis, and derivatization. In addition, the large number of bile acids required the use of more than one column for complete separation. A complete list of the natural and derived bile acids has been published [44].

A. Extraction Procedures

The extraction procedures used [45] have been mainly used in assays where packed columns were employed. In most cases these can be directly applied to capillary columns.

1. *Extraction from Serum*

Most of the early extraction methods use the "hot ethanol" method [46]. One of the many modifications which have been developed is as follows: Plasma is mixed with 10 volumes of ethanol and heated to boiling. The precipitate is removed by centrifugation, and the supernatant is diluted with water to about 70% ethanol. Neutral lipids and cholesterol are removed by extraction with petroleum ether. After evaporation of the ethanol, the bile acids and these conjugated are dissolved in water and extracted with butanol at pH 1.

In most laboratories, the use of adsorbent columns has replaced this method. This method, which makes the removal of interfering substances such as lipids and cholesterol much easier, was first described by Danielson and co-workers in 1963 [48].

Two years later, the use of Dowex columns was reported by Kuksis [49]. In this method a 24-hr continuous extraction with diethyl ether was preceded by hot ethanol extraction. A Dowex-50 cation-exchange column was used for further purification and desalting. The glycine and taurine conjugates were then separated on a Dowex-1 anion-exchange column using a HCl gradient in aqueous alcohol or acetone [50].

Several other methods followed [51,52], each using different adsorbent columns. In the method reported by Sandberg et al. [52], an Amberlyst XN-1006 anion-exchange resin in the chloride form was used. This column was first washed with aqueous organic solvent, converted into the bicarbonate form, and then to the hydroxide form. The bicarbonate form was washed first with 50 ml of 1 N NaOH, followed by 50 ml of 1 N NaOH in 80% aqueous ethanol, and then with distilled water until the eluant was neutral. The serum sample (5–10 ml) was diluted 1:1 with distilled water, and after adjusting to pH 11 with 1 N NaOH, it was placed on the anion-exchange column. After working with water until the washings were neutral, the column was washed with 20 ml of 95% ethanol, then 20 ml ethanol:ethylene chloride (1:1 v/v), and finally 20 ml of 80% ethanol. The bile acids and conjugates could then be eluted with 150 ml of 0.2 M ammonium carbonate in 80% ethanol [53] with an overall recovery of added ^{14}C-labeled deoxycholic, chenodeoxycholic, cholic, and taurocholic acid of between 82% and 96%.

In 1972, Makino and Sjövall [54] presented the method which has gained the greatest acceptance to date. This method uses the Amerlite XAD-2 resin, as suggested by Bradlow [55]. The resin is washed before use with water, methanol, acetone, and then water before storing in water until used. The serum or plasma sample (2 ml) is diluted with 18 ml of 0.1 M NaOH in saline. After 15 min in an ultrasonic bath, it is transferred to a 1 g Amberlyte XAD-2 column and allowed to pass through at a rate of 0.1 ml/min. After washing the column with water until neutral, the bile acids are eluted with 3 ml of ethanol. After alkaline hydrolysis, the hydrolyzate is acidified with 4 N HCl and diluted to 30 ml with distilled water. This sample is then passed through a second XAD-2 column. After washing the column with water until neutral, the excess water is removed from the column by wasing with 5 ml hexane. The adsorbed free bile acids are then eluted with 3 ml of a mixture of hexane:chloroform:methanol (1:1:1, v/v/v). This procedure gave recoveries of 85–95% of added lithocholic acid, while the recoveries of glycocholic, chenodeoxycholic, and cholic acid were 80%, 103%, and 93%, respectively. The recoveries were about the same as those reported by Sandberg et al. [52].

Makino and co-workers [56] have shown that the sulfated bile acids are also separated by the Makino and Sjövall method [54]. The separation of these compounds from the free and conjugated bile acids can be accomplished as follows: After drying the eluate from the Amberlyte XAD-2 column and dissolving the residue in 2 ml of a chloroform:methanol (1:1, v/v) mixture, the sample is passed through a Sephadex LH-20 column (4 g). The free and conjugated bile acids elute with 70 ml of the chloroform-methanol mixture while the sulfated fraction is eluted with 50 ml of methanol.

In our laboratory, we have modified this method [54] slightly. Prior to use, the XAD-2 column is washed with the above solvents plus a final wash of 20 ml of 0.1 M NaOH in saline. This wash using the sample solvent is important since the pore size of the Amberlyte XAD-2 varies with the solvent used [57]. For the same reason, the first wash after adding the sample was saline followed by a water wash until the eluant was neutral. The recovery of the bile acids and their conjugates was improved by washing the column twice with 5 ml of methanol each. The recovery after the first wash was 92% and after both washes was 96% for radioactivelabeled glycocholic and taurocholic acid. The recovery of cholic acid was 98% after the initial 5-ml wash. The methanol washes were pooled, evaporated to dryness, and hydrolyzed enzymatically (see later section). After acidification, the free bile acids were extracted with three 5-ml portions of ether. The first portion extracted 39.6%; the second, 40.1%; and the third, 6%; for a total recovery of 85.7%. Additional extractions did not improve the recovery. The combined ethereal extracts were dried over sodium sulfate (5 g) which removed both the water and water-soluble compounds. Without this step, a chromatogram on capillary column suitable for peak area measurement with a computing integrator could not be obtained. See Figure 13.1 for a summary of this extraction procedure.

2. *Extraction from Bile*

Since the amounts of bile acid in bile from the gallbladder is so high, the need for extensive purification is reduced. Klaassen [58] has published a method based on the work of Sandberg et al. [52] and Okishio et al. [59]. In this procedure the bile sample is diluted, and 1 ml of 0.2 M phosphate buffer (pH 5.8) is added. After enzymatic hydrolysis the internal standard is added, the sample is acidified with 1 ml of 6 M HCl, and the free bile acids are then extracted with two portions of 5 ml of diethyl ether. The method of Makino and Sjövall [54] can also be applied to bile samples without additional modifications.

3. *Extraction from Fecal and Duodenal Samples*

The bile acids in feces are a mixture of primary and secondary acids [60]. There have been several reports on the analysis of single bile acids [61–63], but Eneroth and co-workers have made an attempt to completely characterize the bile acids in feces [64]. The fecal sample is homogenized in 1.5 liters of chloroform-methanol (1:1, v/v) and extracted for 48 hr in a Soxhlet apparatus with hot chloroform-methanol (1:1). After hydrolysis, the acids are extracted with diethyl ether for 16 hr. The ether extract is, in turn, washed with water. After methylation, TLC is used to preseparate the bile acids [65]. This method was used to characterize the human fecal samples in several studies [66,67]. Evrard and Janssen [68] described a method whereby the fecal sample was freeze-dried and then heated with glacial acetic acid with internal standard added for 1 hr at 120°C. An acetic acid-toluene extract was prepared and evaporated to dryness. After alkaline hydrolysis, the unsaponifiable material was removed by extraction with petroleum ether. The bile acids were then extracted after acidification with diethyl ether. The recovery was about 80%.

4. *Extraction from Urine*

The method generally used is a modification of the Makino-Sjövall method [54–56]. The urine sample (30 ml) is passed through a 10-g Amberlyte XAD-2 column. Elution with 60 ml of methanol followed by separation of the sulfated fraction as described above completes the isolation procedure.

5. *Extraction from Tissue Samples*

Two methods have been reported. In the first method [69], cylinders of liver tissue were solubilized with the addition of 50 μl of 3.4 N NaOH, then heated in a boiling water bath. The result was neutralized, and after enzymatic hydrolysis the bile acids were extracted with diethyl ether. After evaporation of the ether, the residue was again extracted and the acids were then derivatized prior to analysis. The second method [70] required a freeze-dried sample which was then extracted twice with 100 ml of 95% ethanol containing 0.1 ml of aqueous ammonia. After working the residue with an additional 50 ml of solvent, the extracts were pooled, concentrated, and chromatographed on Amberlyst A-26. The recovered acids were hydrolyzed enzymatically, and after acidification they were extracted into diethyl ether.

6. *Extraction from Gallstones*

Nakayama [71] powdered the gallstones and extracted the acids with acidic chloroform-methanol (2:1, v/v). Other methods of similar nature have been reported [72].

B. Hydrolysis and Solvolysis

Bile acids are most often found as their glycine or taurine conjugates. These two types can be separated chromatographically by ion-exchange methods [50, 73] or by extraction, since the glycine conjugates can be recovered by extraction with ethyl acetate, while the taurine conjugates are extractable only with butanol [74]. In general, both types are hydrolyzed together under alkaline conditions, which in some cases have been reported to lead to destruction of the acids [52,75,76], mainly due to adsorption onto silicates released from the glass during the hydrolysis. One could use nickel or stainless steel bombs [44], but the best results were obtained by heating with 1 M NaOH at 115°C for 6 hr or by autoclaving in 1.24 M NaHO for 3 hr at 15 psi. Even then, some of the ketocholanic acids decomposed [44].

As a result of these problems, an enzymatic hydrolysis was developed by Nair and co-workers [77]. However, they found that the conjugates were not hydrolyzed at the same rate or extent when working with a cell-free extract of *Clostridium perfringens*. An extract of *Bacillus sphericus* preferentially hydrolyzed taurocholate [78]. After additional purification attempts [79–81], the enzymatic hydrolysis was found to give improved recoveries compared to the alkaline hydrolysis described above [81]. As a result, the enzymatic method using cholylglycyl hydrolase has become the most widely used method for hydrolysis of bile acid conjugates although others have been described [82].

In our laboratory, we have carried out the method of Nair et al. [77] as follows: After elution of the XAD-2 column with methanol, the solvent is evaporated under nitrogen on a water bath at 45°C. The bile acids are dissolved in 2 ml of 0.1 M acetate buffer (pH 5.6), and 25 μl of cholylglycyl hydrolase (specific activity not less than 100 μmol/mg protein) are added along with 0.5 ml of 0.2 M EDTA and 0.2 ml of 0.2 M β-mercaptoethanol. Instead of the original procedure, we incubate this mixture at 37°C overnight. After acidification with 6 N HCl to pH 1, the free bile acids are extractable into diethyl ether as described above.

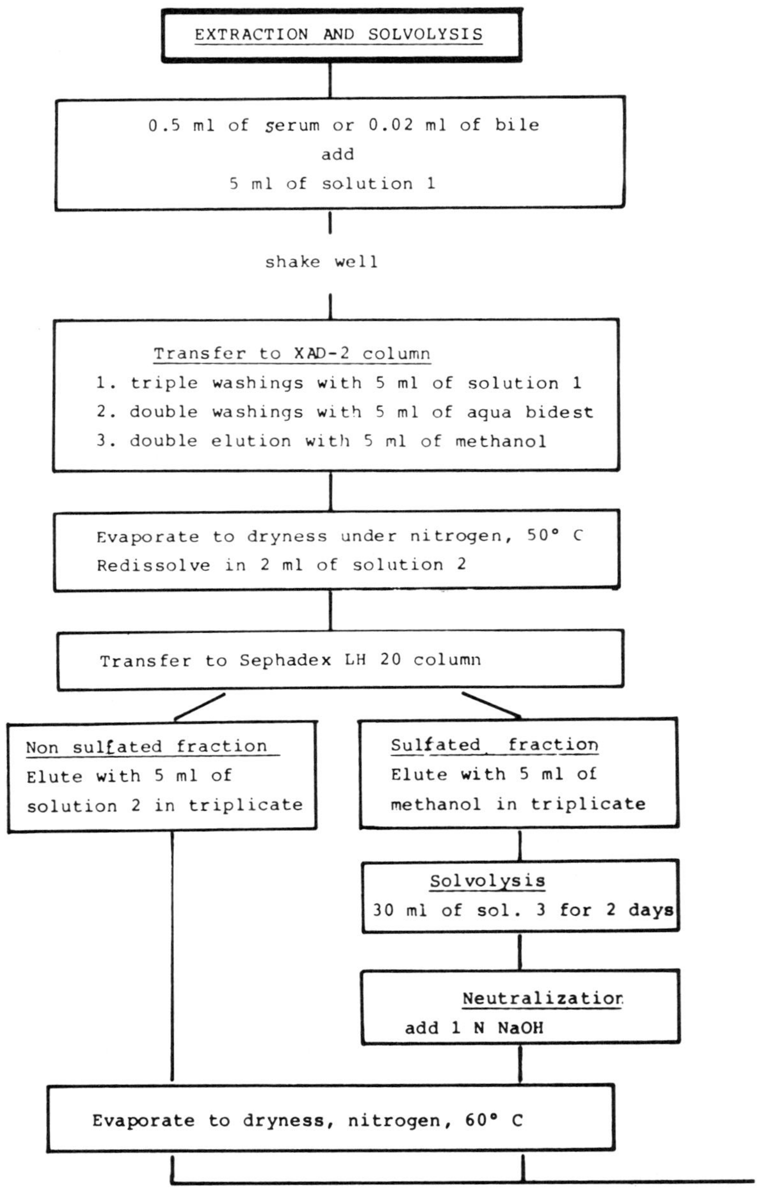

Figure 13.1 Flowsheet diagram of bile acid extraction and derivatization. Solutions: 1, 0.1 M sodium hydroxide in saline; 2, chloroform/methanol 1:1 (v/v), containing 0.01 M sodium chloride; 3, ethanol/acetone 1:9 (v/v), acidified with 2 N HCl to pH 1; 4, 5 ml of aqua bidest/0.25 ml of EDTA buffer/ 0.5 ml of acetate buffer pH 5.6/50 μl cholylglycyl hydrolase/200 μl trifluoroacetic anhydride.

The separation of the sulfated and nonsulfated fractions is best carried out on Sephadex LH-20, as described above [56]. The sulfated fraction can be hydrolyzed by standing for 2 days at room temperature in a solution of 30 ml of ethanol-acetone (1:9, v/v) which has been acidified with a few drops of 2 N HCl to pH 1. After neutralization, the solvent is removed as suggested by Palmer [83] and the residue further hydrolyzed by alkaline and then extracted into diethyl ether.

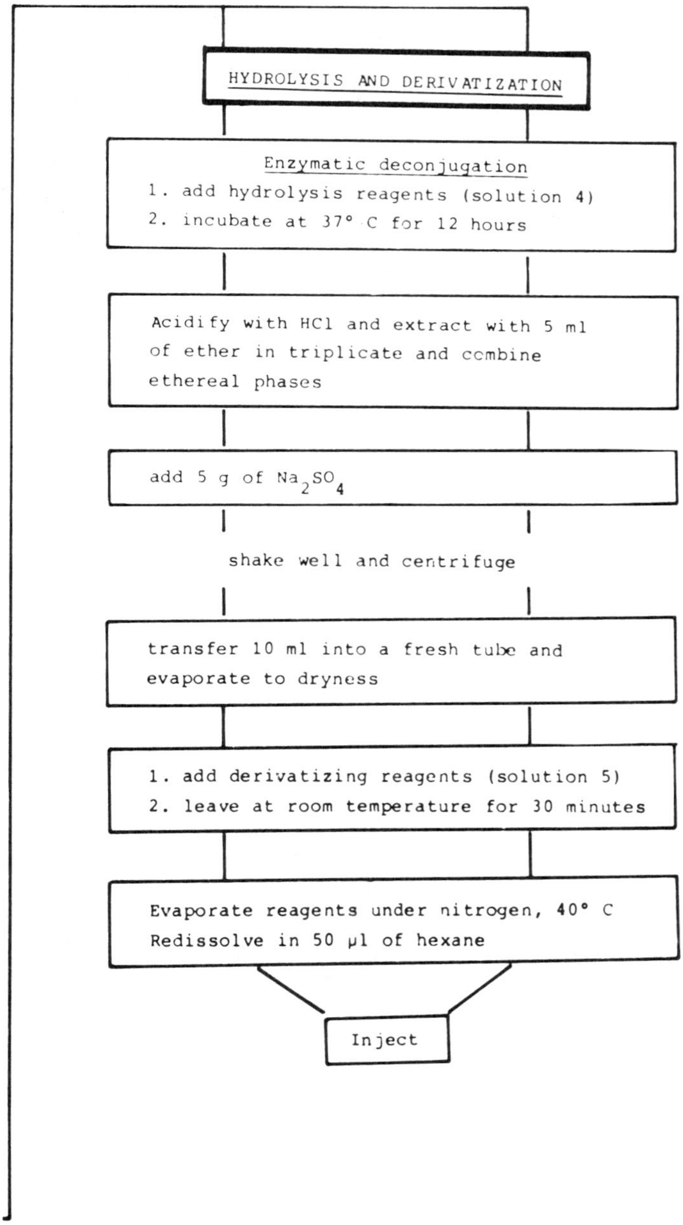

Figure 13.1 (continued)

C. Derivatization

Since bile acids are polyfunctional, they cannot be chromatographed without conversion to a suitable derivative. Early methods chromatographed the bile acids as their methyl esters [42], which because of their extensive polarity tended to be adsorbed irreversibly on the column [84]. As a result, a double derivatization is now used whereby the methyl esters are trimethylsilyated or trifluoroacetylated.

1. *Alkyl Esters*

The most frequently used esters are the methyl esters, which are generally prepared by reaction with diazomethane by the procedure developed for fatty acids by Schlenk and Gellerman [85]: The acids are dissolved in methanol and an ice-cold solution of freshly prepared diazomethane is added until the yellow color persists. The diazomethane is generally prepared from *N*-methyl-*N*-nitroso-*p*-toluenesulfonamide [86]. The method suffers from the disadvantages that diazomethane is explosive and is believed to be carcinogenic [87]. An alternate procedure for preparation of the methyl esters is reaction with methanolic HCl (5%) which can lead to dehydration reactions, complexation, and color formation. Esters of other alcohols have been prepared by Hofman and Mosbach [88]. They have made the propyl esters using propanol and sulfuric acid (99:1, v/v). In our laboratory, we have found that the method described by Imai et al. [90] gives excellent results without the disadvantages associated with diazomethane.

The combination of hexafluoroisopropanol with trifluoroacetic anhydride has been used to derivatize the acid catecholamine metabolites [91] as follows: Hexafluoroisopropanol (100 μl) and trifluoroacetic anhydride (200 μl) are added to the bile acid sample and heated to 37°C for 30 min. After evaporation of the solvent, the residue is taken up in acetonitrile for analysis by gas-liquid chromatography. Analysis by GLC-MS showed that the carboxylic acid groups were esterified while all hydroxyls were trifluoroacetylated, as shown in Figure 13.1. The derivatives are stable for at least 5 days if stored cold. No change in the chromatogram was found over this period, and on 10 repeat runs of a sample containing 10 bile acids plus cholesterol, the mean coefficient of variation was 5–7% for the individual bile acids. Since this method gives such excellent, stable results—no variation in peak areas over 5 days—it is remarkable that the diazomethane method is still used in some laboratories.

2. *Acyl Esters*

Several groups have [43,92] used trifluoroacetic anhydride to derivatize the hydroxyl groups of bile acids. Sjövall's method [92] is as follows: the bile acids are dissolved in 0.2–0.5 ml of trifluoroacetic anhydride at 30–35°C for 15 min. The excess reagent is evaporated and the residue is taken up in acetone for analysis. When pyridine is used, the reaction yield can be improved [93]. Even though there are reports that the trifluoroacetates may undergo thermal decomposition, they still are the derivative of choice since they give the complete resolution of the natural bile acids when subjected to gas-liquid chromatography or gas-liquid chromatography-mass spectroscopy [94]. The agreement in relative retention times from different laboratories has made this the method of choice [95].

Donike [96] has described the acetylation of steroids with *N*-methyl-bis (trifluoroacetamide) (MBTFA), and bis(trifluoroacetamide) (BTFA); however we have tried both reagents for derivatizing bile acid methyl and propyl esters and find no advantage over the use of trifluoroacetic anhydride.

3. *Ethers*

A mixture of dry pyridine:hexamethyldisilazane:trimethylchlorosilane (9:3:1, v/v/v) has been used to derivatize bile acid methyl esters to the corresponding trimethylsilyl ethers [51]. Supina and co-workers [97] have found that this procedure leads to formation of some secondary products from the keto acids. This reaction can be avoided if dimethylformamide is used instead of pyridine. In a comparison of the trimethylsilyl, trifluoroacetyl, and heptafluorobutyryl derivatives, the trifluoroacetates were shown to have slightly better chromatographic properties. The trimethylsilyl derivatives, in general, gave higher response factors when a flame detector was used, while the trifluoroacetates and heptafluorobutyrates were best suited for electron capture detectors.

D. Detectors and Columns

When packed columns are used, the support of choice is QF-1 since it gives the optimal separation of the hydroxy bile acid from the corresponding keto acid. Capillary columns have not been used as extensively but they offer many advantages, as will be seen in the next section. Both flame detectors and electron capture detectors can be used. The fluorinated derivatives are best suited for the latter.

III. APPLICATIONS

The first group to use capillary columns extensively for bile acid analysis was Laatikainen and Hesso [99]. The bile acids were extracted using the method of Makino and Sjövall [54] followed by enzymatic hydrolysis by the Nair and Garcia procedure [81] or by alkaline hydrolysis. The free bile acids were extracted with ethyl acetate and treated with diazomethane after 7-ketodeoxycholic acid was added as an internal standard. The methyl esters were further purified on aluminum oxide and converted into their corresponding trimethylsilyl ethers prior to analysis on a 50 m glass capillary column coated with OV-101 prepared as described by Grob [100,101]. The carrier gas was nitrogen, split 1:10. The column was programmed from 150° to 225°C at 8°C/min. A flame ionization detector was used and the analysis time was 40 min. As noted above, the quantitation was by the internal standard method. In Figure 13.2, a standard bile acid mixture (A), bile acids extracted from serum (B), and from gallbladder bile (C) were chromatographed as the trimethylsilyl ethers of the methyl esters. The peaks due to cholic, chenodeoxycholic, deoxycholic, and lithocholic acid are completely separated. The overall recovery rates of glycocholic and taurocholic acids after enzymatic hydrolysis were 68–80% (mean 81%) and 87–100% (mean 97%), respectively. The same column was used for 300 assays with no change in properties. The overall quantitation gave good agreement with the results of Sandberg et al. [52]. The principal disadvantage of this procedure is that the internal standard was not added prior to extraction, and therefore it could not correct for losses during that step. A second drawback is that lithocholic acid and cholesterol were not separated. Even the time-consuming purification procedures could not assure the absence of all cholesterol, which makes the assay meaningless for lithocholic acid. For quantitative analysis a stream splitter is not advisable.

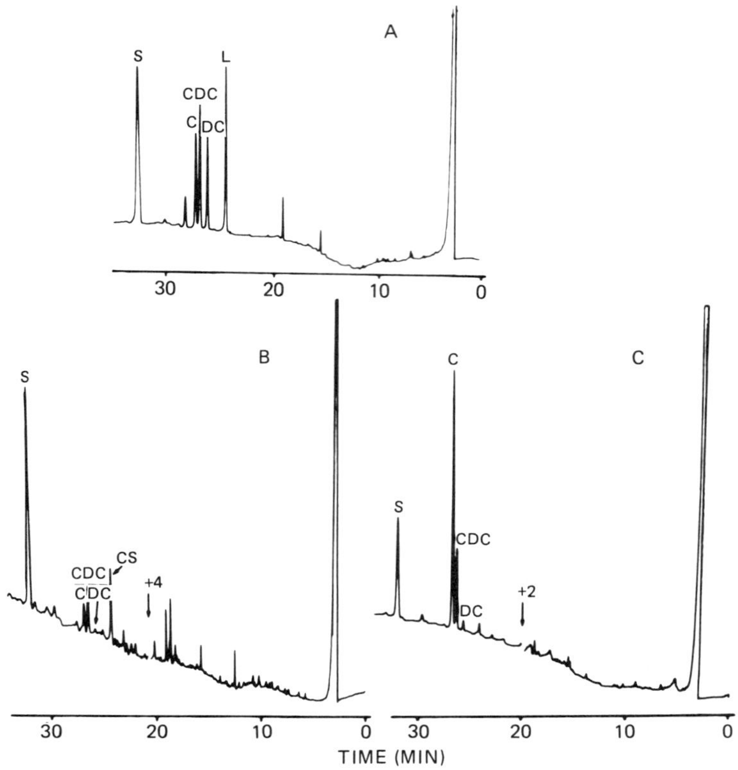

Figure 13.2 Glass capillary gas chromatograms of trimethylsilyl ethers of bile acid methyl esters. A 50-m OV-101 column programmed with 8°C min^{-1} from 150° to 255°C. (a) Reference cholic, C; chenodeoxycholic, CDC; deoxycholic, DC; and lithocholic acid LC; and the internal standard, S. (b) A serum sample from a healthy female. (c) A serum sample from a patient with cholostasis of pregnancy; CS, cholesterol. (From Ref. 99. Reproduced by permission of Elsevier/North Holland Biomedical Press, Amsterdam.)

A second procedure presented by Karlaganis and Paumgartner [102,103] overcomes many of the above disadvantages. The internal standard hyodeoxycholic acid is added to 2 ml serum or 20 μl of bile, and the result is extracted by the Makino and Sjövall method [54] as modified by Schwarz et al. [104], which includes hydrolysis of the sulfated fraction. After alkaline hydrolysis and extraction with ethyl acetate, the free acids are esterified with diazomethane and then converted into their trimethylsilyl derivatives, as described by Makita and Wells [105]. The analysis was carried out on a 20 m × 0.3 mm capillary column coated with a crystal layer of barium carbonate, which in turn was coated with polyethylene glycol 20,000 in methylene chloride by the static coating method [108]. A flame ionization detector was used and the injection port was modified as described by Grob and Grob [106]. The injection port temperature was 300°C and the column temperature was 230°C. Hydrogen was the carrier gas with a flow-rate of 4.4 ml/min and a split ratio of 1:5. This column gave excellent separation of the major bile acids and cholesterol, as can be seen in Figure 13.3. A

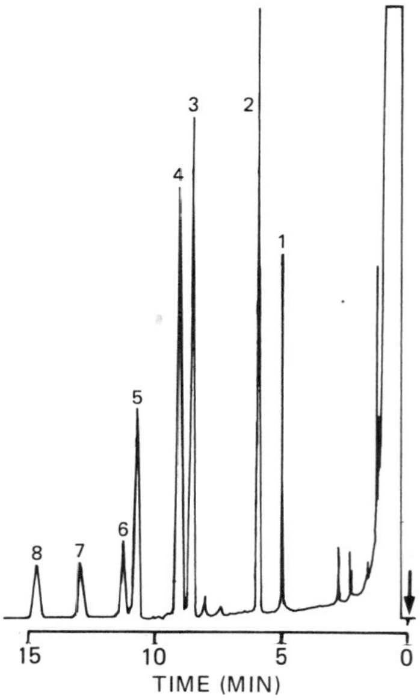

Figure 13.3 Capillary gas chromatogram of TMS ethers of methyl esters of a standard bile acid mixture containing cholesterol (1), cholic (2), deoxycholic (3), chenodeoxycholic (4), hyodeoxycholic (internal standard, 250 pmol per injection) (5), lithocholic (6), ursodeoxycholic (7), and 3-β-hydroxy-5-cholenoic acid (8). A 20-m PG 20 M glass capillary column was used at 230°C. (From Ref. 102. Reproduced by permission of *Journal of Lipid Research.*)

base-line separation of cholesterol, cholic, deoxycholic, chenodeoxycholic, hydrodeoxycholic, lithocholic, ursodeoxycholic, and 3β-hydroxy-5-cholenic acid within a 15-min analysis was achieved [109].

The standard curves are linear and the variation is rather low. The application of this method to the analysis of bile acids in serum and gallbladder bile is shown in Figures 13.4 and 13.5, respectively. There are several problems with this method. As seen in Figure 13.4, almost all of the peaks are on the tail of the solvent front, which makes the use of automatic integrators almost impossible. The broad solvent front can be due to impurities in the sample or to the silylation procedure. The other major problems include the use of diazomethane (as described above) and the short life-time of the column. The same authors [110] have reported the use of on-column injection as described by Grob and Grob [111] to improve the sensitivity and reproducibility.

The first results from our laboratory [112] used the extraction method of Makino and Sjövall [54] and conversion of the methyl esters with diazomethane. The analysis was carried out on a 20 m × 0.32 mm glass capillary column coated with SE-54. Using hydrogen as a carrier gas at a flow-rate of 3.2 ml/min, injection port temperature of 275°C, and a temperature program from 180 to 250°C at 1.2°C/min, a separation of cholanic, cholic, lithocholic, chenodeoxycholic, hyodeoxycholic, 3,12-diketocholic, 7,12-diketolithocholic, and dihydrocholic acid was obtained (Figure 13.6). Because the separation was not the best and because we wanted to avoid the use of diazomethane, we developed the following method: The extraction and derivatization procedures were as

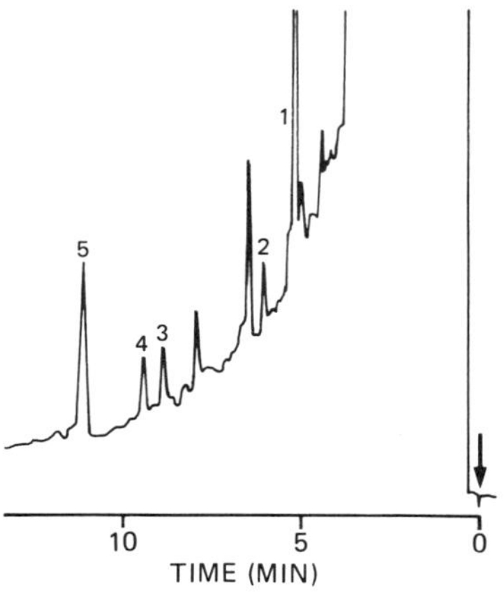

Figure 13.4 Capillary gas chromatogram of TMS ethers of methyl esters of bile acids extracted with XAD-2 from 2 ml of normal human serum containing 6.36 nmol hyodeoxycholic acid as internal standard. Peak numbers are the same as in Figure 13.3. A 20-m PG 30 M glass capillary column was used at 230°C. (From Ref. 102. Reproduced by permission of *Journal of Lipid Research*.)

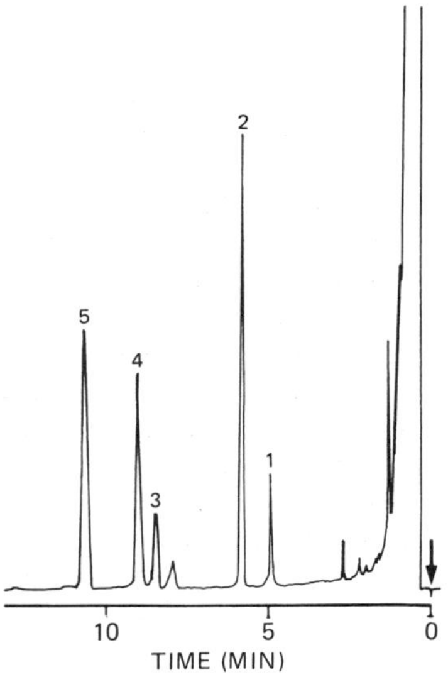

Figure 13.5 Capillary gas chromatogram of TMS ethers of methyl esters of bile acids extracted from 20 μl of human bile containing 127 nmol of hyodeoxycholic acid as internal standard. Peak numbers are the same as in Figure 13.3. A 20-m PG 20 M glass capillary column was used at 230°C. (From Ref. 102. Reproduced by permission of *Journal of Lipid Research*.)

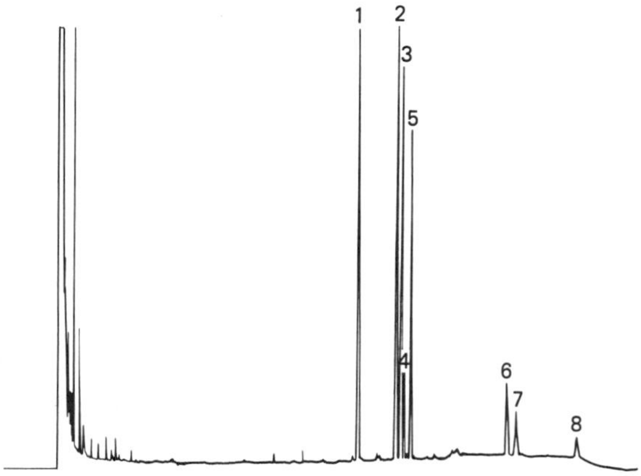

Figure 13.6 Chromatogram of a standard mixture of bile acid methyl ester trifluoroacetates. Column 20 m × 0.32 mm I.D. coated with SE-54. Injection port, 275°C. Program, 180–250°C with 1.2°C min^{-1}. Peak numbers: 1, cholanic acid; 2, cholic acid; 3, lithocholic acid; 4, chenodeoxycholic acid; 5, hyodeoxycholic acid; 6, 3,12-diketocholic acid; 7, 7,12-diketolithocholic acid; 8, dihydrocholic acid.

described in Figure 13.1A 50 m × 0.3 mm glass capillary column coated with OV-1 obtained from H. and G. Jaeggi, Trogen, Switzerland, was used. The column efficiency number (TZ) was 57.1, which gives a TZ/m of 1.14. The equations used are presented in Chap. 10. From this TZ/m value the number of theoretical plates (m) was calculated to be 337,561 or n/m of 6,751. The capacity ratio (k) was found to be 18.9, while the capacity (C) of the column for cholanic acid was 9.7 × 10^{-6} g as calculated from the following equation:

$$C = 0.05\,M\,d^3\,(1 + k)$$

where M is the molecular weight of cholanic acid (360.56) and d is the internal diameter.

The carrier gas was nitrogen, and an electron capture detector was used initially; however, we found that the concentrations of the bile acids in serum were high enough that a flame ionization detector (FID) could be used. The carrier gas was changed to hydrogen to shorten the analysis time. The carrier gas flow-rate was 4.2 ml/min, and flow-rates of the hydrogen and air in the FID were 14 ml/min and 160 ml/min, respectively. The injection port temperature was 275°C, and the analysis program was injection at 70°C, initial temperature 180°C, programmed to 250°C at a rate of 1°C/min. The total analysis time was 70 min. An automatic injector could be used in the analysis. As shown in Figure 13.7, deoxycholic and cholic acid, as well as lithocholic, chenodeoxycholic and Δ5-cholenic acid, were separated by 4,6σ.

Figure 13.8 shows a chromatogram of bile acids extracted from serum as their hexafluoroisopropyl ester trifluoroacetates, and in Figure 13.9 the bile acids extracted from bile as the same derivatives. This general procedure has been used for routine clinical assays [113].

At this time, these last two methods are comparable, each with its own advantages and disadvantages. The major problem with ours is the relatively

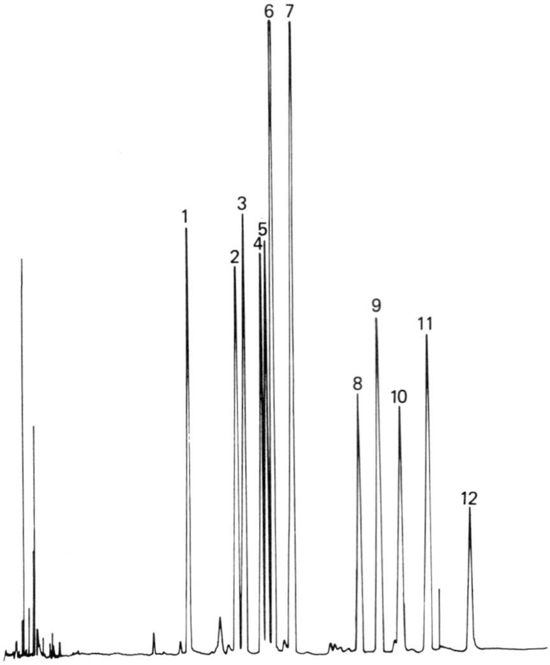

Figure 13.7 Chromatogram of a standard mixture of bile acid hexafluoroisopropyl ester trifluoroacetates. Column: 50 m × 0.33 I.D., coated with OV-1. Injection port, 275°C. Program, 180–250°C with 1.5°C min^{-1}. Peak numbers: 1, cholanic acid; 2, deoxycholic acid; 3, cholic acid; 4, lithocholic acid; 5, chenodeoxycholic acid; 6, Δ5-cholanic acid; 7, ursodeoxycholic acid; 8, cholesterol; 9, 7-ketodeoxycholic acid (internal standard); 10, 3,12-diketocholanic aicd; 11, 7,12-diketolithocholic acid; 12, dihydrocholic acid.

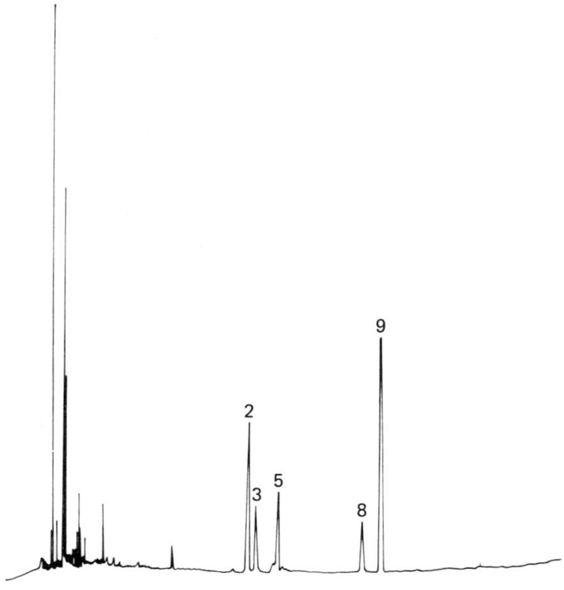

Figure 13.8 Chromatogram of bile acid hexafluoroisopropyl ester trifluoroacetates extracted from human serum. GLC conditions and peak numbers as in Figure 13.7.

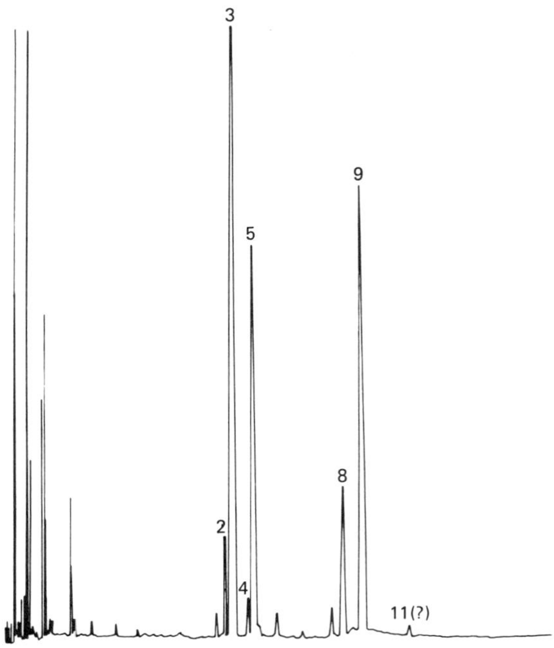

Figure 13.9 Chromatogram of bile acid hexafluoroisopropyl ester trifluoroacetates extracted from human bile. GLC conditions and peak numbers as in Figure 13.7.

long analysis time, 70 min versus 15 min for the Karlaganis et al. method [102,103]. However, our derivatives are stable enough to allow the use of an autoinjector as well as the use of a solid sampler. The use of this equipment makes the length of the analysis less important. The solid sampler also eliminates the solvent front and increases column life. Our present column has been in use for 1 year at 23 hr a day, with no significant change in properties. Which method to use depends on the philosophy of the particular laboratory. Both methods are superior in column efficiency, accuracy, and precision to packed columns, TLC, or HPLC. The use of capillary columns should open up new possibilities in the study of bile acids and their metabolism.

IV. SUMMARY

The development of glass capillary gas-liquid chromatography is only one more step in the logical development of more sophisticated techniques for the study of the physiology and pathophysiology of bile acids. The most widespread methods use the resins such as XAD-2 for extraction. After hydrolysis and/or solvolysis of the sulfated fraction, the isolated bile acids are doubly derivatized, first via esterification and second, acetylation with trifluoroacetic anhydride or trimethylsilylation. No special columns or detectors are needed for the analysis. The trimethylsilyl derivatives have been analyzed on Carbowax 20 M with great success, while the hexafluoroisopropyl ester-trifluoroacetates are best analyzed on OV-1. Both methods now in use have minor disadvantages, but both can be regarded as state-of-the-art techniques. Future efforts will probably focus on increasing the separation efficiencies and adopting these methods to the analysis of any new bile acids and metabolites that are discovered.

REFERENCES

1. B. Josephson, *Physiol. Rev. 21*:463 (1941).
2. M. C. Carey and D. M. Small, *Arch. Intern. Med. 130*:506 (1972).
3. A. F. Hofmann and D. M. Small, *Amer. Rev. Med. 18*:433 (1967).
4. P. P. Nair and D. Kritchevsky, *The Bile Acids: Chemistry, Physiology, and Metabolism*, Plenum Press, New York, 1971.
5. J. Sjövall, *Clin. Chim. Acta 5*:33 (1960).
6. H. Greim, D. Trülzsch, J. Roboz, K. Dressler, P. Czygan, F. Hutterer, F. Schaffner and H. Popper, *Gastroenterology 63*:837 (1972).
7. A. Stiehl, *Europ. J. Clin. Invest. 4*:59 (1974).
8. T. Fedorowski, G. Salen, F. G. Zaki, S. Shefar, and E. H. Mosbach, *Gastroenterology 74*:75 (1978).
9. J. Makino and S. Nakagawa, *J. Lipid Res. 19*:723 (1978).
10. A. F. Hofmann, in *Advances in Internal Medicine* (G. H. Stollermann, ed.), Yearbook Publishers, Chicago, 1976, pp. 501–534.
11. J. A. Summerfield, B. H. Billing and C. H. L. Shackleton, *Biochem. J. 154*:507 (1976).
12. I. Makino, J. Sjövall, A. Norman, and B. Strandwick, *FEBS Lett. 15*: 161 (1971).
13. F. Hutterer, P. G. Bacchin, and H. Denk, *Life Sci. 9*:1159 (1970).
14. H. Schriewer, D. B. von Bassewitz, and H. M. Raven, *Klin Wochenschr. 51*:39 (1973).
15. A. F. Hofmann and G. Paugartner, eds., *Chenodeoxycholic Acid Therapy of Gallstones: Progress Report*, Schattauer, Stuttgart and New York, 1975.
16. H. S. Mekligian, S. F. Phillips, and A. F. Hofmann, *J. Clin. Invest. 50*:1569 (1971).
17. R. H. Palmer, *Arch. Intern. Med. 130*:606 (1972).
18. A. Stiehl, D. L. Earnest, and W. H. Admirand, *Gastroenterology 68*: 534 (1975).
19. A. Stiehl, R. Raedsoh, and B. Kommerell, *Digestion 12*:105 (1975).
20. D. N. Kim, K. T. Lee, J. M. Reiner, and W. A. Thomas, *Exp. Mol. Path. 22*:284 (1975).
21. M. Ogura, Y. Ayaki, and M. Goto, *J. Biochem. 80*:537 (1976).
22. K. A. Mitropoulos, N. B. Myout, G. F. Gibbons, and B. E. H. Balasubramainam, *J. Biol. Chem. 249*:6052 (1974).
23. C. C. Schwartz, Z. R. Vlahcevic, L. G. Halloran, R. Nisman, and L. Swell, *Proc. Soc. Exp. Biol. Med. 156*:261 (1977).
24. D. Mathé, F. d'Hollander, and F. Chevallier, *Biochimie 54*:1479 (1972).
25. J. Ahlberg, B. Angelin, J. Björkhem, and K. Einarsson, *Gastroenterology 73*:1377 (1977).

26. H. J. Mok, K. von Bergmann, and S. M. Grundy, *Gastroenterology 73*:684 (1977).
27. R. G. Knodell, M. D. Kinsey, E. C. Boedeker, and D. P. Collin, *Gastroenterology, 71*:196 (1976).
28. S. J. Mehta, J. E. Struthers, M. D. Kaye, and J. L. Naylor, *Gastroenterology 67*:674 (1974).
29. W. F. Caspary and S. Graf, *Dtsch. Med. Wochenschr. 103*:825 (1978).
30. S. M. Grundy, *J. Clin. Path. 55*:269 (1975).
31. C. R. Pennington, D. E. Ross, M. C. Bateson, and J. A. D. Bouchier, *J. Clin. Path. 31*:58 (1978).
32. K. Einarsson and H. Hellström, *Eur. J. Clin. Invest. 2*:225 (1972).
33. E. C. Osborn, I. D. P. Wootton, L. C. da Silva, and S. Sherlock, *Lancet II*:1049 (1959).
34. J. B. Carey, *J. Clin. Invest. 37*:1494 (1958).
35. Z. R. Vlahcevic, D. Juttijudata, C. C. Bell, and L. Swell, *Gastroenterology 62*:1174 (1972).
36. W. C. McCormick, C. C. Bell, L. Swell, and Z. R. Vlahcevic, *Gut 14*:895 (1970).
37. L. A. Turnberg and G. Grahame, *Gut 11*:126 (1970).
38. Z. R. Vlahcevic, J. Buhac, C. C. Bell, and C. Swell, *Gut 11*:420 (1970).
39. J. E. Struthers, S. J. Mehta, M. D. Kaye, and J. L. Naylor, *Am. J. Dig. Dis. 22*:861 (1977).
40. P. Back, *Z. Gastroenterologie 11*:477 (1973).
41. W. Fröhling and A. Stiehl, *Eur. J. Clin. Invest. 6*:67 (1975).
42. W. J. A. VandenHeuvel, C. C. Sweely, and E. C. Horning, *Biochem. Biophys. Res. Commun. 3*:33 (1960).
43. W. J. A. VandenHeuvel, J. Sjövall, and E. C. Horning, *Biochim. Biophys. Acta 48*:596 (1961).
44. A. Kuksis, in *Lipid Chromatographic Analysis*, Vol. 2 (G. V. Marinetti, ed.), Marcel Dekker, New York, 1969, p. 230.
45. J. Sjövall, in *Methods of Biochemical Analysis* (D. Glick, ed.), Wiely-Interscience, New York, 1964, p. 97.
46. B. Josephson, *Biochem. J. 29*:1519 (1935).
47. A. Norman, *Acta Physiol. Scand. 32*:1 (1954).
48. H. Danielson, P. Eneroth, U. Hellstrom, S. Lindstedt, and J. Sjövall, *J. Biol. Chem. 238*:2299 (1963).
49. A. Kuksis, *J. Amer. Oil Chem. Soc. 42*:276 (1965).
50. B. A. Gordon, A. Kuksis, and J. M. R. Beveridge, *Can. J. Biochem. Physiol. 41*:77 (1963).
51. S. M. Grundy, E. H. Ahrens, Jr., and T. A. Miettinen, *J. Lipid Res. 6*:397 (1965).
52. D. Sandberg, K. Sjövall, J. Sjövall, and D. A. Turner, *J. Lipid Res. 6*:182 (1965).
53. G. W. Kuron and D. W. Tennent, *Fed. Proc. 20*:268 (1961).
54. I. Makino and J. Sjövall, *Anal. Lett. 5*(6):341 (1972).
55. H. Bradlow, *Steroids 11*:265 (1958).
56. I. Makino, K. Shinozaki, S. Nakogawa, and K. Mashimo, *J. Lipid Res. 15*:132 (1974).
57. J. Blome, personal communication.
58. C. D. Klassen, *Clin. Chim. Acta 35*:225 (1971).
59. T. Okishio, P. P. Nair, and M. Gordon, *Biochem. J. 102*:654 (1967).
60. H. Danielson, *Adv. Lipid Res. 1*:335 (1963).

61. M. Jenke and F. Bandow, *Z. Physiol. Chem. 249*:16 (1937).
62. J. B. Carey and C. J. Watson, *J. Biol. Chem. 216*:847 (1963).
63. A. Norman and R. H. Palmer, *J. Lab. Clin. Med. 63*:986 (1964).
64. P. Eneroth, B. Gordon, R. Ryhage, and J. Sjövall, *J. Lipid Res. 7*: 511 (1966).
65. P. Eneroth, *J. Lipid Res. 4*:11 (1963).
66. P. Eneroth, B. Gordon and J. Sjövall, *J. Lipid Res. 7*:524 (1966).
67. P. Eneroth, K. Hellström, and J. Sjövall, *Acta Chem. Scand 22*:1729 (1968).
68. E. Evrard and G. Janssen, *J. Lipid Res. 9*:226 (1968).
69. H. Greim, P. Czygan, F. Schaffner, and H. Popper, *Biochem. Med. 8*:280 (1973).
70. T. Okishio, P. P. Nair, and M. Gordon, *Biochem. J. 102*:654 (1967).
71. F. Nakayama, *J. Lab. Clin. Med. 72*:602 (1968).
72. L. Schoenfield, J. Sjövall, and K. Sjövall, *J. Lab. Clin. Med. 68*:186 (1966).
73. H. S. Wiggins and I. D. P. Wootton, *Biochem. J. 70*:349 (1958).
74. S. J. Levin, G. Johnston, and A. J. Boyle, *Anal. Chem. 33*:1407 (1961).
75. E. M. Mosbach, H. J. Kalinsky, E. Halpern, and F. E. Kendall, *Arch. Biochem. Biophys. 51*:402 (1954).
76. S. J. Levin, J. L. Irvin, and C. G. Johnston, *Anal. Chem. 33*:856 (1961).
77. P. P. Nair, M. Gordon, S. Gordon, J. Reback, and A. I. Mendeloff, *Life Sci. 4*:1877 (1965).
78. R. L. Kelly and E. A. Doisy, Jr., *Fed. Proc 26*:849 (1967).
79. P. P. Nair, M. Gordon, and J. Reback, *J. Biol. Chem. 242*:7 (1967).
80. P. P. Nair, in *Bile Salt Metabolism* (L. Schiff, J. B. Carey, and J. H. Dietchy, eds.), Thomas, Springfield, Ill., 1969, p. 172.
81. P. P. Nair and C. Garcia, *Anal. Biochem. 29*:164 (1969).
82. J. Roovers, E. Evrard, J. Vanderhaeghe, *Clin. Chim. Acta 19*:449 (1968).
83. R. H. Palmer, *Proc. Nat. Acad. Sci. U.S.A. 58*:1047 (1967).
84. D. K. Bloomfield, *Anal. Chem. 34*:737 (1962).
85. H. Schlenk and J. L. Gellerman, *Anal. Chem. 32*:1412 (1960).
86. T. J. de Boer and H. S. Backer, *Rec. Trav. Chim. 73*:229 (1954).
87. C. D. Gutsche, *Org. Reactions 8*:391 (1954).
88. O. Hammarsten, *Z. Physiol. Chem. 61*:459 (1909).
89. A. F. Hofmann and E. H. Mosbach, *J. Biol. Chem. 239*:2813 (1964).
90. K. Imai, Z. Tamura, F. Mashige, and T. Osuga, *J. Chromatogr. 120*: 181 (1976).
91. S. W. Dziedzic, L. B. Dziedzic, and S. E. Githow, *J. Lab. Clin. Med. 82*:829 (1973).
92. J. Sjövall, *Acta Chem. Scand. 16*:1761 (1962).
93. A. Kuksis and B. A. Gordon, *Can. J. Biochem. Physiol. 41*:1355 (1963).
94. G. Jung, J. Pauschmann, W. Voelter, E. Breitmaier, and E. Bayer, *Chromatographia 3*:26 (1970).
95. J. Sjövall, in *Biomedical Applications of Gas Chromatography* (H. A. Szymanski, ed.), Plenum Press, New York, 1964, p. 151.
96. M. Donike, *J. Chromatogr. 78*:273 (1973).
97. W. R. Supina, R. F. Kruppa, and R. S. Henly, *J. Amer. Oil Chem. Soc. 43*:138 (1966).

98. W. J. A. VandenHeuvel, E. O. A. Haahti, and E. C. Horning, *J. Amer. Chem. Soc. 83*:1513 (1961).
99. T. Laatikainen and A. Hesso, *Clin. Chim. Acta 64*:63 (1975).
100. K. Grob, *Helv. Chim. Acta 48*:1362 (1965).
101. K. Grob, *Helv. Chim. Acta 51*:718 (1968).
102. G. Karlaganis and G. Paumgartner, *J. Lipid Res. 19*:771 (1978).
103. G. Karlaganis and G. Paumgartner, *Clin. Chim. Acta*
104. H. P. Schwarz, K. von Bergmann, and G. Paumgartner, *Clin. Chim. Acta 50*:197 (1974).
105. M. Makita and W. W. Wells, *Anal. Biochem. 5*:1 (1972).
106. K. Grob and K. Grob, Jr., *J. Chromatogr. 94*:53 (1974).
107. K. Grob, G. Grob, and K. Grob, Jr., *Chromatographia 10*:181 (1977).
108. K. Grob, G. Grob, and K. Grob, Jr., *J. Chromatogr. 125*:471 (1976).
109. D. C. Jones, G. W. Melchior and M. J. Watkins Reeves, *J. Lipid Res. 17*:273 (1976).
110. G. Karlaganis and G. Paumgartner, *HRC & CC 1*:54 (1978).
111. K. Grob and K. Grob, Jr., *J. Ghromatogr. 151*:311 (1978).
112. H. Jaeger, W. Wagner, J. Homoki, H. U. Klör, and H. Ditschuneit, *Chromatographia 10*:492 (1977).
113. W. Nebelung, H. Jaeger, H. U. Klör, and H. Ditschuneit, 2nd Can. Conf. on Chromatography, Toronto, 1979.

14

Determination of the Absolute Configuration of Chiral Hydroxy Acids by GLC

Johannis P. Kamerling and Johannes F. G. Vliegenthart / *Department of Bio-Organic Chemistry, State University of Utrecht, Utrecht, The Netherlands*

I. INTRODUCTION

In biochemical studies on chiral organic compounds, it is often of great importance to know the absolute configurations of the chiral centers. This information can give insight into the stereospecific course of certain enzymatic reactions. In clinical practice it can indicate specific enzyme deficiencies in inborn errors of metabolism.

The absolute configuration of a chiral compound can be determined in different ways. Generally applied techniques include measurements based on the optical activity (polarimetry, optical rotation dispersion, circular dichroism), enzymatic analysis, nuclear magnetic resonance spectroscopy, and chromatography (gas-liquid chromatography, high-pressure liquid chromatography, etc.).

During the last 20 years significant progress has been made in the application of gas-liquid chromatography (GLC) for the determination of the absolute configuration of biochemically interesting chiral substances. Comprehensive reviews on this subject have been published by Gil-Av and Nurok [1] and Hušek and Macek [2]. The GLC separation of racemic mixtures can be effected in two ways: (1) separation of the enantiomers (if necessary, after protection of polar groups) on a chiral stationary phase; and (2) conversion of enantiomers into diastereomers using a chiral reagent, followed by separation of the diastereomers on a nonchiral stationary phase. After assignment of the GLC peaks by cochromatography with enantiopure substances, the method can be applied for chirality determinations.

The great advantages of the GLC determination of the absolute configurations are:

1. It is a highly sensitive method with a generally applicable detection system.
2. It is not necessary to purify the compound under investigation to a great extent, which facilitates the analysis of material obtained from biological sources.
3. Mixtures of chiral substances can be analyzed without preceding separation.
4. The enantiopurity of chiral compounds can be determined very easily.
5. A coupling with mass spectrometry for identification and verification is possible.

The GLC method has already been applied to a wide range of natural compounds, for example, amino acids [2–9], monosaccharides [10–12], and chiral organic acids. This chapter will deal with the application of capillary and packed column GLC for the analysis of the absolute configuration of chiral hydroxy acids, using diastereomeric derivatives and nonchiral stationary phases. In particular, emphasis is given to chiral hydroxy acids which occur in metabolic pathways.

II. CLINICAL CHEMICAL ASPECTS

In the last decennia, biochemical investigations in relation to inborn errors of metabolism have become very important in human health and disease. Nowadays, many clinical laboratories apply a large number of screening procedures for amino acids, organic acids, steroids, purines, pyrimidines, monosaccharides, oligosaccharides, (glyco)peptides, mucopolysaccharides, etc. In the study of metabolites, GLC in combination with mass spectrometry (MS) has been shown to be a powerful technique for getting definite answers.

GLC-MS analysis of organic acids from physiological fluids has led to the detection and identification of many acids which were not known before [13–15]. A number of new inborn errors of metabolism can be diagnosed, especially in relation with amino acid catabolism [16,17].

The investigation of organic acids from physiological fluids has now reached a stage wherein it has become significant to determine the absolute configuration of the chiral acids as well. Because different stereoisomers will be formed along different metabolic routes, the absolute configuration of the chiral centers might give a clue to the origin of these metabolites.

III. FATTY ACID OXIDATION

In plant and animal systems a wide variety of hydroxy fatty acids may occur [18], which differ with regard to the positions of the hydroxyl groups as well as the chain lengths of these compounds.

A number of metabolic pathways may lead to the formation of hydroxy fatty acids. In the framework of this review the attention has been restricted to the analysis of some products obtained from α and ω2 hydroxylation of a few fatty acids. Furthermore, the determination of the enantiopurity and the absolute configurations of hydroxy fatty acids, which can be prepared through reduction of hydroperoxides formed by the action of lipoxygenases [EC 1.13.-11.12], will be discussed. The latter enzyme is capable of oxygenating polyunsaturated fatty acids containing a 1,4-*cis,cis*-pentadiene system [19]. In this way conjugated hydroperoxy fatty acids are formed. Apparently the source of the enzyme as well as the incubation conditions affect the regio- and stereospecificities of the reaction [20].

IV. CARBOHYDRATES

The determination of the absolute configuration of monosaccharides is required for the structural determination of polysaccharides, oligosaccharides, glycoproteins, glycolipids, and other carbohydrate-containing materials. Also, in several biosynthetic processes dealing with carbohydrates, it can be of importance to know the absolute configuration of the monosaccharides. The introduction of GLC has made it possible to analyze mixtures of monosaccharides directly.

Nowadays, two GLC approaches exist for the determination of the absolute configuration of monosaccharides. The first method, described by Pollock and Jermany [21,22], is based on the oxidation of the aldehyde function and analysis of the formed aldonic acids. In this way the absolute configuration of the substituents at carbon-2 is determined. From the known relative positions of the hydroxyl groups in the parent sugar, it is possible to deduce the absolute configuration of the substituents at the chiral carbon atom adjacent to the primary hydroxyl group, which indicated the D- or L-monosaccharide. The second technique, described by Gerwig et al. [10,11] and Leontein et al. [12], consists of the glycosidation of the monosaccharides with a chiral alcohol and analysis of the formed alkyl glycosides. Although the latter approach is much more convenient, only the analysis of aldonic acids will be included in this review.

V. GAS-LIQUID CHROMATOGRAPHIC DATA

A. Lactic Acid

The enantiomeric forms of lactic acid can be separated on nonchiral stationary phases after acylation with a chiral acid chloride or after esterification with a chiral alcohol. Westley and Halpern [23] have reported the GLC separation of DL-lactic acid on 5% QF-1 (5 ft × 1/8 in.; N_2 flow, 30 ml/min) using the *O*-(-)-menthyloxycarbonyl methyl ester derivatives (R_D/R_L = 1.09; T = 170°C). As shown by Pollock and Jermany [21], D-lactic acid gives rise to two peaks on a Carbowax 20 M capillary column (150 ft × 0.02 in.) after esterification with

(±)-2-butanol, (±)-3-methyl-2-butanol, or (±)-3,3-dimethyl-2-butanol, followed by *O*-acetylation. In relation to these data it has to be noted that on nonchiral stationary phases the combinations D-enantiomer/(-) chiral reagent and L-enantiomer/(+) chiral reagent are eluted together, as are the combinations D-enantiomer/(+) chiral reagent and L-enantiomer/(-) chiral reagent. Kamerling et al. [24] used the *O*-acetylated (-)-menthyl ester derivatives for the resolution of DL-lactic acid.

O-Acetylated L-lactic acid has been employed for the resolution of a large number of (±) chiral alcohols. Capillary columns and packed columns with both polar and nonpolar stationary phases have been applied [1]. Gil-Av et al. [6] demonstrated that increasing chain lengths of 2-*n*-alkanols (4–10 carbon atoms) result in increasing resolution factors. Karger et al. [25] reported on the separation of the L-lactic acid esters of the branched aliphatic alcohols 3-methyl-2-butanol, 3,3-dimethyl-2-butanol, 2-methyl-3-pentanol, and 2,2-dimethyl-3-pentanol. The separation of the L-lactic acid ester of (±)-menthol on a packed column with 1,2,3,4-tetrakis(2-cyanoethoxy)butane has been shown by Nurok [26].

In Figure 14.1a, the separation of the *O*-acetylated DL-lactic acid (-)-menthyl ester on a SP-1000 capillary column is presented [24]. The two peaks were identified by cochromatography with the diastereomeric derivatives of D- or L-lactic acid. To ensure that the menthylation procedure did not lead to racemization, D- and L-alanine were tested. In both cases only the enantiopure substances were found. Commercially available D-lactic acid consisted of

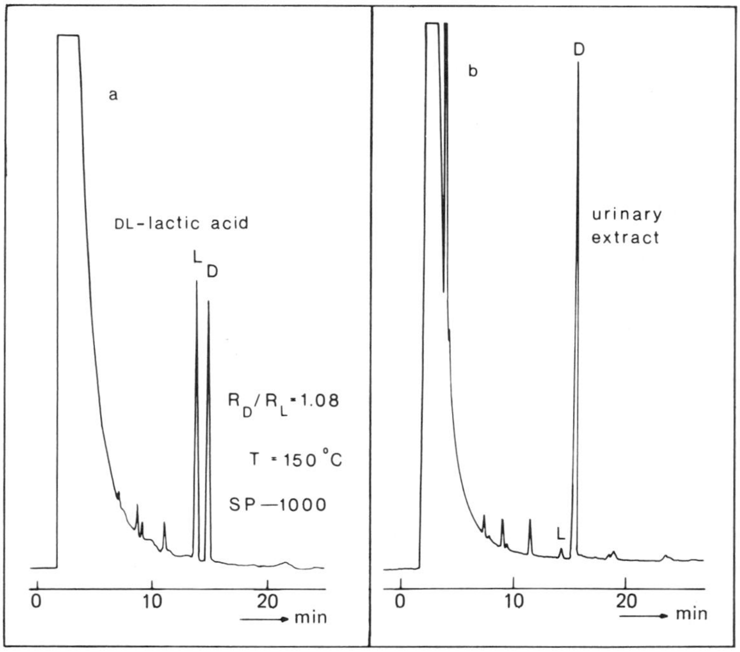

Figure 14.1 Gas chromatograms on a SP-1000 WCOT capillary column (25 m × 0.3 mm I.D.; N_2 flow, 1 ml/min) of the *O*-acetylated (-)-menthyl ester derivatives of (a) commercial DL-lactic acid, and (b) D-lactic acid from an urinary ethyl acetate extract of a patient with D-lactic aciduria. (Data from Refs. 24 and 27.)

98.4% D and 1.6% L, and L-lactic acid of 99.4% L and 0.6% D. Since it was shown that treatment with menthol/HCl did not lead to racemization, the contaminating optical antipodes were considered as intrinsic.

In the screening for inborn errors of metabolism the accumulation of lactic acid in the urine of patients is a relatively often occurring phenomenon. In almost all cases the absolute configuration of the excreted substance has been shown to be L (see Section V.C). However, the excretion of high levels of D-lactic acid, not derived from the intestinal bacterial flora, has been observed by Duran et al. [27]. In the latter case, the absolute configuration was determined using the *O*-acetylated (-)-menthyl ester derivative. Figure 14.1b presents the gas chromatogram of the urinary ethyl acetate extract after derivatization with (-)-menthol/HCl and acetic anhydride. No additional purification was carried out. D-lactic acid was also found to occur in the blood and CSF.

B. Glyceric Acid

Pollock and Jermany [21] reported that the *O*-acetylated (±)-2-butyl, (±)-3-methyl-2-butyl, or (±)-3,3-dimethyl-2-butyl esters of D-glyceric acid give rise to two peaks on a Carbowax 20 M capillary column (150 ft × 0.02 in.). As has been shown by Kamerling et al. [24], the enantiomers of DL-glyceric acid can be resolved as the *O*-acetylated (-)-menthyl ester diastereomeric derivatives. The latter separation on a SP-1000 capillary column is presented in Figure 14.2a. The two peaks were identified by cochromatography with the diastereomeric derivatives of D- or L-glyceric acid.

Patients with hyperoxaluria type II excrete high amounts of oxalic acid and L-glyceric acid [28]. However, glyceric acid accumulation is not restricted to those patients only. It has also been found in patients with other clinical symptoms. Brandt et al. [29] described a patient with D-glyceric acidemia in conjuction with hyperglycinemia. Wadman et al. [30] reported on a patient showing D-glyceric acidemia and aciduria, but without hyperglycinemia. Figures 14.2b and 14.2c show the analyses of the *O*-acetylated (-)-menthyl esters of the urinary organic acids obtained from a patient with L-glyceric aciduria (hyperoxaluria type II), as reported by Kamerling et al. [31], and one with D-glyceric aciduria [30], respectively. In both cases urinary ethyl acetate extracts were used without further purification.

C. Short-Chain Hydroxy Acids

The resolution of the *O*-acetylated (-)-menthyl esters of DL-2-hydroxybutyric acid and DL-3-hydroxybutyric acid have been described by Kamerling et al. [32]. Figure 14.3a shows the separation of these diastereomers on a SE-30 capillary column; the separation of the DL-lactic acid derivatives has also been included. The different peaks were assigned by cochromatography with the derivatives of L-lactic acid, D-2-hydroxybutyric acid, and D-3-hydroxybutyric acid. It has to be mentioned that the diastereomeric derivatives of 3-hydroxybutyric acid elute in the reversed order with respect to those of lactic acid and 2-hydroxybutyric acid.

Patients with lactic acidemia and ketosis regularly show urinary excretion of increased amounts of 2-hydroxybutyric acid. As shown in Figure 14.3b, the absolute configuration of this 2-hydroxy acid and those of lactic acid and 3-hydroxybutyric acid can be determined in one gas chromatographic run. For 2-hydroxybutyric acid, the L configuration was found. Lactic acid occurred in the L configuration, and 3-hydroxybutyric acid in the D configuration.

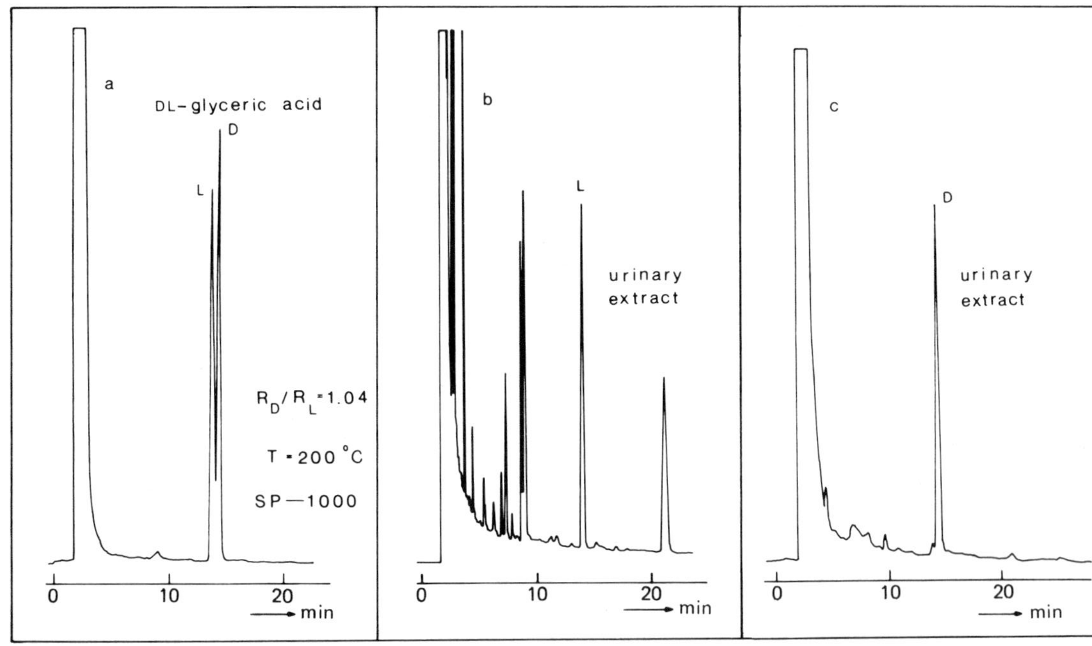

Figure 14.2 Gas chromatograms on a SP-1000 WCOT capillary column (25 m × 0.3 mm I.D.; N_2 flow, 1 ml/min) of the *O*-acetylated (-)-menthyl ester derivatives of (a) commercial DL-glyceric acid, (b) L-glyceric acid from an urinary ethyl acetate extract of a patient with hyperoxaluria type II, and (c) D-glyceric acid from an urinary ethyl acetate extract of a patient with D-glyceric aciduria. (Data from Refs. 24, 30, and 31.)

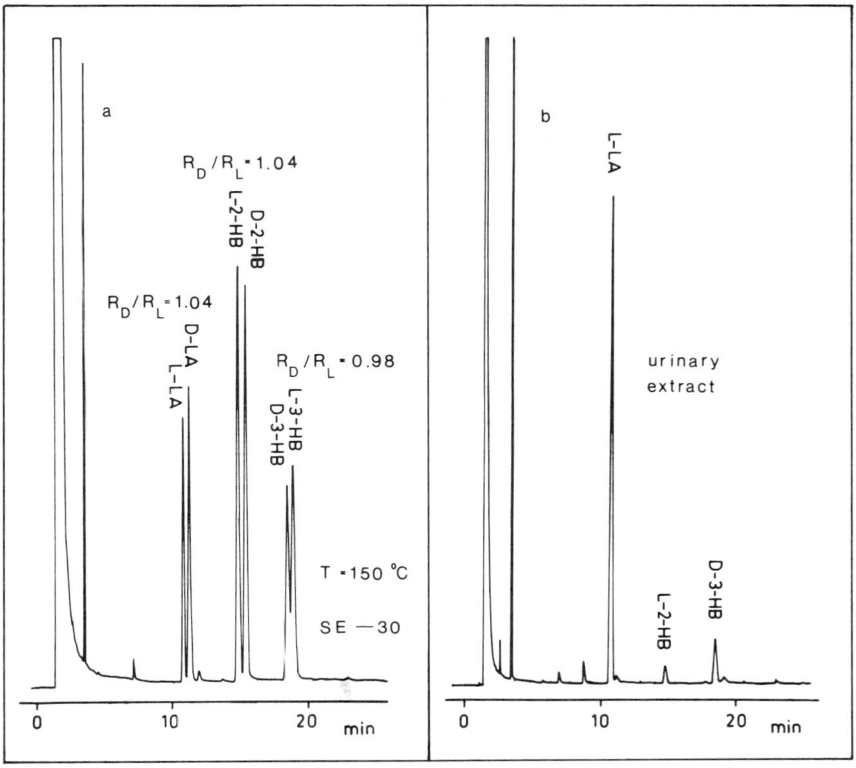

Figure 14.3 Gas chromatograms on a SE-30 WCOT capillary column (25 m × 0.3 mm I.D.; N_2 flow, 1 ml/min) of the *O*-acetylated (-)-menthyl ester derivatives of (a) a mixture of commercial DL-lactic acid (LA), DL-2-hydroxybutyric acid (2-HB), and DL-3-hydroxybutyric acid (3-HB); and (b) the urinary organic acids in a patient with lactic acidemia due to deficiency of cytochromes aa_3 and *b*. (Data from Ref. 32.)

The diastereomeric separation of the *O*-(-)-menthyloxycarbonyl methyl esters of DL-2-hydroxyisovaleric acid and DL-2-hydroxylisocaproic acid on 5% QF-1 (5 ft × 1/8 in.; N_2 flow, 30 ml/min) have been reported by Westley and Halpern [23]. Resolution factors at 170°C were found to be 1.14 and 1.15, respectively. The DL enantiomers of these acids and those of 2-hydroxy-3-methylvaleric acid can also be resolved as the *O*-acetylated (-)-2-butyl ester derivatives (Kamerling and Wadman, unpublished results). In Figure 14.4a, the separation of the *O*-acetylated (-)-2-butyl esters of DL-2-hydroxyisovaleric acid, DL-2-hydroxyisocaproic acid, and DL-2-hydroxy-DL-3-methylvaleric acid on a SP-1000 capillary column are presented. The observed resolution of the diastereomeric derivatives of the latter hydroxy acid was influenced only by the absolute configuration of the substituents at carbon-2. The peak assignments were carried out by cochromatography.

Maple syrup urine disease is attended with the urinary accumulation of the branched-chain amino acids valine, leucine, isoleucine, and *allo*-isoleucine, as well as of the corresponding oxo and hydroxy acids. The concentration of 2-hydroxyisovaleric acid is relatively high in comparison to the other 2-hydroxy acids. The analysis of the urine of such a patient is shown in Figure 14.4b,

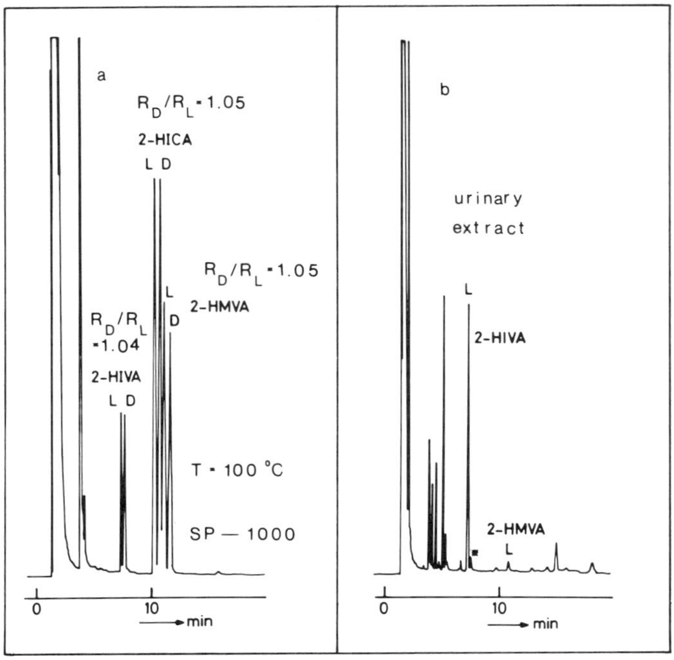

Figure 14.4 Gas chromatograms on a SP-1000 WCOT capillary column (25 m × 0.3 mm I.D.; N_2 flow, 1 ml/min) of the *O*-acetylated (-)-2-butyl esters of (a) a mixture of commercial DL-2-hydroxyisovaleric acid (2-HIVA), DL-2-hydroxy-isocaproic acid (2-HICA), and DL-2-hydroxy-DL-3-methylvaleric acid (2-HMVA); and (b) the urinary organic acids in a patient with maple syrup urine disease (* indicates the small peak obtained from the contaminating (+)-2-butanol in the commercial (-)-2-butanol sample).

demonstrating relatively large amounts of L-2-hydroxyisovaleric acid and smaller amounts of L-2-hydroxy-3-methylvaleric acid (Kamerling and Wadman, unpublished results). Because of the very small quantity of 2-hydroxyiso-caproic acid in the urinary extract, the absolute configuration of the latter acid could not be determined.

The resolution of the *O*-(-)-menthyloxycarbonyl methyl ester derivatives of DL-2-hydroxyheptanoic acid has been described by Hamberg [33] and that of DL-2-hydroxydecanoic acid by Nugteren [34] (see Section V.D.). As reported by van Os et al. [35], the *O*-acetylated (-)-2-butyl ester derivatives of DL-2-hydroxyheptanoic acid could be separated on a SP-1000 capillary column (25 m × 0.3 mm; N_2 flow, 1 ml/min; $R_D/R_L = 1.04$; T = 120°C).

D. Hydroxydicarboxylic Acids

The resolution of the *O*-acetylated di-(±)-2-butyl, (±)-3-methyl-2-butyl and (±)-3,3-dimethyl-2-butyl esters of D-malic acid and D-tartaric acid was carried out by Pollock and Jermany on a Carbowax 20 M capillary column (150 ft × 0.02 in.) [21]. D-malic acid gives rise to two peaks, whereas for D-tartaric acid, a triplet with a ratio 1:2:1 was obtained. As has been reported by Kamerling et al. [36] and van Os et al. [35], the *O*-acetylated di-(-)-2-butyl esters of DL-malic acid, DL-2-hydroxyglutaric acid, DL-2-hydroxyadipic acid, and

DL-2-hydroxysebacic acid can be separated on a SP-1000 capillary column (Figure 14.5). The different peaks were assigned L and D using the L enantiomers as reference compounds. In all cases the L-enantiomer derivative elutes earlier than the D-enantiomer derivative. Esterification of 2-hydroxyglutaric acid yields considerable amounts of the lactone form (see Figure 14.5). Gas-liquid chromatography of the *O*-acetylated di-(±)-2-butyl esters of the L-2-hydroxydicarboxylic acids gives rise to the same chromatographic peak pattern as presented in Figure 14.5. Therefore, the introduced chiral ester function not directly attached at the chiral carbon-2 center of the hydroxy acid does not influence the chromatographic behavior on SP-1000.

Relatively high excretions of malic acid, 2-hydroxyglutaric acid, or 2-hydroxyadipic acid have been observed in patients with different diseases (see Kamerling et al. [36]). From the urine of such patients, Figures 14.6a-e show gas chromatograms of ethyl acetate extracts, treated with (-)-2-butanol/HCl and acetic anhydride. Figure 14.6a shows the chromatogram of a patient with lactic acidemia who excreted among other organic acids L-malic acid [36]; Figure 14.6b represents the chromatogram of a patient with L-2-hydroxyglutaric aciduria [36], described by Duran et al. [37]; Figure 14.6c, the chromatogram of a patient with D-2-hydroxyglutaric aciduria [36], described by Chalmers et al. [38]; Figure 14.6d, the chromatogram of a patient excreting elevated amounts of a mixture of D- and L-2-hydroxyglutaric acid [36]; and

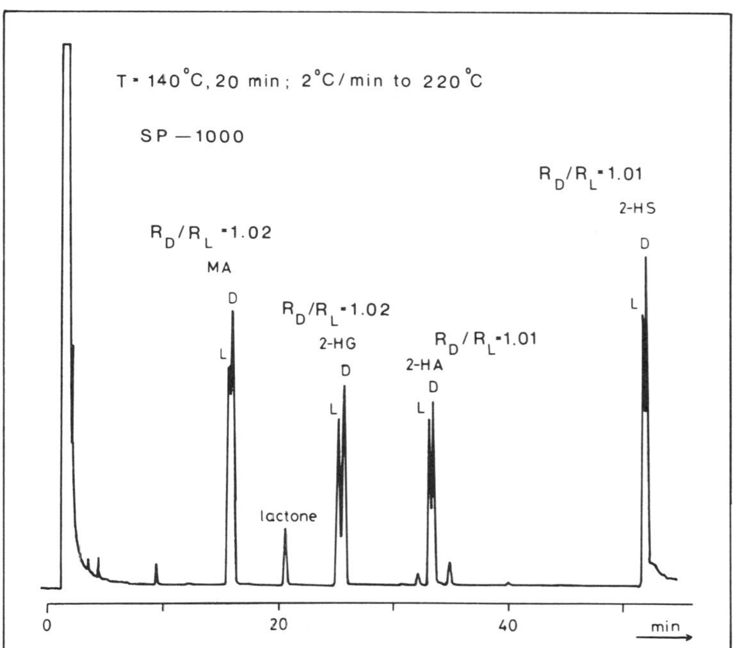

Figure 14.5 Gas chromatogram on a SP-1000 WCOT capillary column (25 m × 0.3 mm I.D.; N_2 flow, 1 ml/min) of the *O*-acetylated di-(-)-2-butyl esters of a mixture of commercial or synthetic DL-malic acid (MA), DL-2-hydroxyglutaric acid (2-HG), DL-2-hydroxyadipic acid (2-HA), and DL-2-hydroxysebacic acid (2-HS). (Data from Refs. 35 and 36.)

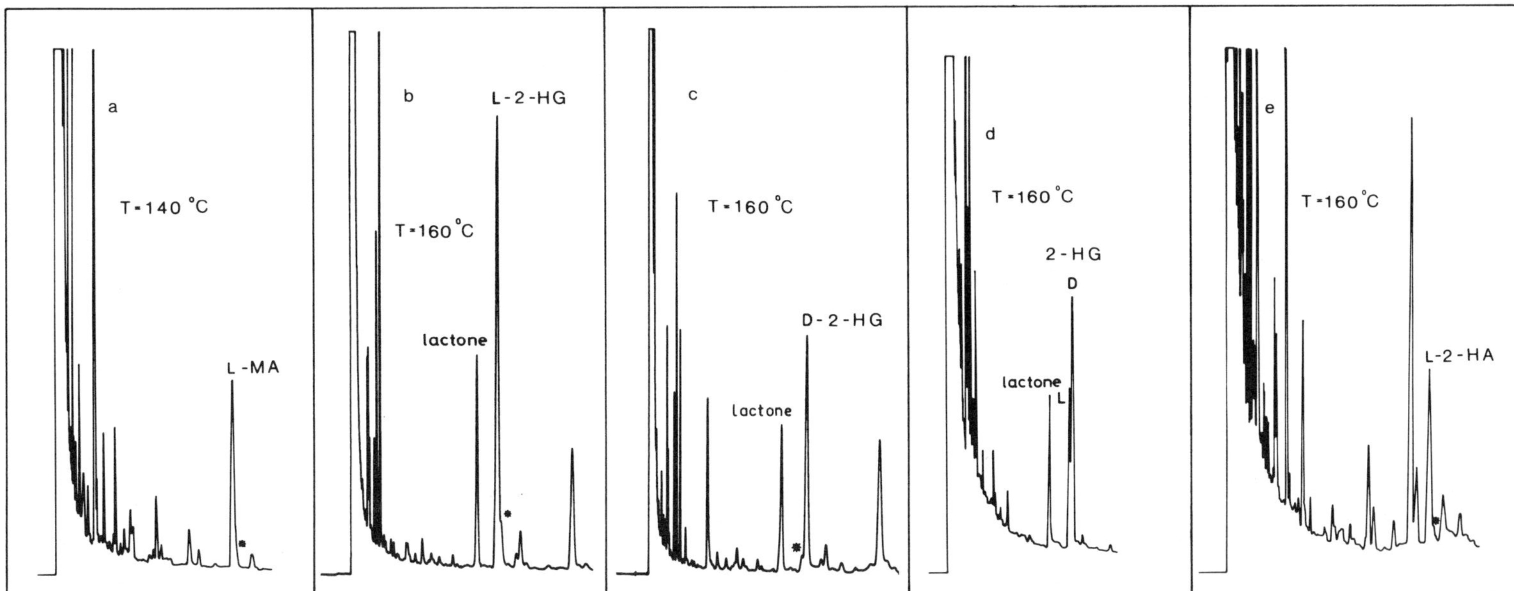

Figure 14.6 Gas chromatograms on a SP-1000 WCOT capillary column (25 m × 0.3 mm I.D.; N_2 flow, 1 ml/min) of urinary extracts of different patients derivatized with (-)-2-butanol and acetic anhydride: (a) L-malic acid; (b) L-2-hydroxyglutaric acid; (c) D-2-hydroxyglutaric acid; (d) D- + L-2-hydroxyglutaric acid; (e) L-2-hydroxyadipic acid (* indicates the shoulder or small peak obtained from the contaminating (+)-2-butanol in the commercial (-)-2-butanol sample). (Data from Refs. 36, 37, and 38.)

Figure 14.6e, the chromatogram of a patient with 2-aminoadipic aciduria, excreting relatively large amounts of L-2-hydroxyadipic acid [36].

Unsaturated fatty acids having a 1,4-*cis,cis*-pentadiene system can be converted into conjugated hydroperoxides by lipoxygenase oxygenation or by autoxidation (see Section III). The stereoisomeric composition of the formed hydroperoxides can not be determined directly by GLC. To obtain suitable 2-hydroxy (di)carboxylic acid derivatives, the hydroperoxides are reduced, and subsequently the prepared *O*-acylated unsaturated hydroxy fatty acid methyl esters are submitted to oxidative ozonolysis. When a chiral *O*-acyl group is present, the 2-hydroxy (di)carboxylic acid methyl ester derivatives can be analyzed directly by GLC. If not, chiral ester groups can be introduced.

Hamberg [33] analyzed the *O*-(-)-menthyloxycarbonyl (di)methyl ester derivatives of 2-hydroxysebacic acid (R_D/R_L = 1.10; T = 200°C) and 2-hydroxyheptanoic acid (R_D/R_L = 1.16; T = 141°C), formed from the *O*-(-)-menthyloxycarbonyl methyl ester derivatives of 9-hydroxy-10,12-octadecadienoic acid and 13-hydroxy-9,11-octadecadienoic acid, respectively (5% QF-1; helium flow, 65 ml/min). In this way it was possible to estimate indirectly the relative amounts of the four isomeric (9-D, 9-L, 13-D, and 13-L) hydroperoxy acids, obtained from linoleic acid by incubation with lipoxygenases from various sources (Table 14.1). Additional examples include the analysis of malic acid, 2-hydroxyadipic acid, 2-hydroxyazelaic acid, and 2-hydroxydecanoic acid.

Borgeat et al. [39,40] reported on the transformation of arachidonic acid into 5-D-hydroxy-6,8,11,14-eicosatetraenoic acid and of homo-γ-linolenic acid into 8-D-hydroxy-9,11,14-eicosatrienoic acid by rabbit polymorphonuclear leukocytes containing lipoxygenase-type enzymatic activities (intermediate hydroperoxides were not detected). GLC analysis of the 2-hydroxyadipic acid derivative (R_D/R_L = 1.10; T = 162°C; 2.2% QF-1), derived from the 5-hydroxy

Table 14.1 Analyses of Hydroperoxyoctadecadienoates Formed from Linoleic Acid on Incubation with Soybean and Corn Lipoxygenases

	Buffer	pH	9-D : 9-L : 13-D : 13-L
A, soybean	0.1 M sodium borate	9.0	4 : 3 : 3 : 90
B, soybean	0.1 M sodium borate	10.5	2 : 2 : 2 : 94
C, soybean	0.1 M sodium borate	9.0	2 : 2 : 3 : 93
D, soybean	0.04 M NH_4OH-NH_4Cl	9.0	6 : 4 : 4 : 86
E, corn	0.1 M potassium phosphate	7.4	86 : 5 : 4.5 : 4.5
F, corn	0.1 M potassium phosphate	6.5	89 : 4 : 3.5 : 3.5
G, autoxidation	—	—	25 : 25 : 25 : 25

A,B: Soybean lipoxygenase from Sigma Chemical Co.
C,D: Soybean lipoxygenase from Nutritional Biochemicals Corp.
E,F: Ammonium sulfate precipate (45-60% saturation) containing corn lipoxygenase.
Source: Data from Ref. 33.

Table 14.2 Comparison of the Enantiomeric Compositions of Hydroperoxides Obtained in Lipoxygenase-Catalyzed Oxygenation of Polyunsaturated Fatty Acids Containing a 1,4-*cis*,*cis*-pentadiene system[a]

Compound[b]	GLC[c] Ozonolysis product(s)	GLC[c] L (hydroperoxide)	GLC[c] D (hydroperoxide)	Chiral center	^{1}H-NMR L (hydroperoxide)	^{1}H-NMR D (hydroperoxide)
I	2-hydroxyheptanoic acid	98	2	C_{13}	97	3
II	2-hydroxysebacic acid	6	94	C_9	6	94
IIa	2-hydroxysebacic acid	71	29	C_9	70	30
III	malic acid	L>92	D< 8	C_{13}	93	7
IV	2-hydroxyheptanoic acid	96	4	C_{15}	97	3
V	2-hydroxyheptanoic acid	94	6	C_{15}	93	7
	malic acid	L< 5	D>95	C_8	7	93
VI	2-hydroxyheptanoic acid	96	4	C_{15}	92	8
	2-hydroxyadipic acid	L< 9	D>91	C_5	10	90

[a]The absolute configurations were determined in two ways: (1) Capillary GLC on SP-1000 of *O*-acetylated (di)-(-)-2-butyl ester derivatives (oxidative ozonolysis of the *O*-acetylated unsaturated hydroxy fatty acid methyl esters; van Os et al. [35]), and (2) ^{1}H-NMR spectroscopy of *O*-α-methoxy-α-trifluoromethylphenylacetylated methyl esters of unsaturated hydroxy fatty acids. The various unsaturated hydroxy fatty acid methyl esters were purified by high-pressure liquid chromatography (see van Os [20]).

[b]I = 13-hydroperoxy 9-*cis*,11-*trans*-octadecadienoic acid (from linleic acid by incubation with soybean lipoxygenase-1, 0.1 M sodium borate pH 9.0, 4°C, O_2 saturated; main product: 13-L_S).

II = 9-hydroperoxy 10-*trans*,12-*cis*-octadecadienoic acid (from linoleic acid by incubation with corn germ lipoxygenase, 0.1 M sodium phosphate pH 6.6, 4°C, O_2 saturated; main product: 9-D_S).

IIa = 9-hydroperoxy 10-*trans*,12-*cis*-octadecadienoic acid (from linoleic acid by incubation with soybean lipoxygenase-2, 0.1 M sodium borate pH 9.0, 4°C, O_2 saturated; main product: 9-L_R).

III = 13-hydroperoxy 9-*cis*,11-*trans*,15-*cis*-octadecatrienoic acid (from α-linolenic acid by incubation with soybean lipoxygenase-1, 0.1 M sodium borate pH 9.0, 4°C, O_2 saturated; main product: 13-L_S).

IV = 15-hydroperoxy 5-*cis*,8-*cis*,11-*cis*,13-*trans*-eicosatetraenoic acid (from arachidonic acid by incubation with soybean lipoxygenase-1, 0.1 M sodium borate pH 9.0, 4°C, O_2 saturated; main product: 15-L_S).

V = 8,15-dihydroperoxy 5-*cis*,9-*trans*,11-*cis*,13-*trans*-eicosatetraenoic acid (from arachidonic acid by incubation with soybean lipoxygenase-1, 0.1 M sodium borate pH 8.75, 4°C, O_2 saturated; main product: 8-D_S,5-L_S).

VI = 5,15-dihydroxy 6-*trans*,8-*cis*,11-*cis*,13-*trans*-eicosatetraenoic acid (from arachidonic acid by incubation with soybean lipoxygenase-1, 0.1 M sodium borate pH 8.75, 4°C, O_2 saturated; main product: 5-D_S,15-L_S).

[c]The values are corrected for the presence of 6% of the (+)-enantiomer in commercially available (-)-2-butanol. SP-1000 capillary column: 25 m × 0.3 mm I.D.; N_2 flow, 1 ml/min. Analysis conditions: 2-hydroxyheptanoic acid, 120°C; malic acid, 140°C; 2-hydroxyadipic acid, 160°C; 2-hydroxysebacic acid, 190°C.

Source: Data from Ref. 35.

unsaturated fatty acid, showed a L:D ratio of 83:17, indicating mainly the D-configuration for the 5-OH group. Analysis of the 2-hydroxyazelaic acid derivative (R_D/R_L = 1.10; T = 180°C; 2.2% QF-1), derived from the 8-hydroxy unsaturated fatty acid, showed a L:D ratio of 82:18, indicating mainly the D-configuration for the 8-OH group. Furthermore, Borgeat and Samuelsson [41] described a second product of arachidonic acid incubation, namely, 5-D,-12-D-dihydroxy-6,8,10,14-eicosatetraenoic acid. A small amount of the 5-L,-12-D stereoisomer appeared to be present. The absolute configurations were assigned on the basis of the GLC data of the degradation products 2-hydroxyadipic acid and malic acid.

As shown by Hamberg and Samuelsson [42], the incubation of arachidonic acid with human blood platelets leads to the formation of 12-L-hydroxy-5,8,10,-14-eicosatetraenoic acid (action of lipoxygenase), and 12-L-hydroxy-5,8,10-heptadecatrienoic acid, and the hemiacetal derivative of 8-(1-hydroxy-3-oxopropyl)-9,12-L-dihydroxy-5,10-heptadecadienoic acid (action of fatty acid cyclooxygenase). The absolute configuration of the first compound was established via the malic acid derivative. The other two compounds yielded 2-hydroxyheptanoic acid.

Nugteren [34] studied the incubation of all-*cis*-8,11-eicosadienoic acid, homo-γ-linolenic acid (all-*cis*-8,11,14-eicosatrienoic acid), and arachidonic acid (all-*cis*-5,8,11,14-eicosatetraenoic acid) with bovine blood platelets. The action of lipoxygenase gave rise to the formation of 12-hydroxy 8-*cis*,10-*trans*-eicosadienoic acid (88% L and 12% D; 2-hydroxydecanoic acid as degradation product), 12-hydroxy 8-*cis*,10-*trans*,14-*cis*-eicosatrienoic acid (91% L and 9% D; malic acid as degradation product), and 12-L-hydroxy 5-*cis*,8-*cis*,10-*trans*,-14-*cis*-eicosatetraenoic acid (see Hamberg and Samuelsson [42]), respectively. For the analysis of the absolute configuration of hydroperoxides a similar approach has been developed by van Os et al. [35], making use of the readily accessible *O*-acetylated (-)-2-butyl ester diastereomers. Table 14.2 summarizes the results of some studies, using the latter derivatives.

The absolute configuration of the nonhydroxy dicarboxylic acid methylsuccinic acid, isolated from normal human urine, has been studied by Zeitman and Lawless [43]. The di-(+)-2-butyl ester diastereomeric derivatives were separated on diethylene glycol succinate as nonchiral stationary phase. The urinary substance was found to have the D-configuration.

E. Long-Chain Hydroxy Acids

Hammarström and Hamberg [44] have reported the GLC resolution of a series of DL-hydroxyoctadecanoic acids with hydroxyl groups at different positions. They studied also the separation of homologous DL-3-hydroxy fatty acids (C_{10}-C_{18}). For this purpose the D-2-phenylpropionate diastereomeric derivatives of the hydroxy acid methyl esters were applied. The results are presented in Table 14.3. As nonchiral stationary phase, 5% QF-1 was employed. The diastereomeric derivatives of the methyl 4-, 7-, and 13-hydroxyoctadecanoates could not be resolved. In view of the latter results, it was stated that it is unlikely that any of the monohydroxyoctadecanoates substituted in the region from carbon-4 to carbon-13 could be resolved in this way. Concerning the homologous series of DL-3-hydroxy acids, it was found that in all cases the diastereomer derived from the D enantiomer has a shorter retention time than that from the L enantiomer (see also D-3-hydroxybutyric acid, Section V.C.).

The separation of the *O*-(-)-menthyloxycarbonyl methyl esters of homologous DL-2-hydroxy fatty acids (C_{14}–C_{26}) on 1.4% OV-210 have been described by

Table 14.3 Resolution of a Series of Long-Chain Hydroxy Fatty Acids on 5% QF-1 (Carrier Gas, He) Using the D-2-phenylpropionate Methyl Ester Derivatives

Hydroxy acid		Resolution factor	T(°C)
DL-17-hydroxy	18:0	1.08 (R_D/R_L)	230
DL-16-hydroxy	18:0	1.08 (R_D/R_L)	230
DL-15-hydroxy	18:0	1.05 (R_D/R_L)	230
DL-14-hydroxy	18:0	1.03	230
DL-13-hydroxy	18:0	—	230
DL-7-hydroxy	18:0	—	230
DL-4-hydroxy	18:0	—	230
DL-3-hydroxy	18:0	1.07 (R_L/R_D)	230
DL-2-hydroxy	18:0	1.03	230
DL-3-hydroxy	10:0	1.06 (R_L/R_D)	200
DL-3-hydroxy	13:0	1.08 (R_L/R_D)	200
DL-3-hydroxy	14:0	1.08 (R_L/R_D)	200
DL-3-hydroxy	16:0	1.08 (R_L/R_D)	200
DL-3-hydroxy	18:0	1.08 (R_L/R_D)	200

Source: Data from Ref. 44.

Hammarström [45] (Figure 14.7). It has been verified that for pure enantiomers the preparation of methyl esters with methanolic HCl (refluxing time, 5 hr) does not give rise to racemization. In all cases, the peak derived from the L enantiomer emerges earlier than that from the D enantiomer. This method has been applied for the determination of the absolute configuration of 2-hydroxy fatty acids from brain cerebrosides. The unsaturated acids were hydrogenated before the analysis was carried out. Only the D configuration was found to occur. Another application was published by Markovetz et al. [46] in relation to studies on fatty acid metabolism (α-oxidation) in peanuts. Enzymatic oxidation of palmitic acid led to the accumulation of D-2-hydroxypalmitic acid. It has to be noted that in contrast to the 2-hydroxy fatty acids, the 3- and ω2-hydroxy fatty acids could not be resolved as their *O*-(-)-menthyloxycarbonyl methyl ester derivatives on QF-1 nor on OV-210 as stationary phases [44].

The determination of the absolute configuration of the ω2-hydroxy acids 17-hydroxyoctadecanoic acid, 19-hydroxyeicosanoic acid, and 21-hydroxydocosanoic acid has been reported by Hamberg [47]. *R*-1-phenylethyl isocyanate was used as chiral reagent. The diastereomeric *N*-(1-phenylethyl)-urethane methyl esters could be separated on 1% QF-1 (1.8 m × 3.5 mm I.D.; carrier gas, helium; resolution factors, 1.08; T = 210°C). Only for the 17-hydroxy acid was the peak sequence estimated. The derivative obtained from the L enantiomer was found to have the smallest retention time. 2-Hydroxy

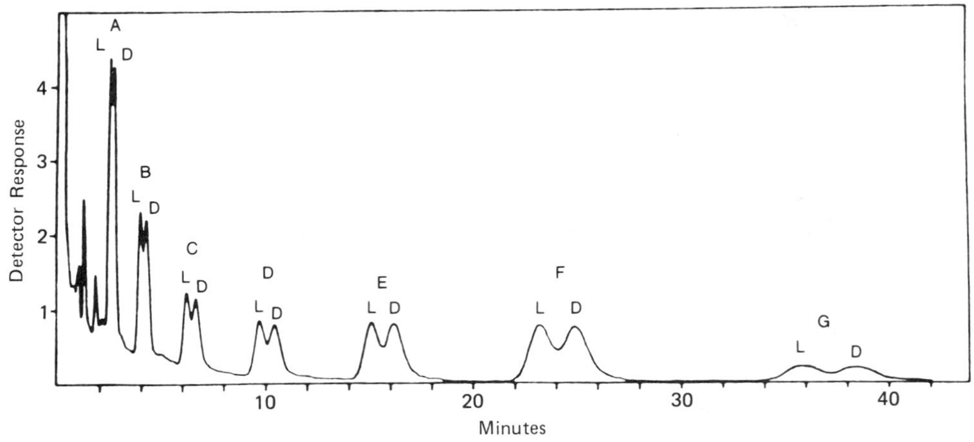

Figure 14.7 Gas chromatogram on 1.4% OV-210 (1.7 m × 3.5 mm I.D.; carrier gas, He) at 250°C of the *O*-(-)-menthyloxycarbonyl methyl ester derivatives of a series of long-chain DL-2-hydroxy fatty acids. A, DL-2-hydroxymyristic acid (C_{14}); B, DL-2-hydroxypalmitic acid (C_{16}); C, DL-2-hydroxystearic acid (C_{18}); D, DL-2-hydroxyarachidic acid (C_{20}); E, DL-2-hydroxybehenic acid (C_{22}); F, DL-2-hydroxylignoceric acid (C_{24}); and G, DL-2-hydroxycerotic acid (C_{26}). (From Ref. 45.)

acid methyl esters and 3-hydroxy acid methyl esters could not be resolved in this way [44]. Applications have been described in studies on the mechanism of microsomal ω2-hydroxylation of fatty acids. Björkhem and Hamberg [48] reported the ω2 hydroxylation of lauric acid to be 40–44% L- and 56–60% D-11-hydroxylauric acid. Hamberg and Björkhem found that with decanoic acid, 75% L- and 25% D-9-hydroxydecanoic acid were formed [49]. The same approach was used by Björkhem and Hamberg in studies on the enzymatic oxidoreduction of 17-hydroxystearic acid [50].

F. Aromatic Hydroxy Acids

Pollock and Jermany [21] reported the occurrence of two peaks for the *O*-acetylated (±)-2-butyl, (±)-3-methyl-2-butyl, and (±)-3,3-dimethyl-2-butyl esters of D-mandelic acid on a Carbowax 20 M capillary column (150 ft × 0.02 in.). Westley and Halpern demonstrated the resolution of DL-3-phenyllactic acid as their *O*-(-)-menthyloxycarbonyl methyl ester derivatives on 5% QF-1 (5 ft × 1/8 in.; nitrogen flow, 30 ml/min; R_D/R_L = 1.14; T = 200°C) [23]. Separations were also obtained for the *O*-acetylated (-)-2-butyl ester derivatives of DL-3-phenyllactic acid and DL-3-(4-hydroxyphenyl)lactic acid on a SP-1000 capillary column (Kamerling and Wadman, unpublished results) (Figure 14.8a,b). The two peaks in Figure 14.8a were assigned by cochromatography with the L enantiomer. The assignment of the peaks in Figure 14.8b was made on the basis of the results obtained with other *O*-acetylated (-)-2-butyl esters of 2-hydroxy acids, described in Section V.

Phenylketonuria and tyrosyluria lead to the urinary accumulation of 3-phenyllactic acid and 3-(4-hydroxyphenyl)lactic acid, respectively. Analysis of the *O*-acetylated (-)-2-butyl esters of the urinary organic acids shows in both cases the presence of only the L enantiomer (Figure 14.8c,d).

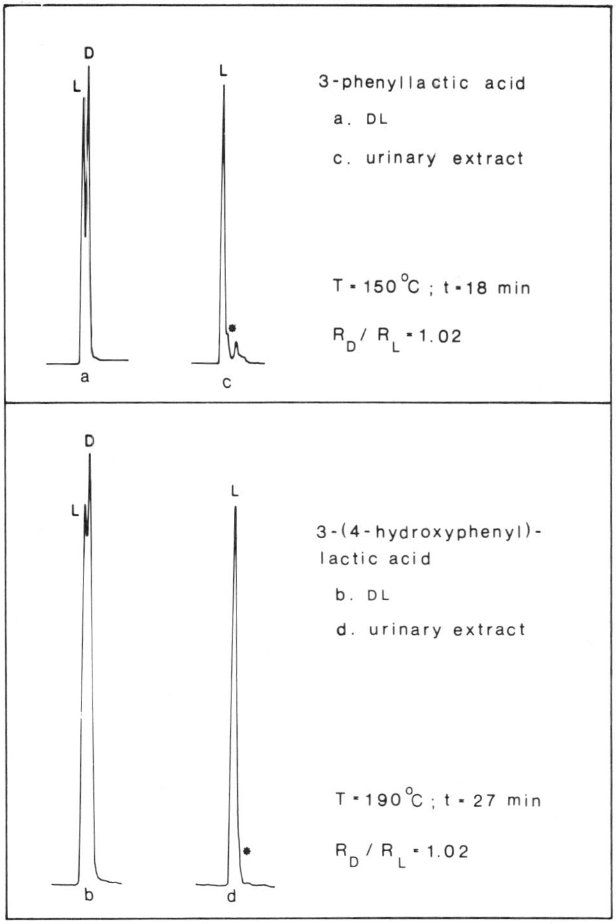

Figure 14.8 Partial gas chromatograms on a SP-1000 WCOT capillary column (25 m × 0.3 mm I.D.; N_2 flow, 1 ml/min) of the *O*-acetylated (-)-2-butyl esters of (a) commercial DL-3-phenyllactic acid, (b) commercial DL-3-(4-hydroxyphenyl)lactic acid, (c) L-3-phenyllactic acid from a urinary ethyl acetate extract of a patient with phenylketonuria, and (d) L-3-(4-hydroxyphenyl)lactic acid from a urinary ethyl acetate extract of a patient with tyrosyluria (* indicates the shoulder obtained from the contaminating (+)-2-butanol in the commercial (-)-2-butanol sample).

G. Aldonic Acids

The separation of the *O*-acetylated (±)-2-butyl, (±)-3-methyl-2-butyl, and (±)-3,3-dimethyl-2-butyl esters of a series of D-aldonic acids (in this case, D according to the sugar nomenclature; see Section IV) on Carbowax 20 M (150 ft × 0.02 in.; helium flow: 6.5–6.8 ml/min) or OV-225 (50 ft × 0.02 in.; helium flow: 6 ml/min) capillary columns has been reported by Pollock and Jermany [21,22]. The underivatized acids were prepared by oxidation of the aldehyde function of tetroses, pentoses, and hexoses. For peak assignments the derivatives obtained with the optically active alcohols were used. The data from the literature presented in Table 14.4 have been arranged in

Table 14.4 Separation Factors of Enantiomeric Aldonic Acids

O-Acetylated aldonic acid (-)-alkyl esters[a]	R_D/R_L[b]	T(°C)	Column material[c]
2B threonic acid	1.03	160	Carbowax 20 M
3M2B threonic acid	1.06	160	Carbowax 20 M
33D2B threonic acid	1.09	160	Carbowax 20 M
2B erythronic acid	1.02	160	Carbowax 20 M
3M2B erythronic acid	1.04	160	Carbowax 20 M
33D2B erythronic acid	1.06	160	Carbowax 20 M
2B xylonic acid	1.03	175	Carbowax 20 M
3M2B xylonic acid	1.05	175	Carbowax 20 M
33D2B xylonic acid	1.09	175	Carbowax 20 M
2B lyxonic acid	1.01	175	Carbowax 20 M
3M2B lyxonic acid	1.03	175	Carbowax 20 M
33D2B lyxonic acid	1.04	175	Carbowax 20 M
2B ribonic acid	1.02	175	Carbowax 20 M
3M2B ribonic acid	1.05	175	Carbowax 20 M
33D2B ribonic acid	1.08	175	Carbowax 20 M
2B arabonic acid	1.03	175	Carbowax 20 M
3M2B arabonic acid	1.06	175	Carbowax 20 M
33D2B arabonic acid	1.13	175	Carbowax 20 M
2B galactonic acid	1.05	180	OV-225
3M2B galactonic acid	1.08	180	OV-225
33D2B galactonic acid	1.16	180	OV-225
2B gluconic acid	1.02	180	OV-225
3M2B gluconic acid	1.04	180	OV-225
33D2B gluconic acid	1.05	180	OV-225

[a] 2B = (-)-2-butanol; 3M2B = (-)-3-methyl-2-butanol; 33D2B = (-)-3,3-dimethyl-2-butanol.
[b] Here D and L indicate the absolute configuration at C-2 and *not* that at the chiral atom adjacent to the primary hydroxyl function (see text).
[c] Carbowax 20M: 150 ft × 0.02 in.; helium flow, 6.5–6.8 ml/min. OV-225: 50 ft × 0.02 in.; helium flow, 6 ml/min.
Source: Data from Refs. 21 and 22.

such a way that D and L indicate the absolute configuration at C-2; the carboxyl function bears only a (-)-2-alkyl group. Using the latter chiral alcohols, it can be seen that L enantiomers are eluted earlier than D enantiomers. Furthermore it was found that a better resolution of diastereomers is obtained with the more sterically hindered alcohols.

ADDENDUM*

Recently, König and Benecke [51] have reported on the gas chromatographic resolution of the *O*-trifluoroacetylated and *O*-trimethylsilylated (+)-3-methyl-2-butyl esters of DL-lactic acid, DL-2-hydroxybutyric acid, DL-2-hydroxyisovaleric acid, DL-2-hydroxyisohexanoic acid, DL-2-hydroxyhexanoic acid, DL-2-hydroxyoctanoic acid, DL-2-hydroxydodecanoic acid, DL-2-hydroxytetradecanoic acid, DL-2-hydroxyhexadecanoic acid, DL-malic acid, DL-2-phenyllactic acid, DL-3-phenyllactic acid, DL-tropaic acid, DL-mandelic acid, DL-4-hydroxy-3-methoxymandelic acid, DL-3-hydroxy-4-methoxymandelic acid, and DL-4-hydroxymandelic acid, and the (+)-3-methyl-2-butyl esters of DL-2-methylsuccinic acid, DL-2-phenyl succinic acid, and DL-2-phenylbutyric acid on SE-30 and OV-17 capillary columns. A typical example of the separation is presented in Figure 14.9. The *O*-trifluoroacetylated derivatives are more

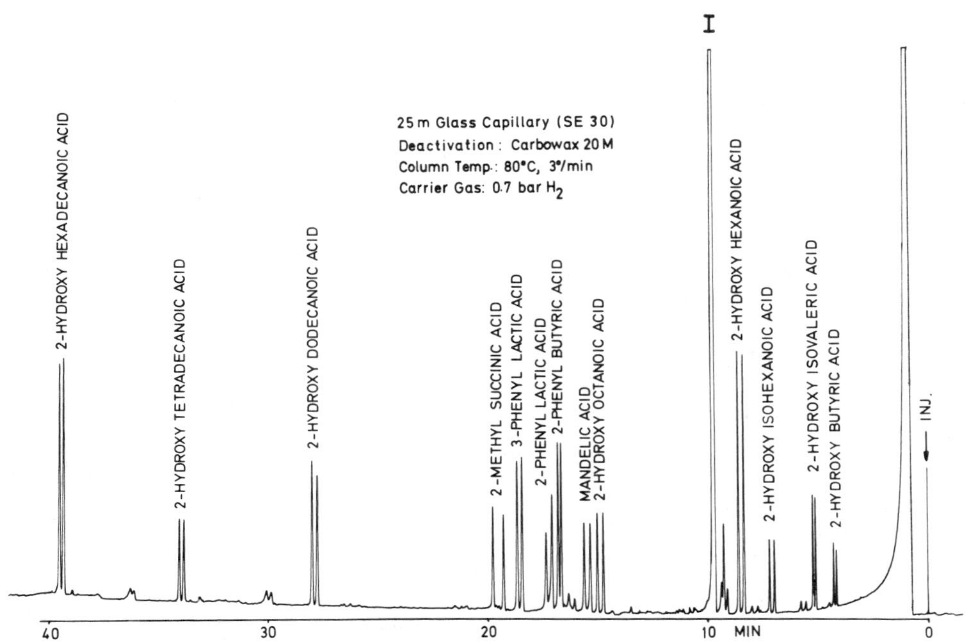

Figure 14.9 Gas chromatogram on a SE-30 WCOT capillary column (25 m × 0.3 mm I.D.) of a series of *O*-trifluoroacetylated (+)-3-methyl-2-butyl esters of DL-2-hydroxycarboxylic acids and (+)-3-methyl-2-butyl esters of DL-2-alkyl(aryl)-carboxylic acids. Internal standard (I): *N*-trifluoroacetyl-L-valine (+)-3-methyl-2-butyl ester. (From Ref. 51.)

---

*After preparation of the manuscript.

volatile and show slightly larger resolution factors than the *O*-trimethylsilylated derivatives. Peak assignments were only made for DL-lactic acid, DL-malic acid, and DL-mandelic acid, showing that the L enantiomers have the longer retention times. It has to be noted that the derivatives of DL-3-hydroxybutyric acid could not be separated (see also Section V.C).

ACKNOWLEDGMENTS

We thank Prof. Dr. S. K. Wadman and Dr. M. Duran (University Children's Hospital "Het Wilhelmina Kinderziekenhuis," Utrecht), and Dr. C. P. A. van Os (Department of Bioorganic Chemistry, State University of Utrecht, Utrecht) for their fruitful collaboration, and Mr. G. J. Gerwig for his technical assistance. The financial support of "Het Praeventiefonds" and the Netherlands Foundation for Chemical Research (SON)/Netherlands Organization for the Advancement of Pure Research (ZWO) is highly acknowledged.

REFERENCES

1. E. Gil-Av and D. Nurok, *Adv. Chromatogr.* *10*:99 (1974).
2. P. Hušek and K. Macek, *J. Chromatogr.* *113*:139 (1975).
3. S. Nakaparksin, P. Birell, E. Gil-Av, and J. Oró, *J. Chromatogr. Sci.* *8*:177 (1970).
4. W. Parr, J. Pleterski, C. Yang, and E. Bayer, *J. Chromatogr. Sci.* *9*:141 (1971).
5. G. E. Bollock and V. I. Oyama, *J. Gas Chromatogr.* *4*:126 (1966).
6. E. Gil-Av, R. Charles-Sigler, G. Fischer, and D. Nurok, *J. Gas Chromatogr.* *4*:51 (1966).
7. W. A. König, W. Rahn, and J. Eyem., *J. Chromatogr.* *133*:141 (1977).
8. M. Hasegawa and I. Matsubara, *Anal. Biochem.* *63*:308 (1975).
9. H. Iwase and A. Murai, *Chem. Pharm. Bull.* *22*:8 (1974).
10. G. J. Gerwig, J. P. Kamerling, and J. F. G. Vliegenthart, *Carbohyd. Res.* *62*:349 (1978).
11. G. J. Gerwig, J. P. Kamerling, and J. F. G. Vliegenthart, *Carbohyd. Res.* *77*:1 (1979).
12. K. Leontein, B. Lindberg, and J. Lönngren, *Carbohyd. Res.* *62*:359 (1978).
13. E. Jellum, O. Stokke, and L. Eldjarn, *Clin. Chem.* *18*:800 (1972).
14. S. C. Gates, C. C. Sweeley, W. Krivit, D. DeWitt, and B. E. Blaisdell, *Clin. Chem.* *24*:1680 (1978).
15. J. Roboz, *Adv. Clin. Chem.* *17*:109 (1975).
16. E. Jellum, *J. Chromatogr.* *143*:427 (1977).
17. M. Duran and S. K. Wadman, *Recent Adv. Clin. Biochem.* *2*:103 (1981).
18. C. R. Smith, Jr., In *Topics in Lipid Chemistry* (F. D. Gunstone, ed.), Logos Press Limited, London, 1970, pp. 277–368.
19. G. A. Veldink, J. F. G. Vliegenthart, and J. Boldingh, *Prog. Chem. Fats Lipids* *15*:131 (1977).
20. C. P. A. van Os, Thesis, State University of Utrecht, The Netherlands, 1980.
21. G. E. Pollock and D. A. Jermany, *J. Gas Chromatography*, *6*:412 (1968).
22. G. E. Pollock and D. Jermany, *J. Chromatogr. Sci.* *8*:296 (1970).
23. J. W. Westley and B. Halpern, *J. Org. Chem.* *33*:3978 (1968).

24. J. P. Kamerling, G. J. Gerwig, J. F. G. Vliegenthart, M. Duran, D. Ketting, and S. K. Wadman, *J. Chromatogr. 143*:117 (1977).
25. B. L. Karger, R. L. Stern, H. C. Rose, and W. Keane, in *Gas Chromatography 1966* (A. B. Littlewood, ed.), Institute of Petroleum, London, 1967, p. 240.
26. D. Nurok, Thesis, University of Capetown, South Africa, 1966.
27. M. Duran, J. P. G. M. Van Biervliet, J. P. Kamerling, and S. K. Wadman, *Clin. Chim. Acta 74*:297 (1977).
28. H. E. Williams and L. H. Smith, *N. Engl. J. Med. 278*:233 (1968).
29. N. J. Brandt, K. Rasmussen, S. Brandt, S. Kølvrda, and F. Schønheyder, *Acta Paediatr. Scand. 65*:17 (1976).
30. S. K. Wadman, M. Duran, D. Ketting, L. Bruinvis, P. K. de Bree, J. P. Kamerling, G. J. Gerwig, J. F. G. Vliegenthart, H. Przyrembel, K. Becker, and H. J. Bremer, *Clin. Chim. Acta 71*:477 (1976).
31. J. P. Kamerling, G. J. Gerwig, J. F. G. Vliegenthart, M. Duran, and S. K. Wadman, *T. Ned. Ver. Klin. Chem.* 5:13 (1980).
32. J. P. Kamerling, G. J. Gerwig, M. Duran, D. Ketting, and S. K. Wadman, *Clin. Chim. Acta 88*:183 (1978).
33. M. Hamberg, *Anal. Biochem. 43*:515 (1971).
34. D. H. Nugteren, *Biochim. Biophys. Acta 380*:299 (1975).
35. C. P. A. van Os, G. P. M. Rijke-Schilder, J. P. Kamerling, G. J. Gerwig, and J. F. G. Vliegenthart, *Biochim. Biophys. Acta, 620*:326 (1980).
36. J. P. Kamerling, M. Duran, G. J. Gerwig, D. Ketting, L. Bruinvis, J. F. G. Vliegenthart, and S. K. Wadman, *J. Chromatogr. 222*:276 (1981).
37. M. Duran, J. P. Kamerling, H. D. Bakker, A. H. van Gennip, and S. K. Wadman, *J. Inher. Metab. Dis. 3*:109 (1980).
38. R. A. Chalmers, A. M. Lawson, R. W. E. Watts, A. S. Tavill, J. P. Kamerling, E. Hey and D. Ogilvie, *J. Inher. Metab. Dis. 3*:11 (1980).
39. P. Borgeat, M. Hamberg, and B. Samuelsson, *J. Biol. Chem. 251*:7816 (1976).
40. P. Borgeat, M. Hamberg, and B. Samuelsson, *J. Biol. Chem. 252*:8772 (1977).
41. P. Borgeat and B. Samuelsson, *J. Biol. Chem. 254*:2643 (1979).
42. M. Hamberg and B. Samuelsson, *Proc. Nat. Acad. Sci., USA 71*:3400 (1974).
43. B. Zeitman and J. G. Lawless, *Biochem. Med. 13*:111 (1975).
44. S. Hammarström and M. Hamberg, *Anal. Biochem. 52*:169 (1973).
45. S. Hammarström, *FEBS Lett. 5*:192 (1969).
46. A. J. Markovetz, P. K. Stumpf, and S. Hammarström, *Lipids 7*:159 (1972).
47. M. Hamberg, *Chem. Phys. Lipids 6*:152 (1971).
48. I. Björkhem and M. Hamberg, *Biochem. Biophys. Res. Comm. 47*:333 (1972).
49. M. Hamberg and I. Björkhem, *J. Biol. Chem. 246*:7411 (1971).
50. I. Björkhem and M. Hamberg, *J. Biol. Chem. 246*:7417 (1971).
51. W. A. König and I. Benecke, *J. Chromatogr. 195*:292 (1980).

15

Capillary GC-MS/SIM Analysis of Serotonin and Metabolites

Cristina Suñol and Emilio Gelpí / *Instituto de Química Bio-Orgánica (CSIC), Barcelona, Spain*

I. INTRODUCTION

The metabolic pathway of the biogenic indolealkylamines starts with tryptophan (TP) and proceeds according to Scheme 1. Of these amines, 5-hydroxytryptamine (5-HT) or serotonin is an established neurotransmitter [1], whereas the role of tryptamine (T), whose levels in rat brain are of the order of 1000 times lower than those of 5-HT, has not been definitively ascertained [2-4]. As for 5-methoxytryptamine (5-MT), its presence and physiological activity [5] in the CNS has been lately questioned [6]. The metabolic fate [7] of these biogenic amines in the central nervous and peripheral systems depends mainly on the action of monoamine oxidase (MAO) systems which convert them to the corresponding acids, 5-hydroxyindoleacetic (5-HIAA), indoleacetic (IAA), 5-methoxyindoleacetic (5-MIAA), and related alcohols forms, 5-hydroxytryptophol (5-HTOL), tryptophol (TOL), and 5-methoxytryptophol (5-MTOL).

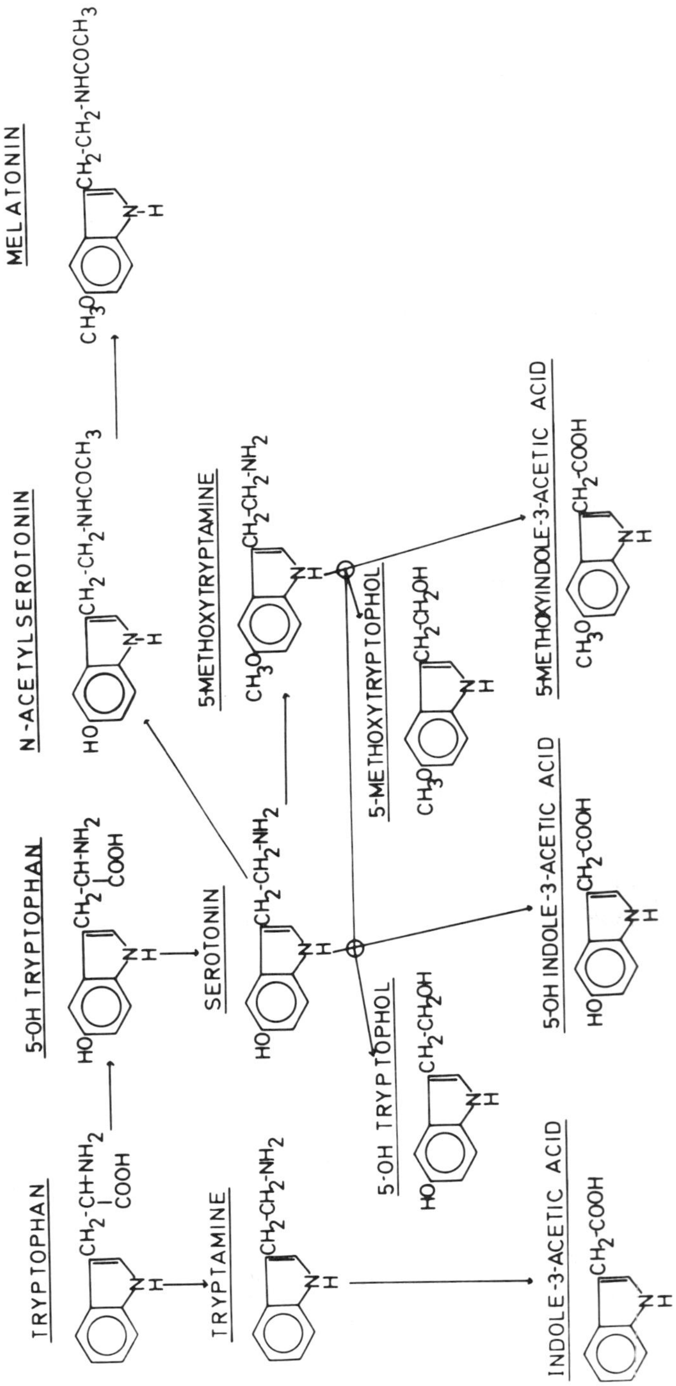
TRYPTOPHAN
5-OH TRYPTOPHAN
N-ACETYLSEROTONIN
MELATONIN
TRYPTAMINE
SEROTONIN
5-METHOXYTRYPTAMINE
5-OH TRYPTOPHOL
5-METHOXYTRYPTOPHOL
INDOLE-3-ACETIC ACID
5-OH INDOLE-3-ACETIC ACID
5-METHOXYINDOLE-3-ACETIC ACID

Scheme 15.1

The alcohols are minor components in this pathway and, as such, are difficult to detect, although 5-HTOL has been detected in human CSF by mass spectrometric techniques [8,9]. Also, 5-HT can undergo a transformation in the pineal gland to *N*-acetylserotonin (NAS) and melatonin (Mel), or to 5-methoxytryptamine (Scheme 15.1).

From various biochemical considerations of the dynamics of such a metabolic system, it follows that the analysis of TP and its indole-related metabolites in human tissue or body fluids could provide important clinical information. For example, it is known that malignant carcinoids secrete large amounts of 5-HT and 5-HIAA [10,11] and that patients with breast cancer show abnormally high levels of TP metabolites in urine [12]. The role of TP metabolism has also been implicated in bladder cancer [13], psychotic disorders, schizophrenia, Hodgkin's disease [14], endogenous depression, Hartnup's disease, phenylketonuria, and Down's syndrome [7], as well as in the mechanism of drugs effective against rheumatoid arthritis [15].

Some of these TP metabolites have been analyzed by various analytical techniques usually limited to the assay of only one compound or a class of compounds (e.g., amines or acids). For instance, 5-HT has been assayed by fluorometric procedures [16,17], biological assay [18], radioimmunoassay [19], isotopic enzymatic assay [20], and high-pressure liquid chromatography (HPLC) with fluorometric [21] or electrochemical detection (ECD) [22].

Nevertheless, the isolated determination of 5-HT by any of these techniques affords a rather restrictive view on the biodynamics of the metabolic pathway as a whole. Furthermore, it has been repeatedly claimed that many of the experimental methods based on most of these techniques are not truly specific (i.e., they are affected by interferences), lack sensitivity, and are excessively time-consuming. In fact, in our experience, only the combined GC-MS/selected-ion-monitoring (SIM) techniques appear to be both sufficiently sensitive and selective to permit the unequivocal identification of most of the sample components eluted from a GC column, providing, at the same time, reliable quantitations of the components. In this regard, although modern HPLC, when coupled to the newer fluorometric and electrochemical detectors, gives detectability limits comparable in some cases to the SIM assays of biogenic amines [23], any attempted identification based only on chromatographic elution volumes, fluorescence spectra, or electrooxidation potential can never be truly unequivocal [23].

For such reasons, mass spectrometry has been widely used for the assay of indoleamines and metabolites in biological samples [2,24–30] with most of the determinations being carried out by previous separation of the sample components in standard GC packed columns. However, as many samples are exceedingly complex, occasionally it may be necessary to resort to higher-efficiency GC systems such as glass capillary column. The relatively higher efficiency and resolution of this type of column provide sufficient power to separate all of the components of the indolealkylamine pathway, thus giving more complete and representative metabolic profiles. Furthermore, glass capillary GC also provides a more inert analytical system, thus reducing the possibilities of sample loss through adsorption and/or degradation processes. However, in practical terms, the use of glass capillary columns requires more attention to technical operating parameters, such as sample introduction and their connection to the mass spectrometer. This is probably one of the reasons why this type of column is not more widely favored relative to the usual GC packed columns. Upon submittal of the manuscript (1980) there is only a preliminary communication on the use of capillary columns in indole amine analysis [31] and an application to 5-MT in human plasma [32].

The low sample capacity of glass capillaries has prompted the introduction of sample-splitting systems, so that only a minor part of the total sample actually enters the column, resulting in a quantitative loss, which in trace analysis may be a serious drawback. This has led to various attempts to bypass the splitter and find practical ways to introduce all of the sample [33-37]. We describe herein a simple method that allows the injection of the sample in the solid state after evaporation of the solvent and which does not require any modification whatsoever of the gas chromatograph [38].

On the other hand, when a capillary GC column, operated under splitless-injection conditions, is connected through a jet separator to a mass spectrometer, a significant part of the sample is lost in the GC-MS interface [38]. For this reason it is advantageous to be able to connect the column to the inlet of the mass spectrometer ion source through direct coupling systems [38-42]. Thus, a truly "splitless" GC-MS system would (1) prevent sample losses at either column inlet or outlet, (2) avoid dead volumes or unnecessary loss of chromatographic performance, (3) permit rapid and easy column interchange, and (4) introduce minimal or no modifications and unnecessary technical sophistication, allowing for maximal simplicity of operation at the same time. The work presented herein gives an account of the capillary GC-MS/SIM system which we have developed along these lines.

II. SAMPLE PREPARATION AND DERIVATIZATION

A. Derivatization of the Samples

The various indolic compounds implicated in the metabolism of tryptophan in the CNS have different functional groups ($-NH_2$, $=NH$, $-OH$, and $-COOH$), as illustrated in Scheme 1. Because of their low volatility and high affinity for other residual polar groups present in the surfaces of the chromatographic systems, with which they easily react (e.g., silanol groups in the glass surfaces of injectors and columns), these compounds become irreversibly immobilized and thus lost in the gas chromatograph; therefore, they cannot be directly analyzed by GC without their previous adequate modification to render them chemically inactive. The usual way to increase their volatility and lower their polarity is to transform them into stable derivatives with good chromatographic characteristics [43]. Although several types of derivatization procedures have been described for biogenic amines in the past [44], the most suitable for the indolealkylamines and their metabolites seem to be those that involve the silylation or acylation of all of the polar groups indicated above, with the exception of the $-COOH$ groups of the acids, which are first methylated when acylation reagents are being used. In this instance, the methylation is carried out by reaction of the acid with acidic methanol, diazomethane, or mixtures of BCl_3/methanol or BF_3/methanol [45], obtaining the corresponding $-COOCH_3$ moiety (subsequently abbreviated herein as $-COOMe$). However, when silylating reagents are used for the derivatization, there is no need for prior methylation of carboxylic acid groups since the silylating agents will readily attack the acid hydrogen, substituting a trimethylsilyl (TMS) group [46].

1. *Silyl Derivatives*

The trimethylsilylation of biogenic indoleamines and related metabolites has been described by various authors [47-52], although their application to practical problems in the metabolism of these amines has been relatively more limited as compared to the corresponding acylated derivatives. The reasons are probably to be found in the fact that the silylated derivatives easily decompose in the presence of traces of water and also tend to undergo structural changes, either soon after their synthesis or during their pass through the gas chromatographic system. An extensive study recently completed in this laboratory [23] indicated that derivative decomposition in the case of silylated species such as 5-hydroxytryptamine and tryptamine is not so much an effect of their inherent chemical stability as it is an effect of irreversible reactions.

A given TMS derivative can react with residual active sites located (1) on the walls or packing of the GC columns, (2) in connecting and transfer lines of the instrumentation, or (3) on the hydrated surfaces of the different pieces of glassware that come in contact with the primarily synthetized products. For instance the formation of silyl derivatives of a compound H—X implies the reaction of this compound with a trimethylsilyl (TMS) donor as shown below:

$$\text{H—X} + \text{R—TMS} \rightarrow \text{X—TMS} + \text{H—R}$$

Thus, in the case of tryptamine, complete silylation can be accomplished by reaction with a mixture of BSTFA-TMSI-pyridine (100:10:4) in a matter of 60 min at 70°C, as previously reported [53]. The derivative thus obtained, the $N^1,N^{\omega},N^{\omega}$-tris(TMS)-tryptamine, can be readily separated on a gas chromatographic column in less than 5 min. Presumably it should be stable, although it could also act as a reagent in the presence of reactive sites with exchangeable hydrogens, as illustrated in Scheme 15.2. The N^1 and N^{ω}, secondary and primary amino groups, respectively, are indicated on the structural formula of tryptamine (I in this scheme).

Evidently these effects lead to a loss of response for the GC peak expected at the retention time of the trisilylated derivative II, since it has been transformed into the N^1,N^{ω}-bis(TMS) derivative III. Derivative III would elute at a different retention time [53] although, having an active N^{ω} hydrogen, it could be lost by adsorption on the chromatographic system.

2. *Acyl Derivatives*

The acylation of indolealkylamines and their metabolites is usually carried out with acid anhydrides or acylimidazole reagents [54]. Biogenic amines, in general,

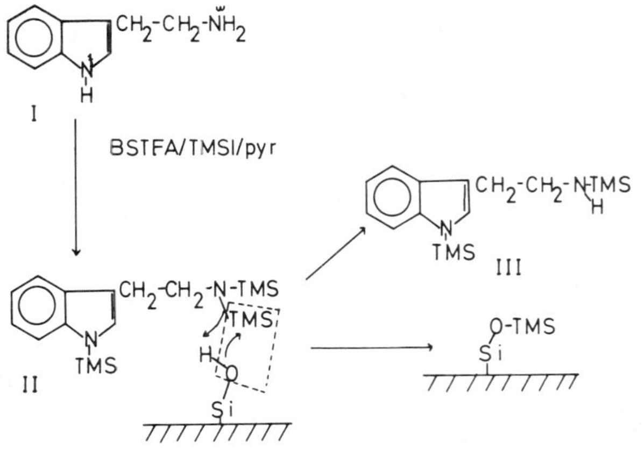

Scheme 15.2

have been converted to propionyl, *n*-butyryl, isobutyryl, pivaloyl, as well as various perfluoroacylated derivatives, such as trifluoroacetyl, heptafluorobutyryl, pentafluoropropionyl, and perfluorobenzoyl [43,54,55], initially developed to enhance the electron-capturing capabilities of non- or low electrophilic organic molecules for their use with the electron capture detector (ECD).

Several years ago, based on our own results and those of other authors, we concluded that in terms of their gas chromatographic characteristics the pentafluoropropionyl (PFP) derivatives gave a more satisfactory performance [43]. Our initial work involved the use of the ECD for biogenic amine profiling, though it was subsequently verified that the specific mass fragmentation patterns of these PFP derivatives were also suitable for combined GC-MS determinations in the multiple-ion detection mode [56] (see below).

In principle, once synthetized according to the general reaction sequence given below for tryptamine as an example,

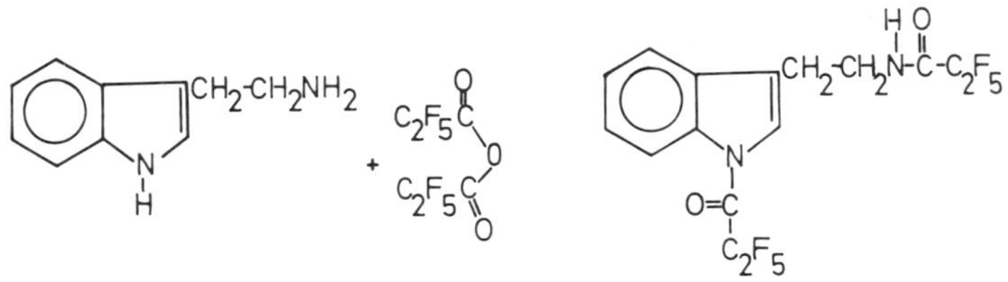

T PFPA T-2PFP

the bis-pentafluoropropionated (2PFP) derivative could be lost before detection either by degradation and/or by irreversible adsorption in much the same manner as described above for the trimethylsilylated derivatives. For example, it has been reported that under certain conditions (e.g., an inadequately inactivated system) a pentafluoropropionyl group (PFP) can be transferred from the β-hydroxyl position of a catecholamine to any of the free silanol groups exposed along the surfaces of the chromatographic system [57]. This process would be equivalent to that depicted in Scheme 15.2 for the TMS derivatives.

In order to drive the reactions to completion, small-scale acylations of biogenic amines are carried out in a relatively large excess of reagent with due care to the selection of the most adequate reaction solvent. According to our own studies the relative yields of reaction can vary significantly, depending on the solvent used [56].

Silylated vs acylated derivatives for biogenic amine determination: From all of the data gathered in the last 7 years we can establish the following practical conclusion regarding the use of silylated vs perfluoroacylated derivatives for tryptophan and its indolic metabolites. Under the usual laboratory conditions, the acylated derivatives are less prone to incur degradative losses and structural changes via sequential loss of acyl groups than their silylated counterparts. Also the silylation procedures often complicate the GC profiles since they are more likely to give rise to multiple derivatives from a single substance (e.g., partially silylated derivatives in different positions). However, if the reactions are properly optimized and if due care is taken in handling both types of derivatives, they seem to be capable of giving comparable detection limits by selected ion monitoring [53]. Another practical consideration to keep in mind is that the perfluoroacylated derivatives are more volatile and thus elute before the corresponding TMS derivatives, thus shortening the analysis time.

Mixed acylation—esterification procedure for biological extracts: For the reasons stated above we have consistently used perfluoropropionated derivatives, prepared according to the following procedure: The methanol eluates obtained from the XAD-2 columns are evaporated to dryness under a stream of purified helium. The dry residue is mixed with 50 μl of a BCl_3/methanol solution to esterify the carboxylic acid groups. After 15 min at a temperature of 85°C the solution is taken again to dryness, adding then to the dry residue (the indole acids have already been methylated) 50 μl of PFPA (pentafluoropropionic anhydride) and 50 μl of AcN (acetonitrile). The AcN has been previously saturated with sodium hydrogen sulfite.

The small vial containing the sample to be acylated is then introduced in an oven at 60°C for 2 hr. At the end of this period the solution is cooled to room temperature, removing the reagent and solvent under a stream of helium. The residue is next taken up in benzene and is then ready for injection into the GC-MS system.

Initially, the benzene used to redissolve the derivatized dry residues contained 5–10% PFPA to further protect the acylated derivatives from decomposition. However, as mentioned previously, these derivatives are, in our experience, stable enough so that recently we have discontinued using these benzene/PFPA mixtures without any appreciable difference in the results.

B. Extraction Procedures

Since the first significant neuroactive indoleamine (serotonin or 5-hydroxytryptamine, 5-HT) was reported by Erspamer et al. [58], various extraction procedures have been described, most of them based on the use of *n*-butanol for 5-HT [59] and ethyl acetate for the acid metabolite 5-HIAA [27,28]. For tryptamine there have been various reported procedures, such as extraction with a mixture of toluene-isoamyl alcohol [60] or with benzene [61]. None of these procedures, however, was suitable for our purposes, as will become clear later on.

In any case it must be taken into consideration that the extraction procedure itself is always implicitly bound to the method of detection and identification of the substances extracted from any given sample. In this sense we have to consider our simplified method of extraction [24] in relation to the aim sought and the capabilities of our analytical system, that is, a combined GC-MS system operated in the SIM mode. Since this detection system is capable of providing a qualitative and quantitative profile of several compounds present in a complex sample, we had the possible option to attempt a concurrent analysis of all or most of the components of the catabolism of tryptophan via the indole pathway [56]. For this reason, if the analytical system is capable of simultaneously analyzing various related metabolites, it is evident that the extraction procedure should provide samples no specifically enriched in only *one* compound or a class of compounds, as most methods do provide (e.g., extraction of an acid fraction), but in *all* of the components of interest. In our case having the SIM system of detection meant that we could process a fraction containing the indole bases, the corresponding acids, and the amphoteric amino acid precursor (tryptophan, TP). Thus, we departed considerably from classical extraction schemes by setting up an original method, finally optimized according to the following overall procedure [24]:

Rat-brain tissue samples with a weight of 1 g or less are homogenized in 10 vol. of an acidic solution of 0.1 N HCl, 1 N KCl, and 0.2% $NaHSO_3$. To avoid contamination of the samples with substances present in the blood, the blood capillaries are blotted with filter paper. Aliquots of the methanolic solutions that contain the deuterated standards (added for quantitative studies) as well as the 5-methoxytryptamine (5-MT, added as "carrier") are then evaporated to dryness in a Potter-Elvehjem homogenizer, under a stream of helium, adding to the dried residues the homogenization medium indicated above and the brain samples. The homogenate is next centrifuged in a Beckman L2-65B ultracentrifuge at 100,000 *g* for 20 min. The supernatant is decanted and the pellet rehomogenized in 10 vol. of the same homogenization medium. This homogenate is centrifuged under the same conditions and the pellet is finally discarded. The two supernatants are pooled and the pH is adjusted to 1.4–1.6. The solution is placed at the top of a short glass column that contains 0.5 g of Amberlite XAD-2 resin. The column is washed with 10 ml of dilute HCl at the same pH of the supernatants, and after discarding this washing, the adsorbed indoles are eluted with 6 ml of distilled methanol. The methanol eluate is collected in a 20-ml pear-shaped glass flask and concentrated to 100–200 μl in a Rotary film evaporator after adding 1 drop of concentrated formic acid and 10 mg of 5-MT. The concentrated sample is next transferred to a Regis 0.5-ml reaction vial, and the dry residue is derivatized by combined methylation-perfluoroacylation as previously described.

The addition of deuterated internal standards at the beginning of the whole process is meant to provide an internal reference for each of the substances under study, thus compensating for the extractive losses of the endogenous counterparts as well as for the different yields and responses of each derivative in the SIM system. On the other hand, the carrier is added to compensate for and prevent such effects as the possible adsorption of the derivatives on the walls of the glassware and/or chromatographic system, as described above. (The carrier, because of its relatively larger proportion with respect to the sample components, blocks any active sites and minimizes the possibilities of degradative reactions and adsorptions.)

According to the results thus far obtained, the method is very simple and efficient, especially if compared to other more elaborate and tedious procedures. For instance, Scheme 15.3 briefly compares in a condensed form the different approaches described for the determination of tryptamine in biological samples. It is evident by inspection of this scheme that the method used in our laboratory is more rapid and simpler and provides more information from any given sample. Multistep extraction and analysis procedures are always prone to significant quantitative losses which in the case of trace analysis may be a serious limitation. For instance, considering the assay of tryptamine as an example, the radioisotopic enzymatic method of Saavedra and Axelrod [60], besides being a two-day procedure, depends on very low blank values for sensitivity reasons and is strongly affected by coextracted interferences due to a lack of enzyme specificity.

This is reflected in the very high values reported for tryptamine in rat brain using this procedure (of the order of 22 ng/g). In contrast, the method of Phillips et al., by virtue of the high specificity of high-resolution mass spectrometry [29], gives values of 0.5 ng/g (400 times less than with the radioisotopic enzymatic method). Nevertheless, it is a rather cumbersome procedure that requires the pooling of two brain samples for optimum results.

While interferences from 5-HT contribute heavily to the reported values of tryptamine according to the radioisotopic enzymatic assay, the coextraction of TP contributes to falsely high values of tryptamine (T) in the spectrofluorometric method of Hess and Udenfriend [66].

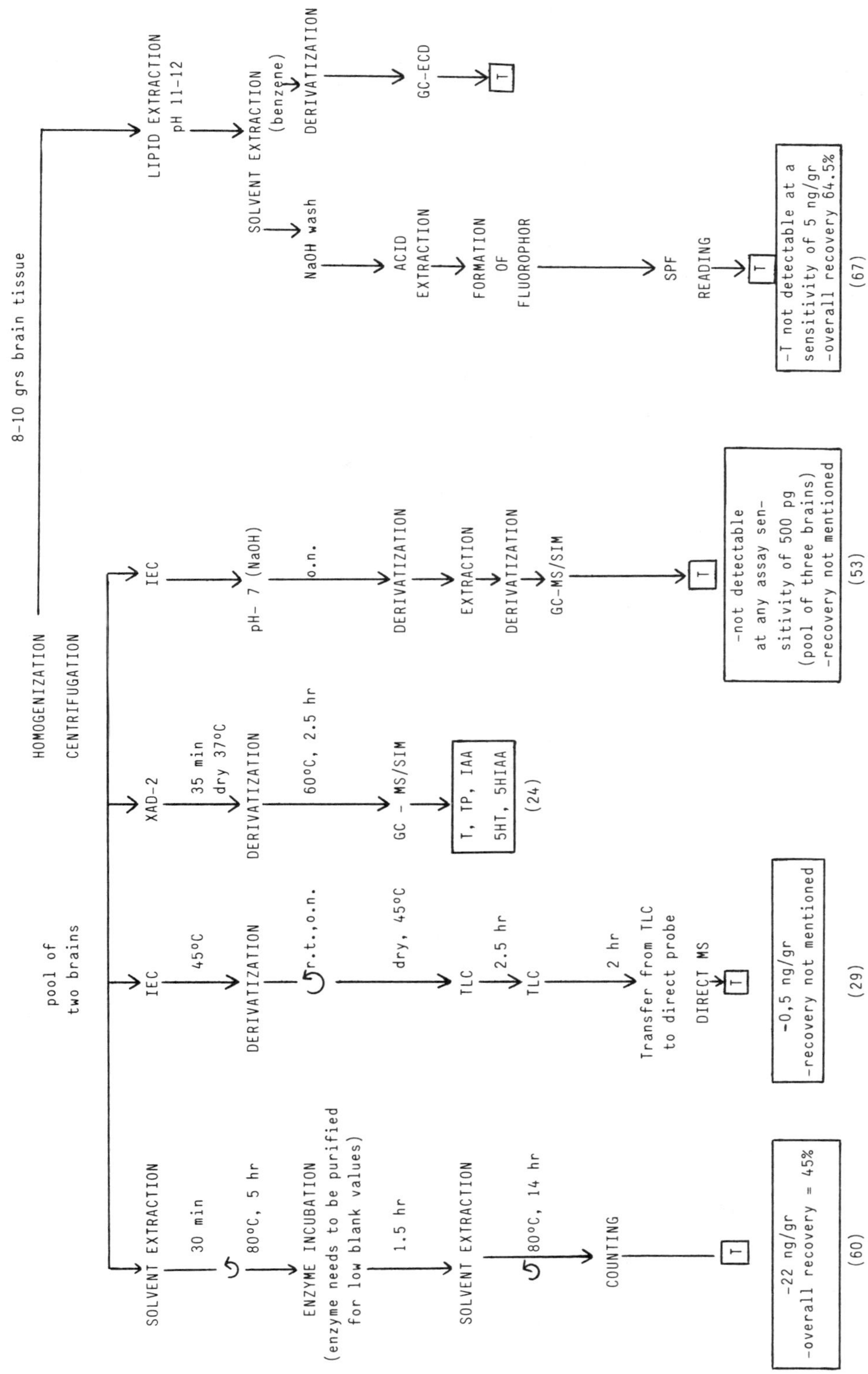
8-10 grs brain tissue
HOMOGENIZATION
CENTRIFUGATION
pool of two brains
SOLVENT EXTRACTION
30 min
80°C, 5 hr
ENZYME INCUBATION
(enzyme needs to be purified for low blank values)
1.5 hr
SOLVENT EXTRACTION
80°C, 14 hr
COUNTING
T
-22 ng/gr
-overall recovery = 45%
(60)
IEC
45°C
DERIVATIZATION
r.t.,o.n.
dry, 45°C
TLC
2.5 hr
TLC
2 hr
Transfer from TLC to direct probe
DIRECT MS
T
-0,5 ng/gr
-recovery not mentioned
(29)
XAD-2
35 min
dry 37°C
DERIVATIZATION
60°C, 2.5 hr
GC - MS/SIM
T, TP, IAA
5HT, 5HIAA
(24)
IEC
pH- 7 (NaOH)
o.n.
DERIVATIZATION
EXTRACTION
DERIVATIZATION
GC-MS/SIM
T
-not detectable
at any assay sensitivity of 500 pg
(pool of three brains)
-recovery not mentioned
(53)
LIPID EXTRACTION
pH 11-12
SOLVENT EXTRACTION
(benzene)
DERIVATIZATION
GC-ECD
T
NaOH wash
ACID EXTRACTION
FORMATION OF FLUOROPHOR
SPF
READING
T
-T not detectable at a sensitivity of 5 ng/gr
-overall recovery 64.5%
(67)

Scheme 15.3

This method, which involves the formation of a fluorophor after acid extraction, has also been used by Martin et al. [67], although these authors were unable to detect tryptamine in rat brain. Finally, the recent method of Warsh et al. [30,53] is the most similar to ours, but it does not allow the concurrent assay of basic and acid TP metabolites.

III. INSTRUMENTATION

A. Capillary Gas Chromatography—Mass Spectrometry

The mass spectrometer Hitachi Model RMU-6H was directly coupled to a Perkin-Elmer Model 3920 gas chromatograph, as illustrated in Figure 15.1. The system operation has been described in detail elsewhere [38]. Briefly, it involves the use of a totally splitless injector and interface connection, both in relation to column inlet as well as outlet. At both ends of the column, vacuum-tight connections can be easily achieved with standard stainless steel Swagelok fittings for chromatography, as shown in inserts *a* and *b* of Figure 15.1. The front and back ferrules regularly used in this type of connection were substituted by a piece of standard high-temperature GC septum disc, cut to fit the inner dimensions of the 1/16-in. Swagelok nuts. This is illustrated in inserts *a* and *b* by the more densely shadowed rectangles which are pierced by the ends of the glass capillary column. This arrangement provides a leak-tight

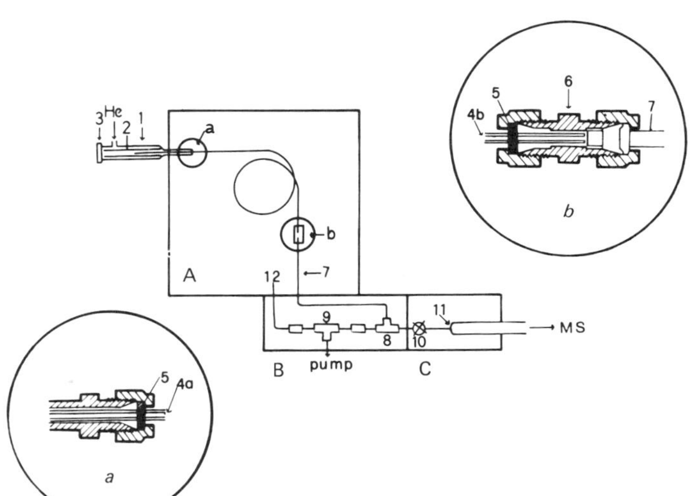

Figure 15.1 Arrangement for direct or indirect connection (through molecular separator) of glass capillary columns to the mass spectrometer. The system is divided into three distincts zones: (A) GLC oven, (B) standard manifold of the Perkin-Elmer model 3920 gas chromatograph, and (C) extra manifold chamber. 1 = standard packed column injector; 2 = 1/8 in. or 1 mm I.D. glass liner; 3 = septum; 4a = capillary column inlet; 4b = capillary column outlet; 5 = septum disc; 6 = 1/16-in. Swagelok union; 7 = stainless-steel capillary (50 cm × 1/16-in. O.D. and 0.15–0.20 mm I.D.) coated with OV-101; 8 = Swagelok "tee'; 9 = single-stage jet separator; 10 - mass spectrometer isolation microcapillary valve (Perkin-Elmer); and 11 = glass-to-metal seal. (From Ref. 38.)

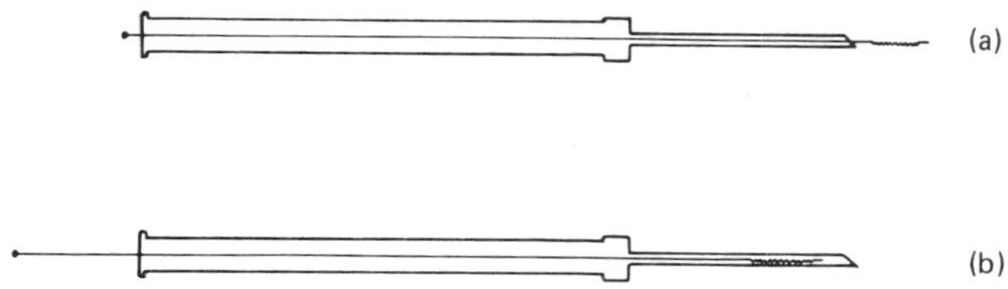

Figure 15.2 Gas chromatographic injector system for solids. Plunger made of inert metal (Au or Pt). (a) loading and desorption position (needle inside injector block); (b) injection position to facilitate piercing the septum disc.

seal when the nuts are manually tightened. The outlet of the capillary column is directly interfaced with the mass spectrometer through a 50-cm-long stainless-steel capillary (7, in Figure 15.1) which ends in a Swagelok "tee" (8, in Figure 15.1). This "tee" is connected to the ion source inlet through a low dead-volume capillary valve (10, in Figure 15.1) manufactured by Perkin-Elmer. The optional connection through a jet separator (9, in Figure 15.1) has been described in a previous publication from this laboratory [38]. The mass spectrometer has been equipped with a four-channel multiple-ion-monitoring system designed and built in or laboratory [62].

The mass spectrometer was operated in the direct connection mode in a pressure range of $1-5 \times 10^{-5}$ mm Hg with a column flow-rate of up to 3 ml/min. The mass spectrometric parameters used throughout this work were set as follows: chamber voltage, 70 eV; emission current, 80 μA; trap current 40–50 μA; accelerating voltage, 2400, 1800, or 1200 V, depending on the ions selected for detection. The temperature of the ion source was kept at 200°C, and that of the connection system at 250°C.

B. Splitless Injection Technique

In order to avoid any unwanted loss of sample at the injection port of the gas chromatograph, by venting off a relatively large proportion of the solution immediately after it is vaporized in the injector, samples were injected with a solids injection syringe equipped with a plunger fitted with a 1-cm-long spiral section supplied by Scientific Glass Engineering, London, Great Britain (order code SI-1RDS). This injector is used according to the following procedure: with the plunger in position a (Figure 15.2), the sample (liquid), previously measured in a standard Hamilton Model 701N or 7001 microsyringe of the type regularly used in gas chromatography, is transferred to the spiral section of the solids injector, where the solvent is allowed to evaporate during a period from 5 to 10 min at room temperature; with the plunger of the solids injector in position b, the needle is next introduced into the injection port of the GC, depressing the plunger to expose the dried sample residue. In this manner, the dried samples are volatilized in less than 2 sec.

IV. APPLICATIONS

The capillary column GC profiles obtained from standard mixtures of the metabolites of tryptophan (Scheme 15.1) are illustrated in Figures 15.3 and 15.4.

For comparative purposes, the corresponding profiles on regularly packed columns are shown next to each of the capillary profiles. The two chromatograms of Figure 15.3 were obtained adjusting the carrier gas flow-rate in both

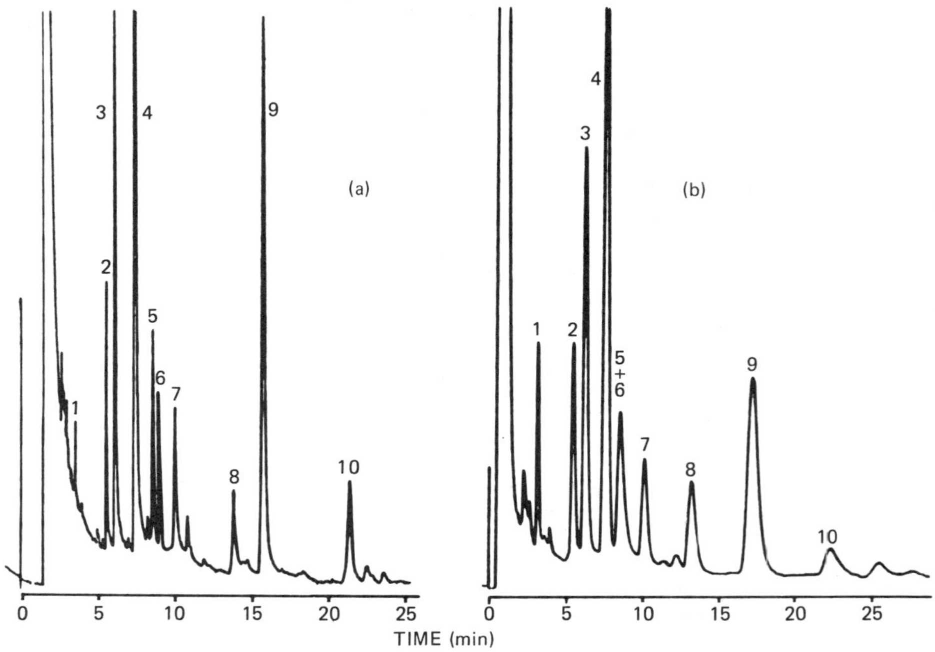

Figure 15.3 Gas chromatographic profiles of a mixture of metabolites of tryptophan separated on OV-17 stationary phase (a) Capillary column, 20 m × 0.37 mm I.D. Column temperature, 150°C. Injection mode, splitter system with a ca. 1:75 ratio. (b) 5% packed column 1.8 m × 2 mm I.D. Column temperature 150°C. 1 = 5-HTOL-3PFP; 2 = 5-HIAA-Me-2PFP; 3 = IAA-Me-PFP; 4 = 5-MTOL-2PFP; 5 = T-2PFP; 6 = 5-HT-3PFP; 7 = NAS-2PFP; 8 = 5-HTOL-2PFP; 9 = 5-MIAA-Me-PFP; 10 = 5-MT-2PFP.

types of columns so that the separation of these compounds takes approximately the same time (ca. 25 min). It is evident, however, that the higher resolution of the capillary columns reflected in the much smaller peak-widths contributes to the total separation of the pentafluoropropionated derivatives of tryptamine (T) and 5-hydroxytryptamine (5-HT), both of which coelute in a single peak on the packed column, approximately at 9 min. The high-resolution power of this capillary column relative to the packed column (both with OV-17 as stationary phase) is practically demonstrated by the higher resolution factors as shown in Table 15.1. For instance, the resolution of the pair of derivatives of 5-HIAA and IAA is 21.8% higher on the capillary column as compared to the packed column. All of the PFP derivatives give only one peak for each substance, except in the case of 5-HTOL, which, under the experimental conditions used for the concurrent derivatization of all of the components of the mixture, gives a response at ca. 3 min for the fully acylated 5-HTOL-3PFP derivative and a second response for a partially acylated (5-HTOL-2PFP) species at ca. 14 min. It must be taken into account that the simultaneous derivatization of substances as chemically diverse as these (substances containing various functional groups of widely different nature) requires the establishment of a suitable compromise in terms of reaction conditions. The optimization of the derivatization reaction for a given substance or class of

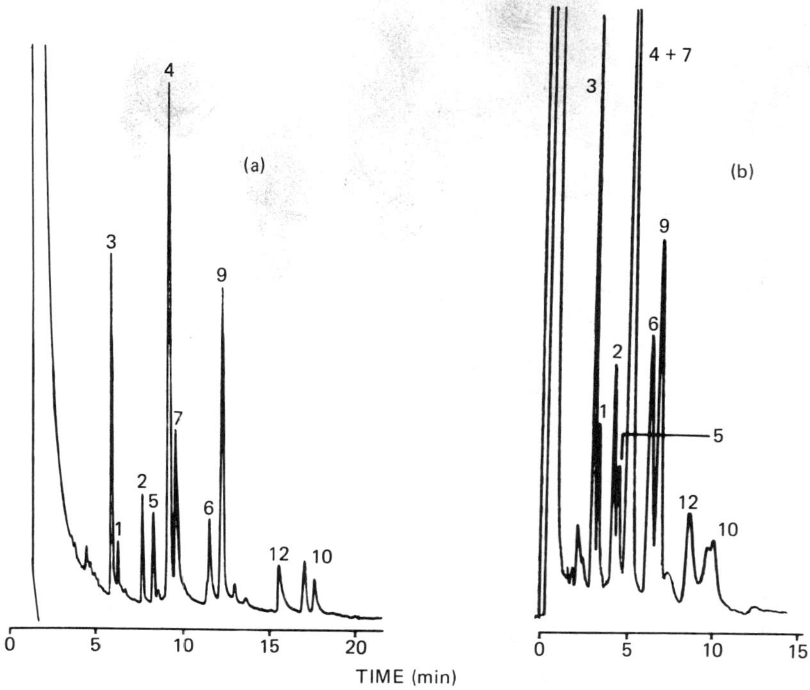

Figure 15.4 Gas chromatographic profiles of a mixture of metabolites of tryptophan separated on an SE-30 stationary phase (a) capillary column, 20 m × 0.3 mm I.D. Column temperature, 150°C. Injection mode, splitter system with ca. 1:75 ratio. (b) 5% packed column 1.8 m × 2 mm I.D. Column temperature, 170°C. Peak identification according to numbering system of Figure 15.3.

substances (e.g., indoleamines vs indoleacids) usually leads to poor results for other substances. Thus, the presence of a peak for 5-HTOL-2PFP indicates that the reaction has not reached the maximum yield under the conditions used.

Table 15.1 Chromatographic Resolution Between Different Pairs of PFP Derivatives

	Capillary column	Packed column
5-HIAA-Me-2PFP/IAA-Me-PFP	1.92	0.88
IAA-Me-PFP/5-MTOL-2PFP	4.19	2.22
5-MTOL-2PFP/T-2PFP	3.55	1.56
T-2PFP/5-HT-3PFP	1.4	0
T-2PFP/NAS-2PFP	6.82	1.63
5-HTOL-2PFP/5-MIAA-Me-PFP	4.35	2.56

The pentafluoropropionated derivative of melatonin, not shown in this capillary GC profile (Figure 15.3), elutes at ca. 27 min.

A similar situation occurs using a less polar stationary phase, such as SE-30, illustrated in Figure 15.4. In this case, the capillary column clearly separates the derivatives of 5-MTOL and *N*-acetylserotonin (NAS), which coelute on the standard packed column.

In practical terms, the increase in chromatographic resolution attainable with capillary columns can be very useful to cope with complex biological extracts since these columns can resolve a much larger number of components than a packed column. In the case of the analysis of tryptophan and its metabolites (see Scheme 15.1), some of which are present in exceedingly small amounts in brain tissue [24], the use of capillary columns, while providing higher resolutions, can enhance the detectability limits, since for a given amount of substance the peak height is always higher in capillary columns as a consequence of their remarkably reduced peak-widths compared to packed columns.

A. Splitless Sample Injection in Capillary Columns

It is known that the low sample capacity of capillary columns requires the provision of splitting systems at the column inlet to avoid overloading the column with large amounts of solvent. However, in practice, the use of a splitting system either prevents the analysis when the sample amount is very low or introduces a significant degree of injection variability since it becomes difficult to repetitively inject the exact, absolute amount of sample. Thus sample splitters are not suitable for reproducible quantitative data. Since we known from experience that the lack of reproducibility at the injection port can be efficiently compensated by the use of deuterated internal standards [24], we were more concerned by the quantitative losses resulting from the sample splitting mode. For these reason we have developed a simple method that requires no modification of either GC column or injector and that allows for a truly splitless operation. The system, which has been described in Sec. III.B., facilitates the introduction of the sample into the capillary column practically free of solvent, as illustrated in Figure 15.5a, compared with the same sample injected on a packed column with the usual microsyringes, shown to the right in Figure 15.5b. The comparison of these two GC profiles also demonstrates that the splitless injection system does not introduce any spurious effects in relative quantitative terms. In other words, the evaporation of the solvent prior to the injection of the sample into the capillary column does not cause or induce any loss of compounds by adsorption and/or degradation, so that the yield of injection for each component is close to 100%. However, as a consequence of the high chemical and photochemical lability of these indole metabolites, the initial injection yields for 5-MTOL-2PFP were of the order of 5% for sample quantities of the order of 40 ng. This effect was traced to the stainless steel of the plunger of the solids injector. The substitution of the original plunger for another made of gold or platinum eliminated completely the processes of adsorption and/or degradation of the sample on the spiral of the plunger [63], with typical results as shown in Figure 15.5a.

The performance of the solids injector fitted with an inert plunger was verified by injection of different amounts of 5-MTOL-2PFP on a packed column [63] with both a Hamilton microsyringe and with the solids injector. From the GC detector response, a relative response ratio was calculated. This study was carried out over a wide range of injected amounts, that is, from 130 pg to

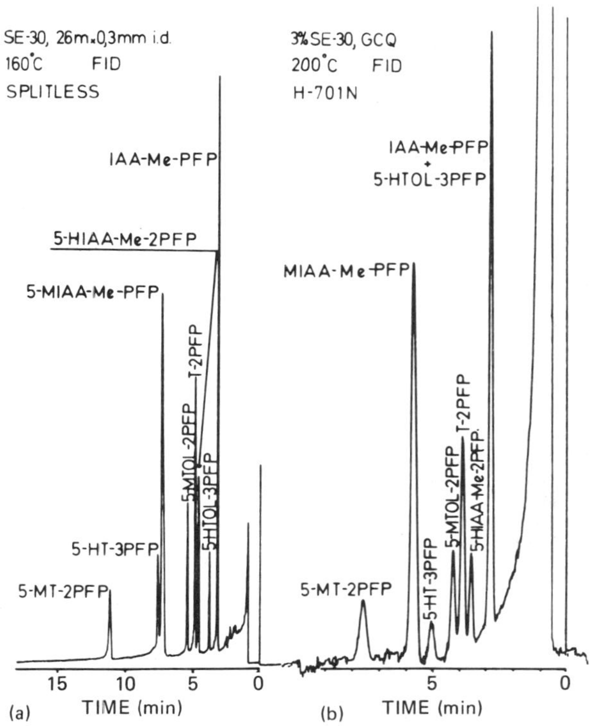

Figure 15.5 (a) Performance of the splitless injection system (solids injector) with a mixture of metabolites of tryptophan separated on an SE-30 capillary column. (b) The same mixture injected in a packed column (with a conventional syringe, Hamilton 701-N). Amounts injected in the range 30–150 ng of each component. Injector temperature held at 250°C.

60 ng. Two types of detection systems were used: the FID for injections of over 5 ng, and the SIM of the mass spectrometer for amounts down to the picogram range, monitoring in this case the fragment ions at *m/e* 306 and 319.

The results obtained over a period of three months are given in Table 15.2. These results indicate that the injection yields for 5-MTOL-2PFP obtained in the splitless mode, as proposed herein for capillary columns, are of the order of 60–70% in the picogram range, 80% in the nanogram range, and 90–100% for amounts injected of the order of 60 ng. Also, as has already been indicated, this method of injection does not introduce any appreciable discrimination. The similarities between the two chromatographic profiles in Figure 15.5 are evident, except for the lack of resolution of the packed column for the 5-HTOL and IAA derivatives and the reversal of the 5-HT derivative, relative to the retention time of the 5-MIAA peak, which is a function of the elution temperature [65].

B. Direct Coupling of GC Capillary Columns to a Mass Spectrometer

The direct coupling system described above is outstanding in its simplicity, as it does not require any instrumental modification and, furthermore, is based on the continued use of regular chromatographic fittings for which no special

Table 15.2 Injection Yields for 5-MTOL-2PFP with Splitless Injection Versus Packed-Column Injection

Amount of 5-MTOL	Percent response[a]	No. of injections[b]
60 ng	90-100	2
5.7 ng	84	10
400 pg	67	4
240 pg	63	5
128 pg	68	5

[a]Ratio of peak heights obtained from solids injector vs conventional packed-column injection system.
[b]Number of injections with each injection system.

training is necessary beyond that required to connect and disconnect a standard packed column. The leak-tight seal achieved with the pieces of septum in place of the usual front and back ferrules of these fittings ensures a working GC-MS connection. In fact, the background pressure inside the mass spectrometer improves as the temperature of the column is raised to 250°C, indicating that most likely, the increase in temperature expands the material of the septum holding the column ends inside the connecting nut, thus improving the tightness of the seal [38].

With this system, all of the effluent of the capillary column is introduced into the ion source of the mass spectrometer, bypassing the separator interface (9, in Figure 15.1). A comparative study carried out by connecting the capillary column to the mass spectrometer in the direct coupling mode as well as through the molecular jet separator (in the latter case make-up gas is added at the column outlet since the separator performs better at high flow-rates) demonstrated that the response increases 100% with the direct connection [38].

The gas chromatograms of a mixture of perfluoroacylated indoles monitored with an FID and with the total ion detector of the mass spectrometer are compared in Figures 15.6a and b. In the first case, the capillary column was connected in the usual fashion to the FID of a Perkin-Elmer 3920 gas chromatograph, while in the second case the column was directly connected to the ion source of the MS. The test mixture contained a representative sample of indoleamines and alcohols, with the amounts injected ranging from 20 to 45 ng per component. With the exception of the large solvent tail recorded by the FID, both profiles are practically identical, thus proving that in practice the direct coupling system herein described does not diminish the operating performance of the capillary column. Nevertheless, in practice the MS vacuum at the outlet of the capillary column reduces the separation efficiency by 25–30% [38,64], although this influence may not be readily appreciated since the initial efficiency of these columns is several orders of magnitude higher than that of packed columns. In other words, even with efficiency losses of 30% it is still possible to obtain chromatograms such as that of Figure 15.6b.

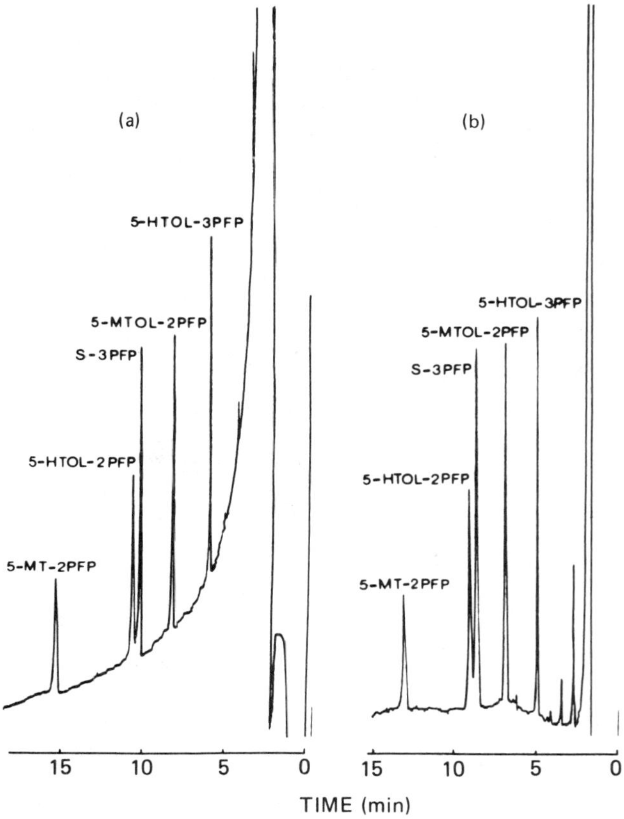

Figure 15.6 (a) Gas chromatographic (FID) profile of a sample of four metabolites of tryptophan (5-HTOL partially and fully perfluoroacylated) separated on an SE-30 glass capillary column (26 m × 0.3 mm I.D.) at 170°C. (b) Total ion monitor (TIM) profile of the same sample under identical GC conditions. The column is directly coupled to the mass spectrometer. Coupling manifold at 250°C. Mass spectrometer conditions: chamber voltage, 70 eV; emission current, 70 μA; trap current, 45–50 μA; and accelerating voltage, 2400 V. Helium carrier gas flow-rate measured at the column outlet at room temperature, 2.6 ml/min. (From Ref. 38.)

The sample was injected in this test using a splitter adjusted to a 1:20 ratio (95% of the sample is vented off to the atmosphere upon injection [38].

C. Capillary Column GC-MS with Selected Ion Monitoring

A detailed study of the mass spectra of each one of the substances of interest enabled us to select the ions most suitable for monitoring with our four-channel SIM unit [62]. Besides the relative abundance of the ion per unit weight of substance, one of the criteria taken into consideration in the process of ion selection was the response of the ion to more than one compound in order to get as much information per analysis as possible from the four-channel system. For instance, by tuning the SIM unit on the ion at a mass-to-charge ratio (*m/e*) of 276, it becomes possible to monitor simultaneously three compounds, namely the PFP derivatives of IAA, T, and TP, given in order of elution, as illustrated in the first row of Scheme 15.4. It has been reported that one of the main fragmentations of perfluoroacylated alkylindoles occurs by cleavage of the α bond in the alkyl chain with retention of the positive charge on the indole nucleus moiety (fragment I in Scheme 15.4). Thus the fragment generated by this process is common to these three derivatives. The same would apply to the simultaneous detection of 5-hydroxyindoles, which can be readily accomplished by tuning the SIM system to monitor the appearance of the fragment ion at *m/e* 438. The increase in 162 mass units for this fragment relative to that at *m/e* 276 is due to the extra perfluoroacylated group in position 5 on the indole ring (see Scheme 15.4).

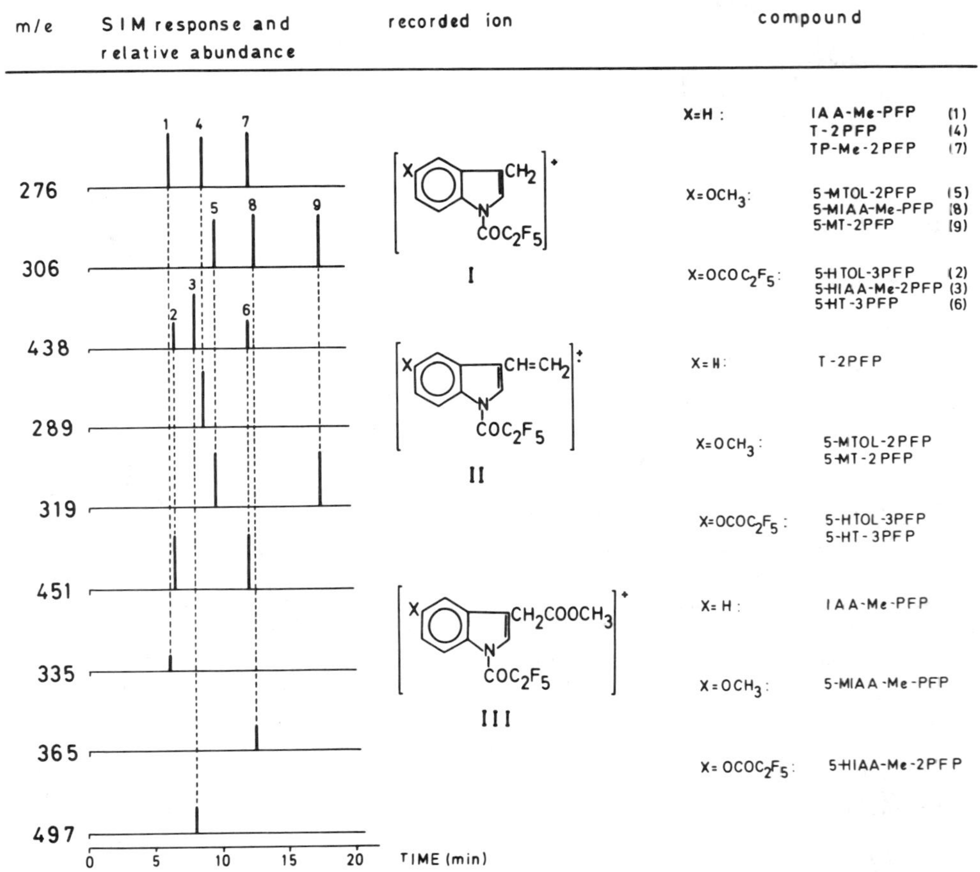

Scheme 15.4

By the same reasoning, monitoring the GC effluent for the fragment ion at mass 306, one would detect the presence of 5-MTOL-2PFP, 5-MIAA-Me-PFP, and 5-MT-2PFP. The other ions selected in this study are also indicated in Scheme 15.4 (see fragments II and III).

In accordance with these considerations the ion current profiles shown in Figure 15.7 illustrate the kind of results obtained by monitoring the ion

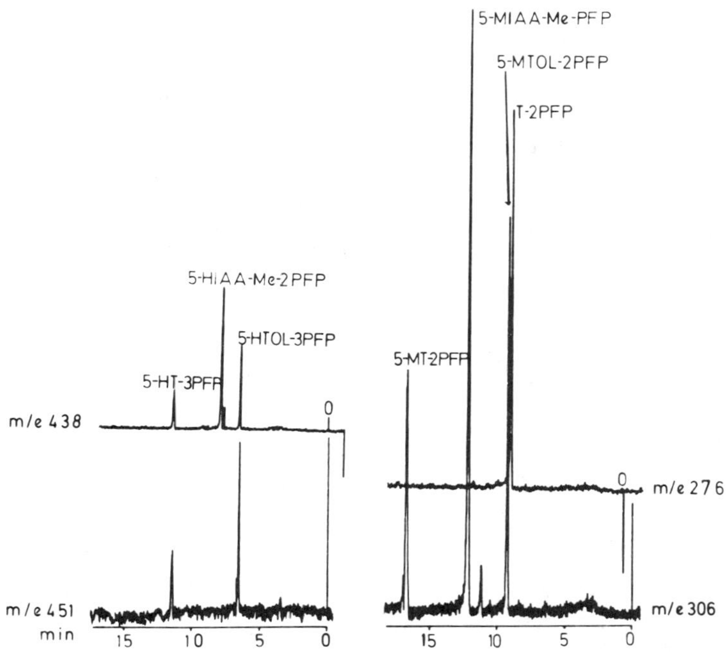

Figure 15.7 Selected ion current profiles of the perfluoroacylated derivatives of a mixture of tryptophan metabolites obtained with an SE-30 capillary column (26 m × 0.3 mm I.D.) directly connected to a mass spectrometer and with the splitless injection system.

currents at *m/e* values of 276, 306, 438, and 451. In this fashion a total of nine compounds could be detected in a single run. Although IAA and TP have not been included in this mixture, if present, they would be detected in the trace of *m/e* 276 at their predicted retention times.

Figure 15.8 shows the representative response curves for 5-MT-2PFP, T-2PFP, and 5-MTOL-2PFP, for amounts ranging from 10 to 300 pg, approximately. As the inset in this figure demonstrates, low picogram-level detection is readily accomplished in these conditions. The corresponding regression coefficients (r^2 = 0.99887 for the 5-MTOL-2PFP, r^2 = 0.99702 for T-2PFP, and r^2 = 0.99734 for 5-MT-2PFP) attest to the linearity between the amount injected in the capillary column and the SIM response. The practical detection limits for the derivatives of 5-MTOL, T, and 5-MT were established at 13, 20, and 30 pg, respectively.

The method can be applied to the study of indole metabolism in samples of extracts of brain tissue or human urine, as shown in the selected ion current profiles given in Figures 15.9–15.11. One of the outstanding advantages of the overall procedure is the possibility of using the deuterated analogue of each one of the compounds of interest as its own internal standard and thus remarkably increasing the precision of the quantitative determinations. For example, Figure 15.9 shows the results obtained with a properly derivatized XAD-2 methanol eluate from a rat brain homogenate analyzed by splitless capillary GC-MS in the selected-ion-monitoring mode. The presence of endogenous TP and IAA is indicated by the response obtained at *m/e* 276, whereas 5-HT

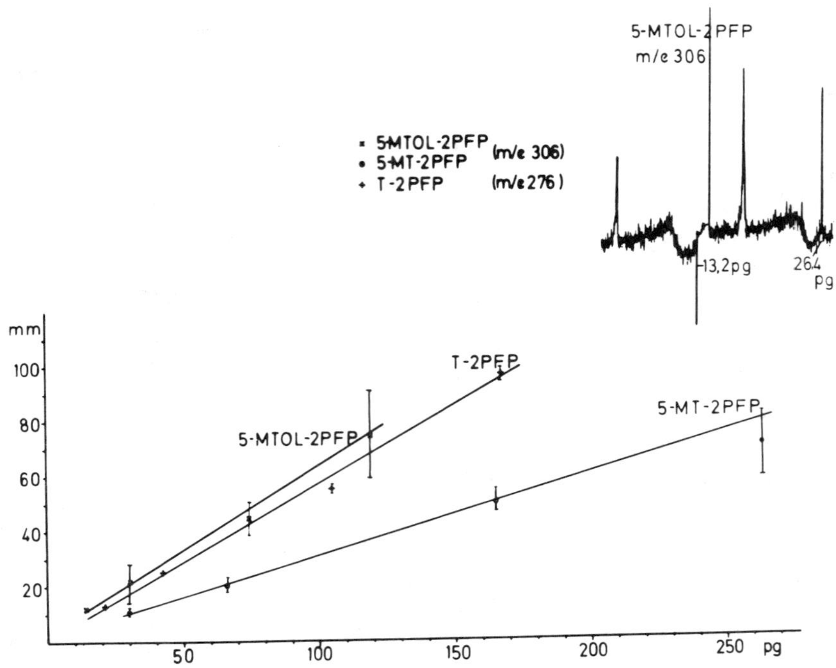

Figure 15.8 SIM response curves for the perfluoroacylated derivatives of 5-MTOL and 5-MT at *m/e* 306 and T at *m/e* 276. The capillary column was directly connected to a mass spectrometer and the sample was injected with the solids injector. The insert shows the response of 10–20 pg of 5-MTOL-2PFP in these conditions.

and 5-HIAA are detected at *m/e* 438. Due to its very low endogenous amounts, the detection of IAA was confirmed by monitoring also the fragment ion at *m/e* 335, obtaining as a further piece of evidence a response at the retention time of IAA. Furthermore, the two ion abundances fit, within the experimental error, the characteristic relative proportions of these two fragments in the mass spectrum of the authentic IAA derivative (Scheme 15.4). For quantitative purposes, the response of these four compounds is directly evaluated relative to the response obtained from a sample spiked with known amounts of the corresponding deuterated counterparts monitored at masses 278 and 440 [24]. This allows an exact quantitation with a degree of reliability thus far unmatched by other techniques. The rest of the peaks appearing on these profiles could be due either to related metabolites or to other substances also responding to these ions, but there is no possibility of mistaken identifications since they do not elute at the predicted retention times, as indicated in Scheme 15.4.

The results obtained with a different rat brain sample analyzed for 5-HT and 5-HIAA are illustrated by the selected ion current profiles of Figure 15.10. Quantitation is also carried out with the corresponding deuterated standards, shown in the lower trace.

Tryptamine was not detected in these samples, as expected, since it has been reported that it is present in rat brain in a concentration of 0.5 ng/g [29] or lower [24,53].

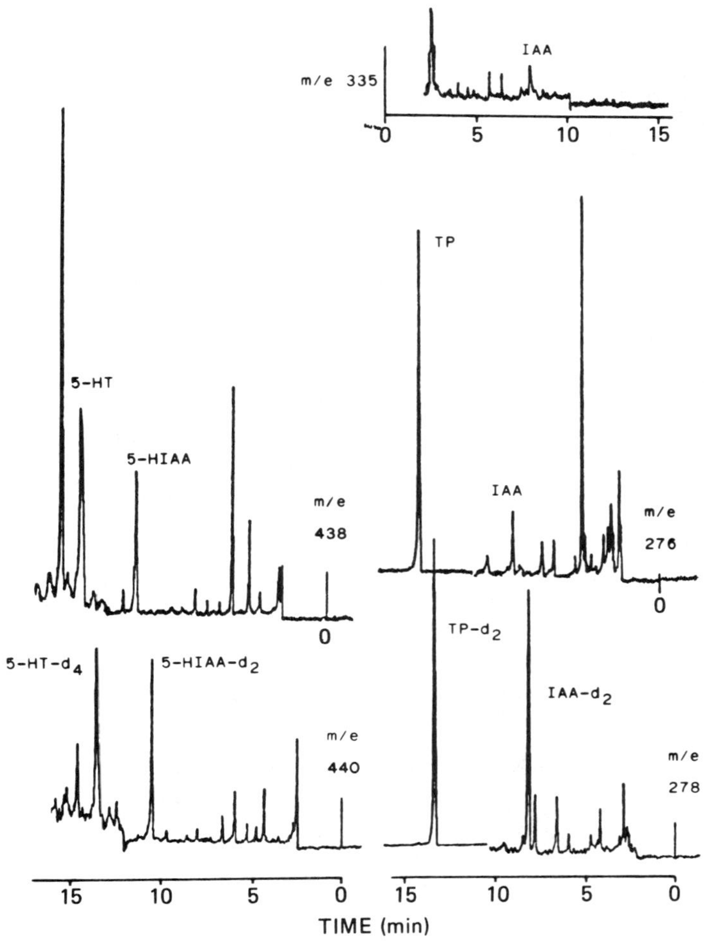

Figure 15.9 Selected ion current profiles from a rat brain extract. SE-30 capillary column (20 m × 0.3 mm I.D.). Column temperature, initially 4 min at 160°C and then programmed to 200°C at 4°C/min. After 10 min the SIM response is attenuated for the ions at *m/e* 335, 276, and 278, and increased for the ions at *m/e* 438, and 440 to compensate for the higher amounts of TP relative to IAA as well as for the different experimental recoveries of 5-HT relative to 5-HIAA.

This method has been also applied to the screening of TP metabolites in samples of human urine from a healthy individual with typical results, as illustrated in the profiles of Figure 15.11. In this manner TP and IAA were detected in the *m/e* 276 ion profile, while the identity of IAA was further confirmed by the response at *m/e* 335. The presence of 5-HIAA in these samples was established in the same way by monitoring the fragment ion characteristic of the 5-hydroxyindoles. A small amount of 5-HT was also detected in these conditions.

A summary of the quantitative data obtained for the rat brain samples is given in Table 15.3. These values compare favorably with the different values reported in the literature, thus proving the reliability of the quantitative

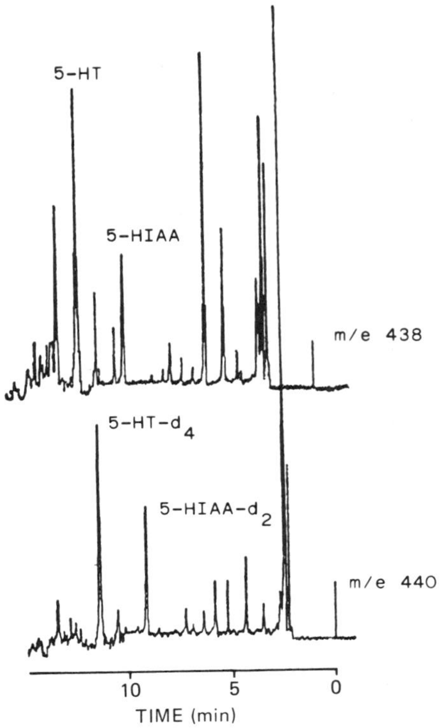

Figure 15.10 Determination of 5-HT and 5-HIAA for a rat brain extract by focusing on the ions at *m/e* 438 (endogenous levels) and 440 (standard levels). Gas chromatographic conditions as in Figure 15.9, except for the program rate which was increased to 8°C/min.

determinations carried out with a combined, high-efficiency glass capillary GC-MS/SIM system, relative to other analytical approaches. In both cases the values obtained for these samples are in agreement with our previous study and also fall within the range of reported rat whole-brain values. The range of concentrations shown in Table 15.3 is derived from the values given in a total of 13 references, as reported recently by Koch and Kissinger [22], which include their own values. In a similar manner, the quantitative data corresponding to these four metabolites in the sample of human urine are given in Table 15.4. The range of reference values is also derived from data given in eight publications from different authors, as compiled by Koch and Kissinger [22]. The only slight discrepancy regarding the somewhat higher level of IAA, both in brain tissue and human urine, is in line with our latest data whose significance will be discussed in more detail elsewhere.

Finally, any suitable biological sample could be screened for other TP metabolites in the same manner and according to the scheme for ion selection indicated above (see Scheme 15.4) (e.g., tryptophols in CSF [8,9] or 5-methoxylated indoles in plasma [32].

In conclusion, the accuracy and precision of the data obtained by selected ion monitoring of the peaks that elute from a high-resolution glass capillary GC column firmly establish this methods as the most reliable to date for metabolic and biomedical studies at low detection levels. In fact, its high

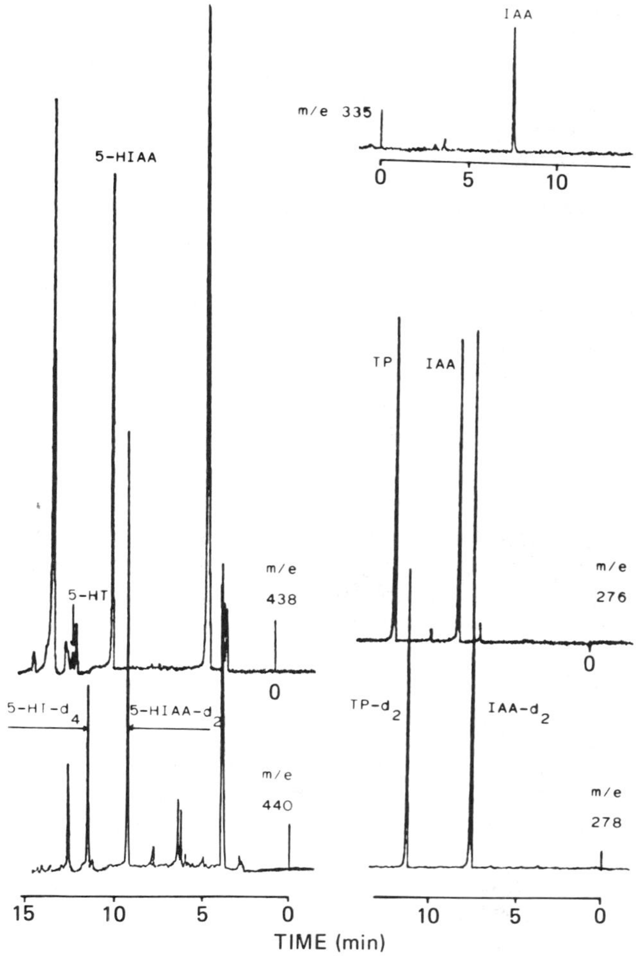

Figure 15.11 Selected ion current profiles from an injection of a human urine extract. GC and MS conditions as in Figure 15.9, except for the program rate (8°C/min) and the lack of any change in sensitivity in the response of the different ion current profiles.

Table 15.3 Quantitation of Tryptophan and Related Indole Metabolites in Rat Brain by Glass Capillary GC-MS/SIM Techniques[a]

	Extract A	Extract B	Literature reference values	
			From this laboratory [24]	Other authors
TP	6		4.16 ± 0.23	2.0-7.91 [22]
5-HT	0.54	0.63	0.53 ± 0.08	0.24-0.71 [22]
5-HIAA	0.52	0.62	0.44 ± 0.02	0.29-0.87 [22]
IAA	0.034		0.013 ± 0.002	0.012 [30]

[a]All values given in μg/g wet tissue.

Table 15.4 Quantitation of Tryptophan and Related Indole Metabolites in Human Urine by Glass Capillary GC-MS/SIM Techniques[a]

		Reference values	
TP	9.7	1.7-27.9	[22]
5-HT	0.16	0.1-0.25	[22]
5-HIAA	2.6	1.1-10	[22]
IAA	13.4	0.3-11	[68,69]

[a]Values given in μg/ml.

degree of specificity is indeed unsurpassed by other conventional procedures. Unequivocal identifications and quantitative analysis of a given compounds rest on a perfect combination of the following factors (in order of increasing individual specificity): (1) enrichment of the compound(s) of interest by selective extraction procedures; (2) selective derivatization techniques tailored to capitalize on the response of the compound(s); (3) characteristic GC retention time; (4) good GC resolution, i.e., compound(s) of interest well separated in time from any possible interferences; (5) measurement of selected ion current profiles at specific *m/e* values; (6) measurement of the abundance ratios of two or more ions characteristic of the spectral fragmentation of the compound(s); and (7) quantitative assay by reference to stable isotope homologues of the compound(s) under study.

Evidently, each one of these operational factors can be individually optimized so that their sum results in the highest possible selectivity and sensitivity. In this regard, one of the most effective ways to optimize a GC separation is by the use of capillary columns. In this work we have paid special attention to the separation of most of the TP metabolites by high-resolution glass capillary columns, demonstrating their advantages over conventional packed columns for the resolution of overlapping peaks. Other factors contributing to the high selectivity of the SIM technique have been previously discussed [24,43, 56]. The results presented demonstrate that the manipulation of this type of column can be as straightforward as for packed columns and also attest to the practical simplicity and feasibility of a splitless capillary GC-MS system.

ACKNOWLEDGMENT

This work was supported by a research grant from the Comisión Asesora de Investigación Científica y Técnica, project no. 1287-76.

REFERENCES

1. A. R. Green and D. G. Grahame-Smith, in *Handbook of Neuropsychopharmacology* (S. H. Snyder, L. I. Iversen, and S. D. Iversen, eds.), Vol. 3, Plenum Press, New York, 1975.

2. A. A. Boulton, in *Trace Amines and the Brain* (E. Usdin and M. Sandler, eds.), Marcel Dekker, New York, 1976.
3. C. A. Marsden and G. Curzon, *Neuropharmacology 18*:159 (1979).
4. W. R. Martin, J. A. Thompson, and M. Nozaki, *Life Sci. 19*:1383 (1976).
5. A. R. Green, S. H. Koslow, and E. Costa, *Brain Res. 51*:371 (1973).
6. T. R. Bosin, G. Jonsson, and O. Beck, *Brain Res. 173*:79 (1979).
7. E. M. Airaksinen and M. M. Airaksinen, in *Serotonin in Mental Abnormalities* (D. J. Boullin, ed.), John Wiley & Sons, New York, 1978.
8. H. Ch. Curtius, M. Wolfensberger, U. Redweik, W. Leimbacher, R. A. Maibach, and W. Isler, *J. Chromatogr. 112*:523 (1975).
9. S. Takahashi, D. D. Godse, A. Naqvi, J. J. Warsh, and H. C. Stancer, *Clin. Chim. Acta 84*:55 (1978).
10. A. Szoerdsma and S. Udenfriend, *J. Clin. Invest. 34*:914 (1955).
11. N. Fornstedt, *Anal. Chem. 50*:1342 (1978).
12. H. L. Davis Jr., R. R. Brown, J. Leklem, and I. H. Carlson, *Cancer 31*:1061 (1973).
13. G. T. Bryan, *Am. J. Clin. Nutr. 24*:841 (1971).
14. V. T. de Vita, B. Z. Chabner, D. M. Livinston, and V. T. Oliverio, *Am. J. Clin. Nutr. 24*:835 (1971).
15. A. M. Krstulovic, P. R. Brown, D. M. Rosie, and P. B. Champlin, *Clin. Chem. 23*:1984 (1977).
16. S. Udenfriend, D. F. Bodganski, and H. Weissbach, *Science 122*:972 (1955).
17. F. P. Miller and R. P. Maickel, *Life Sci. 9*:747 (1970).
18. J. H. Gaddum and K. H. Hameed, *Br. J. Pharmacol. 9*:240 (1954).
19. B. Peskar and S. Spector, *Science 179*:1340 (1973).
20. J. M. Saavedra and J. Axelrod, *J. Pharmacol. Exp. Ther. 186*:508 (1973).
21. A. M. Krstulovic and C. Mataura, *J. Chromatogr. (Biomed. Appl.) 163*:72 (1979).
22. D. D. Koch and P. T. Kissinger, *J. Chromatogr. (Biomed. Appl.) 164*:441 (1979).
23. E. Martínez and E. Gelpí, *J. Chromatogr. 186*:619 (1979).
24. F. Artigas and E. Gelpí, *Anal. Biochem. 92*:233 (1979).
25. O. Beck, I. H. Wiesel, and G. Sedvall, *J. Chromatogr. 134*:407 (1977).
26. F. Cattabeni, S. H. Koslow, and E. Costa, *Science 178*:166 (1972).
27. L. Bertilsson and L. Palmer, *Science 177*:74 (1972).
28. L. Bertilsson, A. J. Atkinson, J. R. Althaus, A. Härfast, J. E. Lindgren, and B. Holmstedt, *Anal. Chem. 44*:1434 (1972).
29. S. R. Phillips, D. A. Durden, and A. A. Boulton, *Can. J. Biochem. 52*:447 (1974).
30. J. J. Warsh, P. W. Chan, D. D. Godse, D. V. Coscina, and H. C. Stancer, *J. Neurochem. 29*:955 (1977).
31. J. J. Franken and M. M. F. Trijbels, *J. Chromatogr. 91*:425 (1974).
32. B. W.Wilson and W. Snedden, *J. Neurochem. 33*:939 (1979).
33. K. Grob and K. Grob, Jr., *J. Chromatogr. 151*:311 (1978).
34. K. Grob and K. Grob, Jr., *J. High Resol. Chromatogr. & Chromatogr. Comm. 1*:57 (1978).
35. P. M. J. van den Berg and Th. P. H. Cox, *Chromatographia* 5:301 (1972).
36. I. Klimes, W. Stünzi, and D. Lamparsky, *J. Chromatogr. 136*:13 (1977).

37. G. Schomburg, H. Behlau, R. Dielmann, F. Weeke, and H. Husmann, *J. Chromatogr. 142*:87 (1977).
38. C. Suñol and E. Gelpí, *J. Chromatogr. 142*:559 (1977).
39. K. Grob and H. Jaeggi, *Anal. Chem. 45*:1788 (1973).
40. N. Neuner-Jehle and F. Etzweiler, *Chromatographia 6*:211 (1973).
41. U. Matthiesen, B. Kump, and W. Staib, *Chromatographia 10*:303 (1977).
42. H. W. Dürbeck, I. Bücker, and W. Leymann, *Chromatographia 11*:295 (1978).
43. E. Gelpí, E. Peralta, and J. Segura, *J. Chromatogr. Sci. 12*:701 (1974).
44. N. Seiler, *J. Chromatogr. (Biomed. Appl.), 143*:221 (1977).
45. G. S. King and K. Blau, in *Handbook of Derivatives for Chromatography* (K. Blau and G. S. King, eds.), Heyden & Sons Ltd., London, 1978.
46. J. Roselló, J. Tusell, and E. Gelpí, *J. Chromatogr. 130*:65 (1977).
47. E. C. Horning, M. G. Horning, W. J. A. Vandenheuvel, K. L. Knox, B. Holmstedt, and C. J. W. Brooks, *Anal. Chem. 36*:1546 (1964).
48. W. J. A. Vandenheuvel, *J. Chromatogr. 36*:354 (1968).
49. P. W. Albro and L. Fishbein, *J. Chromatogr. 55*:297 (1971).
50. W. C. Butts, *Anal. Biochem. 46*:187 (1972).
51. M. Donike, *Chromatographia 9*:440 (1976).
52. F. P. Abramson, M. W. Mc Caman, and R. E. Mc Caman, *Anal. Biochem. 57*:482 (1974).
53. J. J. Warsh, D. D. Godse, H. C. Stancer, P. W. Chan, and D. V. Coscina, *Biochem. Med. 18*:10 (1977).
54. K. Blau and G. S. King, in *Handbook of Derivatives for Chromatography*, (K. Blau and G. S. King, eds.), Heyden & Sons Ltd., London, 1978.
55. D. R. Knapp, in *Handbook of Analytical Derivatization Reaction*, Wiley and Sons, London, 1979, Part II.
56. J. Segura, F. Artigas, E. Martinez, and E. Gelpi, *Biomed. Mass Spectrom. 3*:91 (1976).
57. C. Hiemke, G. Kauert, and D. A. Kalbhen, *J. Chromatogr. 153*:451 (1978).
58. V. Erspamer, *Naunyn-Schmiedebergs Arch. Exp. Pathol. Pharmakol. 196*:343 (1940).
59. D. von Redlich and D. Glick, *Anal. Biochem. 29*:167 (1969).
60. J. M. Saavedra and J. Axelrod, *J. Pharmacol. Exp. Ther. 182*:363 (1972).
61. W. R. Martin, J. W. Sloan, S. T. Christian, and T. H. Clements, *Psychopharm. (Berl.) 24*:331 (1972).
62. F. Artigas, E. Gelpí, M. Prudencio, J. A. Alonso, and J. Baillart, *Anal. Chem. 49*:543 (1977).
63. *C. Suñol and E. Gelpí, 13 International Symposium Advances in Chromatography*, St. Louis, Missouri, October 16–19, 1978.
64. F. Vangaever, P. Sandra, and M. Verzele, *Chromatographia 12*:153 (1979).
65. C. Suñol and E. Gelpí, unpublished results.
66. S. M. Hess and S. Udenfriend, *J. Pharmacol. Exp. Ther. 127*:179 (1959).
67. W. R. Martin, J. W. Sloan, W. F. Buchwald, and S. R. Bridges, *Psychopharmacol (Berl.) 37*:189 (1974).
68. D. J. Byrd, W. Kochen, D. Idzko, and E. Knorr, *J. Chromatogr. 94*:85 (1974).
69. J. A. Hoskins and R. J. Pollit, *J. Chromatogr. 109*:436 (1975).

16

GC Profiling in Metabolic Disorders

Hartmut M. Liebich / *Medizinische Universitätsklinik, Tübingen, Federal Republic of Germany*

I. INTRODUCTION

One of the domains of GC, especially in the form of GC-MS, in biochemistry and clinical chemistry is profile analysis and recognition of normal and abnormal patterns of biological substances in body fluids. Whereas conventional clinical chemical tests determine individual constituents or groups of constituents which are known to be present, profile analysis offers the opportunity to detect and characterize a large number of substances simultaneously, to find new and unexpected substances, to determine changes in the ratios of different components, and to give much more precise information, for instance, on steroid hormones or organic acids, than a single test or a group test could provide.

Abnormalities in urinary or blood steroids may be overlooked by a group test or by radioimmunological determination of one or a few steroids, but are more likely to be detected by profiling. A metabolic acidosis, such as ketoacidosis or lactic acidosis, is described in much more detail by GC profiling of a blood serum or urine sample than by simply measuring the pH value and the lactate concentration in blood. Profile analyses must be expected to add to the understanding of metabolic abnormalities and diseases.

Analysis of metabolic profiles was introduced approximately twelve years ago for urinary steroids, sugars, sugar alcohols, aromatic acids, and serum acids of the Krebs cycle [1]. Especially after the introduction of glass capillary columns, steroid profiles were studied with respect to establishing normal profiles and aberrations due to disease or stress [2–8].

Organic acids are probably that class of low-molecular-weight biological substances for which gas chromatographic methods have contributed most to complement our knowledge of metabolic pathways and to aid, in some cases only to make possible, the clinical diagnosis of a metabolic disease. More than 250 organic acids have been described in the literature. In addition to these, at least the same number of so far unidentified acids have been detected. All of these acids occur in serum and/or urine of healthy individuals, however in a wide range of concentrations. Physiological abnormalities, such as inborn errors of metabolism, endocrinological defects, and diseases of the liver and the kidneys, lead to sometimes very drastic shifts in the concentration of an acid or of one or several classes of acids.

There exists a very large number of publications on organic acids. Many of them are concerned with a limited number or a certain functional class of acids, usually in conjuction with a metabolic abnormality. Some publications deal with a broader spectrum of organic acids, and give detailed information on healthy individuals [9–16] and on patients with metabolic and other diseases [17–20]. In most cases, the acids are analyzed in urine.

A further group of substances, for which profile analyses are applied includes neutral, low-molecular-weight, and volatile constituents in blood [21–26] and in urine [27–32]. These volatile constituents—mainly alcohols and ketones—are metabolic end products. Most of the ketones result from degradation of oxocarboxylic acids. In the following are given examples of metabolic disorders which lead to characteristic alterations in the profiles of the organic acids and the volatile constituents.

II. METHODICAL ASPECTS

A. Isolation of the Volatile Constituents

The isolation of the neutral, low-molecular-weight constituents can be achieved by gas-phase extraction and adsorption at a solid material [22,26,29,31]. An example of the technique is given by the following procedure. To 5% of a 24-hr urine sample an aliquot of 20 g of ammonium sulfate per 100 ml of urine is added in order to increase the volatility of the low-molecular-weight components. As shown schematically in Figure 16.1, the mixture is placed in a 500-ml Erlenmeyer

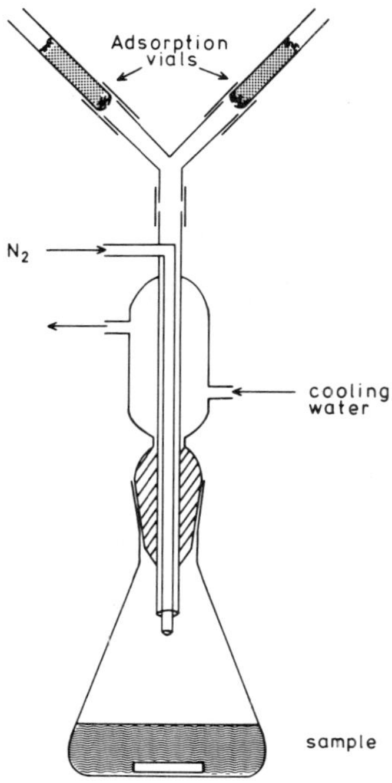

Figure 16.1 Apparatus for the gas phase extraction and adsorption of the volatile constituents.

flask in a water-bath at 90°C and stirred with a magnetic stirrer while nitrogen is passed over the urine at a flow-rate of 20 ml/min for 1 hr. The volatile components are carried with the nitrogen flow through a water condenser into two parallel glass tubes (11 cm × 10 mm O.D.) containing 0.3 g of the adsorbing material, Tenax GC, which is a porous polymer of 2,6-diphenyl-*p*-phenylene oxide (particle size 35–60 mesh). The cooling water of 12 ± 1°C is supplied by a thermostat.

The isolation of the volatile constituents from serum is accomplished according to the same procedure as described above for urine, except that only 5 ml of serum are mixed with 10 g of ammonium sulfate and extracted in a 200-ml Erlenmeyer flask at a temperature of 85°C for 2 hr.

The extraction rate of the gas phase extraction is 95–100% for ketones with five or more carbon atoms. It drops with increasing polarity of the substances and reaches only 50% for acetone and ethanol.

The adsorption rate of the adsorbent Tenax GC is 95–100% for ketones and alcohols with four or more carbon atoms. Strongly polar and hydrophilic substances such as acetone and ethanol are incompletely adsorbed (50–80%).

B. Gas Chromatographic Analysis of the Volatile Constituents

For gas chromatographic analysis the components must be desorbed, by a procedure such as the following: By inserting the adsorbent trap into the GC injector block, the dimensions of which are designed to accept the trap, the components are thermally desorbed at 300°C within 10 min and recondensed in a 1 m × 0.75 mm I.D. precolumn cooled with liquid air. After recondensation, the precolumn and the separating column (100 m × 0.5 mm I.D., stainless steel) are connected. Both precolumn and separating column are coated with Emulphor ON-870. The GC conditions are: carrier gas, nitrogen at 4 ml/min; column temperature, 60°C for 16 min, then programmed to 175°C at 2°C/min and held at that temperature.

Constant sampling temperature, sampling time, and desorption temperature are critical for comparative profile analyses of the volatile constituents. Sampling at 90°C for 1 hr is recommended for urine. Lowering the temperature by 10°C increases the extraction time by 1 hr. The quantitative desorption of the ketones requires a temperature of 300°C. Substances with higher polarity are desorbed at 220°C and below.

The identification of the components is done by GC-MS. The mass spectra may be recorded by a computer, applying the mode of automatic repetitive scanning [31,33].

C. Isolation of the Organic Acids

For the isolation of the acidic components, two principal methods are used: solvent extraction and isolation by anion-exchange chromatography. In solvent extraction, acceptable results are obtained using diethyl ether [17,34,35] or ethyl acetate [36–39], or a mixture of these solvents [9,10]. To improve the extraction, the urine or serum sample can be saturated with salts, such as sodium chloride. The methods are not time-consuming, since normally they are carried out batchwise and only in some cases continuously. However, for many organic acids the procedure is not quantitative; this is especially true for polyhydroxycarboxylic acids and other very polar acids. Protein-containing fluids, especially serum, have to be deproteinized prior to extraction. The removal of the proteins can be achieved either by dialysis or by precipitation. Since protein-bound organic acids are partially lost using the dialysis procedure, the precipitation of the proteins is mostly preferable. However, some organic acids may coprecipitate with the proteins.

Isolation of the organic acids by anion-exchange chromatography is widely used for studying acid profiles. Usually DEAE-Sephadex [11,12,19,40] or Amberlyst A-26 [14,41] are employed. Isolation by anion-exchange is more laborious than solvent extraction procedures. However, it is more quantitative, and it is chosen when detailed profiles of various classes of organic acids, including the hydroxycarboxylic acids, are investigated, and when in comparative studies small changes in the profiles are to be detected. The improvement in the isolation of polyhydroxycarboxylic acids using DEAE-Sephadex instead of diethyl ether/ethyl acetate has been described [42]. Good results are also obtained using Extrelut columns in the isolation procedure [20].

D. Derivatization of the Organic Acids

In the majority of the cases of carboxyl functions of the acids are transformed either into the trimethylsilyl esters or into the methyl esters. Trimethylsilylation is usually achieved with bis(trimethylsilyl)trifluoroacetamide (BSTFA) [9, 11,15,36,43–45] or bis(trimethylsilyl)acetamide (BSA) [19,37–40,46–49]. For the methylation of the organic acids, most laboratories use the reaction with diazomethane [10,14,20,34,41,50–53].

Both the trimethylsilyl esters and the methyl esters have advantages and limitations. Often the methyl esters have better chromatographic properties. Working with GC-MS to identify the acids, the mass spectra of the methyl esters are easier to interpret than the spectra of the trimethylsilyl esters, since the fragmentation of the methyl esters is often simpler and more predictable. When the necessary working precautions are followed, methylation with diazomethane is very simple and rapid. However, the procedure has also some disadvantages. Since the hydroxyl groups of phenolic acids are methylated as well, it is not possible to distinguish between the originally present aromatic —OH groups and $—OCH_3$ groups. Another disadvantage is a possible reaction of diazomethane with double bonds of unsaturated dicarboxylic acids (e.g., fumaric acid) to produce artifacts. Both of these drawbacks are not encountered when working with trimethylsilyl esters.

In addition to the carboxyl group, other functional groups in organic acids are derivatized. Whereas the hydroxyl groups of hydroxycarboxylic acids are transformed into the trimethylsilyl esters when the organic acids are trimethylsilylated, they are usually left underivatized when the organic acids are methylated.

The carbonyl groups of oxocarboxylic acids are transformed into oximes [38,39,49,54], *O*-methyloximes [41], *O*-ethyloximes [40,44], or *O*-trimethylsilylquinoxalinols [55,56]. Without derivatization of the carbonyl groups, many oxocarboxylic acids are not stable for GC analysis.

E. Prefractionation of the Organic Acids

The whole sample of isolated and derivatized organic acids can be analyzed by GC or GC-MS as a total profile [10]. However, because of the high complexity of the mixture of organic acids in serum and in urine, a prefractionation of the total mixture may become essential, when constituents of low concentrations will be studied. Prefractionation is effectively achieved either by preparative TLC [41,51] or by HPLC [51]. An example of a prefractionation by preparative TLC is shown in two schemes, which illustrate the analytical procedures for the analyses of the organic acids in general (Scheme 16.1), and for the oxocarboxylic acids in particular requiring special protective measures at the carbonyl groups (Scheme 16.2). Details of the prefractionation have been described [41]. It is useful to prefractionate into four fractions: fraction 1 contains mainly fatty acids; fraction 2, dicarboxylic acids, the derivatized oxocarboxylic acids, and aromatic acids; fraction 3, hydroxycarboxylic acids, aromatic acids, and acid conjugates; and fraction 4 additional acid conjugates.

F. Separation of the Organic Acids

Packed columns which were formerly used for the GC separation of the organic acids have been usually replaced by glass and more recently fused silica capillary columns. Suitable phases are OV-17, OV-101, SE-30, and OV-1701. Most frequently, columns with 25 m in length are used. Besides the flame ionization detector (FID), specific detectors are chosen, such as the thermionic specific detector for nitrogen containing acids or acid derivatives [41], or the mass spectrometer in the mode of mass fragmentography. The identification of the organic acids is based on their mass spectra and on their GC retention indices.

It must be pointed out that every investigator of organic acids must be aware of possible artifacts as a source of error. These artifacts may be produced at any step of the sample preparation procedure and during the separation. Another source of error is interferences from drugs and medications. Some examples of artifacts and other interferences are described within a review on profiling human body fluids [57]. Many other observations are reported [48,58,59].

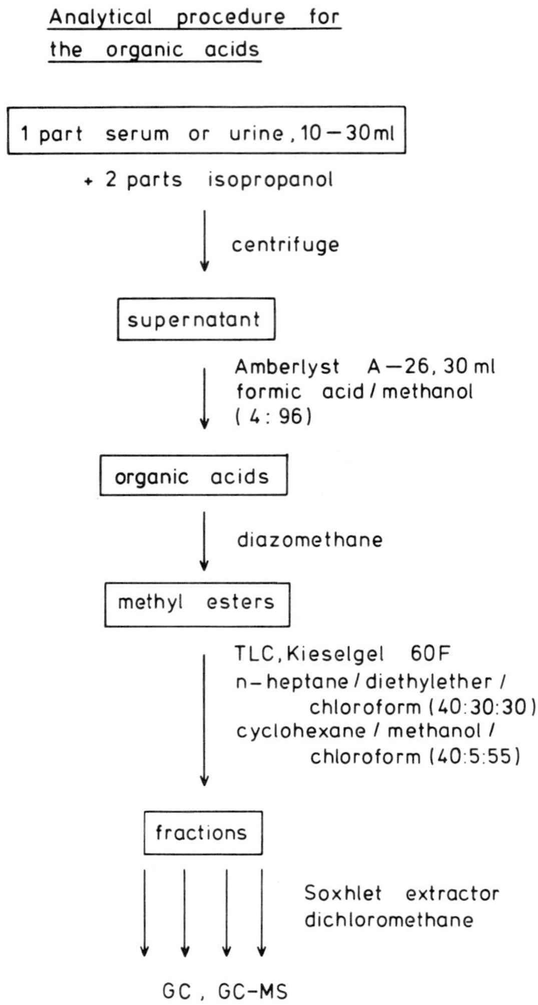

Scheme 16.1

III. INBORN ERRORS OF METABOLISM

An inborn error of metabolism is a congenital defect in metabolism, often in the metabolism of amino acids, caused by an insufficient enzyme activity at a certain step of the biochemical pathway. Quite often, such a defect leads to severe abnormalities in the development of a person, usually a child, and to mental retardation. Since the abnormal development can be prevented by giving a patient suffering from such a defect a special diet, it is essential to detect and to specify the enzyme defect. Many of these defects cause specific organic acidemias and/or acidurias, in some cases in conjunction with aminoacidemias and aminoacidurias. For detection and diagnosis of such defects, GC profiling is quite often the method of choice. In the following some examples of organic acidurias and acidemias shall be given.

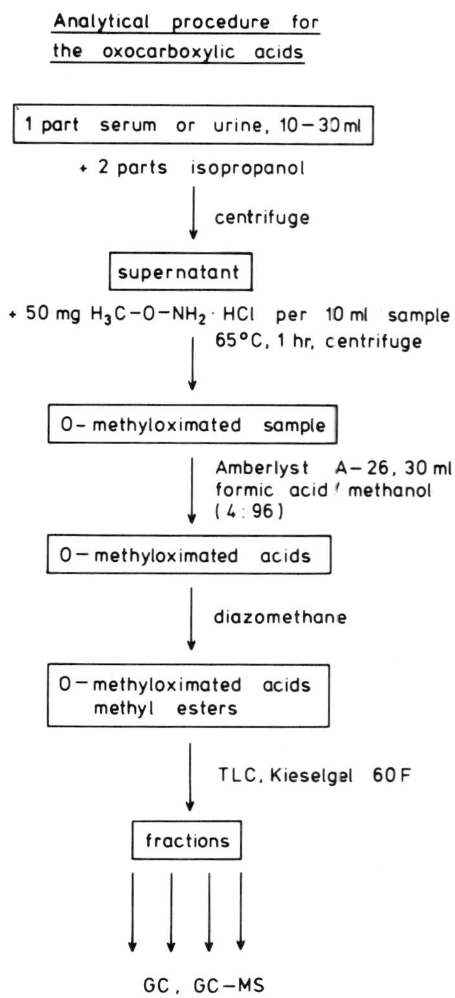

Scheme 16.2

A. Phenylketonuria

Phenylketonuria is the most common disorder in the group of organic acidurias due to an inborn error, and has a frequency in the general population of about 1 per 10,000 live-born infants [60]. Untreated patients suffer from severe mental retardation. In a group of 1778 mentally retarded patients, 1% suffered from phenylketonuria, and approximately 5% had some form of abnormal organic aciduria [61]. The disorder is caused by a deficiency in the activity of phenylalanine 4-hydroxylase. The normal conversion of phenylalanine to tyrosine is prevented, and the alternative pathway to phenylpyruvic acid and its metabolites is utilized, leading to elevated levels of phenylalanine in blood and high excretion of phenylpyruvic acid, phenyllactic acid, and 2-hydroxyphenylacetic acid in urine (Scheme 16.3).

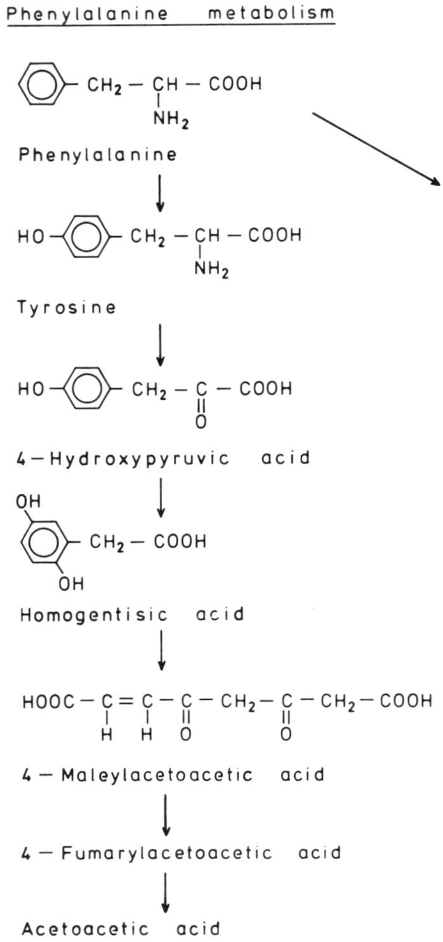

Scheme 16.3

B. Alkaptonuria

Another disorder within the scope of phenylalanine metabolism is alkaptonuria with a frequency of 1 per 100,000. The defect does not lead to mental retardation. It is caused by a deficiency in the activity of homogentisate oxidase and is recognized by a high excretion of homogentisic acid in urine (Scheme 16.3).

C. Maple Syrup Urine Disease

Maple syrup urine disease or branched-chain ketoaciduria is another defect in amino acid metabolism leading to cerebral degeneration and mental retardation. It is caused by a deficiency in the activity of the enzyme which catalyzes the oxidative decarboxylation of the 2-oxocarboxylic acids in the metabolism of valine, leucine, and isoleucine. Consequently, the excretion of the 2-oxocarboxylic acids and their corresponding reduction products is increased in urine, that is, 2-oxoisovaleric acid and 2-hydroxyisovaleric acid (Scheme 16.4), 2-oxoisocaproic acid and 2-hydroxyisocaproic acid (Scheme 16.5), and

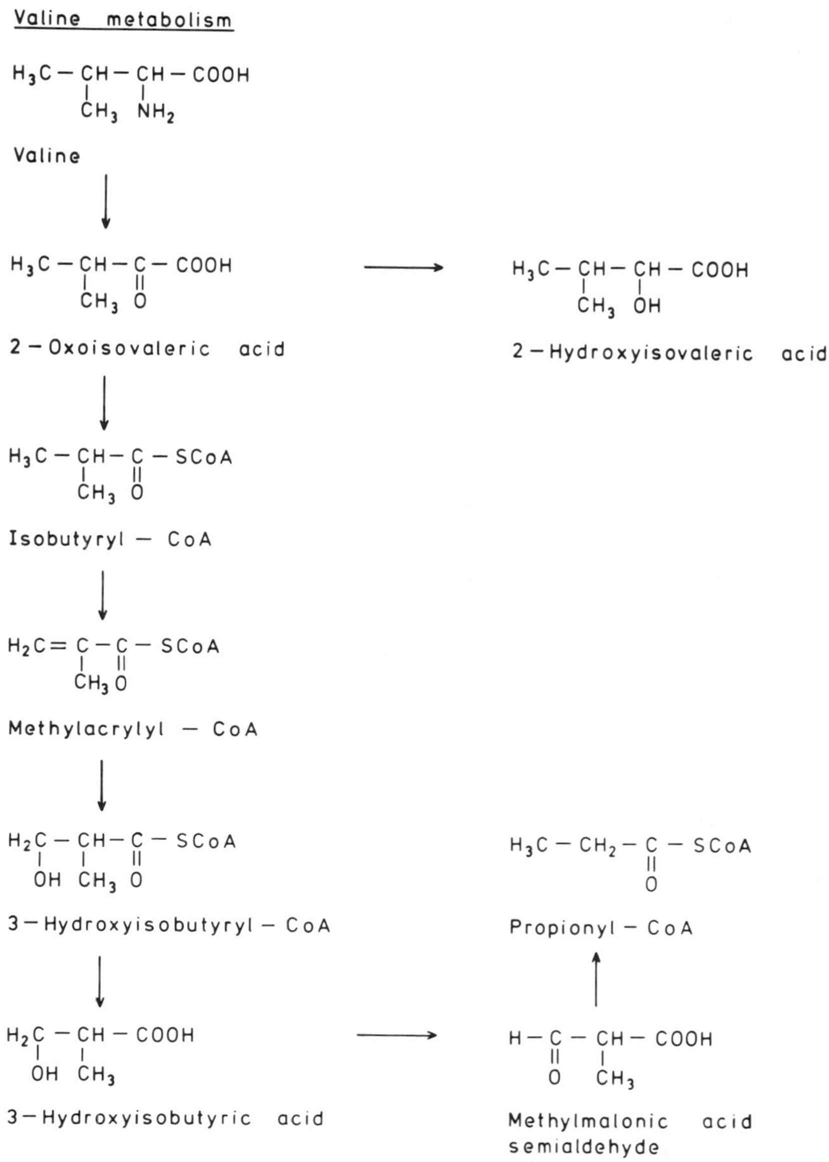

Scheme 16.4

2-oxo-3-methylvaleric acid and 2-hydroxy-3-methylvaleric acid (Scheme 16.6). The amino acids valine, leucine, and isoleucine are elevated in blood and in urine. In addition to the substances whose levels can be directly explained by the enzyme defect, a number of acidic metabolites are excreted in urine which are found in patients with ketoacidosis in general [43].

D. 3-Methylcrotonylglycinuria

Besides maple syrup urine disease there are several other disorders in the metabolism of the branched-chain amino acids. They lead to 3-methylcrotonylglycinuria, propionic acidemia, or methylmalonic acidura, and in some cases to

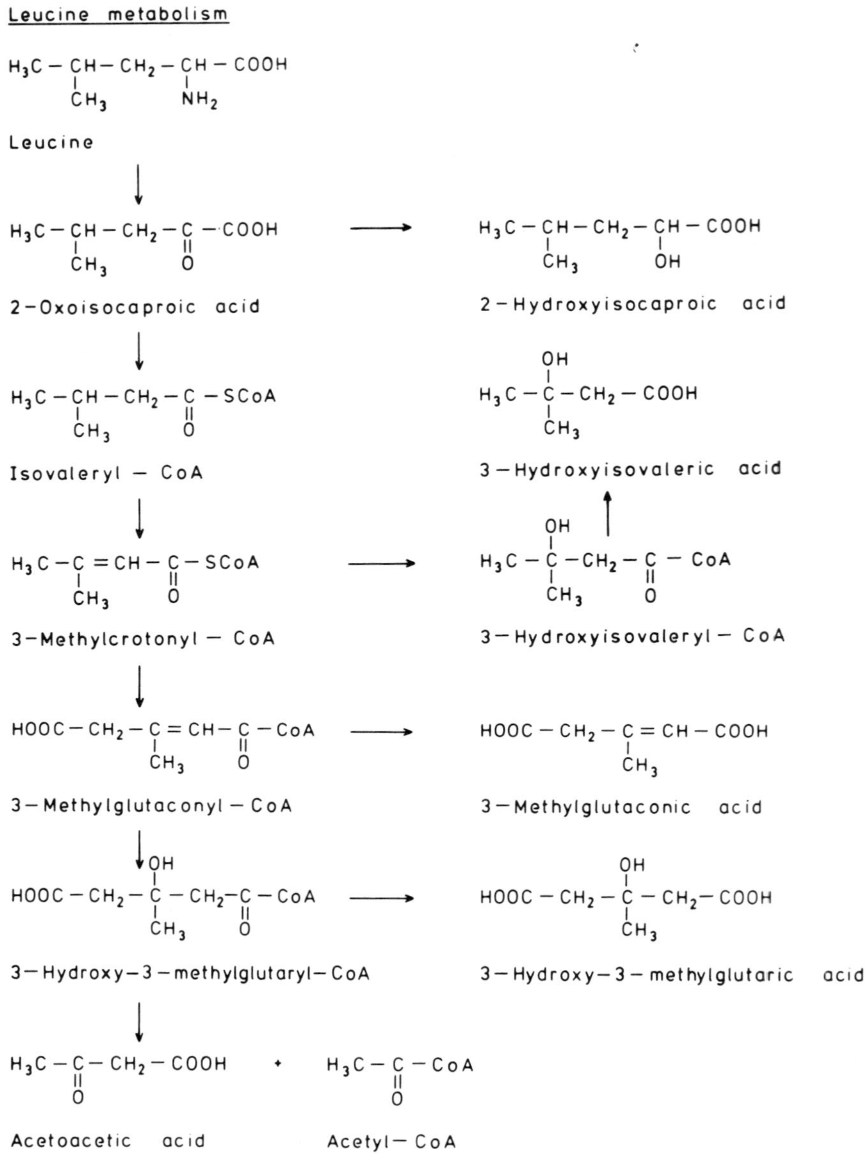

Scheme 16.5

combinations of two or all three of these defects. All three disorders may cause mental retardation.

3-Methylcrotonylglycinuria exists in at least two types and is caused by a deficiency in the activity of methylcrotonylCoA carboxylase in leucine metabolism [57,62]. Due to this enzyme block, 3-methylcrotonyl-CoA (Scheme 16.5) is conjugated with glycine to form 3-methylcrotonylglycine. An additional acid usually formed as a result of the described enzyme block is 3-hydroxyisovaleric acid (Scheme 16.5).

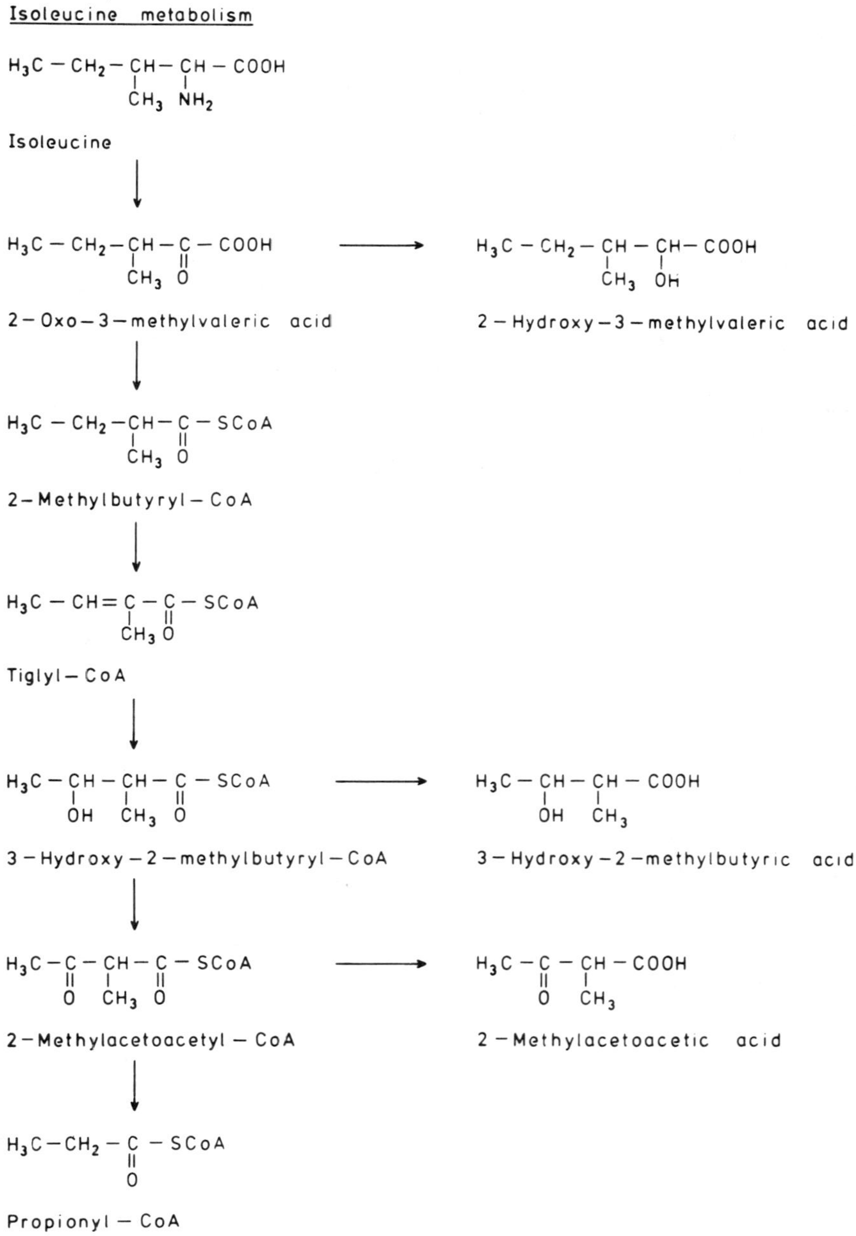

Scheme 16.6

E. Propionic Acidemia

Propionic acidemia is caused by insufficient activity of propionyl-CoA carboxylase, leading to a block in the transformation of propionyl-CoA, derived from valine and isoleucine metabolism (Schemes 16.4 and 16.6), into methylmalonyl-CoA (Scheme 16.7). In serum and urine, elevated levels of propionic acid are generally found. Characteristic metabolites include propionylglycine, 3-hydroxypropionic acid, and several condensation products of acetyl-CoA and propionyl-CoA, such as 3-oxovaleric acid and 3-hydroxyvaleric acid, and 3-oxo-2-methylbutyric acid, 3-hydroxy-2-methylbutyric acid, and 3-oxo-2-methylvaleric acid [63,64].

F. Methylmalonic Aciduria

The defect causing this abnormality lies in the block of the conversion of methylmalonyl-CoA into succinyl-CoA (Scheme 16.7) which is mediated by the enzyme methylmalonyl-CoA mutase. This enzyme requires coenzyme B_{12}, which is derived from vitamin B_{12}. Vitamin B_{12} deficiency leads to a methylmalonic aciduria but is a reversible biochemical disturbance. On the other hand, methylmalonic aciduria can be observed in newborn infants and young children with severe metabolic acidosis who are not vitamin B_{12} deficient [65,66]. This inborn error is very severe and is not vitamin B_{12} dependent. Both forms of methylmalonic aciduria are characterized by excessive amounts of methylmalonic acid in urine (Scheme 16.7).

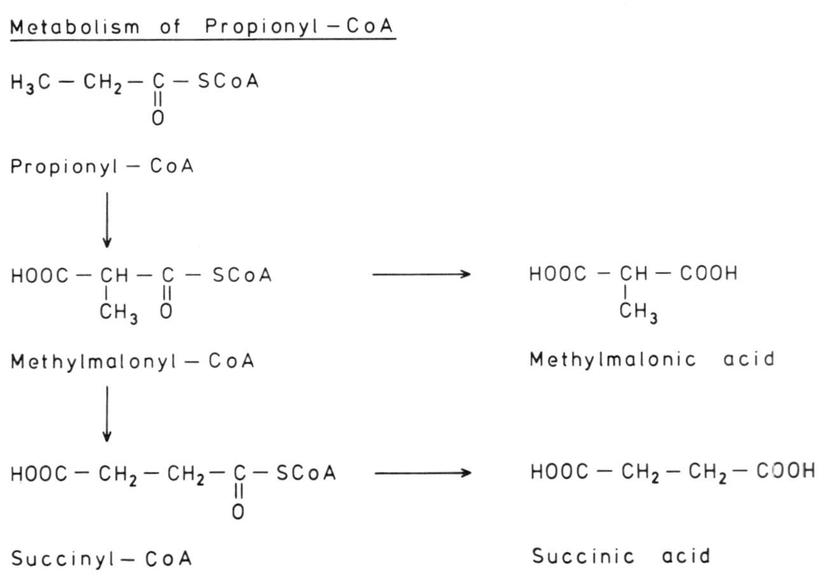

Scheme 16.7

IV. DIABETES MELLITUS

Whereas inborn errors or metabolism are metabolic disorders which occur very seldom in the normal population, diabetes mellitus is a chronic metabolic disease with the very high frequency of 1.5–2.5% in the Western hemisphere [67]. By GC analysis and GC profiling, a number of pathophysiological changes have been discovered in diabetes mellitus. Some of the changes in the concentrations of the metabolites are detectable in blood serum and in urine, even when diabetic patients are well controlled by therapy, and are therefore inherently connected with the disease. Other changes occur in situations of decompensation, that is, during hyperglycemic and ketoacidotic periods.

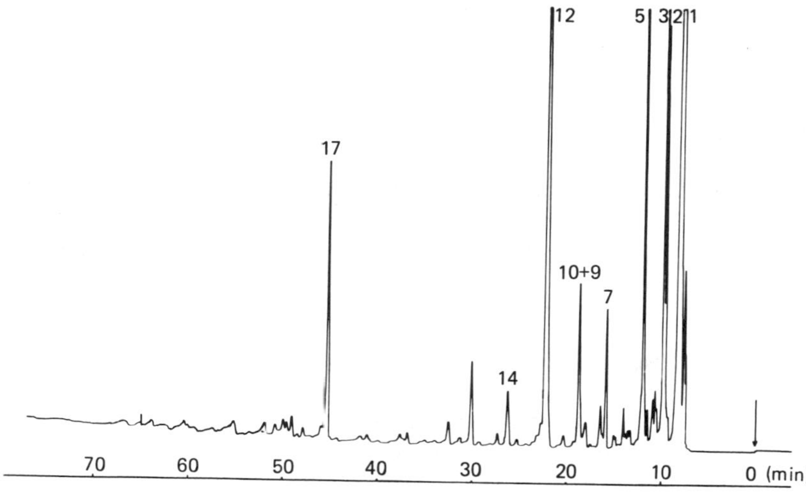

Figure 16.2 Profile of the volatile constituents in urine of a healthy male individual. Peaks 1 = acetone, 2 = butanone, 3 = ethanol, 5 = 2-pentanone, 7 = dimethyldisulfide, 9 = 3-penten-2-one, 10 = *N*-methylpyrrole, 12 = 4-heptanone, 14 = 2-heptanone, 17 = pyrrole.

The aliphatic alcohols belong to the first of the two categories of metabolic changes in diabetes mellitus. The alcohols and their variations in concentration can be observed in the comparative profile analysis of the volatile constituents [28]. In diabetic patients, in many instances elevated levels of ethanol, *n*-propanol, and *n*-butanol are found in serum and in urine (Figures 16.2 and 16.3). By quantitative determination of the alcohols using gas chromatography-mass fragmentography [68], the following normal ranges have been found (zero is assigned to values below the detection limit of the quantitative procedure): ethanol, 0–39 mg/l in serum and 0–46 mg per 24 hr in urine; *n*-propanol, 0–48 μg/l in serum and 0–300 μg per 24 hr in urine; *n*-butanol,

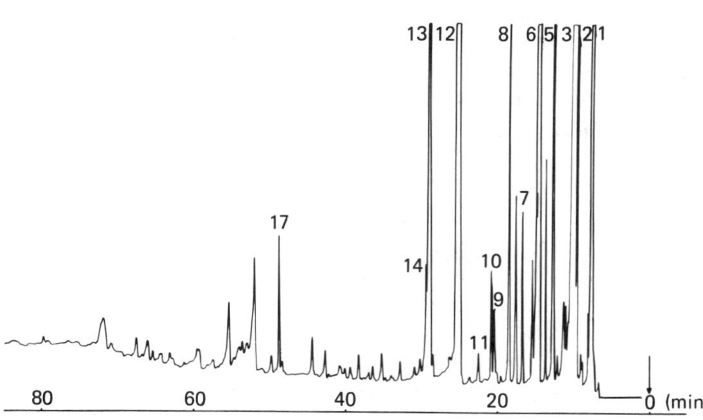

Figure 16.3 Profile of the volatile constituents in urine of an insulin-dependent male diabetic patient. Peak numbers as in Figure 16.2. Additional peaks: 6 = *n*-propanol, 8 = isobutanol, 11 = *n*-butanol, 13 = isopentanol.

0–20 μg/l in serum and 0–18 μg per 24 hr in urine. For diabetic patients, on average the levels are increased by a factor of 5–10.

The methyl-branched alcohols isobutanol and isopentanol, often falling below the detection limit in the serum and urine of normal subjects, are elevated less frequently when the *total* group of diabetics is compared with normal individuals. When, however, only the group of patients with proven diabetic complications such as microangiopathy and neuropathy is considered, in approximately 80% of the cases the excretion of isobutanol and isopentanol is increased. The presence of these two alcohols can be considered to be an indication of the existence of diabetic complications.

V. DIABETIC KETOACIDOSIS

In periods of decompensation of diabetes mellitus, the patients may develop ketoacidosis. During ketoacidosis one encounters very drastic changes in the concentrations of organic acids in serum and in urine. Of the different classes of organic acids, four groups are affected: fatty acids, aliphatic dicarboxylic acids, aliphatic oxocarboxylic acids, and aliphatic hydroxycarboxylic acids.

A. Fatty Acids

The concentrations of the whole spectrum of free fatty acids are elevated in serum and also in urine. This is due to the increase in lipolysis caused by the low insulin activity during ketoacidosis.

B. Aliphatic Dicarboxylic Acids

Of the aliphatic dicarboxylic acids in urine and in serum, mainly the saturated acids with even-numbered carbon chains are elevated, especially succinic acid (C_4), adipic acid (C_6), suberic acid (C_8), and sebacic acid (C_{10}). They are formed from long-chain monocarboxylic acids by ω oxidation followed by β oxidation. The mechanism of the formation of the medium-chain dicarboxylic acids has been proven by feeding ketotic rats with ^{14}C-labeled palmitic acid [69]. The proportion of ω oxidation in fatty acid metabolism in normals is small. It gains importance during ketoacidosis, apparently according to a backup mechanism. It has been found in experiments with rats that between 5 and 20% of the fatty acids are metabolized by initial ω oxidation [70,71].

Methyl-branched dicarboxylic acids, for example, methylsuccinic, methylglutaric, methyladipic, methylpimelic, and methylsuberic, and also some ethyl-branched dicarboxylic acids are very little affected by acidosis. It has been supposed that they are formed by ω oxidation of 3-methyl- or *ante/iso*methyl-branched monocarboxylic acids by the microflora of the intestinal tract [72].

Also very little affected by ketoacidosis are the unsaturated dicarboxylic acids, especially mesaconic acid (trans-methylbutenedioic acid), citraconic acid (*cis*-methylbutenedioic acid), and the two cis/trans isomers of 3-methylglutaconic acid [10].

Profiles of organic acids very similar to those seen in diabetic ketoacidosis are observed in patients who develop ketoacidosis under fasting conditions.

The changes in the serum concentrations and urinary excretion of dicarboxylic acids can be observed already in the total profiles of organic acids (e.g., peaks 12, 19, 27 in Figures 16.4–16.6). More detailed information is obtained when the sample is prefractionated prior to GC analysis. The saturated and

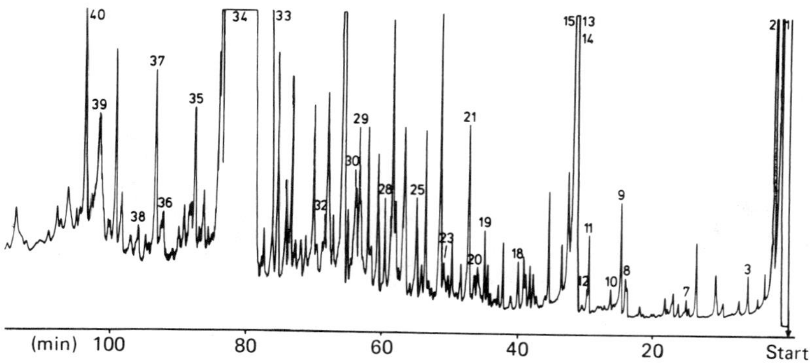

Figure 16.4 Total profile of the methyl esters of the organic acids in urine of a normal individual. Extraction with diethyl ether/ethyl acetate (1:2). GC-conditions: 25-m glass capillary column coated with OV-17; carrier gas, nitrogen at 4 ml/min; column temperature 30°C for 5 min, then programmed to 230°C at 2°C/min and held at that temperature; sample size, 1 μl at a split ratio of 1:20. Peaks: 1 = diethyl ether (solvent), 2 = ethyl acetate (solvent), 3 = 2-hydroxyisobutyric acid, 7 = oxalic acid, 8 = methylmalonic acid, 9 = cresol (derivatized), 10 = phenol (not derivatized), 11 = ethylmalonic acid, 12 = succinic acid, 13 = methylsuccinic acid, 14 = cresol (not derivatized), 15 = benzoic acid, 18 = phenylacetic acid, 19 = adipic acid, 20 = 3-phenylpropionic acid, 21 = 3-methyladipic acid, 23 = 3-hydroxybenzoic acid (as methyl ether), 25 = 4-hydroxybenzoic acid (as methyl ether), 28 = 4-hydroxyphenylacetic acid (as methyl ether), 29 = azelaic acid, 30 = citric acid, 32 = furoylglycine, 33 = saccharin (from diet), 34 = hippuric acid, 35 = indoleacetic acid, 36 = 3-hydroxyhippuric acid (as methyl ether), 37 = 4-hydroxyhippuric acid (as methyl ether), 38 = unknown, 39 = 5-hydroxyindoleacetic acid (as methyl ether), 40 = *N*-phenylacetylglutaminic acid.

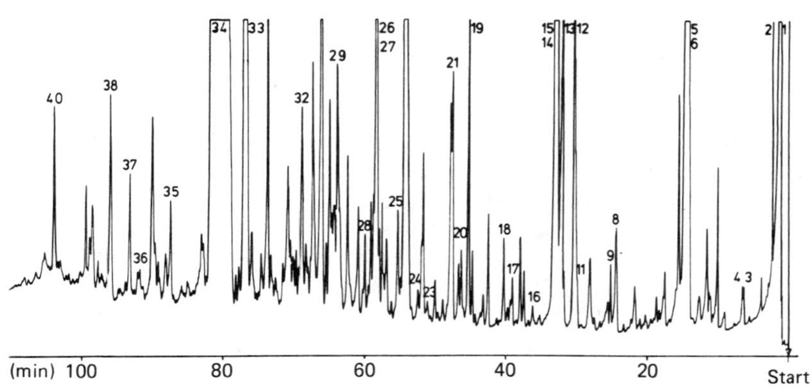

Figure 16.5 Total profile of the methyl esters of the organic acids in urine of a diabetic patient with ketoacidosis. Extraction and GC conditions as in Figure 16.4. Peak numbers as in Figure 16.4. Additional peaks: 4 = pyruvic acid, 5 = 3-hydroxybutyric acid, 6 = acetoacetic acid, 16 = glutaric acid, 17 = 3-methylglutaric acid, 24 = 3-methylpimelic acid, 26 = 2,6-di-*tert*-butylphenol (from diethyl ether), 27 = suberic acid.

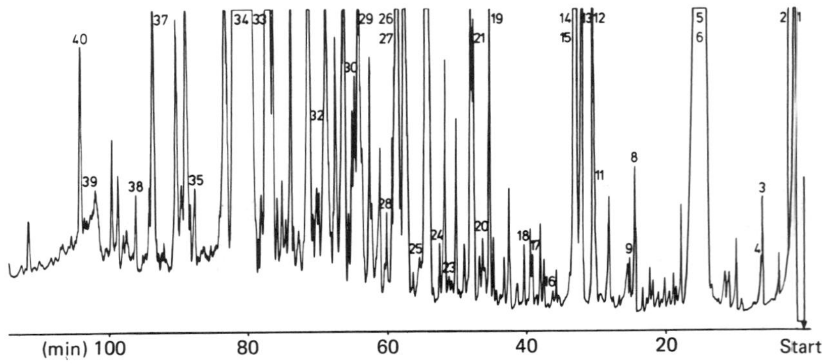

Figure 16.6 Total profile of the methyl esters of the organic acids in urine of a fasting patient with ketoacidosis. Extraction and GC conditions as in Figure 16.4. Peak numbers as in Figures 16.4 and 16.5.

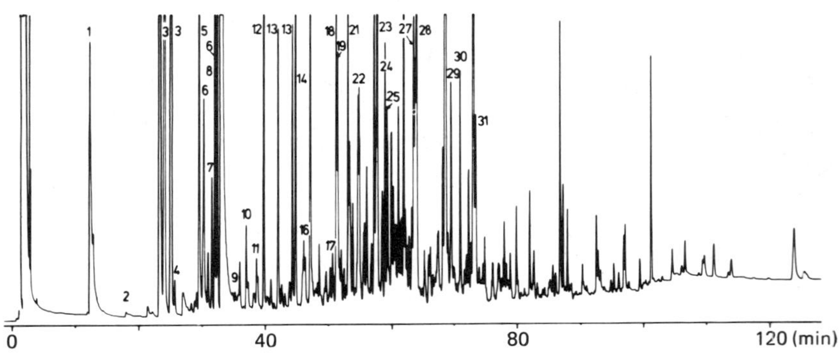

Figure 16.7 Fraction 2 of the methyl esters of the organic acids in urine of a patient with ketoacidosis. Isolation of the acids by anion exchange with Amberlyst A-26, transformation of the oxocarboxylic acids into the *O*-methyloxime derivatives. GC conditions as in Figure 16.4, except for column temperature (40°C for 10 min, then programmed to 230°C at 2°C/min and held at that temperature). Peaks (phenolic acids as methyl esters): 1 = 3-hydroxybutyric acid, 2 = pyruvic acid, 3 = 3-oxobutyric acid, 4 = 2-oxoisovaleric acid, 5 = succinic acid, 6 = 2-oxoisocaproic acid, 7 = 2-oxo-3-methylvaleric acid, 8 = methylsuccinic acid, 9 = citraconic acid, 10 = glutaric acid, 11 = 3-methylglutaric acid, 12 = phenylacetic acid, 13 = 3-methylglutaconic acid, 14 = adipic acid, 15 = phenylpropionic acid, 16 = 3-methyladipic acid, 17 = heptenedioic acid, 18 = heptenedioic acid, 19 = pimelic acid, 20 = hydroxybenzoic acid, 21 = anthranilic acid, 22 = hydroxybenzoic acid, 23 = suberic acid, 24 = hydroxyphenylacetic acid, 25 = 3-methylsuberic acid, 26 = dihydroxybenzoic acid, 27 = phenylpyruvic acid, 28 = azelaic acid, 29 = sebacic acid, 30 = dihydroxybenzoic acid, 31 = 3-carboxy-4-methyl-5-propyl-2-furanpropionic acid.

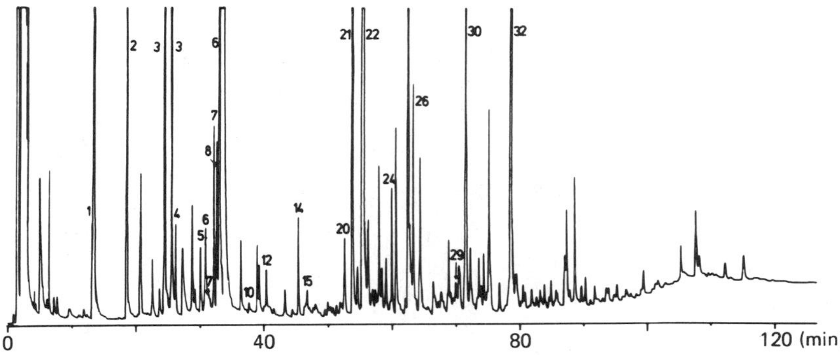

Figure 16.8 Fraction 2 of the methyl esters of the organic acids in serum of the same ketoacidotic patient as in Figure 16.7. Isolation of the acids, GC conditions, and peak numbers as in Figure 16.7.

unsaturated dicarboxylic acids are found in serum and in urine; however, in serum they are found in much smaller concentrations than in urine (Figures 16.7 and 16.8).

C. Aliphatic Oxocarboxylic Acids

Oxocarboxylic acids are formed by ketogenesis or by transamination of amino acids. During ketoacidosis, ketogenesis (Scheme 16.8) is strongly enhanced,

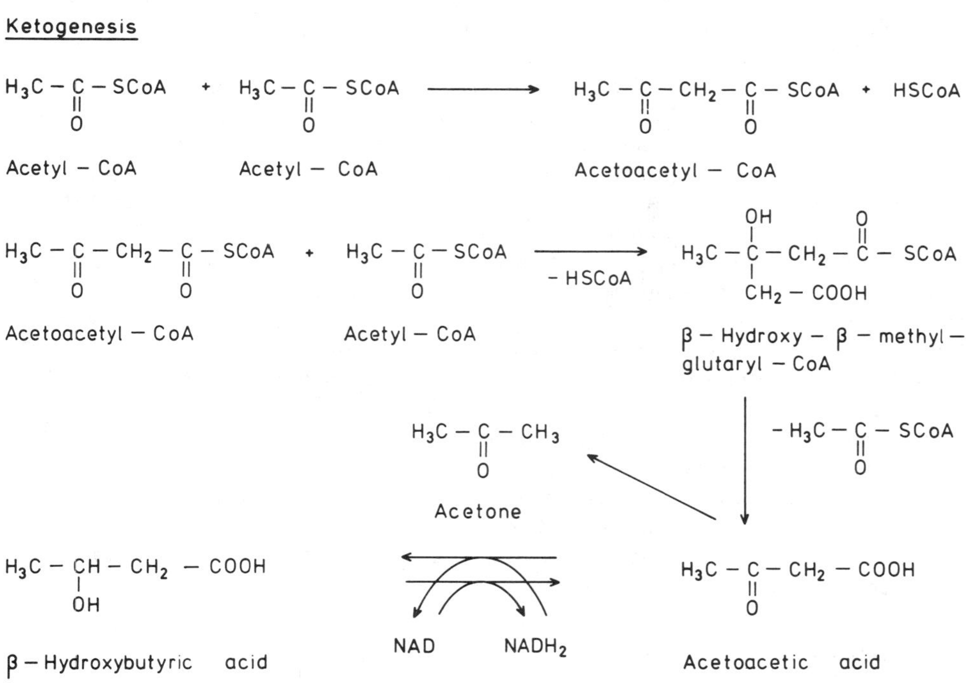

Scheme 16.8

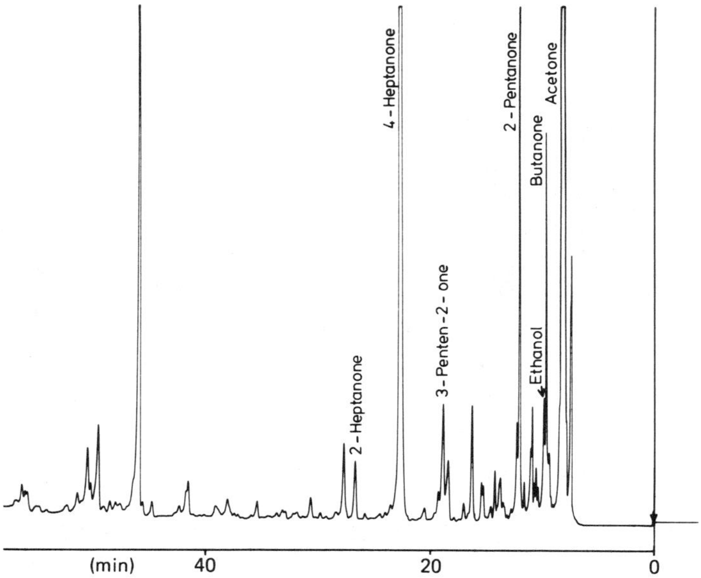

Figure 16.9 Profile of the volatile constituents in urine of a healthy individual.

leading to highly elevated levels of acetoacetic acid and 3-hydroxybutyric acid in serum and in urine (substances 5 and 6 in Figures 16.5 and 16.6; substance 3, acetoacetic acid, in Figures 16.7 and 16.8). The third conventional ketone body, acetone, is detected within the profile of volatile metabolites. Among these volatile metabolites a number of higher-molecular-weight methylketones are found besides acetone, including butanone, 2-pentanone, 3-penten-2-one, and 2-heptanone (Figure 16.9). It is assumed that these methylketones originate from 3-oxocarboxylic acids and are formed from these acids by decarboxylation, analogous to the formation of acetone from acetoacetic acid. During diabetic and fasting ketoacidosis, besides acetone the methylketones 2-pentanone, 3-penten-2-one, and 2-heptanone rise in serum and urine (Figure 16.10) and normalize when the ketoacidosis is overcome.

On the other hand, as a result of the elevated levels of the amino acids valine, leucine, and isoleucine in diabetes [73,74], especially in ketoacidosis, the synthesis of amino acid metabolites is favored. This leads to elevated levels of 2-oxoisovaleric acid (Scheme 16.4), 2-oxoisocaproic acid (Scheme 16.5), and 2-oxo-3-methylvaleric acid (Scheme 16.6), especially in serum.

When the oxocarboxylic acids are analyzed in the form of their *O*-methyloximated acid methyl esters, which contain nitrogen, they can be selectively monitored with the nitrogen-specific detector (Figure 16.11). Depending on the steric conditions, the *O*-methyloximes of the oxocarboxylic acid methyl esters may form syn/anti isomers. The profiles of the oxocarboxylic acids in serum and in urine of normal individuals, diabetic patients under good therapeutical control, and patients with ketoacidosis have been described [41].

D. Aliphatic Hydroxycarboxylic Acids

As pointed out previously, during ketoacidosis there is an increased production of metabolites of the branched-chain amino acids. From valine degradation,

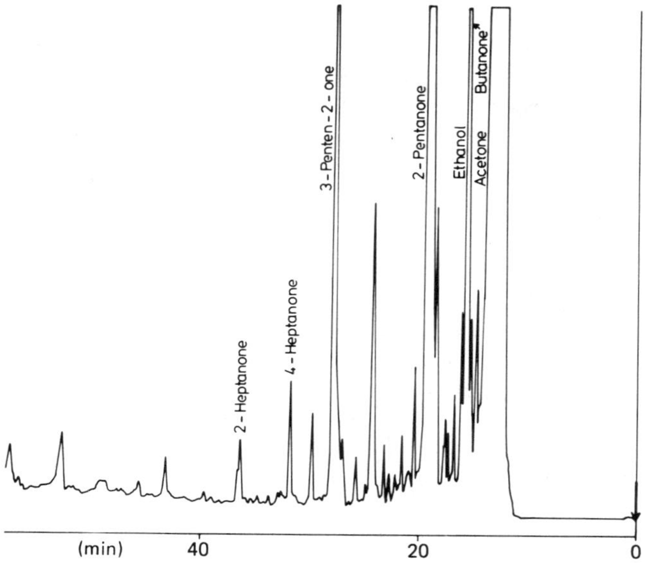

Figure 16.10 Profile of the volatile constituents in urine of a diabetic patient with ketoacidosis.

3-hydroxyisobutyric acid and 2-hydroxyisovaleric acid are formed (Scheme 16.4); from leucine, 3-hydroxyisovaleric acid and 2-hydroxyisocaproic acid (Scheme 16.5); and from isoleucine, 3-hydroxy-2-methylbutyric acid and 2-hydroxy-3-methylvaleric acid (Scheme 16.6) are produced. These acids are elevated in most cases of ketoacidosis [52,75,76]. The hydroxycarboxylic acids that is most drastically increased during ketoacidosis is 3-hydroxybutyric acid, which is formed by reduction of acetoacetic acid in the course of ketogenesis (Scheme 16.8). Besides the described acids, a number of 3-hydroxydicarboxylic acids, probably formed by a combination of ω-oxidation and incomplete β-oxidation have been found in urine [36].

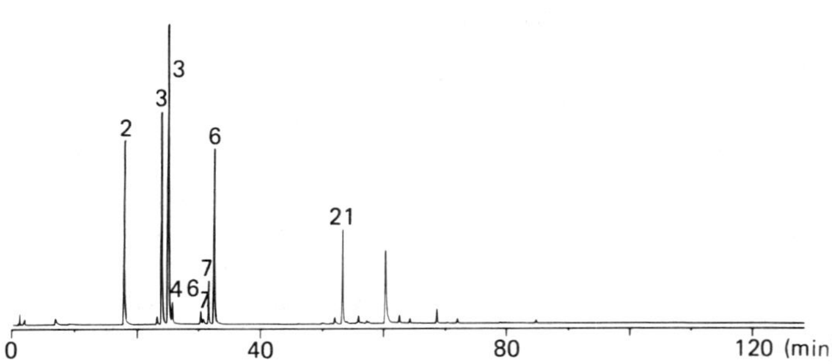

Figure 16.11 Thermionic-specific detection of the O-methyloximated acid methyl esters within fraction 2 of the organic acids in serum of a ketoacidotic patient. Peaks: 2 = pyruvic acid, 3 = acetoacetic acid, 4 = 2-oxoisovaleric acid, 6 = 2-oxoisocaproic acid, 7 = 2-oxo-3-methylvaleric acid, 21 = anthranilic acid.

REFERENCES

1. E. C. Horning and M. G. Horning, in *Advances in Chromatography, 1970* (A. Zlatkis, ed.), Houston 1970, p. 226.
2. C. H. L. Shackleton, J.-A. Gustafsson, and F. L. Mitchell, *Acta Endocrinol.* *74*:157 (1973).
3. A. Luyten and G. A. F. M. Rutten, *J. Chromatogr.* *91*:393 (1974).
4. H. Ludwig, J. Reiner, and G. Spiteller, *Chem. Berichte* *110*:217 (1977).
5. C. D. Pfaffenberger and E. C. Horning, *J. Chromatogr.* *112*:581 (1975).
6. C. D. Pfaffenberger and E. C. Horning, *Anal. Biochem.* *80*:329 (1977).
7. A. L. German, C. D. Pfaffenberger, J.-P Thenot, M. G. Horning, and E. C. Horning, *Anal. Chem.* *45*:930 (1973).
8. G. Spiteller, *Nachr. Chem. Techn. Laborat.* *25*:450 (1977).
9. W. L. Fitch, P. J. Anderson, and D. H. Smith, *J. Chromatogr.* *162*: 249 (1979).
10. H. M. Liebich, A. Pickert, U. Stierle, and J. Wöll, *J. Chromatogr.* *199*:181 (1980).
11. S. C. Gates, N. Dendramis, and C. C. Sweeley, *Clin. Chem.* *24*:1674 (1978).
12. A. M. Lawson, R. A. Chalmers, and R. W. E. Watts, *Clin. Chem.* *22*: 1283 (1976).
13. R. A. Chalmers, M. J. R. Healy, A. M. Lawson, and R. W. E. Watts, *Clin. Chem.* *22*:1288 (1976).
14. M. Spiteller and G. Spiteller, *J. Chromatogr.* *164*:253 (1979).
15. L. Björkman, C. McLean, and G. Steen, *Clin. Chem.* *22*:49 (1976).
16. B. A. Knights, M. Legendre, J. L. Laseter, and J. S. Storer, *Clin. Chem.* *21*:888 (1975).
17. E. Jellum, O. Stokke, and L. Eldjarn, *Clin. Chem.* *18*:800 (1972).
18. E. Jellum, P. Storseth, J. Alexander, P. Helland, O. Stokke, and E. Teig, *J. Chromatogr.* *126*:487 (1976).
19. R. A. Chalmers, A. M. Lawson, and R. W. E. Watts, *Clin. Chim. Acta* *52*:43 (1974).
20. D. Pinkston, G. Spiteller, H. von Henning, and D. Matthaei, *J. Chromatogr.* *223*:1 (1981).
21. B. Dowty, D. Carlisle, J. L. Laseter, and F. M. Gonzales, *Biomed. Mass Spectrom.* *2*:142 (1975).
22. H. M. Liebich and J. Wöll, *J. Chromatogr.* *142*:505 (1977).
23. A. Zlatkis, W. Bertsch, D. A. Bafus, and H. M. Liebich, *J. Chromatogr.* *91*:379 (1974).
24. A. Zlatkis and F. Andrawes, *J. Chromatogr.* *112*:533 (1975).
25. A. Zlatkis and K. Kim, *J. Chromatogr.* *126*:475 (1976).
26. A. Zlatkis, K. Y. Lee, C. F. Poole, and G. Holzer, *J. Chromatogr.* *163*:125 (1979).
27. A. Zlatkis and H. M. Liebich, *Clin. Chem.* *17*:592 (1971).
28. H. M. Liebich, O. Al-Babbili, A. Zlatkis, and K. Kim, *Clin. Chem.* *21*: 1294 (1975).
29. M. Novotny, M. L. McConnell, M. L. Lee, and R. Farlow, *Clin. Chem.* *20*:1105 (1974).
30. A. Zlatkis, W. Bertsch, H. A. Lichtenstein, A. Tishbee, F. Shunbo, H. M. Liebich, A. M. Coscia, and N. Fleischer, *Anal. Chem.* *45*:763 (1973).
31. H. M. Liebich and O. Al-Babbili, *J. Chromatogr.* *112*:539 (1975).

32. H. M. Liebich, O. Al-Babbili, G. Huesgen, and J. Wöll, *Z. Anal. Chem. 279*:148 (1976).
33. H. M. Liebich, *J. Chromatogr. 112*:551 (1975).
34. J. E. Pettersen, E. Jellum, and L. Eldjarn, *Clin. Chim. Acta 38*:17 (1972).
35. J. E. Pettersen and E. Jellum, *Clin. Chim. Acta 41*:199 (1972).
36. J. Greter, S. Lindstedt, H. Seemann, and G. Steen, *Clin. Chem. 26*: 261 (1980).
37. T. Niwa, T. Ohki, K. Maeda, A. Saito, and K. Kobayashi, *Clin. Chim. Acta 99*:71 (1979).
38. H. J. Sternowsky, J. Roboz, F. Hutterer, and G. Gaull, *Clin. Chim. Acta 47*:371 (1973).
39. K. Tanaka, A. West-Dull, D. G. Hine, T. B. Lynn, and T. Lowe, *Clin. Chem. 26*:1847 (1980).
40. R. A. Chalmers, and R. W. E. Watts, *Analyst 97*:958 (1972).
41. H. M.Liebich, A. Pickert, and J. Wöll, *J. Chromatogr. 217*:255 (1981).
42. P. Sims, R. Truscott, and B. Halpern, *J. Chromatogr. 222*:337 (1981).
43. C. Jakobs, E. Solem, J. Ek, K. Halvorsen, and E. Jellum, *J. Chromatogr. 143*:31 (1977).
44. R. A. Chalmers, A. M. Lawson, and O. Borud, *Clin. Chim. Acta 77*: 117 (1977).
45. T. Niwa, K. Maeda, T. Ohki, and J. Sakakibara, *J. Chromatogr. 228*: 59 (1982).
46. F. W. Bultitude and S. L. Newham, *Clin. Chem. 21*:1329 (1975).
47. J. Dosman, L. C. Crawhall, G. A. Klassen, O. A. Mamer, and P. Neumann, *Clin. Chim. Acta 51*:93 (1974).
48. M. Duran, D. Ketting, S. K. Wadman, C. Jakobs, R. B. H. Schutgens, and H. A. Veder, *Clin. Chim. Acta 90*:187 (1978).
49. K. Tanaka, D. G. Hine, A. West-Dull, and T. B. Lynn, *Clin. Chem. 26*:1839 (1980).
50. L. Borg, S. Lindstedt, and G. Steen, *Clin. Chim. Acta 41*:363 (1972).
51. A. Grupe and G. Spiteller, *J. Chromatogr. 226*:301 (1981).
52. S. Landaas, *Clin. Chim. Acta 54*:39 (1974).
53. S. Lindstedt, K. Norberg, G. Steen, and E. Wahl, *Clin. Chem. 22*: 1330 (1976).
54. R. A. Chalmers and R. W. E. Watts, *Analyst 97*:951 (1972).
55. U. Langenbeck, A. Hoinowski, K. Mantel, and H.-U. Möhring, *J. Chromatogr. 143*:39 (1977).
56. U. Langenbeck, A. Mench-Hoinowski, K.-P. Dieckmann, H.-U. Möhring, and M. Petersen, *J. Chromatogr. 145*:185 (1978).
57. E. Jellum, *J. Chromatogr. 143*:427 (1977).
58. C. Jakobs, M. Bojasch, M. Duran, D. Ketting, S. K. Wadmann, and D. Leupold, *Clin. Chim. Acta 106*:85 (1980).
59. M. Ende, P. Pfeifer, and G. Spiteller, *J. Chromatogr. 183*:1 (1980).
60. R. A. Chalmers and A. M. Lawson, *Chem. Britain, 11*:290 (1975).
61. R. W. E. Watts, R. A. Chalmers, and A. M. Lawson, *Lancet I*:368 (1975).
62. L. Eldjarn, E. Jellum, O. Stokke, H. Pande, and P. E. Waaler, *Lancet II*:521 (1970).
63. W. Lehnert and A. Junker, *Clin. Chim. Acta 104*:47 (1980).
64. J. Greter, S. Lindstedt, H. Seeman, and G. Steen, *Clin. Chim. Acta 106*:103 (1980).
65. L. E. Rosenberg, A.-Ch. Lilljeqvist, and Y. E. Hsia, *N. Engl. J. Med. 24*:1319 (1968).

66. S. E. Goodman, E. R. B. McCabe, P. V. Fennessey, B. S. Miles, J. W. Mace, and E. Jellum, *Clin. Chim. Acta 87*:441 (1978).
67. E. Deutsch, and G. Geyer, *Laboratoriumsdiagnostik*, Verlag Brüder Hartmann, Berlin, 1975, p. 698.
68. H. M. Liebich, H. J. Buelow, and R. Kallmayer, *J. Chromatogr. 239*: 343 (1982).
69. J. E. Pettersen, *Clin. Chim. Acta 41*:231 (1972).
70. I. Björkhem, *J. Lipid Res. 19*:585 (1978).
71. W. Kam, K. Kumaran, and B. R. Landau, *J. Lipid Res. 19*:591 (1978).
72. J. E. Pettersen and O. Stokke, *Biochem. Biophys. Acta 304*:316 (1973).
73. P. Felig, E. Marliss, J. L. Ohman, and G. F. Cahill, *Diabetes 19*:727 (1970).
74. H. S. Paul and S. A. Adibi, *Metabolism 27*:185 (1978).
75. S. Landaas, *Clin. Chim. Acta 64*:143 (1975).
76. S. Landaas and C. Jakobs, *Clin. Chim. Acta 78*:489 (1977).

II

PHARMACOLOGY

17

Determination of Bromhexine by Capillary GLC

Dieter Lutz, Werner Gielsdorf, James A. Settlage, Eugen Schlegel, and Halvor Jaeger / *L.A.B. Gesellschaft für pharmakologische Untersuchungen m.b.H. & Co. (Laboratory for Pharmacological Research), Neu-Ulm, Federal Republic of Germany*

I. INTRODUCTION

Bromhexine [*N*-cyclohexyl-*N*-methyl-2-(2-amino-3,5-dibromo)benzylammonium chloride] (Figure 17.1) is a bronchial mucolytic drug widely used in the therapy of hypercrinia, respectively dyscrinia, of the respiratory system. The viscosity of sputum is linked to the presence of acid mucopolysaccharides. Bromhexine causes a fragmentation of acid mucopolysaccharides by stimulating hydrolytic enzyme activity and lysosome formation; due to the alteration of mucopolysaccharide fibers, a decrease in sputum viscosity occurs [1-4].

Pharmacokinetic and metabolic investigations of ^{14}C-labeled bromhexine in man and in animals have shown that bromhexine undergoes extensive and rapid metabolism, yielding mainly dealkylated and hydroxylated metabolites and conjugates [5-8].

One metabolite of bromhexine, ambroxol [*N*-(*trans*-4-hydroxycyclohexyl)-*N*-(2-amino-3,5-dibromobenzyl)amine], is thought to be responsible for the pharmacological actions of bromhexine [9]. The low therapeutic dosage (8-16 mg) of the drug and its extensive biotransformation result in low plasma levels of the unchanged drug (some few nanograms per milliliter). Thus, highly sensitive and specific methods are required for quantitative determination of bromhexine in biological fluids. Until recently, only a few methods existed which fulfilled these requirements.

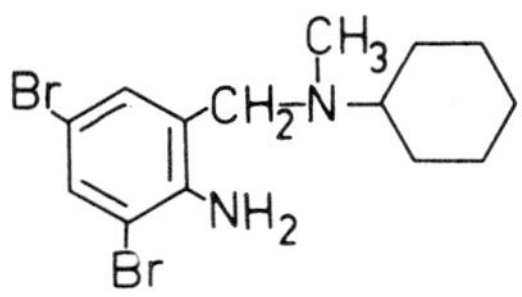

Figure 17.1 Structure of bromhexine.

II. METHODS

A. Comparison of Existing Methods

In 1979 A. P. De Leenher et al. [10] described a gas-liquid chromatographic determination of trifluoroacetylated bromhexine at the nanogram level by means of electron capture detection. The suitability of *N*-ethyl- and *N*-propyl-analogue of bromhexine as internal standards was investigated on packed columns with 11 different liquid stationary phases, with the result that only the *N*-propyl analogue was acceptably resolved from bromhexine on 4% SE-30 as the most suitable phase. In 1980, the same authors [11] adapted this method to a sensitive and reliable electron-capture GLC determination of bromhexine in human plasma: After addition of the internal standard and an aqueous triethanolamine solution to 1 ml of plasma, bromhexine was extracted at alkaline pH into *n*-hexane, transferred to an acidic aqueous solution and back-extracted into *n*-hexane after alkanization; both compounds were derivatized with trifluoroacetic anhydride. The sensitivity of this assay was reported to be 1 ng/ml of plasma. This extraction procedure is used in our laboratory with minor modifications and will be discussed in detail below.

In the same year J. A. A. Jonkheere and co-workers [12] developed a method for quantification of bromhexine in plasma employing gas chromatography-mass spectrometry with selected ion monitoring. Employing $^2H_3$-bromhexine as the internal standard and a sample extraction according to the procedure investigated earlier in the laboratory of these authors [11], a sensitivity to 0.4 ng/ml plasma could be achieved. GC separation could be verified on a 1.8 m × 2 mm silanized glass column packed with 4% SE-30. A comparison of the presented GC-MS method with the electron-capture detection method described earlier [11] showed a good correlation ($r = 0.992$), although selected ion monitoring was found to be superior for levels below 3 ng/ml.

The only method (to our knowledge) until recently for the determination of bromhexine in human plasma using capillary gas chromatography was published by J. Schmid and co-workers in 1982 [13]. As an internal standard they used the structurally very similar *N*-cyclopentyl-*N*-methyl-(2-amino-3,5-dibromobenzyl)amine. The procedure used for the sample cleanup was as follows: 1 ml of plasma was pipetted into a 25-ml centrifuge tube. Then 2 ml of buffer and 10 ml of the internal standard solution (10 ng/ml in hexane) were added. After extraction and phase separation the aqueous layer was discarded and the organic layer extracted with 1 ml of 2 N H_2SO_4. After phase separation the aqueous phase was washed once with 5 ml of hexane, then adjusted to pH 12.5 with 0.5 ml of 5 N NaOH and extracted twice with 5 ml of hexane. The hexane layer was removed into silanized tubes and evaporated; the residue was then dissolved in 25 μl of hexane and transferred into autosample vials; 2 μl were injected into the chromatograph. The chromatographic system was a 25 m × 0.3 mm I.D. WCOT capillary column with OV-17 as the stationary phase in connection with a nitrogen-specific flame ionization detector (N-FID). The authors demonstrated that bromhexine can be well separated from metabolites

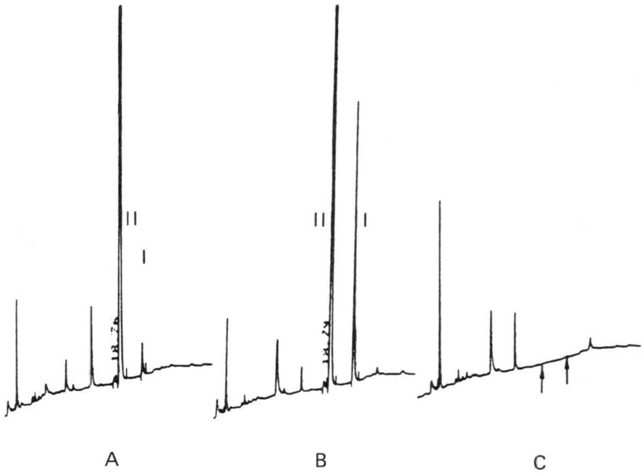

Figure 17.2 Chromatograms of human plasma showing bromhexine at a level of (A) 1.5 ng/ml, (B) 18.7 ng/ml, (C) blank. I = bromhexine, II - internal standard. Column: 25 m × 0.3 mm I.D. WCOT capillary column, OV-17. Detector: N-FID (From Ref. 13. Reproduced by permission of Elsevier Scientific Publishing Company, Amsterdam.)

as well as from endogenous plasma peaks without derivatization to a level of about 0.5 ng/ml, if 2 ml of plasma are extracted. This was only true for OV-17 as the liquid phase; OV-101 was shown to be unsuitable for the separation of bromhexine from the *N*-demethylated bromhexine derivative. The specificity of the assay was excellent as checked by GC-MS; chromatograms of human plasma extracts, obtained with the reported procedure, as shown in Figure 17.2.

One method for the determination of bromhexine in human plasma and urine using a high-performance liquid chromatograph (HPLC) method with UV-detection was described by E. Bechgaard et al. [14] in 1982. Although the assay presented was shown to be suitable for measuring bromhexine plasma levels up to 4 hr after an oral dose of 24 mg some disadvantages still remained compared to the published GLC methods: an internal standard (amitriptyline) was used, which shows no similarity to the structure of bromhexine. Furthermore, a large sample volume (3 ml plasma) is required, and the achieved sensitivity of about 5 ng/ml is not sufficient for more sophisticated pharmacokinetic calculations due to the rapidly declining plasma concentrations.

B. Routine Method Used in the Authors' Laboratory

1. *Extraction and Derivatization*

We prefer the more sensitive and more specific capillary GLC assays for quantitation of bromhexine in plasma. First, a method will be presented which employs a glass capillary column coupled with a nitrogen-specific flame ionization detector. The extraction was performed in the following way: To 2 ml of plasma 100 g of the internal standard orphenadrine, 1 ml of water and 1 ml of 1 N NaOH were added and mixed on a Whirli-mixer for some seconds. After extraction with 7 ml of heptane for 30 min by an inversion shaker and subsequent centrifugation, the upper phase was transferred to another tube and extracted with

2 ml of 1 N H_2SO_4. The upper phase was discarded after centrifugation, the lower phase alkanized with 3 ml of 1 N NaOH and reextracted into 4 ml of heptane; 3 ml of the organic phase were brought to dryness in a stream of nitrogen. The residue was then dissolved in 25 µl of hexane; 1 µl was injected splitless into the chromatograph.

GLC separations were carried out with a Carlo Erba model FV 2150 equipped with a OV-1 30 m × 0.34 mm I.D. glass capillary column (H. a. G. Jaeggy, Trogen, Switzerland) coupled with a thermionic nitrogen-sensitive flame ionization detector (Perkin-Elmer) with an electrically heated rubidium glass bead. Operactions conditions were as follows: injector temperature, 250°C; detector temperature, 275°C. Injections were done in the splitless mode at an oven temperature of 60°C. After 1 min the split was opened and the temperature immediately raised to 160°C (1 min), then programmed at 15°/min to 230°C (3 min)

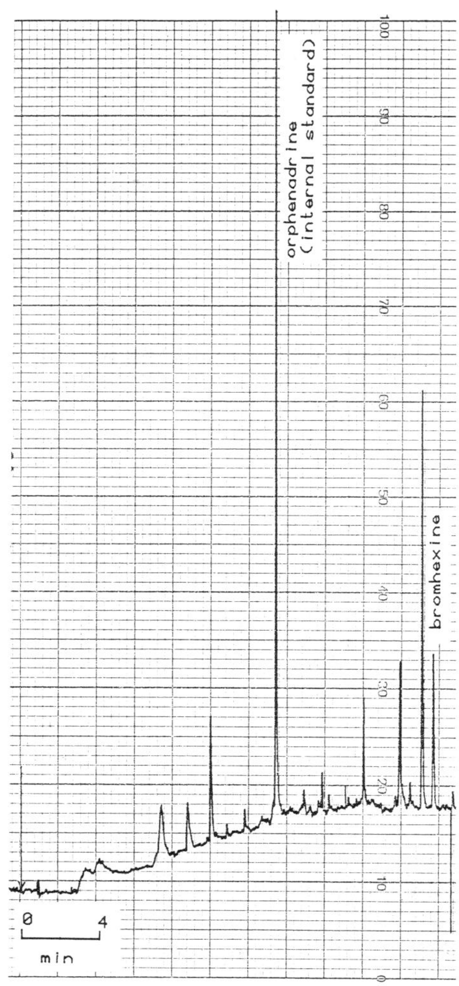

Figure 17.3 Chromatogram of a plasma extract after oral administration of 16 mg bromhexine hydrochloride, representing a level of 10 ng/ml bromhexine. Column: 30 m × 0.34 mm I.D. glass capillary coated with OV-1. Detector: N-FID.

and finally at 5°C/min to 250°C. Gas flow-rates were: for helium (carrier), 2.5 ml/min; for hydrogen, 3.5 ml/min; for air, 70 ml/min; and for nitrogen (makeup gas) 25 ml/min.

Figure 17.3 shows a typical chromatogram obtained by this procedure from human plasma after bromhexine administration. The ideal peak shape of bromhexine makes a derivatization step unnecessary. However, difficulties occurred particularly with the response/reproducibility properties of the N-FID used. During a few days operation period, sensitivity decreased continuously. Consequently, a continuous increase of the heating current of the bead was necessary with the consequence of rapid aging of the rubidium glass bead. This property of the thermionic rubidium-silicate detector, pointed out in detail by J. A. Lubkowitz et al. [15], may be improved by employing bead materials which exhibit better long-term durability. We obtained best results using a method based on the GLC-ECD assay published by A. P. de Leenheer et al. [11].

The use of high-resolution capillary columns in our method allowed the utilization of a structurally analogous internal standard with an *N*-cyclopentyl group instead of the *N*-cyclohexyl moiety in the bromhexine molecule: the *N*-cyclopentyl-*N*-methyl-(2-amino-3,5-dibromobenzyl)amine. This internal standard was synthesized by methods described in the literature [16]; extraction from plasma was performed as follows: All glassware was silanized before use to avoid losses of bromhexine by adsorption on the glass surfaces. To 1 ml of plasma in a 25-ml glass centrifuge tube with a Teflon-lined screw cap, 200 μl of the internal standard solution. *N*-cyclopentyl-*N*-methyl-(2-amino-3,5-dibromobenzyl)ammonium chloride (50 ng) in methanol, 2 ml of a 1 N aqueous triethanol amine solution, and 0.1 ml of a 5 N NaOH solution were added and well mixed. 15 ml of hexane were added and the tube mixed for 30 min on an inversion shaker. After centrifugation, the organic phase was transferred into a glass tube containing 2 ml of 2 N HCl. The tube was shaken for 15 min and the top layer was discarded after phase spearation. The last extraction step was repeated with 5 ml of hexane. The acidic solution was made basic with 1.5 ml of 5 N NaOH and extracted twice with 5 ml of hexane. After centrifugation the combined organic layers were evaporated to dryness under a stream of nitrogen at 45°C.

For derivatization the residue was dissolved in 100 μl of ethyl acetate, and 100 μl of trifluoroacetic acid anhydride (TFAA) were added and mixed. After 5 min, excess trifluoroacetic acid anhydride and ethyl acetate were removed under nitrogen and the residue was dissolved in 100 μl of methanol. 1 μl was injected into the gas chromatograph.

2. *Chromatographic Conditions*

The gas chromatograph used was a Carlo Erba model FV 2150 chromatograph equipped with a 10 mCi ^{63}Ni-electron capture detector and a Grob-type split/splitless injector. The column was a 30 m × 0.32 mm I.D. fused silica capillary with a nonextractable bonded phase DB-5 of 0.25 μm film thickness (J + W Scientific). Carrier gas was hydrogen with a linear velocity of 70 cm/sec. The detector was purged with 28 ml/min of argon/methane (90/10). Injection port and detector temperatures were 200 and 275°C, respectively. Injections (1 μl) were made at an oven temperature of 180°C with the split exit closed during the sampling period of 60 sec. After 3 min the temperature was raised to 265°C at a program rate of 15°C/min.

3. *Quantitative Estimation*

Calibration curves for the determination of bromhexine in plasma were established by preparing a series of blank plasmas spiked with measured, increasing amounts

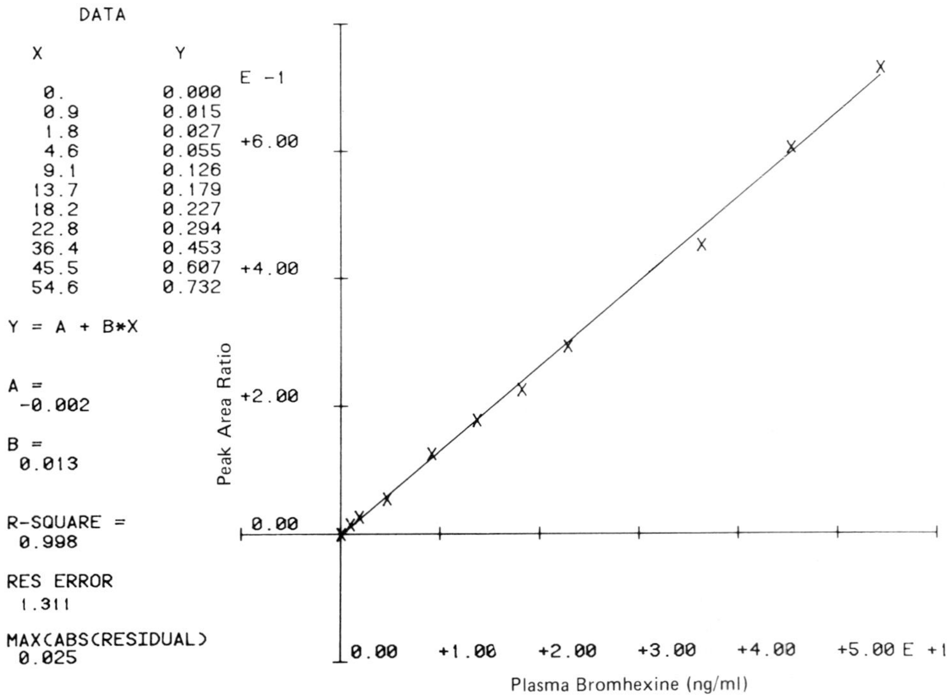

Figure 17.4 Calibration curve for bromhexine in plasma.

of bromhexine (0.9-54.6 ng/ml) in the same way as described above. Linearity of the calibration curve was obtained as indicated in Figure 17.4; bromhexine concentrations were determined by comparing ratios between the peak areas of bromhexine and the internal standard with respect to the standard curve. The extraction recovery of bromhexine from plasma, determined at a concentration of 9.1 ng/ml, was 97 ± 9% (n = 5) compared to water as the sample medium. A within-day precision of the overall procedure (R.S.D.) of 5.5% was obtained by analyzing 5 plasma samples spiked with 9.1 ng/ml bromhexine. The detection limit of the method was about 0.9 ng/ml if 1 ml of plasma was extracted.

4. *Discussion*

Figure 17.5 shows typical chromatograms obtained by the presented method. Derivatization of bromhexine and the internal standard to their corresponding TFA derivatives was quantitative, as demonstrated in Figure 17.6. This increases the ECD response by a factor of 2 compared to the underivatized compounds. A. P. de Leenheer and co-workers [10] reported the TFA derivatives to be stable when kept dry, but relatively unstable in methanol. These observations are confirmed by our investigations. However, if the TFA derivatives, dissolved in methanol, are kept in a refrigerator, no alterations of the chromatograms could be detected within one day, but excessive storage times of the samples should be avoided: for example, Figure 17.6b showed a chromatogram of the same sample as depicted in Figure 17.6a, but after 3-days of storage in the refrigerator. Due to their retention times, the two additional peaks which occur besides the TFA derivatives in the GC trace of Figure 17.6b were identified as underivatized bromhexine and the internal standard, respectively.

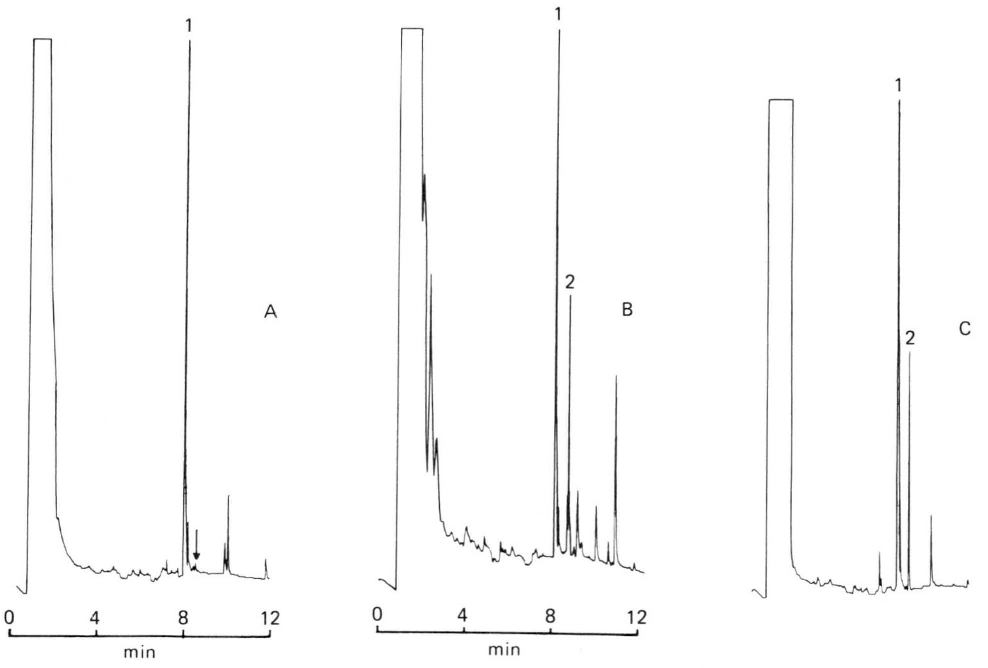

Figure 17.5 A: Chromatogram of a predose plasma extract. B: Chromatogram of a plasma extract after oral administration of 2 × 8 mg bromhexine hydrochloride. C: Chromatogram of a blank plasma spiked with 9.1 ng/ml bromhexine. 1 = Internal standard-TFA, 2 = bromhexine-TFA. Column: 30 m × 0.32 mm I.D. fused silica DB-5. Detector: ^{63}Ni-ECD.

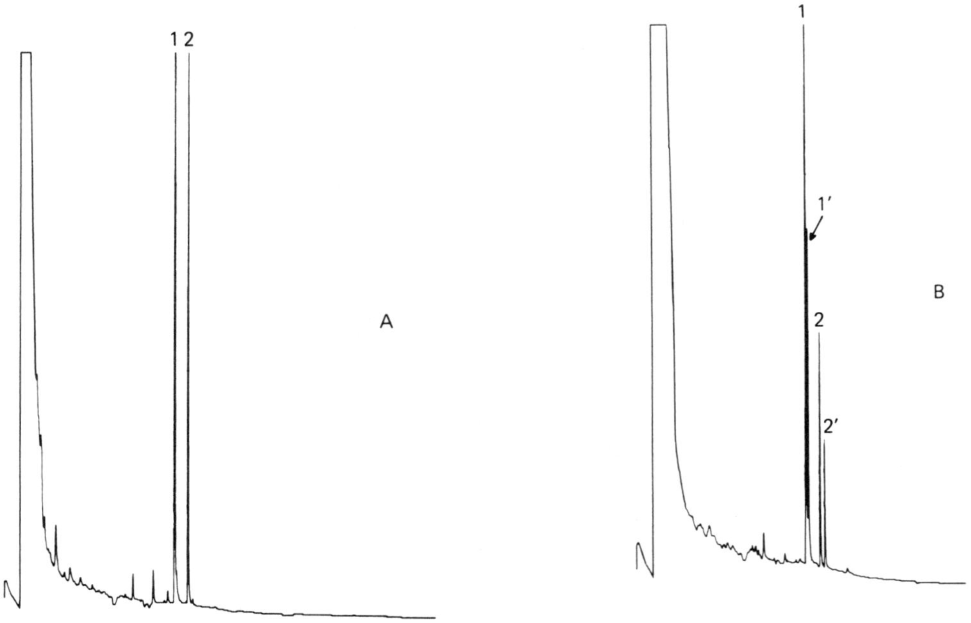

Figure 17.6 Chromatograms of the TFA derivatives of bromhexine (1) and internal standard (2) dissolved in methanol and kept in a refrigerator for different times. Column: same as in Figure 17.5. A: storage time, 1 hr. B: storage time, 3 days. 1' and 2' = underivatized internal standard and bromhexine, respectively.

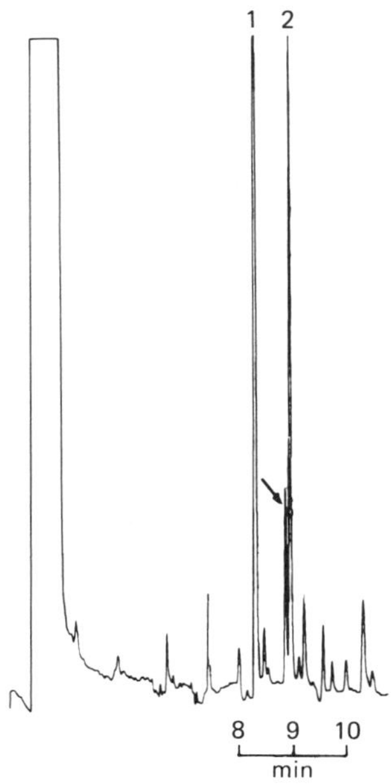

Figure 17.7 Plasma extract after administration of 2 × 8 mg bromhexine hydrochloride. Chromatographic conditions: same as in Figure 17.5. The arrow indicates a metabolite of bromhexine.

5. *Application to a Pharmacokinetic Study*

In contrast to the chromatograms obtained from blank plasma or from spiked plasma, all chromatograms obtained from volunteers after ingestion of bromhexine showed additional peaks due to bromhexine metabolites (Figure 17.5). However, sufficient resolution was achieved for correct quantification. Thus,

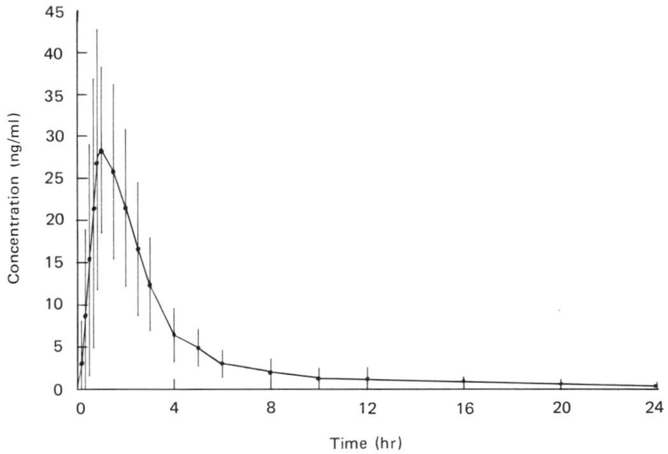

Figure 17.8 Mean plasma bromhexine-levels (ng/ml) of 12 normal human volunteers after oral administration of 2 × 8 mg bromhexine hydrochloride.

the peak (8.85 min) which elutes just before bromhexine (8.94 min) was nearly base-line separated, as illustrated in Figure 17.7. These results as well as those obtained by J. Schmid and co-workers [13] clearly indicate that high-efficiency capillary columns are required to achieve an adequate degree of resolution in order to determine bromhexine in human plasma in the presence of its various metabolites.

The advantages of using capillary columns to an electron capture detector have been demonstrated by several authors [17–19]. A higher sensitivity could be achieved compared to packed columns as a consequence of decreased band spreading and lower column bleed. The usefulness of the presented method in the course of pharmacokinetic studies with bromhexine is demonstrated in Figure 17.8, which shows mean plasma levels of 12 human volunteers after oral administration of the very low dosage of 2 × 8 mg of bromhexine hydrochloride.

REFERENCES

1. H. Buergi, J. C. White, P. C. Elmes, and A. Wash, *Med. Thorac. 21*: 156-167 (1964).
2. H. Buergi, *Praxis 54*:1327-1330 (1965).
3. R. Giesenking and U. Baldamus, *Beitr. Klin. Tuberk. 137*:1-18 (1968).
4. F. Arch, *Arzeim.-Forsch. 13*:480-485 (1963).
5. E. Schraven, F. W. Koss, J. Keck, and G. Beisenherz, *Eur. J. Pharmacol. 1*:445 (1967).
6. Z. Kopitar, R. Jauch, R. Hankwitz, and H. Plezer, *Eur. J. Pharmacol. 21*:6 (1973).
7. R. Jauch and R. Hankwitz, *Arzneim.-Forsch. 25*:1954 (1975).
8. H. T. Nilson, C. G. A. Persson, and T. Tegner, *Acta Pharm. Suecica 13*:241 (1976).
9. E. Mutschler, ed., *Arzneimittelwirkungen*, Wissenschaftl. Verlagsgesellschaft mbH, Stuttgart, 1981.
10. A. P. De Leenheer and L. M. R. Vandecasteele-Thienpont, *J. Chromatogr. 175*:301 (1979).
11. A. P. De Leenheer and L. M. R. Vandecasteele-Thienpont, *J. Pharm. Sci. 69*:99 (1980).
12. J. A. A. Jonkheere, L. M. R. Thienpont, A. P. De Leenheer, P. De Backer, M. Debackere, and F. M. Belpaire, *Biomed. Mass Spectrom. 7*:582 (1980).
13. J. Schmid and F.-W. Koss, *J. Chromatogr. 227*:71 (1982).
14. E. Bechgaard and A. Nielsen, *J. Chromatogr. 228*:392 (1982).
15. J. A. Lubkowitz, J. L. Glajch, B. P. Semonian, and L. B. Rogers, *J. Chromatogr. 133*:37 (1977).
16. J. Keck, *Justus Liebigs Ann. Chem. 662*:171 (1963).
17. P. Devaux and G. Guichon, *Chromatographia 2*:151 (1969).
18. G. Eklund, B. Josefsson, and C. Roos, *J. Chromatogr. 142*:575 (1977).
19. L. Rejhtar and K. Tesarik, *J. Chromatogr. 131*:404 (1977).

18

Analysis of Hydrochlorothiazide by Capillary GLC with Electron Capture Detection

Dieter Lutz, Elsa Ilias, and Halvor Jaeger / *L.A.B. Gesellschaft für pharmakologische Untersuchungen m.b.H. & Co. (Laboratory for Pharmacological Research), Neu-Ulm, Federal Republic of Germany*

I. INTRODUCTION

Hydrochlorothiazide (HCT) is a diuretic widely prescribed for the treatment of hypertension which belongs to the group of the benzothiadiazines.

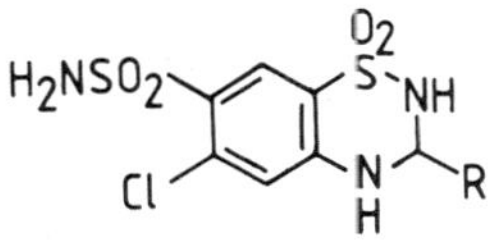

Structure of hydrochlorothiazide (R = H) and butizide [R = $CH_2CH(CH_3)_2$]

Its diuretic effect is based on the inhibition of sodium and chloride resorption in the distal tubules, as well as on an increase of potassium excretion. Following long-term treatment with benzothiadiazines, hypokalemia was observed in 20% of the patients. For this reason hydrochlorothiazide is advantageously combined with potassium-saving diuretics (e.g., triamterene) which nearly compensate for the potassium-eliminating effect of hydrochlorothiazide [1]. In such combination preparations the therapeutic dosage of hydrochlorothiazide is relatively low (approximately 12 mg), requiring very sensitive analytical methods for drug monitoring in plasma samples in the course of pharmacokinetic studies.

II. METHODS

A. Comparison of Existing Methods

In 1975, Vanden Heuvel and co-workers [2] described a gas chromatographic method employing electron capture detection (ECD; 8 mCi ^{63}Ni) and a packed column for hydrochlorothiazide determinations in human plasma; the reported sensitivity was 50 ng/ml. As an internal standard the authors used the bromo analogue of the hydrochlorothiazide. The non-derivatized compounds are sulfonamides with very polar properties and as such are not suitable for analysis by gas-liquid chromatography. Volatility can adequately be increased by formation of the corresponding tetramethyl derivatives by on-column methylation with trimethylanilinium hydroxide (TMAH) in methanol at a flash-heater temperature of 280°C. In this study a glass column packed with a 1% mixture of OV-1 in polycyclohexane-dimethyladipate (6:1) on Gas Chrom P was used.

In spite of the very broad peaks obtained (approximately 2 min at 1/2 peak height for HCT), which are typical for packed columns, the resulting sensitivity is sufficient for determination of the blood level up to 24 hr after oral application of 50 mg hydrochlorothiazide. Furthermore, the analysis can be automated advantageously by the aid of an automatic liquid-sampler; however, the column lifetime was comparatively short (approx. 150 hr).

Walker and Vanden Heuvel [3] modified the packed-column method and used a short glass capillary column (6-m SE-30) with a splitless injection system (dropping glass-needle). However, this arrangement did not yield reproducible results, probably because of insufficient derivatization on the glass-needle. A complete derivatization was obtained by extractive methylation (methyl iodide, sodium hydroxide, tetra-*n*-butyl ammonium hydroxide, methylene chloride). On a 15 m × 0.25 mm I.D. SE-30 glass capillary column with SIM detection they achieved a sensitivity of approximately 2 ng/ml. However, detailed statements about the extraction procedure are not given [4].

In 1975, Lindström and Molander [5] developed an extractive methylation technique for the determination of HCT in plasma, blood corpuscles, and urine, as described previously for the two diuretics chlorthalidone [6] and furosemide [7]. The extraction of HCT from plasma was done at pH 6 with 4-methyl-2-pentanone after addition of the internal standard chlorthalidone. After back-extraction in 0.2 M NaOH, the extractive methylation was carried out at 50°C with methyl iodide in dichloromethane in the presence of tetrahexylammonium hydrogen sulfate. A packed glass column (1.5 m × 2 mm I.D., 1% Sephadex SE-30) with ECD detection served for the gas chromatographic separation. The chromatograms obtained showed the peaks of HCT and the internal standard in the steeply falling shoulder of the solvent peak, making a quantitative determination very difficult. Nevertheless, the minimal HCT concentration which could be determined, was claimed to be 10 ng/ml (2-3 ng/ml if 3 ml of plasma were extracted).

In 1977, Redalieu and collaborators [8] modified the Lindström method in some points: As an internal standard they used the bromo analogue of HCT, and for the chromatographic separation the more polar liquid phase OV-225 was used instead of SE-30. This resulted in a better separation of the HCT peak from the solvent. In spite of relatively broad peak-widths (0.73 min at 1/2 peak height for HCT, Figure 18.1), they reported good quantitation down to 5 ng/ml HCT, if 2 ml of plasma were extracted. An overall recovery rate of 95 ± 8% (SD) over a concentration range of 5-648 ng/ml showed satisfactory accuracy and precision. In our laboratory the best results were obtained with this extraction method and minor modifications in the routine.

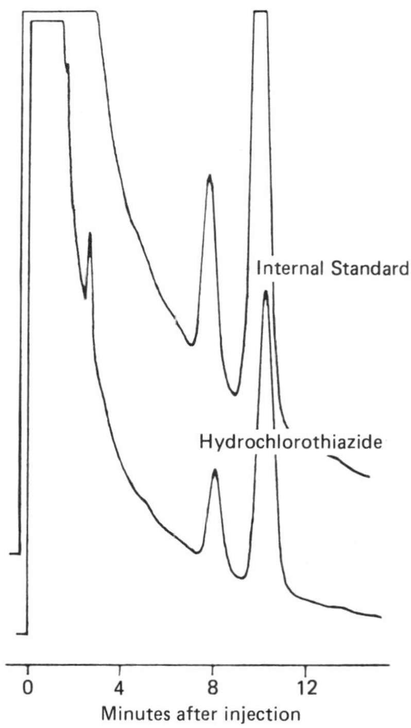

Figure 18.1 Gas chromatogram of human plasma from a subject after administration of 50 mg HCT, obtained on a 0.9 m × 2 mm I.D. glass column, packed with 80-100 mesh 1% OV-225. (From Ref. 8.)

B. Routine Method Used in the Authors' Laboratory

1. Extraction and Derivatization

Compared to the procedure of Redalieu, the most essential modification was the choice of another internal standard: instead of the bromo analogue of HCT we use butizide, the isobutyl homologue of HCT. The gas chromatographic determination of HCT by means of an ECD detection is achieved by a fused silica capillary column instead of a packed column, and therefore only 1 ml of plasma is needed for extraction. The use of an automatic liquid sampler allows a day-and-night operation.

Up to now, HCT determinations by means of capillary gas chromatography in connection with electron capture detection has not been described. In our hands the following procedure yielded the "best" results: To 1 ml of plasma in a glass centrifuge tube with a Teflon-lined screw cap, 100 ng of the internal standard is added and well mixed. After extraction (20 min) with 5 ml isobutyl methylketone, the mixture is centrifuged, and 4 ml of the organic phase are transferred to a clean glass centrifuge tube and reextracted for 20 min with 2.5 ml 0.2 M sodium hydroxide. After centrifugation the organic phase is discarded.

For extractive methylation, 2 ml of the aqueous phase are mixed well with 50 μl of a 0.1 M tetrahexylamminium hydrogen sulfate solution and 5 ml of 0.5 M methyl iodide solution in dichloromethane for 30 min at 50°C in a water-bath.

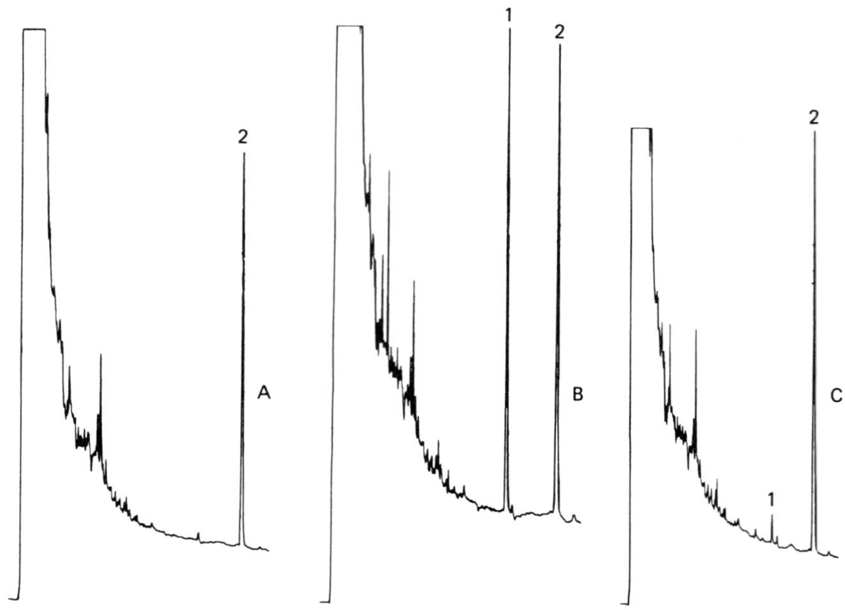

Figure 18.2 GLC analysis of (A) plasma before administration of HCT; (B) plasma 3 hr after administration of 12.5 HCT (73 ng/ml HCT); (C) control plasma spiked with 5 ng/ml HCT. (1 = HCT 8.49 min; 2 = internal standard, 10.83 min (butizide).)

After centrifugation, 4 ml of the organic phase are evaporated to dryness in a conical tube under a nitrogen stream. The residue is then treated with 200 μl cyclohexane for 5 min in an ultrasonic bath and centrifuged. The clear supernatant liquid is transferred into 100-μl sample vials. 1 μl of the sample is injected (splitless) into the gas chromatograph by means of an automatic liquid sampler. Figure 18.2 shows typical gas chromatograms after liquid injection on a 30 m × 0.32 mm I.D. fused silica capillary column DB-5 (J + W Scientific).

2. *Chromatographic Conditions*

The gas chromatograph used was a DANI 3800 chromatograph equipped with a 10 mCi ^{63}Ni electron capture detector and a Grob type split/splitless injector containing a 4 mm × 8 cm glass liner from Carlo Erba.

Injections were made with a DANI automatic liquid sampler, Model ALS 3641, which also controlled the split flow by means of a solenoid valve. Chromatograms were run on a 30 m × 0.32 mm I.D. WCOT fused silica capillary, with a nonextractable stationary phase DB-5 of 0.25-μm film thickness (J + W Scientific).

Carrier gas was hydrogen with a linear velocity of 75 cm/sec. The detector was purged with 28 ml/min of argon/methane (90/10). Injection port and detector temperatures were 242°C and 300°C, respectively. Injection (1 μl) occurred at an oven temperature of 160°C with a closed split exit during the sampling period of 45 sec. Then the temperature was raised to 270°C at a program rate of 30°C/min.

3. *Quantification*

Calibration curves for the determination of HCT in plasma were prepared by assaying a series of blank plasmas spiked with measured amounts of HCT in the same way as described above. Linearity (through the origin) of the calibration curve was obtained between 3 and 300 ng/ml (Figure 18.3). The HCT concentration in a sample was determined by comparing the ratio between the peak of HCT and the internal standard with respect to the established standard curve.

4. *Precision and Sensitivity*

The method allows the quantitative determination of HCT in plasma at concentrations as low as 3 ng/ml. The average recovery for HCT from plasma, determined at a concentration of 25 ng/ml, was 97 ± 4% (R.S.D.), $n = 5$. The precision between days over a one-month period was ±14% R.S.D. ($n = 16$), demonstrated on plasma samples, each spiked with 100 ng/ml of HCT.

The described method for the determination of HCT proved to be appropriate in our laboratory for a large number of samples which usually have to be assayed during the course of pharmacokinetic studies. After more than 500 injections of plasma samples, the capillary column showed no deterioration.

DATA

X	Y
0.	0.000
3.	0.031
3.	0.033
5.	0.053
5.	0.058
10.	0.086
10.	0.098
20.	0.245
50.	0.548
50.	0.534
100.	0.934
100.	1.094
200.	1.915
200.	2.178
300.	3.200

Y = A + B*X

A =
-0.002

B =
0.010

R-SQUARE =
0.996

RES ERROR
0.005

MAX(ABS(RESIDUAL))
0.173

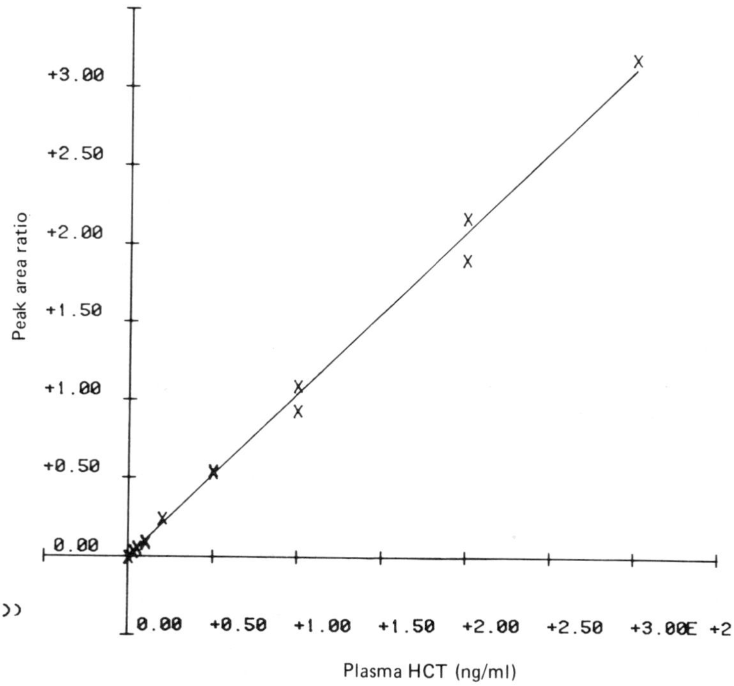

Figure 18.3 Calibration curve for HCT following extraction from plasma samples.

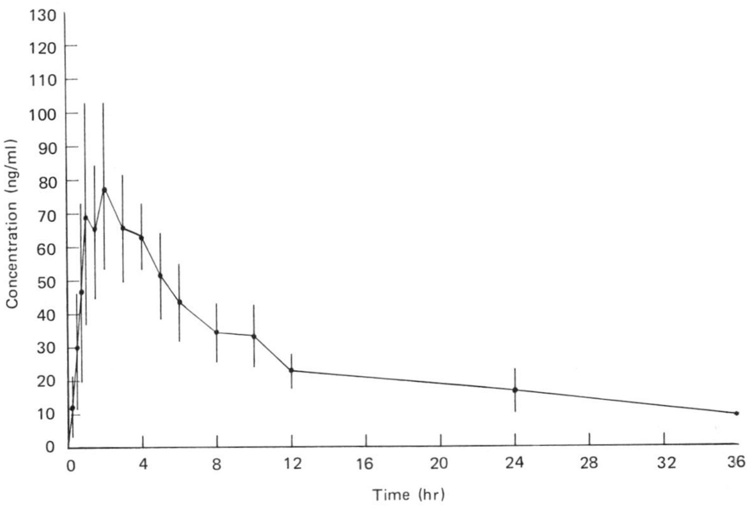

Figure 18.4 Mean HCT plasma levels of 10 volunteers after oral administration of 12.5 mg HCT.

5. *Application*

Figure 18.4 shows the mean plasma HCT levels of 10 volunteers after oral administration of one tablet containing 12.5 mg of HCT. Up to 36 hr after intake of this relatively low dose, HCT levels could be quantitatively determined.

REFERENCES

1. W. Forth, D. Henschler, and W. Rummel, eds., Allgemeine und spezielle Pharmakologie und Toxikologie 2. Auflage, Mannheim, Bibliographisches Institut, Wissenschaftsverlag, 1977.
2. W. J. A. Vanden Heuvel, V. F. Gruber, R. W. Walker, and F. J. Wolf, *J. Pharm. Sci. 64*:1309 (1975).
3. R. W. Walker and W. J. A. Vanden Heuvel, unpublished results.
4. W. J. A. Vanden Heuvel, and J. S. Zweig, *HRC & CC* 3:381 (1980).
5. B. Lindström and M. Molander, *J. Chromatogr. 114*:459 (1975).
6. M. Ervik and K. Gustavii, *Anal. Chem. 46*:39 (1974).
7. B. Lindström and M. Molander, *J. Chromatogr. 101*:219 (1974).
8. E. Redalieu, V. V. Tipuis, and W. E. Wagner, Jr., *J. Pharm. Sci. 67*:726 (1978).

19

Analysis of Clofibrate by Glass Capillary GLC

Halvor Jaeger, Dieter Lutz, Werner Gielsdorf, and James A. Settlage / *L.A.B. Gesellschaft für pharmakologische Untersuchungen m.b.H. & Co. (Laboratory for Pharmacological Research), Neu-Ulm, Federal Republic of Germany*

I. INTRODUCTION

Screening tests in the early sixties by Thorp and Warning [1,2] revealed that a series of aryloxyisobutyric acids effectively reduced plasma concentrations of total lipids and cholesterol in the experimental animal. Since then the efficacy of clofibrate (ethyl *p*-chlorophenoxyisobutyrate, Figure 19.1) in man as well has been demonstrated in the literature many times [3-11]. The drug reduces serum levels of triglycerides up to 22% and those of total cholesterol by 6% [12]. The decrease in serum cholesterol concentrations stems mainly from a decrease in VLDL, whereas LDL cholesterol remains unchanged or even rises.

In another report [13], in asymptomatic men with hypercholesterolemia the lowering effect on serum cholesterol varied between 6 and 11%. Hence in type III hyperlipoproteinemia, plasma concentrations of triglycerides and cholesterol are lowered by as much as 80 and 50%, respectively [14]. In type IIa (isolated elevation of LDL), types IIb, IV and V (elevation of VLDL), the effect is only moderate. In type I (hyperchylomicronemia), no effect could be observed. The mechanism of action is still largely unknown [15-17].

However, the use of the drug has increasingly become circumscribed, as its potential in the prevention of atherosclerosis has come into question [12,13]. Furthermore an increasing number of reports have appeared on undesired side effects, especially gallstone formation, cardiac arrhythmias, and thromboembolic phenomena [18-21].

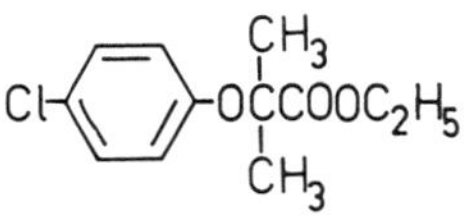

Figure 19.1 Clofibrate (ethyl-*p*-chlorophenoxyisobutyrate).

Upon oral administration, clofibrate is completely absorbed from the intestine and the ester is rapidly hydrolyzed by tissue and serum esterases into the corresponding acid *p*-chlorophenoxyisobutyric acid (CPIB) [22, 23], whose peak concentrations appear in the serum within 4 hr after oral administration [23]. The major fraction of the active metabolite is bound to plasma albumin [24, 25]. Almost the entire administered dose is eliminated within 48 hr after administration [26].

II. METHODS

A. Comparison of Existing Methods

Before publication of the first GC method by Silvestri [31] for analysis of CPIB or its hydrolysis product *p*-chlorophenol, UV absorption measurements were usually applied [22, 27-30], which lack sensitivity and specificity. The extraction of the drug was performed by acidified plasma into a mixture of diethyl ether/light petroleum. The extract was washed, dried over sodium sulfate, and evaporated to dryness. The residue was injected onto a 5% butanediolsuccinate column. Only chromatograms of pure substances were given, not of the substances after extraction from biological material.

Another early method was that of Horning et al. [32] who studied the effect of chronic administration of clofibrate on the excretion of bile lipids in dogs. Sample preparation was as follows: 1 ml of serum was diluted to 20 ml with distilled water, the pH was adjusted to 4.0-4.5 with acetic acid, and the sample was transferred to a DEAE-Sephadex column. Neutral and basic fractions were washed from the column with water, and the acids were eluted with a solution of 150 ml of 1.5 M pyridine-acetic acid. After lyophilization, the residue was transferred with methanol to a graduated centrifuged tube; the methanol was evaporated to approximately 0.2 ml by a nitrogen stream. The solution was then transferred quantitatively to a 1-ml vial to which 10-μl of pyridine was added, and evaporated under a gentle nitrogen stream.

The pyridine salt of CPIBA was redissolved in 50 μl of pyridine, and 100 μl of bis(trimethylsilyl)trifluoroacetamide were added. After mixing in a Vortex mixer, the solution was allowed to stand overnight. As an internal standard methyl myristate was used; the column was a 12-ft 5% SE-30 column with a temperature program of 2°C/min, starting at 100°C. No data evaluating specificity, precision, and accuracy are given.

Although this method seems to be a reliable GC-MS method for the analysis of the active metabolite of clofibrate, it has to be clearly indicated that the extraction is very time-consuming and the chromatogram as well reveals retention times of 50 and 58 min for the trimethylsilyl ester of the chlorophenoxyisobutyric acid and the internal standard, which makes the procedure unsuitable for rapid analysis.

In 1974, two papers on the determination of chlorophenoxyisobutyric acid in biological fluids were published. The first method by Sedaghat et al. [33]

involves saponification, extraction, methylation, and thin-layer chromatography prior to GC analysis. For saponification 1 ml of plasma, urine, or bile, or 1 g of feces was mixed with 10 ml of methanol in a 125-ml glass-stoppered bottle, and 200 μl of the internal recovery standard methylchlorophenoxypropionic acid were added. The mixture was refluxed for 1 hr with 1 ml of 10 N NaOH to hydrolize any conjugated chlorophenoxyisobutyrate possibly present. For extraction, the mixture was acidified with 1 ml of concentrated HCl after cooling, and 20 ml of chloroform and 30 ml of chloroform-methanol (2:1) were added successively. The mixture was shaken for 1 min; 10 ml of water were added and the shaking was continued for 1 min. After centrifugation, the lower phase was transferred to a 500-ml round-bottom flask, and 20 ml of chloroform were added to the upper phase, and the extraction procedure was repeated twice in order to assure a quantitative transfer. The combined lower phases were evaporated to dryness with a rotary evaporator.

Methylation was performed by addition of 5 ml of 5% HCl/methanol to the dried residue. The mixture was left overnight. The methyl esters were concentrated by rotary evaporation at room temperature. Methylchlorophenoxyisobutyrate, methylchlorophenoxypropionate, and the other methyl esters of fatty acids were quantitatively transferred with chloroform-methanol (2:1) to TLC plates for separation from the other fatty acid methyl esters. The respective portion of the TLC plate was eluted into a 100-ml round-bottom flask, and the internal standard methyl arachidonate (C20:0) was added. After evaporation to dryness, the mixture was redissolved in 1 ml of methanol; 5 μl of the sample were injected. The column was a 6 ft × 4 mm I.D. glass column packed with 10% EGSS-X on Gas Chrom Q. Nitrogen was used as a carrier gas at temperatures of 180, 220, and 240°C for column, flash heater, and detector, respectively.

The specificity was found to be good enough, although as an interference peak, methyloleate (C18:0) interfered with the methyl CPIB peak. Therefore, the author stated, that if there was the suspicion of contamination of the sample with fatty acid methyl ester, the TLC procedure has to be repeated.

The sensitivity reported was less than 10 μg/ml of plasma, urine, or bile, or per gram feces. By dissolving the sample in less methanol, sensitivity could be increased to 5 μg/ml. Recovery by means of testing with labeled material was 92% ± 6% from plasma and 91% ± 4% from urine. Reproducibility of the method, expressed as coefficient of variation of 8 sets of replicate samples, was 6%. Although this method obviously yields good quantitative results, the extraction procedure and the cleanup is still very tedious and time-consuming. A chromatogram of an extract from biological origin was not given.

In the same year Knüchel et al. [34] reported that in the fatty acid chromatograms of patients treated with clofibrate a peak was detected that could later be identified as the methyl ester of chlorophenoxyisobutyrate. No quantitation for this method is given and no internal standard is applied, so that quantitation of the drug cannot be readily accomplished.

In 1975, six papers were published on the analysis of clofibric acid in plasma. Berlin [35] used for extraction a method whereby toluene acts as extraction solvent: to 200 μl plasma, 50 μl of 2 M H_3PO_4 and 0.5 ml of the internal standard solution [9 μg/ml 2-(4-chloro-3-methylphenoxy)-2-methylpropionic acid in toluene] were added. After 10 min of shaking, the organic phase was transferred to a 3-ml tapered glass-tube, and the plasma was reextracted with 0.5 ml toluene. 0.5 M Na_2HPO_4 was added to the combined organic phases, and the mixture was shaken for 10 min. The organic phase was removed and the aqueous phase made acidic with 50 μl of 5 M H_3PO_4.

100 μl CH_2Cl_2 were added and the mixture was extracted on a Whirl-mixer for 30 sec. After removal of the aqueous phase, 50 μl of diazomethane in ether were added, and the solution was evaporated under nitrogen to about one-fifth of its volume. 1 μl was injected into the gas chromatograph.

The column was a 6 ft × 3 mm I.D. silanized glass column packed with 3% OV-17 on Gas Chrom Q, 80/100 mesh. Temperatures were 131, 232, and 240°C for the column, the injector, and the detector, respectively. Retention times were 4.3 min for the clofibric acid methyl ester and 6.7 min for the internal standard methyl ester.

The precision (expressed as the variation coefficient of 5 plasma samples at each of 3 different concentrations) was 4.8%. The recovery rate was said to be 98% after the second extraction; the sensitivity, 1 μg of methylchloroisobutyrate per milliliter plasma. The method involves time-consuming re-extraction into phosphate buffer and employs an internal standard which is not commercially available. No chromatogram is given in the paper; furthermore, the use of diazomethane for methylation is regarded to be hazardous, as stated previously [36].

In the same year, Karmen and Haut [37] presented a method that uses a flame ionization detector selectively sensitive to halogen (AFID). The method employs two internal standards, chlorophenoxyacetic acid (CPA) and chlorophenoxypropionic acid (CPP). The extraction was performed by adding 100 μg of CPP and 500 μg of CPA in 0.1 ml acetone to 1 ml serum in a 15-ml test tube. The serum was then acidified with 1 ml 1 N H_2SO_4 and extracted 3 times with 5 ml of diethyl ether. The extracts were combined and dried over 0.1 g anhydrous sodium sulfate and decanted into a second tube. 2 ml of a diazomethane solution were added and the mixture was allowed to stand for 5 min at room temperature. Then 0.2 ml isooctane was added, and the solution was carefully concentrated to 0.2 ml by evaporation with a stream of nitrogen.

The columns used were 183 × 4 mm glass tubes packed with either 10% ethylene glycol adipate polyester (EGA) on Chromosorb W, 80/100 mesh, or 3% SE-30 on Supelcoport, 80/100 mesh. The detector temperatures are not given in this paper. Detectors included a "normal" FID and an AFID detector, recording the chromatograms simultaneously. The AFID record was said to provide more reliable quantification of data because serum lipids interfered with the FID assay. The interference was said to be most significant when the CPIB concentration was low.

The recovery rate of added CPP was said to be close to 90%. More data (showing, for example, precision and accuracy) are not given. According to the chromatogram, CPIB is eluted after 6, CPP after 7, and CPA after 13 min. The chromatograms show considerable tailing, thus making a correct quantification very hard. Furthermore, the method involves the hazardous use of diazomethane and 2 detectors.

In the same year Tuong and Tuong [38] presented their method whereby 1.0 ml of plasma was diluted with distilled water to a final volume of 2 ml. The mixture was then acidified with 1.0 ml of 0.4 M perchloric acid and extracted with 5.0 ml of benzene by gentle shaking for 15 min in an Omni-Shaker at 20 cycles/min. After centrifugation for 15 min at 100 g, 4.0 ml of the organic phase were transferred into a tapered 7-ml sample vial and evaporated to dryness under a stream of nitrogen in a water bath at 40°C. The methylation was performed by using diazomethane; Octadecane in *n*-hexane was used as an external standard.

The column used was a coiled glass column 6 ft × 2 mm I.D. packed with 3% OV-17 on Chromosorb W, 100-120 mesh. The chromatographic conditions were as follows: injector temperature, 180°C; detector temperature, 210°C; column initial temperature 150°C for 6 min, then programmed to 195°C at 10°/min and held at the final temperature for 5 min before recycling. Helium was used as carrier gas at a flow of 25 ml/min. The claimed sensitivity was 250 ng CPIB methyl ester per milliliter plasma, the recovery rate approximately 70%, and the variation coefficient between 5 and 10%. This method also involves use of the hazardous diazomethane; furthermore, a correct quantification by the external standard method in combination with such a low recovery rate cannot be expected.

In the same year Honin and co-workers [26] presented a method that was based on the method of Tuong and Tuong with the exception that after evaporation the residue was taken up with 100 μl of methylated lauric acid in ethyl alcohol (200 μg/ml) as the internal standard. The retention time of the methyl ester of CPIP was 4.5 min and that of lauric acid methyl ester, 3 min. The chromatogram does not show a complete separation of the two compounds; this was achieved on a 2-m column packed with 0.2% OV-17 on GLC-120, 80-100 mesh. The temperatures were: for the column, 140°C; for the injector and detector, 180°C. The accuracy was 10% and the recovery rate 75 ± 5%. The method has the disadvantage of applying an internal standard that occurs naturally in human plasma, although having very low concentrations. Furthermore, the internal standard is added after extraction, thus causing problems in quantification. Again methylation by diazomethane was applied.

In a further publication of the same year, Bruderlein et al. [39] compared their method to the previously published spectrophotometric method of Barrett and Thorp [28] and found that in patients with normal plasma proteins, who did not take any other drugs, the values were comparable, whereas in uremic patients and in patients taking other drugs, the values were not comparable because of interfering substances in the UV spectrum.

The procedure was as follows: to 10 ml of chloroform in a glass-stoppered centrifuge tube are added 0.1 ml of 3-MPA (3-methoxyphenylacetic acid)* as internal standard, 1.0 ml of serum, and 4.0 ml of 0.5 N HCl. The tube was shaken for 10 min and centrifuged for 10 min at 1000 rpm. The aqueous layer was discarded, and 8.0 ml of the chloroform layer transferred to a clean tube. After addition of 5.0 ml of 0.1 N NH_4OH, the tube was shaken for 20 min and recentrifuged. A 4.0-ml aliquot of the aqueous phase was transferred to a third tube containing an equal volume of methanol and evaporated to dryness in vacuo. The residue was transferred in a small volume of acetone to a 0.5-ml Reacti-vial and dried in a stream of nitrogen. Methylation was achieved in 30 min at room temperature by addition of 0.1 ml of methanolic HCl. Excess reagent was removed by a stream of nitrogen. The residue was dissolved in 50 μl of acetone, and an aliquot injected into the gas chromatograph.

The column used as a 6 ft × 2 mm I.D. U-shaped glass column packed with 3.5% SE-30 on 100-120 mesh Gas Chrom Q. The temperatures were 250, 138, and 250°C for injector, column, and detector, respectively. Nitrogen carrier-gas flow was 10 ml/min; the retention times for the methyl esters of 3-MPA and CPIB were 5.5 min and 8.5 min, respectively. The recovery rate of the GLC method was said to be 90 ± 1.3%, independent of the CPIB concentrations. Data on accuracy and precision are not given. Although these data are lacking, the method seems to be quantitative and reliable. The plasma chromatograms are free from interfering peaks and the calibration curves linear.

---

*Internal standard solution: 5 mg 3-MPA in 10 ml methanol.

Also in 1975, Gugler and co-workers [40] presented a paper on the pharmacokinetics of clofibrate. Since the method used in this paper was revised by Gugler and Jensen [41] afterwards, only this revised version is described here:

Venous blood was drawn into heparinized tubes and the plasma was separated by centrifugation. To 1 ml of plasma were added 1.0 ml of 0.4 M hydrochloric acid and 6 ml of toluene containing 120 μl of the internal standard 2-naphthoic acid; the tube was then shaken for 5 min and centrifuged for 3 min at 4000 *g*. A 5-ml portion of the organic phase was transferred to a centrifuge tube, 50 μl of TMA-OH (trimethylanilinium hydroxide) were added, and the mixture extracted on a vortex mixer for 1 min. After a brief centrifugation, 1 μl of the aqueous layer was injected into the gas chromatograph.

The chromatograph separation was carried out on a 6 ft × 4 mm I.D. glass column, packed with 3% SE-30 on 80-100 mesh Gas-Chrom Q. The temperatures were 150, 290, and 270°C for column, injector, and detector (FID), respectively. Retention times were 1.5 min for CPIB and 2.5 min for the internal standard.

The coefficient of variation (six samples each at five concentrations) was 3.3% for plasma and 2.7% for urine. The recovery was 71 ± 2% from plasma and 91 ± 2% from urine. The calibration graphs show excellent linearity, and the chromatograms are of good quality, although they were performed on a packed column. They are given in Figure 19.2.

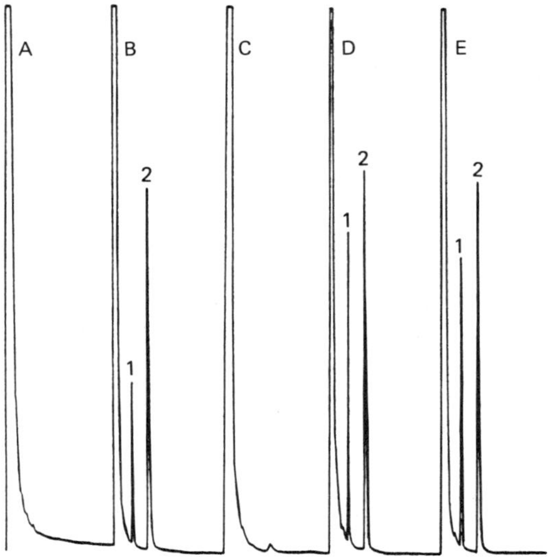

Figure 19.2 Gas chromatograms for: (A) material extracted from blank plasma; (B) material extracted from blank plasma to which CPIB (50 μg/ml) and the internal standard (120 μg) have been added, (C) material extracted from blank urine; (D) material extracted from blank urine to which CPIB (50 μg/ml) and the internal standard (120 μg) have been added; (E) material extracted from plasma obtained from a patient receiving clofibrate. The peaks indicated are: (1) CPIB; (2) the internal standard (2-naphthoic acid). 6 ft × 4 mm I.D. packed column, 3% SE-30 (From Ref. 41, reproduced by permission of Elsevier Scientific Publishing Company, Amsterdam.)

Connor et al. [42] described a method for determination of CPIB, applying two internal standards (*p*-chlorophenoxy)acetic acid and 3-(*p*-chlorophenoxy) propionic acid. A double extraction procedure was carried out and the substances were converted to the corresponding butyl esters by reaction with isobutane in a mixture of methanol and *N*,*N*-dimethylacetamide containing tetramethylammonium hydroxide.

The column used was a 120 cm × 2 mm I.D glass column packed with 3% OV-17 on 100-120 mesh Gas-Chrom Q. The column temperature was programmed from 130 to 240°C at a rate of 16°/min. Injector and detector temperatures were 250 and 300°C, respectively. The retention times for CPIB, CPAA, and CPPA were 2.88, 3.26, and 3.87 min, respectively.

The absolute recoveries from plasma for CPIB, CPAA, and CPPA from serum were 89.1, 95.9, and 99.8%, respectively. The coefficient of variation at a concentration of 79.2 μg/ml of CPIB in plasma was 2.3%; the precision, 4.1%.

The (to our knowledge) latest method for the GC determination of CPIB is that of Wolf and Zimmermann [43]. The extraction procedure is based on a salting-out technique: 1 ml of human plasma, spiked with clofibrate and clofibric acid was added to a 10-ml test tube containing 20 μl of a sodium fluoride solution and 550 mg ammonium sulfate. The mixture was agitated for 30 sec in a Vortex mixer and then centrifuged for 10 min at 3500 rpm for 30 min. The supernate (0.5 ml) was transferred to a fresh test tube, and 100 μg of benzocaine were added. The resulting solution was extracted with 1 ml of ether by agitating for 3 min. Then the sample tubes were sealed and centrifuged at 3500 rpm for 30 min to separate the phases. After centrifugation the ether layer was withdrawn and 1 μl was injected into the gas chromatograph.

Chromatography was performed on a 1 m × 2 mm I.D. glass column packed with 10% Silar-10C on 100-120 mesh Gas-Chrom Q. The injection port, oven, and detector temperatures were 200, 220, and 250°C, respectively. The elution times were 1.8 and 2.5 min for clofibrate and clofibric acid, respectively. The sensitivity was said to be 10 μg/ml with possible decrease to 1 μg/ml. The recovery rate was approximately 100%.

The given chromatogram of the substances after extraction from spiked plasma shows the clofibrate peak in the shoulder of the solvent peak, thus making good quantitative results almost impossible. Furthermore, the simultaneous determination of the two compounds is of questionable value insofar as no clofibrate levels could be detected after injection of the drug.

B. Routine Method Used in the Authors' Laboratory

1. Extraction and Derivatization

None of the above described methods has applied the use of capillary columns. Our own method is according to the extraction procedure based on the paper of Gugler and Jensen [41] with a few minor changes: To 1 ml of serum in a centrifuge tube with a Teflon-lined screw cap, 1 ml of 0.4 M hydrochloric acid and 6 ml of toluene are added. The toluene contains 20 μg/ml of the internal standard 2-naphthoic acid. For samples with an expected concentration of less than 20 μg/ml of CPIB, the concentration of the internal standard is decreased to 2 μg/ml. After a 5-min extraction in the Vortex mixer, the sample is centrifuged for 3 min at 2500 *g* (+4°C). 5 ml of the organic phase are transferred to a conical tube, and 50 μl of diluted Methelute TM (0.1 M trimethylanilinium hydroxide in water-methanol 1:1 v/v) are added with subsequent extraction by vortexing for 1 min. The sample is centrifuged for 3 min at 2000 *g* (+4°C), and 1 μl of the lower phase is injected into the gas chromatograph.

2. Chromatographic Conditions

A Carlo Erba RV 2150 gas chromatograph in connection with an FID was modified for working with capillary columns. A Grob-type split/splitless injector was equipped with a 4-mm I.D. glass liner filled (ca. 2 cm) with silanized glass wool. The chromatograph was equipped with a 50 m × 0.34 mm I.D. OV-1 glass capillary column (H. u. G. Jaeggi, Trogen, Switzerland) with a TZ of 55 for n-C_{13}/n-C_{14}. Carrier gas was hydrogen with a linear velocity of 60 cm/sec. Septum flush and split exit were 4 and 55 ml/min, respectively. Injector and detector temperatures were 300°C. Gas flow-rates for the FID were for hydrogen, 30 ml/min; for air, 350 ml/min; and for makeup gas (nitrogen), 25 ml/min. Injection (1 μl) of the samples was made at 50°C with the split closed for 20 sec. Column temperature was then raised immediately to 170°C. The retention times for the methyl esters of clofibric acid and 2-naphthoic acid were 6.25 and 8.87 min respectively.

Figure 19.3 shows a typical chromatogram of a human serum extract from a volunteer after intake of a therapeutic dose of clofibrate. This chromatogram

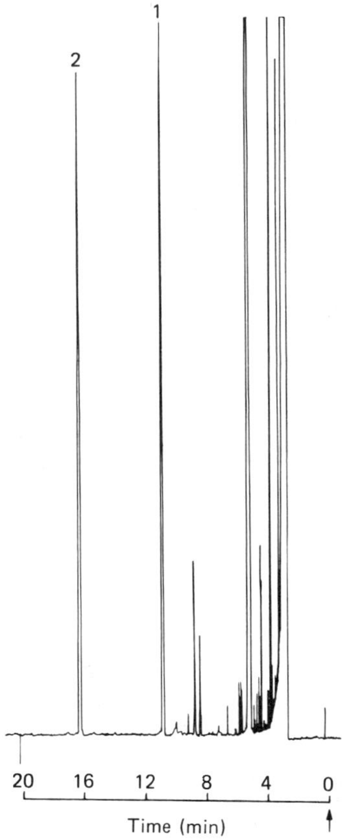

Figure 19.3 Gas chromatogram of a human serum extract after oral administration of 500 mg of clofibrate. Column: 50 m × 0.34 mm I.D. OV-1 glass capillary U_{H_2}: 35 cm/sec. Injection was made into the new column. 1 = Me-CPIB; 2 = internal standard (Me-2-naphthoic acid).

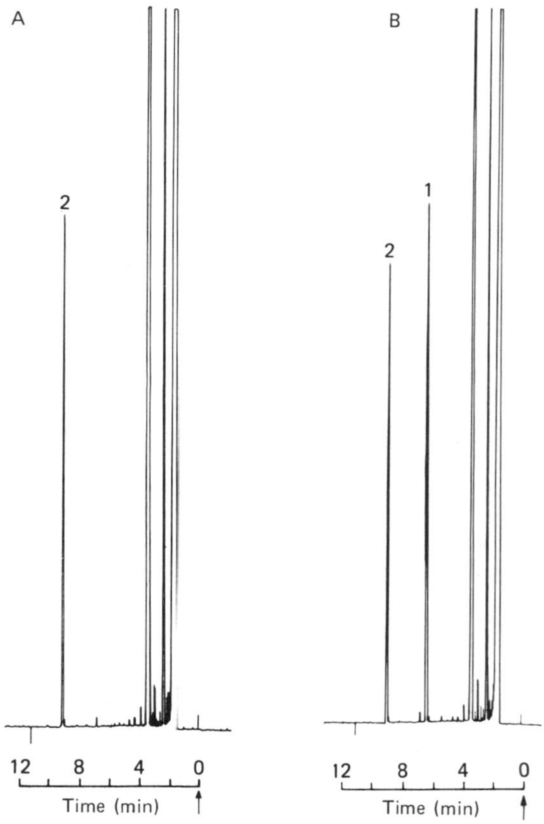

Figure 19.4 Gas chromatogram of human serum extracts (A) before drug administration; (B) after oral administration of 500 mg of clofibrate. Column: same as in Figure 19.3. U_{H2}: 60 cm/sec. GC run after more than 1400 injections. 1 = Me-CPIB; 2 = internal standard (Me-2-naphthoic acid).

was run after only a few times of injections in the new column: A decrease in carrier gas velocity (35 cm/sec) causes an increase of the retention times.

Figure 19.4 shows chromatograms after about 1400 injections on the same column. The excellent properties of the glass capillary remained unchanged. This is due to the fact that nonvolatile material remains in the glass wool of the replaceable inlet liner. In this context, it should be mentioned that the injected samples contained about 50% of water! The absence of any interfering peaks makes it possible to employ shorter columns (e.g., 20 m) in order to obtain shorter analysis times.

3. *Quantification*

Calibration curves for the quantitative determination of clofibric acid in serum have been obtained by addition of known quantities of clofibric acid to blank serum (Figures 19.5 and 19.6). The ratio between the peak area of the methyl esters of clofibric acid and the internal standard were calculated with a Spectra-Physics System 1 Integrator and plotted against the known concentration. Instead of peak areas, peak heights can be determined alternatively.

DATA

X	Y
0.	0.000
1.	0.058
1.	0.058
1.	0.059
2.	0.113
2.	0.113
2.	0.125
5.	0.293
5.	0.288
5.	0.288
10.	0.543
10.	0.547
10.	0.549
20.	1.026
20.	1.035
20.	0.992

Y = A + B*X

A =
0.020

B =
0.051

R-SQUARE =
0.998

RES ERROR
3.394

MAX(ABS(RESIDUAL))
0.039

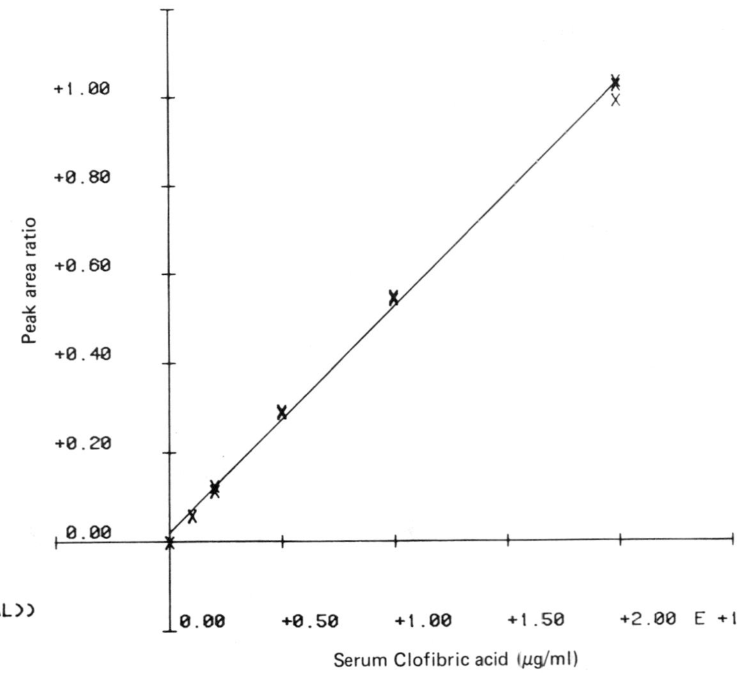

Figure 19.5 Calibration curve for serum CPIB in the concentration range 1-20 μg/ml. Internal standard: 12 μg/ml.

DATA

X	Y
0.	0.000
25.	0.123
25.	0.124
25.	0.127
50.	0.260
50.	0.256
50.	0.263
100.	0.578
100.	0.564
100.	0.591
200.	1.052
200.	1.082
200.	1.052

Y = A + B*X

A =
0.002

B =
0.005

R-SQUARE =
0.996

RES ERROR
6.206

MAX(ABS(RESIDUAL))
0.052

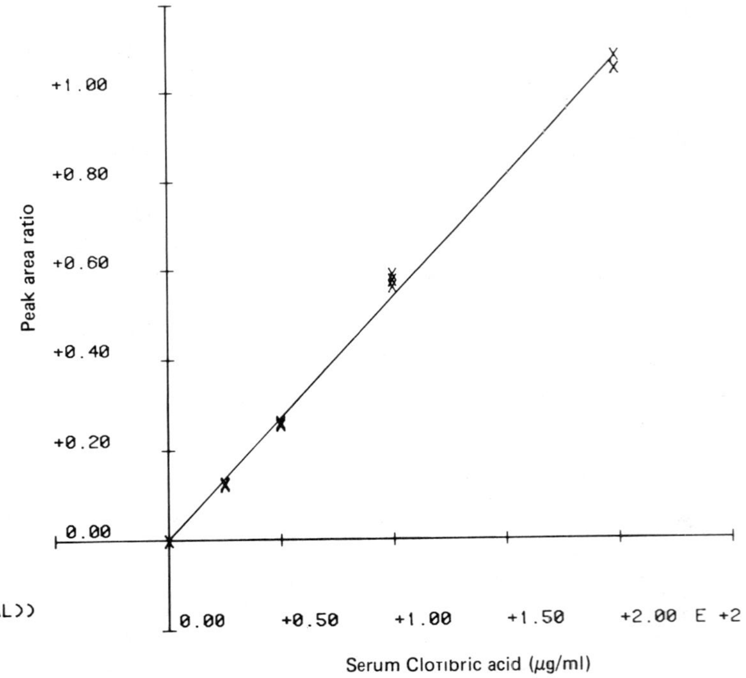

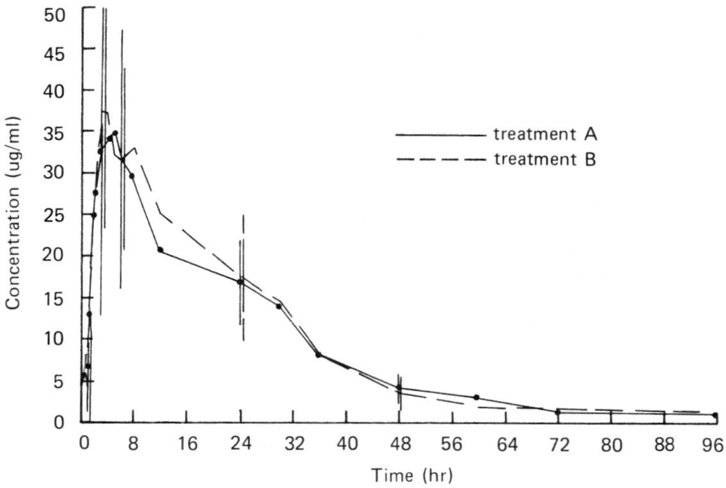

Figure 19.7 Mean serum CPIB levels of 12 normal human subjects after oral administration of one 500-mg clofibrate capsule. Two preparations of different manufacturers are tested in a cross-over study.

4. *Precision*

A within-day precision of the overall procedure (R.S.D.) of 1.5% was obtained on analyzing 5 serum samples spiked with 10 μg/ml of clofibric acid. The day-to-day precision (R.S.D.) over 4 weeks (n = 11) was 10% at a mean concentration of 51 ± 5.6 (S.D.) μg/ml obtained for 11 samples each spiked with 50 μg/ml of clofibric acid.

5. *Recovery*

Known amounts of clofibric acid were added to 1 ml of drug-free serum to give drug concentrations of 20, 50, and 100 μg/ml and extracted in the described way. Percentage recoveries were calculated from a standard curve of the same amounts of clofibric acid, derivatized in the injection port without extraction. A mean recovery of 99 ± 2.5% (S.D.) was found.

6. *Sensitivity*

Concentration of clofibric acid in serum as low as 0.5 μg/ml could be determined quantitatively.

7. *Application*

The utility of the method for the determination of clofibric acid in human serum was demonstrated in a comparison study of serum levels of 12 volunteers after oral administration of 500 mg of clofibrate (equivalent to 422.3 mg clofibric acid) in two different preparations. The mean serum clofibric acid levels of this study are shown in Figure 19.7.

Figure 19.6 Calibration curve for serum CPIB in the concentration range 25-200 μg/ml. Internal standard: 120 μg/ml.

REFERENCES

1. J. M. Thorp and W. S. Waring, *Nature 194*:948 (1962).
2. J. M. Thorp and W. S. Waring, *J. Atheroscler. Res. 3*:351 (1963).
3. M. M. Best and C. H. Duncan, *Amr. J. Cardiol. 15*:230 (1965).
4. E. H. Strisower, *J. Atheroscler. Res. 3*:445 (1963).
5. M. F. Oliver, *J. Atheroscler. Res. 3*:427 (1963).
6. M. F. Oliver, *Progr. Biochem. Pharmacol. 2*:315 (1976).
7. M. F. Oliver, *Br. Med. J. 4*:767 (1971).
8. D. B. Hunninghake and D. R. Tucker, *Circulation 39*:675 (1969).
9. R. I. Levy, D. S. Fredrickson, R. Schuhman, D. W. Bilheimer, J. L. Breslow, N. J. Stone, S. E. Lux, H. R. Sloan, R. M. Chrauss, and D. N. Herbert, *Ann. Intern. Med. 77*:267 (1972).
10. G. M. Grundy, E. M. Ahrens, G. Salen, P. M. Schreibmann, and P. J. Nestel, *J. Lipid Res. 13*:531 (1972).
11. E. Edmond, J. Daver, M. Daver, J. Blanchard, M. Ricoeur, and C. C. Tuong, *Thérapeutique 46*:55 (1970).
12. Coronary Drug Project, *J.A.M.A. 231*:360 (1975).
13. Report from the Committee of Principal Investigators, *Br. Heart. J. 40*: 1069 (1978).
14. R. I. Levy and T. Langer, in *Drugs Affecting Lipid Metabolism: Pharmacological Control of Lipid Metabolism,* Vol. 26 (W. L. S. Holmes, R. Paoletti, and D. Kritchervsky, eds.), Plenum Publishers, New York, 1972, pp. 155-163.
15. S. M. Grundy, E. M. Ahrens, Jr., G. Salen, and E. Quintao, *J. Clin. Invest. 48*:33a (1969).
16. B. J. Kudchodkar, M. S. Sodhi, L. Horlick, and D. T. Mason, *Clin. Pharmacol. Ther. 22*:154 (1977).
17. M. Berman, M. Hall, R. I. Levy, S. Eisenberg, D. W. Bilheimer, D. Phair, and R. M. Goebel, *J. Lipid Res. 19*:38 (1978).
18. M. F. Oliver, *N. Engl. J. Med. 299*:1360 (1978).
19. T. Langer and R. I. Levy, *N. Engl. J. Med. 279*:856 (1968).
20. Coronary Drug Project Research Group, *N. Engl. J. Med. 296*:1185 (1977).
21. J. C. Larosa, W. V. Brown, P. L. Frommer, and R. I. Levy, *Am. J. Cardiol. 23*:266 (1969).
22. J. M. Thorp, *Lancet I*:1323 (1962).
23.
24. A. Sedaghat and E. M. Ahrens, Jr., *Eur. J. Clin. Invest. 5*:177 (1975).
25. J. P. Tillement, R. Zini, P. D'Athis, and G. Vassent, *Eur. J. Clin. Pharmacol. 7*:307 (1974).
26. G. Honin, J. J. Thébault, Ph. D'Athis, J. P. Tillement, and J.-L. Beaumont, *Eur. J. Clin. Pharmacol. 8*:433 (1975).
27. D. B. Hunninghabe and D. L. Azarnoff, *Metabolism 17*:588 (1968).
28. A. M. Barrett and J. M. Thorp, *Br. J. Pharmacol. Chemother. 32*: 381 (1968).
29. L. Almirante, L. Bruseghini, and L. Faci, *Boll. Chim. Pharm. 108*:292 (1969).
30. E. L. Tohman, H. M. Tepperman, and J. Teppeman, *Am. J. Physiol. 218*:1313 (1970).
31. S. Silvestri, *Farmaco Ed. Prat. 25*:197 (1970).
32. M. G. Horning, R. M. Herbert, R. J. Roth, D. L. Davis, E. C. Horning E. P. Fischer, and G. J. Jordan,Jr., *Lipids 7*:114 (1972).

33. A. Dedaghat, H. Nakamura, and E. M. Ahrens, Jr., *J. Lipid Res. 15*:352 (1974).
34. F. Knüchel and M. Ochs, *Arzneim.-Forsch. (Drug Res.) 24*:576 (1974).
35. A. Berlin, *J. Pharm. Pharmac. 27*:54 (1975).
36. H. Jaeger, W. Nebelung, H. U. Klör, H. Ditschuneit, and H. Frank, in this volume, Chap. 13.
37. A. Karmen and H. Haut, *Biochem. Med. 12*:154 (1975).
38. T. C. Tuong and A. Tuong, *J. Chromatogr. 106*:97 (1975).
39. H. Bruderlein, W. T. Robinson, M. Kraml, and D. Dvornik, *Clin. Biochem. 8*:261 (1975).
40. R. Gugler, D. W. Shoeman, D. M. Huffman, J. B. Cohlmia, and D. L. Azarnoff, *J. Clin. Invest. 55*:1182 (1975).
41. R. Gugler and C. Jensen, *J. Chromatogr. 117*:175 (1976).
42. J. N. Connor, G. F. Johnson, and H. M. Solomon, *Clin. Chem. 22*:884 (1976).
43. M. S. Wolf and J. J. Zimmermann, *J. Pharm. Sci. 69*:92 (1980).

20

Glass Capillary Chromatographic Studies of the Metabolism of Drugs and Environmental Chemicals

Marjorie G. Horning, Wanda G. Stillwell*, K. Lertratanangkoon, and K. Halpaap-Wood / *Institute for Lipid Research, Baylor College of Medicine, Houston, Texas*

I. INTRODUCTION

In metabolic studies involving drugs and environmental chemicals, analytical procedures that provide multicomponent analysis are preferred. Most xenobiotics are metabolized by multiple alternate pathways involving several enzyme systems. Since it is not possible to predict which pathways will be affected when experimental parameters are altered, methods that provide quantitative profiles of metabolites rather than data on one or two metabolites are required.

Both glass capillary chromatography (GCC) and high-performance liquid chromatography (HPLC) can be used to profile metabolites in biological fluids and tissues. Because of the sensitivity and selectivity of detection and high

Current affiliation: Delsi/Nermag, Houston, Texas

resolution inherent in these procedures, quantitative multicomponent analyses can be carried out with great precision and accuracy. HPLC methods are most useful for profiling metabolites that can be detected by absorption in the ultraviolet or visible region. Detection is usually not a problem when glass capillary gas chromatography is used; any compound or derivative that can be volatilized can be detected with high sensitivity. However, certain types of reactive metabolites (*N*-oxides, diepoxides) may be unstable under gas chromatographic conditions, even after derivatization, and consequently more definitive profiles may be obtained by HPLC. Conversely, if derivatization stabilizes reactive metabolites, gas chromatographic (GC) analysis may be more useful than HPLC analysis. Fortunately the two methods are complementary; samples which are difficult to analyze by HPLC are frequently easy to analyze by GC and vice versa. The major advantage of gas chromatographic procedures, is that the analytical sample that is used for separation and quantification by glass capillary chromatography can also be used for identification by GC-MS. Although HPLC-MS systems are available, they are not in general use at the present time.

Urinary profiles obtained by either method are reproducible qualitatively and quantitatively for a given species under defined experimental conditions. When the experimental conditions are altered by changing drug dosage, route of administration, or by induction with phenobarbital or β-naphthoflavone, a new set of reproducible profiles is frequently obtained. Profiles provide characteristic patterns that indicate the number and relative amounts of metabolites. Additional valuable information can be obtained when profiles of metabolites are analyzed carefully by glass capillary chromatography and GC-MS. New metabolites are readily identified. Furthermore, the structures of the new metabolites frequently indicate that the compound under study is metabolized by pathways other than those already reported. More rarely, the structures of the metabolites may prove the existence of an entirely new pathway of metabolism.

By comparing profiles of the urinary metabolites excreted by male and female rats and mice, pathways of metabolism influenced by the species and sex of the animal can be defined. Differences in metabolism may be responsible for sex-dependent and interspecies differences in toxicity or pharmacological response.

II. EXPERIMENTAL PROCEDURES

Many specific methods have been developed for the isolation of individual drugs and their metabolites. The procedures described in this section were developed in our laboratory and are considered general procedures for studying the metabolism of a variety of drugs and environmental chemicals.

A. Isolation of Drugs and Their Metabolites from Biological Matrices

1. *Isolation of Nonacidic Drugs and Drug Metabolites by the Use of Salt-Solvent Pairs [1]*

Isolation from plasma or tissue homogenates: An aliquot (0.1-1 ml) of plasma or tissue homogenate (0.2-2.0 ml) was transferred to a 12-ml calibrated screw-capped centrifuge tube, diluted with 4 vols. of glass-distilled water and saturated with powdered anhydrous ammonium carbonate (only a few crystals of undissolved salt remained in the bottom of the tube). After adding 2-5 ml of ethyl acetate, the contents of the tube were mixed thoroughly by inversion and then agitated with a Vortex-type mixer for 20 sec. The phases were separated by centrifugation (15 min), and the organic phase was transferred with a Pasteur pipet to a screw-capped centrifuge (12-ml) tube. The extraction was repeated with 2-5 ml of solvent. The combined solvent extracts containing drugs and their metabolites were evaporated under a nitrogen stream. The residue was dissolved in 0.5 ml of methanol for conversion to derivatives and subsequent analysis by GC and GC-MS.

Isolation from urine or bile: 1-5 ml of urine or bile was transferred to a 12-ml calibrated screw-capped centrifuge tube, and powdered anhydrous ammonium carbonate was added to saturation. Drugs and their metabolites were extracted twice with 5-7 ml of ethyl acetate as described above. The combined extracts were evaporated under nitrogen and redissolved in methanol for subsequent analysis by GC or GC-MS.

2. *Isolation of Acidic Drugs and Drug Metabolites from Urine or Bile*

An aliquot (1-5 ml) of urine or bile was transferred to a 12-ml calibrated screw-capped centrifuge tube. Following acidification of the urine with sulfuric acid to pH 2, 5-7 ml of solvent (ethyl acetate or chloroform) was added, and after mixing thoroughly, the phases were separated by centrifugation. The organic phase was transferred to a screw-capped tube and the extraction repeated with 5-7 ml of solvent. The combined organic extracts were evaporated (nitrogen stream), and the residue of acidic drugs, including many glucuronides, was redissolved in methanol for conversion to derivatives suitable for GC analysis.

3. *Isolation of Drug Metabolites from Urine, Plasma, or Bile after Enzymatic Hydrolysis*

Metabolites present as glucuronide or sulfate conjugates were isolated following enzymatic hydrolysis of the urine, plasma, or bile; acid hydrolysis of conjugates is not recommended. After adding 200 mg of sodium acetate to an aliquot of urine (5-7 ml), plasma (1-2 ml), or bile (2-3 ml) and adjusting the pH to 4.5 with acetic acid, the urine was incubated with 0.1 ml of Glusulase (Endo Labs, Garden City, NY) for 8-20 hr at 37°C. The neutral and basic metabolites including those liberated by enzymatic hydrolysis were extracted by the ammonium carbonate-ethyl acetate procedure.

Acidic metabolites (for example, mercapturic acids), that are not hydrolyzed by Glusulase and remain in the ammonium carbonate-saturated aqueous phase, can also be isolated [2]. After careful acidification (pH 3) with sulfuric acid (3 N), and subsequent saturation with sodium chloride, the acidic metabolites were extracted with 4 × 5 ml of ethyl acetate. The combined ethyl acetate extracts were taken to dryness (nitrogen stream) and redissolved in a small volume of methanol for subsequent derivatization and GC analysis.

4. *Isolation of Drugs and Metabolites from Urine by DEAE-Sephadex Chromatography [3]*

This procedure is used to isolate water-soluble drugs and metabolites which are not readily extracted by organic solvents. The anion exchanger DEAE-Sephadex (A-25, Sigma) was cycled with 0.5 N hydrochloric acid, distilled water, 0.5 N sodium hydroxide solution, and distilled water as recommended by the supplier. An aqueous suspension of the exchanger was poured into a chromatographic column (1.3 cm I.D.) to give a column 9 cm in height. After washing the exchanger with 25 ml of water and with 200 ml of 0.5 M aqueous pyridine acetate solution (39.5 ml of pyridine and 30 ml of glacial acetic acid diluted with distilled water to 1 liter), 5-10 ml of urine (pH 4.5-5.0) were transferred to the column. Basic and neutral compounds were eluted by washing with 100 ml of glass-distilled water. An acidic fraction was eluted with 150 ml of 1.5 M pyridine acetate solution (119 ml of pyridine and 90 ml of glacial acetic acid diluted with glass-distilled water to 1 liter). After lyophilization, the two fractions were transferred with methanol to calibrated screw-capped centrifuge tubes and stored at -20°C until analyzed. The neutral and basic fraction contains urinary sugars and sugar alcohols which may interfere with drug and metabolite analyses. These compounds remain in the aqueous phase when the ammonium carbonate-ethyl acetate procedure is employed. The acidic fraction contains the urinary acids which may interfere with the analysis of some acidic drugs and their metabolites. However, most glucosiduronic acids, mercapturic acids, and related sulfur containing acids have longer retention times (GC) than the endogenous acidic metabolites.

B. Recovery of Drugs and Their Metabolites

To determine the recovery of drug metabolites from urine, bile, homogenates, or plasma, radiolabeled drugs were administered to rats, and appropriate samples of urine, bile, tissue, or blood were collected. The recovery of drugs and their metabolites was followed by comparing the total radioactivity in the biological sample with the radioactivity present in the various fractions.

Table 20.1 illustrates the recovery of phenanthrene metabolites from enzymatically hydrolyzed urine after administration of 10 mg of phenanthrene labeled with 13 μCi of [9-^{14}C] phenanthrene to rats; 0 to 24-, 24 to 48-, and 48 to 72-hr urine samples were collected.

Recovery data of this type provide valuable information about pathways of metabolism. For example, only 58% of the metabolites excreted in rat urine

Table 20.1 Recovery of Radioactive Phenanthrene Metabolites from Enzymatically Hydrolyzed Urine

	0-24 hr		24-48 hr		48-72 hr	
	mg	%[a]	mg	%[a]	mg	%[a]
Recovered as nonacidic metabolites	3.19	(58)	1.73	(61)	0.39	(43)
Recovered as acidic metabolites	1.87	(34)	0.71	(25)	0.26	(29)
Amount remaining in acidified aqueous phase	0.44	(8)	0.40	(14)	0.25	(28)
Total mg/24 hr[b]	5.50		2.84		0.91	

[a]% of total radioactivity in the urine samples.
[b]74% of administered dose was recovered from urine in 72 hr.
Source: K. Lertratanangkoom, M. G. Horning, B. S. Middleditch, W.-S. Tsang, and G. W. Griffin, *Drug Metab. Dispos. 10*:614 (1982).

during 0-24 hr were recovered as neutral metabolites after enzymatic hydrolysis. Acidic metabolites not hydrolyzed by Glusulase and soluble in ethyl acetate were present in all three urine samples and accounted for 25-34% of the urinary metabolites. Since sulfur-containing acids in addition to mercapturic acids have recently been identified as naphthalene metabolites [2], this fraction (which accounts for a third of the urinary metabolites) is of interest. In the 48- to 72-hr urine sample, 28% of the metabolites were not recovered but remained in the acidified aqueous phase after ethyl acetate extraction. However, the 28% was equivalent to only 250 μg, whereas the 14% in the 24- to 48-hr urine was equivalent to 400 μg. With this information, the investigator can decide if the structures of the metabolites remaining in the aqueous phase should be investigated.

Separation of metabolites into nonacidic and acidic metabolites by solvent extraction procedures is usually incomplete. In this example GC-MS analysis proved that traces (1-2%) of neutral metabolites of phenanthrene were present in the acidic extract. However, the procedure outlined provides an accurate and rapid estimation of the quantities of the metabolites present in the individual fractions and indicates the analytical approach that would be most effective in defining the metabolism of the drug under investigation.

C. Derivatization

The chemical properties of the drugs under study determine the type of derivative to be used. For example, we have found that *N,N'*-dimethyl derivatives of barbiturates have better gas chromatographic properties than the *N,N'*-ditrimethylsilyl derivatives. In the study of biphenyl and naphthalene metabolism, silylation with bis(trimethylsilylacetamide) (BSA) was preferred but in the study of quinidine metabolism, silylation with BSA and trimethylchlorosilane (TMCS) or trimethylsilylimidazole (TSIM) was necessary in order to convert the tertiary hydroxyl group of 3-hydroxyquinidine to a silyl derivative quantitatively. Although many different types of derivatives have been used, the procedures described are those found most useful in our laboratory.

1. *Silylation*

Trimethylsilyl (TMS) derivatives were prepared by evaporating the solvent and redissolving the residue of drug and metabolites in a small volume of pyridine. For the urinary aliquots described in the isolation procedures, usually 20-50 μl of pyridine were sufficient. The pyridine solution of metabolites was silylated with 30-60 μl of BSA and then heated at 60°C for 30-60 min. If the metabolites have hindered (tertiary) hydroxyl groups, TSIM or BSA and TMCS was used instead of BSA, and the silylation was carried out at 90°C for 1 hr. TSIM does not form *N*-silyl derivatives readily and can be used when the formation of *N*-silyl derivatives is undesirable.

2. *Methylation*

Methylated derivatives were prepared by dissolving the drug and metabolites in 0.5 ml methanol and adding an excess of ethereal diazomethane (prepared from Diazald). After standing at room temperature for 2-15 min, the excess ether, diazomethane, and methanol were removed with a nitrogen stream and the residue dissolved in pyridine and silylated with BSA.

3. *Ethylation*

If *N*-desmethyl as well as *N*-methyl metabolites are present in the biological sample, the sample should be ethylated instead of methylated. Ethylation converts the *N*-desmethyl metabolites to *N*-ethyl derivatives and phenolic hydroxyl groups to ethyl ethers. Thus it is possible to separate *N*-methyl from *N*-desmethyl metabolites and *O*-methylcatechol from catechol metabolites. This procedure was very useful in studies of the metabolism of methsuximide, mephenytoin, and hexobarbital [4,5].

Ethylated derivatives were prepared by treating an ethanolic solution (0.5 ml) of drugs and drug metabolites with an excess of an ethereal solution of diazoethane (prepared from *N*-ethyl-*N'*-nitro-*N*-nitrosoguanidine) for 15 min. After removal of the excess reagents, the residue was silylated as described above.

The preparation and use of diazoalkanes is hazardous, and for this reason other alkylating procedures are frequently employed. For example, barbiturates can be alkylated with methyl or ethyl iodide in acetone-solid potassium carbonate [6,7]. 5,5-Disubstituted hydantoins can be quantitatively alkylated with methyl or ethyl iodide in the presence of potassium hydroxide [8].

D. Gas Chromatography

Glass open-tubular columns 40-75 m in length were prepared by the method of German et al. [9]. SE-30 capillary columns are prepared by a two-step process on a silanized glass capillary 0.3 mm I.D. and 1.0 mm O.D. In the initial coating containing Silanox (Grade 101; Cabot, Boston, MA), a suspension of Silanox (2 g/100 ml) in carbon tetrachloride containing SE-30 (0.5 g/100 ml) is propelled through the column by a nitrogen stream. After evaporation of the solvent (nitrogen stream) a second coating is applied by passing a 2% solution of SE-30 in isooctane through the column. Columns prepared by this method have approximately 100,000 theoretical plates for tetracosane at 250° (HETP 0.60 mm); the highest value observed for a 60-m column was 150,000 theoretical plates (HETP 0.40 mm).

A precolumn packed with approximately 1 cm of 10% SE-30 on Gas Chrom Q and a gas phase splitter were included in the injection system [9]. The precolumn ensures complete evaporation of the sample before the splitter, and traps nonvolatile components and prevents decomposition of the sample. The life of the column is prolonged if the precolumn is changed frequently, particularly after injection of acidic samples. Capillaries prepared by this procedure are stable up to 325°C and will last 12-15 months.

E. Quantitative Analysis

For quantitative analyses an internal standard should be added to the biological sample preferably before the isolation of drugs and metabolites is initiated. If it is not possible or practical to add the internal standard before isolation, an external standard can be added to the extract or to the sample before or after derivatization. After determining response factors, linearity of response of the instrumental system, and reproducibility of the isolation procedure, quantitative analyses can be carried out.

If the metabolites to be quantified are not available as pure compounds for the determination of response factors, a response factor of 1 with respect to the internal standard is assumed; however, this should be stated clearly when

Table 20.2 Excretion of Biphenyl Metabolites in Noninduced Animals[a]

	Sprague-Dawley rat[b]	C_{57}BL/6Tex mouse[c]
2-Hydroxybiphenyl	1.5 ± 0.2	4.6 ± 0.08
3-Hydroxybiphenyl	1.8 ± 0.3	4.5 ± 0.2
4-Hydroxybiphenyl	28.5 ± 1.6	38.0 ± 4.3
2,3-Dihydroxybiphenyl	trace[d]	0.4 ± 0.06
2,5-Dihydroxybiphenyl	trace	1.9 ± 0.8
3-Methoxy-4-hydroxybiphenyl	<0.5	1.6 ± 0.4
2,4-Dihydroxybiphenyl	trace	1.4 ± 0.1
3,4-Dihydroxybiphenyl	5.1 ± 0.5	29.8 ± 1.9
3,4'-Dihydroxybiphenyl	8.5 ± 0.7	1.1 ± 0.3
4,4'-Dihydroxybiphenyl	44.5 ± 2.5	7.6 ± 1.5
3-Methoxy-4,4'-dihydroxybiphenyl	<1.0	0.9 ± 0.3
3,4,4'-Trihydroxybiphenyl	8.8 ± 2.3	7.4 ± 1.6

[a]Percent of total biphenyl metabolites excreted in urine expressed as the mean ± SE.
[b]n = 4.
[c]n = 5 groups of mice (6 in each group).
[d]less than 0.1%.
Source: Ref. 13.

discussing quantitative results. By assuming a response factor of 1, it is possible to determine if a particular metabolite has increased or decreased significantly in response to a change in one or more experimental parameters; that is, valid ratio measurements can be made.

Results from a quantitative study of biphenyl metabolism carried out in our laboratory are listed in Table 20.2. Most of the metabolites were available as pure compounds so that response factors with respect to the internal standard, 4-bromobiphenyl, could be determined. From the values in Table 20.2, it is clear that not only the animal experiments and isolation procedure but also the gas chromatographic analyses were very reproducible.

III. APPLICATIONS

Gas capillary chromatography can be used to separate and quantitate a number of drugs in a single analysis. Figure 20.1 shows the separation of eleven barbiturates as the *N,N'*-dimethyl derivatives. After selection of an appropriate internal standard, quantification of all 11 barbiturates or any subset of barbiturates can be carried out.

Gas capillary chromatography, however, is most informative when it is used to obtain profiles of drugs and their metabolites present in biological

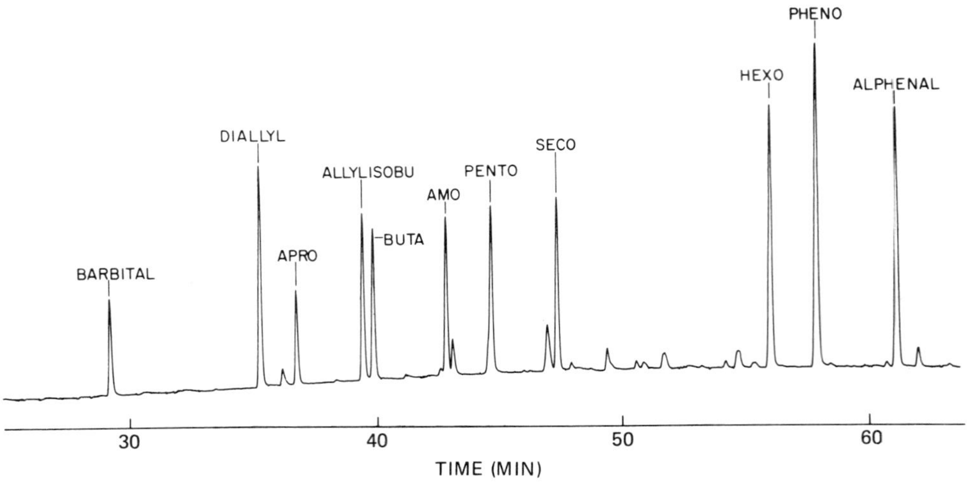

Figure 20.1 Separation of the N,N'-dimethyl derivatives of 11 barbiturates (barbital, diallylbarbituric acid, aprobarbital, allylisobutylbarbituric acid, butabarbital, amobarbital, pentobarbital, secobarbital, hexobarbital, phenobarbital, and alphenal). The analysis was carried out with a 60-M SE-30 glass capillary column by temperature programming at 2°/min form 100°C.

matrices; for example, urine, plasma, bile, or tissue homogenates. Typical profiles of drug metabolites are shown in Figures 20.2-20.4. A urinary profile is shown in Figure 20.2; the 24-hr urine sample was obtained from a patient maintained on phenytoin, phenobarbital, and methsuximide. A profile of naphthalene metabolites excreted in rat bile is shown in Figure 20.3. Figure 20.4 shows a profile of metabolites formed when a 9000 *g* supernatant fraction prepared from mouse liver was incubated with biphenyl. In these three examples, the ammonium carbonate-ethyl acetate procedure was used to isolate the metabolites.

When profiling procedures are employed, it is possible to identify new metabolites, to detect species differences in drug metabolism, to study the effects of inducing agents on drug metabolism, and to define new pathways of metabolism. Examples of each of these applications of profiling by capillary GC will be discussed.

A. Identification of New Metabolites

When urinary metabolites are profiled following administration of a large dose (10-100 mg/kg) of a neutral or basic drug or an environmental chemical to rats or mice, most of the peaks in the capillary GC analysis of the nonacidic fraction represent metabolites. Studies of naphthalene metabolism carried out in our laboratory provide an excellent example [10]. Figure 20.5a shows a gas chromatographic separation of the TMS derivatives of the neutral metabolites of naphthalene isolated from hydrolyzed rat urine after I.P. administration of naphthalene (100 mg/kg).

Of the 21 oxygenated metabolites characterized by capillary GC, 17 were identified as new metabolites. In addition to the previously reported metabolites, we identified two new dihydrodiols, the *cis*-1,2-dihydrodiol and

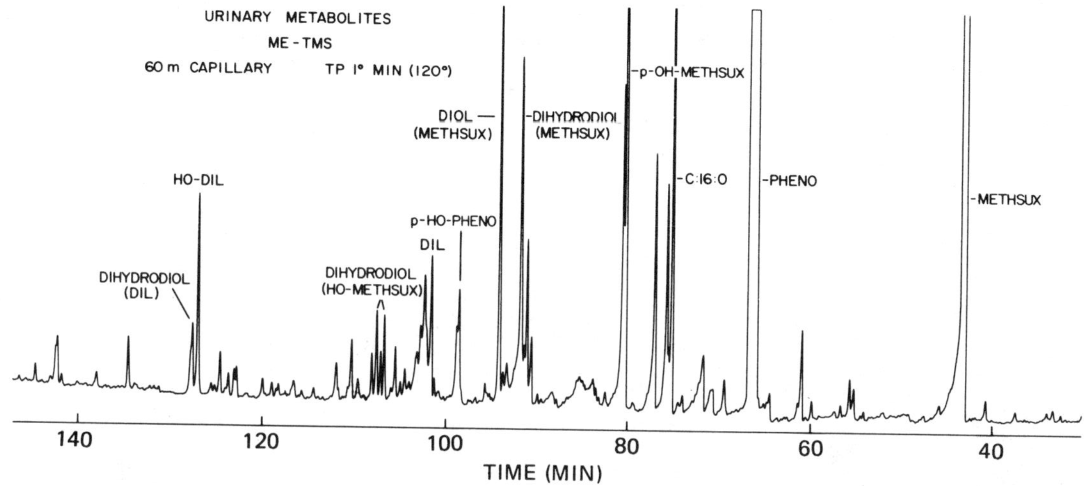

Figure 20.2 Separation of the urinary metabolites excreted by a patient treated with phenobarbital, methsuximide, and phenytoin (Dilantin). The metabolites identified (GC-MS) were methsuximide (METHSUX), phenobarbital (PHENO), *N*,2-dimethyl-2-(4-hydroxyphenyl)succinimide (*p*-OH-METHSUX), *N*,2-dimethyl-2-(3,4-dihydroxy-1,5-cyclohexadien-1-yl)succinimide (DIHYDRODIOL, METHSUX); *N*,2-dimethyl-2-(3,4-dihydroxyphenyl)succinimide (DIOL, METHSUX), 5-ethyl-5-(4-hydroxyphenyl)barbituric acid (*p*-HO-PHENO), phenytoin (DIL), two dihydrodiol metabolites of hydroxymethsuximide (DIHYDRODIOL, HO-METHSUX), 5-phenyl-5-(4-hydroxyphenyl)hydantoin) (HO-DIL), and 5-phenyl-5-(3,4-dihydroxy-1,5-cyclohexadien-1-yl)hydantoin (DIHYDRODIOL, DIL). Palmitic acid (C16:0) was also present in the GC tracing. The separation of the methylated and silylated derivatives was carried out by temperature programming at 1°/min from 120°C with a 60-m SE-30 glass capillary column.

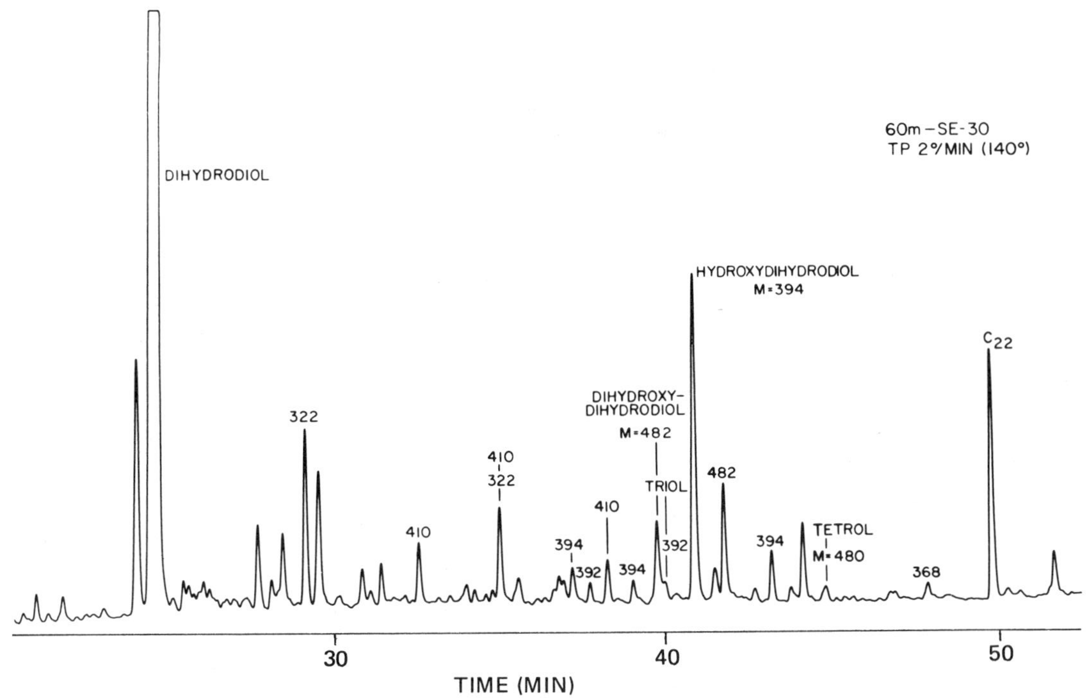

Figure 20.3 Analysis of the TMS derivatives of naphthalene metabolites excreted in rat bile. The metabolites identified were 1β,2α-dihydroxy-1,2-dihydronaphthalene (DIHYDRODIOL), 1,2,3,4-tetrahydroxy-1,2-dihydronaphthalene (DIHYDROXYDIHYDRODIOL), 1,2,4-trihydroxynaphthalene (TRIOL), 1,2,4-trihydroxy-1,2-dihydronaphthalene (HYDROXYDIHYDRODIOL), and 1,2,3,4-tetrahydroxynaphthalene (TETROL). The C-22 hydrocarbon (C_{22}) was added as external standard. The numbers above the peaks refer to the molecular weights of the TMS derivatives of additional naphthalene metabolites in rat bile.

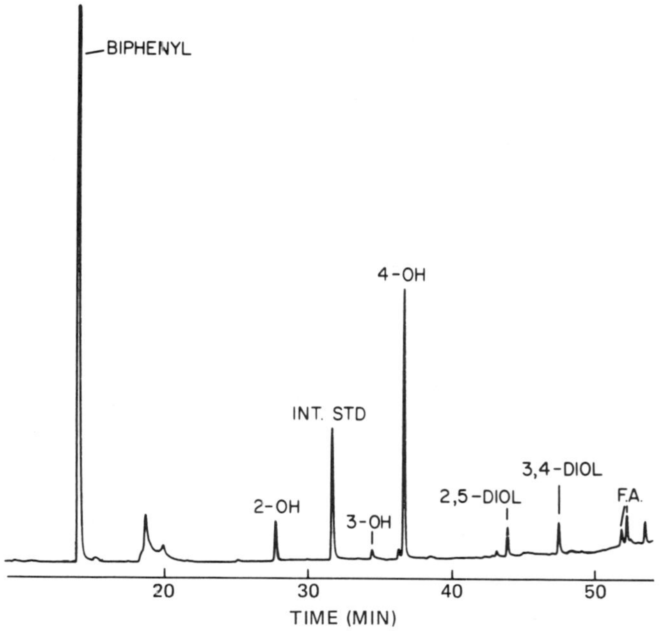

Figure 20.4 Separation of biphenyl metabolites formed by a 9000 *g* supernatant fraction from mouse liver. The metabolites identified were 2-hydroxy- (2-OH), 3-hydroxy- (3-OH), and 4-hydroxybiphenyl (4-OH), 2,5-dihydroxybiphenyl (2,5-DIOL), and 3,4-dihydroxybiphenyl (3,4-DIOL). Fatty acids (FA) were also present. The separation of TMS derivatives was carried out by temperature programming at 2°/min from 140°C on a 60-m SE-30 glass capillary column. (From Ref. 14.)

the *cis*-1,4-dihydrodiol, as well as a dihydrodiol epoxide, and di-, tri-, and tetrahydroxy derivatives of naphthalene. The methylene unit (MU) values for the metabolites are listed in Table 20.3. The mass spectrometric properties of these metabolites have been described [10].

In addition to the oxygenated metabolites, methylthio metabolites of naphthalene were characterized by capillary GC; the peaks in Figure 20.5a labeled with roman numerals were methylthio metabolites. These metabolites, which were excreted in larger quantities in mouse than in rat urine, were identified by GC-MS and by synthesis. Ten methylthio metabolites were isolated from rat urine [11] and six of these were also excreted in mouse urine. A total of 27 new metabolites of naphthalene, 17 in the oxygen series and 10 methylthio derivatives, were separated by capillary GC.

Naphthalene is an unusual example. In our experience quinidine would be considered a more typical example (Figure 20.6). Seven new metabolites in addition to *O*-desmethylquinidine and 3-hydroxyquinidine were detected in the urinary profile of the rat by capillary GC. The structure of the two diastereoisomeric 10,11-dihydrodiols was confirmed by synthesis [12]; the structures of the hydroxylated metabolites in peaks 3-7 are under study. Most quinidine samples also contain 5-20% of 10,11-dihydroquinidine (peak 1), and

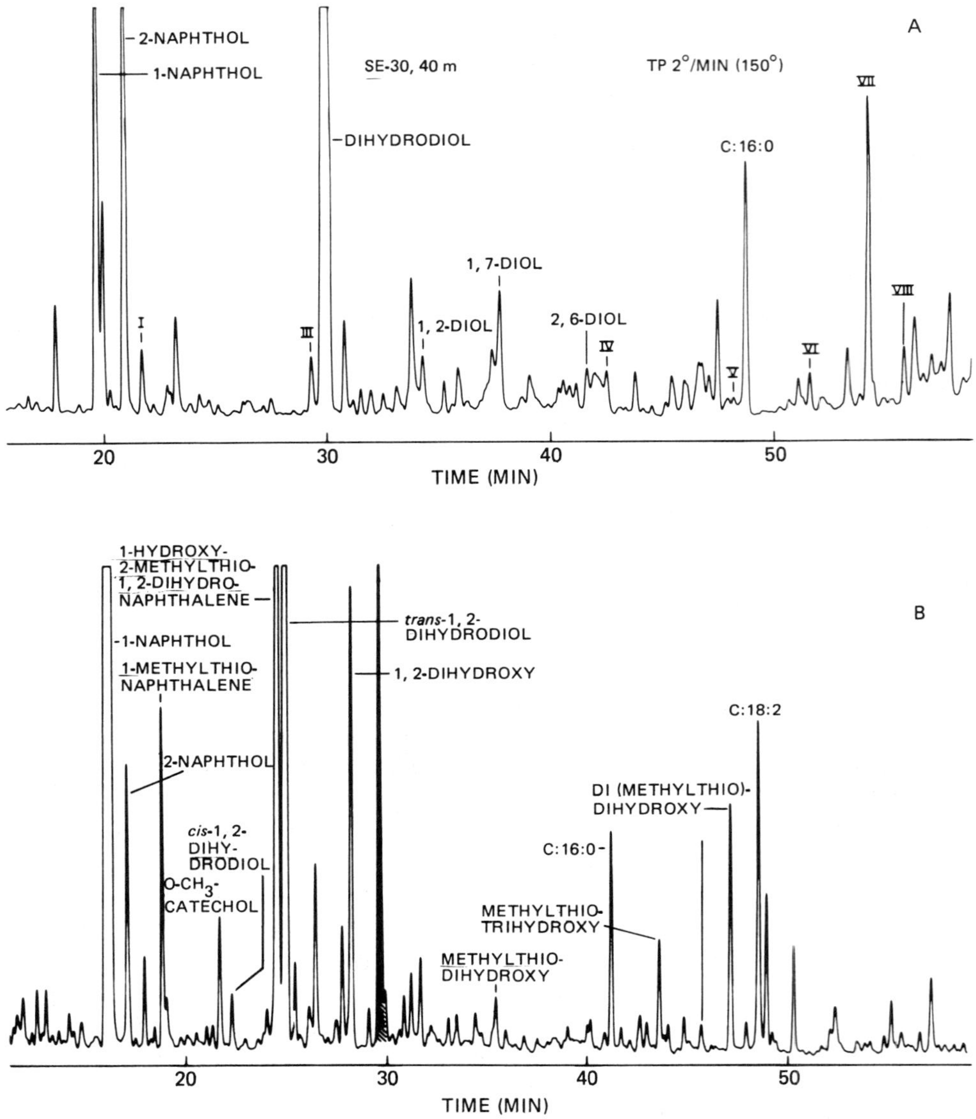

Figure 20.5 (A) Separation of neutral naphthalene metabolites present in hydrolyzed rat urine. The major metabolite is the *trans*-1,2-dihydrodiol. The peaks labeled with roman numerals are methylthio metabolites; VII is a di(methylthio)dihydroxy metabolite, and III is 1-hydroxy-2-methylthio-1,2-dihydronaphthalene. (From Ref. 10.) (B) Separation of the neutral metabolites of naphthalene present in hydrolyzed mouse urine; palmitic (C16:0) and linoleic acid (C18:2) were present in the tracing. The separations of TMS derivatives were carried out on a 60-m SE-30 glass capillary column by temperature programming at 2°/min from 150°C. [From Stillwell, Horning, Griffin and Tsang, *Drug Metab. Dispos. 10*:624 (1982).]

Table 20.3 Urinary Metabolites of Naphthalene

Metabolite	MU[a]
1-Naphthol (I)	15.2
2-Naphthol (II)	15.4
O-Methylcatechol (III)	16.4
1,2-Dihydroxynaphthalene (IV)	17.8
1,7-Dihydroxynaphthalene (V)	18.4
2,7-Dihydroxynaphthalene (VI)	18.9
2,6-Dihydroxynaphthalene (VII)	19.1
cis-1,2-Dihydrodiol (VIII)	16.6
trans-1,2-Dihydrodiol (IX)	17.1
cis-1,4-Dihydrodiol (X)	17.4
Dihydrodiol epoxide (XI) (1β,2α-Dihydroxy-3α,4α-epoxy-1,2,3,4-tetrahydronaphthalene)	19.1
1,2,4-Trihydroxynaphthalene (XII)	20.1
Trihydroxynaphthalene (XIII)	21.0
Trihydroxydihydronaphthalene (XIV)	18.6
Trihydroxydihydronaphthalene (XV)	19.6
Trihydroxydihydronaphthalene (XVI)	19.9
Tetrahydroxynaphthalene (XVII)	21.4
Tetrahydroxydihydronaphthalene (XVIII)	20.0
Tetrahydroxytetrahydronaphthalene (XIX)	19.8
Tetrahydroxytetrahydronaphthalene (XX)	20.25
Tetrahydroxytetrahydronaphthalene (XXI)	20.7

[a]A 60-m SE-30 glass capillary column was used to determine the MU values.
Source: Ref. 10.

the 10,11-dihydro analogues of several of the hydroxylated metabolites were excreted in urine. For example, peak 2 is hydroxydihydroquinidine, a metabolite of 10,11-dihydroquinidine.

A large number of metabolites are usually found for drugs that are *N*-methyl derivatives. During metabolism, extensive hydroxylation and *N*-demethylation occur, and as a result most of the hydroxylated metabolites are excreted as *N*-methyl and *N*-desmethyl derivatives [4]. The two classes of metabolites can be separated easily on a capillary column if the urinary extract is ethylated rather than methylated before silylation. An example of an ethylated and silylated urinary extract is shown in Figure 20.7; the 24-hr

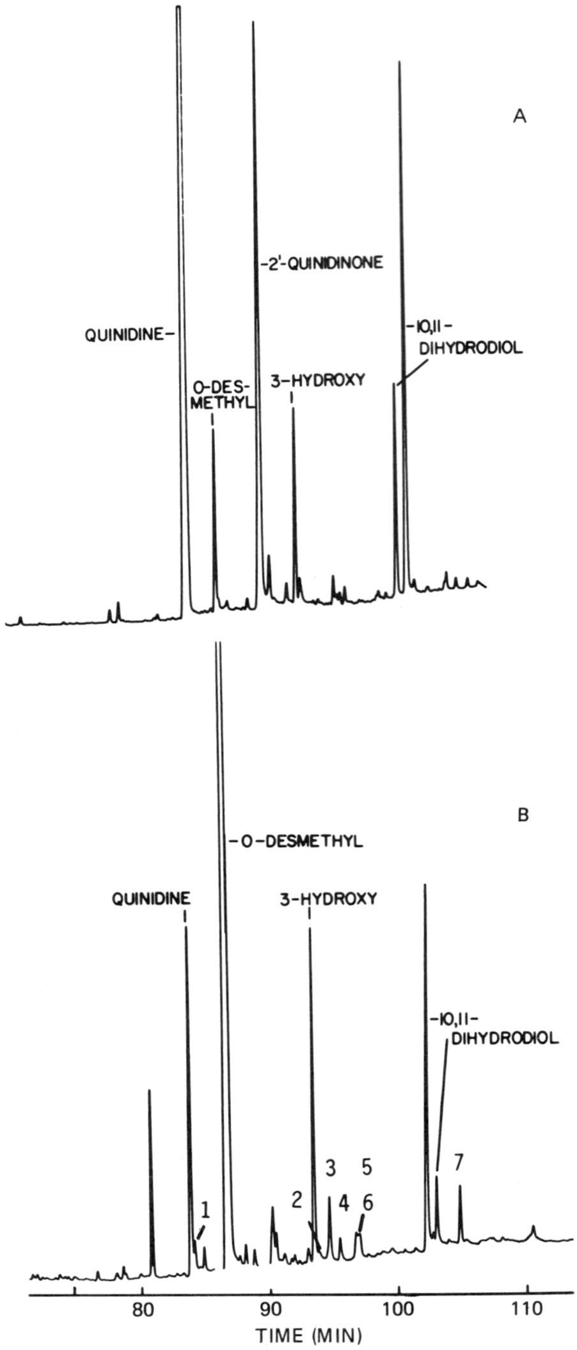

Figure 20.6 (A) Separation of quinidine metabolites present in hydrolyzed human urine. In addition to quinidine and 2'-quinidinone, *O*-desmethylquinidine (O-DESMETHYL), 3-hydroxyquinidine (3-HYDROXY), and the 10,11-dihydrodiol of quinidine were identified. The separation of the TMS derivatives was carried out on a 45-m SE-30 glass capillary column by temperature programming at 2°/min from 120°C. (B) Separation of quinidine metabolites present in hydrolyzed rat urine. The analysis of the TMS derivatives was carried out on a 75-m SE-30 glass capillary column by temperature programming at 2°/min from 120°C.

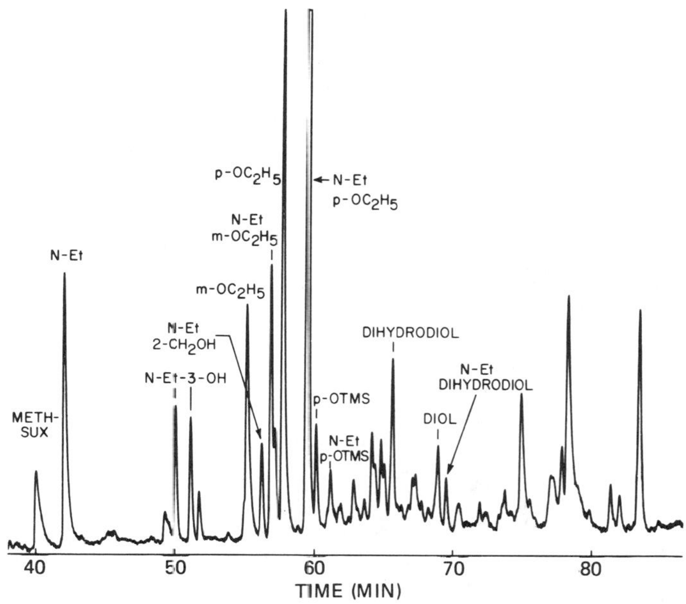

Figure 20.7 Separation of the ethylated and silylated derivatives of methsuximide metabolites present in hydrolyzed guinea pig urine. The demethylated metabolites detected were: 2-methyl-2-phenylsuccinimide (*N*-Et); the trimethylsilyl ethers of two diastereoisomeric 3-hydroxy-2-methyl-2-phenylsuccinimide (*N*-Et-3-OH), 2-hydroxymethyl-2-phenylsuccinimide (*N*-Et-2-CH_2OH), 2-methyl-2-(4-hydroxyphenyl)succinimide (*N*-Et-*p*-OTMS), and 2-methyl-2-(3,4-dihydroxy-1,5-cyclohexadien-1-yl)succinimide (*N*-Et-DIHYDRODIOL); and the ethyl ethers of 2-methyl-2-(3-hydroxyphenyl)succinimide (*N*-Et-*m*-OC_2H_5) and 2-methyl-2-(4-hydroxyphenyl)succinimide (*N*-Et-*p*-OC_2H_5); the trimethylsilyl ether derivatives of *N*,2-dimethyl-2-(4-hydroxyphenyl)succinimide (*p*-OTMS), the dihydrodiol and diol as well as the ethyl ethers of *N*,2-dimethyl-2-(3-hydroxyphenyl)-succinimide (*m*-OC_2H_5) and *N*,2-dimethyl-2-(4-hydroxyphenyl)succinimide (*p*-OC_2H_5). The separation of the ethylated and silylated derivatives was carried out by temperature programming from 100°C at 2°/min on a 60-m SE-30 column.

urine sample was obtained after i.p. administration of methsuximide (*N*,2-dimethyl-2-phenylsuccinimide) to a guinea pig. In addition to *N*-desmethylmethsuximide (2-methyl-2-phenylsuccinimide), the desmethyl analogue of *m*-hydroxy- and *p*-hydroxymethsuximide and the dihydrodiol and several diols of methsuximide were excreted in guinea pig urine. The major urinary metabolite was the *N*-desmethyl metabolite, 2-methyl-2-(4-hydroxyphenyl)-methsuximide (*N*-Et,*p*-OC_2H_5). When ethylated derivatives are prepared, the extent of *N*-demethylation of drugs such as mephobarbital, hexobarbital, mephenytoin, phensuximide, and methsuximide can be quantified.

B. Species Differences in Metabolism

In the naphthalene study, interesting differences in metabolism between the rat and mouse were observed. The *trans*-1,2-dihydrodiol (1β,2α-dihydroxy-1,2-dihydronaphthalene) was the major neutral metabolite in hydrolyzed rat

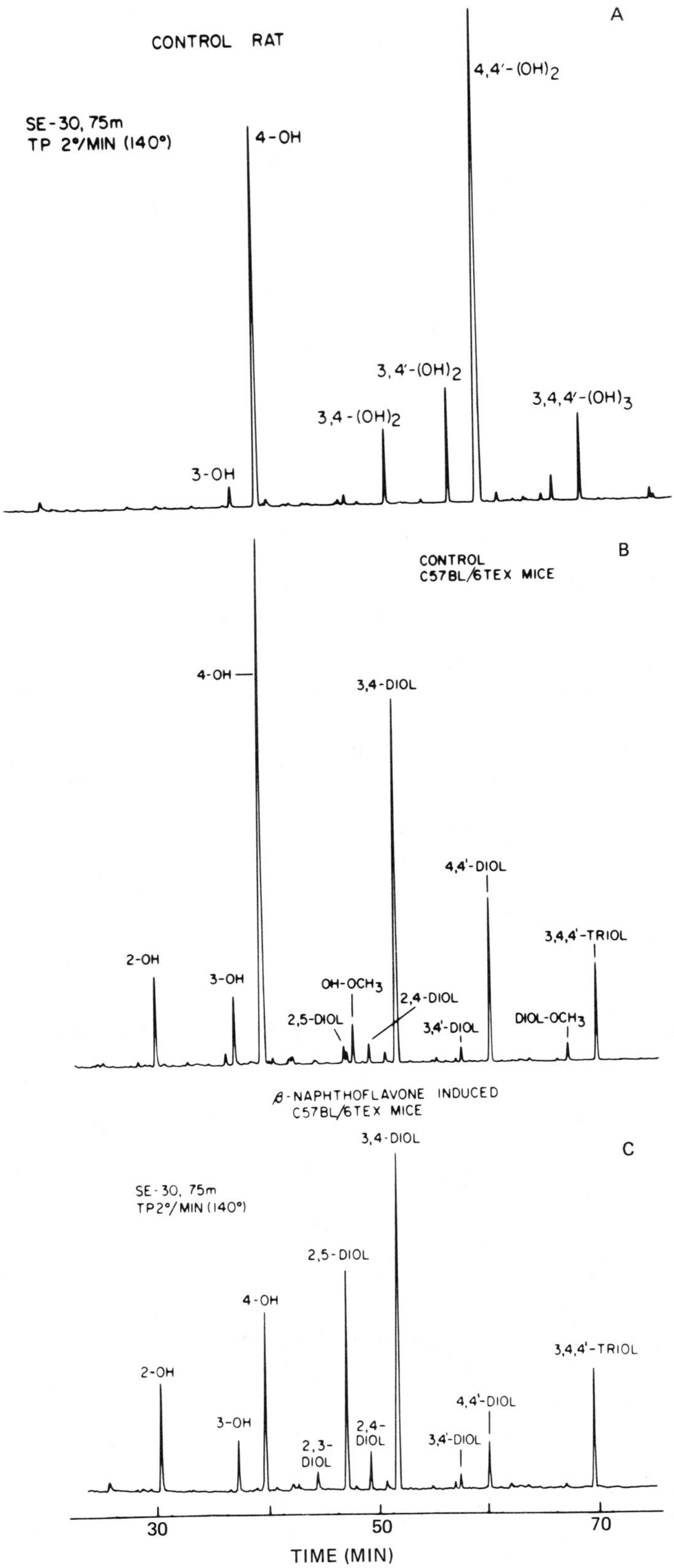

A
CONTROL RAT
SE-30, 75m
TP 2°/MIN (140°)
4-OH
4,4'-(OH)2
3,4'-(OH)2
3,4-(OH)2
3,4,4'-(OH)3
3-OH
B
CONTROL
C57BL/6TEX MICE
4-OH
3,4-DIOL
4,4'-DIOL
3,4,4'-TRIOL
2-OH
3-OH
OH-OCH3
2,5-DIOL
2,4-DIOL
3,4'-DIOL
DIOL-OCH3
β-NAPHTHOFLAVONE INDUCED
C57BL/6TEX MICE
3,4-DIOL
C
SE-30, 75m
TP2°/MIN (140°)
2,5-DIOL
4-OH
3,4,4'-TRIOL
2-OH
3-OH
2,3-DIOL
2,4-DIOL
3,4'-DIOL
4,4'-DIOL
30
50
70
TIME (MIN)

urine; only trace quantities of 1β-hydroxy-2α-methylthio-1,2-dihydronaphthalene were detected. However, in hydrolyzed mouse urine 1β-hydroxy-2α-methylthio-1,2-dihydronaphthalene was a major metabolite in the neutral fraction and was present in amounts comparable to the *trans*-1,2-dihydrodiol. In addition, 1-naphthol was apparently excreted in larger quantities by the mouse than the rat. However, both 1- and 2-naphthol are quite volatile, and for quantitative analysis the isolation procedure would have to be modified to ensure that loss of these volatile metabolites had not occurred. The marked differences in metabolism by the rat and mouse can be seen in Figure 20.5.

The metabolism of biphenyl provides an excellent example of species differences in metabolism (Figure 20.8). The metabolism was studied in Sprague-Dawley rats and in C_{57}Bl/6Tex and DBA/2Tex mice [13]. Both species excreted large quantities of 4-hydroxybiphenyl; this metabolite was the chief product of the first stage of oxidation in both rats and mice. The major product of the second stage of oxidation in the rat was 4,4'-dihydroxybiphenyl, which accounted for 45% of the urinary metabolites. The catechol 3,4-dihydroxybiphenyl was present in much smaller quantities (5% of the total metabolites in urine). In contrast, the catechol was the major diol excreted in the urine of C_{57}Bl/6Tex and DBA/2Tex mice and accounted for 30%, while 4,4'-dihydroxybiphenyl accounted for only 8-10% of the urinary metabolites (Table 20.2). Thus in the rat, second-stage oxidation occurs primarily on the unoxidized ring, while in the mouse second-stage oxidation occurs primarily on the ring already oxidized.

In vitro comparisons were made of the metabolism of biphenyl by 9000 *g* supernatant fractions prepared from livers of Sprague-Dawley rats and Swiss-Webster ICR mice [14]. The major metabolite of biphenyl formed by a 9000 *g* supernatant fraction of liver homogenates from noninduced rats and mice was 4-hydroxybiphenyl. Only very small amounts of dihydroxy metabolites were detected (Figure 20.4). Thus the species differences observed in vivo in noninduced animals were not observed when biphenyl metabolism was studied in vitro.

Quinidine provides another example of species difference (Figure 20.6). The major urinary metabolite of quinidine excreted by the rat following administration of a single acute dose (50 mg/kg, i.p.) was *O*-desmethylquinidine [12]. However, in the human urine sample the major urinary metabolite was 2'-quinidinone which was present only in trace quantities in rat urine. 3-Hydroxyquinidine was also present in human urine in significant quantities.

Figure 20.8 (A) Separation of the TMS derivatives of biphenyl metabolites present in hydrolyzed rat urine. (B) Separation of the TMS derivatives of biphenyl metabolites present in the hydrolyzed urine of C_{57}Bl/6Tex mice. (C) Separation of the TMS derivatives of biphenyl metabolites present in the hydrolyzed urine of C_{57}Bl/6Tex mice induced with β-naphthoflavone. The metabolites identified were 2-hydroxy-(2-OH), 3-hydroxy-(3-OH), and 4-hydroxybiphenyl (4-OH); 2,3-dihydroxy- (2,3-DIOL), 2,4-dihydroxy- (2,4-DIOL), 2,5-dihydroxy- (2,5-DIOL), 3,4-dihydroxy- (3,4-DIOL), 3,4'-dihydroxy-(3,4'-DIOL), and 4,4'-dihydroxybiphenyl (4,4'-DIOL); 3,4,4'-trihydroxybiphenyl (3,4,4'-TRIOL); 3-methoxy-4-hydroxybiphenyl ($OH-OCH_3$); and 3-methoxy-4,4'-dihydroxybiphenyl ($DIOL-OCH_3$). The separations were carried out by temperature programming from 140°C at 2°/min on a 75-m SE-30 glass capillary column. (A and B From Ref. 13.)

Since the three metabolites 3-hydroxyquinidine, O-desmethylquinidine, and 2-quinidinone are pharmacologically active [15], but may elicit different responses, these differences in metabolism are of interest.

The examples cited show that a large number of enzymatic reactions are involved in species differences in metabolism. These differences are most striking for compounds such as quinidine and biphenyl which are metabolized by multiple alternate pathways. Variations in toxicity and pharmacological activity which are observed when two or more species are compared may be due to differences in pathways of metabolism.

C. Induction Studies

The effects of inducing agents such as phenobarbital (a P-450 inducer) and β-naphthoflavone (a P_1-450 inducer) can be studied most effectively by the use of profiling procedures. Again, the most interesting results were obtained with compounds such as biphenyl, quinine, and quinidine which are metabolized by multiple alternate pathways.

Glass capillary chromatography was used to quantitate all of the metabolites in the studies of the effect of inducing agents in biphenyl metabolism [13]. Figure 20.8 compares the GC tracing of the biphenyl metabolites excreted by C_{57}Bl/6Tex control and β-naphthoflavone-induced mice. When quantitative analyses were carried out on 0- to 24-hr samples, it was found that the total quantity of metabolites excreted by control and β-naphthoflavone-induced mice was the same and amounted to 22 μmol/24 hr (40% of the administred dose). The most striking changes after induction were a decreased excretion of 4-hydroxybiphenyl ($p<0.02$) and an increased excretion of 2-hydroxybiphenyl ($p<0.05$) and of 2,5-dihydroxybiphenyl ($p<0.001$). The latter compound was excreted in only trace amounts by control C_{57}Bl/6Tex mice. Thus induction with β-naphthoflavone switched the metabolism from *p*-hydroxylation to *o*-hydroxylation; that is, from 4-hydroxy to 2-hydroxy and 2,5-dihydroxybiphenyl. Since DBA/2Tex mice lack the cytosolic receptor required for induction of P_1-450, they were not induced by β-naphthoflavone and as a result the total amount of urinary metabolites (22 μmol/24 hr) and the urinary profiles were not changed after induction [13].

When rats were induced with β-naphthoflavone, the excretion of metabolites increased from 32 to 54% of the administered dose. The major change was a considerable increase in catechol formation; that is, the metabolism was switched from [4-hydroxy- → 4,4'-dihydroxybiphenyl] to [4-hydroxy- → 3,4-dihydroxy → 3,4,4'-trihydroxybiphenyl]. The effect of β-naphthoflavone on *o*-hydroxylation in the rat was small. The increase in 2-hydroxybiphenyl excretion was only 0.8 μmol/24 hr ($p<0.001$) compared with 3.5 μmol/24 hr for the catechol metabolites ($p<0.01$) [14].

When rats and mice were induced with phenobarbital, the percent of the administered dose of biphenyl excreted in urine did not change significantly and the pattern of metabolites changed only slightly.

D. New Pathways of Metabolism

Frequently the identification of a new metabolite of a drug or environmental chemical proves that the compound is metabolized by additional pathways not considered in earlier studies. For example, the identification of the diastereoisomeric 10,11-dihydroxyquinidines (Figure 20.6) and 10,11-dihydroxyquinines demonstrated that the vinyl side-chains of quinidine and quinine are metabolized to 10,11-diols via 10,11-epoxides [12]. This is a well-established pathway of metabolism for allylic barbiturates [5] but was not a known pathway of metabolism for quinidine and quinine.

In addition, the identification of new metabolites may also indicate the existence of new pathways of metabolism. For example, in the recent studies with naphthalene a large number of new metabolites were identified, and two of the neutral and two of the acidic metabolites indicated that new pathways of metabolism were involved.

One of the neutral metabolites was identified as the *cis*-1,4-dihydrodiol (Table 20.3). The enzymatic reactions involved in the formation of this metabolite are unknown. In an earlier study Sims [16] reported that 1,4-epoxy-1,4-dihydronaphthalene was metabolized by rats to a 1,4-dihydrodiol, but he concluded that naphthalene was not metabolized to the 1,4-epoxide inasmuch as the 1,4-dihydrodiol was not found as a urinary metabolite in his study. As a result of our studies, the in vivo formation of a 1,4-oxide must be considered a very likely possibility.

A second neutral metabolite was identified in our laboratory as 1α,4β-di(methylthio)-2β,3α-dihydroxy-1,2,3,4-tetrahydronaphthalene (Di(methylthio)-dihydroxy, Figure 20.5]. This metabolite could have been formed only from the *anti*-diepoxide of naphthalene by the pathway 1β,2β-epoxy → 1β,2β:3α,4α-diepoxy → 1α-methylthio-2β-hydroxy-3α,4α-epoxy → 1α,4β-di(methylthio)-2β,3α-dihydroxy-1,2,3,4-tetrahydronaphthalene [10]. When the *anti*-diepoxide of naphthalene was administered to rats and mice, the major urinary metabolite was the di(methylthio)dihydroxynaphthalene. The identification of this compound as a urinary metabolite of both naphthalene and the *anti*-diepoxide of naphthalene establishes the diepoxide pathway of metabolism for naphthalene. It seems unlikely that this pathway would be limited to naphthalene, and the diepoxide pathway is probably involved in the metabolism of other polycyclic aromatic hydrocarbons.

A third novel pathway of metabolism was found when the acidic metabolites of naphthalene, isolated after enzymatic hydrolysis of rat and mouse urine, were analyzed by capillary GC (Figure 20.9). In addition to the premercapturic acid (II), which is a known metabolite [17], 1-hydroxy-1,2-dihydro-2-naphthalenyl-*S*-mercaptolactic acid (III) and 1-hydroxy-1,2-dihydro-2-naphthalenyl-*S*-mercaptoacetic acid (IV) were identified (Figure 20.10). These are nonaromatic derivatives of naphthalene.

The pathway outlined in Figure 20.10 is not unique for naphthalene; phenanthrene also is metabolized to a corresponding series of 9-hydroxy-9,10-dihydro-10-mercapturic, mercaptolactic, and mercaptoacetic acids.

In addition to the nonaromatic sulfur-containing acids, the aromatic mercapturic acid (VII), mercaptolactic acid (VIII), and mercaptoacetic acid (IV) derivatives were identified. Compounds VII, VIII, and IX are formed by the facile loss of water and migration of the sulfur-containing moiety under acidic conditions. When the urines were carefully acidified to pH 3, only small amounts of VII, VIII, and IX were present in the GC tracing. Thus, in our study the aromatic sulfur acids VII, VIII, and IX appear to be artifacts produced during isolation rather than metabolites.

In contrast, in a study of halogenated aromatic compounds, Mio [18] has found only the aromatic sulfur-containing acids. For example, 6-mercapturic, 6-mercaptolactic, 6-mercaptopyruvic, and 6-mercaptoacetic acid derivatives of 1-hydroxy-2,4-dichlorobenzene were identified as metabolites of 1,3,5-trichlorobenzene.

The sulfur-containing aromatic acidic metabolites of halogenated compounds appear to be metabolites, whereas the aromatic sulfur-containing acidic

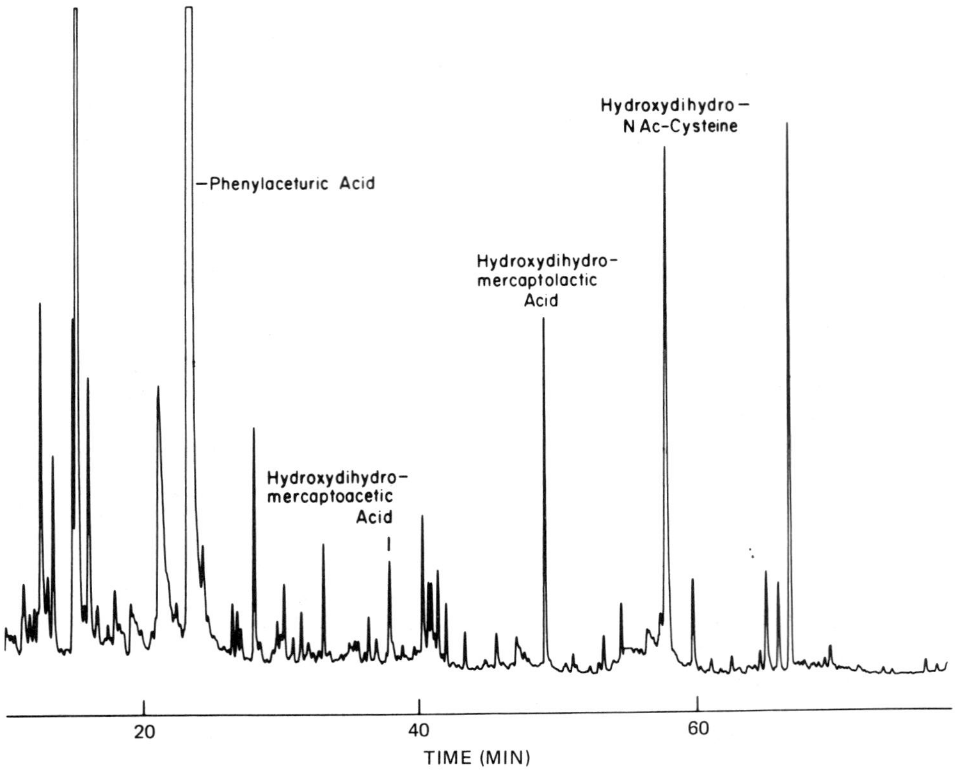

Figure 20.9 Separation of the methyl ester-trimethylsilyl ether (Me-TMS) derivatives of the acidic metabolites of naphthalene present in hydrolyzed mouse urine. The metabolites identified were 1-hydroxy-1,2-dihydro-2-naphthalenyl-*S*-mercaptoacetic acid (HYDROXYDIHYDROMERCAPTOACETIC ACID), 1-hydroxy-1,2-dihydro-2-naphthalenyl-*S*-mercaptolactic acid (HYDROXYDIHYDROMERCAPTOLACTIC ACID), and *N*-acetyl-*S*-(1-hydroxy-1,2-dihydro-2-naphthalenyl)cysteine (HYDROXYDIHYDRO-N-Ac-CYSTEINE). The separation was carried out by temperature programming at 3°/min from 120°C on a 48-m SE-30 glass capillary column. [From G. W. Stilwell, M. G. Horning, G. W. Griffin, and W. E. Tsang, *Drug Metab. Dispos. 10*:614 (1982).]

derivatives fromed via epoxide derivatives of unsaturated hydrocarbons appear to be artifacts formed during isolation. Thus there are two pathways: one involving halogenated aromatic compounds, in which a replacement of a halogen atom by a sulfur-containing nucleophile results in the formation of *aromatic* sulfur-containing acids; and a second pathway defined in our laboratory, in which an epoxide intermediate reacts with a sulfur-containing nucleophile (for example, glutathione, cysteine, or *N*-acetyl cysteine) to form a series of *non-aromatic* sulfur-containing acids.

From these examples it is clear that premercapturic acids and mercapturic acids are not terminal metabolites of glutathione conjugates but that either the cysteine conjugate or premercapturic acid is converted into a series of acidic and neutral sulfur-containing metabolites that account for as much as 50% of the metabolites excreted in urine.

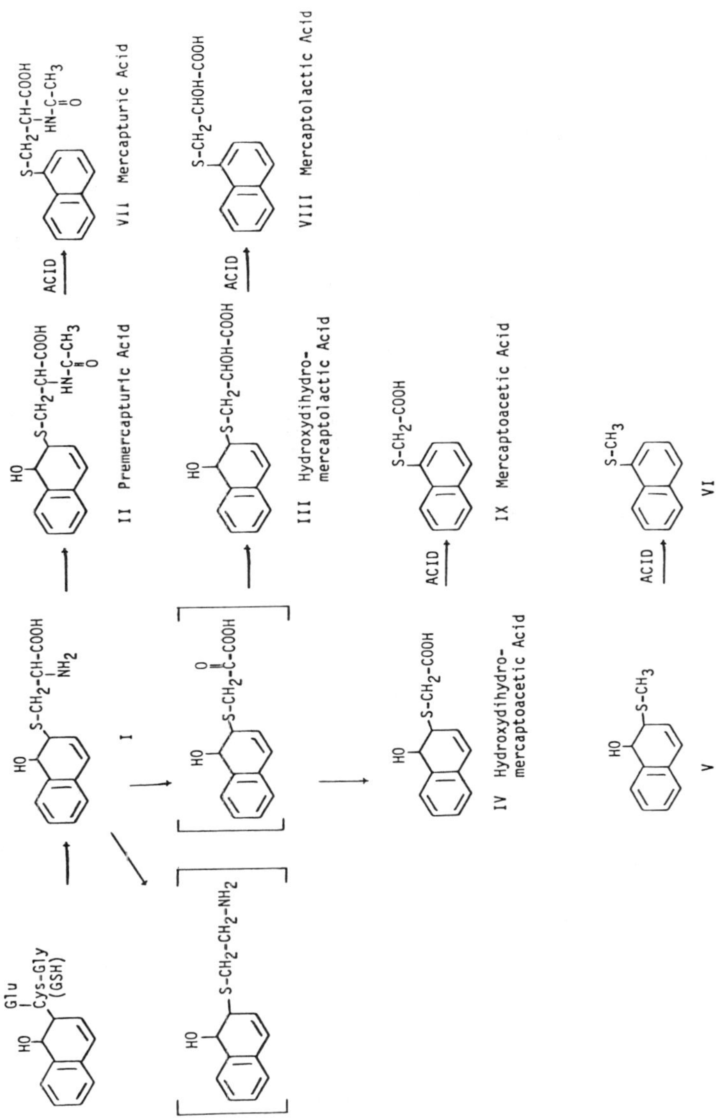

Figure 20.10 Proposed pathway of metabolism of the glutathione conjugate formed from naphthalene epoxide. [From G. W. Stillwell, M. G. Horning, G. W. Griffin, and W. S. Tsang, *Drug Metab. Dispos. 10*:614 (1982).]

IV. SUMMARY

The procedures described for the profiling of metabolites of drugs and environmental chemicals have been in use in our laboratory for over 10 years. The usefulness of the methods, however, increased greatly when stable, reproducible glass capillary columns became readily available. For profile analyses to be meaningful, the resolution provided by capillary columns is essential. The profiles in this chapter were selected to illustrate the effect of species differences and induction on metabolism, and the delineation of new pathways of metabolism. Many additional applications could have been illustrated. Profiles can be used to show sex differences in metabolism, to demonstrate the effect of route of administration on metabolism, and to show the effect of drugs and environmental chemicals on the profiles of specific classes of endogeneous metabolites (sugars, urinary acids, steroids). It was our intent to demonstrate that profile analysis by glass capillary chromatography provides a better and more comprehensive picture of the metabolism of drugs and environmental chemicals than other analytical procedures presently in use.

REFERENCES

1. M. G. Horning, P. Gregory, J. Nowlin, M. Stafford, K. Lertratanangkoon, C. Butler, W. G. Stillwell, and R. M. Hill, *Clin. Chem. 20*: 282 (1974).
2. W. G. Stillwell, M. G. Horning, W.-S. Tsang, and G. W. Griffin, *Pharmacologist 22*:215 (1980).
3. P. I. Jaakonmaki, K-L. Knox, E. C. Horning, and M. G. Horning, *Eur. J. Pharmacol. 1*:63 (1967).
4. M. G. Horning, C. M. Butler, K. Lertratanangkoon, R. M. Hill, T. E. Zion, and P. Kellaway, in *Quantitative Analytic Studies in Epilepsy* (P. Kellaway and I. Petersen, eds.), Raven Press, New York, 1976, p. 95.
5. D. J. Harvey, L. Glazener, D. B. Johnson, C. M. Butler, and M. G. Horning, *Drug Metab. Dispos. 5*:527 (1977).
6. W. Dunges and E. Bergheim-Irps, *Anal. Lett. 6*:185 (1973).
7. W. Dunges, *Chromatographia 6*:196 (1973).
8. R. de Sagher, J. Pocius, and J.-P. Thenot, *J. Chromatogr. 156*:43 (1978).
9. A. L. German, C. D. Pfaffenberger, J.-P. Thenot, M. G. Horning, and E. C. Horning. *Anal. Chem. 45*:930 (1973).
10. M. G. Horning, W. G. Stillwell, G. W. Griffin, and W.-S. Tsang. *Drug Metab. Dispos. 8*:404 (1980).
11. W. G. Stillwell, O. J. Bouwsma, J.-P. Thenot, M. G. Horning, G. W. Griffin, K. Ishikawa, and M. Takaku, *Res. Commun. Chem. Pathol. Pharmacol. 20*:509 (1978).
12. S. E. Barrow, A. A. Taylor, E. C. Horning, and M. G. Horning, *J. Chromatogr. Biomed. Appl. 181*:219 (1980).
13. K. Halpaap-Wood, E. C. Horning, and M. G. Horning. *Drug Metab. Dispos., 9*:97 (1981).
14. K. Halpaap-Wood, E. C. Horning, and M. G. Horning, *Drug Metab. Dispos. 9*:103 (1981).

15. D. E. Drayer, D. T. Lowenthal, K. M. Restivo, A. Schwartz, C. E. Cook, and M. M. Reidenberg, *Clin. Pharmacol. Ther. 24*:31 (1978).
16. P. Sims, *Biochem. J. 95*:608 (1965).
17. E. Boyland, G. S. Ramsay, and P. Sims, *Biochem. J. 78*:376 (1961).
18. T. Mio, Proceedings of the Fourth Meeting of the Japanese Society for Mass Spectrometry, p. 65 (1979).

21

Improved Determination of Isosorbide Dinitrate and Its Metabolites by Capillary Column GC

Dieter Lutz, Johannes Rasper, Werner Gielsdorf, James A. Settlage, and Halvor Jaeger / *L.A.B. Gesellschaft für pharmakologische Untersuchungen m.b.H. & Co. (Laboratory for Pharmacological Research), Neu-Ulm, Federal Republic of Germany*

I. INTRODUCTION

Isosorbide dinitrate (ISDN) is an organic nitrate coronary vasodilator found therapeutically useful in various cardiovascular diseases like angina pectoris and congestive heart failure [1,2].

ISDN is rapidly and extensively metabolized by a glutathione-dependent enzyme system in the liver. This "first-pass" effect leaves only a fraction

---

This chapter is reprinted from *J. High. Resol. Chrom. Chrom. Comm.*, 7: 58-65 (1984).

of the administered oral dose in blood as unchanged ISDN [3-5]. Its metabolites, isosorbide-5-mononitrate (IS-5-MN) and isosorbide-2-mononitrate (IS-2-MN), which both contribute to the pharmacological activity of ISDN, circulate in the blood in much higher concentrations, persist longer and are probably responsible for the role of the drug as a relatively long-acting vasodilator [6,7]. Therefore, it is of great interest to measure plasma levels of all three compounds together, but only a few methods have yet been described [8-14].

In five procedures, conventional packed column GLC was used; however, none of these assays led to satisfactory sensitivity and selectivity. Large sample volumes (3-5 ml) were required [8,11,12], and difficulties occurred with the quantitative mononitrate determination because of peak interferences, mainly by IS-5-MN [11]. In two assays no internal standard was used [10-11].

Two procedures employed capillary gas-liquid chromatography and showed excellent sensitivity and selectivity both for ISDN and the mononitrates [13-14]: In one instance [13], isoidide mononitrate (IIMM) and in the other [14] isomannide mononitrate (IMMN) were used as internal standards for the isosorbide mononitrates determination.

The present study shows, however, that IMMN and IIMN both occur in human plasma per se with peak levels of about 10% of that of IS-2-MN, after oral application of ISDN, particularly in the case of applying large doses of ISDN. Therefore *o*-nitrobenzyl alcohol, which is used in our method, is the better choice as an internal standard for the ISDN metabolites.

The present paper describes an improved simultaneous determination of ISDN and its metabolites in human plasma with an appropriate precision, specifity, and sensitivity in routine analysis of large number of samples. The method was used to obtain more detailed information about the pharmacokinetics of ISDN in humans.

II. EXPERIMENTAL PROCEDURES

A. Chemicals and Reagents

Crystalline isosorbide dinitrate (ISDN), isosorbide-2-mononitrate (IS-2-MN), isosorbide-5-mononitrate (IS-5-MN), isomannide dinitrate (IMDN), and isoidide dinitrate (IIDN) were kindly supplied by Pharma Sanol (Monheim, West Germany) and were homogeneous, checked by GC and TLC. Isomannide mononitrate (IMMN) and isoidide mononitrate (IIMN) were synthesized according to a method described elsewhere [15]. 1,3-Dinitrobenzene (*m*-DNB) was purchased from Serva (Heidelberg, GFR) and was of analytical reagent grade. *o*-Nitrobenzyl alcohol (*o*-NBA) was used as obtained from EGA-Chemie (Steinheim, West Germany). Dichloromethane and ethyl acetate (Nanograde; distilled in glass) were obtained from Mallinkrodt (St. Louis, MO, USA). All glassware was silanized before use with surfasil TM (Pierce, Rockford Ill, USA).

B. Apparatus

The gas chromatograph used was a Fractovap 4160 equipped with a 10-mCi ^{63}Ni-electron capture detector and a Grob-type split/splitless injector, containing a 4 mm × 8 cm glass liner (Carlo Erba, Milan, Italy). Injections were made with an automatic liquid sampler, Model ALS 3641 (DANI, Monza, Italy), which also controlled the split flow by means of a solenoid valve (Brechbühler, Schlieren, Switzerland). The column was a 30 m × 0.32 mm I.D. WCOT fused-silica capillary, with a nonextractable bonded stationary phase DB-5 of 0.25-μm film thickness (J. G. W. Scientific, Rancho Cordova, USA).

Carrier gas was hydrogen, and the detector was purged with 30 ml/min of argon/methane (90/10). Carrier and purge gas were always purified by an Oxisorb kit (Messer-Griesheim, Frankfurt, West Germany). Injector and detector temperatures were 150°C and 220°C, respectively. Injections were made at an oven temperature of 60°C with a closed split exit during the sampling period of 45 sec. Then the temperature was increased ballistically to 120°C with a 3-min hold and subsequent programming at 3°C/min to 195°C.

Measurement of peak areas were done with an integrating printer/plotter (Shimadzu, Model CR 1 B).

C. Extraction

To 1 ml of plasma in a 15-ml glass centrifuge tube with a Teflon-lined screw cap, 50 μl of an ethanolic solution, containing both the internal standards isomannide dinitrate (20 ng) and *o*-nitrobenzyl alcohol (350 ng), were added and the tube was thoroughly mixed for 15 sec. Then 10 ml of dichloromethane were added and the mixture was shaken mechanically for 30 min at 30 rpm. After centrifugation (10 min, 3000 rpm), the top layer was discarded, 8 ml of the organic layer were transferred to a conical glass tube and evaporated to dryness under a stream of nitrogen. The evaporation procedure took place in an ice bath in order to avoid losses of the nitrate esters. The residue was diluted in 300 μl of ethyl acetate, containing *m*-dinitrobenzene (0.6 ng/μl) as a volume-marker and transferred into the autosampler vials; 1 μl was injected into the chromatograph.

D. Quantification

Concentrations of ISDN, IS-2-MN, and IS-5-MN were determined by comparing the peak area ratio (peak area of isosorbide nitrate ester/peak area of internal standard) to the corresponding peak areas of the calibration curves. The two internal standards for ISDN and the mononitrates were isomannide dinitrate and *o*-nitrobenzyl alcohol, respectively. The calibration curves were obtained by extracting plasma samples in the described way, spiked with increasing amounts of ISDN (0.5-75.0 ng/ml), IS-2-MN (2-175 ng/ml), and IS-5-MN (5-450 ng/ml). Calibration curves were prepared every day.

E. Human Study

Twelve healthy volunteers, who had not received any medication for 2 weeks prior to the initiation of the study, and none besides ISDN throughout the duration of the study, were given one tablet of 40 mg isosorbide dinitrate as a sustained release formulation. Blood samples were withdrawn by venipuncture into heparinized tubes before and 0.33, 0.66, 1, 1.5, 2, 3, 4, 6, 8, 12, and 24 hr after administration of the drug. Plasma was immediately separated by centrifugation and stored at -20°C until analysis.

III. RESULTS AND DISCUSSION

A. Extraction and Internal Standards

Table 21.1 presents the absolute extraction recovery with dichloromethane and pentane as the extraction medium. Dichloromethane, as used by M. Ahnoff

Table 21.1 Absolute Extraction Recovery from Plasma with Dichloromethane in Comparison to Pentane as Solvents

Compound	Amount added (ng/ml)	Absolute extraction recovery (%)	
		CH_2Cl_2	Pentane
ISDN	10	89	82
IS-2-MN	50	93	5
IS-5-MN	200	90	5
o-NBA	350	100	n.d.[a]
IMDN	20	94	60

[a]n.d. = not determined.

et al. [14], is a very favorable solvent for all compounds listed, whereas the minor polar pentane is suitable only for ISDN.

The application of a capillary column with its increased separation efficiency, compared to the conventional packed columns, favors the application of internal standards which are structurally very similar to the drug. Hence, for ISDN, which has one nitrate group in exo, the other in endo position, the isomeric IMDN, with both nitrate groups in the endo position, is the internal standard of choice. In the case of IS-2-MN and IS-5-MN, the diastereoisomeric mononitrates IMMN and IIMN, respectively, appear ideal for internal standards and are used by several other authors [13,14,16].

However, after high oral doses of ISDN (at the present time, therapeutical doses of 100-800 mg of ISDN per day are not unusual [17]), two more metabolites appear in the plasma besides unchanged ISDN, IS-2-MN, and IS-5-MN; namely IIMN and IMMN.

Table 21.2 shows plasma levels of IIMN and IMMN of a healthy volunteer after repeated administration of 40 mg ISDN as a standard release formulation under steady-state conditions. IMMN can be determined in plasma up to 12 hr, whereas IIMN, due to its low elimination half-life, already decreases after 9 hr below the detection limit of approximately 2 ng/ml. The peak levels for IMMN and IIMN were 8% and 12%, respectively, of those for IS-2-MN.

Figure 21.1 shows the plasma level curves of the three metabolites IS-2-MN, IMMN, and IIMN. Our results are confirmed by investigations of the ISDN metabolism in rats, performed by Rossel et al. [3]: After oral application of ISDN, considerable amounts of IIMN could be detected in the urine of the animals. Since these authors used a packed column which was unsuitable for the separation of IIMN/IMMN, it is quite possible that, besides IIMN, there was also IMMN! Even with high-resolution capillary GLC this separation could not be achieved in all cases, as mentioned below.

Our results clearly point out that neither IMMN nor IIMN can generally be used as internal standards. However, in two cases these compounds can be recommended as internal standards without scruple:

Table 21.2 Plasma Levels of Isosorbide-2-mononitrate (IS-2-MN), Isomannide Mononitrate (IMMN), and Isoidide Mononitrate (IIMN) in a Subject After Repeated Oral Administration of 40 mg Isosorbide Dinitrate in a Normal Release Formulation

Time	IS-2MN	IIMN	IMMN
0[a]	0	0	0
0[b]	23.8	2.3	3.4
0.33	71.9	9.4	5.8
0.66	113.3	12.3	7.6
1	116.6	11.1	7.7
1.5	112.1	10.9	8.6
2	99.8	7.9	8.1
3.	90.3	6.1	7.2
4	50.5	2.9	5.5
6	30.1	2.0	4.1
9	11.6	n.d.[c]	3.1
12	5.7	n.d.	2.6
24	2.3	n.d.	n.d.

[a]Before the first application of the drug.
[b]Six hours after the last but one application (40 mg ISDN) and immediately before the last application of the drug (40 mg ISDN).
[c]n.d. = not detectable (below 2 ng/ml).

1. The applied dose of ISDN has to be low enough so that the plasma levels of IMMN and IIMN, formed by biotransformation, remain below the detection limit of 2 ng/ml.
2. The mononitrates IS-2-MN and/or IS-5-MN are the parent drugs administered and not formed by ISDN.

For these reasons, we employed the *o*-nitrobenzyl alcohol as an internal standard for the metabolites of ISDN. Smith et al. [18] were the first to use this compound for the determination of the isosorbide mononitrates. But these authors received good results on packed columns only after derivatization of the mononitrates. Furthermore, this method is not suitable for ISDN.

The occurrence of IMMN and IIMN in human plasma after high ISDN doses is of interest not only from the analytical but also from the pharmalogical point of view. Bogaert et al. [6] reported the vasodilatory potency of IIMN in the

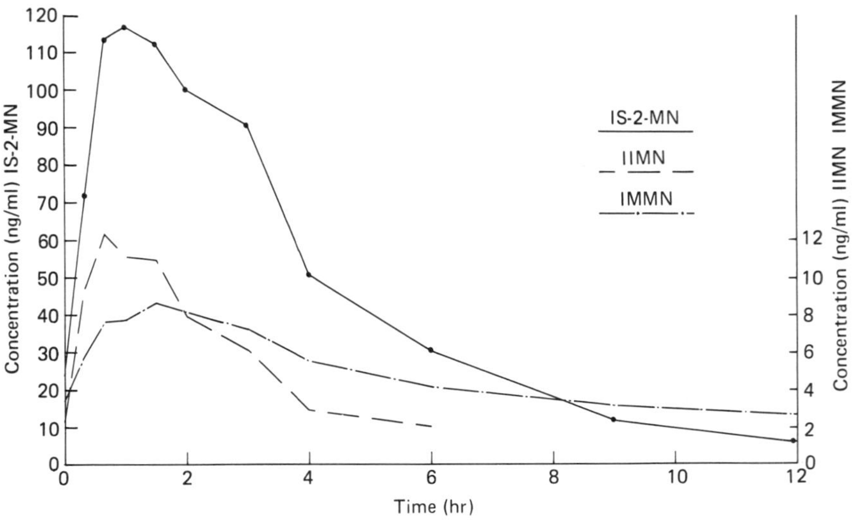

Figure 21.1 Plasma levels of IS-2-MN, IMMN, and IIMN after repeated oral administration of 40 mg ISDN (normal release).

dog, which was approximately equivalent to that of ISDN and corresponds to about two- to threefold that of IS-2-MN at the same dosage. This and our results lead to the suggestion that the pharmacological activity of the ISDN is due not only to the parent drug ISDN, IS-2-MN, and IS-5-MN, but in addition, at least in the case of large ISDN doses, also to IIMN and IMMN!

B. Recovery and Precision

The within-day reproducibility was checked on 5 identical plasma samples, which were analyzed by the described method, each spiked with ISDN, IS-2-MN, and IS-5-MN in various concentrations. The internal standard for ISDN was IMDN, and for the mononitrate metabolites was *o*-nitrobenzyl alcohol. The results are shown in Table 21.3.

The between-days reproducibility was checked during 8 days, on plasma samples spiked with several different amounts of these three compounds. The results are shown in Table 21.4. Tables 21.3 and 21.4 show the excellent reproducibility and accuracy of the method for the simultaneous determination of ISDN and its metabolites. However, these tables also demonstrate the lower day-to-day reproducibility of the assay for IS-5-MN, compared to its satisfactory within-day precision. This might be due to the pronounced adsorption behavior of IS-5-MN towards active glass surfaces. For this reason we strongly recommend daily preparation of calibration curves. In this context, it is of immence importance to keep the glass linear of the injector in a clean condition.

In Table 21.3, the relative standard deviations are given with m-DNB, as an external standard, which was added to the sample after the extraction procedure. In any case the precision is essentially poorer than with the internal standards which were added to the sample before extraction.

Table 21.3 Precision and Recovery in the Within-Day Determinations of ISDN, IS-2-MN, and IS-5MN in Spiked Plasma

Compound	Amount added (ng/ml)	Mean amount found (ng/ml)	No. of determinations	C.V.[a] (±%)	C.V.[b] (±%)	Recovery (%)
ISDN	2	2.1	5	1.7	2.9	105.0
	10	9.8	5	1.8	3.9	98.0
	50	50.1	5	3.6	6.7	100.2
IS-2-MN	10	11.8	5	5.0	7.6	118.0
	50	46.7	5	4.5	9.1	93.4
	100	101.4	5	1.8	8.5	101.4
IS-5-MN	50	54.3	5	3.6	8.3	108.6
	200	192.4	5	2.6	9.8	96.2
	400	403.2	5	2.3	8.7	100.8

[a]Quantification with internal standards.
[b]Quantification with external standard *m*-DNB.

Table 21.4 Precision and Recovery in the Day-to-Day Determinations of ISDN, IS-2-MN, and IS-5-MN in Spiked Plasma

Compound	Amount added (ng/ml)	Mean amount found (ng/ml)	No. of determinations	C.V. (±%)	Recovery (%)
ISDN	5	4.6	8	5.7	92.0
	10	9.7	8	5.6	97.0
	20	19.3	8	3.9	96.3
IS-2-MN	10	11.0	8	3.7	110.0
	25	23.8	8	4.6	95.2
	50	50.2	8	4.1	100.4
IS-5-MN	50	51.1	8	11	102.2
	100	97.4	8	10	97.4
	200	201.5	8	14	100.8

C. Calibration Curves and Sensitivity

The calibration curves are linear in the range 0.5-75 ng/ml ISDN, 2-175 ng/ml IS-2-MN, and 5-450 ng/ml IS-5-MN. They can be described by the equation $y = ax + b$ with a/b for ISDN, IS-2-MN, and IS-5-MN equal to 0.051/-0.013, 0.027/-0.046, and 0.019/-0.015, respectively, and with the coefficient of correlation, r^2, equal to 0.9996, 0.9995, and 0.9988, respectively.

At higher plasma concentrations, a corresponding dilution of the extraction residue with up to 600 μl ethyl acetate has to be done prior to the injection into the gas chromatograph; respectively smaller plasma volumes (<1 ml) are to be extracted in order to stay within the linear part of the calibration curve. This is particularly true for IS-5-MN, which reaches concentrations higher than 450 ng/ml after therapeutical doses of ISDN of 80 mg and more.

The lower limits of detection are 0.5 ng/ml of plasma for ISDN; 1 ng/ml for IS-2-MN, 2 ng/ml for IIMN and IMMN; and 3 ng/ml for IS-5-MN, if 1 ml of plasma is extracted and the final volume is 300 μl.

D. Chromatographic System

The sample injection process is always a critical step in quantitative high-resolution GC analysis. For trace analysis, splitless injection is a sampling technique which can be preformed with an autosampler, and therefore day-and-night operation is possible. Using an autosampler, 36 samples can be analyzed per day.

The chromatograms shown in Figures 21.2, 21.3, and 21.5 were run on a fused-silica capillary column with a TZ of 36 and a coating efficiency of 95.7% for methyl decanoate/methyl undecanoate. At a carrier gas velocity of 38 cm/sec, IMMN and IIMN were not separated, whereas all other compounds were sufficiently separated from endogeneous plasma peaks. IMMN eluted together with IIMN at a temperature of 138°C.

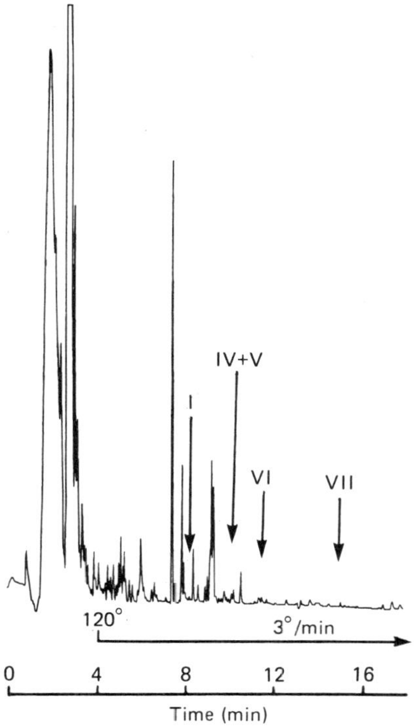

Figure 21.2 Human predose plasma extract. Column 30 m × 0.32 mm I.D. fused silica capillary, ocated with 0.25-μm bonded film of DB-5 (equivalent to SE-54); splitless injection, absence of internal standards. u_{H_2} = 38 cm/sec.

Figure 21.4 shows a chromatogram of the same sample shown in Figure 21.3 on a second column of the same manufacturer with a TZ of 36.6 and a coating efficiency of 97.3%. Increasing the carrier gas velocity to 71 cm/sec caused IMMN and IIMN to elute as well-separated peaks with an elution temperature of 123°C and 124°C, respectively.

A well-known disadvantage associated with packed columns concerns the irreversible adsorption of the mononitrates by active surfaces of the GC Supports [11,14,18]. Efforts were made to overcome this difficulty, in one case by utilizing a highly loaded (30% SE-30) column [11], but problems arose with detector fouling, due to increased bleeding of the liquid phase [18]. In the other case, derivatization of the mononitrates was performed in order to improve their chromatographic properties [18]. One of the essential features of fused silica capillary columns, besides its increased resolution power, is a high degree of inertness. The capillaries applied in the present study are sufficiently inert to detect even a few picograms of underivatized mononitrate injected into the gas chromatograph!

In order to avoid undesired phase stripping at the column entrance using the the splitless injection technique, we recommend capillaries with nonextractable "bonded" phases rather than columns with normal coated phases.

The excellent properties of bonded-phase fused-silica capillaries after more than 400 splitless injections of plasma extracts are demonstrated in Figure 21.6. Only a small increase in tailing of the IS-5-MN peak is detectable. The

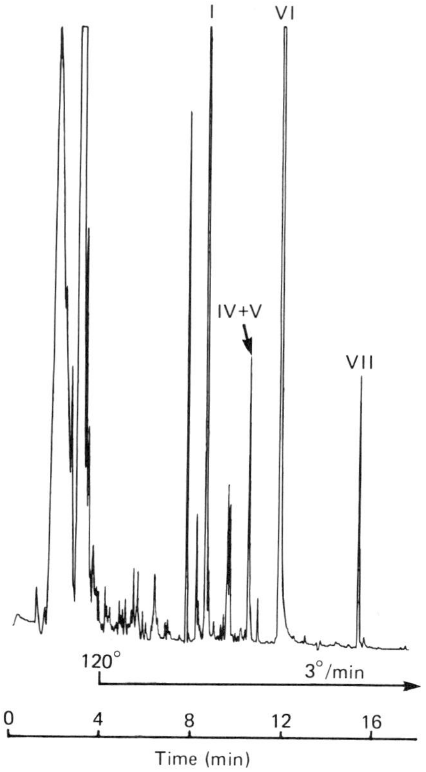

Figure 21.3 Human plasma extract after oral ISDN administration. Absence of internal standards. Column: same as in Figure 21.2, u_{H_2} = 38 cm/sec; IMMN and IIMN not separated. For plasma concentrations of compounds, see Figure 21.5.

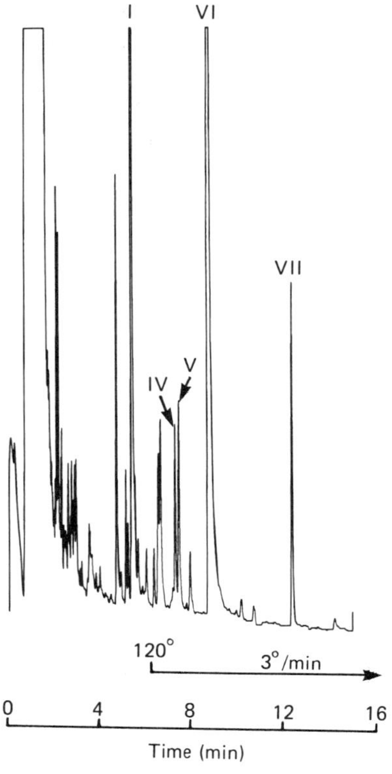

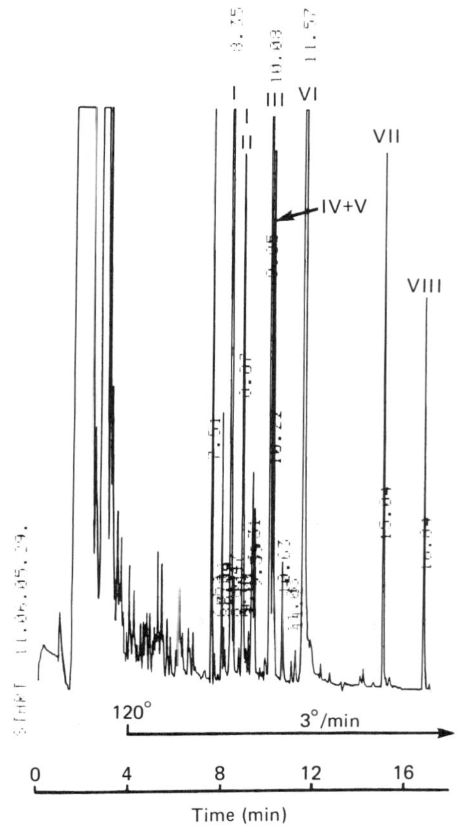

Figure 21.5 Human plasma extract after oral ISDN administration. Plasma sample: same as in Figure 21.3, but with internal standards. Column: same as in Figure 21.2, u_{H_2} = 38 cm/sec. I = IS-2-MN (121.7 ng/ml); II = *o*-NBA (internal standard); III = *m*-DNB (volume marker); IV = IMMN, V = IIMN, VI = IS-5-MN (595.7 ng/ml); VII = ISDN (25.9 ng/ml); VIII = IMDN (internal standard). IMMN and IIMN not separated (21.7 ng/ml both together).

Figure 21.4 Human plasma extract after oral ISDN administration. Sample: same as in Figure 21.3. Column: same as in Figure 21.2 but with an increased separation efficiency for IMMN and IIMN. u_{H_2} = 71 cm/sec. For peak identification, see Figure 21.5.

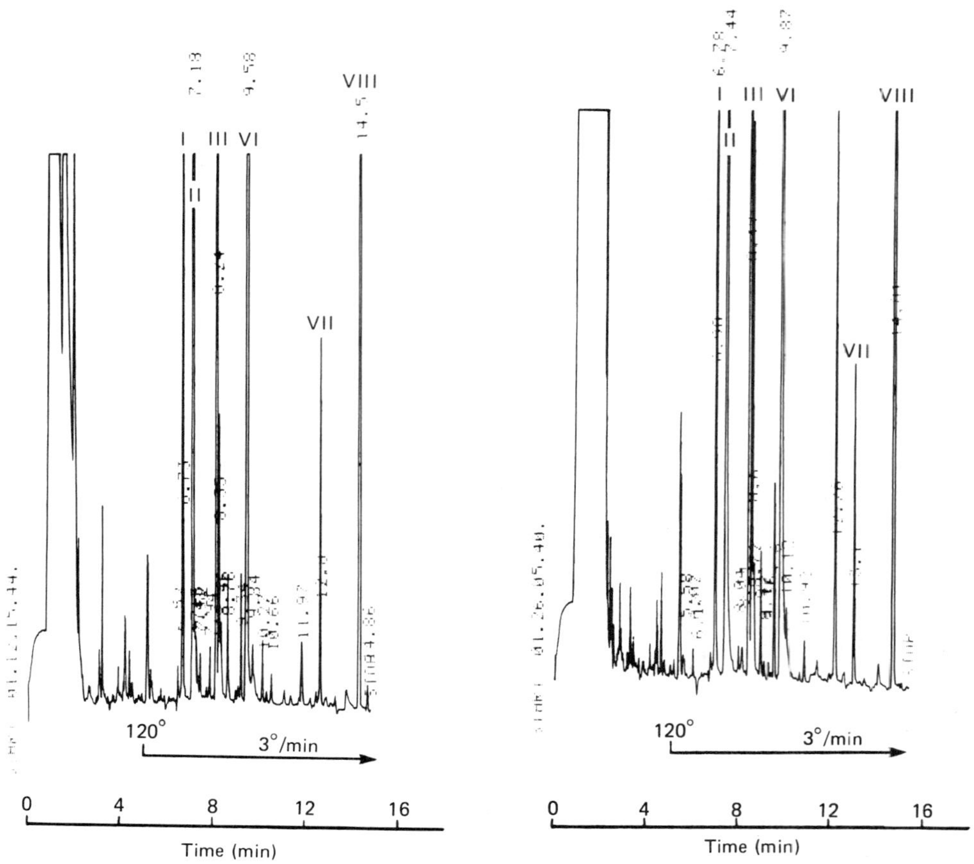

Figure 21.6 Splitless injections of human plasma extracts chromatographed on a 30 m × 0.32 mm I.D. Durabond DB-5 column (J.&W. Scientific). u_{H_2} = 56 cm/sec. Internal standard IMDN, 40 ng/ml. For peak identification, see Figure 21.5. (a) GC run on a freshly installed new column. (b) GC run after more than 400 injections of plasma extracts.

only necessary manipulation of the chromatographic system is a daily cleaning of the glass insert of the injector.

E. Application to a Pharmacokinetic Study in Humans

Table 21.5 shows the results of plasma level determinations of ISDN, IS-2-MN, and IS-5-MN in human volunteers according to the method described. The sensitivity of the method allowed detection of ISDN in plasma even 8-12 hr after oral application of the drug. The mean plasma levels of all 12 volunteers are depicted in Figure 21.7.

Table 21.5 Mean Plasma Levels of Isosorbide Dinitrate (ISDN) and Its Metabolites (IS-2-MN, IS-5-MN) after Single Oral Application of 40 mg Isosorbide Dinitrate in a Sustained-Release Formulation

Time (hr)	ISDN[a] (ng/ml)	IS-2-MN[a] (ng/ml)	IS-5-MN[a] (ng/ml)
0	0	0	0
0.33	2.0 ± 0.9	3.4 ± 1.1	7.7 ± 3.4
0.66	3.6 ± 1.4	7.5 ± 2.0	19.7 ± 6.8
1	5.3 ± 2.6	13.4 ± 5.5	38.1 ± 11.6
1.5	6.6 ± 5.5	21.1 ± 9.5	69.6 ± 18.3
2	8.6 ± 6.2	29.9 ± 12.1	111.4 ± 36.5
3	9.0 ± 5.2	41.3 ± 12.5	165.4 ± 40.0
4	9.2 ± 6.7	49.0 ± 13.5	239.1 ± 86.2
6	6.4 ± 3.7	59.9 ± 14.9	267.5 ± 75.2
8	2.1 ± 0.7	36.1 ± 7.2	232.9 ± 61.4
12	0.4 ± 0.3	12.4 ± 2.6	142.3 ± 44.2
24	0	1.7 ± 1.0	36.8 ± 18.4

[a]Mean of 12 human subjects (mean ± S.D., n = 12).

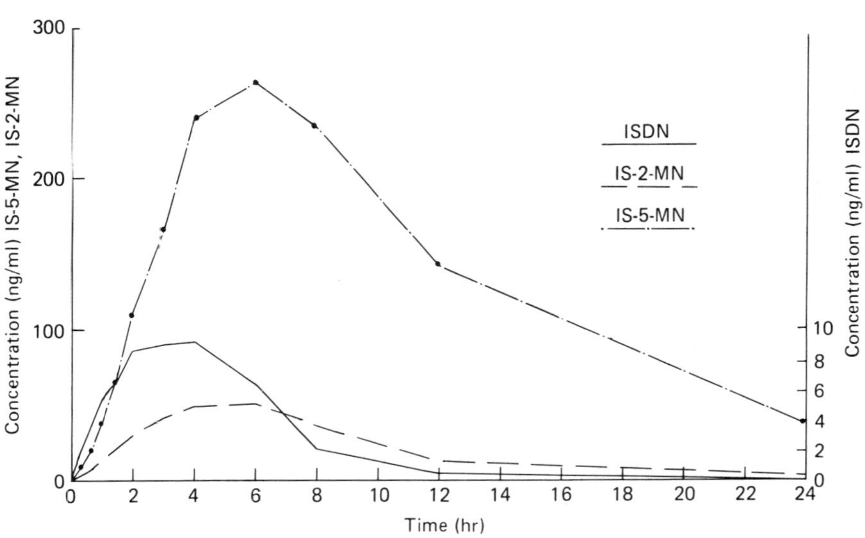

Figure 21.7 Mean plasma levels of ISDN, IS-2-MN, and IS-5MN after single oral administration of 40 mg ISDN (sustained release) to 12 volunteers.

IV. SUMMARY

An automated, specific, and sensitive method for the simultaneous determination of isosorbide dinitrate and its metabolites isosorbide-2-mononitrate, isosorbide-5-mononitrate, isomannide mononitrate, and isoidide mononitrate at concentrations down to 0.5, 1, 3, 3, and 2 ng/ml, respectively, in human plasma is described. The procedure involves the extraction of the drug and its metabolites together with their internal standards isomannide dinitrate and *o*-nitrobenzyl alcohol from plasma with dichloromethane. The analysis is carried out by fused-silica capillary gas chromatography using a ^{63}Ni-electron capture detector. The method is applied to the pharmacokinetics of isosorbide dinitrate in humans.

ACKNOWLEDGMENTS

We thank Mr. H. Maier and Ms. H. Krazeisen for their excellent technical assistance.

REFERENCES

1. W. S. Aronow, in *Organic Nitrates*, (P. Needleman, ed.) Springer-Verlag, Berlin, 1975, p. 163.
2. P. R. Lichtlen, H. J. C. Swan, A. Schrey, and H.-J Engel, eds. Proceedings of the 3rd Nitrate Symposium, Monte Carlo, 1980. Springer-Verlag, Berlin, 1981.
3. M. T. Rosseel and J. G. Bogaert, *Biochem. Pharmacol.* *22*:67 (1973).
4. S. A. Sisenwine and H. W. Ruelius, *J. Pharmacol. Exp. Ther.* *176*: 296 (1971).
5. W. H. Down, L. F. Chasseaud, and R. K. Grundy, *J. Pharm. Sci.* *63*: 1147 (1974).
6. M. G. Bogaert and M. T. Rosseel, *Arch. Pharmacol.* *275*:339 (1972).
7. R. L. Wendt, *J. Pharmacol. Exp. Ther.* *180*:732 (1972).
8. M. T. Rosseel and M. G. Bogaert, *J. Pharm. Sci.* *62*:754 (1973).
9. L. Richard, G. Klein, and J. M. Orr, *Clin. Chem.* *22*:2060 (1976).
10. A. Geigenberger, J. Degen, and H. Maier-Lenz, *Drug Res.* *32*:1138 (1982).
11. D. A. Chin, D. G. Prue, J. Michelucci, B. T. Kho, and C. R. Warner, *J. Pharm. Sci.* *66*:1143 (1977).
12. S. Spörl-Radun, G. Betzien, B. Kaufmann, V. Liede, and U. Abshagen, *Eur. J. Clin. Pharmacol.* *18*:237 (1980).
13. M. T. Rosseel and M. G. Bogaert, *J. Pharm. Sci.* *68*:659 (1979).
14. M. Ahnoff and G. Holm, 4th Symposium on Capillary Chromatography (1981 Hindelang Symposium; R. E. Kaiser, ed), Institute of Chromatography, Bad Dürkheim, 1981, p. 673.
15. M. Anteunis, S. Verhegghe, and T. Rosseel, *Org. Magn. Resonance* *3*:693 (1971).
16. M. G. Bogaert, M. T. Rosseel, and O. Teirlynck, in Proceedings of the 3rd Nitrate Symposium, Monte Carlo, 1980. Nitrates III (P. R. Lichtlen, H. J. C. Swan, A. Schrey, and H.-J. Engel, eds.), Springer-Verlag, Berlin, 1981, p. 23.
17. W. Schneider, *Dtsch. Ärztebl.* *12*:563 (1981).
18. R. V. Smith and J. Besic, *Microchem. J.* *23*:185 (1978).

22

Determination of Bencyclane by GC-MS with Chemical Ionization

James A. Settlage, Werner Gielsdorf, and Halvor Jaeger / *L.A.B. Gesellschaft für pharmakologische Untersuchungen m.b.H. & Co. (Laboratory for Pharmacological Research), Neu-Ulm, Federal Republic of Germany*

I. INTRODUCTION

N-[3-(1-Benzylcycloheptyloxy)propyl]-*N*,*N'*-dimethylammonium hydrogen fumarate (bencyclane; see Structure 1*) represents one compound of the series of cycloalkane ethers which were synthesized in the middle of the

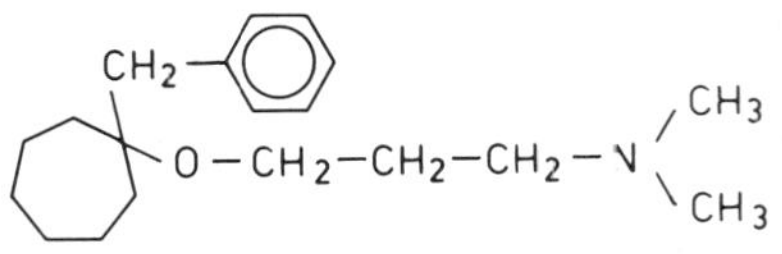

Structure 1

*Dr. Thieman GmbH, Lünen, West Germany.
This chapter is reprinted from *Arzneim.-Forsch./Drug Res.* *34*(1), 3:290-292 (1984).

1960s by Palos et al. [1] and thoroughly tested for their pharmacological/toxicological activities by Komlos et al. [2]. A series of papers on the pharmacology, pharmacokinetics, pharmacodynamics, and uses of bencyclane with an extensive bibliography were published in 1970 [3].

Bencyclane is medically used as a vasodilator, having a direct effect on peripheral blood vessels. Suppression of the Ca^{2+} flow into the cell leads to the electromechanic decoupling at the smooth muscles; furthermore, the general flow characteristics of the blood are enhanced, including cerebral blood flow. Thus, the pharmacological activities of bencyclane affect flow characteristics and cell metabolism of vascular membranes [3,4]. After oral administration of 200 mg, Farago and Simon [5] found peak levels in plasma of about 600 ng/ml and an elimination half-life of approximately 6 hr; Bock [6] measured maximum plasma concentrations of about 2 μg/ml with a tubular maximum (T_{max}) at 2 hr after a single dose of 200 mg (orally given).

A second peak maximum in plasma after 6-7 hr indicates a multicompartment distribution model. The protein binding was determined for about 30%. After a lethal intoxication of 45 Fludilat-Dragées (100 mg each), Erkens [7] measured blood levels of up to 10 μg/ml by GC-MS, but failed to detect the metabolites claimed in the literature (3- and 4-hydroxybencyclane, 4-oxobencyclane, and noroxobencyclane) [3,8]. Since the molecule rapidly undergoes ether cleavage, which leads to bencyclanole and dimethylaminopropanol formation [9], detection in biological material presents some difficulties [3,5-7,10-12].

The published methods for the determination of (1), for example, TLC, GC, UV, and HPLC, are not specific and/or sensitive enough for determinations in biological fluids after *therapeutic* dosages [3,5,6,8,10-12]; most of the cited methods are not able to distinguish between the unchanged drug and its degradation/biotransformation products.

To overcome these obstacles we developed a reliable method for quantitative determination in human plasma. This includes a capillary GC-MS assay employing ammonia as a selective reagent gas for chemical ionization (CI) and registration of the characteristic ions (SIR) of bencyclane.

II. ANALYTICAL APPROACH

A. Extraction of Plasma

One milliliter of plasma is spiked with 100 μl of the internal standard solution (corresponding to 50 ng doxylamine); then 1 ml of buffer solution at pH 9.6 and some grains of sodium chloride are added, well shaken (vortexed), and allowed to stand for 15 min for equilibration.

The mixture is then extracted with toluene/heptane/isoamyl alcohol (70:20:10; v/v) and vortexed for some minutes. After centrifugation (4000 rpm, 10 min) the organic phase is transferred to another tube, where while vortexing for 20 sec, 100 μl hydrochloric acid are added. After centrifugation the organic layer is discarded, the aqueous phase washed with 1 ml pentane, then 1 ml benzene and 10 μl of a 0.5 N sodium hydroxide solution are added.

After vortexing (20 sec) and centrifugation (4000 rpm, 5 min), the organic layer is transferred to another tube and evaporated under a gentle nitrogen flow to dryness. The residue is dissolved in 5 μl benzene; 1 μl of the resulting solution is injected (on-column) in the GC.

B. Preparation of the Standard Solution

Approximately 10 mg doxylamine are dissolved in 100 ml methanol; 0.5 ml of this stock solution is diluted with methanol to 100 ml, resulting in a solution of 0.5 ng/ml.

C. Preparation of the Buffer Solution

500 g of potassium hydrogen phosphate are dissolved in 500 ml distilled water and after cooling, diluted to 1000 ml.

D. Conditions for the GC-MS Combination

All analyses are performed on a VG-Micromass 7035, double-focusing mass spectrometer.

1. *Gas Chromatography*

Conditions

Column:	30 m × 0.25 mm DB-5 (SE-54) fused-silica, bonded phase capillary, (J + W Scientific)
Inlet:	cold
Interface:	220°C
Oven start temp.:	70°C
Oven start delay:	2 min
Program rate:	5°C/min
Final temp.:	320°C
Final delay:	2 min
Reset temp.:	60°C
Net cycle time:	20 min
Carrier gas:	He
Flow rate:	35 cm/sec at 200°C
Injection:	on-column
Injection volume:	1 μl
Injection rate:	2-3 sec
Retention times:	Internal standard, 7 min 14 sec Benzyclane, 9 min 6 sec

2. *Mass Spectrometry*

Basic technique:		Selected Ion Recording (SIR)	
Reagent gas:		Ammonia	
Ions monitored	(*m/z*)		
(A) Ref I (mass)	(269)	Dwell time:	10 msec PFK
(B) Measured mass	(271)	Dwell time:	100 msec doxylamine
(C) Measured mass	(290)	Dwell time:	100 msec bencyclane
(D) Ref II (mass)	(293)	Dwell time:	10 msec PFK
Resolution:		100-flat top peaks	
Source temp.:		150°C	
Multiplier voltage:		3.0 kV	
Emission:		100 μA	
Trap.:		0.2 mv	
eV:		70	
Source pressure with CI:		2×10^{-5} m bar	

E. Recovery

The extraction efficiency (recovery) of bencyclane from plasma was assessed by comparing peak area ratios of bencyclane and doxylamine in a solution containing 30% of the maximum determined bencyclane level. An extraction efficiency of approximately 55% was found.

F. Sensitivity

The assay has proved sensitive down to 500 pg bencyclane per ml plasma. Absolute sensitivity is 100 pg.

G. Reproducibility

The reproducibility of the method was always better than 8% from the mean value determined by five replicated assays.

H. Linearity

A linear relationship between the peak area ratio and plasma concentration of bencyclane exists in the range of 2-100 ng/ml plasma.

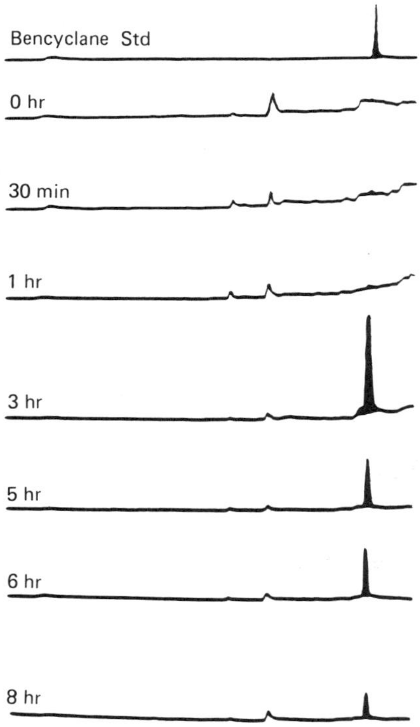

Figure 22.1 Integrated ion-current profiles of bencyclane at various sampling times.

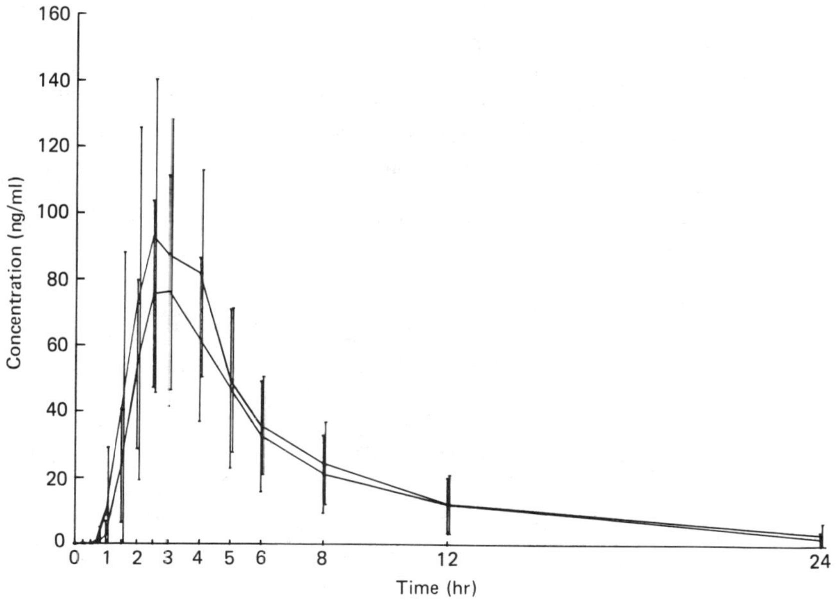

Figure 22.2 Plasma concentrations of two bencyclane preparations after oral application of 100 mg. Upper curve: test preparation. Lower curve: standard preparation.

III. CONCLUSIONS

It is apparent from these results (Figure 22.1) that the proposed assay, employing capillary column GC in combination with a mass spectrometer capable of monitoring reliably preselected characteristic ions by rapid voltage switching, and employing ammonia as a *selective* (!) reagent gas, offers the best method available for the unequivocal quantitative determination of drugs in biological fluids.

This highly sophisticated method (in terms of sensitivity, selectivity, and accuracy) may be the reason why our measured bencyclane levels in human plasma are much lower than the literature claims: most of the published assays can not distinguish between the unchanged drug and its degradation/biotransformation products!

In a single-dose, cross-over trial, we tested two different formulations of bencyclane for their bioequivalency; Figure 22.1 shows the integrated ion-current profiles of bencyclane at various sampling times. Figure 22.2 shows the plasma concentrations of two bencyclane preparations after oral application of 100 mg.

REFERENCES

1. (a) L. Palos, Z. Budai, G. Zólyomi, *Hung. Pat. 151*:865 (1965); *Chemical Abstracts 62*:16125b (1965). (b) *Therapiewoche 25*:2825 (1974).
2. E. Komlos and L. E. Petöcz, *Arzneim.-Forsch. (Drug Res.) 20*:1338 (1970).
3. *Arzneim.-Forsch (Drug Res.)* 20:10a (1970).
4. *Physician's Desk Reference*, 35th ed., Medical Economics Comp., Oradell, N.J., 1981.

5. E. Farago and J. Simon, *Arzneim.-Forsch (Drug Res.)* *20*:1380 (1970).
6. P. R. Bock, *Int. J. Clin. Pharmacol.* *13*:246 (1976).
7. M. Erkens, Speech given at the Int. Symposium, Entwicklung und Fortschritte der forensischen Chemie, Hamburg, March 19-20, 1982.
8. K. Kimura, A. Nagata, and H. Miyawaki, *Xenobiotica* *9*:119 (1979).
9. K. Kigasawa, H. Shimidzu, S. Hayashida, M. Sujino, *Chem. Pharm. Bull.* *29*:1398 (1981).
10. J. Monforte, R. J. Bath, and I. Sunshine, *Clin. Chem.* *18*:1329 (1972).
11. J. Reisch, H.-J. Kommert, H. Alfes, and H. Möllmann, *Z. Anal. Chem.* *247*:56 (1969).
12. K. U. Usbaev, Gyogyszereszet (Budapest) 23(2):54 (1979).

23

GC-MS Determination of Codeine

Werner Gielsdorf, James A. Settlage, and Halvor Jaeger / *L.A.B. Gesellschaft für pharmakologische Untersuchungen m.b.H. & Co. (Laboratory for Pharmacological Research), Neu-Ulm, Federal Republic of Germany*

I. INTRODUCTION

Codeine (1), the methyl ether of morphine, has milder analgesic and respiratory depressant effects than morphine and is therefore widely used as an effective cough suppressant [1,2].

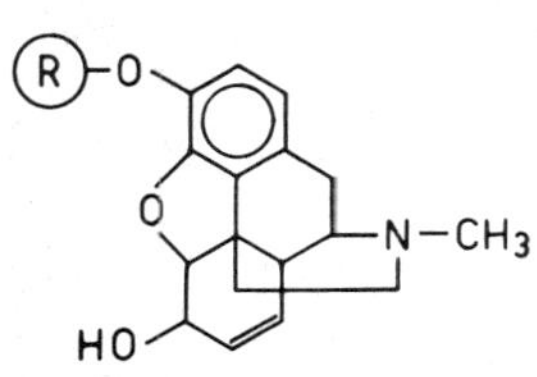

R = CH_3 CODEINE
R = C_2H_5 DIONINE

Structure 1

It is one of the common ingredients of antitussives such as Tylenol w/Codeine Elixir (McNeil Pharmaceutical), Phenergan Expectorant w/Codeine (Wyeth), and Actifed-C Expectorant (Burroughs-Wellcome) [3]. There are more than 30 preparations listed in the *Physician's Desk Reference* which contain codeine [4]; it is the most prescribed drug in the United States]5]. Average doses are sedative, analgestic, and antitussive; indeed, large amounts may cause excitement. When administered orally, 30-60 mg of codeine are approximately equivalent in analgesic effectiveness to 650 mg of aspirin.

Furthermore, it is a known drug of abuse, especially when it is administered intraveneously; both tolerance and addiction can occur [6,7]. Thus this compound raised considerable interest especially among analytical /forensic toxicologists, in drug abuse prevention programs [6]. It is a Schedule II drug under the Controlled Substances Act.

Codeine is well absorbed after both oral and parenteral application; after a single oral dose of 65 mg codeine, maximum plasma levels of approximately 250 ng/ml are reached after $1\frac{1}{2}$ hr [8,9].

After injection of 30 mg codeine, the following excretion products could be identified in human urine within 24 hr: morphine (5-17% of the administered drug), norcodeine (10-21%), conjugated codeine (32-51%), and unchanged codeine (5-10%) [6]. As with most opioids, including morphine, the effect of a given dose is considerably less after oral than after parenteral administration; however, the shape of the time-effect curve also varies with the route of administration.

II. ANALYTICAL METHODOLOGY

Foltz et al. [7] presented a thorough and well documented review of analytical methods for quantitative measurements of some drugs (and their metabolites) which are commonly abused. In principle, the methods described there for the determination of morphine in biological fluids can be used also for codeine determinations; this applies especially for the forensic toxicology, since after codeine injection the morphine has to be quantitatively determined, too.

The numerous methods proposed in the literature for the quantitative determination of codeine in biological fluids are either too nonspecific (RIA, EMIT, FRAT) or insensitive (HPLC, partial GC) for determinations of the extremely low plasma levels (a few ng/ml) of *unchanged* codeine in human plasma after *therapeutic* doses [7]. The specific detection of therapeutic doses in plasma usually is therefore performed with the tritiated substance [10,11]. v. Unruh recently has presented a report on the problems associated with such trials [12].

For the alternative method, therefore, the combination gas chromatography/ mass spectrometry (GC-MS) in connection with registration of specific ions (selected *i*on *r*ecording, SIR) [13,14], chemical ionization (CI) with a selective reactant gas [15,16], and the application of a fused-silica, bonded phase capillary [17] offers a combination which complies with all criteria (sensitivity, selectivity, etc.) for a method for detection of trace levels [18,19]. With respect to the method described by Foltz et al. [7] for the quantitative determination of morphine in biologic fluids, a GC-MS assay for the determination of codeine has been developed.

The high polarity of the compound, which becomes manifest in the strong adsorption and long retention times, makes it advantageous to include a derivatization step on preparing the samples. The derivatization with trifluoroacetic acid anhydride (TFAA), which proceeds nearly twice as fast as the trimethylsilylation, has proven to be especially effective. Besides increasing the transistoriness and sensitivity, this derivatization results in a better selectivity because it is possible to measure in a higher, and therefore cleaner, mass range.

The TFAA derivatization is particularly appropriate for the application of a fused-silica, bonded phase capillary column, whose performance has not been impaired by using TFAA. A further increase in selectivity is achieved by employing ammonia as selective reactant gas for cheimcal ionization:

Compared to the frequently used, relatively acidic reactant gases methane and isobutane, the $(NH_3)_nH^+$(n=1,2,3) protons led to more intensive quasimolecular ions (MH^+) of the two substances, whereas ionization of the coextracted contaminants and bulk materials was supressed [14-16].

Of course, at the required sensitivity of <1 ng/ml, the mass spectrometric measuring technique has to be focused for the registration of specific single ions (SIR). The MH^+ ions of the TFA derivatives of codeine and dionine were *m/e* 396 and 410, respectively. The limit for detection in cyclic scanning has been at 50 ng, but using SIR measurement technique, at 100 pg (absolute).

Dionine, the methyl homologue of codeine, has been chosen as internal standard, since deuterated codeine is not commercially available, whereas high-purity dionine is (E. Merck, Darmstadt).

A. Extraction of Plasma

One milliliter of plasma is spiked with 100 μl of the internal standard solution (corresponding to 30 ng/ml), shaken for some minutes with a vibromixer (Vortex) and allowed to stand for 15 min for equilibration. After adding some grains of sodium chloride and 1 ml phosphate buffer (pH 9.6), the mixture is extracted by vortexing several minutes with toluole/heptane/isoamyl alochol (70:20:10). After centrifugation the organic phase is evaporated at 70°C in an aluminium block under a gentle nitrogen flow.

B. Derivatization

To the extraction residue, 200 μl chloroform and 100 μl trifluoroacetic acid anhydride (TFAA) are added and heated at 70°C for 10 min (aluminium block). After cooling to room temperature, the reaction mixture is evaporated to dryness at 40°C (under a nitrogen flow). This residue is dissolved immediately in 5 μl chloroform, and 2 μl are immediately injected into the GC.

C. Preparing the Solutions

1. *Internal Standard (Ethylmorphine)*

Approximately 13 mg dionine salt are dissolved in 100 ml methanol; 300 μl of this solution are diluted to 100 ml methanol

2. *Buffer System (pH = 9.6)*

500 g potassium hydrogen phosphate are dissolved in 500 ml distilled water and after cooling, diluted to 1 liter.

D. Conditions for the GC-MS Combination

The analyses have been carried out with a double focusing, high-resolution VG Micromass 7035.

1. *Gas Chromatography*

Column:	30-m fused silica capillary DB-1 (SE-30) bonded phase (J and W Scientific) 0.25-mm I.D., 0.25 μm film
Temp.:	Injector: 250°C
	Oven: 200-350°C with 20°C/min
	Interface: 250°C
Carrier gas:	He 33 cm/sec
Split:	1:10

2. *Mass Spectrometry*

Ion-source temp.:	150°C
Ionization energy:	70 eV
Emission:	0.5 mA
Multiplier	3.0 kV
Reactant gas:	Ammonia
Selected ions:	*m/e* 396 (codeine-TFA-derivative)
	m/e 410 (dionine-TFA-derivative)
Dwell times:	200 msec/mass

E. Recovery

The extraction efficiency (recovery) of codeine from plasma was assessed by comparing peak area ratios of codeine and dionine of a solution containing 30% of the maximum determinated codeine level. An extraction efficiency of approximately 76% [1-4] was found.

F. Sensitivity (Relative)

The assay has proved sensitive down to 500 pg codeine per milliliter plasma.

G. Reproducibility

The reproducibility of the method was always better than 8% from the mean value determined by five replicated assays.

H. Linearity

A linear relationship between the peak area ratio and plasma concentration of codeine exists in the range of 2-100 ng/ml plasma (Figure 23.1).

III. CONCLUSIONS

This study summarizes our experiences in the determination of the unchanged drug in human plasma in a single-dose, cross-over trial in order to make statements on the bioavailability [20,21]. We tested two different galenic codeine formulations:

Preparation 1: codeine syrup (2.6 mg codeine/ml)
Preparation 2: codeine soft gelatine capsules (13 mg codeine per capsule).

Two capsules or 10 ml syrup were taken by 12 volunteers, corresponding to 26 mg of the free base. Figure 23.2 shows the mean plasma levels of the unchanged drug.

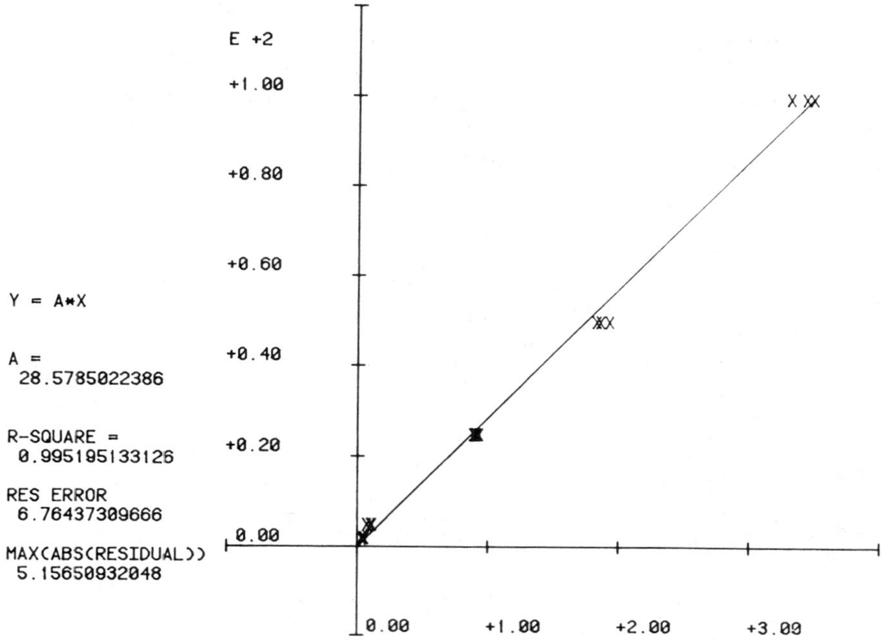

Figure 23.1 Calibration curve of codeine.

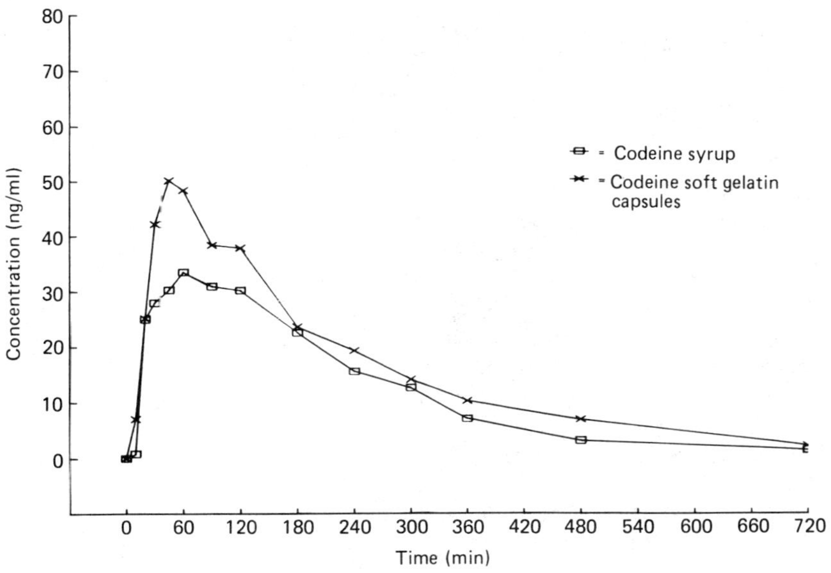

Figure 23.2 Means of plasma concentrations of codeine.

This study clearly demonstrates the benefits of capillary GC in combination with a mass spectrometer serving as an unrivaled sensitive and selective "detector" for the unequivocal determination of drugs in body fluids, for example, in studies for evaluating pharmacokinetic parameters in the course of bioavailability studies. The assay is also applicable to other physiological fluids and can serve as a check on results of less sensitive/specific methods.

REFERENCES

1. E. G. C. Clarke, *Isolation and Identification of Drugs*, The Pharmaceutical Press, London, 1974.
2. *Remington's Pharmaceutical Sciences*, 16th ed., Mack Publ. Com., Easton, Pa., 1980.
3. L. S. Goodman and A. Gilman, eds., *The Pharmacological Basis of Therapeutics*, 5th ed., Macmillian Publishing Co., New York, 1975.
4. *Physician's Desk Reference* (P.D.R.), 35th ed., Medical Economics Comp., Oradell, N.J., 1981.
5. SCRIP No. 664, PJB Publ. Ltd., Richmond, England, February 3, 1982, p. 7.
6. K. K. Kaistha, *Chromatographic Rev. 21*(2):141,151 (1977).
7. R. L. Foltz, A. F. Fentiman, and R. B. Foltz, *GC-MS Assays for Abused Drugs in Body Fluids*, NIDA Research Monograph Series No. 32 (1980).
8. M. K. Brunson and J. F. Nash, *Clin. Chem.* 21:1956 (1975).
9. R. A. Zweidinger, F. M. Weinberg, and R. W. Handy, *J. Pharm. Sci.* 65:427 (1976).
10. W. D. Bechtel and K. Sinterhauf, *Arzneim.-Forsch (Drug Res.)* 28:308 (1978).
11. K. Sinterhauf and W. D. Bechtel, *Arneim.-Forsch. (Drug Res.)* 30: 1012 (1980).
12. G. v. Unruh, in *Bioverfügbarkeit von Arzneimitteln* (N. Rietbrock and B. Schnieders, eds.), G. Fischer Verlag, Stuttgart/New York, 1979.
13. E. L. Ghisalberti, Mass Fragmentography, in *Drug Fate and Metabolism* (E. R. Garett and J. L. Hirtz, eds.), Marcel Dekker, New York/Basel, 1979, 1-67.
14. W. D. Lehmann and H.-R. Schulten, *Angew. Chem.* 90:233 (1978).
15. W. Gielsdorf and P. Toffel-Nadolny, *J. Clin. Chem. Clin. Biochem. 19*: 25 (1981).
16. W. Gielsdorf and E. Klug, *Dtsch. Apoth. Ztg.* 121:1003 (1981).
17. W. Jennings, *Gas Chromatography with Glass Capillary Columns*, Academic Press, New York (1980).
18. B. J. Millard, *Quantitative Mass Spectrometry*, Heyden, London (1978).
19. C. Fenselau, Selectivity in Biomedical Applications, in *High Performance Mass Spectrometry: Chemical Applications* (M. L. Gross, ed.), ACS Series 70, Washington, 1978. 188-208.
20. H. Jaeger, Bioverfügbarkeit und Zulassung von Arzneimitteln, Symposium: Probleme bei der Zulassung von Arzneimitteln, Eschborn, October 10-19, 1981.
21. N. Rietbrock and B. Schnieders, eds., *Bioverfügbarkeit von Arzneimitteln G. Fischer Verlag, Stuttgart/New York, 1979.*

24

Determination of Diphenhydramine and Orphenadrine by GC

Dieter Lutz, Werner Gielsdorf, and Halvor Jaeger / *L.A.B. Gesellschaft für pharmakologische Untersuchungen m.b.H. & Co. (Laboratory for Pharmacological Research), Neu-Ulm, Federal Republic of Germany*

I. INTRODUCTION

For well over 35 years diphenhydramine hydrochloride (1)* has been used medically as an antihistamine and hypnotic [1,2].

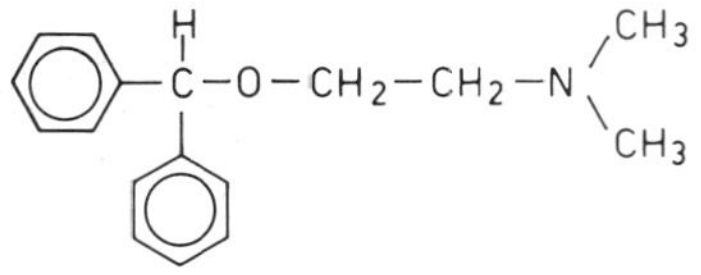

DIPHENHYDRAMINE

Structure 1

*In the United States: Benadryl (Parke-Davis). In Germany: Betadorm-A (Woelm Pharma). In Switzerland: Detensor (Sandoz).

This chapter is reprinted from *J. Clin. Chem. Clin. Biochem.* *21*(10):595-597 (1983).

Chemically it is the 2-diphenylmethoxy-*N*,*N*-dimethylethanamine and possesses anticholinergic, antitussive, antiemetic, and sedative properties, the last mentioned leading to its widespread use as a drug for inducing sleep [3-5].

Since it is a potent and effective H_1-receptor blocker, it is also used for the symptomatic treatment of various allergic diseases, in which its usefulness is clearly attributable to its antagonism of histamine: it inhibits the effects of histamine on capillary permeability and on vascular, bronchial, and other smooth muscles. Even in therapeutic doses the drug elicits side effects, which are rarely serious, but sometimes lead to the withdrawl of the drug [2].

Diphenhydramine is rapidly and well adsorbed with a systemic bioavailability of at least 50% after oral application [6,7]; maximum plasma concentrations are reached after 1-4 hr [8-11]. Further pharmacokinetic measurements include the following: $k_{el} = 0.21\ hr^{-1}$, clearance = 899 ml/min, $V_d = 238$ l, and $AUC_{0\to\infty} = 403$ ng ml^{-1} hr [7]; an apparent plasma half-life of about 3-10 hr is reported [6,10,11].

After per oral administration, the drug is effective for 4-6 hr, with a maximum around 1 hr. Average peak plasma levels (after therapeutic dosage of 50 mg) are about 80-100 ng/ml [4-9]. Plasma protein binding of the drug is approximately 98-99% [6,12].

Less than 3% of a therapeutic dose are excreted unchanged with the urine [11-14]. Biotransformation of diphenhydramine [11,12,15-18] in man takes place by the following main routes:

1. Successive dealkylations and the nitrogen atom, e.g., giving *N*-desmethyldiphenhydramine and *N*,*N*-didesmethyldiphenhydramine.
2. Cleavage at the benzhydrylether function results in the formation of benzhydrol and dimethylaminoethanol.

These degradation routes are well documented in the literature and apply as well to several related colamine/methylcolamine derivatives, e.g., bromazine, brompheniramine, mephenhydramine, chlorphenoxamine, and doxylamine [19]. However, the major metabolite in human plasma and urine has been identified as diphenylmethoxyacetic acid [16].

II. ANALYTICAL METHODS

The physical properties (e.g., UV, IR, NMR, and mass spectral) of the pure substance are summed up by Holcombe and Fusari [20], Clarke [18], Müller [15], and Raeder-Schikorr [17]. In an alkaline medium the compound is fairly stable, but in an acidic medium it is rapidly hydrolyzed to its corresponding benzhydrol and 2-(dimethylamino)ethanol, respectively; a reaction which also takes place under ultraviolet irradiation [20].

There has been published a lot of analytical approaches for the identification/detection of diphenhydramine, for example, paper [21], thin-layer [14], and column [22] chromatography, electrophoresis [21], colorimetric/fluorometric [13], and titrimetric [23] analysis. These, however, are not specific and sensitive enough for determinations in biological fluids (urine, plasma/serum).

Obviously more reliable are the published GC, GC-MS, and HPLC methods: Unfortunately they often include extensive, time-consuming preparations (e.g., special sample clean-up); involving conversion or formation of derivatives [24] and/or operate near the limit of sensitivity [10,11]. Also, some method do not distinguish between the unchanged drug and its metabolites or are prone to interference [8].

In all of the published GC methods, packed columns (with SE-30, PDEAS, Hi-Eff-8B, OV-17, Apiezon-L, and DC-200, or mixtures of them as liquid phases [20], and flame ionization detectors (FID) are used which limit sensitivity to about 10-20 μg/ml [3,6,9,10,17,18,25]; recently published method employ more selective detectors, e.g., the nitrogen-specific detector (N-FID) [8]. Vessman et al. [24] successfully used the electron capture detector (ECD), but their sample clean-up includes, for example, conversion of diphenhydramine to benzophenone, a rather time-consuming and inconvenient procedure.

Liquid-chromatographic assays have been applied only to aqueous samples and/or are not applicable for rapid (screening) procedures [26,27].

The published gas chromatographic-mass fragmentographic method using single ion recording (SIR) of the characteristic ion at m/z 165 fulfills the requirements for a reliable assay but is less suitable for some metabolites (e.g., the *N*-oxide and the diphenylmethoxy acetic acid) which are either thermolabile or not volatile enough [28,29]; furthermore, this method is relatively expensive and often not readily available in a clinical laboratory.

III. RESULTS AND DISCUSSION OF OUR OWN INVESTIGATIONS

With some exceptions we have followed the GC method published by Rough and Calvert [9]:

A. Extraction

To 3 ml of serum in a centrifuge tube (fitted with a Teflon-lined cap) 10 μl of a methanolic solution of the internal standard (14 ng/μl) are added and thoroughly mixed (Vortex) for 10 sec. After mixing 1 ml of sodium hydroxide (1 N) and 7 ml *n*-heptane, the mixture is extracted for 1 hr in an inversion shaker. After centrifuging, the organic layer is transferred to a conical tube and evaporated to dryness under nitrogen at 55°C. The residue is reconstituted in 40 μl acetone, concentrated to approximately 10 μl, and 1 μl is injected in the GC.

B. GC Conditions

The GC analysis is carried out using a Carlo Erba FV 2150 chromatograph equipped with an nitrogen-selective detector (N-FID).

Column:	30 m × 0.25 mm fused silica, bonded phase DB-1 capillary column (J + W Scientific)
Carrier gas:	He 1.2 bar
Make-up gas:	N_2 0.4 bar
Hydrogen:	3 ml/min
Air:	100 ml/min
Temperatures	
Injector:	200°C
Oven:	50°C (1 min) to 180°C for 5 min, then at a rate of 6°C/min to 210°C

Under the described conditions the retention times are:

Diphenhydramine:	10.7 min
Orphenadrine:	12.4 min

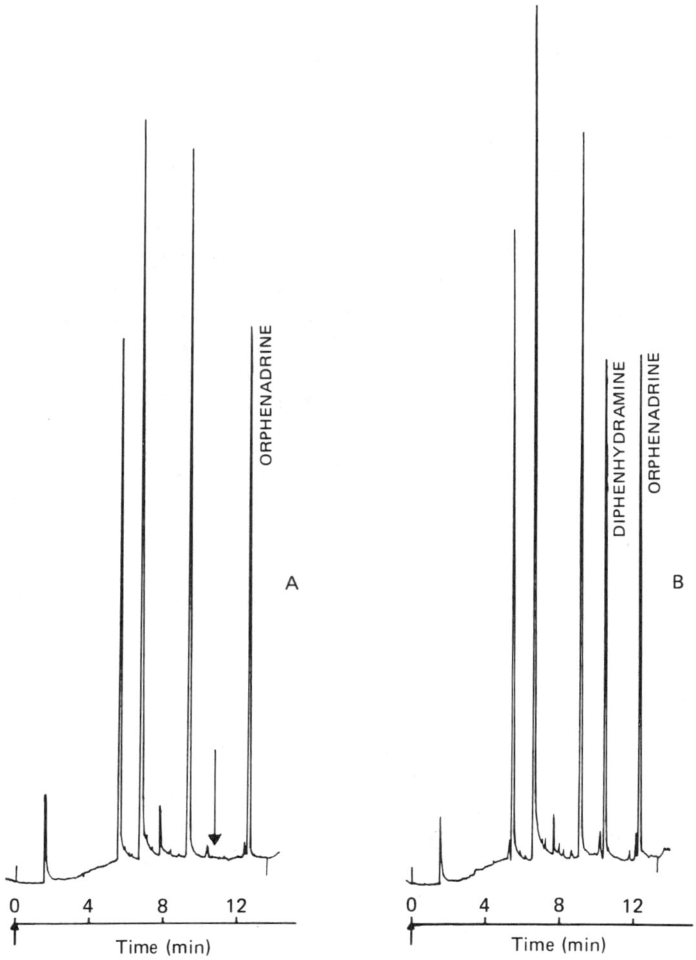

Figure 24.1 Chromatograms of serum extracts. (A) Drug-free serum without internal standard. (B) Four hours after receiving 31 mg diphenhydramine. 1 = diphenhydramine; 2 = orphenadrine (Internal standard).

Figure 24.1A,B shows two representative chromatograms of a serum extract before and after drug treatment. Figure 24.2 shows the serum concentration-time curve of diphenhydramine in 10 volunteers following a single oral dose of 31 mg.

IV. CONCLUSIONS

The presented capillary GC method allows the fast and accurate quantitative determination of either diphenhydramine or orphenadrine and other ethanolamines in human serum. It is also clear from the chromatogram (Figure 24.1) that capillary columns provide more than adequate efficiency and sensitivity.

In combination of a rather nonpolar extraction solvent and a selective detector eliminates the need for a back extraction or further steps in sample clean-up.

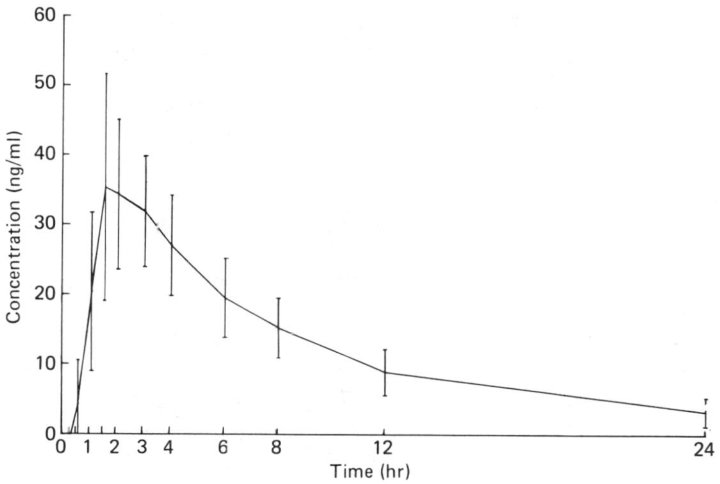

Figure 24.2 Serum concentrations of diphenhydramine in 10 volunteers after ingestion of 31 mg.

A calibration curve was established and checked to be linear between 2.5 and 120 ng/ml serum; the detection limit of 2 ng/ml clearly shows the reliability of the assay.

Average recovery based on 5 determinations was about 100.1% (c.v. = 4.9%); reproducibility between duplicate injections was always better than 3.6%.

REFERENCES

1. *Physician's Desk Reference*, 35th ed., 1981, Medical Economics Co., Oradell, N.J. 1981.
2. *The Pharmacological Basis of Therapeutics*, 5th ed., L. S. Goodman and A. Gilman, eds., Macmillian Publishing Co., New York, 1975.
3. W. R. Külpmann and M. Schreiber, *J. Clin. Chem. Clin. Biochem. 19*: 740 (1981).
4. S. G. Carruthers, D. W. Showman, D. E. Hignite, and D. L. Azarnoff, *Clin. Pharmacol. Ther.* 23:375 (1978).
5. W. Bilzer, U. Gundert-Remy, and E. Weber, *Eur. J. Clin. Pharmacol.* 7:393 (1974).
6. K. S. Albert, M. R. Hallmark, E. Sakmar, D. J. Weidler, and J. G. Wagner, *J. Pharmacolinet. Biopharmaceut.* 3:159 (1975).
7. S. G. Carruthers, D. W. Shoeman, and D. L. Azarnoff, *Clin. Res. 24*: 562A (1976).
8. W. Bilzer and V. Gundert-Remy, *Eur. J. Clin. Pharmacol. 6*:268 (1973).
9. R. Baugh and R. T. Calvert, *Br. J. Clin. Pharm. 3*:1062 (1976).
10. R. S. Albert, E. Sakmar, J. A. Morais, M. R. Hallmark, and J. G. Wagner, *Res. Commun. Chem. Pathol. Pharmacol.* 7:95 (1974).
11. A. J. Glazko, W. A. Dill, R. M. Young, T. C. Smith, and R. I. Ogilvie, *Clin. Pharmacol. Ther. 16*:1066 (1974).
12. S. Pfeifer, *Biotransformation von Arzneimitteln*, Verlag Chemie, Weinheim-New York, 1979.

13. J. Hald, *Acta Pharmacol.* *3*:296 (1947).
14. K. W. Hindmarsh, N. W. Hamon, and D. F. LeGatt, *Clin. Toxicol.* *11*: 245 (1977).
15. R. K. Müller, *Die toxikologisch-chemische Analyse*, Verlag Chemie, Weinheim-New York, 1976.
16. T. Chang, R. A. Okerholm, and A. J. Glazko, *Chem. Pathol. Pharmacol.* *9*:391 (1974).
17. M. Raeder-Schikoor, in *Drug Level Monitoring*, W. Sadée and G. C. M. Beelen, eds., J. Wiley, New York, 1980.
18. E. G. C. Clarke, *Isolation and Identification of Drugs*, The Pharmaceutical Press, London, 1974.
19. W. Gielsdorf and K. Schubert, *J. Clin. Chem. Clin. Biochem.* *19*:485 (1981).
20. I. J. Holcombe and S. A. Fusari, in *Analytical Profiles of Drug Substances*, Vol. 3 (K. Florey, ed.), Academic Press, New York, 1974.
21. A. Baruffini, *Farmaco (Pract. Ed.)* *13*:459 (1958).
22. T. D. Doyle and J. Levine, *J. Assoc. Offic. Anal. Chem.* *51*:191 (1968).
23. *The United States Pharmacopeia* (USP XVIII), U. S. Pharmacopeial Convention, Rockville, MD, 1970, p. 207.
24. J. Vessman, P. Hartvig, and S. Stromberg, *Acta Pharm. Suecia* *7*:373 (1970).
25. T. J. Siek, in I. Sunshine, *Methodology for Analytical Toxicology*, CRC Press Inc., Cleveland, 1975.
26. M. L. E. Bergh and J. de Vries, *J. Liq. Chromatogr.* *3*:1173 (1980).
27. F. L. Vandemark, R. F. Adams, and G. J. Schmidt, *Clin. Chem.* *24*: 87 (1978).
28. R. Eckard, R. Ening, and W. Ening, *Arch. Pharmacol.* *302*:14 (1978).
29. A. Cailleux, A. Turcant, A. Premel-Cabic, and P. Allican, *J. Chrom. Sci.* *19*:163 (1981).

25

Glass Capillary GLC of Nucleosides

C. F. Gelijkens, Andre P. De Leenheer, and Pat. J. F. Sandra / *Rijksuniversiteit Gent, Gent, Belgium*

I. INTRODUCTION

The importance of the nucleosides as constituents of the nucleic acids RNA and DNA is well established [1-3]; in addition, these molecules exert a regulatory activity in cellular processes involving nucleic acids, proteins, carbohydrates, and lipids. As a consequence of this fundamental role in cellular metabolism, much attention has been paid to the nucleosides as potential drugs, and at present, a very wide range of nucleoside analogues have been synthesized and tested for biological activity [4]. They include compounds with antitumor [5], antimicrobial [6], antiprotozoal [7], antiviral [8,9], cardiovascular [10], immunosuppressive [11], and cytokinetic [12] properties.

The synthesis of 5-fluorouracil [13] and its nucleosides by Heidelberger in 1957 may be considered as an example of rational drug design; indeed, the substitution of the fluorine atom in the 5-position of the pyrimidine ring was based upon the finding that tumors preferentially utilize uracil for nucleic acid biosynthesis. Furthermore, because of the small size of the fluorine atom (1.35 Å, compared to 1.20 Å for H), it was anticipated that the 5-fluoro analogue could inhibit the biotransformation of 2'-deoxyuridylate to thymidylate, an essential subunit of DNA [14]. This would then lead to "thymineless" death of the cell, a condition characterized by a continuing RNA and protein synthesis, while DNA synthesis is inhibited [15]. The fluorinated pyrimidines have since become established agents in cancer chemotherapy [16] and have been extensively used as tools in biochemical research [17,18]. The antitumor activity of these drugs depends on the intracellular formation of 5-fluoro-2'-deoxyuridine-5'-monophosphate, a potent inhibitor of thymidylate synthetase [19]. This metabolic activation occurs via pathways also used by the normal endogenous substrates [20]; activity loss of the necessary enzyme systems involved may contribute to the development of resistance of the tumor cell to the fluorinated pyrimidines [16]. Another major difficulty encountered in the clinical application of the nucleoside analogues is their relative lack of selectivity. Thus, they not only inhibit the growth of the rapidly dividing tumor cells, but affect other rapidly growing tissues as well; the resulting toxic effect include leukopenia, stomatitis, diarrhea, and less frequently, thrombocytopenia [16].

In order to avoid unacceptable toxicity and still achieve a satisfactory therapeutic effect, knowledge of the actual levels of the drugs and their intermediate and active metabolites in different biological fluids would be useful [22]. Since most nucleoside analogues have only slightly different structures as compared to the naturally occurring nucleosides [4], highly specific analytical methods have to be used for measurement. Chromatographic techniques have been widely used in the nucleoside field [23]. In the present chapter, only the capillary GC aspects will be discussed. Specialized readers may feel that particular points have not received sufficient attention. Many interesting packed-column GC results in nucleoside analysis are even not mentioned. The aim of this contribution is mainly to make it clear that capillary GC has already contributed much to this difficult technology of nucleoside analysis. The "quantitative" determination of some pharmacologically interesting nucleosides will be emphasized. The compounds studied (Table 25.1) include naturally occurring (VIII-XI, XIII), synthetic (IV-VI), and nucleosides of medicinal importance, used in cancer chemotherapy (I-III) and in treatment of viral infections (VII, XII) [24,25].

Table 25.1 Structural Formulas of Nucleosides

	Nucleoside	R_1	R_2	R_3	R_4	R_5	
I	5-Fluoro-2'-deoxyuridine	F	OH	H	H	–	FUDR
II	5-Fluorouridine	F	OH	OH	H	–	FURD
III	1-(2-Deoxy-β-D-lyxofuranosyl)-5-fluorouracil	F	H	H	OH	–	3'-*epi*-FUDR
IV	5-Chloro-2'-deoxyuridine	Cl	OH	H	H	–	ClUDR
V	5-Chlorouridine	Cl	OH	OH	H	–	ClURD
VI	5-Bromo-2'-deoxyuridine	Br	OH	H	H	–	BrUDR
VII	5-Iodo-2'-deoxyuridine	I	OH	H	H	–	IUDR
VIII	2'-Deoxyuridine	H	OH	H	H	–	UDR
IX	Uridine	H	OH	OH	H	–	URD
X	Thymidine	CH_3	OH	H	H	–	THYM
XI	2'-Deoxyadenosine	–	–	H	–	H	dADO
XII	9β-D-arabinofuranosyladenine	–	–	H	–	OH	Ara-A
XIII	Adenosine	–	–	OH	–	H	Ado

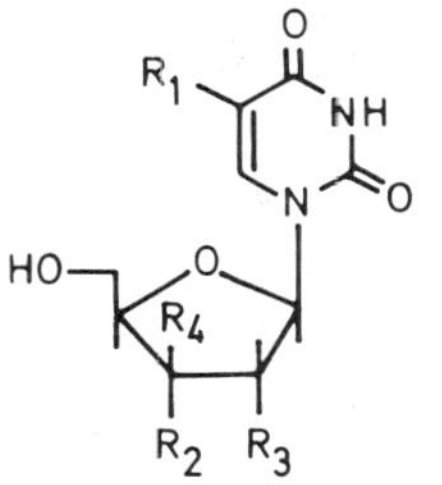

I - X

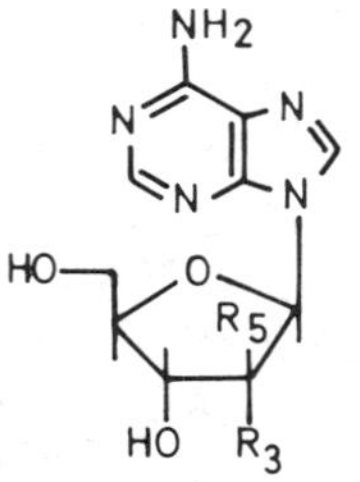

XI - XIII

II. SAMPLE PREPARATION AND DERIVATIZATION

A. Sample Preparation

One of the most tedious aspects of the quantitative analysis of nucleosides is the enrichment process required to isolate them from the biological fluids. Due to their highly polar character, the nucleosides are poorly extracted from aqueous solutions (serum, plasma, or urine) by double-phase extraction procedures. In addition, the required high polarity of the organic solvent always causes the coextraction of many endogenous impurities, resulting in interfering peaks. This is most pronounced in urine samples in which drug levels are usually higher compared to plasma samples but where the biological matrix is much more complex. Additional clean-up procedures are required to remove urinary salts (phosphates, sulfates) and urinary pigments. The drawbacks of solvent extraction [poor yield (recovery ∼60%) and low specificity due to the highly polar organic phase] resulted in the development of on-column extraction techniques. Due to the different clean-up procedures for plasma and urine samples, they are treated separately.

1. *Serum or Plasma Samples*

Two-phase extraction after salting out with ammonium sulfate using diethyl ether-isopropanol (4:1) was applied for the analysis of the nucleosides of uracil and thymine [26]. The same technique was used for the determination of 5-fluorouracil [27]. Solvent extraction of mice serum samples with *n*-butanol was described by Pantarotto et al. [28]. The extraction recoveries from serum were established and were, for example, 58 ± 8% for 5-FUDR and 60 ± 5% for ARA-C (arabinosyl cytosine). On-column extraction was developed by De Leenheer and Gelijkens [29]; 5-fluoro-2'-deoxyuridine (5-FUDR), being a weak acid (pK_a = 7.66) is present almost entirely as a monoanionic species (99.55%) at pH = 10. It is therefore quantitatively adsorbed on an anion-exchange resin of the quaternary ammonium type, AG1—X4(Cl^-). Other plasma constituents are removed with a water-methanol wash. Subsequent quantitative elution of the nucleosides (5-FUDR) is performed with 0.3 M acetic acid in methanol. The extraction recovery of 5-FUDR from serum in the 5 ng-1 μg/ml range was established by using 6-[^{3}H]FUDR and showed a mean value of 93.2 ± 2.1% (S.D.). By this simple and rapid technique, relatively pure extracts are obtained. Moreover, large sample series can be handled simultaneously. The technique was used for the measurement of 5-FUDR serum levels in cancer-treated patients (see Section IV).

Typical procedure: Mix a 1-ml serum sample and 100 μl internal standard (IS) solution with 5 ml carbonate buffer (pH 10 and ionic strength 0.1) and apply to an extraction column. Allow the diluted sample to pass through the resin bed, and wash the column consecutively with 10 ml H_2O and 10 ml methanol. Elute the drug and IS with 10 ml of 0.3 M acetic acid in methanol. Evaporate the collected eluate to dryness under a nitrogen stream at 50°C.

Comments:

1. The internal standard, a structurally related product, must be added as soon as possible to compensate for losses further on.
2. Extraction columns (2.0 cm × 6 mm I.D.) are prepared with AG1-X4 (Cl^-) resin (100-200 mesh, Bio-Rad, Richmond, Va.) using glass wool-plugged glass columns fitted with a Teflon tap and a 15-ml reservoir on top.

2. *Urine Samples*

The anion-exchange technique developed for serum samples cannot be directly applied to urine samples. The urinary inorganic salts are coextracted and saturate the resin by their high concentration and great affinity. The nucleosides are not quantitatively retained and, moreover, the salts coelute and inhibit the derivatization reaction. Urinary pigments are also adsorbed and interfere in the chromatographic analysis. The problem has been studied [30], and a clean-up procedure based on $Ba(OH)_2$ treatment and Sephadex LH-20 chromatography has been developed. In this procedure the inorganic salts are precipitated as their Ba salts. Nucleosides and pigments are extracted by anion-exchange on AG1-X4 (Cl^-), and the pigments are removed by chromatography on a Sephadex LH-20 column. Details of the extraction and clean-up procedure are shown in Scheme 1.

Urine (1 ml)

↓ Add: 100 μl internal standard solution.
1 ml of 10^{-1} M $Ba(OH)_2$.
Mix and centrifuge

Supernatant

↓ Apply to AG1-X4 (Cl^-) resin column (2.0 cm × 6 mm I.D.).
Wash with 10 ml H_2O, 10 ml CH_3OH.
Elute with 10 ml 0.3 M HAc in CH_3OH.

Crude Extract

↓ Evaporate to dryness at 50°C under N_2.

Residue

↓ Dissolve in 200 μl CH_2Cl_2:CH_3OH (93:7).
Apply to Sephadex LH-20 column (6.0 cm × 6 mm I.D.).
Wash with 10 ml CH_2Cl_2:CH_3OH (97:3).
Elute with 10 ml CH_2Cl_2:CH3OH (94:6).

Eluate

↓ Evaporate to dryness at 50°C under N_2.

Dry Extract ⟶ Derivatize ⟶ CGC analysis

Scheme 1 Nucleoside extraction.

If ribonucleosides have to be analyzed, an alternative procedure can be used. Ribonucleosides are selectively adsorbed under certain conditions on columns containing immobilized phenylboronic acid gel [31]. *Cis*-diol boronate complexes are formed while the urinary pigments and salts are not retained by the boric acid gel and can be removed with a buffer washing. The ribonucleosides are then quantitatively eluted under acidic conditions (formic acid). This technique was applied to the quantitative analysis of 5-fluorouridine (FURD) in human urine. The extraction procedure is outlined in Scheme 2.

B. Derivatization

The application of gas-liquid chromatography to nucleoside analysis presents several problems which are essentially due to the high polarity and low volatility of these compounds. The study of nucleosides by GC therefore comprises the preparation of derivatives on an analytical scale (microgram range) in a controllable and reproducible way. In the first report of the GC analysis of nucleosides, Miles and Fales [32] mentioned the preparation of acetyl, methyl, and isopropylidene derivatives. Most of the subsequent work with nucleosides, however, has involved silyl derivatives. Trimethylsilyl (TMS) derivatives were first reported by Hancock and Coleman [33]. The derivatives were prepared using hexamethyldisilazane (HMDS) and trimethylchlorosilane (TMCS) in pyridine; method based on the work of Sweeley et al. [34]. The same silylating mixture was employed by Sasaki and Hashizume [35,36]. Improved results were

Urine (1 ml)

↓ Add: 100 μl internal standard solution.
100 μl 5 M $NH_4Ac/MgAc_2$ (pH 9).
Mix and centrifuge.

Supernatant

↓ Apply to Boric Acid Gel column (4.0 cm × 6 mm I.D.).
Wash with 9 ml 0.5 M $NH_4Ac/MgAc_2$ (pH 9).
Elute with 5 ml 0.1 M HCOOH.

Eluate

↓ Add 300 μl 0.5 M carbonate buffer (pH 10).
Apply to AG1-X4 (Cl^-) resin column (2.0 cm × 6 mm I.D.).
Wash with 10 ml H_2O, 10 ml CH_3OH.
Elute with 10 ml 0.3 M HAc in CH_3OH.

Extract

↓ Evaporate to dryness at 50°C under N_2.

Dry Extract ⟶ Derivatize ⟶ CGC analysis

Scheme 2 Ribonucleoside extraction.

obtained with bis(trimethylsilyl)trifluoroacetamide (BSTFA) in acetonitrile [37-39]. This reagent, sometimes catalyzed by trimethylchlorosilane (TMCS) [40], is mostly used in recent work. Acetyl [41], trifluoroacetyl [42,43], phenylboronate/TMS [44], and isopropylidene/TMS [41] derivatives have also been used in nucleoside analysis.

Multiple derivative formation for some nucleosides, using silylating agents, has been observed. The preparation of mixed derivatives, e.g., methoxime/TMS derivatives [45] and trifluoroacetyl/TMS derivatives [46], eliminates multiple derivative formation. Some investigators have studied the formation of permethyl derivatives. Their low molecular weight and high degree of stability make methyl derivatives very useful in gas chromatography [41,47,48] and gas chromatography-mass spectrometry [49]. To prepare permethyl derivatives, several methods have been proposed. Dolhun and Wiebers [41] used silver oxide and methyl iodide, whereas Baker [47] employed trimethylphenyl ammonium hydroxide (TMPAH) to methylate nucleosides and nucleotides. von Minden and McCloskey [50] prepared methyl derivatives with methylsulfinyl carbanion [51,52], generated by heating a NaH/oil suspension in DMSO, and methyl iodide as reagents. The methylsulfinyl carbanion can also be obtained by dissolving potassium *tert*-butoxide in dry DMSO [53,54]. This procedure is less hazardous and yields a more stable reagent. This technique was used by De Leenheer and Gelijkens [48], and the GC behavior for the derivatives of 10 pyrimidine and 3 purine nucleosides on 13 stationary phases was studied.

The mass spectral characteristics of the derivatives were identical to those found by von Minden and McCloskey [50] and indicate the number of $-CH_3$ groups introduced in each nucleoside. For the compounds studied, the ribosyl or 2'-deoxyribosyl moiety was completely *O*-methylated. The position of the $-CH_3$ group(s) in the pyrimidine moiety was studied by UV spectrophotometry, which supported the location of the methyl group at *N*-3, rather than an *O*-4 methylation. For adenine nucleosides, UV spectrophotometry suggested the generation of a N^6,N^6-dimethyl group. The derivative of uridine is shown in Figure 25.1.

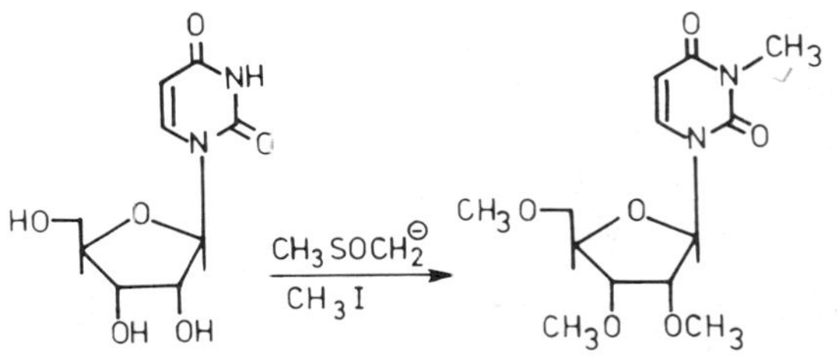

Figure 25.1 Derivative formation (*N,O*-permethyl) of uridine.

1. *Typical Procedure*

Preparation of the methylsulfinyl carbanion reagent: Six grams of potassium *tert*-butoxide in 100 ml of dry DMSO are heated to 70°C with mechanical stirring. After 2 hr, the solution is filtered and 10-ml portions are stored at -18°C. Prior to use, this stock solution is diluted 4 times with dry DMSO to yield the reagent. An aliquot of the reagent is tested by mixing a 50-μl aliquot with 100 μl of a solution containing 1 mg triphenylmethane in DMSO. An intense red color should develop immediately.

Derivatization: An amount in the range 0-100 μg nucleoside is dissolved in a mixture of 100 μl DMSO and 50 μl of the diluted methylsulfinyl carbanion reagent. After a reaction time of 1 min, 100 μl of methyl iodide is added and the mixture allowed to stand for 1 hr. The reaction is stopped by addition of 5 ml of H_2O, and the *N,O*-permethylated nucleosides are extracted into 5 ml $CHCl_3$. The organic layer is washed 3 times with 5 ml H_2O, dried over Na_2SO_4, and evaporated to dryness under a nitrogen stream at 60°C. The residue obtained is dissolved in $CHCl_3$, amount depending on the application (see further).

2. *Comments*

The *N,O*-permethylated derivatives are more stable with respect to hydrolysis than the TMS derivatives, and their low molecular weight is favorable for MS work. The derivatives have excellent chromatographic properties on different column systems. They possess a well-defined structure, and the reaction scheme followed allows the preparation of a diversity of alkyl derivatives [CH_3-, $CH_3(CH_2)_n-$, CD_3 etc.]. The reproducibility of the procedure was found to be excellent. As has been pointed out by De Leenheer and Gelijkens [48], there is a remarkable influence of the methylsulfinyl carbanion reaction time on the derivatization process. Better yields are obtained at reaction times less than 1 min. For quantitative work, shorter reaction times can be used (higher sensitivity); this does not matter if the internal standard is added to the sample (plasma or urine) before the derivatization [48].

III. INSTRUMENTATION

An analytical method developed for the determination of nucleosides in urine and plasma must fulfill two requirements: sensitivity and selectivity. Although the sample preparation technique can partly separate the compounds of interest from the biological fluid, the choice of the chromatographic separation technique and the detection method are in this respect much more important.

A. Capillary Gas Chromatography (CGC)

In GC the use of glass capillary columns offers several well-known advantages over packed columns: better separation, increased sensitivity (lower detection limit due to less peak broadening), better detector stability, shorter analysis time, improvement of the quantitative data, etc. Another point which must be emphasized is that by the high efficiency of a capillary column, the selectivity of the stationary phase is less important than in packed column chromatography. In the laboratory of one of the authors (P.S.), which is involved with all kinds of analyses, 95% of the applications can be performed on two stationary phases: an apolar gum pase (OV-1) and a polar gum phase (Superox) [55].

The advantages of capillary GC over packed column GC in nucleoside analysis are clearly illustrated in Figure 25.2. The analysis of some pyrimidine nucleosides is compared on a packed column (GasChrom Q 100-120 mesh, 2.3 m × 2.0 mm I.D.) coated with 3% SP 2250 (a methyl phenyl silicone polymer) and on

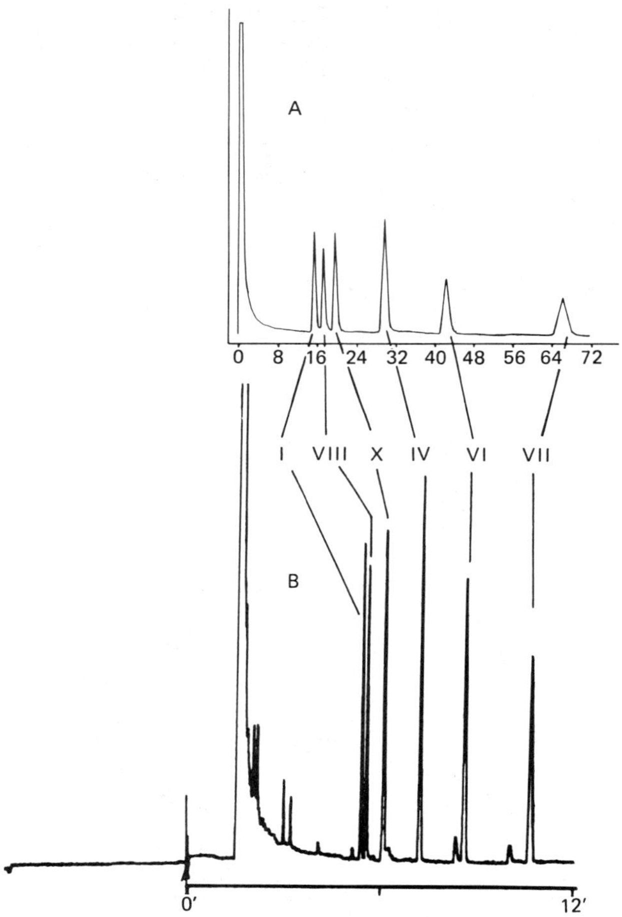

Figure 25.2 Comparison of GC (A) and CGC (B) of some nucleosides as their *N,O*-permethyl derivatives. Conditions: see text; for peak numbering see Table 25.1.

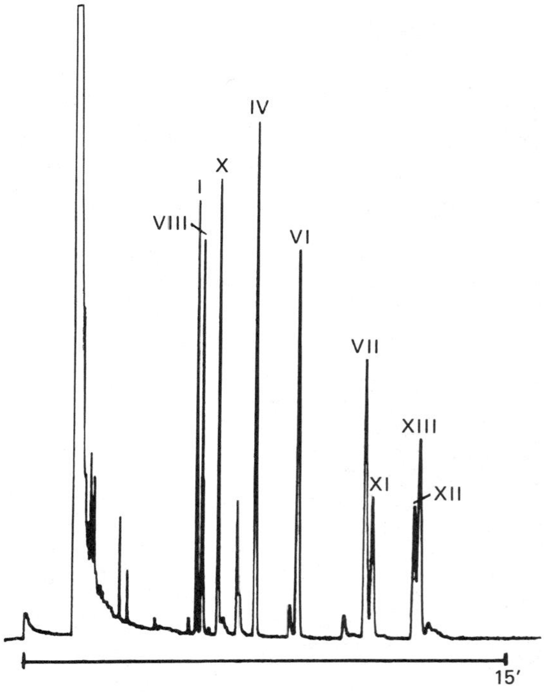

Figure 25.3 CGC separation of pyrimidine and purine nucleosides as their *N,O*-permethyl derivatives. Column: 20 m × 0.5 mm I.D. deactivated by HTS treatment and coated with OV-1 (df, 0.3 μm). Temperature: 200° isothermally, cold on-column injection (Carlo Erba Fractovap 4160). For peak numbering, see Table 25.1.

a capillary column (20 m × 0.5 mm I.D.) deactivated by HTS (*h*igh-*t*emperature-silylation) treatment [56,57] and coated with OV-1 (df, 0.3 μm). Both analyses were run isothermally at 200°C.

For packed-column GC it was established [48] that the best results in terms of separation, selectivity, and good peak symmetry were obtained on medium polar methyl phenyl silicone phases (SP-2230, OV-17, OV-25). Poor resolution and tailing factors up to 33 were noted on OV-1 [58]. As can be deduced from Figure 25.2b, separation on the HTS-OV-1 capillary column is excellent, and the analysis time is reduced by a factor of 6. Moreover, the amount injected on the packed column was 500 ng and on the capillary column 5 ng per nucleoside. Baseline separation of epimers such as FUDR/3'-*epi*-FUDR and Ado/Ara-A (Figure 25.3) is, however, not possible on apolar phases. On Superox columns, good resolution is obtained as illustrated in Figure 25.4, showing the analysis of 2'-deoxyadenosine, 9β-D-arabinofuranosyladenine, and adenosine on a 15 m × 0.3 mm column, coated with Superox-4 and operated isothermally at 230°C.

B. Injection Technique

The two preferred injection techniques for high-temperature quantitative capillary gas chromatography are the falling needle injector [59] and the cold

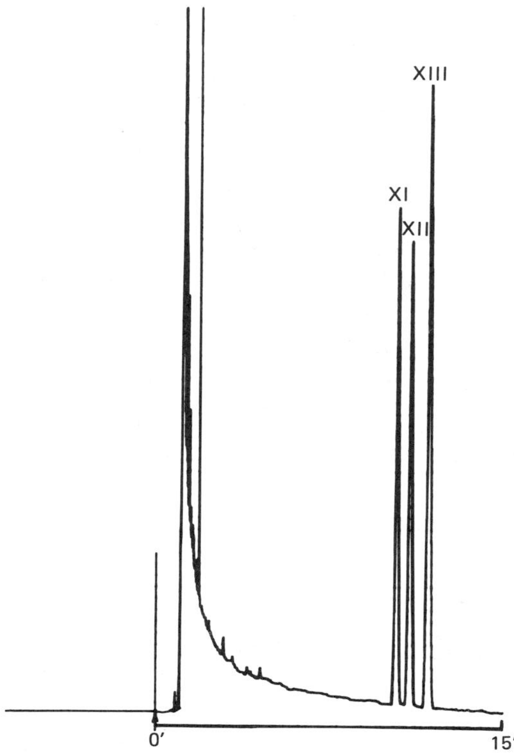

Figure 25.4 CGC separation of some purine nucleosides as their *N,O*-permethyl derivatives. Column: 15 m × 0.3 mm I.D. coated with Superox-4 (df, 0.2 μm). Temperature, 230° isothermally. For peak numbering, see Table 25.1.

on-column injector [60,61]. They have both been applied to nucleoside analysis. The potentials of both systems were recently discussed [62]. The possibility of introducing large injection volumes (1-3 μl) into the capillary column contributes to a greater overall sensitivity.

C. Detection Technique

Sensitivity and selectivity are furthermore enhanced if specific detectors are used. The thermionic nitrogen detector (NP detector) provides high sensitivity and specificity for nucleosides, and detection limits of 10 ng/ml biological fluid have been reported [31]. Best results, however, can be achieved with mass spectrometry, and more especially, with selected ion monitoring (SIM). The electron impact [EI or CE (He)] mass spectra of *N,O*-permethylated nucleosides show molecular ions of low relative intensity (<1%) due to the intense fragmentation of the molecule. This is illustrated by the CE (He) mass spectra of *N,O*-permethyl-5-FUDR (Figure 25.5). Moreover, the fragmentation pattern is complex, and the electron impact mode can hardly be applied to quantitative measurements of nucleosides in the low concentration range such as found in biological fluids. Mass spectra recorded under chemical ionization (CI) conditions show much less fragmentation. The mass spectra of permethylated 5-FUDR in the CI (CH_4), CI(iC_4H_{10}), and CI(NH_3) modes are shown in

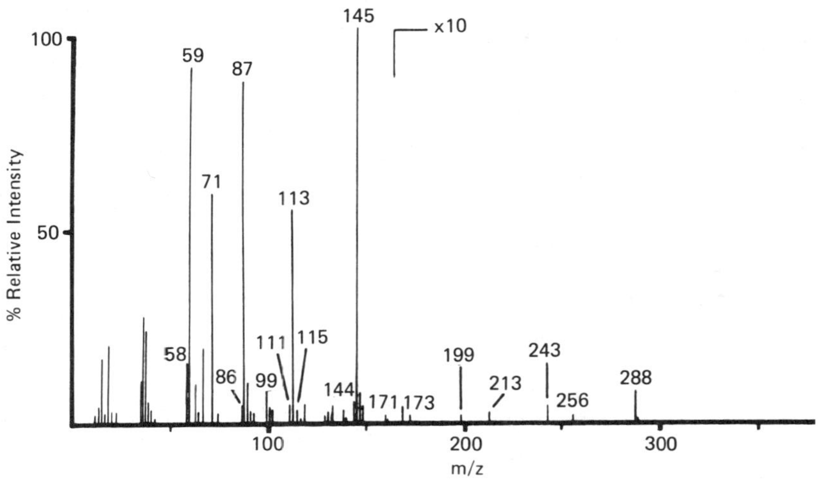

Figure 25.5 CE-He (electron impact) mass spectrum of *N,O*-permethyl 5-FUDR (Finnigan 3200 quadrupole GC-MS).

Figure 25.6A,B, and C. The use of methane or isobutane as reagent gas gave less favorable results as far as sensitivity for selected ion monitoring is concerned. Ammonia as reagent gas yields less fragmentation; an intense quasi-molecular ion $[MH]^+$ as well as a characteristic adduct ion $[M.NH_4]^+$ is obtained and it is therefore the best choice [63]. For the highest sensitivity, working conditions and internal standard are selected in such a way that single-ion monitoring can be applied. Focusing the mass spectrometer alternatively on two or more *m/z* values is associated with sensitivity loss. Nucleoside serum concentrations down to 1 ng/ml have been measured with a satisfactory signal-to-noise ratio, provided a clean ion source was available. This sensitivity is reached only if the capillary column is directly coupled with the ion source [64], preventing sample loss in a separator. It is our experience that separator efficiency (LKB 9000S MS) can be as low as 14%.

IV. APPLICATIONS

A. Determination of Uracil and Thymine Nucleosides in Picomole Amounts by CGC-MS (SIM) [26]

C. Finn, H. J. Schwandt, and W. Sadée described a method for the determination, with high precision and sensitivity, of free uracil (Ura) and thymine (Thy) at a picomol level in plasma samples, utilizing stable isotope dilution and capillary gas chromatography electron-impact mass spectrometry in the selected ion monitoring mode. $[^{15}N_2]$Ura and $[^{2}H_3]$Thy served as internal standards. This assay has been extended to the analysis of very low levels of nucleosides by developing a method for quantitative hydrolysis to the respective bases.

Samples containing 10-1000 pmol of Ura and Thy nucleosides in aqueous solution and 200 pmol $[^{15}N_2]$Ura and $[^{2}H_3]$Thy as internal standards in methanolic solution were transferred to 2-ml glass ampules. The liquid was evaporated with nitrogen and 0.1 ml 6 N HCl was added. The ampules were flame sealed and

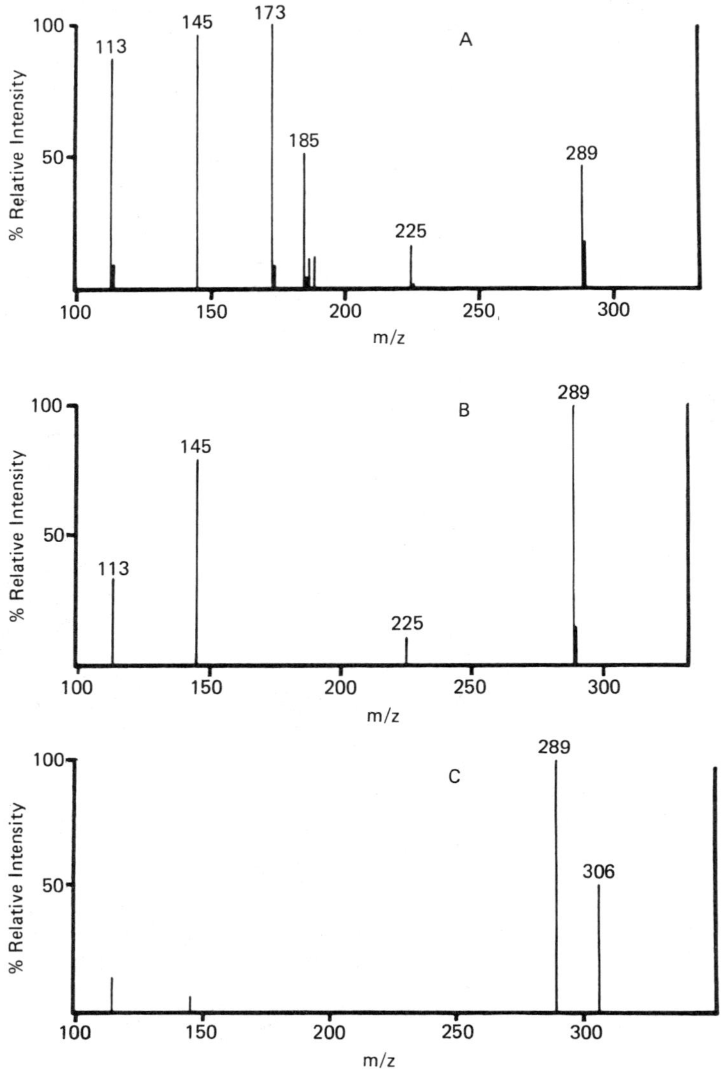

Figure 25.6 (A) CI(CH_4) (chemical ionization) mass spectrum of *N,O*-permethyl 5-FUDR. (B) CI(iC_4H_{10}) (chemical ionization) mass spectrum of *N,O*-permethyl 5-FUDR. (C) CI(NH_3) (chemical ionization) mass spectrum of *N,O*-permethyl 5-FUDR. (Finnigan 3200 qudrupole GC-MS).

heated in an aluminum block for 2 hr at 180°. The ampules were then opened, the HCl evaporated with nitrogen, and the residue transferred to a microcentrifuge tube using a freshly prepared mixture of ethyl acetate:methanol:conc. ammonium hydroxide (18:1:1). The overall yield of Ura and Thy formed by this method and transferred was about 60% and 80%, respectively. Both Ura and Thy were chemically stable under the hydrolysis conditions and no exchange of the isotopic labels was observed. The solutions containing Ura and Thy and [$^{15}N_2$]Ura and [$^{2}H_3$]Thy were evaporated in a vacuum centrifuge and the residue was derivatized using a mixture of BSTFA:pyridine (1:3) at 120° for 30 min giving the di-trimethylsilyl derivatives.

The samples were injected as well on packed as on capillary columns. The authors noted that "the major advantage of capillary columns lies in the clear separation of Ura-TMS and Thy-TMS from the solvent peak and in a greater number of theoretical plates of the column."

The capillary column (10 m × 0.35 mm I.D. coated with SE-52) was connected via an open split separator to the ion source of a Hitachi MS-52 mass spectrometer working in the electron impact mode. Detection was performed either by repetitive scanning and plotting reconstructed mass chromatograms or by SIM of low quantities of Ura and Thy (≪ 10 ng injected). The following ions were selectively monitored: *m/z* 255, $[M\text{-}CH_3]^+$ of $Thy(TMS)_2$; 256, $[M]^{+\cdot}$ of $Ura(TMS)_2$; and 258, $[M]^{+\cdot}$ of $[^{15}N_2]$-$Ura(TMS)_2$, and $[M\text{-}CH_3]^+$ of $[^2H_3]$-$Thy(TMS)_2$. A mass chromatogram of Ura-TMS and Thy-TMS is shown in Figure 25.7. Levels of free Ura and Thy were measured in human and rat plasma with a sensitivity of 20-40 pmol ml^{-1}. Determination of Ura and Thy nucleoside levels in biological fluids was not reported.

B. Quantitation of 5-Fluorouridine in Human Urine by CGC with a Nitrogen-Selective Detector [31]

The ribonucleoside 5-fluorouridine is assumed [14] to be an important intermediate in the formation of 5-fluoro-2'-deoxyuridine-5'-monophosphate, which is a strong inhibitor of thymidylate synthetase [19] and the active metabolite of the chemotherapeutic agent 5-fluorouracil [13]. Chadwick et al. demonstrated

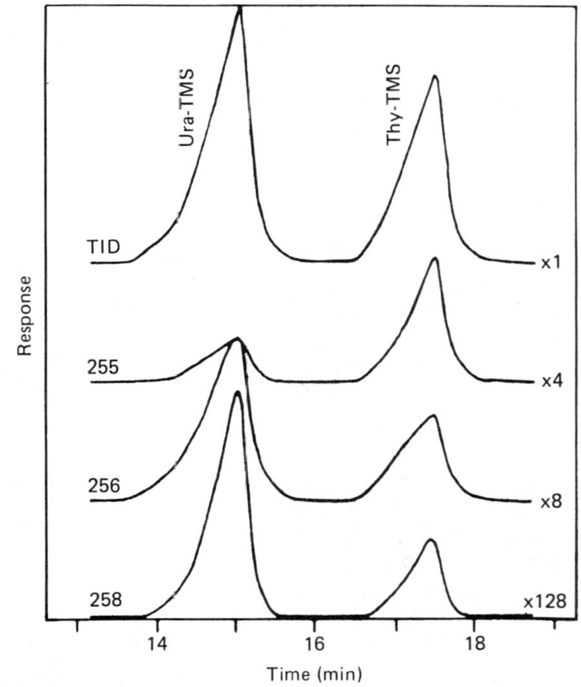

Figure 25.7 Mass chromatogram of Ura-TMS and Thy-TMS obtained by capillary GC-MS under electron impact. (From Ref. 26.)

the presence of 5-fluorouridine in blood, urine, and tissues of mice bearing solid L 1210 lymphocytic leukemia after i.v. administration of ^{14}C-labeled 5-fluorouracil [65] and ^{14}C-labeled 5-fluoro-2'-deoxyuridine [66]. For pharmacokinetic studies of 5-fluorouracil and its metabolites in humans, a method was developed for the determination of 5-fluorouridine in urine samples [31]. 5-Chlorouridine was added as internal standard to the urine sample, and the ribonucleosides were isolated by application of the on-column extraction procedure with immobilized phenylboronic acid. Due to the complex nature of the urine sample, an additional clean-up step by anion exchange on AG1-X4(Cl^-) was found necessary to remove coextracted endogenous impurities that could interfere with the gas chromatographic analysis using the thermionic nitrogen detector. The isolated nucleosides were transformed into the *N,O*-permethyl derivatives and injected on a glass capillary whisker-type column, 15 m × 0.25 mm I.D., coated with FFAP (Carbowax 20 M modified with 2-nitroterephthalic acid anhydride). The column was installed in an HP 5710 A gas chromatograph fitted with a nitrogen-phosphorus thermionic detector. Linear calibration curves were obtained by plotting the peak height ratios of 5-fluorouridine/internal standard versus the urinary concentrations of standardization samples prepared by supplementing blank urine with known amounts of 5-fluorouridine.

Figure 25.8 shows the chromatogram of a urine extract containing 0.270 μg/ml of 5-fluorouridine and 1.0 μg/ml of 5-chlorouridine (IS). The precision of the technique was excellent since the internal standard is added directly to the urine, thus compensating for eventual losses during analysis. The within-run precision (CV) of the technique was 5% (n = 10) at the 200 ng/ml concentration level, while the minimal detectable amount was found to be 10 ng of 5-fluorouridine/ml when a 1-ml urine sample was analyzed as described. The analytical recovery of 5-fluorouridine was established by the external standard method in the 0-1 μg/ml concentration range and was found to be 68 ± 6% (S.D.).

C. Identification and Quantitation of 5-Fluorouridine in Rat Urine After Administration of 5-FURD or 5-FUDR [67]

The previous study led to the estimation of 5-fluorouridine formed in rats after administration of 5-fluorouracil (5-FURD) or 5-fluoro-2'-deoxyuridine (5-FUDR) as reflected by its urinary excretion. The nucleosides were isolated by anion-exchange extraction (u_A) as well as by selective adsorption on Boric Acid Gel (u_B) (Section II.A.2). The extracts were permethylated using potassium *tert*-butoxide in dimethyl sulfoxide and methyl iodide as previously described (Section II.B.1).

The extracts were subjected to TLC, HPLC, CGC/MS, and CGC on two different columns: (1) 12 m × 0.25 mm (I.D.) whisker-type, statically coated with OV-25; (2) 15 m × 0.25 mm (I.D.), directly coated with Superox-FA (a thermostable FFAP-like material; RSL, Eke, Belgium). These analyses revealed the presence of 2'-deoxyuridine, thymidine, and 5-fluorouridine.

The 5-fluorouridine levels were measured by CGC with nitrogen-selective detection. Some representative results are summarized in Table 25.2. Figure 25.9A shows the chromatogram of a blank rat urine extract and Figure 25.9B a chromatogram of rat urine extract after i.p. administration of 5-fluorouracil

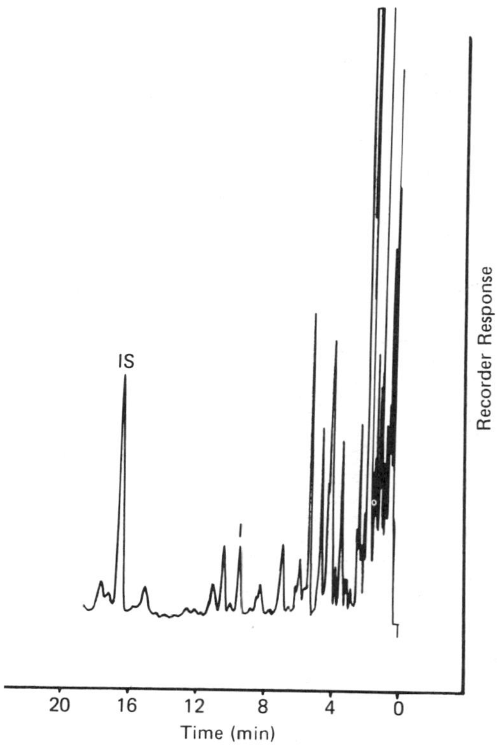

Figure 25.8 CGC chromatogram of a derivatized urine extract containing 0.270 μg/ml of 5-fluorouridine (I) and 1.0 μg/ml of internal standard (IS), 5-chlorouridine. Column: 15 m × 0.25 mm with FFAP; oven temperature, 220°; carrier gas, He + H_2 (9 + 1); flow-rate, 3 ml/min. NP thermionic detector (Model 18789A Hewlett-Packard). (From Ref. 31.)

Table 25.2 Urinary 5-Fluorouridine Excretion in Rats as Determined by CGC

Series 1 (5-fluorouracil, 200 mg/kg, i.p.)		Series 2 (5-fluoro-2'-deoxyuridine, 380 mg/kg, i.p.)	
Rat	μg/kg/24 hr	Rat	μg/kg/24 hr
1	4.6	13	167
2	7.6	14	120
3	2.8	15	184
4	5.9	16	245
5	7.0	17	99
6	8.9	18	407

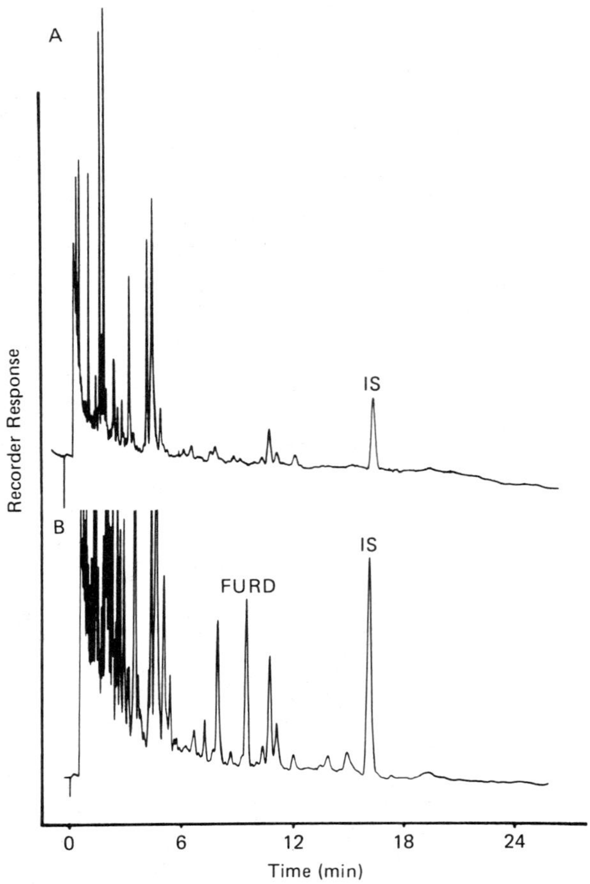

Figure 25.9 (A) CGC of a permethylated blank rat urine extract (u_B) (B) CGC of a permethylated rat urine extract (u_B) obtained after i.p. administration of 5-fluorouracil (200 mg/kg). Column: 15 m × 0.25 mm I.D. coated with Superox-FA (RSL Belgium), 220°C, NP detection. FURD = 5-fluorouridine, IS = internal standard (5-chlorouridine). (From Ref. 67.)

(200 mg/kg). The urinary 5-fluorouridine levels were determined in samples collected between 0 and 24 hr after drug administration. Samples collected between 24 and 48 hr were close to the detection limit (10 ng/ml). These data indicate that only a small amount of the dose (5-fluorouracil or 5-fluoro-2'-deoxyuridine) was excreted as 5-fluorourudine.

D. Measurement of 5-Fluoro-2'-deoxyuridine Serum Levels by CGC Chemical-Ionization Mass Spectrometry [68]

The 2'-deoxynucleoside 5-fluoro-2'-deoxyuridine (5-FUDR) is a useful antineoplastic agent [69,70]. Although the mode of action of 5-FUDR is established, there is a lack of understanding of the pharmacokinetics of the drug [22]. Monitoring 5-FUDR concentrations in body fluids in order to achieve an adequate therapeutic effect and avoid unacceptable toxicity has been studied by Gelijkens et al. [68]. After addition of 3'-*epi*-FUDR (IS), the drug was isolated from the serum sample by anion exchange [AG1-X4(Cl^-)] chromatography. The *N,O*-permethylated derivatives were prepared and the samples were analyzed on a capillary column (10 m × 0.25 mm I.D.) coated with Superox-4 [55]. The column was coupled directly to the ion source [64] of a quadrupole mass spectrometer operating in the CI(NH_3) mode; CI-SIM measurements were performed by focusing the instrument on *m/z* 289 ($[MH]^+$ of the *N,O*-permethyl derivatives of 5-FUDR and 3'-*epi*-FUDR). The epimers (5-FUDR and 3'-*epi*-FUDR are well separated on the capillary column (Figure 25.10), permitting single-ion-monitoring measurements by focusing on one ion. The loss in sensitivity associated with focusing alternatively on two or more *m/z* values is thus eliminated.

A representative selected-ion-monitoring recording is shown in Figure 25.11. Recovery of 5-FUDR from human serum averaged 93 ± 2% (S.D.) in the 0-1μg/ml range; using single ion monitoring, a linear detector response was obtained for concentrations up to 50 ng 5-FUDR/ml, while a detection limit of about 1 ng/ml was attained. The method was applied to in vivo studies by measuring the disappearance curves of 5-FUDR in serum of treated cancer patients. The time course of 5-FUDR in serum levels for a patient who had received 25 mg kg^{-1} i.v., is shown in Figure 25.12. A linear elimination phase is observed that seems to be leveling off after some time.

E. Analysis of Purinyl Nucleoside Isomer Composition by CGC [71]

CGC is also very useful to check the purity of nucleosides used for drug administration [67] or formed by synthetic procedures. To obtain the nucleosides in a pure form, reaction products are mostly subjected to preparative liquid chromatography. The advantage of capillary gas chromatography-mass spectrometry over other identification techniques is the small amount required (μg). The possibilities of CGC in this respect have been recently demonstrated by P. Fischer, G. R. Lösch, and D. Müller [71].

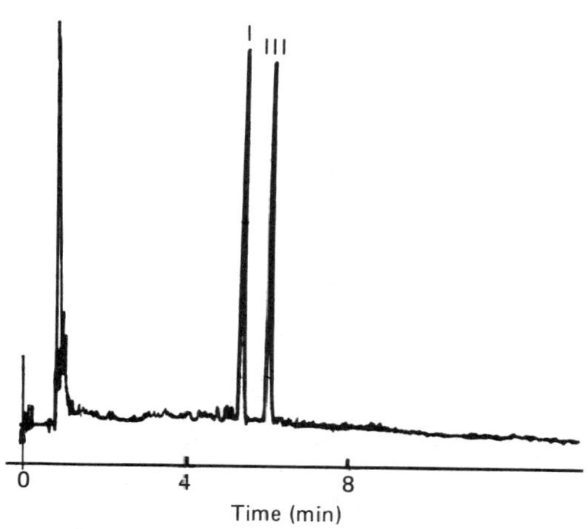

Figure 25.10 CGC chromatogram of the epimers 5-FUDR and 3'-*epi*-FUDR on a capillary column (10 m × 0.25 mm) coated with Superox-4, operated at 200°C.

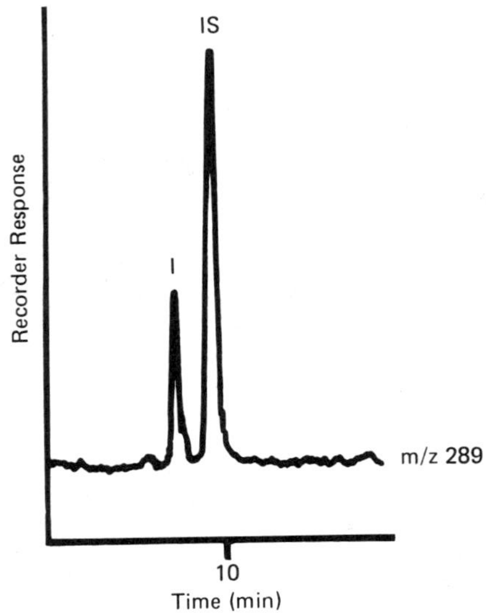

Figure 25.11 CI(NH_3) selected ion monitoring of a derivatized serum sample containing 15 ng 5-FUDR (I)/ml, IS = internal standard, 3'-*epi*-FUDR. A Finnigan 3200 MS was operated under the following conditions: multiplier voltage, 2.4 kV; reagent gas, NH_3 (0.75 torr). (From Ref. 68.)

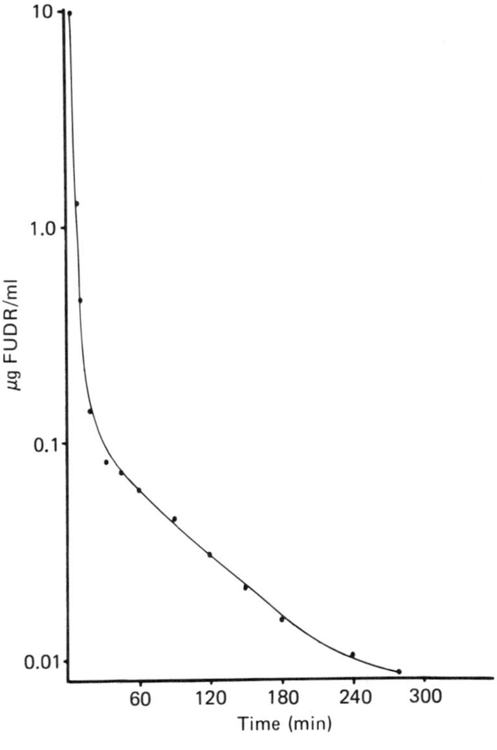

Figure 25.12 5-FUDR serum levels of a patient at different times after i.v. administration of a 25 mg kg^{-1} dose. (From Ref. 68.)

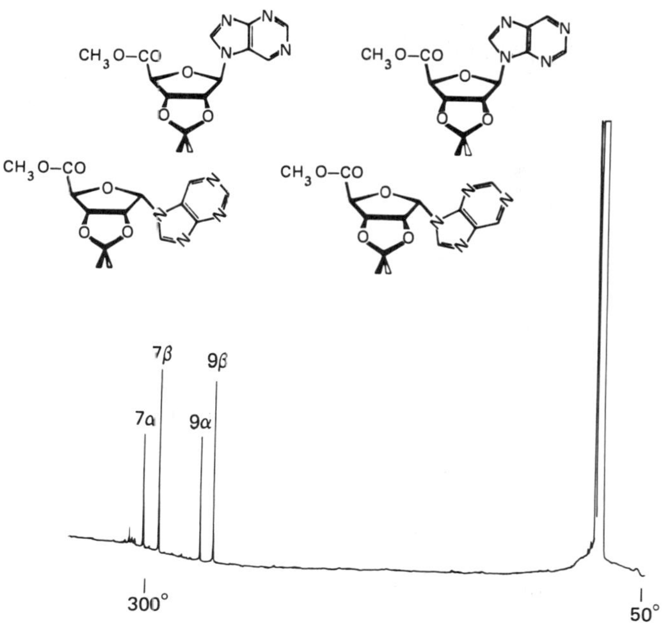

Figure 25.13 GC of an artificial mixture of 9α-, 9β-, 7α-, and 7β-purinyl methyl-2',3'-*O*-isopropylidene-D-ribofuranuronates dissolved in CH_2Cl_2 (on-column injection at 50°; temperature-programmed 5°/min to 300°; carrier He, 0.8 bar). (From Ref. 71.)

The four linkage isomers of purinyl nucleosides were well separated on a 20 m × 0.3 mm column treated with HMDS and coated with SE-52. A CGC trace of the separation of an artificial mixture of 9α-, 9β-, 7α-, and 7β-purinyl methyl 2',3'-*O*-isopropylidene-D-ribofuranuronates is shown in Figure 25.13.

V. SUMMARY

Some recent methods in instrumentation, sample preparation, and derivatization for the analysis of nucleosides by capillary gas-liquid chromatography have been presented, as well as several applications of quantitative measurements in biological fluids, using different detection systems (flame ionization detector, thermionic nitrogen-selective detector, mass spectrometer). It is evident that glass capillary columns offer several major advantages over analytically packed columns for qualitative and quantitative measurements in the nucleoside field.

Future developments and refinements of the analytical techniques may include use of isotopically (^{13}C, ^{2}H, ^{15}N, ^{18}O) labeled nucleosides [72] as internal standards for increased precision in selected-ion-monitoring work. Additionally, the possible use of combined glass capillary gas chromatography-negative-ion chemical ionization-mass spectrometry [73,74] should be mentioned. As a result of the greatly increased sensitivity associated with this technique, detection of femtogram amounts of phenylethyl amines (as pentafluorobenzoyl derivatives) has already been achieved [75]. The application in the nucleoside field might require development of suitable derivatives with satisfactory electron-capture cross section; pilot studies using the electron capture detector may be useful for this purpose.

Also, attempts could be made to synthesize phosphorous-containing nucleoside derivatives; indeed, the thermionic nitrogen-selective detector has an even greater sensitivity for phosphorus compounds. In this way, even lower detection limits could be attained [76,77].

ACKNOWLEDGMENTS

Thanks are due to Prof. Dr. M. Verzele for the facilities extended in preparing this contribution and for his interest in this work.

REFERENCES

1. P. A. Levene and W. A. Jacobs, *Chem. Ber. 42*:2474 (1909).
2. P. A. Levene, L. A. Mikeska, and T. Mori, *J. Biol. Chem. 85*:785 (1930).
3. J. D. Watson and F. H. Crick, *Nature 171*:737 (1953).
4. A. Bloch in *Drug Design* (E. J. Ariens, ed.), Academic Press, New York and London, 1973, vol. IV, p. 285.
5. J. P. Howard, N. Cevik, and M. L. Murphy, *Cancer Chemother. Rep. 50*:287 (1966).
6. J. J. Fox, K. A. Watanabe, and A. Bloch, *Prog. Nucleic Acids Res. Mol. Biol. 5*:251 (1966).
7. L. E. Stephen and A. R. Gray, *J. Parasitol. 46*:509 (1960).
8. H. J. Eggers and I. Tamm, *Ann. Rev. Pharmacol. 6*:231 (1966).
9. F. M. Schabel, Jr., *Ann. N.Y. Acad. Sci. 173*:215 (1970)
10. M. Rockwell and M. H. Maguire, *Mol. Pharmacol. 2*:574 (1966).
11. R. S. Schwartz in *Human Transplantation* (F. T. Rapaport and J. Dausset, eds.), Grune and Stratton, New York, 1968, p. 440.
12. B. J. Srivastava, *Int. Rev. Cytol. 22*:349 (1967).
13. R. Duschinsky, E. Pleven, and C. Heidelberger, *J. Am. Chem. Soc. 79*:4559 (1957).
14. C. Heidelberger, *Prog. Nucleic Acids Res. Mol. Biol. 4*:1 (1965).
15. P. V. Danenberg, *Biochim. Biophys. Acta 473*:73 (1977).
16. C. Heidelberger, in *Handbook of Experimental Pharmacology* (A. C. Sartorelli and D. G. Johns, eds.), Springer-Verlag, Berlin, Heidelberg, 1975, Vol. 38/2, p. 193.
17. G. B. Grindey and C. A. Nichol, *Cancer Res. 32*:527 (1972).
18. R. C. Jackson, *J. Biol. Chem. 253*:7440 (1978).
19. K. U. Hartmann and C. Heidelberger, *J. Biol. Chem. 236*:3006 (1961).
20. D. G. Johns, in *Handbook of Experimental Pharmacology* (A. C. Sartorelli and D. G. Johns, eds.), Springer-Verlag, Berlin, Heidelberg, 1974, Vol. 38/1, p. 270.
21. D. Kessel and T. C. Hall, *Cancer Res. 29*:1749 (1969).
22. C. E. Myers, R. Diasio, H. M. Eliot, and B. A. Chabner, *Cancer Treatment Rev. 3*:175 (1976).
23. K. H. Schram and J. A. McCloskey in *GLC and HPLC Determination of Therapeutic Agents* (K. Tsuji, ed.), Marcel Dekker, Inc., New York and Basel, 1979, part 3, p. 1149.
24. H. E. Kaufman, *Proc. Soc. Exp. Biol. Med. 109*:251 (1962).
25. H. E. Kaufman, E. D. Ellison, and W. M. Townsend, *Arch. Opththalmol. 84*:783 (1970).

26. C. Finn, H. J. Schwandt, and W. Sadée, *Biomed. Mass Spectrom.* *6*:194 (1979).
27. C. Finn and W. Sadée, *Cancer Chemother. Rep.* *59*:279 (1975).
28. C. Pantarotto, A. Martini, G. Belvedere, A. Bossi, M. G. Donelli, and A. Frigerio, *J. Chromatogr.* *99*:519 (1974).
29. A. P. De Leenheer and C. F. Gelijkens, *J. Pharm. Sci.* *67*:417 (1978).
30. C. F. Gelijkens and A. P. De Leenheer, *Abstracts,* 27th IUPAC Congress Helsinki, n° 217 (1979).
31. A. P. De Leenheer and C. F. Gelijkens, *J. Chromatogr. Sci.* *16*:552 (1978).
32. H. T. Miles and H. M. Fales, *Anal. Chem.* *34*:860 (1962).
33. R. L. Hancock and D. L. Coleman, *Anal. Biochem.* *10*:365 (1965).
34. C. Sweeley, R. Bentley, M. Makita, and W. Wells, *J. Am. Chem. Soc.* *20*:2497 (1963).
35. Y. Sasaki and T. Hashizume, *Anal. Biochem.* *16*:1 (1966).
36. T. Hashizume and Y. Sasaki, *Anal. Biochem.* *24*:232 (1968).
37. C. W. Gehrke and A. B. Patel, *J. Chromatogr.* *123*:335 (1976).
38. C. W. Gehrke and A. B. Patel, *J. Chromatogr.* *130*:103 (1977).
39. A. B. Patel and C. W. Gehrke, *J. Chromatogr.* *130*:115 (1977).
40. J. K. MacLeod, *J. Org. Chem.* *41*:3959 (1976).
41. J. J. Dolhun and J. L. Wiebers, *Org. Mass Spectrom.* *3*:669 (1970).
42. W. A. Koenig, L. C. Smith, P. F. Crain, and J. A. McCloskey, *Biochemistry* *10*:3968 (1971).
43. W. A. Koenig, L. C. Smith, P. F. Crain, and J. A. McCloskey, *Chem. Ber.* *105*:262 (1972).
44. J. J. Dolhun and J. L. Wiebers, *J. Am. Chem. Soc.* *91*:7755 (1969).
45. W. C. Butts, *J. Chromatogr. Sci.* *8*:474 (1970).
46. U. I. Krahmer, J. G. Liehr, K. J. Lyman, E. A. Orr, R. N. Stillwell, and J. A. McCloskey, *Anal. Biochem.* *82*:217 (1977).
47. K. M. Baker, *Adv. Mass Spectrom. Biochem. Med.* *1*:231 (1976).
48. A. P. De Leenheer and C. F. Gelijkens, *Anal. Chem.* *48*:2203 (1976).
49. B. L. Hillcoat, M. Kawai, P. B. McCulloch, J. Rosenfeld, and D. K. O. Williams, *Br. J. Clin. Pharm.* *3*:135 (1976).
50. D. L. von Minden and J. A. McCloskey, *J. Am. Chem. Soc.* *95*:7480 (1973).
51. P. A. Leclercq and D. M. Desiderio, Jr., *Anal. Lett.* *4*:305 (1971).
52. E. J. Corey and M. Chaykovsky, *J. Am. Chem. Soc.* *87*:1345 (1965).
53. J. Eagles, W. M. Laird, R. Self, and R. L. M. Synge, *Biomed. Mass Spectrom.* *1*:43 (1974).
54. J. I. Brauman, J. A. Bryson, D. C. Kahl, and N. J. Nelson, *J. Am. Chem. Soc.* *92*:6679 (1970).
55. M. Verzele and P. Sandra, *J. Chromatogr.* *158*:111 (1978).
56. K. Grob, G. Grob, and K. Grob, Jr., *HRC & CC* *2*:31 (1979).
57. M. Godefroot, M. Van Roelenbosch, M. Verstappe, P. Sandra, and M. Verzele, *HRC & CC* *3*:337 (1980).
58. C. F. Gelijkens, Ph.D. thesis, University of Ghent (1979).
59. P. M. J. Van den Berg and Th. P. H. Cox, *Chromatographia* 5:301 (1972).
60. K. Grob and K. Grob, Jr., *J. Chromatogr.* *151*:731 (1978).
61. M. Galli, S. Trestianu, and K. Grob, Jr., *HRC & CC* *2*:366 (1979).
62. M. Verzele, G. Redant, S. Qureshi, and P. Sandra, *J. Chromatogr.*, *199*:105 (1980).
63. M. S. Wilson and J. A. McCloskey, *J. Am. Chem. Soc.* *97*:3436 (1975).

64. F. Vangaever, P. Sandra, and M. Verzele, *Chromatographia 12*:153 (1979).
65. M. Chadwick and W. I. Rogers, *Cancer Res. 32*:1045 (1972).
66. M. Chadwick and C. Chang, *Cancer Treat. Rep. 60*:845 (1976).
67. C. F. Gelijkens, P. Sandra, F. Belpaire, and A. P. De Leenheer, *Drug Metabolism and Diposition 8*:363 (1980).
68. C. F. Gelijkens, P. Sandra, and A. P. De Leenheer, *Biomed. Mass Spectrom. 7*:572 (1980).
69. T. Buroker, M. Samson, J. Correa, R. Fraile, and V. K. Vaitkevicus, *Cancer Treat. Rep. 60*:1277 (1976).
70. B. Cady and R. A. Oberfield, *Am. J. Surg. 127*:220 (1974).
71. P. Fischer, G. R. Lösch, and D. Müller, *HRC & CC 3*:161 (1980).
72. B. H. Min and W. A. Garland, *Res. Comm. Chem. Pathol. Pharmacol. 22*:145 (1978).
73. D. F. Hunt and F. W. Crow, *Anal. Chem. 50*:1781 (1978).
74. W. A. Garland and B. H. Min, *J. Chromatogr. Sci. 17*:279 (1979).
75. D. F. Hunt and S. K. Sethi, in *High Performance Mass Spectrometry, Chemical Applications* (M. L. Gross, ed.), ACS Symposium Series, no. 70, 1978, p. 150.
76. K. Jacob and W. Vogt, *J. Chromatogr. 150*:339 (1978).
77. K. Jacob C. Falkner, and W. Vogt, *J. Chromatogr. 167*:67 (1978).

26

Analysis of Peptide Antibiotics by Glass Capillary GC-MS

Wilfried A. Koenig / *Institut für Organische Chemie, Universität Hamburg, Hamburg, Federal Republic of Germany*

I. INTRODUCTION

Peptide sequence analysis is a challenging task in the field of structure identification of natural products. Efficient microchemical methods such as the Edman degradation techniques have been elaborated for elucidating the primary structure of a peptide chain. Only occasionally have gas phase analytical procedures been applied to solve these structural problems, mainly because of the high polarity and low volatility of peptides and the limited amount of information obtained from small peptide fragments as compared to the large number of amino acids in a protein. Nevertheless, gas chromatography, combined with mass spectrometry as a sensitive and highly specific separation and detecting system, has been frequently used for structural studies of smaller peptides, particularly peptide antibiotics and peptide hormones.

The main problems encountered in determining the structure of peptide antibiotics arise from the great variety of structural peculiarities of this compound class:

1. Microorganisms generally do not produce only one peptide chain, but a variety of homologues or analogues, whereby amino acids are exchanged by others. This microheterogeneity causes problems for the isolation of pure compounds.
2. Many peptide antibiotics contain both D- and L-amino acid enantiomers , indicating that biosynthesis of peptide antibiotics occurs by a mechanism different from protein biosynthesis.
3. A large number of unusual amino acids have been detected in peptide antiobiotics. *N*-methylated or α-C-alkylated amino acids are quite common. Also amino acids with additional hydroxy or amino groups in different positions have been identified.
4. Peptide antibiotics usually contain constituents which are not amino acids, such as amino alcohols, fatty acids, β-keto acids, or hydroxy acids. In many cases, the end groups are blocked by substitution with these molecules or by intramolecular ring formation.

This property and the presence of D enantiomers and of unusual amino acids causes resistance against proteolytic enzymes. It is also the main reason why conventional analytical methods, stepwise degradation, and end-group determination techniques cannot routinely be applied to the investigation of peptide antibiotics.

Comparatively few peptide antibiotics are therapeutically used today; the penicillins, cephalosporins, bacitracin, gramicidin A, tyrocidine, polymyxin B, and the actinomycins are the most noteworthy among these. A review article on the properties of peptide antibiotics was published by Bodanszky and Perlman [1].

II. SAMPLE PREPARATION

Most peptides isolated from natural sources are too large for direct investigation by gas chromatography or mass spectrometry, even after derivatization. For this reason, partial cleavage of the peptide chain and formation of smaller peptide fragments are necessary before volatile derivatives are formed and the analysis by GC or GC-MS can be started (Scheme 1). From the fragments

Peptide Antibiotic

↓ conc. HCl/37°C/12-24 hr

Peptide Fragments (2-6 amino acids)

↓ volatile derivatives
↓

Separation on Glass Capillary Column
El and Cl Mass Spectra

Scheme 1

identified by mass spectrometry after gas chromatographic separation, the original total sequence may be derived, if a sufficient number of overlapping fragments can be identified. For an unambiguous arrangement of the partial sequences, the knowledge of at least one end-group of the peptide chain is necessary. Difficulties may arise when the amino end-group is blocked by small acyl groups (formyl, acetyl), or when the C-terminal group is an amino alcohol, or when a cyclic structure is formed by a peptide bond between the amino and carboxylic end groups.

Free end-groups may be detected by conventional methods such as dinitrophenyl (DNP) derivatization according to Sanger [2], or by dansylation [3], or by hydrazinolysis [4]. Gas chromatography can also be used for end-group determination in combination with sequential studies [5,6].

Several procedures have been suggested for the formation of volatile derivatives of peptides. Only two methods have been successfully applied in sequence analysis of peptides by GC-MS: the formation of trifluoroacetylated peptide esters according to Weygand and co-workers [7] and the reduction of *N*-acetyl peptides to *N*-ethyl polyamino alcohols and the investigation of the *O*-trimethylsilylated derivatives according to Biemann and his associates [8-10]. Permethylated peptide derivatives [11], although frequently used for mass spectrometric sequence analysis, have not been adopted for GC investigations.

A. Partial Cleavage of Peptides

Mild acid hydrolysis of peptides may be achieved by using concentrated hydrochloric acid at 37° for 12 hr or with various concentrations of HCl and temperatures and hydrolysis times between a few hours and several days. Nonaqueous trifluoroacetic acid may also be used as an especially mild reagent for selective cleavage of labile peptide bonds. Acidic hydrolysis is rather unspecific, although certain peptide bonds are very sensitive against acidic conditions, particularly when aspartic acid, serine, threonine, and glycine are involved; both amino acids participating in a peptide bond naturally influence their strength, however. In other cases, high stability due to steric hindrance has been observed, especially when valine or isoleucine is forming the peptide bond. Unpredictable stabilities may occur in peptide antibiotics with uncommon amino acids. As an example, the extreme lability of the peptide bond between α-aminoisobutyric acid (Aib) and proline in alamethicins and related peptide antibiotics should be mentioned [12]. A review article on protein hydrolysis has been published by R. Hill [13].

Only in rare cases do proteolytic enzymes display activity against peptide antibiotics. Exo- and endopeptidases, however, may be used to cleave other peptides specifically at certain peptide bonds. Particularly useful for sequencing peptides by GC-MS is the application of dipeptidylaminopeptidases (cathepsins), as demonstrated by Caprioli et al. [14].

B. Formation of Derivatives for GC-MS

As already mentioned, two procedures are of major interest in GC-MS applications.

1. *Acylated Peptide Esters (Scheme 2)*

In this type of derivative, the high polarity is reduced by substitution of free amino, hydroxy, and carboxyl groups with retention of the intact peptide

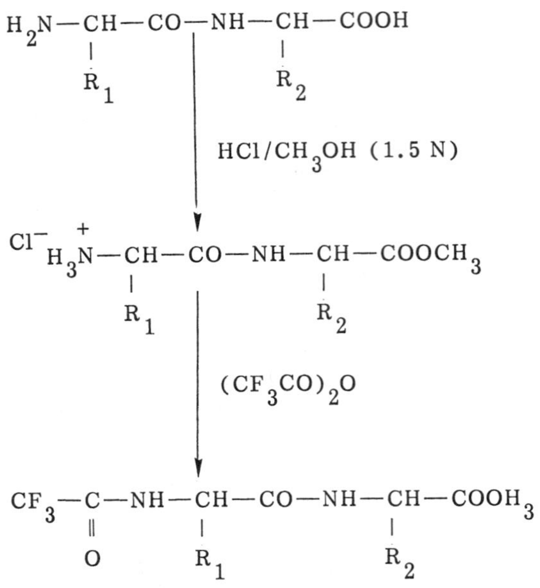

Scheme 2

bond. As acyl groups, the trifluoroacetyl (TFA) and, to a minor degree, the pentafluoropropionyl groups have come in use. The TFA group was introduced by Weygand and his associates as an amino-protecting group in peptide synthesis. In the following years Weygand and Prox and other groups made use of the high volatility of TFA amino acid and peptide esters in gas chromatography and mass spectrometry [15]. They could demonstrate, that *N*-TFA peptide methyl esters up to pentapeptides can be eluted from packed GC colunms [16], and that the mass spectra show distinct sequence ions by fragmentation of the amide bonds.

The derivatization proceeds in two steps. In the primary step, esterification of the free carboxyl end-group and simultaneously of the side-chain carboxyl functions of aspartic acid and glutamic acid is brought about by using a solution of dry HCl gas in methanol (∿1.5 N) at room temperature or, alternatively, at 100°C for 30-60 min, which gives higher yields but with possible cleavage of sensitive peptide bonds. Also, diazomethane in methanol solution may be used for esterification, but this reagent gives only poor yields of esters and occasionally results in N-methylation.

In the second step, N-trifluoroacetylation of amino and imino groups and O-trifluoroacetylation of hydroxy groups are achieved using trifluoroacetic acid anhydride [16] in dichloromethane solution at room temperature or at higher temperatures. Other trifluoroacetylating agents such as methyltrifluoroacetic acid thioesters, and *N*-methyl-*N*-trifluoroacetyl-trifluoroacetic acid amide [17] react only with primary and secondary amino groups, leaving the hydroxy groups of serine, threonine, and tyrosine unaltered. The yields of N-acylation are not always satisfactory with these latter reagents.

Side reactions occurring during this derivatization procedure may arise through acidic hydrolysis of peptide bonds; elimination of hydroxyl or thiol functions in serine, threonine, and cysteine; cyclization of glutamic acid to pyrrolidone carboxylic acid; formation of cyclic 2-trifluoromethyl-Δ^3-oxazolidine-5-ones [18] or diketopiperazines [19]; and the hydrolysis of amide functions of asparagine or glutamine during the esterification reaction.

2. *O-Trimethylsilylated Polyamino Alcohols (Scheme 3)*

To increase further the volatility of peptides, Biemann et al. [8] in 1959 suggested the reduction of the carbonyl groups of acetylated peptide esters to $-CH_2$ groups by reaction with lithium aluminium hydride. This results in the formation of *N*-ω-ethyl polyamino alcohols, which exhibit high volatility and excellent mass spectrometric fragmentation properties. Still higher volatility can be achieved by using *N*-TFA or *N*-PFP derivatives as starting material; after reduction, partially fluorinated *N*-alkyl derivatives are obtained.

The derivatization procedure starts with the esterification of the peptide partial hydrolyzate, usually by using a solution of HCl gas in methanol, as described above. Esterification is not necessary, however, if diborane is used as a reducing agent [9,20]. In this case, the first step of the derivatization procedure is the acylation step. Acetylation of a peptide mixture may be brought about by dissolving the sample in a 1:1 mixture of acetic acid anhydride and dry pyridine and keeping the reaction mixture at room temperature for 45 min, or by treating the peptides with a mixture of 200 µl of acetic acid, 200 µl of methanol, and 75 µl of acetic acid anhydride at room temperature for 15 hr.

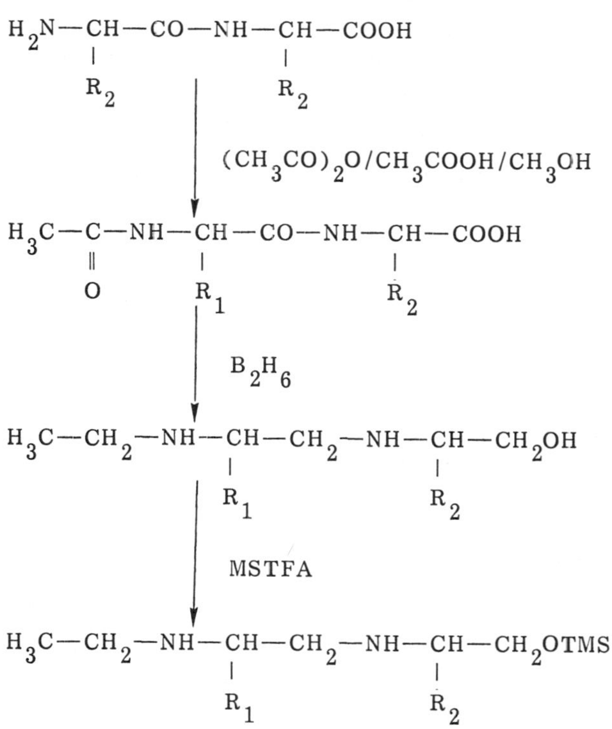

Scheme 3

Trifluoroacetylation proceeds as described above. The acylation, particularly trifluoroacetylation of free peptides, however, may cause racemization of the C-terminal amino acid by formation of an oxazolinone derivative [18]. As a result of the racemization, diastereoisomeric peptide derivatives are generated, which are separated on good GC columns and may complicate the analysis.

Reduction of acylated peptides may be performed by an excess of a solution of diborane in tetrahydrofurane and refluxing the reaction mixture for 4 hr. The reaction is stopped by adding a small amount of HCl solution in methanol. To destroy borane complexes, the mixture has to be heated for 30 min at 90°C, and the solvent and boric acid methyl ester evaporated by a stream of nitrogen [9]. Repeated addition of methanol and evaporation to dryness seem to be of advantage for the removal of side products.

In the final step, *O*-trimethylsilyl (TMS) derivatives of the polyamino alcohols are obtained by addition of *N*-methyl-*N*-trimethylsilyl trifluoroacetamide at room temperature or by reaction with a mixture of trimethylsilyl diethylamine and pyridine at 55°C.

III. TECHNIQUES AND INSTRUMENTATION

A. Gas Chromatography

Gas chromatography and electron-impact mass spectrometry have been used repeatedly for sequence analysis of peptide antibiotics [21-24]. However, most of the work was done with packed columns and the application of glass capillary columns has never been described before in this field. This may partly be due to the fact that high-temperature stable glass capillaries with efficiently deactivated inner surfaces are very important for the separation of polar peptide derivatives. Glass capillaries with such properties are still not generally available.

Our own investigations [25] of the properties of glass capillary columns, prepared according to different procedures which are described in detail in the literature, show us that high-temperature gas-phase Carbowax 1500 deactivation [26] before coating of the columns is advantageous. Also the precipitation of a layer of colloidal silicic acid, as proposed by Schulte [27], seems to support the formation of a steady and relatively inert inner surface for the film of the stationary phase.

B. Mass Spectrometry

Mass spectral fragmentation behavior of peptide derivatives has been discussed in detail in several articles in the literature [28,29]. The occurrence of sequence-indicating ions A_1, A_2, A_3..., B_1, B_2. B_3..., or Z_1, Z_2, Z_3... for each amino acid in the peptide chain is of major importance. These ions are produced by α cleavage as a result of charge localization at a heteroatom:

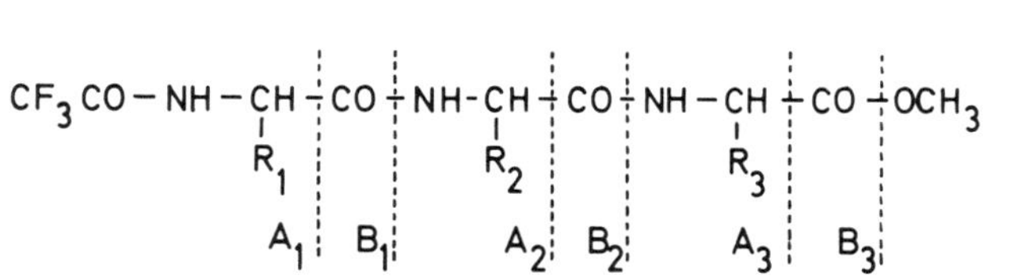

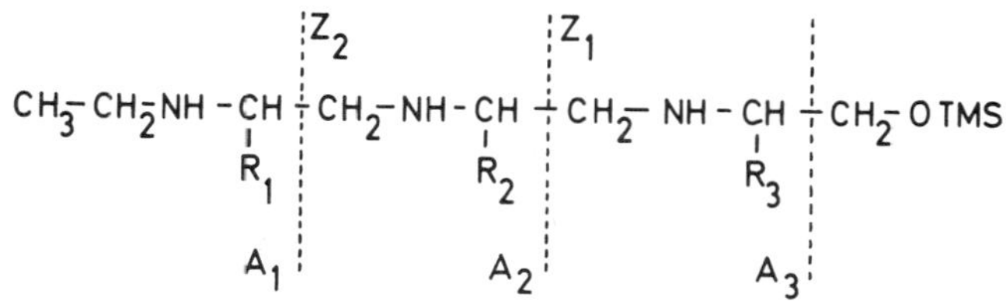

In addition to sequence information, the identification of the molecular ion is highly appreciable. It is a general experience that in either type of derivative, the relative intensity of sequence-indicating ions decreases towards the high-mass end of the spectra.

In the electron impact (EI) mass spectra of peptide derivatives, molecular ions above 1% relative intensity are rarely observed. This makes the identification of the C-terminal amino acid in peptides larger than dipeptides very difficult. In recent years, however, chemical ionization (CI) mass spectrometry has become an important complementary tool [30,31]. The main difference between CI and EI mass spectra is the presence of very intense protonated molecular ions and low-intensity fragment ions in CI spectra. Ionization of sample molecules is brought about by proton transfer from ions of a reactant gas (methane or isobutane are most commonly used), yielding even electron $[MH]^+$ ions in contrast to highly labile $M^{\dagger}$ radical ions produced by electron impact, which have a much higher excitation energy and tend to decompose into fragment ions. In previous investigations we have already shown that trifluoroacetylated peptide esters give very informative CI mass spectra [6,25].

For all investigations described in the following section, a Hewlett-Packard gas chromatograph 5840 with a 22 m × 0.4 mm SE-30 glass capillary (pyrex glass, Carbowax 1500 gas-phase deactivated at 280°C, colloidal silicic acid interlayer) was used in combination with a Hewlett-Packard 5985 A quadrupole mass spectrometer. The sample was introduced in the splitless mode onto the cooled capillary (40°C for 1 min, then heated to 120°C at 30°/min and temperature programmed from 120 to 230°C at 6°/min). The column had a separation number of 22, and helium of 0.7 bar was used as a carrier gas, resulting in a flow-rate of 4-8 ml/min. The end of the glass capillary was attached to a platinum/iridium capillary outlet splitting device going to the ion source of the mass spectrometer and to a flame ionization detector (FID). A makeup gas-flow of 30 ml/min of helium was used to increase the flow to the FID. This technique allows the monitoring of the FID signal in parallel to the total-ion-current signal.

IV. APPLICATIONS

In the following, the results of sequence investigations of some membrane-modifying peptide antibiotics are presented. These compounds have aroused considerable interest as agents with the ability to produce fluctuating, voltage-dependent ionic currents in lipid membranes, and therefore may serve as model systems for nerve conduction.

A. Sequence Studies of Alamethicin

The first antibiotic of this group we have studied is the nonadecapeptide alamethicin. As a characteristic property of peptide antibiotics, natural alamethicin contains several analogous sequences due to the exchange of amino acids in different positions by homologues. The acetylation of the N-terminal amino acid, an amino alcohol L-phenylalanol at the C-terminus, and the occurrence of α-amino isobutyric acid (AIB) are further structural peculiarities of this group of peptides. Previous results of sequence studies by Ovchinnikov and co-workers [32] and by Martin and Williams [33] were revised in new investigations by Rinehart et al. [24].

After treatment of a 2.5-mg sample of alamethicin for 12 hr in concentrated HCl at 37°C, the dried residue of peptide fragments was submitted to acetylation, diborane reduction, and trimethylsilylation. The gas chromatographic separation (Figure 26.1) and investigation by mass spectrometry yielded a list of di- to pentapeptides (Table 26.1). As an example of the mass spectrometric identification of a partial sequence, the EI and CI mass spectra of peak 27 of Figure 26.1 are given in Figure 26.2. It may be noted that all the sequence ions of the A and Z series are present in the EI spectrum, while the CI spectrum displays an intense $[MH]^+$ ion.

The overlapping partial sequences were sufficient to derive the complete structure. Moreover, it can be realized from the identified fragments that there are at least seven analogous sequences present. This is in accordance with earlier high-performance liquid chromatographic investigations of natural alamethicin by G. Jung and co-workers [34]. The presence of N-terminal *N*-acetyl Aib could be rationalized by GC/MS investigations, performed also by G. Jung et al. [12].

Our investigations show, in addition to complete sequence information, the position of glutamine and glutamic acid in the peptide chain. The identification of asparagine and glutamine has always been a big problem because of hydrolysis of the amide side-chain to aspartic acid and glutamic acid during partial acid cleavage of the peptide or during the esterification step. By using mild

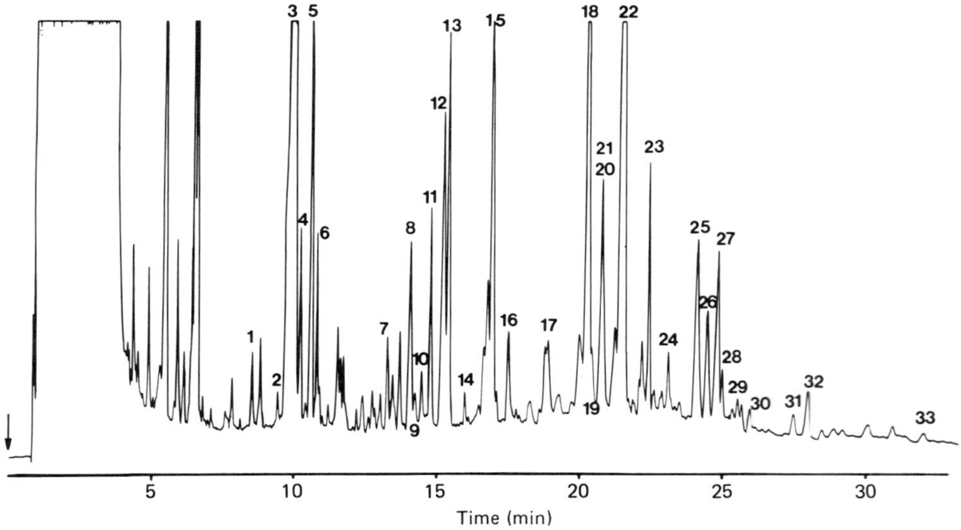

Figure 26.1 Glass capillary gas chromatogram of *N*-ethyl,*O*-trimethylsilyl-polyamino alcohol derivatives from a partial hydrolyzate of alamethicin. (Adapted from Ref. 37.)

Table 26.1 Partial Sequences from Alamethicin Hydrolyzate Identified by GC-MS

Pro-Aib [4]
Pro-Aib-Ala [13]
Pro-Aib-Ala-Aib [18]
Pro-aib-Ala-Aib-Ala [26] Pro-Aib-Ala-Aib-*Aib* [27] Pro-Aib-*Aib*-Aib-Ala [31] Pro-Aib-*Aib*-Aib-*Aib* [32]
Aib-Ala-Aib [9] Ala-Aib-*Aib* [8]
Ala-Aib [1]
Ala-Gln-Aib [17]
Gln-Aib [7]
Glu-Aib-Val [21]
Glu-Aib-Val-Aib [28] Ala-Gln-*Aib*-Aib-Aib [25]
Val-Aib [2]
Val-Aib-Gly [10]
Val-Aib-Gly-Leu-Aib [30]
Gly-Leu-Aib [12]
Leu-Aib [5]
Leu-Aib-Pro-Val [29]
Pro-Val-Aib [15]
Pro-Val-Aib-Aib [22] Pro-Val-Aib-*Val* [23] Pro-Val-*Val*-Aib [24]
Val-Aib-Aib [11] Val-Aib-*Val* [14]
Aib-Aib [3]
Glu-Gln [19]
Glu-Gln-Phol [33]
Phol [6]

1	2	3	4	5	6	7	8	9	10	11	12	13	14	15	16	17	18	19	20
(Ac-Aib-)	Pro-	Aib-	Ala-	Aib-	Ala-	Gln-	Aib-	Val-	Aib-	Gly-	Leu-	Aib-	Pro-	Val-	Aib-	Aib-	Glu-	Gln-	Phol
			Aib	Ala	Aib			Aib							Val	Val			
															(Iva)	(Iva)			

Source: Adapted from Ref. 37

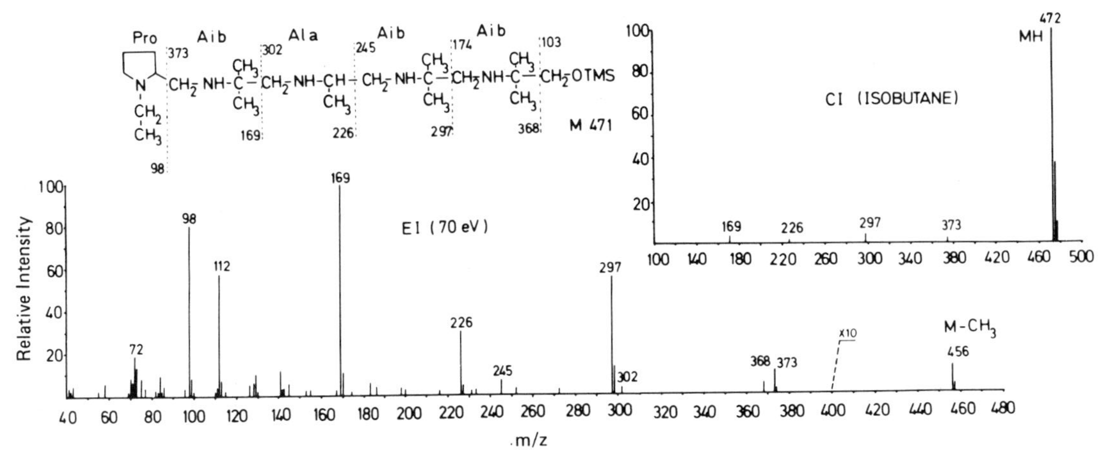

Figure 26.2 Electron impact (70 eV) and isobutane chemical ionization mass spectra of N-ethyl,O-trimethylsilyl-polyamino alcohol derivative of a pentapeptide fragment from partial hydrolyzate of alamethicin. (Adapted from Ref. 37.)

conditions for partial hydrolysis and omitting esterification in the derivatization procedure, it is possible to obtain different derivatives for glutamine and glutamic acid. While the glutamic acid side-chain forms a hydroxypropyl residue during reduction with B_2H_6, the amide residue of glutamine is acetylated in the aceylation step and reduced to an *N*-ethylaminopropyl group. This is demonstrated in the mass spectrum of the derivative of Glu-Gln (Figure 26.3).

B. Sequence Studies of Trichotoxin A 40

Trichotoxin A 40 [35] is a peptide antibiotic with similar biological and structural properties as alamethicin. As unusual constituents, trichotoxin contains L-valinol at the C-terminus, D-α-ethylalanine (D-isovaline) [36], and also a high content of Aib. By mild hydrolysis with trifluoroacetic acid, trichotoxin could be split in an N-acetylated dodecapeptide and a C-terminal hexapeptide residue starting with L-proline and ending with L-valinol. The N-terminal peptide chain could further be separated into three homologues by silica gel chromatography and countercurrent distribution. The C-terminus correspondingly consisted of two homologous chains. All these partial sequences were investigated separately.

For GC-MS investigations the N-trifluoroacetylated peptide methyl esters of a 12-hr partial hydrolyzate were analyzed under identical conditions as described for alamethicin. Figure 26.4 shows the gas chroamtographic separation of one of the dodecapeptide partial hydrolyzates. The fragments, identified by their EI and CI mass spectra, are listed in Table 26.2. Sufficient overlapping allows an unambiguous arrangement of the fragment peptides to reconstruct the original structure. Acetylated aib could be identified in a separate analysis [12].

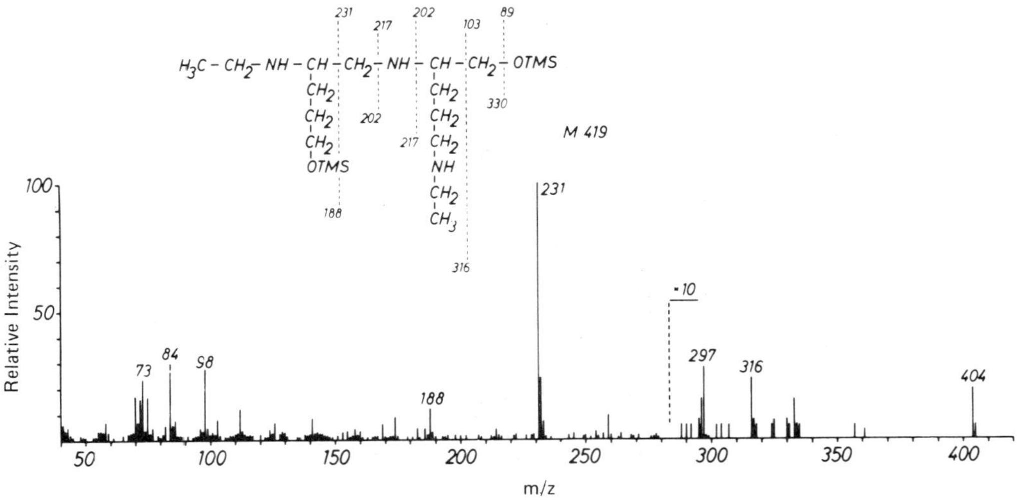

Figure 26.3 Electron impact (70 eV) mass spectrum of Glu-Gln derivative from partial hydrolyzate of alamethicin. (Adapted from Ref. 37.)

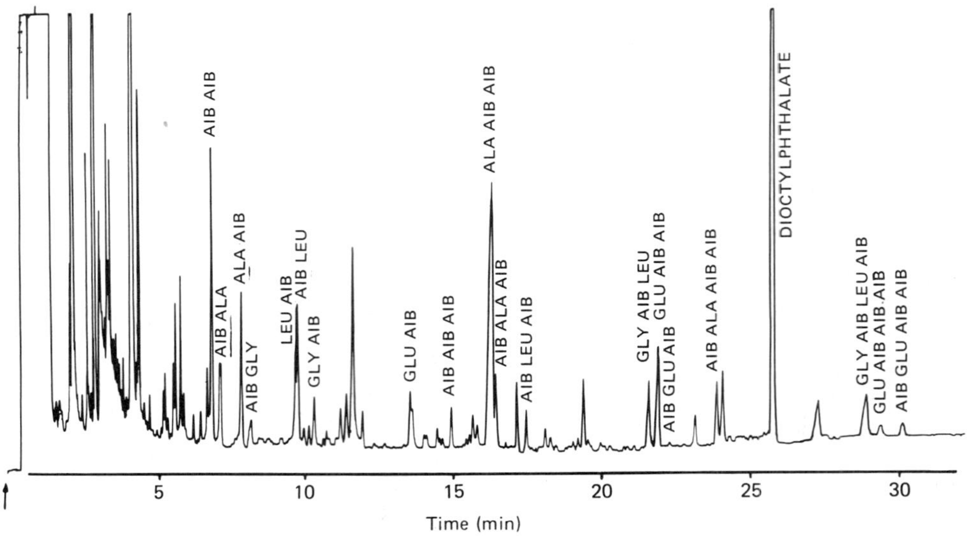

Figure 26.4 Glass capillary gas chromatogram of *N*-TFA-peptide methyl esters from partial hydrolyzate of the N-terminal dodecapeptide fragment of trichotoxin A 40 (Adapted from Ref. 37.)

Table 26.2 Partial Sequences from a Hydrolyzate of an N-Terminal Dodecapeptide from Trichotoxin A 40

Partial sequences
Aib-Gly
Gly-Aib
Gly-Aib-Leu
Gly-Aib-Leu-Aib
Aib-Leu
Aib-Leu-Aib
Aib-Glu-Aib
Aib-Glu-Aib-Aib
Glu-Aib
Glu-Aib-Aib
Glu-Aib-Aib-Aib
Aib-Aib
Aib-Aib-Aib
Aib-Ala
Aib-Ala-Aib
Aib-Ala-Aib-Aib
Ala-Aib
Ala-Aib-Aib

(Ac)-Aib-Gly-Aib-Leu-Aib-Glu-Aib-Aib-Aib-Ala-Aib-Aib

Source: Adapted from Ref. 37.

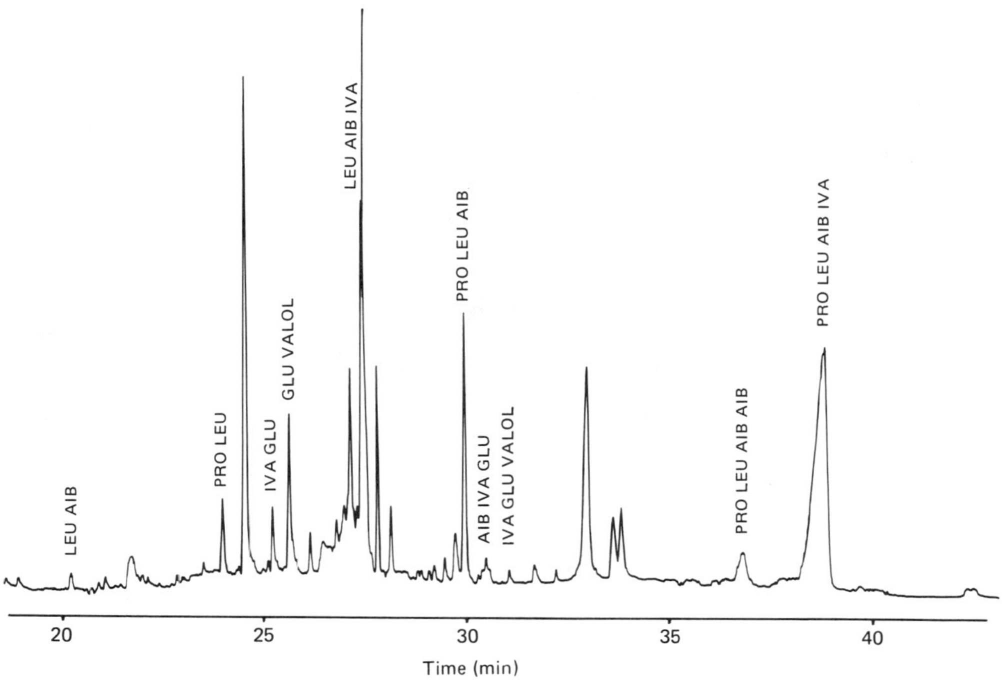

Figure 26.5 Glass capillary gas chromatogram of *N*-TFA-peptide methyl esters from a partial hydrolyzate of the C-terminal pseudohexapeptide of trichotoxin A 40 (Adapted from Ref. 38.)

By the same procedure both of the C-terminal hexapeptides were analyzed. They differ only by the exchange of one Aib residue against D-isovaline. The GC of the partial hydrolyzate is shown in Figure 26.5 and in Table 26.3 the partial sequences as identified by GC-MS are listed. Figure 26.6 once more points out the advantage of CI mass spectrometry as a complementary technique in peptide identification. In the 70 eV EI spectrum of the tetrapeptide

Table 26.3 Partial sequences from a Hydrolyzate of a C-Terminal Hexapeptide from Trichotoxin A 40

Pro-Leu
Pro-Leu-Aib
Pro-Leu-Aib-Iva
Leu-Aib
Leu-Aib-Iva
Aib-Iva-Glu
Iva-Glu
Iva-Glu-Valol
Glu-Valol
Pro-Leu-Aib-Iva-Glu-Valol

Source: Adapted from Ref. 37.

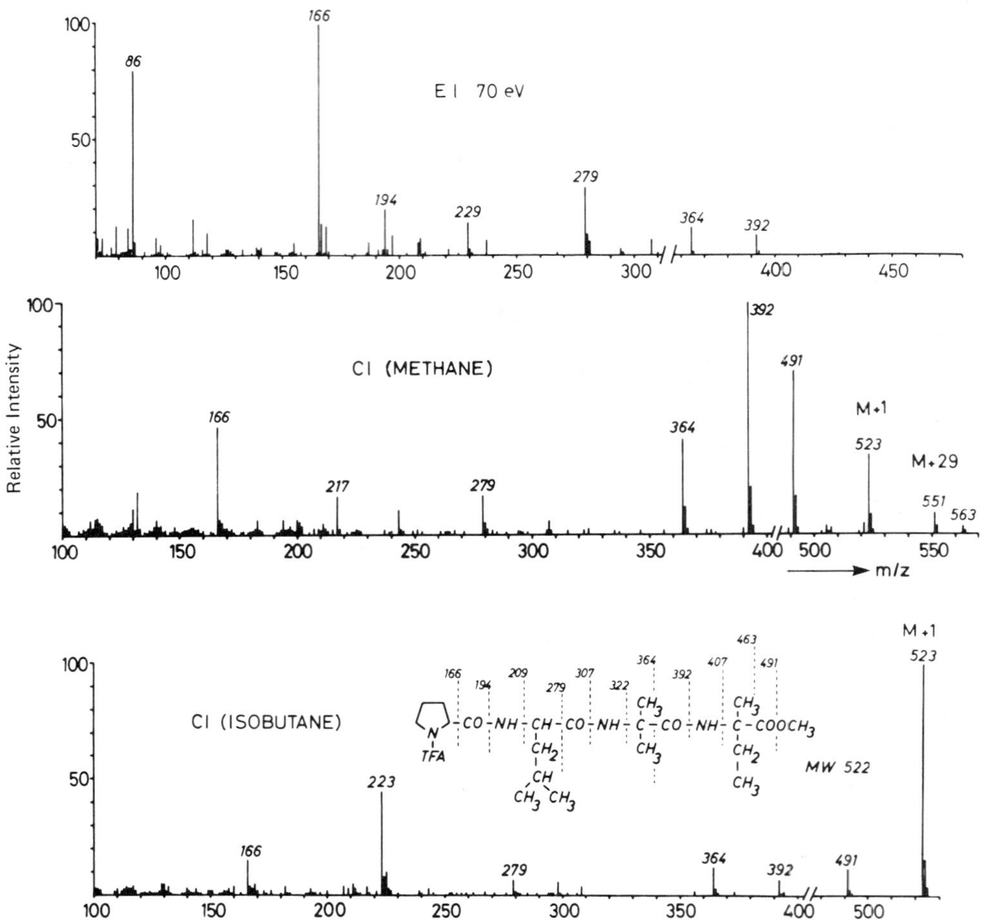

Figure 26.6 Electron impact (70 eV) and chemical ionization mass spectra (methane and isobutane) of N-TFA-peptide methyl ester from partial hydrolyzate of the C-terminal pseudohexapeptide of trichotoxin A 40. (Adapted from Ref. 37.)

Pro-Leu-Aib-Iva, there is no indication of the C-terminal amino acid. The isobutane CI mass spectrum reveals an intense $[MH]^+$ ion and still a complete set of sufficiently intense sequence-indicating ions. The methane CI spectrum also shows a moderately intense $[MH]^+$ ion, but more intense fragmentation and additional ion signals in the mass range above the molecular weight due to ion-molecule reactions between substrate and reactant gas ions. This clearly demonstrates the higher energy transfer from protonated methane to the sample molecules as compared to isobutane. The energy difference can be estimated from the difference in proton affinities and is responsible for the degree of fragmentation of the peptide molecule.

V. SUMMARY

Glass capillary gas chromatography proves to be an efficient technique for the separation of volatile derivatives of di- to hexapeptides as obtained by partial hydrolysis of peptide antibiotics. At present, two types of derivatives are in use: *N*-TFA-peptide methyl esters and *N*-ω-ethyl-*O*-TMS-polyamino alcohols. The latter derivatives are somewhat more volatile. Both types of derivatives are suitable for elution from carefully deactivated glass capillary columns as symmetrical peaks. For the identification of the peptide sequence, electron impact and isobutane chemical ionization mass spectra exhibit complementary properties.

A crucial point in the combined technique is the thermostability of the column deactivation. Future work should also concentrate on extending the range of application to longer peptide chains and to peptides with more trifunctional amino acids.

ACKNOWLEDGMENTS

The author is indebted to W. Guenther and M. Aydin, Universität Hamburg, who who did some of the exterimental work, to Prof. G. Jung and Dr. H. Brueckner, Universität Tübingen, for their cooperation and for peptide samples, and to the Bundesministerium für Forschung und Technologie for supplying the GC-MS system.

REFERENCES

1. M. Bodanszky and D. Perlman, *Science 163*:352 (1969).
2. F. Sanger, *Biochem. J. 39*:507 (1945).
3. W. R. Gray, in *Methods in Enzymology 25B* (C. H. W. Hirs and S. N. Timasheff, eds.), Academic Press, New York, 1972, p. 121.
4. S. Akabori, K. Narita, H. Hanabusa, and A. Tsugita, *Nippon Kagaku Zasshi 76*:225 (1955).
5. W. A. Koenig, C. Engelfried, H. Hagenmaier, and H. Kneifel, *Liebigs Ann. Chem. 1976*:2011.
6. W. A. Koenig, H. Krohn and C. Engelfried, in *Peptides 1978* (T. Z. Siemion and G. Kupryszewski, eds.), Wroclaw University Press, Poland, 1979, p. 255.
7. F. Weygand, A. Prox, H. H. Fessel, and K. KunSun, *Z. Naturforsch. 20b*:1169 (1965).
8. K. Biemann, F. Gapp, and J. Seibl, *J. Am. Chem. Soc. 81*:2274 (1959).
9. J. A. Kelly, H. Nau, H. J. Foerster, and K. Biemann, *Biomed. Mass Spectrom 2*:313 (1975).
10. H. Nau, H. J. Foerster, J. A. Kelley, and K. Biemann, *Biomed. Mass Spectrom. 2*:326 (1975).
11. B. C. Das, S. D. Géro, and E. Lederer, *Biochem. Biophys. Res. Commun. 29*:211 (1967).
12. H. Brueckner and G. Jung, *Chromatographia 13*:170 (1980).
13. R. Hill, *Adv. Protein Chem. 20*:37 (1965).
14. R. M. Caprioli, W. E. Seifert, and D. E. Sutherland, *Biochem. Biophys. Res. Commun. 55*:67 (1973).
15. E. Bayer and W. A. Koenig, *J. Chromatogr. Sci. 7*:95 (1969).

16. A. Prox and F. Weygand, in *Peptides* (H. C. Beyermann, A. van de Linde, and M. van der Brink, eds.), North Holland Publ. Co., Amsterdam, 1967, p. 158.
17. M. Donike, *J. Chromatogr. 78*:273 (1973).
18. H. Matsuo, Y. Fujimoto, and T. Tatsuo, *Tetrahedron 39*:3465 (1965).
19. E. Fischer, *Ber. Deut. Chem. Ges. 39*:2893 (1906).
20. H. Frank and D. M. Desiderio, *Anal. Biochem. 90*:413 (1978).
21. E. Bayer, K. H. Gugel, K. Haegele, H. Hagenmaier, S. Jessipow, W. A. Koenig, and H. Zaehner, *Helv. Chim. Acta. 55*:244 (1972).
22. W. A. Koenig, W. Loeffler, W. H. Meyer, and R. Uhmann, *Chem. Ber. 106*:816 (1973).
23. G. Jung, W. A. Koenig, D. Leibfritz, T. Ooka, K. Janko, and G. Boheim, *Biochim, Biophys. Acta 433*:164 (1976).
24. R. C. Pandey, J. Co. Cook, and K. L. Rinehart, *J. Am. Chem. Soc. 99*:8469 (1977).
25. W. A. Koenig and W. Guenther, *HRC & CC 2*:378 (1979).
26. D. Cronin, *J. Chromatogr. 97*:263 (1974).
27. E. Schulte, *Chromatographia 9*:315 (1976).
28. K. Biemann, in *Biochemical Applications of Mass Spectrometry* (G. R. Waller, ed.), Wiley, New York, 1972, p. 405.
29. H. Nau, *Angew. Chem. 88*:74 (1976).
30. M. S. B. Munson and F. H. Field, *J. Am. Chem. Soc. 88*:2621 (1966).
31. W. J. Richter and H. Schwarz, *Angew.Chem. 90*:449 (1978).
32. Y. A. Ovchinnikov, A. A. Kiryushkin, and I. V. Kozhevnikova, *J. Gen. Chem. USSR 41*:2105 (1971).
33. D. R. Martin and R. J. P. Williams, *Biochem. Soc. Trans 3*:166 (1975).
34. G. Jung and H. Brueckner, private communication.
35. H. Brueckner, W. A. Koenig, M. Greiner, and G. Jung, *Angew. Chem. 91*:508 (1979) and *Angew. Chem. Int. Ed. Eng. 18*:476 (1979).
36. H. Brueckner, G. J. Nicholson, G. Jung, K. Kruse, and W. A. Koenig, *Chromatographia 13*:209 (1980).
37. W. A. Koenig and M. Aydin, in *Peptides 1980* (K. Brunfeldt, ed.), Scriptor, Copenhagen, 1981, p. 711.
38. W. A. Koenig and M. Aydin, in *Chemistry of Peptides and Proteins*, Vol. 1 (W. Voelter, E. Wünsch, J. Ovchinnikov, and V. Ivanov, eds.), de Gruyter & Co., Berlin, New York, 1982, p. 173.

27

Internal Standards in Drug Analysis by Glass Capillary GLC-MS

W. J. A. VandenHeuvel, J. R. Carlin, and R. W. Walker / *Merck Sharp & Dohme Research Laboratories, Rahway, New Jersey*

I. INTRODUCTION

Faced with the need for specific and sensitive assays for drugs, their metabolites, and endogeneous compounds (the levels of which are affected by drug administration), research workers in drug metabolism and related fields during the 1960s turned to GLC systems employing packed columns and flame ionization detectors. Electron capture detectors were used for compounds which contain elements and/or functional groups of high electron affinity. Derivatization was utilized in some cases to introduce the potential for detection via electron capture, and this approach also served on occasion to improve separations. Nonselective or nonpolar polymers such as SE-30 and OV-1 were the most commonly employed liquid phases, but greater column selectivity was frequently achieved by use of the polar stationary phases OV-17 and OV-210 (QF-1). Enhanced selectivity was achieved, notably during the 1970s, by the use of element-specific detectors (for sulfur, phosphorus, and especially nitrogen) and the combined gas-liquid chromatography-mass spectrometric (GLC-MS) technique known as selected ion monitoring or mass fragmentography [1].

A few papers report on the use of longer than normal packed columns in drug analysis, but in general little attempt was made to improve selectivity by employing packed columns of greater than 2 m in length [2,3]. Furthermore, it is worthy to note that increases in assay specificity (and sensitivity) by the application of high-resolution glass capillary column GLC technology to drug analysis did not occur until the mid to late 1970s, even though the feasibility of steroid analysis using capillary columns was demonstrated nearly a half dozen years earlier. Whatever the reasons for this significant time lag, the benefits of capillary column GLC are now widely recognized by workers in drug metabolism and related fields [4].

II. CHOICE OF INTERNAL STANDARD

In addition to specificity and sensitivity, another important component for establishing a useful assay is the proper choice of an internal standard. Although occasionally added to the final isolate prior to the GLC step, an internal standard is far more properly introduced into the biological specimen prior to the initiation of the isolation procedure. Ideally an internal standard for a GLC-based assay should mimic the behavior of the drug or metabolite of interest during the isolation procedure which leads to the sample to be analyzed, but the two compounds ultimately must be distinguishable on the basis of GLC behavior or by the detection device employed. This requirement means that for GLC assays involving nearly all the commonly utilized detection systems, the internal standard and the drug must exhibit different retention times. Thus the chemical analogue of the drug (e.g., *tert*-butyl vs. isopropyl) is frequently employed as the internal standard [5,6]. Not only do such pairs of compounds usually exhibit very similar properties with respect to an isolation procedure and yet are separable on the GLC column, but the analogue may well be readily available from the laboratory which synthesized the drug. Detection via selected-ion-monitoring mass spectrometry is of course amenable to such approach, since the analogue usually produces ions of m/e values differing in mass by the difference in molecular weight between the drug and internal standard [5]. The ideal internal standard, however, is a compound as close in structure to the drug as possible, but which can still be differentiated from the drug by the detection system. A stable isotope-labeled analogue should thus be the internal standard of choice concerning the isolation procedure, and such compounds can be used when selected ion monitoring (SIM) is the mode of detection.* Selected ion monitoring with a stable isotope-labeled internal standard is thus a powerful combination for selective and sensitive detection and assay of drugs and related compounds, and this approach has found great favor in recent years [7-11]. Very few papers have appeared in the literature describing further enhancement of this specificity (and sensitivity) by use of glass capillary columns with selected ion monitoring [12-16], although reports on methods using packed columns abound.

*Radiolabeled analogues (e.g., ^{14}C at high specific activity, ⩾55 mCi/mM) can also be used as internal standards. These compounds may be available because of their use in radiodistribution studies, but there is, with good reason, a hesitancy to use radioactivity in this manner. There is no difference between monitoring stable or radioactive isotope-labeled species with respect to the mass spectrometric technique.

III. APPLICATIONS

A. Measurement of Prostaglandins

In 1976 Maclauf and co-workers [17] reported the application of glass capillary column GLC to the measurement of prostaglandins. Compared to packed column GLC, the separations were much improved; also of critical importance, because of the reduced sample adsorption and the sharper peaks obtained with capillary columns, sensitivity (detection via flame ionization) was increased by a factor of 100. These workers succeeded in coupling their column with a mass spectrometer, and 2 years later Sors et al. [12] reported on a collaborative study in which selected ion monitoring was used to measure the in vitro biosynthesis of the arachidonic acid metabolite, thromboxane B_2 or TXB_2. The capillary column

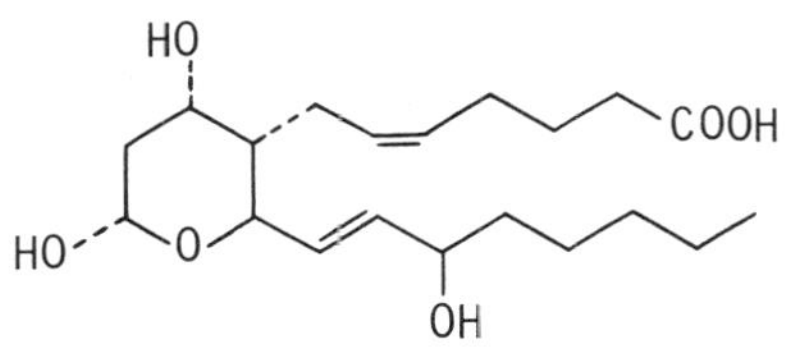

TXB_2

GLC-mass spectrometric approach, which employed the closely related PGF_2-d_4 as the internal standard, served to confirm the validity of a radioimmunoassay method for quantitation of TXB_2. The two compounds are analyzed as their methyl ester-trimethylsilyl ether derivatives. It was demonstrated that the formation of TXB_2 during arachidonic acid-induced aggregation of both human platelet-rich plasma and washed human platelets is strongly inhibited by the antiinflammatory agent indomethacin (see Table 27.1).

Other, more recent, examples of the use of capillary columns in the GLC analysis of prostaglandins and related compounds are found in the papers by Fitzpatrick [18], Rigaud and co-workers [19], Erlenmaier et al. [13], and Walker and associates [14]. Derivatization was employed in all cases. The report of Erlenmaier et al. [13] discusses the greater accuracy and sensitivity observed for the quantitation of urinary prostaglandins via capillary rather than packed column GLC-mass spectrometry. These desirable qualities were proposed to result from the much greater resolving power of the capillary column. For the determination of the urinary concentrations of PGE_2, PGF_2, and 7-hydroxy-5,11-diketotetranorprostane-1,16-dioic acid (PGM), the major

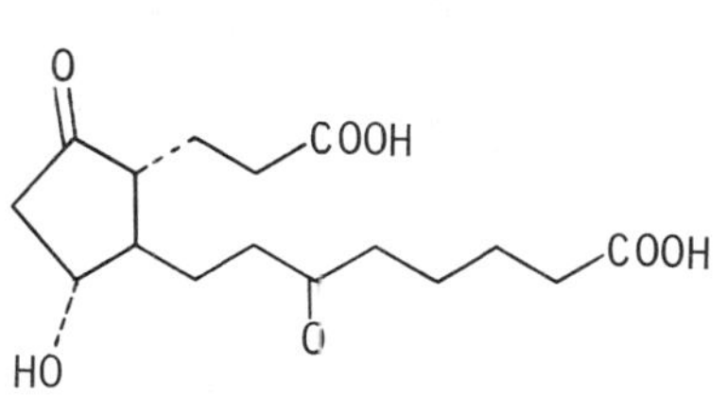

PGM

human urinary metabolite of PGE_1 and PGE_2, deuterium-containing analogues of the naturally occurring prostaglandins were used as internal standards. The lower detection limit (S/N of 2:1) for PGE_2, using an SE-30 30-m glass capillary column, was found to be 50 pg (using splitless injection) via selected

Table 27.1 TXB_2 Formed During Arachidonic Acid-Induced Aggregation of Washed Human Platelets

	TXB_2, μg/ml[a]	
Sample	RIA[b]	SIM[c]
1	1.85	2.02
2	1.91	1.76
3	2.18	1.73
4	2.08	1.58
5	2.41	2.90
6	2.18	1.48
7	1.63	2.18
8[d]	0.027	0.04
9[d]	0.025	0.03

[a]Concentration of TXB_2 in an ethanol extract of incubation mixture (which contained 3.2×10^9 platelets).
[b]Radioimmunoassay.
[c]Selected ion monitoring.
[d]Incubation carried out in the presence of indomethacin (1.4×10^{-4}M).
Source: Ref. 12.

ion monitoring, whereas with a packed column (1% Dexsil 300, 1.8 m) the minimum amount which could be detected quantitatively was 1 ng (50 ng of PGE_2-d_4 coinjected in each case).

Walker et al. [14] described an isotope-dilution capillary column GLC-mass spectrometric procedure for the analysis of PGM in the urine of patients treated with indomethacin [20]. As in the Erlenmaier paper, PGM-d_7 was used as the internal standard, and the compounds were analyzed as their methyl ester-methoxime-trimethylsilyl ethers. Selected ion monitoring in both the electron impact and the chemical ionization modes was employed. Capillary column GLC, with its inherently greater resolving power, was used to reduce the incidence of impurity interference. A detection limit of approximately 0.5 ng of PGM/ml of urine (20-ml aliquot) was found. The results (see Table 27.2) from a series of rheumatoid arthritis patients confirmed previous reports that administration of indomethacin significantly reduces the urinary output of PGM (presumably by inhibition of PGE_1 and PGE_2 biosynthesis).

B. Assay of β-Blockers

β-Adrenergic receptor blocking agents or β-blockers have become widely used drugs during the past decade. GLC-based methods have been employed frequently in studying their metabolism and determining their concentrations in

Table 27.2 Urinary Excretion (μg/24 hr)[a] of PGM (7α-Hydroxy-5,11-diketotetranorprostane-1,16-dioic acid) in Patients with Rheumatoid Arthritis Treated with Indomethacin via Oral[b] and Oral/Suppository[c] Routes

		Before drug		Oral		Oral/ suppository	
Patient	Sex	EI[d]	CI[e]	EI	CI	EI	CI
A	M	5.34	5.60	2.76	2.66	2.34	2.41
B	F	2.40	2.68	0.95	0.77	0.81	0.89
C	F	3.73	4.64	1.50	1.59	0.96	1.18
D	M	15.3	15.5	6.20	7.03	4.56	6.42
E	F	7.18	7.46	1.08	1.07	3.36	3.26
F	F	12.2	13.5	8.68	8.96	2.33	2.83

[a]Averaged result from 2 injections.
[b]25 mg 3 times daily for 3 weeks.
[c]100 mg nightly for 3 weeks.
[d]Electron impact ionization.
[e]Chemical ionization.
Source: Ref. 14.

body fluids [5-7,16,21-23]. The Medical University of South Carolina group has been particularly active in studying propranolol [7,24], and recently Walle

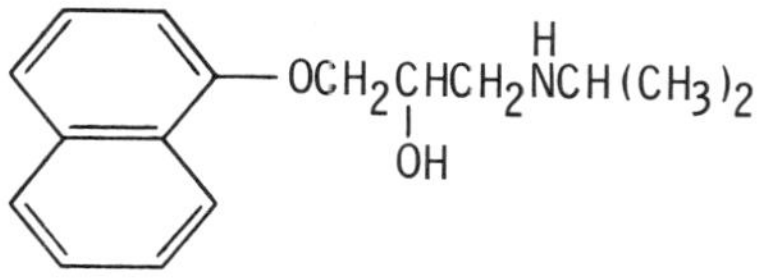

Propranolol

and Walle [7] reported on a packed column GLC-MS method for determining the bioavailability of propranolol and propranolol-d_2 in the dog. These authors used propranolol-d_6 as the stable isotope-labeled internal standard in their selected ion monitoring assay. Funke et al. [5] have utilized the selected-ion-monitoring approach for determining the serum levels of another β-blocker, nadolol, but these authors used a chemical, rather than an isotopically labeled analogue as their internal standard.

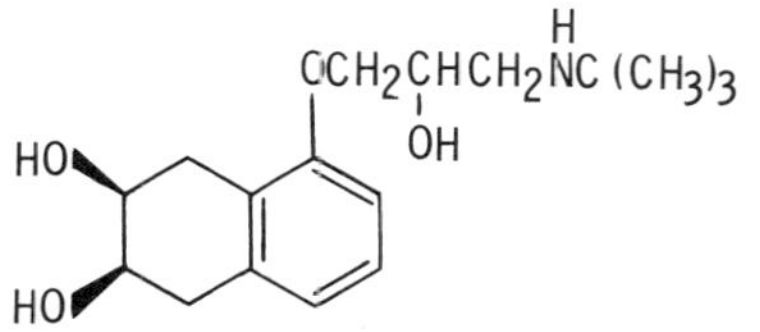

Nadolol

Capillary column GLC has been used in the assay of β-blockers. Thus DeBruyne and associates [22] developed an assay (electron capture detection of heptafluorobutyryl derivatives) for alprenolol and oxprenolol in 0.1-0.2 ml

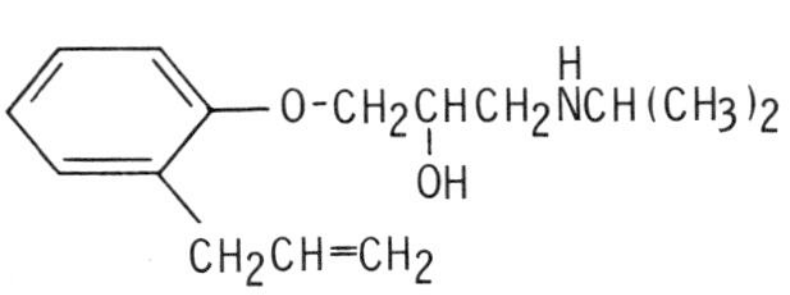

Alprenolol

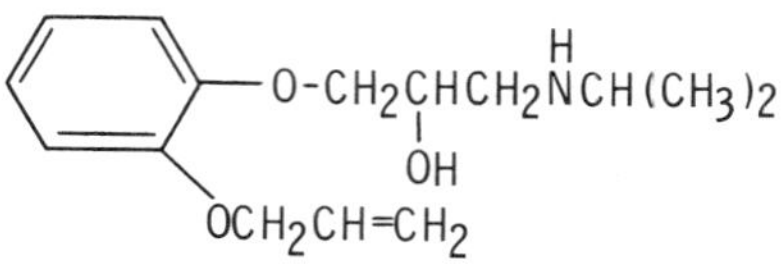

Oxprenolol

of serum. Not only are these volumes only one-tenth those necessary for an earlier packed column assay for these drugs, but the capillary column approach uses a simplified extraction procedure and does not require redistilled solvents. The authors credit this to the high resolving power of capillary columns (25-m OV-101) relative to that of previously used packed columns (2-m OV-101). Peak interference is much reduced, leading to increased accuracy and sensitivity with the capillary column approach. The lowest detectable amount of the derivatized β-blocker was 2 pg (injected with a split ratio of 1:12), with a standard deviation of 3% at 25 ng/ml over a period of several months. The capillary column GLC separation of a number of β-blocker enantiomers as their diastereoisomeric *N*-trifluoroacetyl-1-prolyl and *N*-heptafluorobutyryl-1-prolyl amides has been studied by Caccia et al. [25]. These authors found both better separations and shorter retention times with the capillary (60-m SCOT column, OV-225) than with the packed (2-m, 3% OV-225) column.

Conversion from packed to capillary column GLC usually improves drug assay sensitivity because with the latter, adsorption losses are minimized and the greater column efficiency leads to greater peak height per unit mass. The higher resolving power of capillary columns reduces the possibility of impurity interference because of its enhanced separation over packed column GLC. Selected ion monitoring also offers high specificity and sensitivity approaching that of electron capture detection. Capillary column GLC in combination with selected ion monitoring should possess exceptional specificity and sensitivity, both very necessary in the assay of drugs possessing pharmacological activity at the low ng/ml level. The β-adrenoreceptor blocking agent timolol is such a drug. An electron-capture detection packed-column assay developed by Tocco

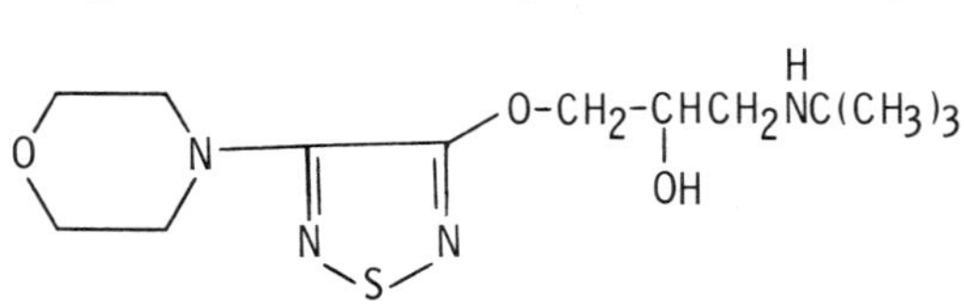

Timolol

$^{13}C_3$-timolol, $C(^{13}CH_3)_3$; timolol-d_9, $C(C^2H_3)_3$

et al. [6] possesses a sensitivity approaching 2 ng/ml human plasma, and a nitrogen-selective detection-based GLC assay capable of measuring plasma levels 10 hr post-dose has been reported by Else et al. [23].

Carlin et al. [16] have devised a selected-ion-monitoring capillary column GLC-mass spectrometric assay for timolol and [$^{13}C_3$]timolol in human plasma, timolol-d_9 serving as the internal standard. The study was designed to demonstrate that an assay possessing the requisite sensitivity and specificity could be established and that no significant metabolic kinetic isotope effect exists with [$^{13}C_3$]timolol. A key step in the isolation procedure involves use of a rapid, simple, reverse phase-type separation procedure utilizing a cartridge (SEP-PAK) containing C_{18} packing. The final isolate is trimethylsilylated prior to the introduction using splitless injection (dropping glass needle) into the GLC system. The ions monitored are *m/e* 86 (timolol), 89 [$^{13}C_3$]timolol) and 95 (timolol-d_9) which arise from the side chains of these compounds, e.g., $[CH_2NHC(CH_3)_3]^+$. A very small portion of the *m/e* 86 ion arises from the fragment ion:

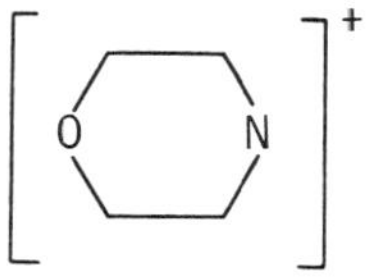

The method is capable of assaying for timolol in plasma at ∿0.5 ng/ml; at 7 ng/ml the CV% values (over a 1-month period) for timolol and [$^{13}C_3$]timolol are 3.7 and 1.4, respectively. Thanks to the specificity resulting from use of capillary column GLC and selected ion monitoring, isolates from control plasma yield essentially no response at the retention times of the three compounds of interest. Two selected-ion chromatograms from plasma isolates are presented in Figures 27.1 and 27.2 (2 hr and 12 hr post-dose, respectively). These are from a volunteer given 5 mg of timolol maleate and 5 mg of [$^{13}C_3$]timolol maleate. Plasma levels from a volunteer given 10 mg of timolol maleate and 10 mg of [$^{13}C_3$]timolol maleate are reported in Table 27.3. It is evident from these data that the concentrations of the two forms of the drug are equivalent, and that there is no significant metabolic kinetic isotope effect. Assays such as this one, with demonstrated specificity and sensitivity, should be useful in generating data for determining bioavailability via coadministration of drug and stable isotope-labeled drug (see, for example, Wolen et al. [26]).

C. Quantification of Ketobemidone

Another example of using stable isotope-labeled internal standards in capillary column selected-ion-monitoring assays was reported recently by Bondesson and Hartvig [15]. Their work relates to the quantification of the narcotic analgesic ketobemidone in human plasma. Ketobemidone (as its *O*-acetyl

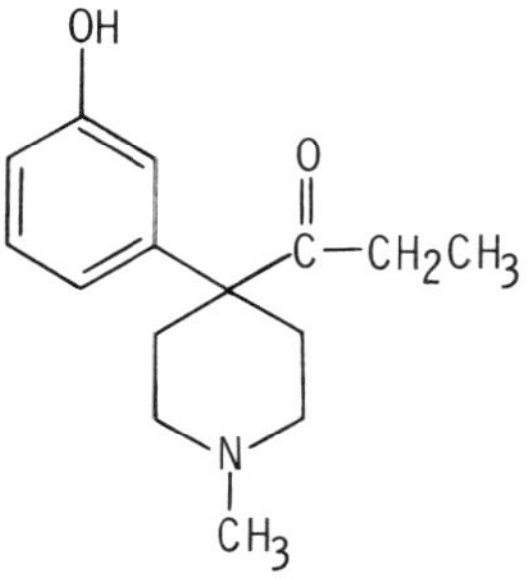

Ketobemidone

Table 27.3 Plasma Levels (ng/ml) of Free Drug Following a Single Oral Dose of Timolols[a]

Time (hr)	Unlabeled timolol	$[^{13}C_3]$Timolol
0.5	9.4	9.1
1	41.8	40.3
1.5	40.9	39.9
2	25.6	26.0
3	22.9	22.8
4	19.5	19.8
6	13.2	12.8
8	7.3	7.0
10	5.5	5.3
12	3.3	3.5

[a]Unlabeled timolol maleate (10 mg) and $[^{13}C_3]$timolol maleate (10 mg).
Source: Ref. 16.

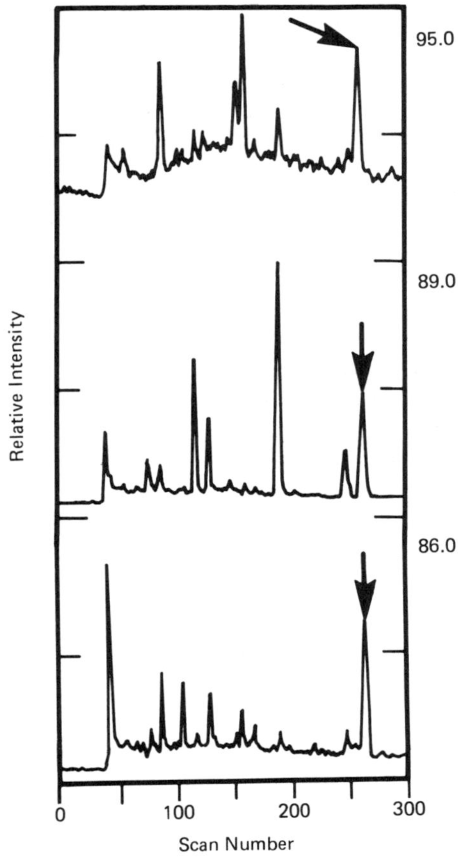

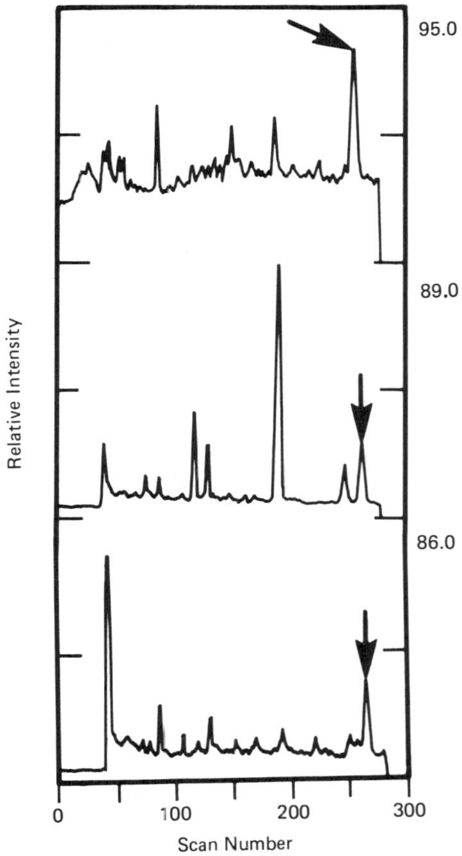

Figure 27.2 Selected-ion-monitoring response plots (*m/e* 86, 89, 95) resulting from assay of the isolate from 1 ml of plasma (12-hr post dose) from a volunteer given 5 mg of timolol maleate and 5 mg of [$^{13}C_3$]timolol maleate. Conditions same as for Figure 27.1. (From Ref. 16, and reproduced with permission of the copyright owner.)

Figure 27.1 Selected-ion-monitoring response plots (*m/e* 86, 89, 95) resulting from assay of 4% of the isolate from 1 ml of control human plasma (to which had been added 7.29 ng each of timolol and [$^{13}C_3$]timolol and 7.38 ng of [$^{2}H_9$]-timolol prior to the initiation of the isolated procedure). The peak heights represent ∿200 pg injected of each compound. The sample was trimethylsilylated and injected (dropping glass needle) into a 16-m SE-30 glass capillary column at 225°C; retention times for the compounds are ∿4 min. (From Ref. 16, and reproduced with permission of the copyright owner.)

derivative) levels were initially measured using packed column GLC, but because of adsorptive losses during chromatography and to enhance selectivity (and also sensitivity), the authors switched to use of a capillary column (18 m, SP-1000) approach with acetylated isolates from plasma containing low levels of drug. This method permits quantification of plasma concentrations down to 2 ng/ml (relative standard deviation of 4% at 5 ng/ml) and involves monitoring the molecular ions of the acetylated drug (*m/e* 289) and internal standard ketobemidone-d_4 (*m/e* 293).

Although few articles on the use of stable isotope-labeled internal standards in drug analysis by glass capillary column GLC-MS methods have appeared, there is every reason to believe that this situation will change soon. Once it becomes apparent that by switching from packed to capillary column selected-ion-monitoring, the selectivity and sensitivity of this latter technique are enhanced (as they are with other GLC detectors), the conversion rate will increase markedly. The ultimate technique could be high-resolution GLC combined with high-resolution or exact-mass (rather than low-resolution or nominal-mass) selected ion monitoring, but that is an additional (and more costly) step away.

REFERENCES

1. C.-G. Hammar, B. Holmstedt, and R. Ryhage, *Anal. Biochem.* *25*:337 (1968).
2. W. J. A. VandenHeuvel, *Separation Sci.* *3*:151 (1968).
3. J. J. DeRidder, P. C. J. M. Koppens, and H. J. M. VanHal, *J. Chromatogr. (Biomed. Applics.)* *143*:289 (1977).
4. W. J. A. VandenHeuvel and J. S. Zweig, *HRC & CC* *3*:381 (1980).
5. P. T. Funke, M. F. Malley, E. Ivashkiv, and A. I. Cohen, *J. Pharm. Sci.* *67*:568 (1978).
6. D. J. Tocco, F. A. deLuna, and A. E. W. Duncan, *J. Pharm. Sci.* *64*:1879 (1975).
7. T. Walle and U. K. Walle, *Res. Commun. Chem. Pathol. Pharmacol.* *23*:453 (1979).
8. W. A. Garland, B. J. Hodshon, G. Chen, G. Weiss, N. R. Felicito, and A. MacDonald, *J. Agric. Food Chem.* *28*:273 (1980).
9. T. E. Gaffney, C.-G. Hammar, B. Holmstedt, and R. E. McMahon, *Anal. Chem.* *43*:307 (1971).
10. R. W. Walker, H.-S. Ahn, G. Albers-Schönberg, L. R. Mandel, and W. J. A. VandenHeuvel, *Biochem. Med.* *8*:105 (1973).
11. J. T. Briggs, W. H. Holland, S. Chang, P. P. Hipps, and W. R. Sherman, *J. Pharm. Sci.* *65*:261 (1976).
12. H. Sors, P. Pradelles, F. Dray, M. Rigaud, J. Maclouf, and P. Bernard, *Prostaglandins* *16*:277 (1978).
13. T. Erlenmaier, H. Müller, and H. W. Seyberth, *J. Chromatogr. (Biomed. Applics.)* *163*:289 (1979).
14. R. W. Walker, V. F. Gruber, J. Pile, K. Yabumoto, A. Rosegay, D. Taub, M. L'E. Orme, F. J. Wolf, and W. J. A. VandenHeuvel, *J. Chromatogr. (Biomed. Applics.)* *181*:85 (1980).
15. U. Bondesson and P. Hartvig, *J. Chromatogr.* *179*:207 (1979).
16. J. R. Carlin, R. W.Walker, R.O. Davies, R. T. Ferguson, and W. J. A. VandenHeuvel, *J. Pharm. Sci.* *69*:1111 (1980).

17. H. Maclouf, M. Rigaud, J. Durand, and P. Chebroux, *Prostaglandins* *11*:999 (1976).
18. F. A. Fitzpatrick, *Anal. Chem.* *50*:47 (1978).
19. M. Rigaud, P. Chebroux, A. Soustre, J. Durand, H. Rabinovitch, and J. C. Breton, *Adv. Chromatogr.* *00*:615 (1979).
20. N. Baber, L. D. C. Halliday, W. J. A. VandenHeuvel, R. W. Walker, R. Sibeon, J. P. Keenan, T. Littler, and M. Orme, *Ann. Rheum. Dis.* *38*:128 (1979).
21. D. J. Tocco, A. E. W. Duncan, F. A. deLuna, J. L. Smith, R. W. Walker, and W. J. A. VandenHeuvel, *Drug Metab. Dispo.* *8*:236 (1980).
22. D. DeBruyne, H. Kinsun, M. A. Moulin, and M. C. Bigot, *J. Pharm. Sci.* *68*:511 (1979).
23. O. F. Else, H. Sorenson, and I. R. Edwards, *Eur. J. Clin. Pharm.* *14*:431 (1978).
24. T. Walle, E. C. Conradi, U. K. Walle, T. C. Fagan, and T. E. Gaffney, *Clin. Pharmacol. Ther.* *26*:686 (1979); *27*:22 (1980).
25. S. Caccia, C. Chiabrando, P. DePonte, and R. Fanelli, *J. Chromatogr. Sci.* *16*:543 (1978).
26. R. L. Wolen, R. H. Carmichael, A. S. Ridolfi, L. Thompkins, and E. A. Ziege, *Biomed. Mass Spectrom.* *6*:173 (1979).

28

Analysis of Cannabis

Neil K. McCallum / *Chemistry Division, Department of Scientific and Industrial Research, Petone, New Zealand*

I. INTRODUCTION

Cannabis has been used in some parts of the world for thousands of years, although it is the last decade which has seen its most spectacular gains in the West. The first chemical examination of *Cannabis sativa* was made by Tscheepe in 1821 [1]. A considerable body of further work was completed by others in the middle and late nineteenth century [1], but it was not until 1964 that the major active constituent, Δ^1-tetrahydrocannabinol (Δ^1-THC; see Figure 28.1), was isolated in its pure form [2]. The surge in scientific interest brought about by the recent great increase in the use of cannabis has produced a vast proliferation of literature, but real progress in the areas of chemistry, biochemistry, and metabolism has been slow. Most of the impediments to this progress have been analytical. They arise from the chemical and chromatographic similarity of many of the compounds occurring in the plant, and from the problems of detecting the minute amounts of the compounds that often have to be dealt with. Much of the work has been directed towards forensic needs, and therefore there has been considerable emphasis on highly refined methods required for discriminating and identifying these compounds.

The earliest forensic tests entailed microscopic examinations of plant material [3], simple color tests [4], and thin-layer chromatography [5]. These techniques have been largely superseded by the much higher resolution methods of high-pressure liquid chromatography (HPLC) and gas-liquid chromatography (GLC)—the latter in particular being used very widely [6]—although even with these methods, obtaining adequate resolution for the complex mixtures encountered is often a problem. Recent developments favoring the very high resolving power of capillary GLC columns, often coupled with selective detector systems such as electron capture, nitrogen/phosphorus, and mass fragmentometry of selected ions, are a logical step to overcome these problems. Up to now the main body of gas chromatographic literature has been concerned with conventional columns, and this review will therefore address itself to the literature considered potentially useful to future workers with capillary columns as well as the capillary column work already reported.

II. INSTRUMENTATION

The most scientific interest is centered upon the cannabinoids which are a large group of compounds including the main active constituent of cannabis and several chemically unstable compounds. The possibility of double bond isomerization [7], decarboxylation of the acids [8], and oxidation [9] requires that all surfaces that these compounds come in contact with are inert, and thus glass columns, glass injector liners, etc., should always be used. Further, it has been demonstrated that low levels of the nonpolar cannabinoids can be adsorbed from solution with great facility and be tightly bound. Thus all the glass surfaces should be silanized [10], and contact with plastics (except Teflon) and rubber should be excluded [10]. Oxygen should be rigorously excluded from the carrier gases to avoid the possibility of oxidation of these compounds at the elevated column temperatures.

Analyses of compounds in other groups of naturally occurring compounds which are also found elsewhere in nature (e.g., the terpenes, waxes, etc.), may not require such rigid analytical conditions [11] and are covered adequately in the general literature on these compounds.

The maximum size of sample to be introduced onto the capillary column is a problem, particularly in low-level analyses. There are a number of injection devices which are commercially available and whose theory of operation is discussed elsewhere in this book. The splitless devices are to be preferred where quantitative precision is important, and only glass or glass and gold systems should be used. In addition to these, a splitless device utilizing the Grob effect [12] but without back-venting excess solvent vapor, has been reported for use in analyzing underivatized cannabinoids. Relatively large volumes (approximately 10 μl) can be chromatographed [13].

The mixtures encountered in cannabinoid analysis are often so complex that even the high resolution provided by capillary columns is not adequate. In these instances, preliminary purification of the extract and/or selective detection is needed. A wise choice of extraction solvent and chemical partition are the simplest approaches, and these will be discussed later under the appropriate headings. Preliminary chromatography of the extract can also be employed [14], but the most elegant method uses a conventional chromatographic column with the facility for trapping any desired fraction of the effluent gases. The trapped fraction is then diverted onto a capillary column for further gas chromatography [15]. In this way, considerable refinement of separation and a desirable increase in loading capacity are produced.

Selective detection has the advantage of avoiding time-consuming preliminary purifications, and for biological samples in particular, it is valued for enhanced sensitivity of detection. Formation of an electron-capturing derivative [16] followed by gas chromatography with electron capture detection is the least selective method and generally requires a preliminary purification of the extract in all but the least demanding situations. Nevertheless, this approach does have some following because of the very high sensitivities of detection which can be achieved. Formation of phosphate derivatives followed by specific phosphorus detection (flame photometric or alkali bead) offers considerably improved selectivity and almost as good sensitivity of detection for the phenols [17]. Mass fragmentometry (electron impact) is another very selective and sensitive (albeit expensive) type of detection. The latter detector has the advantage that, in conjuction with gas chromatographic retention times, a judicious choice of the ions being monitored [18] can provide good confirmation of the identity of the compound [14,18]. In some situations, metabolites of a compound with the same fragments in common can also be followed. Chemical-ionization mass spectrometric detection has the advantage that it provides an intense molecular ion, but it has not been used extensively [19], and for cannabinoids electron-impact mass spectrometry is to be preferred because of the more complex fragmentation patterns.

III. THE CONSTITUENTS OF CANNABIS

The analytical chemistry of *Canabis sativa* has been largely concerned with the family of compounds known as "cannabinoids" since it is this group which is unique to cannabis and which includes the important biologically active compounds found in the plant. However, many groups of other compounds are also present, and the use of capillary column gas chromatography has been very useful in their investigation.

A. Noncannabinoid Constituents

1. *Basic Constituents*

Compared to the effort which has gone into the other areas of cannabis chemistry, the basic constituents have been rather neglected. Those compounds which have been isolated and identified by methods other than GLC are reviewed elsewhere [1,20].

Haq et al., following reported tumor-accelerating characteristics of some *N*-alkylated carbazoles and indoles, did find and quantify carbazole, indole, and skatole in marihuana smoke condensate. They used conventional GLC columns (Silar 5-CP) [21], but their analytical conditions may not be transferable to capillary column work. Lousberg and Salemink have reported the use of capillary gas chromatography for identification of a number of naturally occurring amines [20b].

2. *Neutral Constituents*

The plant waxes have partition properties similar to those of many of the cannabinoids and are generally coextractives with this other more important group of compounds. The composition of the wax is complex, containing many parafinic materials [21,22]. Gas chromatographic data using conventional columns with OV-25, OV-17, QF-1, and SE-30 stationary phases have been reported for these compounds, and presumably this will be of some use in future high-resolution work [23].

Cannabis has a characteristic, slightly sweet odor, and there has been considerable interest in the volatile constituents contributing to this smell. Hood et al. compared head-space components with those of the essential oil of cannabis using conventional GLC columns and OV-101 and Reoplex 400 as stationary phases. Twenty-seven known compounds, mainly monoterpenes and sesquiterpenes, were found in the essential oil of Mexican marihuana, and a proportion of these also occurred in the head space [24]. Bercht et al. [11] prepared samples from cannabis from a variety of sources both by passing a continuous flow of nitrogen through the plant material and collecting the distillate in a cold trap at -80°C, and by hydrodistillation. They reported the presence of at least 23 monoterpenes when the extracts were analyzed on a stainless-steel WCOT column (100 m × 0.25 mm, polyethylene glycol 1450 + 1% alkaterge). Analyses were also done using conventional columns with SE-33 and Reoplex-400 stationary phases, following that of earlier work [25], but the capillary column appeared to be markedly superior in all respects. Stahl and Kunde also have utilized capillary columns (stationary phase not mentioned) for analyzing the steam distillate of Lebanese hashish: 230 components were observed, including several not previously found in cannabis [26].

B. Cannabinoid Constituents

The cannabinoids are a large group of compounds unique to cannabis and its extracts. Δ^1-Tetrahydrocannabinol (Δ^1-THC, see Figure 28.1) is the compound responsible for the majority of psychotomimetic action of the plant. Other constituents and metabolites have different degrees and types of biological activity [27]. Formulas and abbreviations for the more important cannabinoids are given in Figure 28.1.

1. *Neutral Constituents*

The more important of the "neutral" constituents are shown in Figure 28.1. They possess a free phenolic but no carboxylic acid grouping. They are not extractable into aqueous base but may be separated from genuinely neutral constituents such as terpenes, steroids, etc., by extracting into Claisen's alkali [28]. They are very nonpolar [29] and therefore are extractable by light petroleum as well as by more polar solvents. A large number of stationary phases (SE-30, OV-1, OV-17, SE-52, OV-7, OV-210, OV-225, XE-60, QF-1, Dexsil 300 GC, and Carbowax 20 M) have been used in conventional columns for gas chromatography of these compounds [30,31]. Most of these systems gave adequate separations of the major constituents, but some overlaps were found to occur. Thus it has been suggested that some naturally occurring alkanes may interfere with cannabinoid analyses [23], that cannabidiol (CBD, see Figure 28.1) may be confused with cannabichromen [32], and methyl and propyl homologues may be confused with the amyl compounds [33] unless the retention times of the silylated derivatives are also checked. All these observations support the use of capillary in preference to conventional gas chromatographic columns.

Remarkably little capillary column work has been done with these constituents. McCallum reports using a SCOT column (70 m × 0.5 mm, coated with SE-30) for the analysis of Δ^1-THC, cannabinol (CBN), and cannabidiol (CBD) (see Figure 28.1) in conjuction with the splitless injection system, discussed in Section II [13]. Spiteller and co-workers, using glass capillary columns coated with OV-101 and SE-30 and in conjuction with a mass spectrometer, have proved the potency of this approach by isolating a number of new

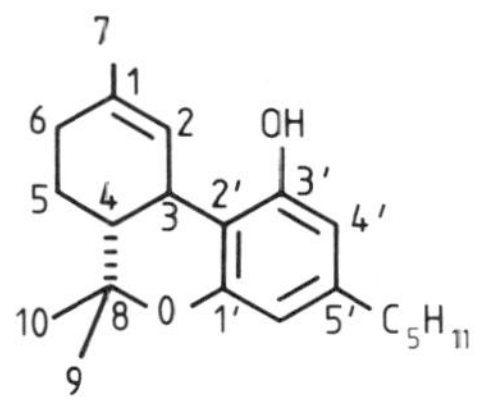

1. Δ^1-Tetrahydrocannabinol
Δ^1-THC

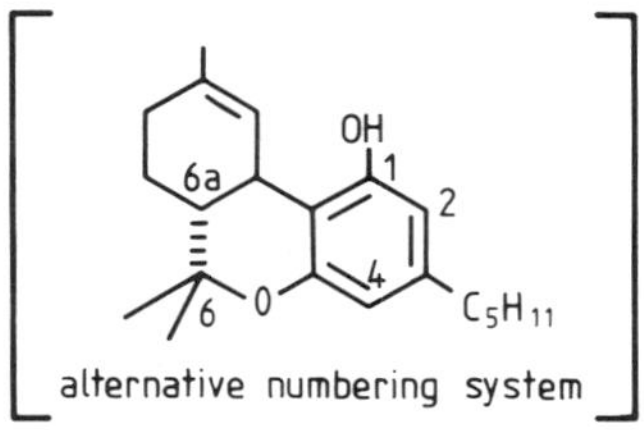

[Δ^9-THC]

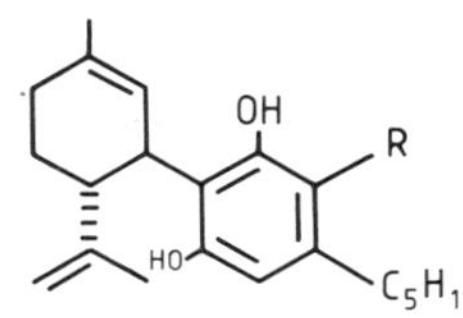

2. R = H, Cannabidiol (CBD)
R = COOH cannabidiolic acid

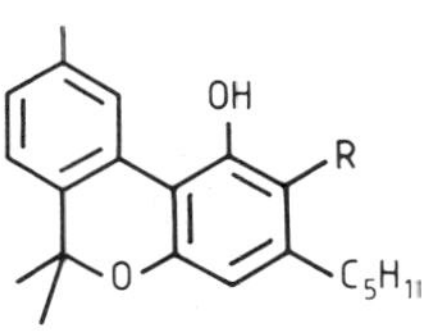

3. R = H, Cannabinol (CBN)
R = COOH Cannabinolic acid

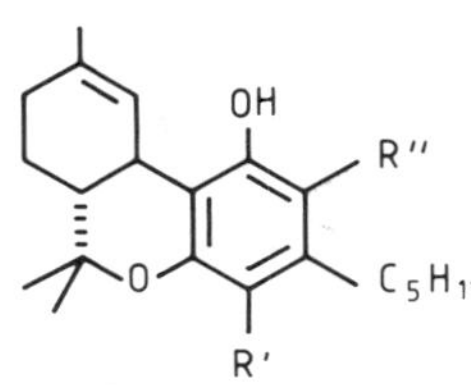

4. R′=H R″= COOH;
Δ^1 –THC acid A.
R′=COOH R″=H;
Δ^1 –THC acid B.

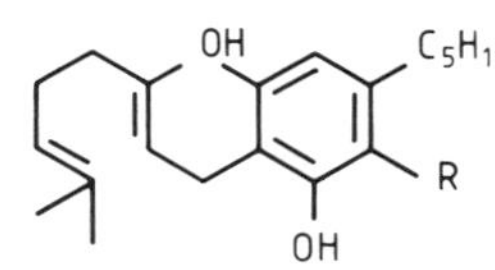

5. R = H, Cannabigerol
R = COOH cannabigerolic acid

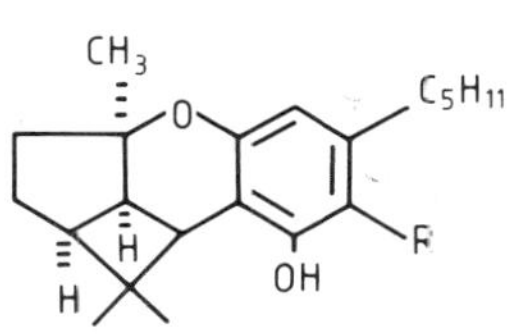

6. R=H cannabicyclol
R=COOH cannabicyclolic acid

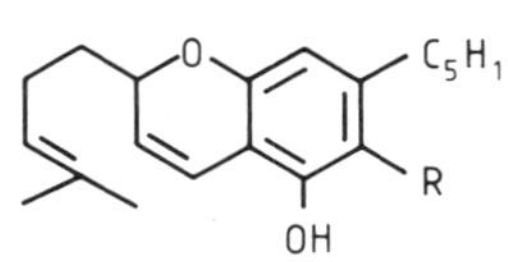

7. R = H Cannabichromen
R = COOH cannabichromenic acid

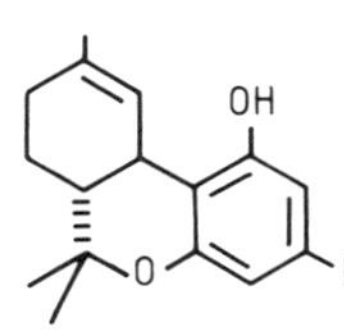

8. R = C_3H_7;
Δ^1- tetrahydrocannabivarol
(propyl - Δ^1 - THC).
R = CH_3;
Δ^1 - tetrahydrocannabiorcol
(methyl – Δ^1 – THC).

Figure 28.1 Common, naturally occurring cannabinoids.

compounds from the very complex mixtures obtained as the neutral extractives of hashish [34-36]. Crombie and Crombie have published an extensive investigation into the gas chromatography of neutral cannabinoids. After a survey of the most suitable stationary phases, gas chromatography of these compounds was carried out on a conventional column (carbowax 20 M) and as their trimethylsilyl ethers using a SCOT column (OV-225). It was reported that between these two columns it was possible accurately to analyze the 28 natural and potentially natural cannabinoids. Plots of the side-chain length against log of retention times were linear for the C_3-C_7 compounds, but not for C_1 [37].

This is in accordance with earlier work by Bailey [38] and should be of considerable value, particularly for identification and quantification of minor constituents. They also report simple "miniaturized syntheses" for eight natural cannabinoids (CBD, Δ^1-THC, Δ^6-THC, CBN, cannabichromen, cannabicyclol, and cannabicitran) and their C_1, C_3 and C_7 homologues [37].

2. Acidic Constituents

In general, these are the "neutral" cannabinoids with a carboxylic acid moiety substituted in the aromatic ring. The more important of these are shown in Figure 28.1. These compounds are relatively nonpolar and can be separated from neutral cannabinoids in a light petroleum extract by partition with 1 N NaOH. Use of weaker bases such as aqueous sodium carbonate are not as efficient [39]. Above 80°C [40] the acids can decarboxylate with ease unless protected as methyl esters or by silylation [41]. Crombie and Crombie report "miniaturized syntheses" of these compounds and their homologues which are suitable for use by the analytical chemist, and separations using a conventional OV-17 column and a SCOT OV-225 for 34 compounds are reported. In this work it was found that the methyl esters of the compounds gave poorer separations and required higher column temperatures than when the phenolic moiety was also trimethylsilylated. In a few cases, overlaps between compounds were observed but these could be avoided by a preliminary, simple thin-layer chromatographic separation [41]. Grote and Spiteller [35] report the isolation of water-soluble cannabinoids by absorption on XAD resins and also separation by forming trimethylsilyl methyl esters for gas chromatography using a 25 m × 0.3 mm SE-30 capillary column.

3. Cannabis Constituents After Smoking

Cannabis products are often administered by smoking, and it is necessary to understand the effect of the smoking process on the constituents before considering the significance of their levels in the material to be smoked. Fentiman et al condensed marihuana smoke in cooled traps and analyzed the condensate using GLC-MS (chemical ionization as well as electron impact) with conventional columns. A number of noncannabinoid acids and phenols were identified [19]. It seems unlikely that any of these compounds make a significant contribution to the "cannabis effect," although contribution to cocarcinogenesis and ciliostasis remains a separate question [42]. Novotny and Lee trapped marihuana smoke from the atmosphere in a room onto the polymer, poly-*p*-2,6-diphenylphenylene oxide, and with the aid of GLC and using a capillary column (SF-96; 60 m × 0.4 mm, I.D.), found an extremely complex mixture with hundreds of components which was quite distinguishable from the normal room background and from tobacco smoke. Only the major cannabinoids were identified [43]. Later work from the same authors, comparing marihuana, tobacco, and Δ^1-THC in model pyrolysis experiments, and using GLC with a capillary column (SE-52; 11 m × 0.26 mm, I.D.), resulted in the identification of over 150 polynuclear hydrocarbons in the smoke. Again these compounds would be expected to make no contribution to the "cannabis effect," but they were shown to arise in much higher concentrations from the cannabinoids in marihauana than from tobacco, and they are of particular interest for their potential carcinogenicity as well as being important decomposition products of the cannabinoids [44]. The pyrolysis of CBD has also been studied in some detail using conventional GLC-MS. A number of new cannabinoids and cracking products were identified, but it is unlikely that any of these compounds have a significant influence on the overall "cannabis effect" [45].

The cannabinoid acids are known to decarboxylate easily and in high yield to the parent cannabinoid [46] during the smoking process and also during gas chromatography of the free acid. This parallel can be used to advantage when analyzing marihuana (which is to be used for smoking) for available Δ^1-THC. Considerable amounts of unchanged cannabinoids can be inhaled via smoking. The proportion of Δ^1-THC reported to survive smoking varies widely from as little as 20% [47] to the more commonly accepted range of 50-60° [33]. Clearly a major factor is the combustion temperature (and therefore, indirectly, moisture content of the material) as well as the medium in which the cannabinoids are pyrolyzed. Thus, the proportions of cannabinoids delivered to the subject, when marihuana is smoked, normally show little change [48,49], but when tobacco cigarettes (which are more acidic than marihuana [50]) containing cannabinoids are smoked, the proportions delivered can vary. Under these conditions CBD can undergo an acid-catalyzed cyclization to Δ^1-THC [50], and the oxidation of Δ^1-THC to CBN is also enhanced (2.7-6.8%) [49]. The use of dried tobacco (and therefore high combustion temperatures) increases the latter transformation still further to approximately 40% of the Δ^1-THC [17]. The smoking of preextracted hemp straw with cannabinoids spiked onto it may allow isomerization of some of the Δ^1- to Δ^6-THC [48]. Over 80% of the cannabinoids in the smoke are absorbed by experienced users [47].

4. *Analytical Procedures*

Quantitative gas chromatographic analysis is best carried out relative to internal standards. Comparison between the cannabinoid peak in question and the internal standard can be done either by peak area or peak height provided the peaks show no tailing. In our work and in that of others [30], we have found acceptable precision and greater convenience with the latter method. A large number of internal standards have been used; methadone, cocaine, androst-4-ene-3,17-dione, dibenzyl phthalate, methyl oleate, *n*-eicosane, *n*-docosane [51], and other *n*-alkanes [30]. All of these are probably suitable provided that certain criteria are met. These criteria include the following: (1) the cannabinoid in question is efficiently extracted (the internal standard is normally added after the extraction); (2) very low levels are not being quantified; (3) the internal standard peak does not overlap any naturally occurring peak; and (4) in those instances where cannabinoids are being derivatized, the standard contains the same functional groups of a similar reactivity to the compound in question. In the analysis of levels of cannabinoids less than about 1 μg/ml, absorption of the cannabinoids onto glass can be a problem [10] unless the glassware is silanized (most conveniently by the method outlined in [17]). The situation can further be assisted by choosing an unnatural cannabinoid of similar structure as internal standard (e.g., for the analysis of Δ^1-THC, the *n*-heptyl homologue of Δ^6-THC [17] or a deuterated form of the compound if detection is by MS [14]) and using it at high concentrations so that it acts as a "carrier," that is, it can compete successfully for the available absorption sites. The stability of the internal standard solution during storage is of importance to the accuracy of the final results. With the noncannabinoids, in some cases it is documented [52], and in others it may be self-evident or have to be checked. Cannabinoids may be stored at -20°C in nonpolar solvents or in ethanol, but chloroform in particular should be avoided [52]. Calibration curves made from comparing standard solutions of each cannabinoid against internal standard are required before the unknown solutions can be quantified. Most cannabinoids are oils and are therefore likely to retain some solvent. The oils should therefore be prepared as standard solutions utilizing ultraviolet absorbance data [53] rather than by gravimetric methods, with the proviso that sufficient care be taken to purify the cannabinoids from others which may interfere with this estimation. The most common example of this is the removal of CBN and also Δ^6-THC from Δ^1-THC standard solutions by TLC [17] or HPLC [54].

Important considerations in the analysis of cannabis material is the method of storage and extraction prior to the estimation. The plant material is best stored at -20°C [55,56] if it is not be be analyzed soon, and it should be air-dried at room temperature prior to solvent extraction [46]. Light petroleum and ethanol have been used as extraction solvents, but the work of Fairbairn and Liebmann indicates that chloroform gives greater extraction efficiency than light petroleum (two 30-min chloroform extractions give >99% recovery) and many less noncannabinoid coextractives than ethanol [46]. The appropriate internal standard is then added, and unless the solution is to be analyzed immediately, it is preferable to evaporate the extract and reconstitute in another solvent such as light petroleum or ethanol to avoid decomposition of the cannabinoids [52]. Depending on the success of the gas chromatographic conditions in separating the required compound, it is possible that separation of some of the noncannabinoids in the extract prior to analysis may be needed [23]. If this is so, column chromatography on Florisil [57], alumina [23], and molecular sieves [14] is possible. However, this should not be necessary with a high-resolution capillary column and/or the gas chromatographic conditions suggested by Crombie and Crombie [37]. It should be borne in mind that the extract will also contain the carboxylic acids corresponding to the parent cannabinoids, and these will decarboxylate under gas chromatographic conditions, elevating the apparent level of cannabinoid. Such a level is relevant to the cannabinoid available from the smoking process (with the appropriate adjustment for pyrolysis) rather than the amount actually in the plant [46]. If the acids and neutrals have to be analyzed separately, a preliminary purification by back-extraction into 1 N NaOH [58,59], column chromatography [23], or derivatization of the extract [41,60] is necessary. Formation of silyl ethers has also been recommended for resolving some cannabinoids on low-resolution columns [32], and separations of the stable silyl ether or methyl ester derivatives of acids have also been reported [37,41,60]. These derivatives are easily formed, but a reduction in quantitative precision is reportedly produced [32]. In many instances, careful choice of high-resolution columns should be sufficient to avoid cannabinoid overlap and therefore the need for silylation, and the cannabinoid acids may be quantifiable with greater precision from the difference of two analyses: the first, including the decarboxylated acid and the second, after the acids have been removed by one of the above methods. It should also be noted that under certain conditions, losses of cannabinoid on the gas chromatographic column are possible at lower concentrations of stationary phase [30], and in the case of CBD and cannabigerol, by pyrolysis in the injector port over 150° [61]. This will be evident when a calibration curve is constructed. The former case should be able to be corrected by silylation of the column or by increasing the stationary phase loading.

C. Cannabinoids in Biological Samples

1. *Analysis for Cannabinoids*

Introduction: Most of the interest in the analysis of cannabinoids in biological samples has centered on the three most abundant cannabinoids (Δ^1-THC, CBN, and CBD) and also Δ^6-THC which is often used as a more stable model for Δ^1-THC. Δ^1-THC is considered the major active component of cannabis [27], and for this reason most of the work has dealt with it. Δ^1-THC is not normally found in urine [62,63], although Green has reported its presence at low levels [64], and it is the blood samples which are usually analyzed because of their greater relevance (such as it may be) to brain levels. In the case of post-mortem analyses, a brain sample may well give more information.

Most of the analytical problems encountered arise from the very low levels of cannabinoid which have to be determined (down to 1 ng/ml of blood being typical levels 4-6 hr after smoking (cf. Figure 28.2)). At these levels, conventional gas chromatographic detection systems are no longer sufficiently sensitive, and even when a detection system is sufficiently sensitive, it is probable that detection will be frustrated by the vast proliferation of other substances in the extract which obscure the compound in question. Therefore, sufficient selectivity of detection is as important a criterion as sensitivity, and the poorer the selectivity the more complicated and tedious the method will become with the preliminary purification steps that have to be introduced. The polarity of the extraction solvent (and therefore the number of coextractives) as well as the possibility of contaminants (from detergents used on glassware, etc.,) also have bearing on gaining sufficient analytical specificity. Another problem arising from the low levels of the drug is that at these levels, binding of the drug to glassware and other materials must also be prevented [10].

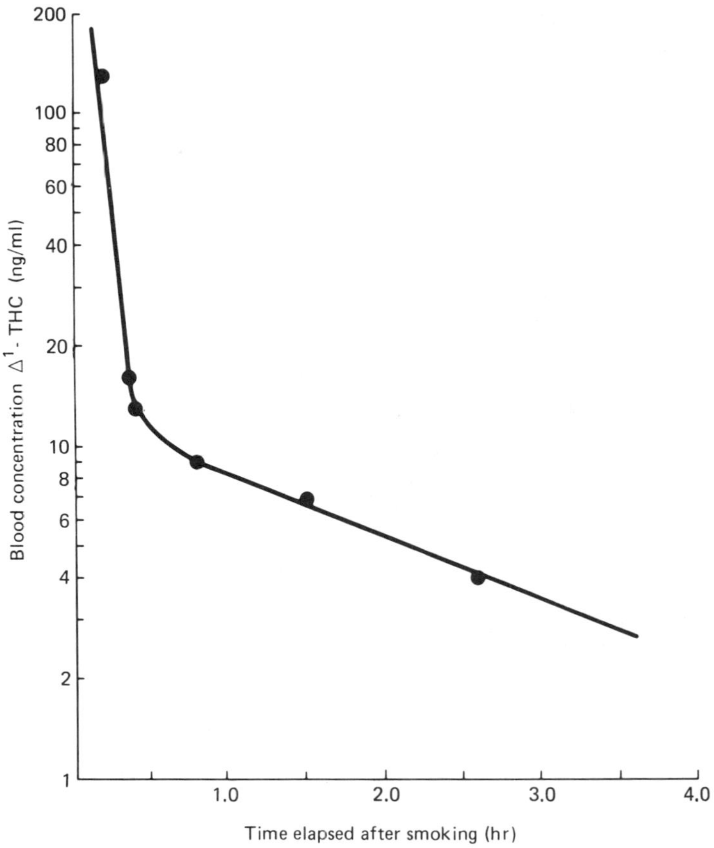

Figure 28.2 Blood levels of Δ^1-THC during the period of intoxication following the smoking of a "joint" containing 8 mg Δ^1-THC and 1.6 mg CBN.

The glassware: Clearly, different methods of detection will be sensitive to different types of contamination, and therefore the adopted methods for cleaning the glassware should normally be those specified in the literature for that particular method. If none is specified, the proposed detergent should be blanked before being used routinely. We recommend avoiding the alternative of chromic acid cleaning. Several methods for silanizing glassware exist. Most of these use silanizing agent in a solvent, and they are both slow and inconvenient for large quantities of glassware. Exposure of glassware to an atmosphere of silanizing agent in a sealed container (such as a desiccator with a small beaker of silanizing agent contained in it), and then transferring the glassware to another container where it is exposed to methanol vapor, is a rapid and very convenient alternative [17].

Extraction solvents: The most frequently used extraction solvent is light petroleum (or pentane or hexane) by itself [17,65-67] or containing 1.5% isoamyl alcohol [14,15,68,69]. Choice of the latter is generally made on the grounds of improved extraction efficiency, which is normally quoted as being greater than 90%. Reports of extraction efficiencies for the former are generally lower, varying between 68 and 92%. In our work, we have found that use of 1.5% isoamyl alcohol/heptane for extraction introduces many co-extractives that an extensive purification prior to GLC analysis would be necessary. On the other hand, using the same analytical techniques with pentane or hexane as extraction solvent, only the cannabinoids are seen in an otherwise very clean extract [17] (see Figure 28.3). Lower extraction

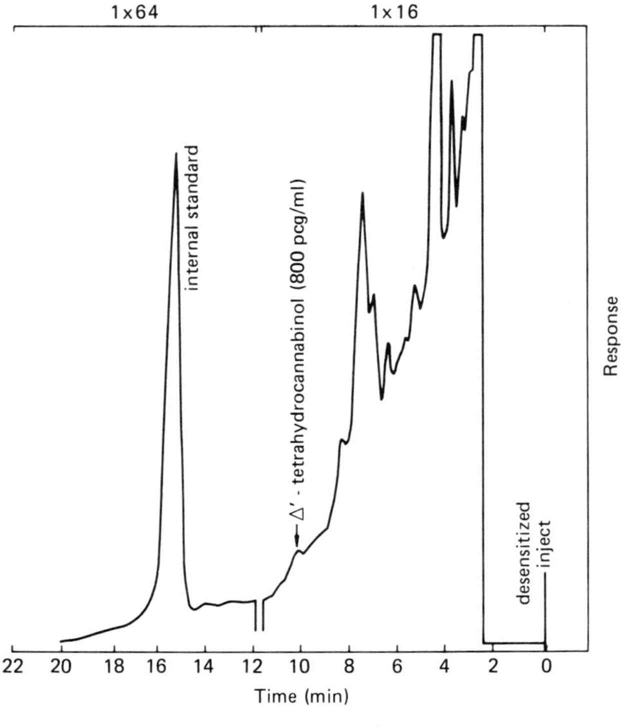

Figure 28.3 Gas chromatogram from the analysis of 1 ml of blood taken from a known drug user who was found three days after being drowned. The Δ^1-THC level in this blood was 800 pg/ml.

efficiencies in the latter situation need not be a problem either. The underlying reason for the widely different quoted extraction efficiencies for cannabinoids in this case appears to be that pentane and hexane extract the compounds at a much slower rate than when isoamyl alcohol is present. Therefore, if it is found that extraction efficiencies are not high enough, it is better to extract more vigorously or for longer periods than to change solvents. If the internal standard added to the blood prior to extraction is very similar to the compound being analyzed, and if the detector has adequate sensitivity, efficiency of extraction may be sacrificed to speed of analysis without untoward effects on accuracy of results [17].

Internal standards: Internal standard (usually in ethanol) is added to the blood prior to extraction and compared to the cannabinoid peak in the final chromatogram (ratios of peak heights or peak areas). The ratio of the two peaks is then referred to a graph previously constructed from analyses of blank bloods to which known amounts of cannabinoid had been added (see Figure 28.4). The role of the internal standard is two-fold. First, its traditional function is to ensure that mechanical losses or dilutions and losses by binding to glassware, etc., have no effect on the result although they do have an effect on the overall yield of compounds for analysis. Second, the internal standard may be used as a "carrier" [17,70]. Losses of cannabinoids to glassware, etc., are a problem only in the lower nanogram range and, although the right internal standard will allow accurate analyses, it may still be

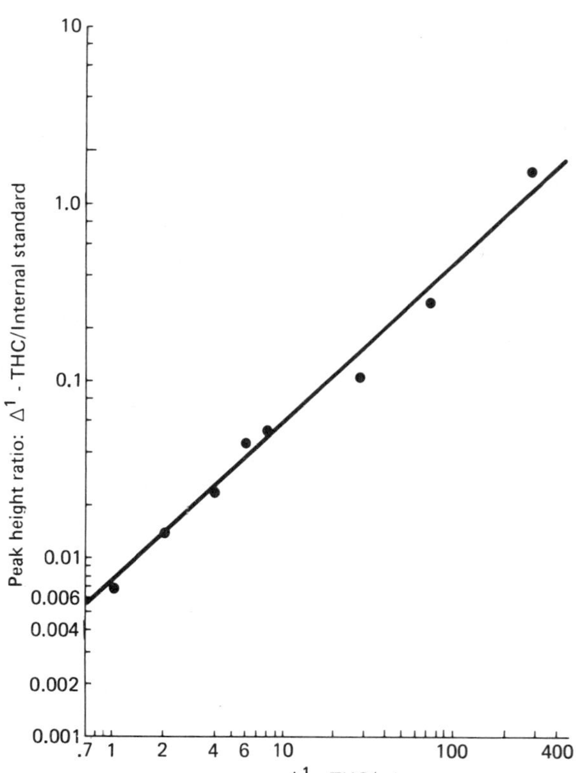

Figure 28.4 Standard curve for Δ^1-THC in blood.

desirable to increase the overall cannabinoid yield to extend the sensitivity of the method being used. If the internal standard is an unnatural homologue of the cannabinoid analyzed for, and is used at relatively high concentrations (e.g., 0.1-1 μg for Δ^1-THC internal standard), the majority of cannabinoids occupying available binding sites will be internal standard (if the Δ^1-THC concentration is low), and the overall proportion of cannabinoid lost to such binding will be significantly reduced.

The choice of internal standard to some extent is dictated by the analytical technique being used. Clearly, the ideal internal standard is deuterated Δ^1-THC for Δ^1-THC analyses, and this is used when mass fragmentometry is the detection system [14a,28,67]. In practice, interference from exchange reactions with the internal standard is a problem and may limit the lowest quantifiable levels [70] of an otherwise very sensitive technique. With other detectors, the criterion for a good internal standard is close chemical similarity, and preferably, a retention time slightly longer than the cannabinoid so that large concentrations of internal standard will not swamp small concentrations of the cannabinoid to be measured. In our analyses we use the *n*-heptyl homologue of Δ^6-THC. This was chosen instead of the Δ^1-THC homologue on grounds of better chemical stability [17]. Other workers have used hexahydrocannabinol [71,72] and the naturally occurring CBD and CBN, the latter two having limited utility for obvious reasons [73].

The methods: The method of Fenimore et al. [15,71] is the only routine gas chromatographic method which has been adapted for use with capillary columns. The method involves extraction of blood serum with 1.5% isoamyl alcohol in hexane, derivatization of the extract with pentafluoropropionic anhydride, and ultimately, separation by GLC using a capillary column with electron capture detection. Unfortunately the derivatization and detection lack sufficient specificity, and a preliminary purification is necessary. This is achieved by introducing the extract into a conventional packed GLC column with facilities to trap the required fraction and to introduce it directly into a capillary column for further chromatography. This reduces the number of interfering coextractives to the point where Δ^1-THC levels down to 0.5 ng/ml may be quantified; however, compounds with greater retention times than the Δ^1-THC are also present, and as a result, the total gas chromatographic analysis time is approximately 45 min [71]. Disadvantages of this technique, therefore, appear to be analysis time and to a lesser extent the need for specialized equipment. It is quite possible that changing the extraction solvent to hexane could enable the analysis time to be considerably improved.

Garrett and Hunt report a method based on the conventional column chromatography of the pentafluorobenzoates with electron capture detection. Again a preliminary purification of the extract is required, but this time it is achieved by the use of HPLC. Presumably, capillary columns similar to those above could be substituted in this method, providing an increase in resolution and sensitivity, and again the use of hexane as extraction solvent may have its advantages. It should also be pointed out that, thus far, this method has only been tested for dog plasma, and it is possible that analyses of human blood could present further problems [73].

The use of mass spectrometric detection has enjoyed the widest success to date. Skinner reported its use for the detection of Δ^1-THC in human urine and rat blood and urine, and good sensitivity of detection was achieved [69]. However, with the more difficult medium, human blood, the method of Agurell et al. was one of the first methods of analysis for Δ^1-THC, and it is probably still one of the most widely used GLC methods [14a,70]. With this technique (also called multiple ion detection) the ion intensities of characteristic molecular fragments from the GLC column eluent are monitored in the form of a chromatogram. With Δ^1-THC, for example, m/e = 299, 314 can be monitored, and similarly for the deuterated internal standard. The relative peak heights of Δ^1-THC and internal standard provide the means for calculating the concentration in the original sample. Unfortunately in this case, the selectivity of detection and the resolution of the columns used were not sufficient to separate endogenous plasma constituents, and a preliminary liquid chromatographic cleanup was therefore necessary [14b,70]. The sensitivity of detection of this method is good (0.3 ng Δ^1-THC/ml plasma) and only limited by the extent of exchange of the deuterated Δ^1-THC internal standard (usually d_3 or d_7) [70]. Other workers have developed similar approaches, and Agurell reports that the use of silyl derivatives provides a small increase in sensitivity of detection [70b]. Rosenfeld et al. recommend an on-column methylation of the Δ^1-THC and the addition of the trideuterio derivative of this compound (Δ^1-THC-OCD_3) as internal standard, to the extract immediately prior to gas chromatography [28].

This method does appear to be much shorter than that of Agurell, and it manages to bypass the need for preliminary chromatography by this combination of derivatization and a selective extraction of the interfering lipids using dilute Claisen's alkali [28]. A disadvantage of the method is its reliance on the reproducibility of extraction yields. These workers report a 5.5% standard deviation in their results, but we have found in routine work with a range of bloods of different ages and condition and differences in the condition of glassware, that variations in extraction efficiencies may be rather higher.

The use of mass spectrometry in the chemical ionization mode (CI) has also proved to be of value. Detrick and Foltz have developed a method in which they add deuterated internal standard, extract the blood with hexane, wash the extract with acid and base, form the trimethylsily derivatives, and quantify by GLC using a conventional OV-17 column and CI mass spectrometry detection, concentrations down to ca. 0.5 ng/ml [67]. Wall and coworkers use a similar approach, but with a 1.5% isoamyl alcohol/light petroleum extraction, hexahydrocannabinol internal standard, and pentafluoropropionate derivatives. No preliminary purification was necessary [72].

The final approach to the analyses of human bloods involves the use of specific phosphorus detection. Addition of the heptyl homologue of Δ^6-THC as internal standard, hexane or pentane extraction of the blood, followed by derivatization with diethyl phosphorochloridate, and gas chromatography with specific phosphorus detection (flame photometric or alkali bead) offers outstanding selectivity and limits of detection about 0.5 ng Δ^1-THC/ml blood. Figure 28.3 illustrates such a chromatogram, and it can be seen that in the area where cannabinoids are expected to occur, no significant concentrations of other coextractives can be seen [17]. Because of the lack of preliminary purification this is a very rapid technique by comparison to other available methods. The method has also been adapted to the analysis of brain samples [17], and it is one of only two available methods for this purpose [74].

General comments: The field of detection of cannabinoids in biological fluids is expanding rapidly, and as yet there is no "accepted" method to be used. Of all the methods described above, only two utilize the superior resolving power of capillary columns. Clearly, with the remaining methods there is scope for development using any of the current techniques for sample concentration combined with capillary columns. Such work should be well rewarded by fewer problems with coextracted endogenous material and greater sensitivity of detection. It may also be that some of these methods can be improved by changing extraction solvents.

The choice of what provides the basis of a satisfactory method will depend on the aims and circumstances of the worker. For epidemiological studies speed is the essence, and the worker may well choose an immunoassay technique for screening large numbers of samples [76] and a more precise method for confirmation of those which appear to be positive. Criteria as to what constitutes an unequivocal identification are difficult to define. Mass spectrometric detection has the advantage of characteristic mass fragments as well as retention times, and probability-based matching (PBM) of a greater number of peaks than what is normally used [64] may well prove to be a valuable refinement to this technique. All other techniques have only retention times as evidence and as a minimum, confirmation on a nonpolar column (e.g., OV-1) and a polar column (e.g., Hi Eff-9BP) would be necessary. However, in the final consideration, only large numbers of analyses of known blank bloods and positive bloods will give the operator an indication of minimum requirements.

In conclusion, it must be stressed that cannabinoids, and in particular Δ^1-THC, are among the most difficult of drugs to analyze for. Many problems are generated by the very low levels of drugs normally analyzed for, from their binding properties, and from the large number of compounds (naturally occurring and otherwise) which can interfere with the analyses. For these reasons particular attention must be paid to preparation of glassware, redistillation of all solvents, excluding contact with all materials except glass and Teflon, keeping the analytical set-up completely separate from all other cannabis products and manipulations. In general, the literature method should be followed scrupulously unless the change is carefully considered and checked.

2. *Analysis for Cannabinoid Metabolites*

Neutral metabolites: Cannabinol (Figure 28.1) occupies a unique position as a naturally occurring compound which is also a metabolite of Δ^1- and Δ^6-THC. It has been found as a metabolite in the blood of animals [76-79] and in humans [64], although Wall et al dispute the latter result. It is possible that the mode of Δ^1-THC administration [72] used by Wall et al. (in serum albumin, intravenously) was responsible for their negative results [80]. Cannabinol may be analyzed for, by the methods already discussed above.

The majority of the remaining "neutral" metabolites are cannabinoids which have been hydroxylated at saturated carbons in any or several of the positions shown (see Figure 28.5). The most important of the hydroxylated metabolites is undoubtedly 7-hydroxy-Δ^1-THC which is reported to be considerably more potent than the parent compound [81]. Presumably this compounds plays some

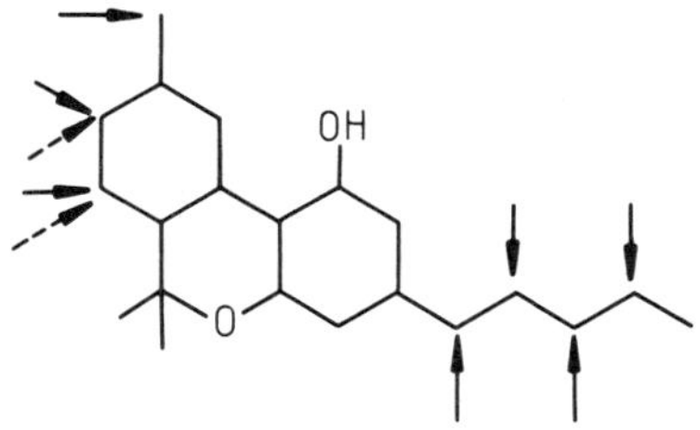

Figure 28.5 Possible positions for hydroxylation in the cannabinoid nucleus.

role in the overall "cannabis effect" although it is probable that it is not as great [82] as first suggested in the "active metabolite" hypothesis [83]. Wall and co-workers have developed a GLC/mass spectrometric method for the analysis of this compound. It involves a petroleum ether/1.5% isoamyl alcohol extraction, purification using liquid chromatography on Sephadex LH 20 (c.f., [14b]), and silylation of the 7-hydroxy-Δ^1-THC before GLC-MS in the EI mode. Δ^1-THC and CBN can be quantitated from the same extract. The liquid chromatographic elution volumes are documented for a very large number of metabolites [14b], and these may well be analyzable using similar techniques. In an alternative approach, Harvey and Paton have found that formation of the tri-*n*-butylsilyl derivatives of extracts containing cannabinoids plus mono- and dihydroxycannabinoids enables a gas chromatographic separation of all these compounds [84]. This may prove valuable in an alternative analytical method based on GLC-MS. Foltz et al. report a method similar to that of Wall's, but using GLC-MS in the CI mode, preliminary purification of the extract is avoided. He claims twice the sensitivity compared to EI mass spectrometric detection with greater selectivity [85].

Acidic metabolites: Agurell and co-workers have pointed out that Δ^1-THC-7-oic acid is the major metabolite for Δ^1-THC, having concentrations almost equal to Δ^1-THC within 50 min of smoking and then slightly higher for more than 24 hr. They propose this as the basis for identifying cannabis intoxication and present a method [86] similar to their previous method for quantifying Δ^1-THC in the blood [70]. The metabolite is extracted from buffered blood with diethyl ether, converted to the methyl ester using diazomethane, separated from other impurities using liquid chromatography, and silylated prior to analysis by GLC-MS in the EI mode. The results are described as "semi-quantitative," and the detection limits have not been established. The authors point out there may well be other acidic metabolites at relatively high levels in human blood, corresponding to those already found in other animals [87,88] and which may also be analyzed by this method.

Green reports using a similar method for Δ^1-THC-7-oic acid in urine with probability-based matching (PBM) of between 10 and 15 mass fragments in the standard compound to those detected in the GLC eluent. The compound can be identified even in relatively complex mixtures, and this eliminates the need for a preliminary cleanup of the extract [89].

Wall et al. suggest that the Δ^1-THC-7-oic acid forms a stable protein complex which needs to be broken by an initial acetone extraction. Removal of solids, concentration of the aqueous acetone to an aqueous extract, which is then extracted with diethyl ether and purified by HPLC, followed by silylation and quantification with GLC-MS as in his previous work. Detection limits are ca. 0.5 ng/ml plasma [90]. A most interesting report is that of the method of Foltz et al. still under development. This technique involves direct injection of body fluids onto reverse-phase HPLC, derivatization of the appropriate fractions, and analysis of the extracts on capillary column GLC-MS, with mass spectrometric detection in the CI mode. It is claimed that Δ^1-THC, 7-OH-THC, and Δ^1-THC-7-oic acid can be analyzed for simultaneously and down to 10 pg/ml. Foltz stresses that in this method, considerable advantage is gained from the capillary column because of better resolution, much improved sensitivity of detection, and less chance of loss of sample from adsorption and decomposition of the sample because of the lower operating temperatures enjoyed by these columns [85].

General comments: It will be seen that little work has been done on the analysis of cannabinoid metabolites in human blood. Methods for the two Δ^1-THC metabolites should be valuable for the large number of other metabolites which have yet to be investigated. The current methodology uses conventional GLC columns, and if this is modified to take advantage of the improved sensitivity and resolution of capillary columns, significant advantages would accrue.

REFERENCES

1. R. Mechoulam, in *Marijuana: Chemistry, Pharmacology, Metabolism, and Clinical Effects* (R. Mechoulam, ed.), Academic Press, New York, 1973, pp. 4-7.
2. (a) Y. Gaoni and R. Mechoulam, *J. Am. Chem. Soc. 86*:1646 (1964); (b) Y. Gaoni and R. Mechoulam, *J. Am. Chem. Soc. 93*:217 (1971).
3. J. W. Fairbairn, *Bull. Narcotics 24*:29 (1972).
4. L. J. Grlić, *Bull. Narcotics 16*:29 (1964).
5. R. Mechoulam, N. K. McCallum, and S. Burstein, *Chem. Rev. 76*:75 (1976).
6. W. M. L. Crombie, in *Cannabis and Health* (J. D. P. Graham, ed.), Academic Press, New York, 1976, p. 25.
7. E. C. Taylor, K. Lenard, and Y. Shvo, *J. Am. Chem. Soc. 88*:367 (1966).
8. R. Mechoulam, Z. Ben-Zvi, B. Yagnitinsky, and A. Shani, *Tetrahedron Lett.* 2339 (1969).
9. R. K. Razdan, A. J. Puttick, B. A. Zitko, and G. R. Handrick, *Experientia, 28*:121 (1972).
10. (a) E. R. Garrett and C. A. Hunt, *J. Pharm. Sci. 62*:1211 (1973); (b) E. R. Garrett and C. A. Hunt, *J. Pharm. Sci. 63*:1056 (1974).
11. C. A. L. Bercht, F. J. E. M. Küppers, R. J. J. Ch. Lousberg, C. A. Salemink, A. Baerheim Svendsen, and J. Karlsen, *U.N. Secretariat Document*, ST/SOA/SER.S/29 (1971).
12. K. Grob and K. Grob, Jr., *J. Chromatogr. 94*:53 (1974).
13. N. K. McCallum and E. R. Cairns, *J. Pharm. Sci. 66*:114 (1977).
14. (a) S. Agurell, B. Gustaffson, B. Holmstedt, K. Leander, J.-E. Lindgren, I. Nilsson, F. Sandberg, and M. Asberg, *J. Pharm. Pharmacol. 25*:554 (1973); (b) K. Fonseka, M. Widman, and S. Agurell, *J. Chromatogr. 120*:343 (1976).
15. D. C. Fenimore, R. R. Freeman, and P. R. Loy, *Anal. Chem. 45*: 2331 (1973).
16. N. K. McCallum and R. J. Armstrong, *J. Chromatogr. 78*:303 (1973).
17. N. K. McCallum and E. R. Cairns, *J. Anal. Toxicol. 2*:89 (1976).
18. T. B. Vree, D. D. Breimer, C. A. M. van Ginneken, J. M. van Rossum, R. A. de Zeeuw, and A. H. Witte, *Clin. Chim. Acta 34*:365 (1971).
19. A. F. Fentiman, Jr., R. L. Foltz, and G. W. Kinzer, *Anal. Chem. 45*: 580 (1973).
20. M. D. Willinsky, in *Marijauana: Chemistry, Pharmacology, Metabolism, and Clinical Effects* (R. Mechoulam, ed.), Academic Press, New York, 1973, p. 139. (b) R. J. J. Ch. Lousberg and C. A. Salemink, *Pharm. Weekbl. 108*:1 (1973)
21. M. Z. Haq, M. Zamir-ul, S. J. Rose, L. R. Deiderich, and A. R. Patel, *Anal. Chem. 46*:1781 (1974).
22. R. Mechoulam, N. K. McCallum, and S. Burnstein, *Chem. Rev. 76*:80 (1976).
23. R. A. de Zeeuw, J. Wijsbeek, and Th. M. Malingré, *J. Pharm. Pharmacol. 25*:21 (1973).

24. L. V. S. Hood, M. E. Dames, and G. T. Barry, *Nature 242*:402 (1973).
25. M. C. Nigam, K. L. Handa, I. C. Nigam, and L. Levi, *Can. J. Chem. 43*:3372 (1965).
26. E. Stahl and R. Kunde, *Tetrahedron Lett. 30*:2841 (1973).
27. R. Mechoulam and H. Edery, in *Marijuana: Chemistry, Pharmacology, Metabolism, and Clinical Effects* (R. Mechoulam, ed.), Academic Press, New York, 1973, p. 116.
28. J. J. Rosenfeld, B. Bowins, J. Roberts, J. Perkins, and A. S. Macpherson, *Anal. Chem. 46*:2232 (1974).
29. E. W. Gill, G. Jones, and D. K. Lawrence, *Biochem. Pharmacol. 22*:175 (1973).
30. N. E. Hoffman and R. K. Yang, *Anal. Lett. 5*:7 (1972).
31. W. L. Crombie, in *Cannabis and Health* (J. D. P. Graham, ed.), Academic Press, New York, 1976, p. 25.
32. C. E. Turner, K. W. Hadley, J. Henry, and M. L. Mole, *J. Pharm. Sci. 63*:1872 (1974).
33. W. M. L. Crombie, in *Cannabis and Health* (J. D. P. Graham, ed.), Academic Press, New York, 1976, p. 26.
34. J. Friedrich-Fiechtl and G. Spiteller, *Tetrahedron 31*:479 (1975).
35. H. Grote and G. Spiteller, *J. Chromatogr. 154*:13 (1978).
36. H. Grote and G. Spiteller, *Tetrahedron 34*:3207 (1978).
37. L. Crombie and W. M. L. Crombie, *Phytochemistry 14*:213 (1975).
38. K. Bailey, D. Legault, and D. Verner, *J. Chromatogr. 87*:263 (1973).
39. M. D. Willinsky, Ph.D. dissertation, University of Toronto, 1971.
40. N. J. Doorenbos, P. S.Fetterman, M. W. Quimby, and C. E. Turner, *Ann. N.Y. Acad. Sci. 191*:3 (1971).
41. L. Crombie and W. M.L. Crombie, *Phytochemistry 16*:1413 (1977).
42. E. L. Wynder and D. Hoffmann, *Adv. Cancer Res. 8*:249 (1964).
43. M. Novotny and M. L. Lee, *Experientia 29*:1038 (1973).
44. M. Novotny, M. L. Lee, and K. D. Bartle, *Experientia 32*:280 (1976).
45. H. J. W. Spronck and C. A. Salemink, *Recl. Trav. Chim. Pays-Bas 97*:185 (1978).
46. J. W. Fairbairn and J. A. Liebmann, *J. Pharm. Pharmacol. 25*:150 (1973).
47. S. Agurell and K. Leander, *Acta Pharm. Suecica 8*:391 (1971).
48. R. Mechoulam, in *Marijuana: Chemistry, Pharmacology, Metabolism, and Clinical Effects* (R. Mechoulam, ed.), Academic Press, New York, 1973, p. 70.
49. K. O. Fehr and H. Kalant, *Can. J. Physiol. Pharmacol. 50*:761 (1972).
50. W. Quarles, G. Ellmanand, and R. Jones, *Clin. Toxicol. 6*:211 (1973).
51. W. M. L. Crombie, in *Cannabis and Health* (J. D. P. Graham, ed.), Academic Press, New York, 1976, p. 29.
52. C. M. Bonuccelli, *J. Pharm. Sci. 68*:262 (1979).
53. R. Mechoulam, in *Marijuana: Chemistry, Pharmacology, Metabolism, and Clinical Effects* (R. Mechoulam, ed.), Academic Press, New York, 1973 p. 149.
54. E. E. Knaus, R. T. Coutts, and C. W. Kazakoff, *J. Chrom. Sci. 14*:525 (1976).
55. C. E. Turner, K. W. Hadley, P. S. Fetterman, N. J. Doorenbos, M. W. Quimby, and C. Waller, *J. Pharm. Sci. 62*:1601 (1973).
56. G. S. Lewis and C. E. Turner, *J. Pharm. Sci. 67*:876 (1978).
57. M. D. Willinsky, in *Marijuana: Chemistry, Pharmacology, Metabolism, and Clinical Effects* (R. Mechoulam, ed.), Academic Press, New York, 1973, p. 141.

58. R. Mechoulam and Y. Gaoni, *Tetrahedron 21*:1223 (1965).
59. R. Mechoulam and Y. Gaoni, *Fortschr. Chem. Org. Naturst. 25*:175 (1967).
60. P. S. Fetterman, N. J. Doorenbos, E. S. Keith, and M. W. Quimby, *Experientia 27*:988 (1971).
61. K. Bailey, *J. Chromatogr. 160*:288 (1978).
62. S. H. Burnstein, in *Marijuana: Chemistry, Pharmacology, Metabolism, and Clinical Effects* (R. Mechoulam, ed.), Academic Press, New York, 1973, p. 180.
63. L. E. Hollister, S. L. Kanter, R. D. Board, and D. E. Green, *Res. Commun. Chem. Pathol. Pharmacol. 8*:579 (1974).
64. D. E. Green, in *Cannabinoid Assays in Humans* (R. E. Willette, ed.), NIDA Research Monograph 7, National Institute on Drug Abuse, Maryland, 1976, p. 70.
65. D. S. Kreuz and J. Axelrod, *Science 179*:391 (1973).
66. S. Agurell, I. M. Nilsson, A. Ohlsson, and F. Sandberg, *Biochem. Pharmacol. 19*:1333 (1970).
67. R. Detrick and R. Foltz, in *Cannabinoid Assays in Humans* (R. E. Willette, ed.), NIDA Research Monograph 7, National Institute on Drug Abuse, Maryland, 1976, p. 88.
68. L. Lemberger, S. D. Silberstein, J. Axelrod, and I. J. Kopin, *Science 170*:1320 (1970).
69. R. F. Skinner, *Proc. West. Pharmacol. Soc. 15*:136 (1972).
70. (a) A. Ohlsson, J.-E. Lindgren, K. Leander, and S. Agurell, in *Cannabinoid Assays in Humans* (R. E. Willette, ed.), NIDA Research Monograph 7, National Institute on Drug Abuse, Maryland, 1976, p. 48 (b) A. Ohlsson, S. Agurell, J.-E. Lindgren, and K. Leander, in *Cannabinoid Analysis in Physiological Fluids* (J. A. Vinson, ed.), ACS Symposium Series 98, American Chemical Society, Washington D.C., 1979, p. 73.
71. D. C. Fenimore, C. M. Davis, and A. H. Horn, in *Cannabinoid Assays in Humans* (R. E. Willette, ed.), NIDA Research Monograph 7, National Institute on Drug Abuse, Maryland, 1976, p. 42.
72. M. E. Wall, T. M. Harvey, J. T. Bursey, D. R. Brine, and D. Rosenthal, in *Cannabinoid Assays in Humans* (R. E. Willette, ed.), NIDA Research Monograph 7, National Institute on Drug Abuse, Maryland, 1976, p. 107.
73. E. R. Garrett and C. A. Hunt, in *Cannabinoid Assays in Humans* (R. E. Willette, ed.), NIDA Research Monograph 7, National Institute on Drug Abuse, Maryland, 1976, p. 33.
74. J. L. Valentine, P. J. Bryant, P. L. Gutshall, O. H. M. Gan, E. D. Thompson, H. C. Niu, in *Cannabinoid Assays in Humans* (R. E. Willette, ed.), NIDA Research Monograph 7, National Institute on Drug Abuse, Maryland, 1976, p. 96.
75. A. R. Chase, P. R. Kelley, A. Taunton-Rigby, R. T. Jones, and T. Harwood, in *Cannabinoid Assays in Humans* (R. E. Willette, ed.), NIDA Research Monograph 7, National Institute on Drug Abuse, Maryland, 1976, p. 1, et loc. cit.
76. E. W. Gill and D. K. Lawrence, *Biochem. Pharmacol. 23*:1140 (1974).
77. M. Widman, M. Nordqvist, S. Agurell, J.-E. Lindgren, and F. Sandberg, *Biochem. Pharmacol. 23*:1163 (1974).
78. Z. Ben-Zvi, J. R. Bergen, and S. Burnstein, *Res. Commun. Chem. Pathol. Pharmacol. 7*:201 (1974).

79. N. K. McCallum, B. Yagen, S. Levy, and R. Mechoulam, *Experientia 31*:520 (1975).
80. N. K. McCallum, A. Gugelmann, C. A. M. Brenninkmeijer, and R. Mechoulam, *Experientia 33*:1012 (1977).
81. I. M. Nilsson, S. Agurell, J. L. G. Nilsson, A. Ohlsson, F. Sandberg, and M. Wahlqvist, *Science 168*:1228 (1970).
82. E. W. Gill, G. Jones, and D. K. Lawrence, *Biochem. Pharmacol. 22*: 175 (1973).
83. R. Mechoulam, *Science 168*:1159 (1970).
84. D. J. Harvey and W. D. M. Paton, *J. Chromatogr. 109*:73 (1975).
85. R. L. Foltz, P. A. Clarke, B. J. Hidy, D. C. K. Lin, A. P. Graffeo, and B. A. Petersen, in *Cannabinoid Analysis in Physiological Fluids* (J. A. Vinson, ed.), ACS Symposium. Series 98, American Chemical Society, Washington, D.C., 1979, p. 59.
86. M. Nordqvist, J.-E. Lindgren, and S. Agurell, in *Cannabinoid Assays in Humans*, (R. E. Willette, ed.), NIDA Research Monograph 7, National Institute on Drug Abuse, Maryland, 1976, p. 64.
87. S. Burstein, J. Rosenfeld, and T. Wittstruck, *Science 176*:422 (1972).
88. M. Nordqvist, S. Agurell, M. Binder, and I. M. Nilsson, *J. Pharmac. Pharmacol. 6*:471 (1974).
89. D. E. Green, F. Chao, K. O. Loeffler, and S. L. Kanter, in *Cannabinoid Analysis in Physiological Fluids* (J. A. Vinson, ed.), ACS Symposium Series 98, American Chemical Society, Washington D.C., 1979, p. 93.
90. M. E. Wall, D. R. Brine, J. T. Bursey, and D. Rosenthal, in *Cannabinoid Analysis in Physiological Fluids* (J. A. Vinson, ed.), ACS Symposium Series 98, American Chemical Society, Washington, D.C., 1979, p. 39.

29

Capillary GLC Determinations of Barbiturates and Diphenylhydantoin

Wolfgang Dünges / *Fachbereich Geowissenschaften der Universität, Mainz, Federal Republic of Germany*

I. INTRODUCTION

This chapter deals with (WCOT and SCOT) capillary gas chromatography of barbiturates and of their metabolites and also of diphenylhydantoin (DPH) and its metabolites since this drug is often used in combination with phenobarbital in the treatment of epilepsy. Their combined capillary GC determination is, therefore, important but is still somewhat problematic. The scope of the survey will not allow the discussion of other chemically or pharmacologically related substances. Trace analyses of barbiturates and of diphenylhydantoin can be carried out only after derivatization*, which has to meet several requirements.

*For one of the rare exceptions, see [1].

1. The reaction has to be complete.
2. The reaction should lead only to one product.
3. The derivatives should be stable before and during the chromatographic analysis.
4. Reagents and solvents should not affect the liquid phase of the GC columns.

The last two points are of greater importance here than in work with packed columns.

The paper is arranged in two main parts: Under the heading of "Literature" (Section II) a survey of the state of the art is given, while new results from our lab will be presented under the title of "Experiments" (Section III). The literature has been scanned through June of 1981. An excellent review by Marjorie G. Horning covers the publications concerning GC (i.e., mainly packed-column GC) of barbiturates and DPH until 1978 [2].

II. LITERATURE

A. Silylation

N,N-Trimethylsilyl derivatives of barbiturates [Figure 29.1C, X = $Si(CH_3)_3$] were introduced for packed column GC quite early. Because of their lability (see, e.g., [2]), however, they are not used for the capillary GC of barbiturates. M. C. Horning [2] uses silylation in extracts from biological material after alkylation of the barbiturate ring with diazomethane in order to convert free OH groups or other reactive moieties. As a consequence of this procedure, quantitation is improved, lifetime of the column is prolonged, and cleaner chromatograms are provided. Horning has not made clear, however, whether these points are all valid for capillary GC as well.

H.-J. Egger et al. [3] have determined DPH and its phenolic dihydrodiol metabolites in plasma, urine, and homogenized tissue using WCOT glass capillaries (SE-30, 25 m), either with FID or with MS detection. The heterocyclic rings were nearly quantitatively *N,N*-bis-silylated with bis(trimethylsilyl)trifluoroacetamide in dry pyridine (analogue C, Figure 29.1), only 1-2% of the *N*-mono product were reported. The different hydroxy groups of the metabolites were derivatized as well. It was reported that (1) use of other silylation

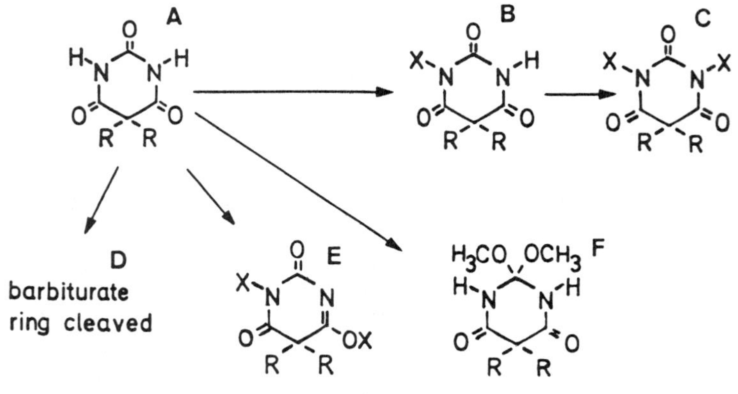

Figure 29.1 Reaction products from barbiturates. X = CH_3, or other alkyl, $Si(CH_3)_3$.

agents or conditions, (2) reaction with diazomethane or with perfluoroacetanhydride, and (3) permethylation under alkaline conditions led to the degradation of the metabolites or to other side reactions.

B. Pyrolytic Alkylation

Pyrolytic alkylation with quaternary ammonium compounds is used widely for barbiturate analyses with packed columns. Cleavage of the barbiturate ring [2] and other side reactions [4] have been described. With a special technique, pure trimethylanilinium hydroxide was coinjected with a chloroform solution of the barbiturates [5]. Thus J. De Graeve and Vanroy were able to avoid side reactions. Twelve different barbiturates and diphenylhydantoin were determined from plasma using a 20-m SCOT glass capillary column with SE-30. Negative effects on the column were not found even after prolonged use. Brallobarbital [5-allyl-5-(2-bromoallyl)barbituric acid] was decomposed; no by-products from other barbiturates or diphenylhydantoin were reported.

C. Alkylation with Dimethylformamide-Dimethylacetal

Alkylation with dimethylformamide-dimethylacetal has been introduced for packed column GC of barbiturates by V. S. Venturella et al. [6]. These authors postulate that compound F (Figure 29.1) is the product, whereas M. G. Horning et al. [7] report that the *N,N'*-bismethyl-product C is formed as well. O. Mathiesen [9] has shown the usefulness of the reagent for capillary GC (15-m WCOT column with OV-101) analyses of barbiturates. Without any discussion of [6,7], this author proposes structure C (Figure 29.1) for the derivatives on the basis of MS with chemical ionization.

A. S. Christophersen et al. [10] have analyzed similar alkylation mixtures using 20-m WCOT glass capillaries with OV-1 and MS detection. Relying on Venturella's paper [6], and without citing [7] or [9], they come to the conclusion that their product has structure F (Figure 29.1). Applying special injection techniques and other optimized conditions, they were able to achieve quantitative conversions of the investigated barbiturates. Plasma extracts were similarly analyzed. However, structure F appears to be somewhat improbable, since the peaks in the published chromatograms show no tailing, which should be noticeable if compounds with two free NH groups being analyzed.

D. Extractive Alkylation

M. Garle and Petters [11] determined 15 barbiturates from plasma after extraction with charcoal and extractive alkylation of the first extract with tetrabutylammonium hydroxide and ethyl iodide in dichloromethane. Degradation products were not described, but the recovery values varied from the more volatile compounds (70%) to those with higher boiling points (110%), as the solutions containing *N,N'*-bis-alkylbarbiturates were taken to dryness. Separation was carried out with a SCOT glass capillary column (45-m, SE-30).

E. Methylation with Diazomethane

Methylation with diazomethane is the least complicated alkylation technique for barbiturates, but it leads to *N,O* by-products, such as E of Figure 29.1. M. G. Horning [2] have used this procedure for the analysis of barbiturates with a WCOT glass capillary column (60 m with SE-30). Using selective-ion

MS, and with carbon-13 isomeric barbiturates as internal standards, it was possible to quantitate the procedure despite side reactions. 95% of the diphenylhydantoin was converted to the monomethyl product (analogue to B, Figure 29.1) by diazomethane [2]. Quantitation was done as with barbiturates.

F. Claisen's Carbonate Method

Claisen's carbonate method is remarkably well suited to the alkylation of barbiturates. It consists of heating the acetone solutions of alkyl halogenids and barbiturates with solid potassium carbonate [12]. The reaction mixtures can be injected directly into the gas chromatograph. Wu and Pearson [13] have shown the applicability of methylation of barbiturates with the carbonate method to their work with WCOT steel capillary columns. These columns (15 m, with OV-17) did not deteriorate after prolonged usage. The method was applied to blood serum analyses as well.

Own experiments with glass capillaries [14] demonstrated the advantages of the carbonate method: no by-products were observed, and there were no negative effects on columns or detectors; this was also confirmed by others [15,16]. Best separations for the *N,N'*-bis-alkyl products were obtained with a WCOT glass capillary coated with SF-96 (40 m). Methyl, allyl, and benzyl products were prepared in 7-μl reactions with the aid of a microrefluxer [17]. The retention indices of the products of 27 barbiturates were given, and blood level determinations were shown.

With WCOT glass capillary columns (37 m, with SE-30), R. De Sagher et al. [15] have investigated the alkylation of diphenylhydantoin under the conditions of the carbonate method. Using acetone, and with methyl iodide and potassium carbonate, they obtained 85% of the *N,N'*-bis-methyl product. After addition of methanol, the yield was 98%, but 2% of a *N,O* side product was observed. Nearly complete reaction with a decreased content of side products was achieved with ethyl iodide in acetone/methanol after the addition of potassium hydroxide in water; here, the reaction mixture had to be taken to dryness and the residue was reconstituted with benzene, aliquots of which were analyzed.

III. EXPERIMENTS

It was shown in our laboratory [12,14] as well as those of others [13,15,16] that Claisen's carbonate method is the method of choice for the alkylation of barbituric acids, especially if capillary GC is used. It was thus far not satisfactory, however, for the alkylation of diphenylhydatoin (see Figure 29.2A and [15]), and of phenolic metabolites. Derivatization techniques which had been proposed for diphenylhydantoin (namely, silylation [3] and ethylation under aqueous alkaline conditions [15]) are likewise not applicable for barbiturates. Thus no method was available which could lead exclusively to *N,N'*-bis-alkyl products both of barbiturates and of diphenylhydantoin, and whereby the reaction mixtures could be directly injected without purification steps into GC using capillaries.

For this reason a further investigation of the conditions of Claisen's carbonate method was carried out. Two approaches seemed promising: (1) crown ether catalysis, and (2) optimization of the reactivity of the alkali carbonate.

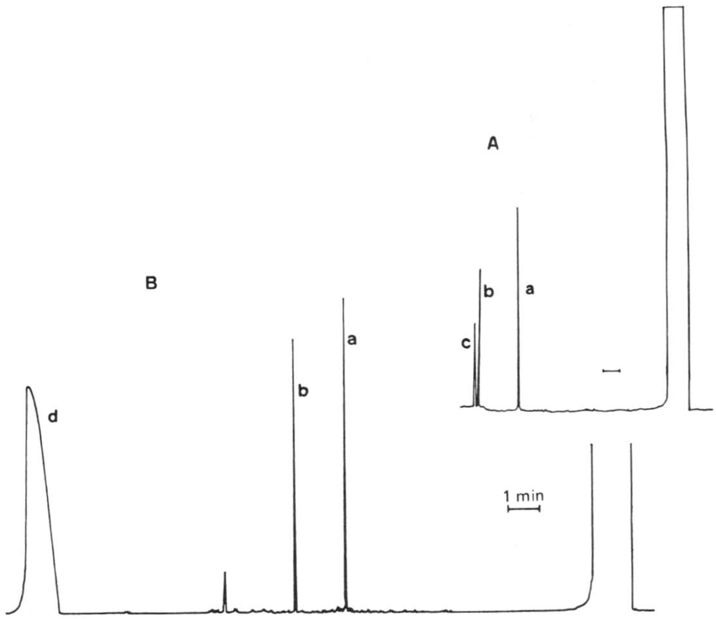

Figure 29.2 Gas chromatograms from reflux reactions of phenobarbital and of diphenylhydantoin with methyl iodide, nondried K_2CO_3 in presence (B) or absence (A) of dibenzo-18-crown-6 (in acetone). Crown ether leads to complete methylation of diphenylhydantoin (b); monomethylhydantoin (c) is no longer visible. (a) *N,N'*-dimethylphenobarbital; (d) dibenzo-18-crown-6. For chemical conditions, see text. GC was performed with a 20-m glass capillary column with OV-17. Temperature, 50°C, 3-min hold, then 18°/min to 270°C. Splitless injection. GC, Carlo Erba, type 2100.

A. Crown Ether Catalysis

Using crown ether catalysis, we improved to 100% the yields to the reaction of 4-bromomethyl-7-methoxycoumarin (Br-Mmc) with organic acids under conditions of the carbonate [18] (see also [19]). In addition to these experiments with the fluoroescence label, we were able to achieve quantitative esterification of fatty acids with methyl iodide if crown ethers were present in high concentrations [20]. We therefore used the following conditions: To 0.5 mg of finely powdered potassium carbonate (p. A, E. Merck, Darmstadt, W. Germany) placed in a 7-μl vessel were added 5 μl of dry acetone (see [19]) containing phenobarbital and diphenylhydantoin and 2 μl of methyl iodide which was saturated with dibenzene-18-crown-6 (about 15%). After refluxing the mixture for 30 min with a microrefluxer (see Figure 29.3 [21]), 1 μl was injected into a GC with a WCOT glass capillary.

Chromatogram B (Figure 29.2) of such an analysis shows two sharp peaks of *N,N'*-bismethylphenobarbital (a) and of *N,N'*-bismethyldiphenylhydantoin (b), as confirmed with authentic compounds.* As compared to chromatogram

***N,N'*-bismethyldiphenylhydantoin was a gift from H. Schäfer, Desitin-Werke, Hamburg, W. Germany.

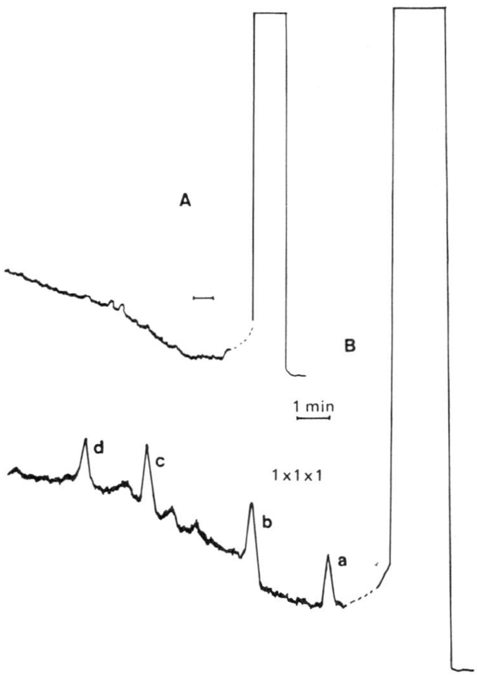

Figure 29.3 Gas chromatograms of (B) dilute reaction mixtures of phenobarbital (a), diphenylhydantoin (b), "hydroxyphenobarbital" (c) and "hydroxydiphenylhydantoin" (d). (A) Blank. Permethylation of all products can be assumed. 2 μl of 7-μl reactions were injected. Each peak corresponds to about 4 ng. GC conditions were as in Figure 29.2, but temperature was increased ballistically. For chemical conditions, see text. Dry Cs_2CO_3 was used under controlled conditions; see also [21].

A, peak c—which is most probably *N*-monomethyldiphenylhydantoin—is absent here. In experiment A, no crown ether was used, under otherwise identical conditions.

Although chromatogram B indicates that 100% methylation was achieved in both cases, the results seemed not totally satisfactory because of the presence of peak d from the crown ether. While here the GC determinations are not affected, this might be the case if solutes with greater retention times are present, such as hydroxy metabolites. In addition, the time of analysis is increased considerably because of the late peak.

B. Optimization of Alkali Carbonate Reactivity

Reactions with alkali carbonates of different water content seemed interesting since in a two-phase reaction like the carbonate method, surface conditions play a crucial role; therefore, these conditions might be influenced by crystal water. Thus in the carbonate reaction of barbiturates with Br-Mmc, a marked influence of the water content on Mmc side-reactions was in fact observed [19]. Methylation reactions with different salts were carried out: 0.5 mg of alkali carbonates with a very high or very low water content were added to 7-μl vessels using a microhood which was described earlier [19]. During these manipulations, 5-10% air humidity was maintained (see also [21]). Other conditions were as described above; crown ethers were not present.

From A and B in Figure 5.4 [21], it is obvious that use of dry Cs_2CO_3 and K_2CO_3 results in 100% *N,N'*-bismethylation for phenobarbital and for diphenylhydantoin. With moist cesium and potassium carbonate, the monomethyl product from diphenylhydantoin (c) can be discerned in the chromatograms A' and B'. With dry cesium carbonate and under similar experimental conditions, the metabolites 5(*p*-hydroxyphenyl)-phenylhydantoin and 5(*p*-hydroxyphenyl) 5-ethylbarbiturate ("hydroxyphenobarbital") were reacted together with their mother compounds. Figure 29.3 shows the chromatogram of a highly diluted reaction mixture. Again, only four peaks are present, indicating the completeness of reaction without by-products.

IV. CONCLUSIONS

From the literature and from experiments, it is clear that the success of capillary GC analysis depends to high degree on the derivatization process. It can now be reported that:

1. Barbituric acids and diphenylhydantoin can be determined with capillary GC, using an optimized version of the carbonate method.
2. The reaction of the two hydroxy metabolites seems quantitative as well. This has to be confirmed by further experiments.

A systematic investigation of the parameters of the carbonate method should be carried out. A stepwise investigation of the reaction of simple compounds (barbiturates, phenols, diols) should lead to the development of optimal conditions for the alkylation of more complex substances such as hydroxy metabolites. First results are encouraging. It is hoped that this path can be followed.

ACKNOWLEDGMENTS

The experiments have been carried out in the laboratories of W. Dosch, Fachbereich Geowissenschaften der Universität Mainz, to whom I am greatly indebted.

I thank F. P. Ott, Norheim, and W. Christmann, Mainz for helpful discussions, and B. Dünges, P. Dünges, and B. Soltau for excellent technical assistance.

This work has been generously supported from the Doktor-Robert-Pfleger-Stiftung, Bamberg, W. Germany.

REFERENCES

1. C. A. Cramers, E. A. Vermeer, L. G. van Kuik, J. A. Hulsman, and C. A. Meijers, *Clinica Chim. Acta 73*:97-107 (1976).
2. M. G. Horning, in GLC and HPLC determinations of therapeutic agents, Pt. 2 (K. Tsuji, ed.), M. Dekker, New York, 1978, 477-518.
3. H.-J. Egger, W. Wittfoht, and H. Nau, in *Quant. Mass Spectrometry in Life Sciences*, (Proceedings of the 2nd International Symposium, Ghent,

1978 (A. P. DeLenheer, R. R. Roncucci, and C. P. van Peteghem, eds.), Elsevier, Amsterdam, 1978, 303-314.
4. W. C. Kossa, J. McGee, S. Ramachandran, and A. J. Webber, *J. Chromatogr. Sci. 17:* 177-187 (1979).
5. J. DeGraeve and J. Vanroy, *J. Chromatogr. 129*:171-179 (1976).
6. V. S. Venturella, V. M. Gualario, and R. Elang, *J. Pharm. Sci. 62*: 662-668 (1973).
7. M. G. Horning, L. Glazener, and E. C. Horning, unpublished results, cited in [8].
8. D. J. Harvey, J. Nowlin, P. Hickert, C. Butler, O. Gansow, and M. G. Horning, *Biomed. Mass Spectrom. 1*:340-344 (1974)
9. U. Matthiesen, *Beitr. Ger. Med. 37*:337-345 (1979).
10. A. S. Christophersen and K. E. Rasmussen, *J. Chromatogr. 192*:363-374 (1980).
11. M. Garle and I. Petters, *J. Chromatogr. 140*:165-169 (1977).
12. W. Dünges and E. Bergheim-Irps, *Analyt. Lett. 6*:185-195 (1973).
13. A. Wu and M. L. Pearson, *Analyt Lett. 10*:381-393 (1977).
14. W. Dünges, R. Langlais, and R. Schlenkermann, *J. HRC & CC 2*:361-365 (1979).
15. R. DeSagher, J. Pocius, and J.-P Thenot, *J. Chromatogr. 156*:43-53 (1978).
16. M. Donike, private communication
17. W. Dünges, *Analyt. Chem. 95*:963-964 (1973).
18. W. Dünges, A. Meyer, K. E. Müller, M. Müller, R. Pietschmann, C. Plachetta, R. Sehr, and H. Tuss, *Z. Analyt. Chem. 288*:361-368 (1977).
19. W. Dünges, *Prä-chromatographische Mikromethoden*, A. Hüthig, Heidelberg, 312 pp. (1979).
20. M. Höckel, W. Dünges, A. Holzer, P. Brockerhoff, and G. H. Rathgen, *J. Chromatogr. 221*:205-214 (1980).
21. W. Dünges, Chapter 5, this volume.

30

Capillary GLC in Pharmacokinetics and Drug Metabolism Studies

Albertus G. de Boer*, Nicholas P. E. Vermeulen, and Douwe D. Breimer
/ *University of Leiden, Leiden, The Netherlands*

I. INTRODUCTION

During the past decade, gas chromatography has been developed into a powerful technique for the separation and quantitative determination of drugs and metabolites in biological fluids [65,89,110]. This has to be partly attributed to the considerable progress in the technology of preparing stable and high-resolution capillary columns [34,52,98]. Improvements in the injection and detection techniques, however, are equally well responsible for the increased analytical power of gas chromatography. With regard to sample introduction into the capillary column, a better understanding of the critical factors has led to the development of more precise and accurate sample introduction systems

* *Current affiliation*: Department of Pharmaceutical Technology and Biopharmacy, University of Leiden, Leiden, The Netherlands.

[10,53,99]. Improvements in detection are mainly the result of advances in the selectivity and sensitivity of available detection techniques [26,28,72].

In this chapter, a short survey is given of sample introduction systems for capillary columns. A relatively simple injector device for solids, recently developed in our laboratory, is described in more detail. Next, the gas chromatographic procedures for the quantitative analysis of etomidate, lidocaine, propranolol, pemoline, nitrazepam, hexobarbital metabolites, and cyclohexylmercapturic acids in plasma or urine are reviewed. These assays were recently developed for pharmacokinetic and metabolic studies, and they are based on the combined use of solid injection, selective detection, and short capillary columns.

II. SAMPLE INTRODUCTION INTO A CAPILLARY COLUMN

One of the critical factors in operating capillary gas chromatography is the selection of a proper way of sample introduction into the capillary column. Ideally, the sample is introduced into the first plates of the column. According to Harris [57], the volume of the injected sample should not exceed $V/2\sqrt{N}$, where V is the retention volume and N is the plate number. To fulfill this prerequisite, various injection systems have been developed and described in the literature. They can be subdivided into head space, liquid, and solid injection devices.

A. Head Space Injection

The head space injection technique has achieved considerable importance in trace analysis of volatile compounds in complex mixtures [54]. For example, it has been applied in the analysis of ethanol in blood [54], trichloroethanol in blood [27], and various inhalation anesthetics [9]. Recently, Ettre et al. have shown that head space injection is also applicable in gas chromatography with open-tubular capillary columns [44]. The main advantage of head-space injection in comparison with other injection techniques is that rather clean samples can be obtained from the biological samples in a relatively simple way.

B. Liquid Injection

Direct introduction of relatively large volumes of liquid samples into capillary columns by means of a syringe is a matter of great delicacy. Generally, only limited volumes of liquid can be injected to avoid deterioration, overload, and loss of efficiency of the column. To meet this purpose, various types of liquid injection systems were developed.

1. *Injection with Splitting*

One of the first techniques employed in capillary gas chromatography included a sample injection splitter. It rejects 90-99% of the sample and introduces only a small percentage into the column. The splitting technique is subject to a number of critical factors, such as molecular weight, polarity, and concentration of the sample components, geometry and materials of the splitter, and the temperature regulation of the whole splitting device [100]. The linearity of the splitter and the injection temperature are especially important with respect to the accuracy and precision of quantitative assays. Although commercial splitters are available which seem to satisfy these demands, the use of this type of injector is limited to the quantitative assay of low concentrations of drugs or metabolites in biological fluids: only small fractions of the biological sample can be introduced into the capillary column.

2. *Splitless Injection*

Splitless injection, according to Grob [50,51], is quite suitable for trace analysis. This injection technique, however, requires time-consuming oven temperature programming to accomplish the thermal desorption of the liquid samples which have been trapped in the first part of the column. With a packed precolumn, it allows injection of liquid volumes up to 20 μl [83]. This kind of injection always has to be optimized by trial and error with respect to solvent, desorption time, and temperature [100]. In addition, the column performance may be influenced negatively by solvent overload and pollution.

Alternatively, the selective sampling technique, being also a kind of splitless injection, makes use of a double column system and is operated isothermally [99,100].

Generally speaking, splitless injecting is time-consuming and therefore not well suited for the assay of drugs or metabolites in biological fluids, at least in pharmacokinetic and drug metabolism studies, where large numbers of samples have to be assayed.

3. *On-Column or Direct Injection*

On-column injection of a sample into a capillary column, whereby a liquid is injected directly without previous evaporation of the solvent, into a cold area of the column, generally yields accurate and reproducible results [4,53]. The direct injection technique, as described by Schomburg et al. [100] allows injection of liquid volumes up to 10 μl into a capillary column (0.25 mm I.D.). Direct injection into a wider-bore capillary column (0.5-1.0 mm I.D.) was found to be rather promising [119]. Inherent drawbacks of on-column or direct injection, however, are rapid pollution of the capillary column and short span of life owing to breaking of the liquid-phase film, which is mainly due to the large amounts of solvent injected.

C. Solid Injection

Solid injection has proved to be an excellent method for sample introduction into a capillary column. This is particularly true in the quantitative analysis of low drug and metabolite concentrations. Since the solvent is evaporated completely before injection, relatively large fractions of extracts of biological fluids can be introduced. The method, however, can be applied only with compounds having boiling points higher than 200-250°C. Although various solid injection systems have been described in the literature [5,10,70,77,88,100, 107], many of them are not suitable for routine analysis of large numbers of samples. The best known in this respect is the all glass "falling needle" solid injector [10]. It exhibits excellent injection properties, but it is rather vulnerable, and its glass needle is easily polluted on repeated injection.

In our laboratory we now use a new solid injector, the "ball-valve" solid injector. It is a modified version of one previously introduced by Driessen and Emonds [39]. The "ball-valve" solid injector is composed of three main parts (Figure 30.1). The lower part is a stainless-steel adapter necessary to connect the injector to the injection port of the gas chromatograph. This adapter is provided with cooling ribs in order to avoid thermal deformation of the Teflon seal of the ball valve (middle part) by overheating. The

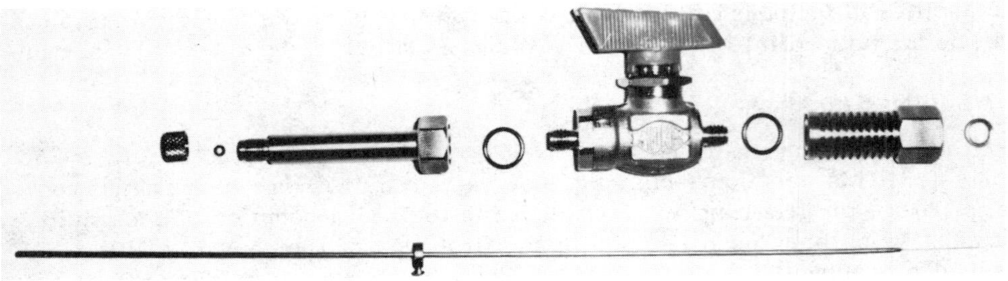

Figure 30.1 Various parts of the "ball-valve" solid injector. From top to bottom: adjustable screw, small O-ring, upper stainless-steel part for guiding the needle, a larger O-ring, ball-valve, another large O-ring, lower stainless-steel part with cooling ribs and a copper O-ring. Left: stainless-steel needle with adjustable resistance and thin stainless-steel tip.

ball-valve* is a 1/8-inch stainless-steel valve (Hoke, Gerolock, cat. no. 7122 G 2 S, Gommer, Voorburg, The Netherlands). The upper stainless-steel part is used as a guide for the stainless-steel needle (length, 30 cm; diameter, 1.9 mm; length tip, 2.5 cm; diameter tip, 0.3 mm) and for the gastight shutoff of the whole injection system after the needle has been introduced into the upper part of the solid injector. This shutoff is achieved using a small O-ring (ARP no. .003½; material, Buna S or natural rubber, shore 70A, Eriks, Alkmaar, The Netherlands). By means of a copper O-ring, the connection between the injection port of the gas chromatograph and the "ball-valve" solid injector is made gastight. A detailed drawing of the upper and lower stainless-steel parts of the solid injection system is shown in Figure 30.2.

The following procedure is applied for the introduction of a sample into the capillary column using the "ball-valve" solid injector. Several microliters of a solution are brought onto the tip of the stainless-steel needle by means of a 10-μl syringe. After evaporation of the solvent, the needle is introduced into the upper part of the "ball-valve" solid injector. By means of the adjustable screw and the O-ring at the top, a gastight shutoff is effected. Next the ball valve is opened, and the needle is rapidly injected into the injection port of the gas chromatograph. The depth of the needle into the injection port can be regulated using the adjustable resistance on the needle (Figure 30.1). After evaporation of the sample for 10 sec, the needle is lifted so far that the ball valve can be closed. This may be verified by a mark on the needle. Next the adjustable screw on the top is loosened and the needle can be removed.

The advantage of this "ball-valve" solid injector in comparison with the "falling-needle" injector is that more needles can be used at the same time. In the case of very short retention times (e.g., 2-4 min), this allows rapid sampling. Another advantage is that nonvolatile residues on the needle after injection can be easily removed. In those cases where metal-catalyzed decomposition of compounds was observed, a glass-coated tip of the injection needle was used successfully.

---

*We now use a Nupro plug valve: 1/8-inch Swagelock, cat. no. ss-2P4T, Nupro Company, Willoughby, Ohio, U.S.A.

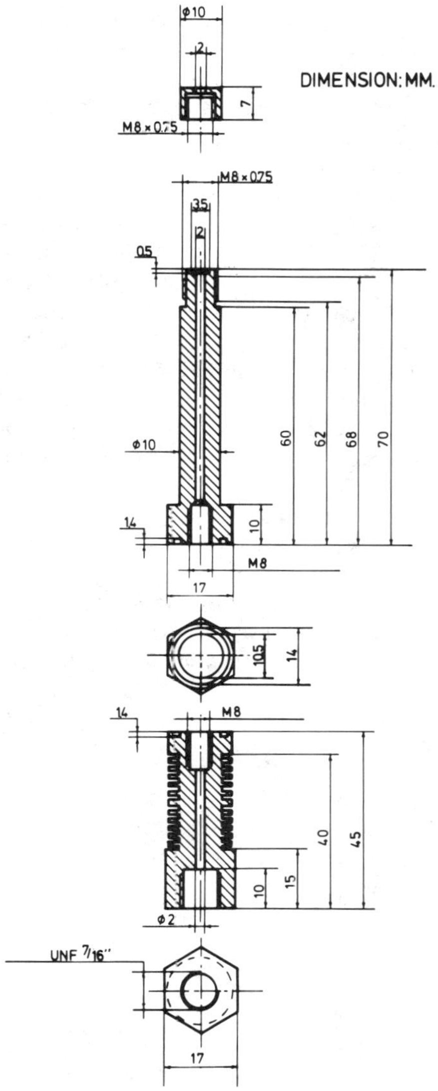

Figure 30.2 Upper and lower stainless-steel parts of the "ball-valve" solid injector in detail.

The above-described "ball-valve" solid injector has a compact and solid construction and is less vulnerable than the "falling-needle" device. Using our solid solid injector, various compounds were and still are assayed quantitatively, such as pemoline [113], theophylline [16], lidocaine [18,20], salicylamide [19], nitrazepam [17], clonazepam [17], propranolol [22], etomidate [21], antipyrine [69], phenacetin [69], hexobarbital and metabolites [114], vinylbital and metabolites [117], mercapturic acids [118], pentobarbital [15], the laxative 1,8-dihydroxyanthraquinone [3], and 3-carboxyantipyrine [35].

Properties of the "Ball-Valve" Solid Injector

The properties of the "ball-valve" solid injector can be illustrated by comparing its results with those obtained after liquid injection. In the following examples an extract was used of human plasma, spiked with lidocaine (L) and *N*-methylhexobarbital (MeH). In each example 0.8 μl of this extract was either injected or brought on the tip of the needle of the solid injection system, corresponding with about 3 ng of lidocaine and 4 ng of *N*-methylhexobarbital. The SCOT column (Carbowax 20 M KOH; length, 10 m; I.D., 0.4 mm), flows, and temperatures were the same in each example.

In the first example, detection was carried out with a flame ionization detector (FID, Figure 30.3). The left part of this figure shows a gas chromatogram after liquid injection of the sample. A large solvent peak is seen, whereas lidocaine (L) and *N*-methylhexobarbital (MeH) are hard to distinguish. Decreasing the attenuation in order to quantify lidocaine does not improve the situation in this case. Upon solid injection of the plasma extract with the "ball-valve" solid injector, a gas chromatogram is obtained, as depicted in the middle part of Figure 30.3). The solvent peak has almost completely disappeared and the lidocaine peak and the *N*-methylhexobarbital peak are well separated. Upon decreasing the attenuation (right part of Figure 30.3), lidocaine and *N*-methylhexobarbital can be measured quantitatively.

In the second example, detection was carried out with a nitrogen-selective detector (N/P-FID). From Figure 30.4, at first sight it seems that solid injection is of no great advantage in this case, because after liquid as well as solid injection the "solvent peak" is relatively small. However, if we compare the relative and the absolute peak heights of lidocaine and *N*-methylhexobarbital, it becomes clear that both the relative and the absolute peak heights are

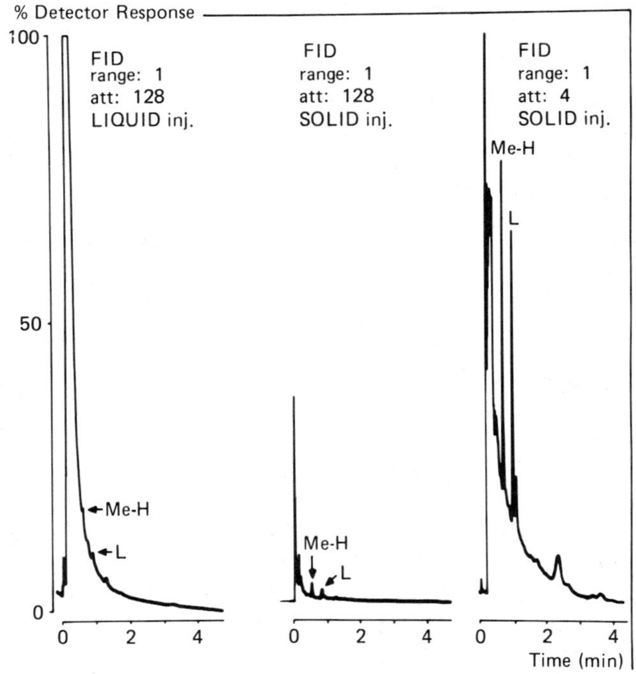

Figure 30.3 Gas chromatograms obtained after liquid and solid injection ("ball-valve" solid injector) of about 3 ng lidocaine (L) and 4 ng *N*-methylhexobarbital (MeH) into a Carbowax 20 M KOH SCOT column with flame ionization detection.

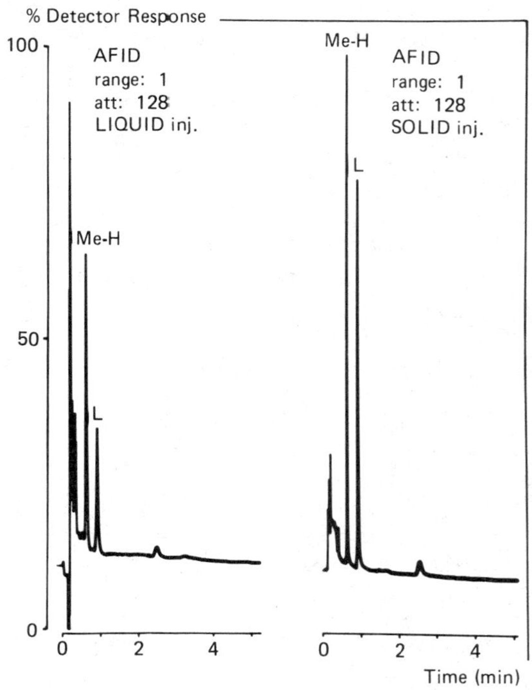

Figure 30.4 Gas chromatograms obtained after liquid and solid injection ("ball-valve" solid injector) of about 3 ng lidocaine (L) and 4 ng *N*-methylhexobarbital (MeH) into a Carbowax 20 M KOH SCOT column with nitrogen selective detection. From left to right: liquid injection (0.8 μl), attenuation 128; solid injection, attenuation, 128.

negatively influenced when liquid injection is applied. Comparison of Figures 30.3 and 30.4 also shows that the nitrogen-selective detector is far more sensitive (about 70 times) that the flame ionization detector.

With respect to the accuracy of the assays when using the solid injector, it was found that the coefficient of variation of the peak height ratio of lidocaine/*N*-methylhexobarbital following consecutive injections of ethanol solutions containing 100 ng lidocaine + 500 ng *N*-methylhexobarbital and 2000 ng lidocaine + 500 ng *N*-methylhexobarbital was 3.5 and 0.6% (n = 5), respectively.

It may be concluded that the "ball-valve" solid injector is a valuable and accurate tool, particularly in quantitative gas chromatographic assays of high as well as of low concentrations of lidocaine in plasma. In the last part of this chapter, this conclusion will further be substantiated by the results of the assays of other compounds.

III. APPLICATIONS OF CAPILLARY GAS CHROMATOGRAPHY WITH "BALL-VALVE" SOLID INJECTION AND SELECTIVE DETECTION

A. Instrumentation

Except for pemoline, all experiments were performed on Hewlett-Packard gas chromatographs (model 5710), equipped with a N/P-selective detector (model

18789A) or with a pulse-modified EC detector (constant current, model 5713A). Pemoline was assayed with a Pye GCV gas chromatograph, equipped with a rubidium chloride nitrogen-selective detector (model 795014). The GC-MS analyses were performed on an LKB 2091/2130 (EI/CI) gas chromatograph-mass spectrometer-computer system. Ion source temperature was 200°, separator, 210°C; electron energy, 70 eV; trap current 30 μA; and accelerating voltage, 3500 V. Chemical ionization was carried out with ammonia as a reactant gas. For the mass chromatographic studies, mass spectra were recorded continuously and stored in memory of a DEC PDP 11/05 computer. The mass fragmentographic measurements were performed, computer-controlled, by measuring selected ions continuously.

The glass capillary columns were drawn from borosilicate glass with a Hupe and Bush glass-drawing machine. The capillary columns (length, 8-10 m; I.D., 0.4-0.5 mm) were coated according to the dynamic procedure described by Cramers et al. [34]. In the first step, a fumed silanized silica (Tullanox, 10 μm) deactivated with benzyltriphenylphosphonium chloride (BTPPC) and suspended in dichloromethane/toluene (3:1) was used. In the second step, the stationary phase was deposited on the support layer in a dynamic way: a 1% filtered solution of Carbowax 20 M in dichloromethane; a 1% filtered solution of KOH in ethanol, followed by a 1% filtered solution of Carbowax 20 M solution in dichloromethane; 1.6% solution of PPE-21 in dichloromethane/toluene (1:), followed by a solution of 3% OV-17 in dichloromethane. The carrier gas (helium) was used at a flow-rate of 5-10 ml/min.

In all experiments injection was carried out with the "ball-valve" solid injector as described before in this chapter. The temperatures of the column, injector, and detector are specified in the legends of the respective figures.

B. Nitrogen-Selective Detection

The nitrogen-selective detector contains an alkali-metal salt, which has to be heated for the thermionic emision of alkali particles. In the earlier nitrogen-selective detectors, heating took place by means of a flame. At present, however, in most detectors this is achieved electrically, whereby stability and performance have improved considerably [28,68]. The chemical mechanism and the nature of the response, being enhanced selectively for nitrogen- and phosphorous-containing compound, have not yet beene elucidated completely [2,24, 28,73,102]. Nevertheless, there are several advantages in using a nitrogen-selective detector compared to the normal flame ionization detector, especially when drugs or metabolites have to be analyzed in biological fluids.

Owing to the enhanced selectivity of the nitrogen-selective detector, in general, less attention is required as concerns the selectivity of the extraction procedure. Consequently, extraction is less time-consuming and laborious. The great selectivity of the detector also reduces the possible interference of endogenous substances. In many cases, the detection limit of the nitrogen-selective detector will be lower than that of the normal flame ionization detector. This, however, greatly depends on the number and the position of the nitrogen atoms in a drug molecule.

1. *Etomidate [21]*

In the literature two procedures have been described for the determination of the intraveneous anesthetic agent etomidate (1) in plasma:

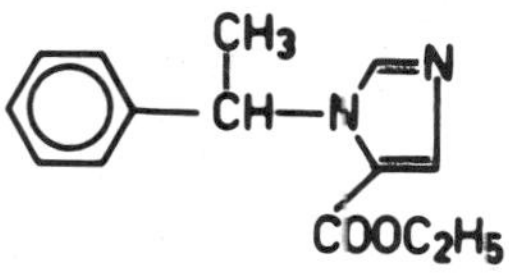

Structure (1)

1. A gas chromatographic method which requires a laborious extraction procedure and which is not sensitive enough to determine very low plasma concentrations [124].
2. A mass fragmentographic method, which requires a gas chromatograph-mass spectrometer and which is therefore not suitable for routine analysis [56].

We developed a method for the assay of etomidate in human plasma which is more sensitive and less laborious than the GC method published previously [124]. Our method was successfully applied to pharmacokinetic studies in man [21].

The extraction of etomidate from 1.0-ml portions of human plasma was carried out by a single extraction step with *n*-pentane during 15 sec.

In Figure 30.5 are shown the gas chromatograms of plasma extracts from a surgical patient before, and 8 and 240 min after intraveneous injection of 17 mg of etomidate. Propoxate was chosen as an internal standard because it is an analogue of etomidate. (In general, analogues exhibit similar physicochemical and chromatographic properties, which as a consequence, may positively influence the reproducibility of a quantitative assay.) The reproducibility of the whole analytical procedure for the determination of etomidate is reflected in the coefficient of variation at different concentrations, being 6.1% ($n = 5$). There was a linear relationship between the detector response of etomidate and propoxate in the etomidate concentration range of 12.5-375 ng/ml of plasma. The extraction yield in this range was 77%. The detection limit was about 5 ng etomidate per ml of plasma.

By means of GC-MS it was verified that etomidate as well as proxate is eluted from the gas chromatograph unchanged.

In conclusion, a simple single extraction, a short capillary Carbowax 20 M column, a nitrogen-selective detecotr, and a "ball-valve" solid injector allow the rapid determination of therapeutic concentrations of etomidate in human plasma. Propoxate can be determined in a similar manner.

2. Lidocaine [18, 20]

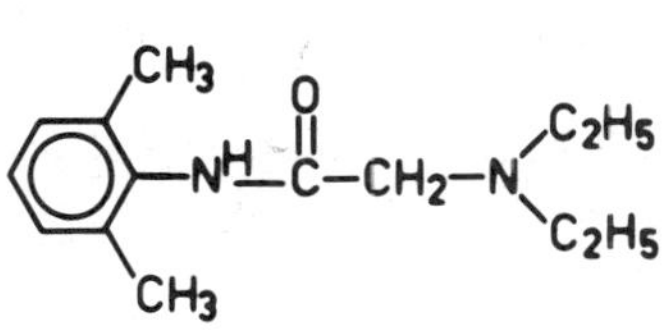

Structure (2)

In the next example, a relatively simple, sensitive, and selective assay is discussed for the determination of lidocaine (2) in human whole blood, plasma, and saliva, and rat blood by utilizing gas chromatography with a capillary Carbowax 20 M KOH SCOT-column, a "ball-valve" solid injector, and a nitrogen-selective detector.

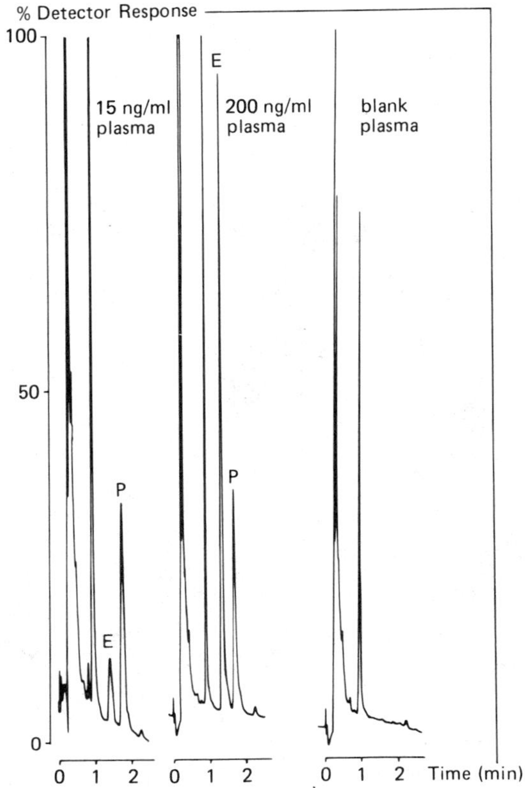

Figure 30.5 Gas chromatograms of a plasma extract obtained from a male surgical patient immediately before (right) and 8 min (center; 200 ng etomidate per ml plasma) and 240 min (left; 15 ng etomidate per ml plasma) after receiving 17 mg etomidate i.v.; E = etomidate, P = propoxate (internal standard; 125 ng/ml plasma). Temperatures: column, 170; detector, 300; injection port, 200°C. (From Ref. 21.)

Various gas chromatographic assay methods have been described for the determination of lidocaine in biological fluids. These methods were based on the use of mass spectrometric detection [82,105], flame ionization detection [6,7,11,14,30,37,62,66,67,75,79,80,90,97,106], and nitrogen-selective detection [1,60,61,96]. The methods using mass spectrometric detection, however, are less suitable for routine determination of large numbers of samples. Of the other methods, those using nitrogen-selective detection are to be preferred because of the increased selectivity and sensitivity. It is especially the increased selectivity of this detector that makes it possible to use relatively simple extraction procedures, however, can be augmented by using a capillary column instead of a packed one. In order to avoid overloading and too rapid deterioration of the capillary column, it is generally necessary to use splitting techniques for the introduction of the samples into the column [96]. On the other hand, it is possible to apply solid injection, by which the introduction of large amounts of solvents is prevented.

In our assay, lidocaine was extracted twice from plasma and whole blood on a whirlmixer with *n*-pentane during 10 sec.

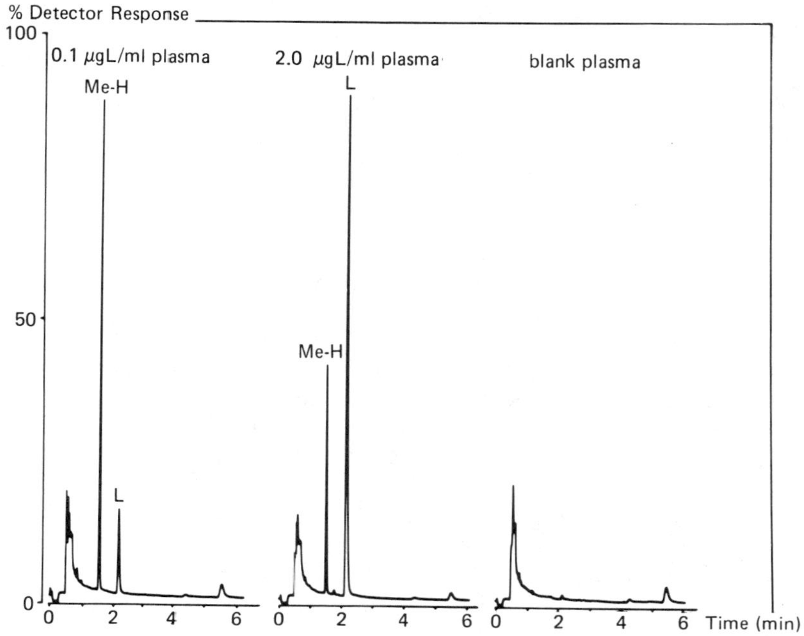

Figure 30.6 Gas chromatograms of extracts of 1 ml blank plasma and plasma samples containing 0.1 and 2.0 μg lidocaine (L)/ml, obtained after solid injection into a Carbowax 20 M KOH SCOT column with nitrogen-selective detection. *N*-methylhexobarbital (MeH) was used as internal standard (0.5 μg/ml plasma). Temperatures: column, 160; detector, 300; injection port, 250°C. (From Ref. 20.)

Figure 30.6 shows typical gas chromatograms of the extracts obtained from human plasma, containing a relatively low and high concentration of lidocaine. In addition, the figure shows a gas chromatogram of a blank plasma extract. From the latter chromatograms it is clear that there is no interference of endogenous components with lidocaine, nor with the internal standard *N*-methylhexobarbital. The same chromatograms were obtained when human whole blood, saliva, and rat whole blood samples were analyzed. The peak eluting with a retention time of 5.5 min (Figure 30.6) was found to originate from caffeine. Linear calibration curves were obtained in a lidocaine concentration range of 10-2000 ng/ml of plasma. The coefficient of variation in this concentration range was always smaller than 5% ($n = 5$). The recovery of lidocaine from rat whole blood, human whole blood, plasma and saliva was 81.1, 75.3, 72.0, and 74.0%, respectively. The detection limit of the assay was about 1 ng/ml sample.

Regarding the selectivity of this analytical procedure of lidocaine, attention should be paid to the fact that an alkaline Carbowax 20 M KOH capillary column was chosen diliberately. In general, an alkaline stationary phase improves the selectivity of the assay of amines, mainly because acidic compounds will be trapped in the column. The almost complete absence of solvent upon introduction of the sample with the aid of the solid injector allows optimal performance of the capillary column by reducing tailing effects and deterioration of the stationary phase, which improves the sensitivity of the assay. The short retention times and the short extraction time (10 sec) with a proper solvent permit the assay of many samples per day.

3. *Propranolol* [22]

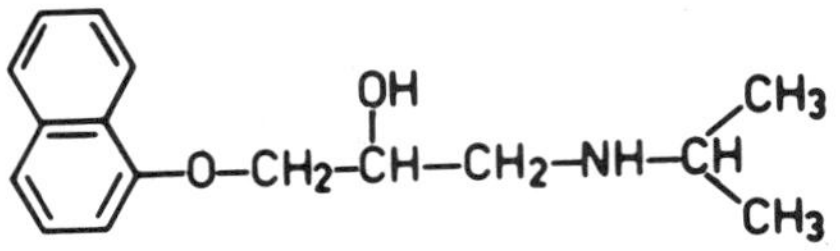

Structure (3)

Several methods have been described for the determination of propranolol in plasma. They comprise thin-layer chromatography (TLC) [46,58], high-performance liquid chromatography (HPLC) [81], gas chromatography-mass spectrometry (GC-MS) [42,121], gas chromatography (GC) [36,38,64,111,120], and fluorimetry [12,103]. The TLC methods are not sensitive enough to determine low propranolol concentrations [58] or require rather large plasma samples [46]. The fluorimetric method of Black et al. [12], as modified by Shand et al. [103], is of limited specificity and sensitivity [32]. In contrast to this, the HPLC method which uses fluorimetric detection is rather sensitive and selective [81]. The GC-MS methods [42,121] are very sensitive and specific, but they are not very suitable to analyze large amounts of samples. The GC methods all require derivatization of propranolol, resulting in the formation of di-trifluoroacetic [36,11,120], dipentafluoropropionic [64], or di-heptafluorobutyric [38] acid derivatives. They comprise the introduction of halogen into the molecule in order to make the determination of low concentrations of propranolol in plasma possible by electron capture detection.

We have developed a relatively simple, selective, and sensitive method for the determination of propranolol in human or rat plasma, which is based upon derivatization with acetic acid anhydride followed by capillary gas chromatography and nitrogen-selective detection [22].

Propranolol was extracted twice from alkalinized plasma (pH 13) with 3 ml freshly distilled diethyl ether on a whirlmixer during 5 sec. After evaporation of the solvent, derivatization was carried out in dichloromethane using acetic acid anhydride. Subsequently the mixture was shaken with a tertiary sodium phosphate solution.

In Figure 30.7, gas chromatograms are shown of extracts of human plasma, containing high and low propranolol concentrations, as well as the gas chromatogram of a blank plasma extract. Obviously, there is no interference of the peaks of the derivatives of propranolol or the internal standard with those of endogenous plasma substances or metabolites. As internal standard, oxprenolol was chosen because of its structural resemblance to propranolol. It allows internal standardization of the extraction as well as the derivatization step of the analytical procedure. In principle, this should improve the reproducibility of the overall analysis. The reproducibility of the assay of propranolol according to our assay is reflected by the coefficient of variation determined at the concentrations of 12.5 and 100 ng/ml plasma, being 15 and 4.8% (n = 5) respectively. In addition, the calibration curves were linear in the concentration range of 5-100 ng/propranolol per ml plasma. The detection limit was about 1 ng/ml, which is sufficiently low for pharmacokinetic and bioavailability studies.

By means of combined GC-MS it was confirmed that the derivatized compounds of propranolol and oxprenolol elute unchanged from the capillary column.

When evaluating this method and comparing it to other gas chromatographic procedures, one of the major advantages is that the extraction procedures, one of the major advantages is that the extraction procedure is relatively simple.

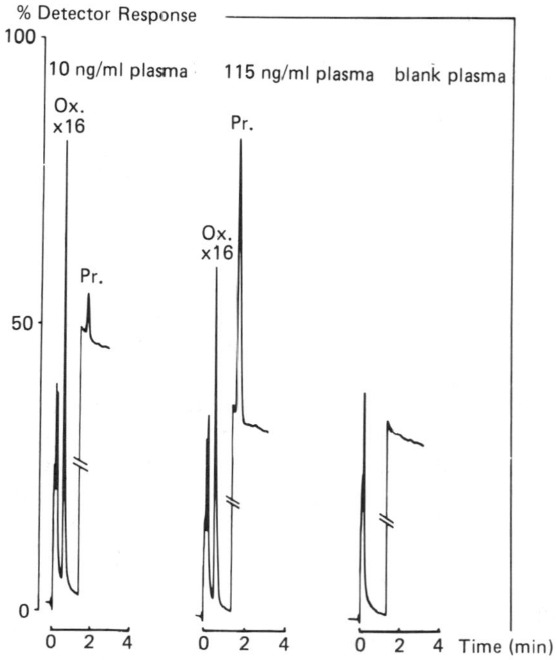

Figure 30.7 Gas chromatograms of extracts containing 10 and 115 ng propranolol/ml human plasma and blank human plasma, obtained following solid injection into a Carbowax 20 M SCOT column with nitrogen-selective detection. Oxprenolol was used as internal standard (250 ng/ml plasma). Temperatures: column, 210; detector, 300; injection port, 300°C. (From Ref. 22.)

The final step of shaking the derivatized drugs with a tertiary sodium phosphate solution is necessary for neutralizing the residual acetic acid and acetic acid anhydride. Once the samples were derivatized, they were stable for at least one week. The derivatization procedure with acetic anhydride improves the gas chromatographic properties of propranolol and oxprenolol. This is also the case in the procedures where esterification is performed with halogen-containing derivatization agents, being required for electron capture detection [36,38,64,111,120]. Due to halogen introduction, however, the selectivity of the analytical procedure may be negatively influenced because endogenous compounds may be derivatized and detected as well. The use of the capillary Carbowax 20 M column also appears to be an improvement in the analysis of propranolol. Many samples can be analyzed in a relatively short time. Solid injection is required in order to detect the low concentrations of propranolol and to prevent deterioration of the capillary column by the organic solvent.

4. *Pemoline [113]*

Pemoline (2-imino-5-phenyl-4-oxazolidinone, see Figure 30.8, I) is a stimulant of the central nervous system. Because of its low lipophilicity and its instability under acidic and alkaline conditions, it is difficult to extract pemoline from biological fluids. As a consequence, methods such as thin-layer chromatography (TLC) [40] and high performance liquid chromatography (HPLC) [31], which employs direct extraction of pemoline from human urine, lack the sensitivity

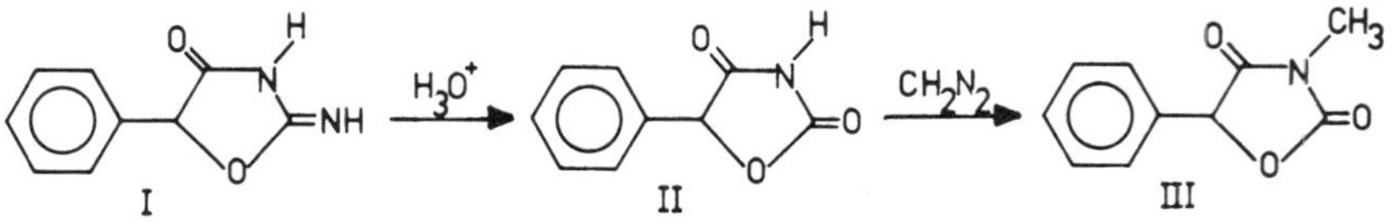

Figure 30.8 Derivatization scheme of pemoline (I). First, acidic hydrolysis of pemoline to 5-phenyl-2,4-oxazolidinedione (II) followed by methylation to *N*-methyl-5-phenyl-2,4-oxazolidinedione (III).

and selectivity required for quantitative determination. Gas-liquid chromatography (GLC) cannot be applied because pemoline is not eluted unchanged, and derivatization yields a variety of different compounds. For these reasons, several authors have described methods whereby pemoline is hydrolyzed by acidic treatment to 5-phenyl-2,4-oxazolidinedione (Figure 30.8, II) prior to extraction [23,33,71,112]. This compound is easily extracted and determined by gas chromatography. However, all these methods neglect the occurrence of compound II as a metabolite of pemoline. Moreover, they are not sufficiently sensitive for the assay of therapeutic concentrations of pemoline in plasma and saliva.

We developed a quantitative analytical procedure for pemoline in plasma, saliva, and urine, which is based on capillary gas chromatography with solid injection and nitrogen-selective detection. The method takes account of the occurrence of the oxazolidinedione as a metabolite of pemoline in urine.

The hydrolysis of pemoline in human urine is rapid and essentially complete within 15 min at 80°C in 1 N hydrochloric acid [112]. Under the same conditions, complete conversion of pemoline is achieved in human plasma and saliva.

The extraction of the hydrolysis product of pemoline from portions of 3 ml of plasma or 2 ml of saliva was carried out with 5 ml of dichloromethane on a whirlmixer, during 10 sec. Fensuximide was used as an internal standard. Upon evaporization of the organic extraction layer, the residue was methylated with diazomethane.

In Figures 30.9 and 30.10, the gas chromatograms are shown of plasma, containing 0.61 μg pemoline per ml, and of saliva, containing 0.63 μg/ml, respectively. The samples were obtained from a human volunteer who received 50 mg of pemoline orally. From the blank chromatograms, it can be seen that there is no interference with endogenous plasma or saliva constituents.

By means of GC-MS analysis it was shown that the methylated oxazolidinedione (III) and the internal standard fensuximide (I.S.) are eluted from the gas chromatograph without chemical alteration. Furthermore, it was found that the peaks, having retention times of 2.1 and 4.8 min, correspond with methylated hippuric acid and with caffeine, respectively.

According to standard curves there was a linear relationship between the ratio of the peak height of the methylated oxazolidinedione (III) to that of fensuximide and the concentration of pemoline between 0.05 and 1.0 μg/ml of plasma and saliva. The reliability and accuracy of the assay procedure was assessed with the aid of five calibration curves for plasma and saliva each. The corresponding correlation coefficients were in all cases better than 0.999, clearly establishing the linearity of the calibration curves. The coefficient of variation at all concentrations was always lower than 5% (n = 5). For plasma as well as saliva, extraction times of 10 sec gave satisfactory and reproducible extraction yields of 82 and 100%, with a coefficient of variation at each concentration of <3% (n = 5). The detection limit was about 1 ng per single injection with a signal-to-noise ratio of 1:10.

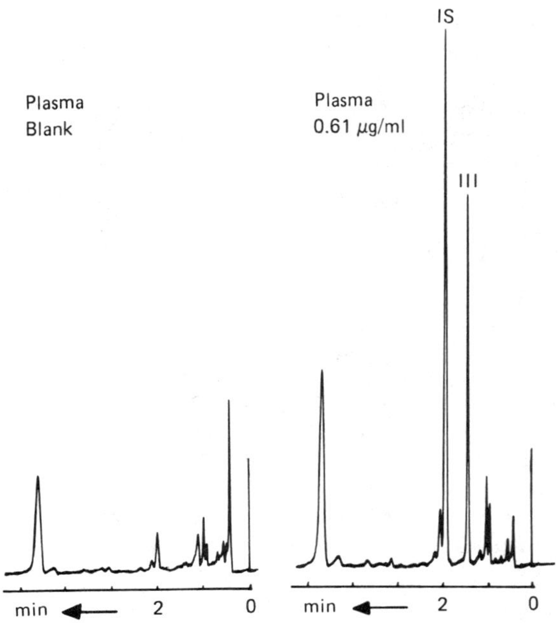

Figure 30.9 Gas chromatogram of a 2.0-ml blank extract of deproteinized human plasma and of an extract of plasma containing 0.61 μg/ml pemoline. Temperatures: column, 190; detector, 260; injection port, 250°C. (From Ref. 113.)

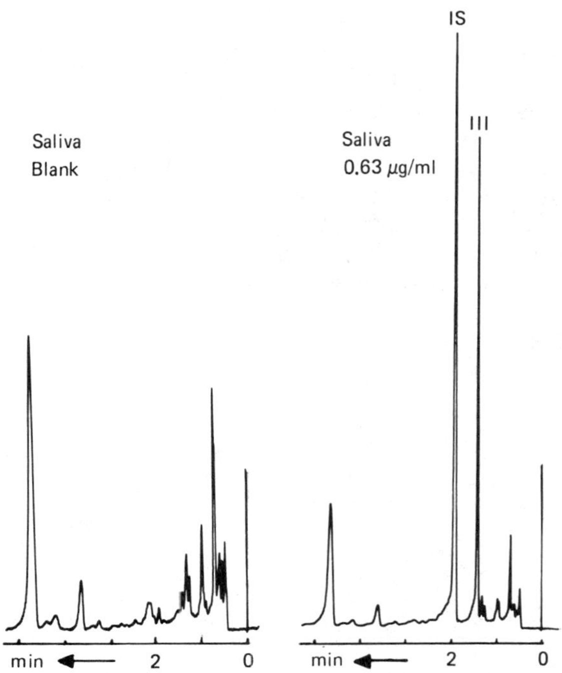

Figure 30.10 Gas chromatogram of a 2.0 ml blank extract of human saliva and of saliva containing 0.63 μg/ml pemoline. Temperatures: see Figure 30.9. (From Ref. 113.)

With respect to the analysis of pemoline in human urine, a similar analytical procedure as described for plasma and saliva was followed except that a short extraction procedure with dichloromethane/*n*-pentane (1:1) was used, preceding the acidic hydrolysis of pemoline to the oxazolidinedione (II). The preextraction step was necessary, since the oxazolidinedione (II) also appeared to be a metabolite of pemoline. The lowest detectable concentration of pemoline in urine was about 100 ng/ml, which is about five times lower than can be achieved with the method described previously [112]. This higher sensitivity is a direct consequence of the use of a capillary SCOT column with a total number of 10,000 theoretical plates (calculated for the methylated oxazolidinedione III), compared with a micropacked column which has a considerably lower total number of theoretical plates (N = 1000 for the same compound).

It was stated [115] that the assay described for the quantitative determination of pemoline in human urine, plasma, and saliva is rapid and relatively simple. It takes into account the occurrence of 5-phenyl-2,4-oxazolidinedione itself as a metabolite of pemoline, while mandelic acid, another metabolite of pemoline, or other endogenous substances do not interfere with the method. The method has been applied to study the pharmacokinetics of pemoline and its "oxazolidinedione metabolite" in human urine, plasma, and saliva [115].

C. Electron-Capture Detection

The electron capture detector (ECD) is a selective detector which is especially sensitive to electrophilic compounds such as halogen-containing compounds. Its sensitivity greatly depends on the number and the nature of electrophilic atoms in a particular molecule. The ECD is 10-100 times as sensitive as the N/P detector, and thus quantitative determination of drug concentrations lower than 1 ng/ml should be possible. The detection limit of an assay can sometimes be substantially lowered by derivatization of drugs or metabolites with halogen-containing reagents. However, the selectivity of an assay can be negatively influenced by such derivatization procedures, because of coderivatization of background substances.

The early EC detectors generally were rather sensitive to contamination, which especially in routine operation proved to be a serious drawback. With the recent development of pulse-modified EC detectors [72], this drawback is largely overcome. Nevertheless, large amounts of solvent still can better be avoided in routine analyses. Therefore, solid injection is preferable for sample introduction. It will protect the detector as well as the capillary column from contamination and thereby maintain optimal performance of the gas chromatographic system.

1. *Nitrazepam [17]*

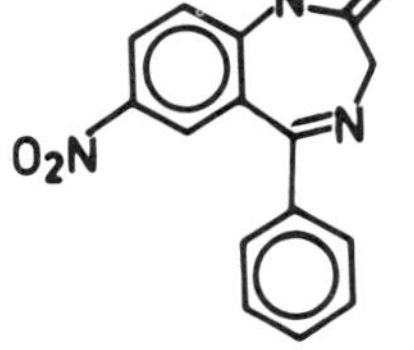

Structure (4)

For nitrazepam, various analytical methods have been described in the literature, comprising thin-layer chromatography (TLC) [84-87,92], photometry

[91,108], fluorometry [94,95], and radiometry [93], high-performance liquid chromatography [101,123], and gas chromatography [8,41,48,55,63,74,78,104]. Because of lack of sensitivity of specificity, however, several of these methods cannot be applied to pharmacokinetic investigations when therapeutic doses are used. Some of the gas chromatographic methods determine the benzodiazepines after acidic hydrolysis as benzophenones [8,74,78,104], which gives rise to loss of specificity. Others require derivatization, e.g., methylation [41,63] and trimethylsilylation [48]. Some methods require laborious extraction procedures for the isolation of the compounds from blood or plasma which, in addition to chemical derivatization and long retention times, may render such methods very time-consuming.

It was our aim to develop a method for the assay of nitrazepam in plasma that could be applied to pharmacokinetic and bioavailability studies in man. The method should therefore be specific and sensitive and also rapid because of the great number of samples to be analyzed. The use of a SCOT capillary column, together with solid injection and electron capture detection, proved to be suitable for this purpose.

The extraction of nitrazepam from plasma was carried out on a whirlmixer during 20 sec, using twice 5 ml of a mixture of light petroleum (b.p., 40-60°C)/ dichloromethane (1:1).

In Figure 30.11 are shown the gas chromatograms of the extracts of plasma 1.5 and 80 hr after oral administration of 5 mg of nitrazepam to a human

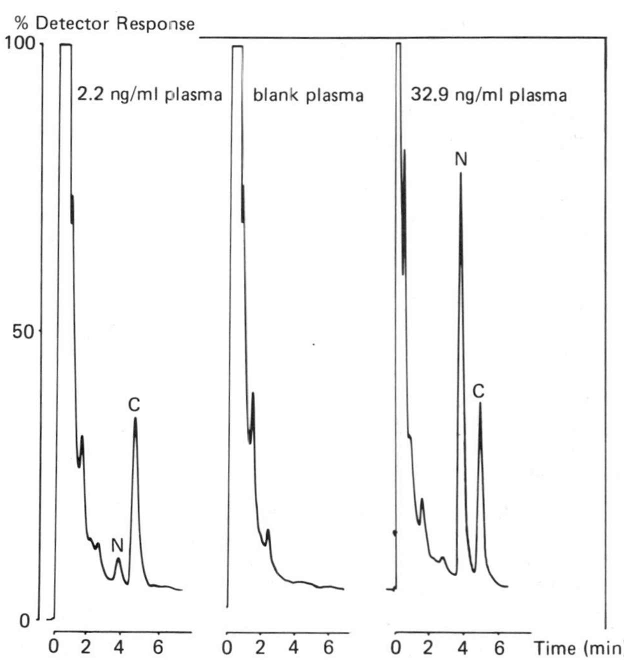

Figure 30.11 Gas chromatograms of a 1-ml plasma extract obtained from a volunteer immediately before (middle) and 1.5 hr (right) and 80 hr (left) after, receiving 5 mg nitrazepam orally. N = nitrazepam, C = clonazepam (internal standard, 25.0 ng/ml plasma). Temperatures: column, 230; detector, 300; injection port, 250°C. (From Ref. 17.)

volunteer. From the blank chromatogram, also shown in Figure 30.11, it is clear that there is no interference from endogenous plasma constituents or metabolites. Clonazepam was used as internal standard. Although the retention time of clonazepam is slightly longer than that of nitrazepam, the overall time of one gas chromatographic run is still within 6 min. There was a linearity between the detector response (peak area nitrazepam/peak area clonazepam) and the concentration of nitrazepam between 10 and 100 ng/ml of plasma. The recovery of nitrazepam in this range was 92% (n = 4, coefficient of variation <3.8%) and for clonazepam 75% (n = 4, coefficient of variation <6%). The reproducibility of the assay of nitrazepam is reflected in the coefficient of variation in this concentration range, being about 5%. The detection limit was about 1 ng of nitrazepam per ml of plasma.

By means of combined GC-MS, it was confirmed that both compounds left the gas chromatographic column unchanged.

In conclusion, the present method allows the accurate and specific determination of nitrazepam (and of clonazepam) in plasma in relatively low concentrations. The use of a capillary SCOT column appeared to be a definite improvement in the analysis of theese benzodiazepines. An additional advantage of using capillary columns is the low column bleeding, which is particularly important when electron capture detection is employed. Interference with other benzodiazepines, such as oxazepam, lorazepam, flurazepam, chlordiazepoxide, nordiazepam, diazepam, hydroxydiazepam, and medazepam, does not occur. Preliminary studies on the metabolism of nitrazepam have indicated that its major metabolites do not interfere with the assay of the parent compound.

D. Mass Spectrometric Detection

The mass spectrometer is one of the most generally applicable detectors for gas chromatography. This is mainly due to its high sensitivity and selectivity. In addition, however, the mass spectrometer may provide a wealth of structural information. The introduction of capillary columns in GC-MS has even increased the power and versatility of this analytical technique. The low flow-rate of carrier gas into the mass spectrometer and the high separation power of capillary columns have been recognized as distinct advantages for the gas chromatograph mass spectrometer combination [76,122]. Other important features of combining capillary columns with GC-MS include faster analyses, higher sensitivity of the assays, and a lower background [45].

In this section, three examples will be discussed to illustrate specific advantages of using mass spectrometry in combination with capillary gas chromatography and solid injection.

1. Hexobarbital Metabolites: Mass Chromatography [116]

The next two examples deal with the metabolism of hexobarbital [Structure I of (5)] a compound which in many studies has been used as a model substrate to measure the activity of oxidative microsomal enzymes in the liver. Monooxygenase reactions are known to give 3'-ketohexobarbital and 3'-hydroxyhexobarbital as major, and N-demethylated products as minor metabolites [29,116]. Qualitative as well as quantitative analyses of these metabolites were mainly performed with the aid of thin-layer chromatography (TLC), radioisotopic techniques, and countercurrent distribution [59,116]. In some studies, mass spectrometry whether or not combined with gas chromatography, was successfully used [25, 47,109,125]. Recently we developed a quantitative procedure for the assessment of 3'-hydroxy- and 3'-ketohexobarbital in urine of rat and man, which was based

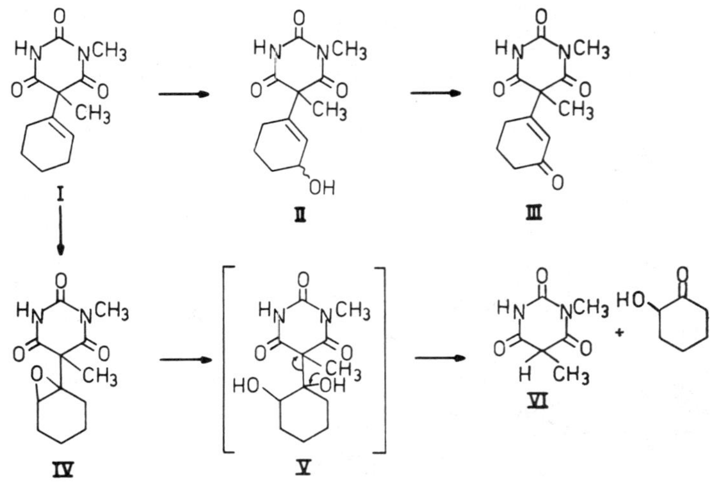

Structure Group (5)

on capillary gas chromatography with solid injection and nitrogen-selective detection [114]. In addition, 1,5-dimethylbarbituric acid, being the end product of a new type of epoxide-diol pathway, was identified with the aid of mass chromatography [114].

Figure 30.12 represents part of the total-ion-current chromatogram and the mass chromatogram of the ions 56 and 156 of an extract of rat urine; after the rat had been treated intraperitoneally with hexobarbital, 0.5 ml of urine was extracted once with 5 ml ethyl acetate at pH 0. Since the retention times and the relative intensities of the ion at m/z 56 and m/z 156 exactly correspond to those of synthetic 1,5-dimethylbarbituric acid, this mass chromatogram is to be considered as a positive identification of this compound in urine. Similarly, 1,5-dimethylbarbituric acid was identified in urine of man treated orally with hexobarbital.

The molecular weight of 1,5-dimethylbarbituric acid is 156 amu., which is considerably lower than that of the parent compound (236 amu). Its relative retention time appeared to be 0.4, compared to hexobarbital. As a consequence, upon liquid injection, 1,5-dimethylbarbituric acid was hidden under the solvent peak, at least when the same isothermal chromatographic ocnditions were applied, which are normal for hexobarbital and its known metabolites. Using the mass spectrometer as detector, the solvent peak is normally vented off. Therefore, the introduction of the sample into the capillary column via solid injection was especially appropriate in this analysis, because it enabled the mass spectrometer to take spectra even of that part which, following liquid injection, would have been vented off.

2. *Hexobarbital Metabolites: Mass Fragmentography [116]*

Using the mass spectrometer for quantitative analyses, the instrument is preferably operated in the mass fragmentographic mode. In this mode, certain ions, which are more or less characteristic for the compounds to be measured, are preselected and monitored continuously.

In the following example, the quantitative assay of 3'-hydroxy- and 3'-ketohexobarbital in the urine of a patient with cholestasis is described [116]: 0.5 ml of urine was adjusted to pH 5.4 with 0.5 ml of a sodium phosphate buffer and subsequently extracted twice with a mixture of ether/n-pentane/1-pentanol (60:40:2). The combined organic layers were evaporated to dryness

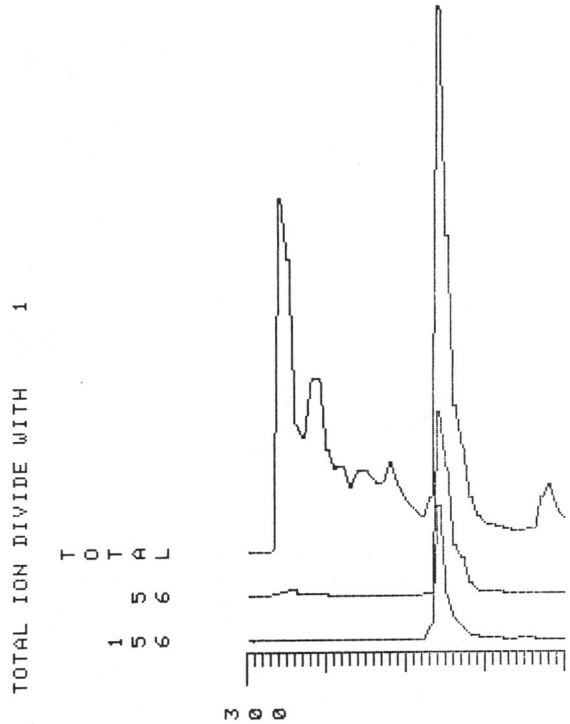

Figure 30.12 Total ion current and mass chromatogram of the ions at m/z 56 and 156 of an extract of rat urine, following treatment of the rat with hexobarbital. Temperatures: column, 180; injector, 250°C.

and the residue was redissolved in 300 μl of a mixture of absolute ethanol and acetone (80:20).

Figure 30.13 presents a single-trace mass fragmentogram of such an extract of a patient treated intraveneously with hexobarbital. By means of mass chromatography the following peaks were identified: I, unchanged hexobarbital; II, 3'-hydroxyhexobarbital; III, 3'-ketohexobarbital; I.S., the internal standard 1,5-dimethyl-5-(2'-ketocyclohexyl)barbituric acid; and VI, 1,5-dimethylbarbituric acid. The shoulder marked with I' was identified as 1',3'-dienehexobarbital. 1,5-Dimethylbarbituric acid is present in the mass fragmentogram of the urinary extract at pH 5.4, but its recovery was substantially higher at pH 0, because of its pK_a value of 4.0. Therefore, this compound was extracted at pH 0, as was described previously [114]. The detection limits for the measurement of the metabolites II and III with this mass fragmentographic method were sufficiently low to measure urinary excretion up to 72 hr, whereas at all concentrations, the precision of this method was within the limits of ±7% (n = 5).

For the measurement of 3'hydroxy- and 3'-ketohexobarbital in the urine of these patients, mass fragmentographic detection was preferred to the previously applied nitrogen-selective detection [114], since sometimes nitrogen-containing substances, most likely originating from comedicated drugs, interfered with the metabolites to be measured. From the mass fragmentogram shown, it is obvious that by measuring m/z 156, which is the ion corresponding

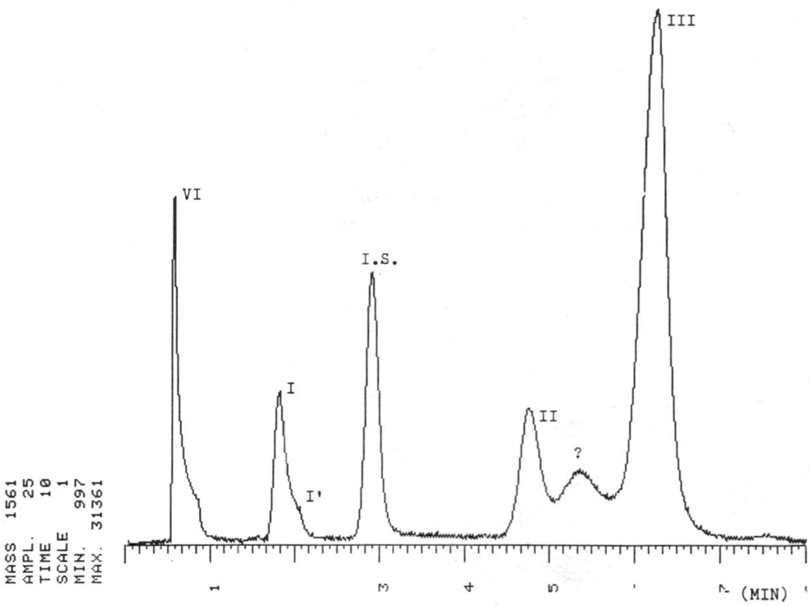

Figure 30.13 Mass fragmentogram (m/z 156) of a urinary extract obtained from a patient treated with hexobarbital; it contains unchanged hexobarbital (I), 1',3'-dienehexobarbital (I'), 3'-hydroxyhexobarbital, 3'-ketohexobarbital (III), 1,5-dimethylbarbituric acid, the internal standard (I.S.) and an unknown metabolite. Temperatures: column, 220; injector, 250°C.

to 1,5-dimethylbarbituric acid, practically a barbiturate-selective detector is obtained. Moreover, it is worth noting that the use of well-deactivated and short capillary SCOT columns permit the assessment of the metabolite of hexobarbital without chemical derivatization.

3. *Cyclohexylmercapturic Acids* (6) *[118]*

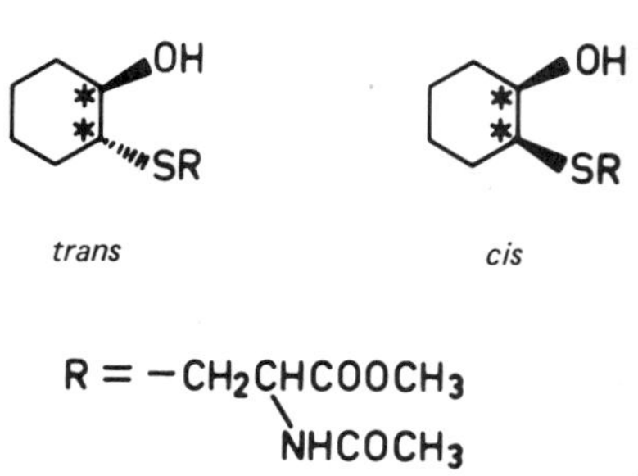

Structure Group (6)

The last application described in this chapter deals with the ability of chemical ionization gas chromatography-mass spectrometry to estimate ratios of *cis*- and *trans*-mercapturic acids of cyclohexene oxide (S-substituted *N*-acetyl-(1)-cysteine derivatives) [118].

Mercapturic acids are ultimate products of an important biotransformation reaction of epoxides, i.e., glutathione conjugation. The stereospecificity of this type of enzymatic conjugation reaction was recently studied with the aid of cyclohexene oxide as a simple model substrate [13]. For this study an analytical method was needed which would provide the possibility of assessing the ratio of the cis- and trans stereomeric mercapturic acids in biological fluids. With capillary SCOT gas chromatography it was not possible to separate these stereoisomers, nor with Carbowax 20 M HP, nor with a mixture of OV-17/PPE-21 as stationary phase. Derivatization of the free hydroxyl group in both stereoisomers appeared to be rather complicated and not suitable for solving the problem.

Previously, chemical ionization and electron-impact ionization mass spectrometry have proved to be suitable to demonstrate stereochemical differences between bifunctionally substituted compounds [49]. Therefore the electron impact and chemical ionization behavior of the cis and trans stereoisomeric mercapturic acids of cyclohexene oxide was studied. With ammonia as a reactant gas, it was possible to discriminate between these two stereoisomers on the basis of the apparent stereoselectivity in the loss of H_2O from abundant protonated molecular ions. Figures 30.14 and 30.15 show the mass fragmentograms of m/z 176 [MH^+], 258 [MH^+-H_2O], and 257 obtained from pure cis and trans isomers. The cis stereoisomer, in fact consisting of two diastereoisomers, coincides with the two diastereomeric trans compounds, which in their turn are

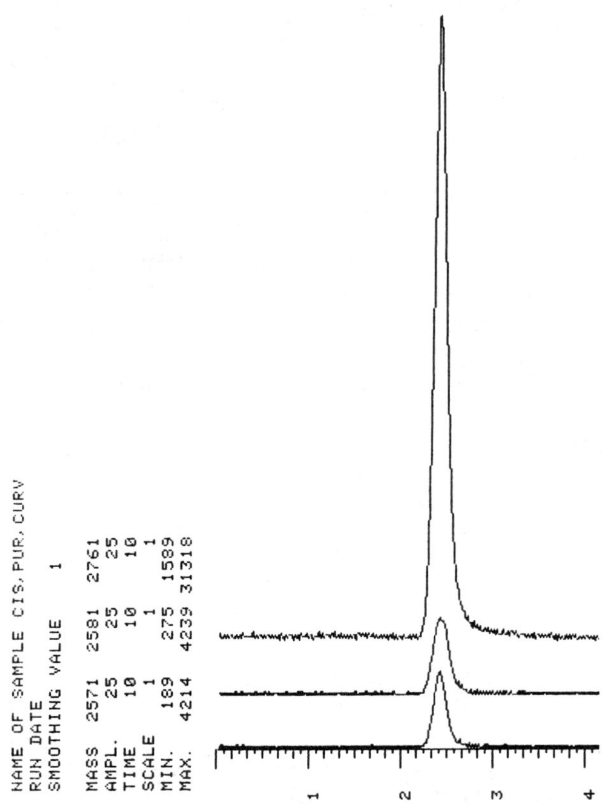

Figure 30.14 Mass fragmentogram (m/z 257, 258, and 276) of the *cis*-mercapturic acids of cyclohexene oxide upon ammonia chemical ionization. (From Ref. 118.)

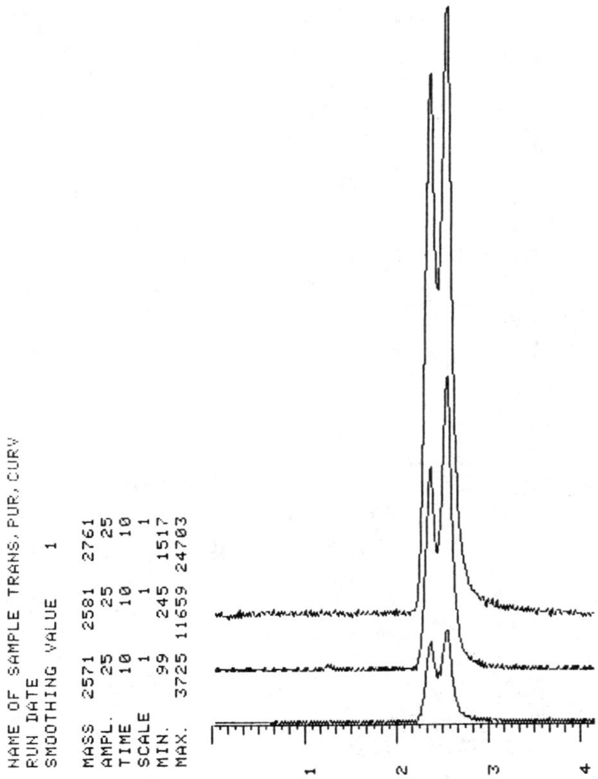

Figure 30.15 Mass fragmentogram (*m/z* 257, 258, and 276) of the *trans*-mercapturic acids of cyclohexene oxide upon ammonia chemical ionization. Temperatures: column, 190; injector, 250°C. (From Ref. 118.)

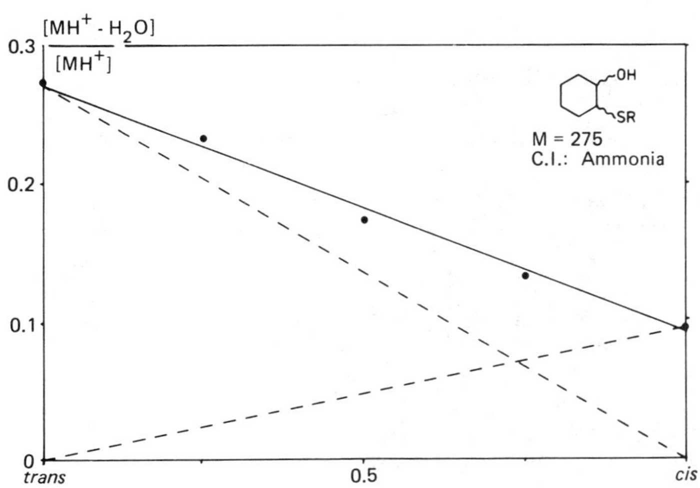

Figure 30.16 Calibration curve of samples containing different relative amounts of *cis*- and *trans*-mercapturic acids of cyclohexene oxide used to determine the ratio of the cis and trans isomers in urine of rats treated with cyclohexene oxide. (From Ref. 118.)

eluted slightly separated. By running samples containing different amounts of cis and trans isomers and measuring peak areas of the *m/z* values of 276, 258, and 257, a calibration curve was obtained in which after correction of the natural ^{13}C contribution of *m/z* 257 to *m/z* 258, the ratio 258:276 is plotted against the relative amounts of cis and trans compounds. As can be seen in Figure 30.16, such a calibration curve is linear and is in fact the sum of the individual calibration curves of the cis and trans isomers. The standard deviation of 5 measurements of single ratios of cis and trans was about 4%. Using this type of calibration curve it was found that within the experimental error, only *trans-S*-(2-hydroxycyclohexyl)-*N*-acetyl-(1)-cysteine was excreted in the urine of rats treated with cyclohexene oxide [13].

IV. CONCLUDING REMARKS

From the brief description of solid injection and the applications discussed in the previous sections, it is evident that solid injection into capillary columns has a number of distinct advantages. First, because no solvent is injected upon sample introduction into a capillary column, stripping off and breaking of the liquid film in the column is greatly reduced. Secondly, pollution of the column due to involatile sample components is largely avoided. As a consequence of both features, column lifetime is extended considerably. Thirdly, it is obvious that in the case of the analysis of compounds with very short retention times, smaller amounts can be detected, because of the absence of a solvent peak. Ultimately, since no sample splitting is needed, larger parts of samples sometimes laboriously prepared, can be introduced into the capillary column. In many cases, this may result in the detectability of lower levels of compounds.

With respect to the use of capillary columns in comparison with packed columns, it seems evident that in the analysis of complex mixtures, the enhanced selectivity and sensitivity are of major importance [110]. In particular cases this may result in faster analysis as well as analyses at lower temperatures. The latter minimizes effects of column bleeding and sometimes favors the analysis of thermolabile compounds.

Ettre et al. [43] have shown that the resolution of compounds with very short retention times is better on SCOT columns than on WCOT columns. Realizing this and the fact that SCOT columns often possess a relatively high sample-capacity, in some cases (e.g., routine analysis in pharmacokinetic studies) SCOT columns are to be preferred. Another evident advantage referred to in this chapter, includes the favorable adsorption properties of capillary columns. As a consequence, very polar compounds such as the metabolites of hexobarbital as well as subnanogram levels of compounds such as therapeutic concentrations of nitrazepam can be assayed routinely.

The selective gas chromatographic detectors, whose applications were discussed in this chapter, intrinsically permit the detection of subnanogram amounts of compounds (low electronic noise level). This also holds true for the analysis of drugs, metabolites, and other compounds in complex biological mixtures, since, as a consequence of their apparent selectivity, interference of endogenous constituents is greatly reduced (low "biological" noise level). In practice, reasonably easy and rapid sample preparation steps can therefore be applied.

This chapter was meant to illustrate that in gas chromatography the choice of a proper combination of injection system, column, and detector provides a powerful analytical technique in the analysis of complex mixtures. For the quantitative determination of xenobiotics required in pharmacokinetic and drug metabolism studies, the choice of solid injection, and short, medium polar, or polar SCOT capillary columns and selective detectors have proved to be very successful.

REFERENCES

1. K. K. Adjepon-Yamoah and L. F. Prescott, *J. Pharmacol. 26*:889 (1974).
2. W. A. Aue, C. W. Gehrke, R. C. Tindle, D. L. Stalling, and C. D. Ruyle, *J. Gas Chromatogr.* 5:381 (1967).
3. A. J. Baars, R. J. Vermeulen, and D. D. Breimer, *J. Chromatogr. 120*:217 (1976).
4. H. T. Badings, J. J. G. van der Pool, and J. G. Wassink, *Chromatographia 8*:440 (1976).
5. E. Bailey, M. Fenoughty, and J. R. Chapman, *J. Chromatogr. 96*:33 (1974).
6. A. H. Beckett, R. H. Boyes, and J. B. R. Parker, *Anaesthesiology 20*:294 (1965).
7. A. H. Beckett, R. N. Boyes, and P. J. Appleton, *J. Pharm. Pharmacol. 18*:76S (1966).
8. G. P. Beharrell, D. M. Hailey, and M. K. McLaurin, *J. Chromatogr. 70*:45 (1972).
9. H. H. Beneker Kolmer, A. G. Burm, C. A. Cramers, J. M. Ramakers, and H. L. Vader, *J. Anaesthesiol. 47*:1049 (1975).
10. P. M. J. van den Berg and Th. P. H. Cox, *Chromatographia 5*:301 (1972).
11. N. Benowitz and M. Rowland, *Anaesthesiology 39*:639 (1973).
12. J. W. Black, W. A. M. Duncan, and R. G. Shanks, *Br. J. Pharmacol. 25*:577 (1965).
13. P. J. van Bladeren, D. D. Breimer, C. J. R. Seghers, N. P. E. Vermeulen, and A. van der Gen, *Drug.Metab. 9*:207 (1981).
14. W. L. Blankenberger, C. A. Di Fazio, and F. A. Berry, *Anaesthesiology 42*:325 (1975).
15. A. G. de Boer and D. D. Breimer, unpublished results.
16. A. G. de Boer and D. D. Breimer, *Pharm. Weekbl. 112*:764 (1977).
17. A. G. de Boer, J. Röst-Kaiser, H. Bracht, and D. D. Breimer, *J. Chromatogr. Biomed. Appl. 145*:105 (1978).
18. A. G. de Boer, Ph.D. Thesis, University of Leiden, 1979.
19. A. G. de Boer, J. M. Gubbens-Stibbe, F. H. de Koning, A. Bosma, and D. D. Breimer, *J. Chromatogr. 162*:457 (1979).
20. A. G. de Boer, D. D. Breimer, H. Mattie, J. Pronk, and J. M. Gubbens-Stibbe, *Clin. Pharmacol. Ther. 26*:701 (1979).
21. A. G. de Boer, J. B. Smeekens, and D. D. Breimer, *J. Chromatogr. 162*:591 (1979).
22. A. G. de Boer, D. D. Breimer, and J. M. Gubbens-Stibbe, *Pharm. Weekbl. Sci. Ed. 2*:101 (1980).
23. M. van Boven and P. Daenens, *J. Chromatogr. 134*:415 (1977).
24. V. V. Brazhnikov, M. V. Gur'ev, and K. I. Sakodynsky, *Chromatogr. Rev. 12*:1 (1970).
25. D. D. Breimer, Ph.D. Thesis, University of Nijmegen, 1974.
26. D. D. Breimer, in *Therapeutic Relevance of Drug Assays*, Boerhaave Series, Vol 14, (F. A. de Wolff, H. Mattie, and D. D. Breimer, editors) University Press, Leiden, 1979.

27. D. D. Breimer, H. C. J. Ketelaars, and J. M. van Rossum, *J. Chromatogr. 88*:55 (1974).
28. C. A. Burgett, D. H. Smith, and H. B. Bente, *J. Chromatogr. 134*:57 (1977).
29. M. T. Bush and W. L. Weller, *Drug Metab. Rev. 1*:249 (1972).
30. G. Caille, J. Lelorier, Y. Latour, and J. G. Besner, *J. Pharm. Sci. 66*:1383 (1977).
31. G. P. Cartoni and F. Natalizia, *J. Chromatogr. 134*:57
32. C. A. Chidsey, P. Morselli, G. Bianchetti, A. Morganti, G. Leonetti, and A. Zanchetti, *Circulation 52*:313 (1975).
33. S. Y. Chu and L. T. Sennello, *J. Chroamtogr 137*:343 (1977).
34. C. A. Cramers, E. A. Vermeer, L. G. van Kuik, J. A. Hulsman, and C. A. Meyers, *Clin. Chim. Acta 73*:97 (1976).
35. M. Danhof, A. G. de Boer, E. de Groot-van der Vis, and D. D. Breimer, *Pharmacology 19*:215 (1979).
36. P. N. Degen and W. Riess, *J. Chromatogr. 121*:72 (1976).
37. C. A. Di Fazio and R. E. Brown, *Anaesthesiology 34*:86 (1971).
38. E. Di Salle, K. M. Baker, S. R. Bareggi, W. D. Watkins, C. A. Chidsey, S. Frigerio, and P. L. Morselli, *J. Chromatogr. 84*:347 (1973).
39. O. Driessen and A. Emonds, *Proc. Kon. Ned. Acad. Wetensch.*, Ser. C., *77*:171 (1974).
40. H. Eberhardt and M. Debackere, *Arzneim. Forsch. 15*:929 (1965).
41. H. Ehrsson and A. Tilly, *Anal. Lett. 6*:197 (1973).
42. H. Ehrsson, *J. Pharm. Pharmacol. 28*:662 (1976).
43. L. S.Ettre and E. W. March, *J. Chromatogr. 91*:5 (1974).
44. L. S. Ettre, J. E. Purcell, J. Widomski, B. Kolb, and P. Pospisil, *J. Chromatogr. Sci. 18*:116 (1980).
45. J. Eyem, *Chromatographia 8*:456 (1975).
46. Y. Garceau, I. Davis, and J. Hasegawa, *J. Pharm. Sci. 67*:826 (1978).
47. N. Gerber, R. Lynn, R. Holcomb, W. L. Weller, and M. T. Bush, *J. Pharmacol. Exp. Ther. 177*:234 (1971).
48. M. S. Greaves, *Clin. Chem. 20*:141 (1974).
49. M. M. Green, in *Topics in Stereochemistry*, (E. L. Eliel and N. L. Allinger, eds.), Wiley Interscience, New York, 1977.
50. K. Grob and G. Grob, *J. Chromatogr. Sci. 7*:584 (1969).
51. K. Grob and K. Grob, Jr., *J. Chromatogr. 94*:53 (1974).
52. K. Grob and K. Grob, Jr., *J. Chromatogr. 125*:471 (1976).
53. K. Grob and K. Grob, Jr., *J. Chromatogr. 151*:311 (1978).
54. H. Hachenberg and A. P. Schmidt, in *Gas Chromatographic Head Space Analysis*. Heyden and Son, Ltd., London, 1977.
55. D. M. Hailey, *J. Chromatogr. 98*:527 (1974).
56. M. J. van Hamme, J. J. Ambre, and M. M. Ghoneim, *J. Pharm. Sci. 66*:1344 (1977).
57. W. E. Harris, *J. Chromatogr. Sci. 11*:184 (1973).
58. B. W. Hadzija and A. M. Mattocks, *J. Pharm. Sci. 67*:1307 (1978).
59. R. R. Holcomb, N. Gerber, and M. T. Bush, *J. Pharmacol. Exp. Ther. 188*:15 (1974).
60. H. B. Hucker and S. C. Stauffer, *J. Pharm. Sci. 65*:926 (1976).
61. T. R. Irgens, W. M. Henderson, and W. H. Shelver, *J. Pharm. Sci. 65*:608 (1976).
62. T. Ishikawa, *Jap. J. Anaesthesiol. 23*:246 (1974).
63. L. Kangas, *J. Chromatogr. 136*:259 (1977).
64. R. E. Kates and C. L. Jones, *J. Pharm. Sci. 66*:1490 (1977).

65. C. H. Kaye, in *Progress in Drug Metabolism*, Vol. 4, (J. W. Bridges and L. F. Chasseaud, eds.), John Wiley and Sons, New York, 1980.
66. J. B. Keeneghan, *Anaesthesiology 29*:110 (1968).
67. B. J. Kline and M. F. Martin, *J. Pharm. Sci. 67*:887 (1978).
68. B. Kolb and J. Bischoff, *J. Chromatogr. Sci. 12*:625 (1974).
69. L. G. J. de Leede, A. G. de Boer, and D. D. Breimer, *Biopharm. Drug. Dispos. 2*:131 (1981).
70. J. W. de Leeuw, W. L. Maters, D. v.d. Meent, and J. J. Boon, *Anal. Chem. 49*:1881 (1977).
71. J. C. Libeer and P. Schepens, *Pharm. Weekbl. 112*:210 (1977).
72. R. J. Maggs, P. L. Jaynes, A. J. Davies, and J. E. Lovelock, *Anal. Chem. 43*:1966 (1971).
73. H. Maier-Bode and M. Riedman, *Residue Reviews*, Springer, New York, 1975.
74. Y. Masuda, *Jap. J. Leg. Med. 25*:445 (1971).
75. L. E. Mather and G. T. Tucker, *J. Pharm. Sci. 63*:306 (1974).
76. W. McFadden, *Techniques of Combined Gas Chromatography/Mass Spectrometry*, Wiley Interscience, New York, 1973.
77. O. Mlejnek and J. Blatnicky, *J. Chromatogr. 105*:382 (1975).
78. K. Møller-Jensen, *J. Chromatogr. 111*:389 (1975).
79. E. Naito, M. Matsuki, and K. Shimoji, *Anaesthesiology 47*:466 (1977).
80. R. L. Nation, E. J. Triggs, and M. Selig, *J. Chromatogr. 116*:188 (1976).
81. R. L. Nation, C. W. Peng, and W. L. Chiou, *J. Chromatogr. Biomed. Appl. 145*:429 (1978).
82. S. D. Nelson, W. A. Garland, G. D. Breck, and W. F. Trager, *J. Pharm. Sci. 66*:1180 (1977).
83. M. Novotny and R. Farlow, *J. Chromatogr. 103*:1 (1975).
84. H. Oelschlager, J. Volke, and G. T. Lim, *Arch. Pharm. 298*:213 (1965).
85. H. Oelschlager, J. Volke, G. T. Lim, and U. Frank, *Arzneim. Forsch. 16*:82 (1966).
86. H. Oelschlager, J. Volke, and G. T. Lim, *Arzneim. Forsch. 17*:637 (1967).
87. O. Pribilla, *Arzneim. Forsch. 10*:1148 (1965).
88. K. E. Rasmussen and J. Karlsen, *J. Chromatogr. Sci. 14*:67 (1976).
89. E. Reid, Blood, *Drugs and Other Analytical Challenges*, John Wiley & Sons, New York, 1978, Vol. 7.
90. F. Reynolds and A. H. Beckett, *J. Pharm. Pharmacol. 20*:704 (1968).
91. J. Rieder, *Arzneim. Forsch. 10*:1134 (1965).
92. J. Rieder, *Arzneim. Forsch. 10*:1140 (1965).
93. J. Rieder, *Arzneim. Forsch. 10*:1142 (1965).
94. J. Rieder, *Arzneim. Forsch. 23*:207 (1973).
95. J. Rieder, *Arzneim. Forsch. 23*:212 (1973).
96. M. T. Rosseel and M. G. Bogaert, *J. Chromatogr. 154*:88 (1978).
97. M. Rowland, P. D. Thomson, A. Guichard, and K. L. Melmon, *Ann. N.Y. Acad. Sci. 179*:383 (1971).
98. G. Schomburg, H. Husmann, and F. Weeke, *J. Chromatogr. 99*:63 (1974).
99. G. Schomburg, R. Dielman, H. Husmann, and F. Weeke, *J. Chromatogr. 122*:55 (1976).
100. G. Schomburg, H. Gehlau, R. Dielman, F. Weeke, and H. Husmann, *J. Chromatogr. 142*:87 (1977).

101. C. G. Scott and P. Bommer, *J. Chromatogr. Sci. 8*:446 (1970).
102. J. Sevick, *Chromatographia 6*:139 (1973).
103. D. G. Shand, E. M. Nuckolls, and J. A. Oates, *Clin. Pharmacol. Ther. 11*:112 (1970).
104. J. A. F. de Silva, I. Behersky, C. V. Publisi, M. A. Brooks, and R. E. Weinfeld, *Anal. Chem. 48*:10 (1976).
105. J. M. Strong and A. J. Atkonson, *Anal. Chem. 44*:2287 (1972).
106. G. Svinhufvud, B. Ortegren, and S.-E. Jacobsson, *Scan. J. Clin. Lab. Invest. 17*:162 (1965).
107. F. J. Tehan and J. L. Dye, *Anal. Chem. 47*:1876 (1975).
108. S. L. Tompsett, *J. Clin. Pathol. 21*:366 (1968).
109. R. M. Thompson, N. Gerber, R. A. Seibert, and D. M. Desiderio, *Drug Metab. Disp. 1*:489 (1973).
110. W. J. A. VandenHeuvel and A. G. Zacchei, in *Drugs, Fate and Metabolism: Methods and Techniques*, Vol. 11 (E. R. Garret and J. L. Hirtz, eds.), Marcel Dekker, New York, 1978.
111. P. Vermeij, Thesis, University of Amsterdam, 1977, p. 35.
112. N. P. E. Vermeulen, D. de Roode, and D. D. Breimer, *J. Chromatogr. 137*:333 (1977).
113. N. P. E. Vermeulen, M. W. E. Teunissen, and D. D. Breimer, *J. Chromatogr. 157*:133 (1978).
114. N. P. E. Vermeulen, B. H. Bakker, J. Schultink, A. van der Gen, and D. D. Breimer, *Xenobiotica 9*:289 (1979).
115. N. P. E. Vermeulen, M. W. E. Teunissen, and D. D. Breimer, *Br. J. Clin. Pharmacol. 8*:459 (1979).
116. N. P. E. Vermeulen, Ph.D. Thesis, University of Leiden, 1980.
117. N. P. E. Vermeulen, B. H. Bakker, D. Eylers, and D. D. Breimer, *Xenobiotica 10*:159 (1980).
118. N. P. E. Vermeulen, J. Cauvet, W. C. H. H. Luyten, and P. J. van Bladeren, *Biomed. Mass Spectrom. 7*:413 (1980).
119. M. Verzele, M. Verstappe, P. Sandra, E. van Luchene, and A. Vuye, *J. Chromatogr. Sci. 10*:668 (1972).
120. T. Walle, *J. Pharm. Sci. 63*:1885 (1974).
121. T. Walle, J. Morrison, K. Walle, and E. Conradi, *J. Chromatogr. 114*: 351 (1975).
122. J. T. Watson, *Introduction to Mass Spectrometry: Biomedical, Environmental, and Forensic Applications*, Raven Press, New York, 1976.
123. D. J. Weber, *J. Pharm. Sci. 61*:1797 (1972).
124. J. Wynants, R. Woestenborghs, and J. Heykants, Janssen Research Product Information Service, Section II: Biological Research Reports, Serial no. R26490/13, Janssen Pharmaceutica, B-2340 Beerse, Belgium, 1974.
125. T. D. Yih, Ph.D. Thesis, University of Nijmegen, 1976.

Index